THE GAZETTEER OF SCOTLAND

W. & A. K. JOHNSTON'S
GAZETTEER OF SCOTLAND

Including a Glossary of the Most Common
Gaelic and Norse Names

W. & A. K. JOHNSTON, LIMITED
EDINBURGH AND LONDON
1937

W. & A.K. Johnston's *Gazetteer of Scotland*
was first published in Edinburgh in 1937.

Some twenty years later, in 1958, a revised edition appeared
under the imprint of W. & A.K. Johnston & G.W. Bacon Ltd.
Johnston & Bacon, as the company became known, brought
out another major revision, the third edition, in 1973.

In 2001 Johnston & Bacon was acquired by
Famedram Publishers, who are maintaining the imprint.

This is a facsimile reprint of the original 1937 edition.
A fully revised fourth edition of the gazetteer is in
preparation.

ISBN 07179 4608 8
© Copyright 2004 Famedram Publishers Ltd.
Printed by Thomson Press (I) Ltd., C-35 Phase II, Noida.

CONTENTS

plain and up the valleys of the longer rivers life is a little easier, for the soil is richer and deeper and the climate is a little more genial. Oats and barley are successfully cultivated and beef-cattle reared. It should be noted that the Rowett Institute in Aberdeen is one of the principal centres in the British Isles for the study of cattle-rearing methods, and the region is famous for its black-polled Aberdeen-Angus breed. Distilling is an important industry in this area, and along the north-east coast are to be found the principal fishing centres in Scotland. The busiest season is from June till August when the boats are engaged in the herring fishing. In the latter part of the year cod, haddock, whiting, and other fish are caught. On the west Stornoway is noteworthy, while the chief centres on the east are Wick, Fraserburgh, Lerwick, Peterhead, and Aberdeen.

The Highlands are poor in minerals, but hydro-electric power has been developed in many parts (see introductory tables), and the smelting of aluminium is carried on at Foyers, Kinlochleven, and Fort William.

The Midland Valley.

This is often called the Central Lowlands, but is only so in comparison with the higher land of the regions to the north and south of it. To the north, separated from the Highland edge by a well-marked valley, lie the Sidlaw Hills, the Ochil Hills, the Campsie Fells, and the hills of Renfrew and Ayrshire. Through the gaps between these hills break the rivers Tay, Forth, and Clyde. To the south moorland offshoots of the Southern Uplands penetrate the Midland Valley and the only really distinctive range of hills is the Pentlands. For the rest, the region is mainly undulating lowland. Even before the period known as the Industrial Revolution, this was the most densely peopled region in the whole country, for it contains the most valuable agricultural land. Whereas cattle-rearing and dairying are the dominant features of farming in the west, cereal and fruit cultivation are more important in the drier east, particularly in Strathmore, the Carse of Gowrie, and the Lothians. An enlightened attitude to farming on the latest scientific lines has enabled farmers to wrest from the soil large yields per acre. Agriculture alone, however, cannot account for the high density of population in this part of the country where three-quarters of the total population of Scotland live.

It is only since the Industrial Revolution that such a great concentration of population became a feature of the Midland Valley, and we have to look for an explanation in the existence of minerals and in the easy means of communication in the region. The coalfields have given rise to highly industrialised and densely peopled areas. In the east, the Fife and Lothians fields are connected by seams running under the Firth of Forth. In the centre, the Lanarkshire field extends through East Stirlingshire to Clackmannan; and in the west, coming close to the sea like the Fife and Lothians fields, is the N. Ayrshire field. The richer iron ore that occurred with the coal, especially in the central field, is practically exhausted, and the iron industry now relies mainly on imported ores. The first ironworks on any extensive scale in Scotland were built at Carron, near Falkirk, in 1759, but the smelting industry soon spread into Lanarkshire (Coatbridge, Airdrie, Wishaw, Glasgow, etc.) and Ayrshire. The Lanarkshire area is now the principal seat of the heavy iron and steel industries, and machinery of all kinds is made and engineering in all its branches practised.

The Clyde area is the largest shipbuilding district in the world. The cotton industry is centred chiefly in Paisley and Glasgow, and the latter has also woollen mills, chemical works, soap works, etc. Greenock refines sugar. The Ayrshire field exports coal to Ireland from Ardrossan, Irvine, Troon, and Ayr, and supports at Kilmarnock the industries of engineering, woollens, carpets, and a boot and shoe factory. The Fife and Lothians fields export coal, chiefly to the Baltic, from Burntisland, Methil, and Leith, and support at Dunfermline, Kirkcaldy, and Burntisland such industries as linen, linoleum, and shipbuilding and aluminium smelting. Edinburgh has the largest printing industry outside London, besides a large brewing and distilling trade. In the outskirts of the city there are paper mills. Oil is distilled from shale at Broxburn in West Lothian, and the refining of imported crude oil is also carried on. While linen manufacturing is still practised in the county of Angus, the manufacture of jute sacking is the principal industry at Dundee. Here, and at Perth, preserves are made, and the water in the neighbourhood is suitable for the dyeing industry. Communication within the Midland Valley is very easy, and the estuaries of the Forth and Clyde approach to within 24 miles of each other.

THE SOUTHERN UPLANDS.

This region has not the grandeur of the Highlands, for the hills are much lower and present more rounded contours. Here too, however, the rocks are old and hard so that the hillier regions have but a poor covering of soil and are clad with coarse grass and some heather. The flatter areas are confined to the shores of the Solway Firth and the lower part of the Tweed valley. Mount Merrick (2764 ft.), in the south-west, is the highest peak, but the main hill group, Hart Fell, Broad Law, etc., lie in the centre. The rivers Dee, Nith, and Annan flow south-eastwards to the Solway Firth, the Tweed flows north and east to the North Sea, and the Clyde north and north-west to the Atlantic.

As in all parts of Britain, the west is wetter than the east, and, broadly, the south-west coast is a region the valleys and lowlands of which are important for dairy-farming. Stranraer, at the sheltered head of Loch Ryan, is the port for the shortest sea route to Ireland, since Portpatrick on the open sea has no protection from storms. Dumfries, besides being a railway junction and market-town, manufactures hosiery. The eastern half of the Southern Uplands lies in the basin of the Tweed, and the drier climate renders this *the* sheep-farming region of Scotland and accounts for the manufacture of woollen goods in such towns as Galashiels, Hawick, Peebles, Selkirk, etc. The bulk of the raw wool is now imported from Australia, New Zealand, and South Africa. The lower part of the valley—the Merse—is admirably suited for agriculture.

The only mining activity in the Southern Uplands is a little coal in the neighbourhood of Sanquhar in the Upper Nith and some lead from Leadhills and Wanlockhead. Hydro-electric power is being developed in Kirkcudbright-shire on the River Dee.

COMMUNICATIONS.

The country is served by the L.M.S. and L.N.E. railways, the former entering from Carlisle and the latter from Berwick. From Carlisle a line crosses into Scotland and proceeds westward to Stranraer *via* Annan and Dumfries ; another makes its way to Glasgow by Annan, Dumfries, the Nith valley, Kilmarnock, and the Barrhead gap in Renfrewshire ; a third crosses the border from Carlisle, and by means of the Annan valley and the upper Clyde reaches Carstairs where the line forks, one branch going to Edinburgh and another to Glasgow.   From Carlisle a L.N.E. line runs up the Liddel Water, cuts across to Hawick on the Teviot, and proceeds northwards by Galashiels and Gala Water to Edinburgh. The East Coast route from London comes into the country *via* Berwick, skirts the east end of the Lammermuirs to reach Dunbar, and then strikes west to reach Edinburgh.   The L.M.S. route to the north goes by Stirling, Perth, the Tay and Garry valleys, and then by Drumochter Pass reaches the valley of the Spey.   At Aviemore the line divides into two branches, one going direct to Inverness *via* Carrbridge and the other to Inverness *via* Forres.   Beyond Inverness the line divides again at Dingwall, one portion going west to Kyle of Lochalsh and the other northwards to Thurso and Wick.   At Dunblane a branch line proceeds by Callander, Lochearnhead, Crianlarich, and Dalmally to Oban. From Perth a L.M.S. line runs through Strathmore to join the L.N.E. line just outside Montrose, from which place both proceed northward to Aberdeen.   The L.N.E. route to the north from Edinburgh crosses the Forth and Tay bridges and then proceeds *via* Dundee north to Aberdeen, from which point it runs through Keith, Elgin, Forres, and Nairn to Inverness.   From Glasgow the L.N.E. runs north by Crianlarich and Tyndrum and Fort William to Mallaig. Both L.M.S and L.N.E. run lines between Glasgow and Edinburgh.

The foreign trade of Scotland is carried on mainly through the ports in the Central Lowlands.   Those on the west side trade chiefly with America, the West of England, and Ireland, while those on the east side exchange goods with the countries bordering the North and Baltic Seas.

COUNTIES OF SCOTLAND

COUNTIES.	Area in Square Miles.	Population Census, 1931.	Density in Square Miles.	% increase (+) or decrease (−) on 1921 Population.
AREA AND POPULATION.				
Aberdeen . . .	1,955	300,430	153·6	− 0·2
Angus	875	270,190	308·7	− 0·3
Argyll . . .	3,213	63,014	19·6	−18·0
Ayr	1,128	285,182	252·8	− 4·7
Banff	641	54,835	85·5	− 4·3
Berwick	461	26,601	57·7	− 5·8
Bute	218	18,822	86·3	−44·2
Caithness . . .	686	25,656	37·4	− 9·3
Clackmannan . .	55	31,947	580·8	− 1·8
Dunbarton . . .	241	147,751	613·1	− 1·9
Dumfries . . .	1,063	81,060	76·3	− 0·5
East Lothian . .	271	47,369	174·8	− 0·2
Fife	492	276,261	561·5	− 5·7
Inverness . . .	4,088	82,082	20·1	− 0·5
Kincardine . . .	383	39,864	104·1	− 4·6
Kinross . . .	82	7,454	90·9	− 6·4
Kirkcudbright . .	898	30,341	33·8	− 2·3
Lanark . . .	882	1,585,968	1798·1	+ 2·2
Midlothian . . .	362	526,277	1453·8	+ 3·9
Moray . . .	476	40,805	85·7	− 1·8
Nairn	164	8,294	50·6	− 5·6
Orkney . . .	376	22,075	58·7	− 8·4
Peebles . . .	355	15,050	42·4	− 1·8
Perth . . .	2,528	120,772	47·8	− 3·8
Renfrew . . .	245	288,575	1177·8	+ 0·7
Ross and Cromarty .	3,078	62,802	20·4	−11·3
Roxburgh . . .	665	45,787	68·9	+ 1·8
Selkirk . . .	257	22,608	87·9	0·0
Shetland or Zetland .	551	21,410	38·8	−16 1
Stirling . . .	447	166,447	372·3	+ 2·9
Sutherland . . .	2,028	16,100	7·9	− 9·6
West Lothian . .	120	81,426	678·6	− 3·0
Wigtown . . .	486	29,299	60·3	− 4·8
	29,785	4,842,554	162·3	− 0·8

SCOTTISH PARLIAMENTARY CONSTITUENCIES

Constituency.	ELECTORATE IN 1935.		
	Men.	Women.	Total.
ABERDEEN CITY—			
North	25,252	27,606	52,858
South	23,531	32,788	56,319
ABERDEEN ANE KINCARDINE—			
Central	18,892	21,092	39,984
Eastern	17,317	18,522	35,839
Kincardine and Western	13,978	15,550	29,528
ANGUS	14,703	16,678	31,381
ARGYLLSHIRE	20,486	23,230	43,716
AYR BURGHS	23,976	29,686	53,662
AYR AND BUTE—			
Bute and Northern	24,062	29,620	53,682
Kilmarnock	21,187	26,547	47,734
South	20,546	19,919	40,465
BANFFSHIRE	15,600	16,918	32,518
BERWICK AND HADDINGTON	22,684	24,763	47,447
CAITHNESS AND SUTHERLAND	13,383	14,410	27,793
DUNBARTONSHIRE	28,310	32,906	61,216
DUMBARTON BURGHS	19,965	19,779	39,744
DUMFRIESSHIRE	21,266	24,581	45,847
DUNDEE (2 seats)	48,380	64,081	112,461
DUNFERMLINE BURGHS	18,272	19,862	38,134
EDINBURGH—			
Central	16,688	19,513	36,201
East	20,989	24,008	44,997
North	20,777	26,009	46,786
South	17,973	30,311	48,284
West	26,716	34,687	61,403
FIFE—			
East	21,043	26,764	47,807
West	23,791	22,490	46,281
GALLOWAY	19,619	22,460	42,079
GLASGOW—			
Bridgeton	18,700	18,837	37,537
Camlachie	20,204	22,543	42,747
Cathcart	20,309	25,863	46,172
Central	22,652	21,852	44,504
Gorbals	22,785	23,292	46,077
Carried Forward	684,036	797,167	1,481,203

Constituency.	ELECTORATE IN 1935.		
	Men.	Women.	Total.
Brought Forward	684,036	797,167	1,481,203
GLASGOW—(*continued*)			
Govan .	20,419	21,025	41,444
Hillhead	14,518	22,254	36,772
Kelvingrove	19,844	29,993	49,837
Maryhill	25,446	28,371	53,817
Partick .	17,328	19,671	36,999
Pollock .	18,751	25,597	44,348
St. Rollox	19,677	20,309	39,986
Shettleston	22,476	24,220	46,696
Springburn	21,953	23,237	45,190
Tradeston	18,655	18,927	37,582
GREENOCK .	22,294	23,374	45,668
INVERNESS AND ROSS AND CROMARTY—			
Inverness	21,910	24,020	45,930
Ross and Cromarty	13,302	14,452	27,754
Western Isles	13,546	14,642	28,188
KIRKCALDY BURGHS	20,620	22,612	43,232
LANARKSHIRE—			
Bothwell	22,367	21,713	44,080
Coatbridge	19,920	19,102	39,022
Hamilton	17,674	17,367	35,041
Lanark .	19,728	21,469	41,197
Motherwell	19,549	18,771	38,320
Northern	29,880	29,427	59,307
Rutherglen	24,483	26,580	51,063
LEITH .	23,606	26,327	49,933
LINLITHGOWSHIRE	24,870	22,907	47,777
MIDLOTHIAN AND PEEBLES—			
Northern	23,532	27,155	50,687
Peebles and Southern	16,754	17,782	34,536
MONTROSE BURGHS	13,109	17,832	30,941
MORAY AND NAIRN	13,960	16,510	30,470
ORKNEY AND SHETLAND	14,501	15,979	30,480
PAISLEY	25,217	30,236	55,453
PERTH AND KINROSS—			
Kinross and Western	15,541	18,691	34,232
Perth	21,868	26,939	48,807
RENFREWSHIRE—			
East	38,177	45,024	83,201
West	18,400	20,996	39,396
ROXBURGH AND SELKIRK	20,636	26,381	47,017
SCOTTISH UNIVERSITIES (3 Seats)	38,231	14,750	52,981
STIRLING AND CLACKMANNAN—			
Clackmannan and East	22,553	23,650	46,203
West	17,476	17,860	35,336
STIRLING AND FALKIRK BURGHS	21,789	22,913	44,702
	1,498,596	1,676,232	3,174,828

THE SCOTTISH PEERAGE

Scottish Representative Peers are marked with a †.

Those marked * sit in the House of Lords either as British Peers or Scottish Representative Peers. The others are not members of the House of Peers.

*Abercorn, Earl (and 3rd Duke) of, James Albert Edward Hamilton, K.G. Seats, Duddingston House, Midlothian ; Baron's Court, Newtown Stewart, Co. Tyrone, Ireland.—Government House, Hillsborough, Co. Down.—Residence, 68 Mount Street, London, W.1.—Club, Carlton.

*Aberdeen and Temair, 8th Earl (2nd Marquess) of, George G. Gordon, O.B.E., Deputy - Lieutenant for Aberdeen, Haddo House, Aberdeenshire and 16 Westbourne Street, W.2.

Ailsa, Marquess of (Earl Cassillis), Archibald Kennedy, Lord-Lieutenant of Ayrshire. Seats, Culzean Castle, Maybole.—London residence, 65 Lancaster Gate, W.2.

†Airlie, 11th Earl of, David Lyulph Gore Wolseley Ogilvy, Cortachy Castle, Kirriemuir. — London residence, 48 Bryanston Square, W.1.

Arbuthnott, 14th Viscount, John Ogilvy, Lord-Lieutenant of Kincardineshire, Arbuthnott House, Fordoun, Kincardineshire.

*Argyll, 10th Duke of, Niall Diarmaid Campbell, Hereditary Master of the Household in Scotland, Lord-Lieutenant of Argyllshire, Inveraray Castle, Inveraray, Argyllshire.—2 Observatory Gardens, Campden Hill, London, W.8.

*Atholl, 8th Duke of, John George Stewart Murray, K.T., P.C., G.C.V.O., C.B., D.S.O., Lord-Lieutenant of Perthshire. Seats, Blair Castle, Blair-Atholl, and Dunkeld House, Dunkeld.—98 Elm Park Gardens. London, S.W.10.

†Balfour of Burleigh, 7th Baron, George John Gordon Bruce, a Deputy-Lieutenant of Clackmannanshire, Kennet House, Alloa, Clackmannanshire.—31 Bedford Gardens, London, W.8.

Balfour, 2nd Earl of, Gerald William Balfour, Fishers' Hill, Woking, Surrey.—Clubs, Carlton and Athenæum.

†Belhaven and Stenton, 11th Baron, Lieut.-Col. Robert Edward Archibald Hamilton, C.I.E., a Deputy-Lieutenant of Lanarkshire, Wishaw House, Wishaw.—41 Lennox Gardens, London, S.W.1.—Clubs, Carlton and U.S.

†Breadalbane and Holland, Earl of, Sir Charles William Campbell, M.C., D.L., J.P., West Monkton, Taunton.

*Buccleuch and Queensberry, Duke of, Walter John Montagu - Douglas - Scott. Seats, Dalkeith House, Midlothian ; Bowhill, Selkirkshire ; Eildon Hall, St. Boswells, Roxburghshire ; Drumlanrig Castle, Thornhill ; and the Lodge, Langholm.—London residence, 2 Grosvenor Place, S.W.1.—Clubs, Carlton, Guards, and New, Edinburgh.

Buchan, 15th Earl of, Ronald Douglas Stuart Mar Erskine, Amondell House, Mid-Calder.—Club, Bachelors'.

Bute, Marquess of. See Dumfries, Earl of.

†Caithness, 18th Earl of, Norman Macleod Buchan, C.B.E., Auchmacoy, Aberdeenshire.

Carnwath, 13th Earl of, Sir Ronald Arthur Dalzell, 119 Gloucester Terrace, Hyde Park, London, W.2.—Club, Carlton, London.

Cassillis, Earl. See Marquess of Ailsa.

Cathcart, 6th Baron (Earl), Alan Cathcart, 39 Bryanston Square, London, W.1.—Clubs, Bachelors' and Carlton.

*Colville of Culross, Baron (Viscount), Charles Alexander Colville.—Residence, Point Colville, near Victoria, British Columbia ; Danegate House, Eridge Green, Sussex.

*Crawford, 26th Earl of (Premier Earl of Scotland), David Alexander Edward Lindsay, P.C., K.T. Seats, Balcarres, Colinsburgh, Fife ; Haigh Hall, Wigan, Lancashire.—7 Audley Square, London, W.1.—Clubs, Carlton and Burlington Fine Arts.

Dalhousie, 15th Earl of, John Gilbert Maule Ramsay. Seats, Dalhousie Castle, Midlothian ; Brechin Castle and Panmure House, Angus.

Dingwall, Baroness, Nan Ino Cooper (Baroness Lucas). Struy Lodge, Beauly ; The Hall, Horsey, Norfolk.—22 Sussex Square, London, W.2.

Dumfries, Earl of, and Bute (Marquess of Bute), John Crichton-Stuart, K.T. Seats, Mount Stuart, Isle of Bute ; Dumfries House, Cumnock, Ayrshire ; The Old Place of Mochrum, Wigtownshire ; Cardiff Castle, Glamorganshire ; Edinburgh, 5 Charlotte Square. — London, 26 Queen Anne's Gate, S.W.1.

*Dunblane, Viscount (Duke of Leeds), John Francis Godolphin Osborne, Hornby Castle, Bedale, Yorkshire.

Dundonald, 13th Earl of, Thomas Hesketh Douglas Blair Cochrane, Lochnell Castle, Argyle ; 101 Mount Street, London, W. ; and Gwrych Castle, Abergele, North Wales.—Clubs, Turf and Guards.

*Dunmore, Earl of, Alexander Edward Murray, V.C., D.S.O., M.V.O., a Deputy-Lieutenant of Inverness-shire. Seats, Borve Lodge, South Harris, Inverness-shire ; Breton Hill, Weybridge.—17 Cleveland Gardens, W.2.

Dysart, Countess of, Wenefryde Agatha Tollemache, Greaves.

*Eglinton and Winton, 16th Earl of, Archibald Seton Montgomerie. Seat, Eglinton Castle, Irvine.—70 Cadogan Square, S.W.1.

*Elgin and Kincardine, 10th Earl of, Sir Edward James Bruce, Bt., C.M.G., Deputy-Lieutenant for Fifeshire. Seat, Broomhall, Dunfermline, Fifeshire.

*Elibank, 2nd Baron (Viscount), Charles Gideon Murray, Lord Lieutenant of Peeblesshire. Seats, Darn Hall, Eddleston, Peeblesshire ; Elibank Cottage, Selkirkshire.—238 St. James's Court, Buckingham Gate, S.W.1.

*Elphinstone, 16th Baron, Sidney Herbert Elphinstone, K.T., D.L., J.P., Carberry Tower, Musselburgh, Midlothian ; Maryland, Worplesdon, Surrey.—6 Upper Brook Street, W.1.—Carlton and Marlborough Clubs, London, S.W.

*Erroll, 22nd Earl of, Josslyn Victor Hay, Hereditary Lord High Constable of Scotland, Walls, Ravenglass, Cumberland.—Club, Marlborough.

†Fairfax, 12th Baron, Albert Kirby Fairfax, 10 Stanhope Street, Hyde Park, W.2.—Clubs Carlton, Bachelors', Brooks, London, and New, Edinburgh.

†Falkland, 13th Viscount, Lucius, Plantagenet Cary, O.B.E., Carlton Club (Premier Viscount of Scotland).

Forbes, 22nd Baron (Premier Baron of Scotland), Atholl Laurence Cunyngham, a Deputy-Lieutenant of Aberdeenshire. Seats, Castle Forbes, Keig, Aberdeen ; Brux Lodge, Kildrummy, Aberdeenshire.

*Forrester, 4th Baron (Earl of Verulam), James Grimston. Seat, Gorhambury, St. Albans, Herts.

*Galloway, 12th Earl of, Randolph Algernon Ronald Stewart. Seat, Cumloden, Newton-Stewart.—Clubs, Army, Navy, and Bachelors'.

*Glasgow, 8th Earl of, Patrick James Boyle, D.S.O., Kelburn Castle, Fairlie, Ayrshire.

Gray, Baroness, Ethel Eveleen Gray Campbell ; Brownswood, Enniscorthy, Co. Wexford, Ireland. Residence, 73 Pont Street, London, S.W.1.

†Haddington, 12th Earl of, George Baillie-Hamilton, M.C. Seat, Tyninghame House, East Lothian.

*Hamilton and Brandon, 13th Duke of, Alfred Douglas Douglas-Hamilton (Premier Peer of Scotland), a Deputy-Lieutenant of Lanarkshire. Seats, Dungavel, Strathaven ; Ferne House, Shaftesbury, Dorset.

Herries, Baroness, Gwendolen (Duchess of Norfolk), Mary Constable-Maxwell. Seats, Everingham Park, Yorkshire; Kinharvie,

Dumfries ; Arundell Castle, Sussex ; Norfolk House, St. James's Square, London, S.W.1.

*Home, 13th Earl of, Charles Cospatrick-Archibald Douglas-Home, Lord-Lieutenant of Lanarkshire and a Deputy-Lieutenant of the County of Berwick and of the County of the City of Glasgow. Seats, The Hirsel, Coldstream, Berwickshire ; Douglas Castle and Bothwell Castle, Lanarkshire.—Clubs, Travellers' and Carlton, London.

Hopetoun, Earl of (Marquess of Linlithgow), Victor Alexander John Hope, K.T., O.B.E., D.L. Seats, Hopetoun House, South Queensferry ; Ormiston Hall, East Lothian.

Huntley, 11th Marquess of, Charles Gordon (Premier Marquess of Scotland), P.C., J.P., a Deputy-Lieutenant of Aberdeenshire. Seats, Aboyne Castle, Aberdeenshire ; Orton Longueville, Peterborough.—24 Grosvenor Square, W.1.—Clubs, Royal Societies, London.

†Kellie, Earl of. See Mar and Kellie.

*Kincardine, Earl of. See Elgin and Kincardine, Earl of.

Kinloss, Baroness, Mary Morgan-Grenville, C.I.—Moreton Lodge, Maids Moreton, Buckingham.

*Kinnaird, 12th Baron, Kenneth FitzGerald, a Deputy-Lieutenant of Perthshire. Seat, Rossie Priory, Inchture, Perthshire.

*Kinnoull, 14th Earl of, George Harley Hay, Balhousie Castle, Perthshire.

*Kintore, 10th Earl of, Arthur George Keith-Falconer, K.T., P.C., G.C.M.G. Seats, Keith Hall, Inverurie, Aberdeenshire ; 105 Mount Street, London, W.1.—Clubs, Carlton, Marlborough, Brooks' and Turf, London.

Lauderdale, 15th Earl of, Sir Ian Colin Maitland. Seats, Thirlestane Castle, Lauder, Berwickshire.—Clubs, Bachelors', London ; and New, Edinburgh.

*Leeds, Duke of. See Dunblane, Viscount of.

*Lennox, Duke of, Frederick Charles Gordon-Lennox (Duke of Richmond and Gordon). Seat, Gordon Castle, Fochaber, Morayshire.—Goodwood, Chichester.

†Leven and Melville, Earl of, Archibald Alexander Leslie-Melville. Seat, Glenferness, Nairn.—1 Sussex Square, London, W.2.

†Lindsay, Earl of, Reginald Lindesay-Bethune. Seats, Kilconquhar House and Wormistone House, Fife.—25 Berkeley Square, London, W.1.

*Linlithgow, Marquess of. See Hopetoun, Earl of.

*Lothian, 11th Marquess of, Philip-Henry Kerr, C.H. Seats, Newbattle Abbey, Dalkeith, Midlothian ; Mounteviot, Ancrum, Roxburghshire ; and Blickling Hall, Aylsham, Norfolk.—Club, Travellers'.

*Loudoun, 12th Countess of, Edith Maud Abney Hastings. Seats, Loudoun Castle, Galston ; Manor House, Ashby-de-la-Zouche, Leicestershire.

*Lovat, 15th Baron, Simon Christopher Joseph Fraser. Seat, Beaufort Castle, Beauly, Inverness-shire.

Mansfield, Earl of. See Stormount, Viscount.

Mar, 35th Earl of, Lionel Walter Young (Scottish Premier Earldom), 12 Churchfield Road, Ealing, London, W.1.

†Mar and Kellie, Earl of, Walter John Francis Erskine, K.T. (Premier Viscount of Scotland), Convener and Lord-Lieutenant of Clackmannanshire, Alloa House, Alloa; Kellie Castle, Fife.—Marlborough and Carlton Clubs, London, S.W.; New Club, Edinburgh.

*March, Earl of. See Wemyss and March, Earl of.

*Melville, Earl of. See Leven and Melville, Earl of.

Midlothian, Earl of. See Rosebery, Earl of.

*Montrose, 6th Duke of, James Graham, C.V.O., C.B., Lord-Lieutenant of Bute, Deputy Lieutenant of Stirlingshire. Seats, Buchanan Castle, Drymen, Stirlingshire; Brodick Castle, Isle of Arran.

*Moray, 18th Earl of, Francis Douglas Stuart, M.C. Seats, Doune Lodge, Perthshire; Darnaway Castle, Forres; Castle-Stuart, Inverness-shire; Donibristle, Fife; and Kinfauns Castle, Perthshire.

†Morton, 21st Earl of, Sholto Charles John Hay Douglas. Seats, Conaglen, Ardgour; Dalmahoy, Ratho, Midlothian; Aberdour Castle, Fife.

*Nairne, Baron, Henry William Edmond Petty-Fitzmaurice (Marquess of Lansdowne), D.S.O., M.V.O. Seats, Bowood Park, Caine, Wilts; Derreen, Kenmore, Co. Kerry.—20 Mansfield Street, W.1.

*Napier and Ettrick, Baron, Sir Francis Edward Basil Napier. Seat, Thirlestane Castle, Selkirk.

Newburgh, Earl of, Carlo Giustiniani-Bandini.— 1 Via Virginio, Orsini, Rome.

Northesk, 11th Earl of, David Ludovic George Hopetoun Carnegie, a Deputy-Lieutenant for Angus. Seat, Ethie Castle, Arbroath.—6 Hans Crescent, S.W.1.

Orkney, 7th Earl of, Edmond Walter Fitz-maurice. Seats, Glanmore, Templemore; Tythe House, Stewkley, Bucks.

Perth, 15th Earl of, William Huntly Drummond, Machany, Perthshire.—Scottish Club, London.

Polwarth, 9th Baron, Walter George Hepburne-Scott, C.B.E., V.D., Lord-Lieutenant of Selkirkshire, and a Deputy-Lieutenant and Convener of Roxburghshire. Seat, Humbie House, East Lothian.

*Queensberry, Duke of. See Buccleuch and Queensberry, Duke of.

†Queensberry, 10th Marquis of, Francis Archibald Kelhead Douglas, 67 Tufton Street, S.W.1.

*Reay, 13th Baron, Æneas Alexander Mackay. Seat, Arnheim, Holland.

Richmond and Gordon, Duke of. See Lennox, Duke of.

*Rollo, 11th Baron, William Charles Wordsworth Rollo, C.B. Seat, Duncrub Park, Dunning, Perthshire.

*Rosebery, Earl of (Earl of Midlothian), Albert Edward Primrose, H.M.A., D.S.O., M.C., Lord-Lieutenant of Midlothian and of Linlithgowshire; Durdans, Epsom. Seat, Dalmeny Park, Linlithgowshire.—38 Berkeley Square, London, W.1.

Rothes, 20th Earl of, Malcolm George Dyer Edwardes Leslie. Seat, Leslie House, Leslie, Fife.—Junior Carlton Club, London, S.W.

*Roxburghe, 9th Duke of, Sir George Victor Robert John Innes-Kerr. Seat, Floors Castle, Kelso; Broxmouth Park, East Lothian.— 2 Carlton House Terrace, London, S.W.1.

Ruthven, 9th Baron, Walter Patrick Hore-Ruthven, C.B., C.M.G., D.S.O., a Deputy-Lieutenant of Perthshire.—Government House, Guernsey.

†Saltoun, 19th Baron, Alexander Arthur Fraser, M.C., Aberdeenshire. Seat, Philorth, Fraserburgh, Aberdeenshire. Cross Deep, Twickenham, 1.—Turf and Pratt's, London; and New, Edinburgh.

Seafield, 13th Countess of, Nina-Caroline Ogilvie Grant of Grant. Seats, Cullen House, Cullen; Castle Grant, Grantown, Morayshire; Balmacan, Inverness-shire.

†Sempill, 19th Baron, William Francis Sempill, Seats, Fintray House and Cragievar Castle, Aberdeenshire.

†Sinclair, 16th Baron, Archibald James Murray Saint-Clair, M.V.O., a Deputy-Lieutenant of Berwickshire. Seats, Nisbet House, Duns; Herdmanston, Haddington; and Milton Park, Dalry, Kirkcudbrightshire.

*Southesk, 10th Earl of, Sir Charles Noel Carnegie, a Deputy-Lieutenant of the Counties of Aberdeen, Kincardineshire, and Forfar. Seats Kinnaird Castle, Brechin; Crimonmogate, Lonmay, Aberdeenshire; and Elsick, Kincardineshire.—Clubs, Carlton, Bachelors', and Royal Yacht Squadron, London.

*Stair, 12th Earl of, John James Dalrymple, D.S.O., Lord -Lieutenant of Wigtown. Seats, Lochinch, Wigtownshire; Oxenfoord Castle, Midlothian.—17 Eaton Square, London, S.W.1

*Stormont, 11th Viscount, Mungo David Malcolm Murray (Earl of Mansfield), B.A. Seat, Scone Palace, Perth.

Strathallan, Viscount. See Perth, Earl of.

*Strathmore and Kinghorn, Earl of, Claude George Bowes-Lyon, K.T., G.C.V.O., Lord-Lieutenant of Angus, a Deputy-Lieutenant of County of the City of Dundee. Seats, Glamis Castle, Forfar; Streatham Castle, Darlington.—84 Eaton Square, S.W.1.

*Sutherland, 5th Duke of, George Granville Sutherland-Leveson-Gower, Lord Lieutenant of Sutherlandshire. Seat, Dunrobin Castle, Golspie.

Torphichen, 13th Baron, John Gordon Sandilands, Calder House, Mid-Calder, Midlothian.

Tweeddale, 11th Marquis of, William George Montagu Hay, Yester House and Hope's House, Gifford, East Lothian.—New Club, Edinburgh.

Verulam, Earl of. See Forrester, Baron.

Wemyss and March, 11th Earl of, Hugo Richard Wemyss-Charteris-Douglas, a Lord-Lieutenant of the County of East Lothian. Seats, Gosford House, Seton, and Amisfield, East Lothian; Elcho Castle, Perthshire; Neidpath Castle, Barns, Peeblesshire; Stanway, Gloucestershire.—3 Wyndham Place, W.1.

FERRIES

ALTHOUGH Scotland is on the whole a mountainous country, it has a high reputation for good roads. These naturally tend to follow the river valleys, particularly in the more highland areas, and travellers by road may obtain a wide variety of scenery in a very short distance. Close to the shores of many of the sea lochs in the west run first-class roads, and although the route between given points may be somewhat circuitous, the scenic beauty is ample compensation for any loss of time involved. Naturally, on such a coast-line, ferries play an important part, and below we give a list of the principal ones, together with the scale of charges.

FERRIES (Courtesy of Automobile Association)

1. Erskine to Old Kilpatrick, across river Clyde.

Daily steamboat service, 6 a.m. to 10 p.m.; Saturdays, 11 p.m.; Sundays, April to September, 10 a.m. to 11 p.m.; October to March, 10 a.m. to 10 p.m. Easy embarkation.

Charges.—Motor car, 9d.; motor cycle, 3d.; motor cycle and side-car, 6d.; motor caravan, 10d.; trailer caravan, 6d.; small luggage trailer, 6d. Driver's fare included in above. Additional passengers, 1d. each.

2. Renfrew to Yoker, across river Clyde.

Steamboat service continuously day and night. Double rates midnight to 5 a.m. Easy embarkation.

Charges.—Motor car and occupants, 6d.; tri-car, 3d.; motor cycle, 2d.; motor cycle and side-car, 3d.; motor caravan, 8d.; trailer caravan, 6d.; luggage trailer. 4d. Driver's fare included in above. Additional passengers, ½d. each.

5. Govan to Partick, across river Clyde in Glasgow.

Continuous service on weekdays from Monday to Saturday inclusive, 6 a.m. to 10 p.m. No Sunday service. Free ferry.

4. Connel Bridge, across Loch Etive.

L.M.S. railway bridge, with track for cars alongside line. Open daily, including Sundays, from 7 a.m. to 10 p.m. The bridge will be opened specially between the closing and opening hours on previous notice being given to the Stationmaster at Connel Ferry, stating date and time of proposed crossing and in which direction. In addition to the usual charges an extra fee of 2/- between 10 p.m. and opening time will be payable for this service.

Charges.—Motor car, 10/-; two-seater, 7/6; tri-car, 7/6; motor cycle, 2/-; motor cycle and side-car, 4/-; motor caravan, 10/-; trailer caravan, 10/-; luggage trailer, 5/-; passengers, 2d.

Commercial vehicles and charabancs are not allowed to cross the bridge. Hackney carriages and hiring cars of a seating capacity not exceeding five (including driver) may use the bridge at the discretion of the Company.

5. Ballachulish, across Loch Leven.

Charges.—Motor cars, under 9 h.p., 2/6; 9 h.p. to 16 h.p., 3/6; 16 h.p. and over, 5/-. Buses and charabancs capable of seating 8 persons and driver, 10/-; motor lorries and vans: 2 tons, 5/-; 3 tons, 7/6; 3 to 5 tons (limit), 10/-; cycles, 6d.; motor cycle, 1/-; motor cycle and side-car, 2/-; passengers, 4d.; when two passengers and over, including driver of a car, 2d. each.

6. Dornie to Ardelve Hotel, across Long Long.

Continuous service from 8 a.m. to 8 p.m., Sundays and Bank Holidays included. Weather and light permitting, cars will be ferried after 8 p.m. if necessary. Wire if after dark. Delay of up to two hours at low spring tides. Care necessary on approach from south.

Charges.—Motor car, 6/-; motor cycle, 1/-; motor cycle and side-car, 1/6; tri-car, 5/-; motor caravan, 6/-; trailer caravan, 5/-; luggage trailer, 3/-; maximum weight, 2 tons 15 cwt. Overall length of vehicle, 18 ft.; width, 6 ft. 6 in. Passengers, 2d. Fare and a half after 10 p.m.

7. Kyle of Lochalsh to Kyleakin, across Kyleakin to Isle of Skye.

Motor ferry boat (during summer months an extra motor ferry boat is in service). Continuous service during daylight for 3 hours before and 3 hours after spring (high) tides; 4 to 4½ hours after neap (high) tides. Fairly easy embarkation. Advisable to wire in advance to " Ferry Lessee, Kyle of Lochalsh." No Sunday service.

Charges.—Motor car up to 12 h.p., 8/-; over 12 h.p., 12/6; tri-car, 8/-; motor cycle, 2/6; motor cycle and side-car, 5/-; motor caravan up to 12 h.p., 8/-; over 12 h.p., 12/6; trailer caravan, 8/-; luggage trailer, 4/-; maximum weight, 3 tons 10 cwt. Passengers, 6d. each.

8. Strome (Ross-shire), across Loch Carron.

Motor ferry. Continuous service during daylight. Occasional delay of 1 hour before and 1 hour after very low spring tide. Care necessary on approach from south. No Sunday service.

Charges.—Motor car, 10/-; tri-car, 7/-; motor cycle, 2/-; motor cycle and side-car, 3/-; motor caravan, 10/-; trailer caravan, 5/-; luggage trailer, 5/-; maximum weight, 3 tons; maximum wheel base, 14 ft.; overall length, 18 ft.

9. Kylesku (Sutherland).

Motor ferry boat. Continuous service during daylight with slight delay at certain tides. Service from about beginning of May to middle of October. Sundays included.

Charges.—Motor car up to 10 h.p., 6/- (return 9/-); from 10 h.p. to 20 h.p., 8/- (return 12/-); exceeding 20 h.p., 10/- (return 15/-); motor cycle, 2/6 (return 5/-); motor cycle and sidecar, 3/6 (return 7/-); trailers, 6/- (return 12/-).

10. Kessock (Inverness), across Beauly Firth.

Steamboat carrying one large car only. Summer: from April to June inclusive, also September and October. Weekdays 7.45 a.m. and hourly to 7.45 p.m. Sundays 9.45 a.m. and hourly to 5.45 p.m. 1st July to 31st August, weekdays 7.45 a.m. and hourly to 9.45 p.m. Sundays 9.45 a.m. and hourly to 9.45 p.m. Winter: from October to April. Weekdays 7.45 a.m. and hourly to 5.45 p.m. Sundays 9.45 a.m., 1.45, 4.45 p.m. (weather and tide permitting). Steamer leaves north side at the hour. Embarkation rather awkward.

Charges.—Motor car, two-seaters, 2/6; four-seaters, 4/-; large cars, 4/6; tri-car, 1/6; motor cycles, 9d.; motor cycle and side-car, 1/3; motor caravan, 4/- to 6/-; trailer caravan, 1/6 to 2/-; luggage trailer, 1/6; passengers, 3d.

11. Dundee to Newport, across Firth of Tay.

Steamboat service. Weather permitting, hours of sailing:
From Dundee.—Weekdays: 1st May to 30th September, 7 a.m., 8 a.m., thereafter half-hourly till 9.30 p.m., then 10.15 p.m. Sundays: hourly 9 a.m. to 9 p.m., then 10.15 p.m.

From Newport.—Weekdays: 1st May to 30th September, 7.30 a.m., 8.30 a.m. thereafter half-hourly till 9.30 p.m., then 10.15 and 10.45 p.m. Sundays: hourly 9.30 a.m. till 9.30 p.m., then 10.45 p.m.

Services slightly curtailed during winter months. Embarkation easy.

Charges.—Motor cars, 4 seats, 3/6 ; 2 seats, 3/- ; tri-car, 2/6 ; motor cycle, 8d. ; motor cycle and side-car, 1/4 ; charabancs (maximum 20 ft.), 7/- ; charabancs (maximum, 30 ft.), 10/- ; caravans (12 ft.), 6/6 ; passengers, 4d.

12. Granton to Burntisland, across Firth of Forth.

Steamboat service.

From Granton.—Weekdays: frequent service from 8 a.m. to 5.15 p.m. (April to September inclusive) and 6.55 p.m. (June to September inclusive). Sundays: 9 a.m., 1.0, 3.10, 5.5 p.m.

From Burntisland.—Weekdays: frequent service 8.45 a.m. to 7.45 p.m. (June to September inclusive). Sundays: 9.45 a.m., 2.25, 4.0, 6.0 p.m.

Charges.—Motor car, 15/- ; 2-seater, 10/- ; tri-car, 7/6 ; motor cycle, 2/6 ; motor cycle and side-car, 5/- ; motor caravan, 10/- per ton (minimum 10/-) ; trailer caravan, 7/6 per ton or fraction (minimum charge, 15/-) ; luggage trailer, 5/-.

13. Queensferry (North and South), across Firth of Forth.

Steamboat service.

From North Queensferry every half-hour commencing at 7.30 a.m. till 11 p.m., both weekdays and Sundays.

From South Queensferry every half-hour commencing at 8 a.m. till 11.30 p.m., both weekdays and Sundays.

	Single.	Return.
Charges.—Motor cycle (excluding driver) . . .	1/2	2/–
Motor cycle and side car (excluding driver) .	2/2	3/8
Cars up to 8 h.p. (including driver) . .	3/–	5/–
8 h.p. to 13 h.p. (including driver) .	3/9	6/–
13 h.p. to 18 h.p. (,,) .	4/6	7/6
18 h.p. to 29 h.p. (,,) .	5/6	9/–
over 29 h.p. (,,) .	6/6	11/–
Trailers per foot of length . . .	9d.	1/4
Passenger	4d.	7d.

14. Alloa and South Alloa, across river Forth.

Ferry leaves Alloa every hour from 7 a.m. ; from South Alloa 20 minutes after each hour. From April to September last ferry from Alloa at 7 p.m. South Alloa, 7.30 p.m. From November to February, last ferry from Alloa at 5 p.m. South Alloa, 5.30 p.m. During March and October last ferry from Alloa at 6 p.m. South Alloa, 6.30 p.m.

Sundays: April to September, ferry leaves Alloa hourly, 8 a.m. to 7 p.m. ; South Alloa, 8.20 a.m. to 7.30 p.m. March and October ferry leaves Alloa hourly, 8 a.m. to 6 p.m. South Alloa, 8.20 a.m. to 6.30 p.m. November to February ferry leaves Alloa hourly, 8 a.m. to 5 p.m. ; South Alloa, 8.20 a.m. to 5.30 p.m.

Charges.—Motor car up to 10 h.p., 1/6 ; over 10 h.p., 2/- ; tri-car, 1/6 ; motor cycle, 8d. ; motor cycle and sidecar, 11d. ; motor caravan, 2/- ; trailer caravan, 1/6 ; luggage trailer, 6d. ; passengers, 3d.

15. Mallaig to Armadale (Skye).

Starting from Mallaig at 8 a.m., continuous service either way till a late hour at night. Crossing takes 40 minutes No Sunday crossing. Motorists advised to wire approximate hour of arrival to "Isles Ferry, Mallaig."

Charges.—Cars up to 14 h.p., £1 (single), 35/- (return) ; over 14 h.p., 25/- and £2 motor cycle, 5/- ; motor cycle with sidecar, 7/6 ; passengers, 1/6 each.

SHIPPING INFORMATION

COASTAL PASSENGER SERVICE

London & Edinburgh Shipping Co.—
  8 Commercial Street, Leith, and 126 Hope Street, Glasgow.
  Regular Passenger Service :
    Leith to London—Mondays, Wednesdays, and Saturdays.
    London to Leith—Tuesdays, Thursdays, and Saturdays.
    The Company has combined steamer and motor coach tours.

The Dundee, Perth & London Shipping Co.—
  62 Robertson Street Glasgow ; East Dock Street, Dundee ; and 18 Mincing
  Lane, London.
  Regular Passenger Service :
    Dundee to London—Wednesdays and Saturdays.
    London to Dundee—Wednesdays and Saturdays.
    Particulars of Tours in Company's Booklet.

The Aberdeen, Newcastle & Hull Steamship Co. Ltd.—
  Humber Dock, Hull ; Quayside, Newcastle ; Regent and Waterloo Quays,
  Aberdeen.
  June to September : Hull to Newcastle and Aberdeen, Wednesdays.
                     Newcastle to Aberdeen, Thursdays.
                     Aberdeen to Newcastle and Hull, Saturdays.
                     Newcastle to Hull, Mondays.
  Particulars of Tours in Company's booklet.

The Aberdeen Steam Navigation Co. Ltd.—
  Waterloo Quay, Aberdeen Wharf, Limehouse, London, E.
  Operates a service between Aberdeen and London.

Clyde Shipping Co. Ltd.—
  78 Carlton Place, Glasgow.
  Every Monday, *via* Belfast for Waterford, Plymouth, and London.
    „ Tuesday, *via* Belfast for Southampton and London.
    „ Thursday, *via* Dublin for Waterford and Cork.
    „ Friday, *via* Belfast for London.

Messrs M'Callum, Orme & Co. Ltd.—
  45 Union Street, Glasgow, C.1.
  Circular Tours to Western Isles twice every ten days.
  Special Cruises to St. Kilda and round Isle of Skye.
  The regular service to Skye is operated by the L.N.E.R.

North of Scotland, Orkney & Shetland Steam Navigation Co. Ltd.—
  St. Matthew's Quay, Aberdeen.
  Steamers leave Leith at frequent intervals for Stromness, Scalloway, Walls, Aith,
  Hillswick, Lerwick, Wick, Thurso, Kirkwall.
  Direct sailings between Leith, Aberdeen, and Lerwick.

Williamson-Buchanan Steamers Ltd.—

308 Clyde Street, Glasgow, C.1.

May to September.—Daily sailings between Glasgow (Broomielaw) and Dunoon, Rothesay, Largs, etc.

Daily sailings between Greenock and Gourock ; to Crarae and Inveraray *via* Kyles of Bute ; Gourock and Campbeltown *via* Fairlie and Lochranza (Arran).

Daily excursions to Lochgoilhead, Arrochar, etc.

Campbeltown & Glasgow Steam Packet Co.—

Old Quay Head, Campbeltown, and 5 Bridge Wharf, Broomielaw, Glasgow.

Daily sailings between Glasgow and Campbeltown.

David M'Brayne Ltd.—

44 Robertson Street, Glasgow.

All Year Round Services.—Daily except Sunday : Glasgow, Loch Fyne, Islay, Oban ; Tarbert and Islay ; Lochgoilhead service ; Oban, Sound of Mull, and Tobermory ; Oban and Lismore ; Oban, Ballachulish, and Fort-William ; Mallaig, Kyle of Lochalsh, and Portree ; Mallaig, Kyle of Lochalsh, and Stornoway.

Tuesdays, Thursdays, and Saturdays : Fort-William, Inverness, and intermediate ports.

Mondays, Wednesdays, and Fridays : Oban, Castlebay, and Lochboisdale ; Kyle of Lochalsh ; Harris, Lochmaddy, Lochboisdale, and Mallaig ; Inverness, Loch Ness, Fort Augustus, and Bonavie.

Mid June to Mid September.—Oban, Staffa, and Iona.

Burns & Laird Lines Ltd.—

52 Robertson Street, Glasgow.

Belfast—Daylight Service from Ardrossan.
Belfast—Night Mail Service from Glasgow.
Regular sailings to Dublin and Londonderry from Greenock.

AIR LINES OPERATING IN SCOTLAND

(1) Inverness—Wick—Kirkwall.
(2) Aberdeen—Wick—Dingwall.
(Operated by Highland Airways Ltd.)

Booking Offices—

Inverness : Messrs. Macrae & Dick, 36 Academy Street.
Wick : Messrs. Robertson & Sons, Bridge Street.
Kirkwall : The " Orcadian " News Office.

(1) Aberdeen—Wick—Thurso—Kirkwall.
(2) Aberdeen—Edinburgh.
(3) Aberdeen—Hull—London
(Provisionally Suspended).
(Operated by Aberdeen Airways Ltd.)

Booking Offices—

Aberdeen : Airport, Dyce.   Caledonian Hotel, Union Terrace.
Thurso : Royal Hotel.
Kirkwall : Kirkwall Hotel.
Edinburgh : Rossleigh Motors Ltd., Shandwick Place.

Edinburgh—Newcastle-on-Tyne—Leeds—London.
Service Suspended.
(Operated by North-Eastern Airways Ltd.)

Booking Offices—

Edinburgh : Messrs. A. Watt & Sons, 1 South St. David Street.
Newcastle : Messrs. Ace Motors Ltd., Barras Bridge.
Leeds :   Wallace Arnold Tours, Corn Exchange.
London : North-Eastern Airways Ltd., Heston Airport, Hounslow.

Glasgow—Belfast—Liverpool—London.
(Operated by British Airways Ltd.)

Booking Offices—

Glasgow : Union Transit Co., 381 Argyle Street.
Belfast : Grand Central Hotel, 18 Royal Avenue.
Liverpool : Messrs. Crosthwaite & Co., Speke Airport.
London : Hillman's Airways Ltd., Essex Airport, Stapleford, Essex.

(1) Glasgow—Campbeltown—Islay.
(2) Glasgow—Skye.
(3) Glasgow—Isle of Man.
(Operated by Northern & Scottish Airways Ltd.)

Booking Offices—

Glasgow : Renfrew Airport.
Campbeltown : Campbeltown Airport.
Islay : Duich Aerodrome.

Glasgow—Belfast—Liverpool—Manchester—Birmingham—London.
(Operated by Railway Air Services.)

Booking Offices—

Glasgow : Central Station.
Belfast : York Road Station.
Liverpool : Lime Street Station.
Manchester : L.M.S. Office, 47 Piccadilly.
Birmingham : New Street Station.
London : Euston House, Seymour St., N.W. 1.

LIST OF OWNERS OF STATIONS GENERATING ELECTRIC POWER IN SCOTLAND

SCOTLAND'S industrial progress has always been largely based on coal and even in the generating of electricity it is still a very serious rival to water power, for in the last fifteen years or so there has been a large reduction in the cost of producing electric power from coal. There is no doubt that the water power resources exist. The question is one of relative costs.

In the subjoined table we give a list of stations producing electricity, distinguishing the hydro-electric ones, and indicating those which supply power to the grid of the Central Electricity Board. Many of the stations given are comparatively small, and the tendency is for those to be closed down and supplies taken in bulk either directly or indirectly from the Central Electricity Board.

NORTH SCOTLAND

Aberdeen Corporation.
Beauly Electric Supply Co. Ltd.
Buckie Corporation.
Duncan's Electricity Supply Corporation (Ballater, Ellon, and Kintore *).
Dunoon & District Electricity Supply Co. Ltd.
Elgin Electric Supply Co. Ltd.
Gordon-Richmond Estates Co. (Fochabers *).
Grampian Electricity Supply Co. (Kingussie, Rannoch * †, and Tummel * †).
Grantown-on-Spey Electricity Supply Co.
Inverness Corporation (Waterloo Place, Bucht,* and Inshes).

Kirkwall Corporation.
Lairg Electricity Supply Co. Ltd.
Lerwick Corporation.
Lochaber Power Co.*
L.N.E.R. (Mallaig).
Lossiemouth Corporation.
Oban Corporation.
Ross-shire Electricity Supply Co. Ltd. (Lochluichart *).
Rothesay Corporation.
Rothesay Tramways Co. Ltd.
Stornoway Electric Supply Co. Ltd.
Tobermory Corporation.*
Wick Corporation.

* Water Power Stations.   † Stations supplying electricity to the Grid.

CENTRAL SCOTLAND

Ayrshire Electricity Board (Kilmarnock†).
Clyde Valley Electric Power Co. (Clyde's Mill,† and Yoker †).
Crieff Electric Supply Co. Ltd.
Dundee Corporation (Carolina Port †).
Edinburgh Corporation (Portobello †).
Electric Supply Corporation Ltd. (St. Andrews).
Falkirk Corporation.

Glasgow Corporation (Dalmarnock,† Pinkston, and Govan Destructor Station).
Greenock Corporation.
Kirkcaldy Corporation.
Lanarkshire Hydro-Electric Power Co. (Corra Linn * †, and Stonebyres * †).
Scottish Central Electric Power Co. (Bonnybridge †).
Stirling Corporation.

* Water Power Stations.   † Grid selected Stations.

SOUTH SCOTLAND

Dumfries (Maxwelltown *).   Scottish Southern Electric Supply Co. Ltd. (Galashiels †).
Urban Electric Supply Co. Ltd. (Hawick).

* Water Power Station.   † Grid selected Station.

NOTE.—The Hydro-electric stations of the Galloway Water Power Company are also Grid Selected Stations, namely, Tongland, Glenlee, Kendoon, Carsfad, and Earlstoun. Tongland and Glenlee will begin operations this year, and the other three are now under construction.

BRITISH BROADCASTING CORPORATION

(Courtesy of B.B.C.)

The broadacsting service to Scotland is at present provided by the following transmitters :—

National Programme—			Wave-length Metres.	Frequency Kilocycles per sec.	Power Kilowatts.
Droitwich National	.	.	1500	200	150
Scottish National .	.	.	285.7	1050	50
Scottish Programme—					
Scottish Regional .	.	.	391.1	767	50
Composite Programme—					
Aberdeen .	.	.	.	233.5	1285	1

Situation of Transmitter :

Droitwich National	.	.	2° 6′ 30″ W. Long., 52° 17′ 40″ N. Lat.
Scottish Regional and National (Westerglen, Falkirk)	.	.	3° 48′ 55″ W. Long., 55° 58′ 17″ N. Lat.
Aberdeen	.	.	2° 06′ 00″ W. Long., 57° 09′ 00″ N. Lat.

With regard to the distribution of the National Programme, the greater part of Scotland derives this from the Droitwich transmitter, but the Scottish National provides a service of this programme at a relatively high level of field strength to the heavily populated areas round Edinburgh and Glasgow. This is particularly necessary in the case of Glasgow due to the prevalence of electrical interference, which is more noticeable on the longer wave-length of Droitwich.

LIST OF WIRELESS STATIONS IN SCOTLAND FOR COMMUNICATING WITH SHIPS AT SEA BY TELEGRAPHY AND TELEPHONY

(Courtesy of Marconi House)

Port Patrick Radio, W. Wigtownshire. | Wick Radio, Caithness.

LIST OF DIRECTION FINDING BEACONS

Butt of Lewis, Outer Hebrides.
Cumbrae, Firth of Clyde.
Isle of May, Firth of Forth.
Kinnaird Head, N.E. Aberdeenshire.
Mull of Kintyre, W. Argyll.

North Ronaldsay Lighthouse, Orkney Islands.

Sule Skerry Lighthouse, W. of Orkney Islands.

NO-LICENCE AND LIMITATION

AREAS IN SCOTLAND WITH NO LICENCE FOR THE SALE OF LIQUOR BY LOCAL OPTION

1. Burghs.

Kirkintilloch (Dumbartonshire).
Kilsyth (Stirlingshire).
Lerwick (Zetland).
Cullen (Banffshire).
Findochty (Banffshire).

Stromness (Orkney).
St. Monance (Fifeshire).
Stewarton (Ayrshire).
Wick (Caithness).

2. Wards of Larger Burghs.

Glasgow—
   Cathcart Ward.
   Pollokshields Ward.
   Whiteinch Ward.
Rutherglen—
   Crosshill Ward.

   Gallowflat Ward.
   Greenhill Ward.
   Stonelaw Ward.
Kilmarnock—
   Grange Ward.
*Greenock—*8th Ward.

3. Parish Areas.

Rathen (Aberdeenshire).
Fenwick (Ayrshire).
Kilmory (Buteshire).
Watten (Caithness).
Holm (Orkney).
Barvas (Ross and Cromarty).

Balfron (Stirlingshire).
Cathcart (Renfrewshire).
Kilmacolm (Renfrewshire).
Sandsting (Zetland).
Tingwall (Zetland).
Unst (Zetland.

Total : 30 No-Licence Areas.

AREAS IN SCOTLAND IN WHICH THE NUMBER OF LICENCES FOR THE SALE OF LIQUOR IS LIMITED BY LOCAL OPTION VOTE

1. Burghs.

*Peterhead (Aberdeenshire).
*Eyemouth (Berwickshire).
 Milngavie (Dumbartonshire).

Newport (Fifeshire).
Airdrie (Lanarkshire).

2. Wards of Larger Burghs.

Glasgow—
   Dennistoun Ward.
   Govanhill Ward.
   Kelvinside Ward.
   Langside Ward.
   North Kelvin Ward.

   Partick West Ward.
   *Pollokshaws Ward.
   *Hamilton—3rd Ward.
   *Motherwell and Wishaw—5th Ward.
   *Greenock—1st Ward.
   *Paisley—7th Ward.

3. Parishes.

 Kilbirnie (Ayrshire).
*Avondale (Lanarkshire).
 Erskine (Renfrewshire).

*Mearns (Renfrewshire).
 Killearn (Stirlingshire).
 Uphall (West Lothian).

Total : 22 Limitation Areas.

* Denotes areas in which Further Limitation Resolution is in force.

GOLF

THIS game is a favourite and democratic sport in Scotland, and has been played in the country from the middle of the fifteenth century. Although it has spread to all parts of the world, the headquarters of the game is still at St. Andrews and out of the multitude of courses in the country the following world-famous courses may be specially mentioned—St. Andrews, Muirfield, Prestwick, Troon, Turnberry, Gleneagles, North Berwick.

SCOTTISH GOLF COURSES

	Holes
Aberdeen—	
Municipal Courses :	Holes
King's Links . 6 (*auxiliary*) and 18	
Hazlehead . . . 9 and 18	
Victoria 18	
Others :	
Royal Aberdeen G.C., Bridge of Don 9 and 18	
Balnagask G.C., Balnagask . . 18	
Muscar G.C., Bridge of Don 9 and 18	
Northern G.C., Golf Road . . 18	
Aberdour (Fife) 18	
Aberfeldy (Perthshire) . . 9	
Aberfoyle (Perthshire) . . 9	
Aberlady (East Lothian)—	
Kilspindie 18	
New Luffness . . . 18	
Aberlour (Banffshire) . . 9	
Aboyne (Aberdeenshire) . . 18	
Airdrie (Lanarkshire)—	
Airdrie G.C. 18	
Easter Moffat . . . 18	
Alexandria (Dunbartonshire) . 18	
Alloa (Clackmannanshire) . . 9	
Alness (Ross-shire) . . . 9	
Alva (Clackmannanshire) . . 9	
Alyth (Perthshire) . . . 13	
Annan (Dumfriesshire) . . 18	
Anstruther (Fife) . . . 9	
Arbroath (Angus) . . . 18	
Archerfield (East Lothian), strictly private 18	
Arran—	
Brodick 18	
Corrie 9	
Corriecravie 9	
Lamlash 18	
Lochranza 9	
Machrie Bay 9	
Pirnmill 9	
Shiskine 12	
Whiting Bay 18	
Auchinblae (Kincardineshire) . 18	
Auchterarder (Perthshire) . . 9	
Auchtermuchty (Fife) . . 9	
Aviemore (Inverness-shire) . 9	

	Holes
Avonbridge (Stirlingshire) . . 9	
Ayr, two courses, each . . . 18	
Balfron (Stirlingshire) . . 9	
Ballantrae (Ayrshire) . . . 18	
Ballater (Aberdeenshire) . . 18	
Ballingry (Fife) . . . 9	
Ballinluig (Perthshire) . . 9	
Banchory (Kincardineshire) . . 18	
Banff (Banffshire) . . . 18	
Barrhead (Renfrewshire) . . 18	
Barry (Angus) 18	
Bathgate (West Lothian) . . . 18	
Bearsden (Dunbartonshire)—	
Bearsden G.C. . . . 9	
Douglas Park G.C. . . 18	
Windyhills G.C. . . . 18	
Beith (Ayrshire) . . . 9	
Bellshill (Lanarkshire) . . 18	
Biggar (Lanarkshire) . . 18	
Bishopriggs (Lanarkshire) . . 18	
Bishopton (Renfrewshire) . . 18	
Blackford (Perthshire) . . 9	
Blair Atholl (Perthshire) . . 9	
Blairgowrie (Perthshire) . . 18	
Blairmore (Argyllshire) . . 9	
Boat of Garten (Inverness-shire) . 9	
Bonar Bridge (Sutherland) . . 9	
Bo'ness (West Lothian) . . 18	
Bonnybridge (Stirlingshire) . . 9	
Bothwell (Lanarkshire) . . 18	
Braemar (Aberdeenshire) . . 18	
Brechin (Angus) . . . 18	
Bridge of Allan (Stirlingshire) . 9	
Bridge of Weir (Renfrewshire)—	
Old Course, Ranfurly G.C. . . 18	
Ranfurly Castle G.C. . . 18	
Broadford (Skye) . . . 9	
Broomieknowe (Midlothian) . . 18	
Brora (Sutherland) . . . 18	
Broxburn (West Lothian) . . 9	
Buchlyvie (Stirlingshire) . . 9	
Buckie (Banffshire) . . . 9	
Burghead (Elgin) . . . 9	
Burntisland (Fife) . . . 18	
Busby (Renfrewshire) . . 9	

	Holes
Giffnock (Renfrewshire)	18
Gifford (East Lothian)	9
Girvan (Ayrshire)	18
Glasgow—	
Municipal Courses :	
Alexandra Park	9
Bellahouston Park	9
Blackhill	18
Deaconsbank	18
Knightswood	9
Littlehill	18
Linn Park	18
Ruchill	9
Others :	
Glasgow G.C., two courses, each	18
Alexandra G.C.	18
Balmore G.C.	18
Cathcart Castle G.C.	18
Cathkin Braes G.C.	18
Cow Glen G.C.	18
Crow Wood G.C.	18
East Renfrewshire G.C.	18
Haggs Castle G.C.	18
Mount Ellen G.C.	18
Pollock G.C.	18
Ralston G.C.	18
Sandyhills G.C.	18
Toryglen G.C.	9
Williamwood G.C.	9 and 18
Glencorse (Midlothian)	18
Gleneagles (Perthshire), two courses, each	18 and 1 of 9
Glenfarg (Perthshire)	9
Glenluce (Wigtownshire)	9
Golspie (Sutherlandshire)	18
Gourock (Renfrewshire)	18
Grangemouth (Stirlingshire)	18
Grantown-on-Spey (Moray)	18
Greenlaw (Berwickshire)	9
Greenock (Renfrewshire)	9 and 18
Gullane (East Lothian), three courses, each 18 holes, and 1 course free for children, 9 holes	
Hamilton (Lanarkshire)	9 and 18
Harburn (Midlothian)	18
Hawick (Roxburghshire)	18
Helensburgh (Dunbartonshire)	18
Helmsdale (Sutherlandshire)	9
Hopeman (Morayshire)	9
Huntly (Aberdeenshire)	9
Innellan (Argyllshire)	9
Innerleithen (Peeblesshire)	9
Insch (Aberdeenshire)	9
Inveraray (Argyllshire)	9
Inverbervie (Kincardineshire)	9
Invergordon (Ross-shire)	9
Inverness	18
Inverurie (Aberdeenshire)	9
Irvine (Ayrshire)—	
Irvine G.C.	18

	Holes
Irvine (*continued*)—	
Irvine Municipal	18
Ravenspark G.C.	18
Isle of Islay (Argyllshire)—	
Islay G.C., Machrie	18
Gartmain G.C., Bowmore	9
Isle of Tiree (Argyllshire)	18
Jedburgh (Roxburghshire)	9
Johnstone (Renfrewshire)	18
Keith (Banffshire)	9
Kelso (Roxburghshire)	9
Kelty (Fife)	9
Kemnay (Aberdeenshire)	9
Kenmore (Perthshire)	18
Kilbirnie (Ayrshire)	9
Killin (Perthshire)	9
Kilmacolm (Renfrewshire)	18
Kilmarnock (Ayrshire)—	
Kilmarnock (Barassie G.C.)	18
Kilmarnock Municipal	9 and 18
Kilsyth (Stirlingshire)	9
Kincardine-on-Forth (Fife)	9
Kincraig (Inverness-shire)	9
Kinghorn (Fife), Municipal course	18
Kingsbarns (Fife)	9
Kingussie (Inverness-shire)	18
Kinloch-Rannoch (Perthshire)	9
Kinross (Kinross-shire)	9
Kintore (Aberdeenshire)	9
Kippford (Kirkcudbrightshire)	9
Kirkcaldy (Fife)	18
Kirkcudbright	9
Kirkintilloch (Dunbartonshire), two courses, each	18
Kirkwall (Orkney)	18
Kirn (Argyllshire)	18
Kirriemuir (Angus)	18
Kyle (Ross-shire)	9
Ladybank (Fife)	9
Laggan (Inverness-shire)	9
Lanark	9 and 18
Langholm (Dumfries-shire)	9
Largs (Ayrshire), two courses	9 and 18
Lauder (Berwickshire)	9
Laurencekirk (Kincardineshire)	9
Leadhills (Lanarkshire)	9
Lennoxtown (Stirlingshire)	9
Lenzie (Lanarkshire)	18
Lerwick (Shetland)	9
Leslie (Fife)	9
Lesmahagow (Lanarkshire)	9
Leuchars (Fife)	9
Leven (Fife)	18
Linlithgow (West Lothian)	9
Livingstone (West Lothian)	9
Lochgelly (Fife)	9
Lochgilphead (Argyllshire)	9
Lochmaben (Dumfriesshire)	9
Lochwinnoch (Renfrewshire)	9
Lockerbie (Dumfriesshire)	9

GLOSSARY

OF

THE MOST COMMON GAELIC OR CELTIC AND NORSE PLACE-NAMES IN SCOTTISH TOPOGRAPHY, TOGETHER WITH THE GAELIC NAMES FOR MANY OF THE TOWNS, VILLAGES, AND DISTRICTS IN SCOTLAND.

NO Celtic inhabitant, speaking in his own language, has ever applied the name "Scotland" to his native land; to him it is Alba, the oldest name by which Scotland and the whole of Great Britain was known, and although the Celtic element only occupies what has been termed the "Celtic fringe," they have left an imperishable record of their former possessions in the Celtic names of all the rivers of Great Britain. Scotland was invaded by the Romans, and they occupied the lands south of the Forth and Clyde for nearly four hundred years, and established forts and camps as far north as Kincardineshire. Later the Saxons penetrated as far as the Lothians, while the Danes had settlements along the East coast, from the Firth of Forth to the Moray Firth. Towards the end of the eighth century the Norsemen took possession of and governed the western islands and parts of the mainland from the north of the Moray Firth round Cape Wrath to the Firth of Clyde, the Isle of Man, and a part of Ireland round Dublin. The Norse occupied these lands until they were defeated at the battle of Largs (1230), and the part of Scotland they ruled over for nearly four hundred years gradually returned to the Scottish Crown.

The earliest inhabitants of Scotland were the Picts, a race who spoke a language akin to modern Welsh, and their original habitations in Scotland are approximately determined by the test word "Pit," a place, farm, or town, comparable with Celtic "Baile" and English "Ton." There are still about 315 "Pits" in Scotland, and these are to be found from the south of Sutherland to Stirling, the largest number being in the counties of Aberdeen, Perth, Fife and Kinross, amounting in all to 193, and showing that these districts must have formed the heart of the Pictish kingdom. However, as the modern Celtic name for town is "Baile," many of the former "Pits" are now "Baile." Early in the fifth century there was yet another invasion, this time from Ireland, by a tribe called "Scotti." They occupied at first some of the Western Isles, then the mainland of Argyllshire, and gradually absorbed Pictland, until, at the beginning of the eleventh century, they gave a king and their name to Scotland. Henceforth the country which was first known to historians as Alba or Albion, became in succession Pictovia, Caledonia, and finally Scotland.

It is obvious that so many foreign tribes, speaking different languages, having gained a footing and established settlements, there would be a mixture of languages in the place-names. The Norse influence was so great that in some parts, Lewis for example, the Norse names are in the ratio 4 to 1 of the Celtic names, while many of the domestic words used by the present inhabitants are derived from Norse, but their origin is disguised under Celtic influence. As an example, the Norse "Hop," an inner bay, is on the West of Scotland; "Ob," as in Oban, the little bay; in Lewis it is "Tob," while in the Orkneys and north Sutherland it is "Hope," as Long Hope, Loch Hope, and Ben Hope. In Roxburgh, Peebles, and Selkirk, it is "hope," meaning a land-bay. "Bolstadr," a farm, town, or village is -bol, -pol, or -pool, as in Eirebol, "beach town" (Sutherland), and Barepool, in Tiree. In some cases while the Norse name is that in general commercial use, the local name is Celtic—Tain is in Gaelic "Baile Dhubhaich," St. Duthac's towns; Dingwall, the Norse "Court town," i.e. Tingwall, is in Gaelic "In 'ir-pheofharan," confluence of the Peffer burn; Beauly, "beautiful place," is in Gaelic "A' Mhanachainn," place of Monks; the Norse "Fjallr," a hill, becomes

" Fell " or " Val," as in Goatfell in Arran, Oiseval (E. Fell) in St. Kilda, while Kilda itself is from Norse " Kilde," a well. The Norse " Habest," high stead, becomes Tabost (Lewis) ; " Holmr," an island at high water, is in Gaelic " Tolm or Tuilm."

The oldest names in Scotland, as in most countries, are the river names. Everywhere invaders have retained these, *e.g.* Melbourne is new, but its river Yarra Yarra is the native name ; in U.S.A. we have the Mississippi and Missouri, while in Scotland the Dee, Don, Ness, and others were names given in pagan times and are the names of river gods and goddesses.

The Gaelic " Dail," a field, invariably forms the initial part of a name, *e.g.* Dalkeith, " Che's field." While the Norse " Dal " or " Dale " forms the termination, *e.g.* Liddisdale, sloping or sheltered valley. Swordale which occurs often is " Swardalr," green turf dale. " Fjordr," a firth or bay, in Gaelic takes the form of -ort, -art, -ford, and -ost ; Knoydart is " Knuts' fjord " ; Skiport is " ship fjord " ; Gruinnard, " shallow fjord " ; Laxford, " salmon firth." Innse was the British or Brythonic name for " island," as in Inchkeith, Inchcolm, etc., but in Scottish topography most of the islands have taken the Norse form " Ey-land," which is in Gaelic " Eilean," and both " Innse " and " Eilean " are applied to river bends and meadows near a lake as well as to islands, *e.g.* Perth with its two Inches. Hebrides is a misreading of Hebudes, just as Iona is a misreading of " Ioua " (the local pronunciation is Ee as the two e's in thee). Ptolemy (second century) writes " Aebudae " and says there are five islands, the only one which can now be recognised from his names is Mull. The name is now applied to all the islands from Mull of Kintyre to the Butt of Lewis under *Outer* and *Inner* Hebrides.

The meanings of many of the older names given by a race who have left no written record of their language are now lost, such as Ben Ledi, Firth of Forth, etc. Philologists during recent years have by diligent research added greatly to our knowledge of Scottish names, and foremost among these is Professor W. J. Watson whose invaluable work on Celtic place-names of Scotland is a masterpiece. Others worthy of note are the late Professor Donald Mackinnon, Dr. Alex. MacBain, Dr. Alex. Cameron, and for Norse names Dr. George Henderson, Robert Locke Bremner, and Professor Jacob Jakobsen.

PRONUNCIATION

Initial *Bh* or *Mh* equals *V*, but after a broad vowel equals *W*, as in English " Now."
  ,,   *C* equals *K*.
  ,,   *Fh* is silent.
  ,,   *Ph* equals *F*.
  ,,   *Sh* or *Th* equals *h*.
  ,,   *S* after *An t* is silent.
Final—*aidh* equals *y* as in my.
  ,,   —*idh* equals *y* as in duty.
Th final, or when flanked with vowels, is a strong breathing.
Ch in contact with *a*, *o*, or *u*, is a strong guttural, as in loch.
Ch, in contact with *e* or *i*, is a guttural as in German *ich*.

THE ARTICLE

A' equals the, as in *A' Chreag* : the rock.
Am equals the, as in *Am Boc* : the buck.
An t- equals the, as in *An t-Eilean Sgitheanach* : the isle of Skye.
Na equals the, as in *Na Bruthaichean* : the braes.
A' equals of the, as in *Allt a' choire Dhuibh* : burn of the black corry.
An t- equals of the, as in *Bagh an t-Siosalaich* : Chisholm's bay.
Na equals of the, as in *Slochd na Beinne* : hollow of the mountain.
Nan equals of the, as in *Meall nan Eun* : hill of the birds.
Nam equals of the, as in *Coire nam Bo* : corry of the cows.

Abbreviations : G. for Gaelic ; N. for Norse.

c

Aber : G. Obar and Abar (confluence, as in Aberdeen, Abergeldie, and Abernethy, confluences of the Don, Geldïe, and Nethy; over 70 in Scotland).

Aberdeen : G. Obar Dheathan (confluence of the r. Don).

Aberdour : G. Obair Dhobhair (confluence of the Dour water).

Aberfeldy, G. Obair-pheallaidh (confluence of the Peallaidh with the Tay, now called Moness Burn).

Aberfoyle : G. Obair-phuill (confluence of pools).

Abhainn, Aibhne, and **Amhuinn, Aimhne :** River.

Acair, Acairseid : Anchor and anchorage ; e.g. Bagh na-h-Acair seid (bay of the anchorage).

Achadh, Achaidh : A field, a place.

Achencairn : G. Achadh a' chairn (meadow of the cairn).

Achnahannet : G. Achadh na h-Annaid (field or meadow of the mother Church).

Agh, Aighe, Aighean : Hind, heifer.

Ail, Aileach or **Aillig** (old form): A rock or stony place ; e.g. Ail-Clutha (rock on the Clyde, now Dumbarton Castle).

Ailean, Ailein, or **Ailean, Ailein (à):** Green spot, enclosure, meadow.

Ailsa Craig : G. Creag Ealasaid (from old **Ail,** a rock, and modern **G. Creag,** both meaning rock).

Airidh, Airidhe, or **Airigh, Airighe (à):** Sheiling.

Aisir, Aisre : A pass between rocks.

Ald, Alt, Ault, Auld : G. Allt (a burn, stream ; as in Aldclune, G. Allt Chluain, burn of the meadow ; Altnabreae, G. Allt nam Breac, trout burn ; Aultbea G. Allt Beithe, birch burn ; Auldearn, G. Allt Eireann).

Alltbea or **Aultbea :** G. Allt-Beithe (birch-stream).

Almond River : Abhainn Aman.

Alness : G. Alanais (from Allan, a bog, wet place).[1]

Amat, Amaite : N. Á-mót (river-meet, confluence), e.g. Amat on r. Oykell where the Rappach joins it.

Amulree, G. Àth Maol-Ruibhe (ford of St. Mulrhu).

Annaid, Annaite, and **Annat** (na h-Annaidean, the annats) ; A Mother Church.

Aoineadh, Aoinidh : Steep brae with rocks, moraine ; e.g. Aoinead Mòr, the great steep.

Aonach, Aonaich : Moor or market-place.

Arbhar, Arbhair : Corn.

Arbroath (old form **Aberbrothok**): G. Obar Bhrotháig (confluence of the small stream Brothock with the sea).

Ard, Aird : G. Àrd Àirde (a height, promontory ; e.g. Ardlamont, height of Lamont ; Ardmeanach, mid-height ; Ardrossan height of the little promontory.

Aros : N. ár-ós (river mouth). See Aros, Mull.

Ath, Atha : A ford (mas.); a kiln (fem.); e.g. am Bannath, bottom ford, now Bonar Bridge.

Attow Ben : G. Beinn Fhada (long mountain).

Auch, Ach : G. Achadh (a field ; Achanalt, G. Achadh nan Allt, field of the burns ; Auchnashellach : G. Achadh nan Seileach, field of the willows).

Aven, or **Avon** (Banff): G. Athfhinn (Fingal's ford).

Averon River : G. Abhainn Abharan (Alness).

Aviemore : G. Agaidh-mhòr (the big gap ; i.e. wide valley of the Spey. Agaidh Bheag is a small gap in the Cairngorms).

Avon : G. Abhainn, q.v.

Ayr : G. Inbhir-air (confluence of the r. Ayr, with sea) ; Ayr is N, eyrr (a gravelly beach).

Bac, Bhaic, Bacaichean : Bank, peat bank.

Bad, Bhaid : A tuft, a clump of trees, a place, or shrubs ; common place-name in Sutherland.

Badcall : G. Bada-call (hazel-clump).

Badenoch : G. Bàideanach (marsh or lake land).

Bàgh, Bhàigh : A bay.

Baile, Bhaile : A town or hamlet, homestead.

Bal and **Ball** as a prefix is from G. Baile, q.v.; e.g. Balnagown, G. Bail 'a Ghobhainn (the smith's town).

Balgair, Bhalgair : A fox.

Balindore : G. Bail' an deora (pilgrims' town).

Ballachulish : G. Bail' a' chaolais (town on the channel).

Ballantrae : G. Bail'-an-traigh (town by the shore).

Ballater : G. Beal'tair or Bealadair.

Balloch : G. Bealach (a pass).

Ballochantuy : G. Bealach an-t-Suidhe (W. Kintyre, the resting pass).

Balmoral : G. Baile Mhoireil (large place, a clearing, open space.) [1]

Balnagown : G. Baile nan Gobhainn (town of the smiths).

Balnakeil : G. Baile-na-Cille (town of the church).

Balnakyle : G. Baile na Coille (town of the wood).

Balnespick : G. Baile-an-easbuig (town of the bishop).

Balquhidder : G. Both-fuidir.

Bàn, Bhàn, Bhàin, Bàine, Bana : Fair, white.

Banff : G. Banbh (a pig, an old name for Ireland).

Bannockburn : G. Allt-a-bhonnaich (stream from the peaks, stream gives name to the village).

Ban-righ : Queen.

Barcaldine : G. Am Bana-calltuinn (place of hazel).

Bàrd, Bhàird : (1) A poet ; (2) an enclosed meadow (Scotch ward).

Barpa (Hebrides and Skye) : A rude conical heap of stones, sepulchral ; Eng. barrow ; N. Hvarf (hill on the horizon).

Bàta, Bhàta, Bataichean : A boat, boats.

Bàthaich, Bhathaich : Byre, sanctuary, shelter (in deer forests).

Beag, Bheag, Bhig, Bige, Beaga : Little.

Bealach, Bhealaich : A pass between hills.

Bealaidh, Bhealaidh : Broom.

Bean, Mna (pron. Mra), **Ban :** Wife, woman.

Bearn : A gap.

Beaufort Castle : G. Dùnaidh.

Beauly : Fr. beau lieu (pretty place) ; G. is A' Mhanachain (the monkery).

Beinn, Bheinn, or Beinne : Mountain.

Beith, Bheithe : Birch ; Allt. Beith (birch-burn).

Beithir, Bheithir : e.g. Beinn a' Bheithir (serpent, wild beast, monster).

Ben : G. Beinn (mountain, originally horn, peak), e.g. Ben Wyvis, G. Beinn Uais (high mountain) ; Ben Macdui, G. Beinn Mhic Duibhe (McDuff's mountain).

Ben An' (Trossachs) : is in G. Am Binnean (the little mountain).

Benbecula : Beinn nam faolgha (height of the ford).

Beul, Beòil, Bheòil : Mouth.

Biast, Béiste : Beast, monster.

Binnean, Binnin : (small and peaked mt.), e.g. Ben An' (Trossachs), G. Am Beinnean (the little pinnacle.)

Biolaire, Bhiolaire : Water-cresses.

Biorach, Bioraiche : (1) adj. sharp-pointed ; (2) n. dog-fish.

Black Isle : An t-Eilean Dubh (black isl.).

Blair : G. Blàr, q.v. (peatmoss).

Blair-Atholl : G. Blàr an Athoill.

Blairgowrie : Blàr-goibhre.

Blàr, Bhlàir : Cleared space, plain.

Bó, Bà : cow, cattle ; e.g. Bealach nam Bó (pass of the cattle).

Boat of Garten : Coit Chartain.

Bodach, Bhodaich : Old Man, spectre.

Bog, Bhuig : (1) adj. soft ; (2) n. a soft place.

Bogha, Bhogha, or Bodha Bhodha (pron. Boa and Voa) : N. Bodi (a breaker) ; (sunk rock in sea).

Boisdale : Baghasdal.

Bonar Bridge : Drochaid a' bhanna (bridge at the shallow).

Bonawe : Bun-Atha (foot of the ford).

Bonskied : Bonn-sgaoid, Old Monskeid.

Borg, N. : A fortress.

Both, Bothan or Bothain : A primitive stone or turf house ; bothy.

Bradan, Bhradain : Salmon.

Brae, Bread : G. Braigh (upper part). (Braemar, G. Braigh Mhàr, height of Mar ; Breadalbane, G. Braghad Albainn, height of Alba, i.e. original name for Scotland.)

Braigh, Bhraghad : Upper part.

Brander, Pass of : is in G. Cumhang a' Bhrannraidh.

Breac, Bhreac, Bhric, Brice, Breaca : (1) adj., speckled ; (2) n., trout.

Breadalbane : Bràghad Albainn (the heights of Alba (Scotland)).

Breug, Breugach, or Bréige : false, applied to stone cairns erected on mountain as guides ; e.g. Buachaille Bréige, false shepherd.

Bridge of Allan : Drochaid Ailein.

Bridgend : Ceann drochaid (bridge head). Drumnadrochit, G. Druim na Drochaid, the bridge at the ridge, Loch Ness.

Britain : Breatunn, generally -uinn.

Broc, Bhruic : The badger ; e.g. Allt 'a Bhruic (badger burn).

---

[1] *Place-Names*, by Alex. MacBain, p. 182.

Brochan, Brochain : literally gruel or porridge, but applied in place-names to anything broken up or comminuted, as Coire Brochain (Cairngorms), the corry of the broken stones.

Brodick, Breadhaig, N. Breidha-fjord (broad ford).

Broom, River, and **Loch :** G. Abhainn Bhraoin ; Old Ir. bróen (a drop, shower water.) [1]

Bruach, Bhruaich, or **Bruthach, Bhruthaich :** Bank, brink, steep place, brae ; e.g. Tighnabruaich (Kyles of Bute), house at the brae.

Buachaille Etive : Buachaille Eite (where Eite means a loch goddess). [2]

Buachaille, Bhuachaille : herdsman ; e.g. Buachaille Etive (watchman of Etive).

Buidhe, Bhuidhe (pron. Buie and Vuie) **:** Yellow ; Buidheanaich (yellow place).

Bùirich : Bhùirich roar (bellow, e.g. Meall a' Bhùirich).

Bun : literally root, but in place names generally applied to the mouth of a river or stream, as in Bunaven (G. Bun na h-Aibhne), Bunawe, Bunchrew (Inverness). (G. Bun chraobh, bottom of wood).

Bunchrew : Bun-chraobh (near the tree).

Cadha (pron. Caa) **:** Steep place, a pass.

Cailleach, Caillich : A nun, old woman, hag (cf. bodach) ; e.g. Allt na Caillich (burn of the hag).

Cairn : G. Càrn, (heap of stones, rocky hill, e.g. Cairntoul) ; G. Càrn an t-Sabhail (hill of the barn).

Caithness : Modern, Gallaibh ; old Cataibh (country of the tribe cats).

Cala, or **Caladh :** A harbour.

Calltuinn, Calltuinne : Hazel.

Calman, Calmain, or **Calaman, Calamain, of Columan :** a dove.

Càm, Chaim : Bend, crooked ; Cambusmore (Callander), the big bend.

Camas, Chamais : a channel, a bay, in inland places a bend.

Cambus : G. Camas, q.v., e.g. Cambusmore, G. Camas Mòr (big bend).

Camus nan Gall : Camus nan Gall (channel of the foreigners).

Canach, Chanaich : Cotton grass.

Canna : Canaidh, Eilean Chanfhaidh, an t-Eilean tarsuinn.

Caochan, Chaochain : Wild : A streamlet.

Caol, Caolas, Chaolais : narrow, strait, firth, kyle, e.g. An Caol Arcach (the Pentland Firth).

Caora, Caorach : Sheep.

Caorunn, Chaoruinn : Rowan tree or mountain ash.

Cape Wrath : N. hvarf (a turning point). See WRATH. Cr. Am Parbh.

Capleach, Caiplich : Place of horses.

Capull, Chapuill : Horse or mare.

Car, bend, e.g. **Bealach Carach :** The winding pass.

Càrn, Chùirn, or **Chàirn :** Heap of stones applied to round rocky hills.

Càrn Toul (Cairngorms) : Càrn an t-sabhail (rocky hill of the barn).

Càrr, or **Càthair** (pron. Kāar) **:** Rough or broken mossy ground.

Cas, Caise : Steep.

Cat, Chait : A cat.

Ceann, Chinn, Cinn : Head, a headland ; Kincardine, G. Cinn-chardain, Wood-end (Watson).

Ceapach, Cheapaich : A tillage plot.

Cearc, Circe : a hen ; Cearc-fhraoich (moor hen, grouse).

Cearcall, Chearcaill : a circle, a hoop, e.g. Coire Chearcaill (the circular corrie).

Ceard, Ceardach, Cheardaich : Craftsman, smithy, forge.

Ceò, Cheathaich : mist, e.g. Coire Cheathaich (the misty corry).

Chanonry : A'Chananaich (residence of Canons).

Chicken Head : Rudha na Circe (prom. of the hens, the point of Eye peninsula, Lewis).

Cill, Cille, Ceall : church, burying-place, e.g. Cille Mhuire (St. Mary's Church) ; Baile na Cille (town of the church).

Cioch, Ciche : a pap ; Beinn nan Ciochan (hill of the paps), old name for Lochnagar.

Clach, Cloiche : stone, e.g. Clach Dhion Shelter stone.

Clachan : place of stones, applied primarily to a stone house, especially a cell or church ; secondarily to a " Kirktown " hamlet, e.g. Clachan, W. Kintyre.

Clachnaharry : G. Clach na h-aire (stone of the watching), north of Inverness.

[1] See Watson's *Place-Names of Ross and Cromarty*, p. 241.
[2] Prof. Watson's *Celtic Place-Names of Scotland*, p. 47.

Cladach, Chladaich : Shore, beach.

Claidheamh, Chlaidhimh : Sword.

Claignonn : Skull, head, rounded hillock.

Clamhan, Clamhain : Kite, Buzzard.

Cleit, Chleit : Rocky eminence.

Clibreck : Clibric, Ben Klibreck, Sutherland (3164 ft.)

Cluain, Cluaine : A green plain, pasture.

Clunie, Cluanie : G. Cluainigh (meadowplace), *e.g.* Clunie in Badenoch.

Cnàimh, Cnàmha : Bones.

Cnap, Chnaip (pronounced locally krap) : A hillock.

Cneamh, Chneamh, or Creamh, Chreamha : Wild garlic.

Cnoc, Chnuic, Cnocan (pronounced krock) : A round hill on maps. Knock, as Knockander (Cnocan Dubh), black knoll, vil. near Tomintoul.

Coigach, A'Choigeach (a fifth part ; dist. N.W. Ross and Cromarty).

Coileach, Choilich : A cock ; *e.g.* Coileach-froach (a heath-cock).

Coill, Coille : Wood, forest.

Còinneach, Chòinnich : Moss ; Chòinneachan (moss, place of moss).

Coir, Coire, Choire : A round hollow in mountain side, cirque ; Corryarrick (the short corrie, through which General Wade's road passes and rises to 2507 ft.)

Colintraive : Caol an t-snàimh (the swimming channel, where cattle were swam across in olden times).

Con, Choin : Dog. See Cu.

Coolin Hills : An Cuilionn, An Cuilfhionn ; Cuilioun Rumach (the C. of Rhum) ; an Cuilionn Sgitheanach (the C. of Skye).[1]

Corryarrick : Coire Ghearraig (the short Coire). See Coir.

Craig : G. Creag, q.v., *e.g.* Craighouse, rock house ; Craigendoran, G. Creag an Dobhrain (rock of the otter).

Craigellachie, Creag-Eileachaidh (meaning rock rock, *i.e.* the old form ailich, rock, and the modern creag, a rock).

Craobh, Craoibhe : Tree ; Crieff is in G. Craobh (place of trees).

Crasg, Chraisg, Chroisg : A crossing, generally a watershed.

Crask : Crasg, crossing the watershed on the road from Lairg to Tongue, Sutherland.

Creachann, Creachainn : the bare windswept place about the top of a hill.

[1] *Celtic Place-Names*, Watson, p. 95.

Creag, Chreag, Creige : A crag, rock, or cliff ; Conchreag (combination of rocks).

Crioch, Criche : boundary ; *e.g.* Allt na Criche (burn of the boundary).

Cròcah, Cròcaich : Branched, branching ; full of bends.

Croe, Glen : Gleann crotha.

Crom, Cruime : Crooked.

Cromarty : Cromba bend.

Crodh, Chruidh : Cattle.

Crossg, see Crasg.

Cruach, Chruach, Chruaich : A heap, stack, bold hill.

Cruachan : A haunch ; Cruachan Beann (haunch of peaks).

Cu, Con, or **Choin :** Dog.

Cuach, Chuaich, Cuaiche : A cup-shaped hollow.

Cuilc, Cuilce : A reed.

Cumbrae, Great : Cumradh mór ; N. Kumreyjae ; the isles of Kumr may be Welsh Kymry.

Cumhann, Cumhainn, Cuinge : Narrow, a strait ; *e.g.* Coalais Cumhaun, W. Sutherland (the narrow channel).

Cùil : Nook, recess.

Cuileann, Chuilinn : Holly.

Cuith, Cuithe : Pit, wreath of snow, narrow glen.

Cùl, Chùil, Cùile : Back, hill-back, nook.

Culloden : G. Guil-lodair (nook of the marsh).

Dail : A field, dale, haugh, dalach (dalloch) of a dale.

Dal, Dall : when used as a prefix it is from G. Dail ; as an affix it is generally from N. Dalr, a dale ; *e.g.* Dalnaspidal, G. Dail na Spideil (field of the hostel) ; Netherdale, N. Nedri-dalr (lower dale) ; Liddisdale, sloping valley.

Dalkeith : Dail-ché (Keith generally means wood).

Dalmally : Dail-mhàilidh (a saint named Maillidh).

Dalwhinnie : Dail-chuinnidh (champions dale).[2]

Damh, Daimh : Ox, stag.

Darach, Daraich : Oak.

Dearcag : A berry, or of berries.

Dearg, Dheirg : Red.

Diebidale, N. : Deep dale.

Diollaid, Diollaide : Saddle ; *e.g.* Beinn na Diollaide (hill of the saddle).

Doire, Dhoire : Grove, hollow.

[2] Watson, p. 145.

Doirlinn, Doirlinne, or **Doirling, Doirlinge :** Isthmus (beach).

Don, River : Deathan (gives its name to Aberdeen which is in G. Obair Dheathain, confluence of the Don).

Dornie : An Dòirnidh (Ceannaiche na Dòirnidh, the Dornie merchant). The old name—Bun-dà-loch—is now applied only to the east end of the village.

Dornoch : Dornoch (place of stones).

Douglas : Dùbhghlas (black water).

Doune : An Dunaidh, a port.

Drem, Drom, Drum : G. Druim, q.v., *e.g.* Druimnadrochit, G. Druim na Drochaide (ridge of the bridge), Loch Ness.

Drochaid, Drochaide : Bridge.

Druim, Droma : A ridge, the back.

Drumalbin : Druim-albainn (the ridge of Alba (Scotland)).

Drummond : G. Druiminn (old locative case of drum " at ridge.")

Drumnadrochit : Druim na drochaid (bridge of the ridge).

Drumouchter : Druim-uachdair (ridge of the upper ground).

Dubh, Dhubh, Dhuibhe, Dubha : Black.

Duine, Dhaoine, or **Daoine :** A man, men.

Dumbarton : Dun Breatunn (fort of the Britons).

Dùn, Dùin, or **Dhùin :** Fortress, castle, heap, mound.

Dun : G. Dùn, q.v., *e.g.* Dumbarton ; G. Dùn Breatunn (fortress of the Britons).

Dunbeath : Dun Bheitheadh (hill or fort of the birches).

Dundee : G. Dùn Deagh ("Fort of Daig(h)"). See W., p. 220.

Dunfermline : Dun Pharlain (the fort of [Mac] Farlane).

Dunkeld : Dun Chaillinn (the fort of the Caledonians).

Dunrobin : Dun-robain (Robin's Fort).

Dunvegan : Dun-bheagan (fort of Began).

Each, Eich : Horse.

Eadar : between, appears in place-names as Eadar Da Chaolas (between two kyles) ; Eadar Da Fhaoghail (between two fords) ; Eadar Da Ghobhal (between two forks) ; Eadar Da Chaolas (between two fords) ; parish name in Sutherland and Lewis.

Eaglais, Eaglaise : Church.

Eala, Ealachan : Swan or swans ; Loch na-h Eala (loch of the swans).

Ear : East.

Earn R. : Abhainn Éire.

Eas, Easa, Easan : Waterfall, rough ravine (Perthshire).

Easgaidh, Easgainn : A marsh.

Eccles : G. Eaglais (a church), *e.g.* Ecclefechan (Church of St. Fechan).

Eddrachillis : Eadar-dà-chaolais (between two fords).

Edinburgh : G. Dùn-eideann (the meaning, says Watson, "is quite obscure," p. 341.

Eilean, Eilein, Eileanan : Island or islands ; An-t Eilean Dubh (the black isle).

Eilrig : Deer pass, place where deer were killed or captured.

Embò : Euraboll. See Eribol.

Eribol : G. Euraboll, N. Eyrr-bol (beach-stead).

Eun, Eòin : Bird or birds.

Evanton : Bail' Eòghaiun, am Bail' ùr (Evan's town).

Fad, Fhad, Fhada, Fada : long ; *e.g.* Beinn Fhada (long mountain.).

Faire : Watching ; Cnoc na Faire (hill of the watching).

Falkirk (Fawkirk) : is in G. an Eaglais Bhreac (speckled church).

Falloch, Glen : Gleann Falach.

Fang, Faing, or **Fhaing :** Sheep-pen, fank.

Faoghail, or **Faodhail** (pron. fao-ul) : A ford in sea channel.

Faoileag, Faoileige : Sea gull.

Fasnakyle : Fas na coille (place of the wood).

Feadan, Fheadain : Narrow glen or hollow, streamlet.

Fear, Fir, Fhir : A man.

Fearn, Fhearna : Alder tree.

Féith, Féithe (pron. fæ) : Bog, slowly moving stream (lit. vein) ; Feithe Buedhe (yellow stream), Cairngorms.

Fell : N. Fjall (a rough hill), appears as a termination ; Val, *e.g.* Schreval (scree fell).

Feshie : Feisidh (1230 Cessy) earliest form Vestia, root ved. (wet) (Macbain).

Fiadh, Fhéidh : Deer.

Fife : Fiobha (Fib, one of the sons of Cruithne).

Fillan : Faolan (St. Fillan's, died 777).

Findhorn, River : Uisge Éire.

Forsinard : Forsain-àird, N. fors (waterfall) and G. Aird (a height).

Fort Augustus : Cill Chuimein (St. Quimens church).

Fortrose: A'Chananaich (place of Canons).
Fort William: An Gearasdan, Inbhir-lòchaidh is its oldest name, but it passed from Maryburgh to Gordonsburgh, then to Duncansburgh.
Foula, Shetland: N. Fugey (bird-island).
Freuchie: Fraochaidh (place of heather).
Fuar: Cold, *e.g.* Fuar-ghlaic (cold hollow); Fuaralaich and Fuaralacha (cold place); Fuar-Mhonadh (cold mountain), Loch Ness side.
Fuaran, Fhuarain: Well, spring; secondarily, a green spot.

Gabhar, Gabhair, or **Gaibhre;** also **Gobhar, Gobhair,** or **Goibhre:** Goat.
Gaineamh, Gainimh, or **Gaineamhach, Gaineamhaich:** Sand.
Gairloch: Geàrr-loch (short loch).
Gall, Ghaill: Stranger, lowlander; the Sutherland men called the Caithness Gaillich (strangers).
Gamhainn, Gaimhne, or **Gamhna:** A stirk.
Gàradh, Ghàraidh: Wall or dike, also a garden.
Garbh, Gharbh, Ghairbh, Gairbhe, Garba: Rough.
Garrynahine: Gearraidh na h-aibne (the enclosure at the river).
Garten, Boat of: Coit Ghartain (coit, a boat).
Geadh, Gheòidh: A goose.
Geal, Gheal, Ghil, Gile, Geala: White.
Geàrr: Short; Geàrr and Gearrsaich (a hare); *e.g.* Gareloch (the short loch).
Geàrraidh (Sheiling): Outer pastures (Lewis).
Geodha, Geo, Gio: N. Gja (a chasm, rift).
Geusachan: Giùthsachan (place of firs).
Gigha: Giogha (appears in Sagas as Gudey, *i.e.* god-isle).
Gill: N. A ravine, *e.g.* Trailgil (Troll-ravine, the equivalent in G. is Gleann, a glen).
Giubhas, Giuthas, Ghiubhais and **Ghiubh-sachan:** Fir, place of firs; Kingussie is in G. Ceann Ghiubhaisich, head of the firs.
Glac, Glaic: A hollow.
Glais: A stream.
Glas, Ghlas, Ghlais, Glaise, Glasa: Grey or green.
Glasgow: Glasgu in 1136; G. Glaschú (Dr. Watsons says it probably means "green hollow.")[1]

Gleann, Ghlinne: Narrow valley, dale, glen; Gleann Mòr na-h Alba is the G. for the great glen from Inverness to Fort William.
Glen, G. Gleann: q.v.
Glenbucket: Gleann Buichead (Buichead's Glen).
Glencroe: Gleann crò.
Glenelg: Gleann Eilg. See ELGIN.
Glengyle: Gleann Goill.
Goat Fell: Gaota-bheinn, N. Geitar-fjall.
Gob, Ghuib: Point, beak.
Gobha, Ghobhainn: Blacksmith; *e.g.* Balnagown in Ross is the Smith's town.
Gobhar, see GABHAR.
Gobhlach, Ghoblaich: Forked.
Golspie: Goillspidh; in 1330 Goldespy, in 1448 Golspi; no satisfactory explanation of the first has been given, the termination—pie—is the N. (by a habitation). Its Celtic name is Kilmaly.
Gorm, Ghorm, Ghuirme, Guirme: Green and blue.
Grampians, The: Monadh Dhrum Uachdair, W. of Struan; Monadh Minigeig, E. of Struan.
Grantown: Baile nan Granndaich (the Grants' Town).
Grian, Gréine: The sun.
Gruinnard River: N. Grunnfjördr (shallow firth).
Guala or **Gualann, Ghualainn:** Shoulder of a hill; Gualin House, an old hostel, now a shooting-lodge in Durness, Sutherland.

Hebrides: Ptolemy says " Above Ivernia (Ireland) are the isls. called Ebomdai, or Aiboudai, five in number." Of these only the isl. of Mull (Malaios) has been recognised. The modern application of the name applies to the whole of the Western Isles from the Mull of Kintyre to the Butt of Lewis under Outer and Inner Hebrides.
Holm: G. Tuilm; an isl. in a river or near the shore, from N. Holmr. An isl. joined to the mainland at low water.
Hutick, Ben: Beinn Thutaig, mt. in N. Sutherland.

Iar: West.
Imrich, Imriche (locally pron. Irmich): removing, flitting. Bealach na-h-Imriche (pass of the flitting).

[1] *Celtic Place-Names of Scotland*, p. 386.

Inbhir, Inbhire : place of meeting of rivers, or where a river falls into the sea or lake ; confluence of Aber, Amat. See RIVER.

Inch : G. Innis, q.v.

Inchcape : Off Angus ; Innis ceap (now Bell rock).

Inchkeith : Innis Ché (the isl. of Ché).

Inghean, Inghinn : Daughter, same as Nighean.

Innis, Innse : (1) Isl ; (2) meadow by the side of a river, haugh ; (3) resting-place for cattle, etc.

Inver : G. Inbhir, q.v. Inverness, Inverey, confluences of the Ness and Ey. (Over 400 Invers in Scotland.)

Invergordon : An Rudha, Rudha 'nach Breacaidh ('nach=aonach, market).

Iodhlann, Iodhlainn or **Iolainn, Iolainne :** Corn, barnyard.

Iolair, Iolaire : Eagle.

Iosal, Isle : Low.

Katrine, Loch : Loch Ceiterein (Katrine's Loch).

Ken, Kin : G. Ceann, a head, prom., e.g. Kingussie, G. Ceann a'Ghiuthsaich (head of the fir wood) ; Kintail, G. Ceann an t-sàile (end of the salt water) ; Kendrum (head ridge).

Keppoch : G. Ceapach : A tillage plot.

Kilcalumkill : Cill Chalum Chille (St. Columba's church).

Kilchurn : Caol a' chùirn (strait at the cairn, Loch Awe).

Kildonan : Cill Donnan (St. Donan's Church.

Kinbrace, Cinn a'bhràist.

Kincardine : Cinn Chardainn (head of the wood).

Kincraig : Ceann na creige (head of the rock).

Kinlochbervie : Ceann Loch Biorbhaidh.

Kirkcudbright : Cille Chuithbeirt (St. Cuthbert's Church).

Knock : G. Cnoc (locally pron. Crock) : A round hill, hillock.

Knockantinny : Cnoc an teine (watch-fire hill).

Knockfarrel : Cnoc-fearralaidh (projecting or high).[1]

Kyle : G. Caol and Caolas (a strait).

Lag, Luig : A hollow.

Lagan, Lagain : A little hollow.

Lagg, Laggan : G. Lag and Lagan, q.v ; Loch Lagan, the loch in the hollow.

Làir, Làire : Mare.

Làirig, Làirige : A pass.

Laogh, Laoigh : Calf ; e.g. Beinn Laoigh (calf mountain).

Lawers, Ben : Labhar. The name, says Watson, is derived from the stream rising on Ben Lawers, Uisge Labhar (loud burn).

Leabaidh, Leapa : A bed, a lair, as in Leabaidh an Fheidh (the deer's lair).

Leac, Lice or **Lic :** Flat stones ; Leacach (stony slope).

Learg, Leirge : A plain, hillside.

Leathad, Leathaid : A slope.

Leathann, Leathainn or **Leathan, Leathain :** Broad.

Leith : Half, as in Laith-allt (half-burn, i.e. burn with one steep side).

Leitir, Leitire, Leitreach : Slope, side of a hill.

Lesmahagow : "Lios Mo-Fhéngu, my fechin's enclosure," Watson.

Leum, Leuma : A leap.

Lewis : Eilean Leòdhas, Leòdhas ; N. Ljodhus ; the real meaning, says Watson, "seems best found in Ljodhus—house of songs or lays." [2]

Liath, Leith, or **Leithe :** Grey ; Liathanaich (greyness).

Linlithgow : Gleann Iucha (lake of Lithgow or wet-hollow).[3]

Linn, Linne : A pool, e.g. Braclinn (Callander) ; speckled fall, Linn of Dee (Braemar).

Lismore : Lios mhor (great garden).

Loch, Locha, Lochan, Lochain : Lake, lakelet.

Lochnagar : Loch na gàire.

Logie, see LAG.

Loinn, Loinne : An enclosure, land.

Lomond, Ben : Beinn Laomuinn (beacon hill).[4]

Lòn, Loin (in Galloway, Land) : Marsh, morass (in Skye, a slow stream).

Long, Luinge : A ship.

Losg, Loisgte : Burnt ground.

Losgann, Losgainn : Frog (common Lewis place-names).

[1] Watson, *Place-Names of Ross and Cromarty*, p. 98.
[2] *Place-Names of Ross and Cromarty*, p. 263.
[3] Watson, *Celtic Place-Names of Scotland*, p. 384.
[4] *Celtic Place-Names*, by Prof. W. J. Watson, p. 212.

Luachrach, Luachraich : Place of rushes.
Lùb, Luib, : A bend.
Lui, Ben : Beinn Laoigh (the calf mountain).

Machar, Machair : A plain, *e.g.* A'Mhachar (Durness) ; Machrihanish (Kintyre).
MacLeods of Dunvegan and Glenelg : Siol Thormaid.
Madadh, Mhadaidh : Dog, wolf, fox (appears in Scottish place-names as fox ; generally qualified Madadh-ruadh (red fox)).
Magh, Mhaigh, Mhagha : A plain, a field, *e.g.* Moy in Inverness.
Maigheach, Mhaighiche : A hare.
Màm, Mhàim : lit. a swelling ; in place-names, large, round, or gently rising hill, *e.g.* Màm na Gualainn (rising ground).
Manach, Mhanaich : a monk ; A'Mhanachainn (the monkery). (G. for Beuly and Fearn in E. Ross.)
Maol, Maoile : Bare top.
Maree, Loch : Loch Ma-ruibh (old name, Loch Feadhal feas) ; originally Loch Ewe and its S. end is still known as Ceann Loch Ewe (head of Loch Ewe).
Meadhon, Mheadhoin : Middle.
Mealfourvounie : Meal-fuar-mhonaidh (cold mountain range).
Meall, Mhill : Lit. a lump, applied to a round hill.
Meanabh-Chrodh, Meanabh-chruidh : Small or young cattle (properly Meanbh ; hence Ben Venue, G. A'Bheinn Mheanbh).
Meann, Mhinn : A kid.
Mèinn, Mèinne : Ore, *e.g.* Allt na Meinne (ore-burn).
Meur, Meoir : Finger branches, applied to small streams.
Min, Mine : Smooth.
Mna : wife, *genitive singular* of Bean, q.v.
Mòd, Mhòid : generally applied to a small knoll, where courts of justice sat in ancient times.
Mòine, Mòinteach, or Monadh : Peat, mossy ground (outer pasture in Lewis).
Mol, Mal : Shingly beach ; N. Möl (pebbles).
Monadh : hill, mt., as in Monadh-Ruadh (red mountains), Badenoch.
Mòr, Mhòr, Mhòir, Mòire : Large, great.
Morven : A'Mhorairne (the great land).
Moulinearn : Muileann-fheàrna (the mill at the Alders, a pl. near Pitlochry).

Mousa (Shetland) : N. Moosa (the moory isle).
Muc, Muice : Pig. A'Mhuclach (the piggery).
Muck Island : Eilean nam muc, Tir-chrain (Isle of swine).
Muileann, Mhuilinn : A mill.
Muir, Mhara, or Mara : The sea.
Mult, Mhuilt : Wether.
Muran, Mhurain : Sea bent.

Nairn : Inbhir Narunn (confluence of the r. Nairn).
Nathair, Nathrach, Nathraichean : Serpent.
Naver, River : Nabhuir (? old form Nabaros).
Nead, Nid : A nest.
Ness, Loch : Loch Nis (loch takes its name from the r. Ness).
Nighean, Nighinn : Daughter, young woman.

Ob, Oba, Oban : A bay, from N. Hóp (a bay) whence Longhope (Orkney). Loch Hope (Sutherland) appears as Tob in Lewis.
Oban : N. hóp (an inner bay).
Odhar, Odhair, Uidhre, Idhir : Dun-coloured.
Ogle, Glen : Gleann Oguil.
Oidhche (pron. Oech-a) : night ; *e.g.* Airidh na h-Aon Oidhche (sheiling of the one night (Lewis)) ; Loch na h-Oidhche (night loch, a loch that fishes best at night).
Oitir, Oitire : Sand bank ; oitir (at Otter ferry, Loch Fyne).
Onich : Omhanaich, Othanaich (foam-frothed (MacBain)).
Or, Oir : Gold.
Orchy, Bridge of : Drobhaid Urchaidh (on wood).
Ord, Uird : Round hill ; Ordan (a little round hill).
Orkneys : Old G. Innse (the isls.), Orc modern Arcaibh (whale-isles.)
Os, Ois, or Osa : River mouth, or outlet ; in Lewis (slowly moving water).
Overscaig : N. Ofarsgaig (upper strip).

Pàirc, Pàirce : A park, a field.
Pait, Paite, Paitean : A hump, place, ford or fords.

Pet, or **Pit :** Farm or piece of land, vil., sometimes a hollow, *e.g.* Pitmaduthy), G. Pit 'ic Dhuibh (Macduff's steading); Pitlochry, G. Bailechloichridh (stone-steading).

Pitlochry : Baile Chloichridh (stone steading).

Pittentrail : N. and G. Bail' an tràill (Thrall's stead).

Poll, Phuill : A pool or pit.

Port, Phuirt : Port, harbour, ferry.

Portmahomack : Port-mo-Cholmaig (Colman's Port).

Preas, Phris : Bush, shrub, copse.

Raineach, Rainich : Fern.

Ràmh, Ràimh : An oar.

Rathad, Rathaid : Road, a way.

Reamhar, Reamhair, Reamhra : Thick, fat.

Réidh, Réidhe : Smooth, level, plain.

Rhiconich : Ruighe-chóinnich.

Riabhach, Riabhaich : Brindled, greyish.

Righ, Righe : King (Ban-righ, Queen).

Rispond : Ruspuinn, small harbour E. of Durness, Sutherland.

Rogart : Sgire Rao'ird, ("Raghart, *i.e.* Roghart, Ra-ghart, Big field," (Watson).)

Ròinn, Ròinne : Point, prom.

Ròn, Ròin : A seal.

Rothiemurchus, Ràta mhurchais (Muirgu's place (MacBain)).

Roy, Glen : Gleann Ruaidh Red Glen.

Rudha : A spit, a prom.; Rudha na-h-Aoiriun, prom. of the Mass (Ardnamurchan).

Ruigh, Ruighe, Ruidh, Ruidhe : A run for cattle, sheiling, land sloping up to a hill.

Rum now **Rhum :** Eilean Ruma, "Rìoghachd na Forraiste fiadhaich."

Ruthven : Ruadhainn (the red place).

Sabhal, Sabhail : Barn.

Sac, Saic : Horse load.

Sagart, Sagairt : Priest.

Saidhe, Saigh, Saighe : Bitch.

Sàil, Sàile : (1) heel ; (2) salt water.

Saobhaidh, Saobhaidhe, Saobhaidhean : Fox-den.

Schiehallion : Sidh Chailleann (mt. of the Caledonians).

Sealg, Seilge : Hunt, hunting.

Seamrag, Seamraig : Trefoil or shamrock.

Sean, Seana : Old; Sean chreag (old rock).

Seangan, Seangain and **Sneaghan :** An ant.

Searrach, Searraich : Foal, colt.

Seileach, Seilich : Willow.

Sgadan, Sgadain : Herring.

Sgarbh, Sgairbh : The cormorant.

Sgeir, Sgeire : A sea rock ; N. Sker, an isolated rock in the sea.

Sgoilt, Sgoilte : split, *e.g.* Clach Sgoilte (split stone).

Sgor, Sgòrr or **Sgùrr :** Rocky peak.

Shetland, or **Zetland :** G. Sealtainn; N Hjaltland ; Prof. Jakobsen says, " No sure explanation has yet been offered of the name." [1]

Sian, Sine : Storm.

Sionnach, Sionnaich : Fox.

Sithean, Sithein (pron. Shee-an): A hillock; fairy knoll.

Sleamhuinn, Sleamhna : Slippery.

Sleac, Slic : flat stones, (Badenoch) same as Leac.

Slochd, Sloc, or **Sluichd :** Deep, hollow ; Slochd Mòr, the ravine through which the Highland railway passes between Carrbridge and Inverness.

Sneachd, Sneachda : Snow.

Socach, Socaich : Snout, projecting place, mossy ground between fork of streams, often anglicised as Succoth.

Spean, Glen : Gleann Spiothain (Glen of the r. Spean).

Speireag, Speireig : Sparrow-hawk.

Spey, River : G. Abhainn Spé, hawthorn stream (Watson).

Spréidh, Spréidhe : Cattle.

Srath, Sratha (locally pron. Stra.): A strath, valley.

Sròn, Sròine : A nose or point.

Sruth, Srutha, Sruthan, Sruthain : Current, a stream, streamlet.

Stac, Staca : Steep conical hill.

Stack, Ben, Beinn Stac, a stack.

Stirling : Sruighlea.

Stob : a point; Stob mòr, big point (Collins).

Stornoway : Steornabhadh. N. Stjorn, steering, *i.e.,* Steerage Bay.

Strath : G. Srath, q.v., a flat valley.

Strome Ferry : Port an t-Sròim (port at the prom. or nose).

Stronachlachair : Sròn a' chlachair (prom. of the mason).

Strone : G. Sròn, a nose, prom., *e.g.,* stone (Firth of Clyde).

[1] *Shetland Place-Names,* Jakob Jakobsen p. 119.

Strontian : G. Sròn an t-Sithein (prom. of the knoll).

Struan : G. Sruthan (a stream, current, streamlet).

Struy, Strathglass : G. an t-sùridh, place of streams (Watson).

Stùc : a peak, e.g. Stùc Garbh Mòr (the big, rough peak).

Suidhe : sitting or resting place, level shelf in a hill side; so also Spardan (a roost).

Suilven in Sutherland : G. Sula-bheinn; N. sulr (pillar), i.e. the pillar mt. often called " the sugar loaf."

Sutherland (place) : Cataibh (the country of the Cat Tribe).

Sutors of Cromarty : Na Sùdraichean (the place of tanners (Watson)).

Tabost : G. Tàbost, N. Habost (high stead).

Tain is in G. Baile Dhubhthaich (St. Duthac's town) ; Tain Tene (1227) Dr. Henderson thinks may be from N. eyrar-ping.

Tairbeart, Tairbeirt, Tarbert, Tarbet : A narrow isthmus, hence the Tarberts on the W. of Scotland. Place where boats were pulled across.

Tarbert : An Tairbeart (tarrainn-bat) ; i.e. narrows where small boats were pulled across.

Tarsuinn : Transverse, across.

Tayinloan : Taigh 'n lòn (house by the marsh) on W. Kintyre.

Taynuilt : Taigh an uillt (house by the burn), Argyll.

Tayvallich : Taigh a'bhealaich (house of the pass).

Tayvullin : Taigh a' mhuilinn (house of the mill).

Teampull, Teampuill : A temple or church.

Thurso : G. Inbhir Theorsa, confluence of the r. Horsa ; N. Thiórsá, Bull's Water (W.).

Tigh, Tighe : Tay, Ty

Tigh, Tighe, Tay, Ty : A house, e.g. Tighnabruaich, G. Tigh na Bruaich (house of the bank) ; Tayloin, G. Tigh an Lòin (house by the marsh).

Tipper : G. Tiobair (a well).

Tiree : Tir-iodh (corn-land (Henderson)).

Tòb : N. a bay (Lewis) ; same as Ob, q.v., from N. hóp (an inner bay).

Tobar, Tobair, Tober : A well.

Tobermory : Tobar Mhoire (St. Mary's Well).

Toll, Tuill : Hole.

Tom, Tuim : A round hillock ; e.g. Tomnahurich ; G. Tom na h-Iubhraich (hillock of the yew wood).

Tomintoul : Tom an t-sabhail (hillock of the Barn, the second highest vil. in Scotland, 1160 ft.).

Tomnahurich : Tom na h-iùbhraich (hillock of the yew wood), near Inverness.

Tòn, Tòin : Buttock, haunch.

Torc, Tuirc : A boar.

Torr, Torra : A heap, hill, castle.

Tràigh, Traighe, and Tràghad : (Sand) beach.

Tri-tighearnan : Three lords.

Trossachs, The : Na Tròiseachan (the cross places).

Trotternish : Trondairnis (in G. from the N. Throndar-nes). i.e. Thrond's Ness.

Tulach, Tulaich : A knoll, a hillock.

Tunnag, Tunnaig : A duck.

Tweed, River : Tuaidh.

Tyndrum : G. Tigh an Droma (house of the ridge); G. Druim Alba (ridge or backbone of Scotland).

Uachdar, Uachdair : Top, upper part.

Uaine : Green or blue ; Lochan Uaine (the green loch).

Uam Var : An Uamh mhór (the big cave).

Uamh, Uamha, Uaigh, Uaighe : Cave.

Uchd : Breast, an ascent ; Uchdan : A short, steep bank.

Uidh : Isthmus, slow moving water.

Uig : A nook, hollow, or bay.

Uisg, Uisge : Water.

Ulaidh : Treasure.

Ulbh, Ulbhaidh : N. Ulfr (a wolf).

Ullapool : G. Ullapul ; N. Ulli-bolstador, Uilli's stead.

Uruisg : human monster, goblin, brownie, e.g. Coire an Uruisg (Loch Katrine), corry of the monster.

Uyea (Shetland) : N. Oya (the isle).

Val : (1) N. Fjall (a high hill) ; (2) Hvall (a knoll).

Vat : N. Vatn (water) ; e.g. Loch Langavat (long water (Lewis)).

LIST OF ABBREVIATIONS

ac.	. . . .	acres.
affl.	. . . .	affluent.
alt.	. . . .	altitude.
bur.	. . . .	burgh.
civ.	. . . .	civil.
co.	. . . .	county.
dist.	. . . .	district.
div.	. . . .	division.
eccl.	. . . .	ecclesiastical.
ft.	. . . .	feet.
G.	. . . .	Gaelic.
ham.	. . . .	hamlet.
isl.	. . . .	island.
L.M.S.	. . .	London, Midland and Scottish Railway.
L.N.E.	. . .	London and North Eastern Railway.
m.	. . . .	mile, or miles.
mfrs.	. . . .	manufactures.
mt.	. . . .	mountain.
mun.	. . . .	municipal.
N.	. . . .	Norse.
par. (s.)	. . .	parish (es).
parl.	. . . .	parliamentary.
pl.	. . . .	place.
P.O.	. . . .	Post Office.
pop.	. . . .	population.
prom.	. . . .	promontory.
q.s.	. . . .	quoad sacra.
r.	. . . .	river.
rt.	. . . .	right.
ry. sta.	. . . .	railway station.
sub.	. . . .	suburb.
T.O.	. . . .	Telegraph Office.
Vil.	. . . .	village.

GAZETTEER OF SCOTLAND

A

Abbey, q.s. par., Arbroath town, Angus.

Abbey, q.s. par., E. Edinburgh; ry. sta., L.N.E. (Abbeyhill).

Abbey, q.s. par., Paisley, N.E. Renfrewshire; 15,924 ac. Contains the town of Johnstone and parts of Barrhead and Paisley. Flax, cotton mills, and engineering works.

Abbey, a ham. on the Tyne, 1½ m. E. of Haddington sta., E. Lothian. Has remains of convent. Parliament sat here in 1548.

Abbey Burn, a stream S. Kirkcudbrightshire, enters Solway Firth at harb. of Burnfoot, 6 m. S.E. of Kirkcudbright.

Abbey Craig, a rocky knoll 1¾ m. N.E. of Stirling (ht. 362 ft.). Wallace Monu. on summit, rising to 220 ft., and commands a magnificent view of the Forth Valley.

Abbey Green, see LESMAHAGOW.

Abbey Head, applies to a headland on the Solway, 6 m. S.E. of Kirkcudbright.

Abbey Hill, Berwickshire, a hill (913 ft.) 6 m. N.N.W. of Duns.

Abbey St. Bathans, par. and ham., Berwickshire. Took its name from Cistercian Nunnery founded by a daughter of William the Lion. 4 m. W.S.W. of Grant's House sta. P.O. 5845 ac. Pop. 169.

Abbeythune, a seat in Angus, 1 m. S. of Inverkeilor and 2 m. S.W. of Lunan Bay.

Abbotrule, an ancient par. in Roxburghshire, divided equally in 1777 between the pars. of Hobkirk and Southdean.

Abbotsford, q.s. par., Govan par., Glasgow.

Abbotsford, Roxburghshire, the mansion-house erected by Sir Walter Scott on the right bank of the r. Tweed, 2 m. W. of Melrose. Built in 1817–1821, it stands on a terrace and commands a magnificent view of the surrounding country. It contains Scott's library of over 20,000 volumes, and is open to the public at certain hours during the summer season.

Abbotsford Ferry, Selkirkshire; sta. on L.N.E. ry., 2 m. S.E. of Galashiels.

Abbotshall, former par., now in par. of Kirkcaldy and Dysart, Fife.

Abbotshaugh, Stirlingshire, a quondam abbey now obliterated, near Grangemouth, in Falkirk par., Stirlingshire. Its grange gave the name to Grangeburn, and through that to Grangemouth.

Abbots Isle, small green island near the foot of Loch Etive, Muckairn par., Argyll.

Abbots Tower, ancient ivy-clad tower, 40 ft. high, ½ m. E.N.E. of Sweetheart Abbey in New Abbey par., Kirkcudbrightshire.

Abbotsview, convalescent home of the Scottish Co-operative Wholesale Society, Selkirkshire, opposite Abbotsford.

Abbotswell, seat in the par. of Nigg, Kincardineshire, 1 m. S. of Aberdeen.

Abdie, par., Fife, on Firth of Tay; 4850 ac. Pop. 670. Contains Newburgh, its post town. Limestone, red sandstone, and trap rock occur. Lindores loch, 4 m. in circumference. Vestiges of Macduff Castle and Wallace Camp, ½ m. from the Firth; modern mansions —Incrye Abbey and Lindores House. Ry. sta., Lindores.

Aber, a ham. and islet, Dunbartonshire, S.E. shore of Loch Lomond, par. of Kilmaronock. The islet is 1 m. N. of the ham.

Aberarder, glen in Aberdeenshire, and in the Dee Valley, between Crathie and Invercauld, affords a good view of Ben Avon (3843 ft.).

I

Aberarder, ham. and estate, par. of Daviot and Dunlichty, 12 m. S. of Inverness. P.O., T.O.

Aberarder Lodge, N.E. end of Loch Laggan, par. of Laggan, Inverness-shire.

Aberargie, vil., Perthshire, Abernethy par., 4 m. E.S.E. of Bridge of Earn. P.O.

Abercairney, seat, S.E. Perthshire, 3 m. E. of Crieff; ry. sta., L.M.S., same name, 4¼ m. E. of Crieff.

Aberchalder, a locality on the Caledonian Canal, 5 m. S.W. of Fort Augustus, Inverness-shire. Here Prince Charles mustered 2000 men (1745) before his march south.

Aberchalder, Wester, an estate in Daviot and Dunlichty par., Inverness-shire.

Aberchirder, a police bur., Banffshire, 9¼ m. S.W. of Banff; 63 ac. Pop. 868. P.O., T.O. Bank.

Abercorn, par. and vil., West Lothian; 4508 ac. Pop. 775. Vil., 3 m. N.N.W. of Winchburgh.

Abercorn, ry. sta., L.M.S., Paisley. Called after the Duke of Abercorn, who owns some of the land here.

Abercrombie, a ham., 1½ m. W. of St. Monans, Fifeshire.

Aberdalgie, par., vil., and seat, 4 m. S.W. Perth; par. 4172 ac. Pop. 235. P.O.

Aberdargie, see ABERARGIE.

Aberdeen, co. town of Aberdeenshire, parl., royal, and mun. bur., is the fourth largest town in Scotland. Pop. 167,259. The town lies chiefly between the rivers Dee and Don, with the small district of Torry to the S. of the Dee and in the co. of Kincardineshire; 130½ m. N.E. of Edinburgh by L.N.E., and 158¼ m. by L.M.S. via Stirling. Comprises Old and New Aberdeen; is built almost entirely of granite, and is aptly called " The Granite City." It is the seat of a University formed by union of the University and King's College of Old A., and the University and Marischal College of New A. The Scottish Fishery Board Marine Laboratory and Aquarium is situated at Torry, and the Royal Naval shore wireless sta. at Stoneywood. It has extensive docks, electric trams and buses, G.P.O. with branches, lifeboat sta., and all the Scottish banks are represented. Principal exports—granite, cattle, and fish—

as much as 600 tons of fish being landed daily. Industries include shipbuilding, engineering, fish curing, papermaking, granite polishing, cordage, tarpaulin, and sail-cloth manufacturing. Returns two members to Parliament.

Aberdeenshire, a maritime co. in N.E. Scotland. Area 1971 sq. m. or 1,251,451 ac. Pop. 300,430. Bounded on the N. and E. by the North Sea, on the W. by Banffshire and Inverness-shire, and on the S. by Perthshire, Angus, and Kincardineshire. The coast-line is mostly bold and rocky, inland hilly, and in the S.W. mountainous. The greatest heights are Ben Macdhui 4296 ft., Cairntoul 4141 ft., and Cairngorm 4084 ft. The chief rivers are the Dee, Don, Deveron, and Ythan, all noted salmon rivers. A large portion is under wood, and there are also extensive tracts of moorland, but towards the sea the land is fertile and comparatively level. Large numbers of sheep and cattle are reared, including the famous Polled Angus breed of cattle. It has important fisheries and granite quarries. Balmoral is the King's Highland residence. Combines with Kincardineshire to form three divisions, returning one member to Parliament for each division.

Aberdona House, seat and estate in Clackmannanshire, 5 m. E.N.E. of Alloa.

Aberdour, coast, par., and vil., S.W. Fifeshire, 3 m. W. of Burntisland; 5793 ac. Pop. 2055. P.O., T.O. Vil. pop. 1310. Situated in a sheltered place on the shore of the Firth of Forth, it is a favourite watering place, has good hotels, ry. sta. L.N.E. Contains the vil. of Donibristle Colliery, and includes the island of Inchcolm, 1¼ m. S., and the ruins of a nunnery and ancient castle; also Mansion of Donibristle.

Aberdour, par. and vil., Aberdeenshire; 14,034 ac. Pop. 1180. Lies 8 m. E. of Fraserburgh, the vil. called New Aberdour, founded in 1798. On the Moray Firth, has a rocky seaboard of 6 m. Contains the fishing vil. of Pennan. Inland the surface is level. Vil. has P.O., T.O. Vil. pop. 426.

Aberdour House, Aberdeenshire, 1½ m. N.E. of New Aberdour.

Aberdour House, girls' private school.

Aberfeldy, G. *obar Pheallaidh* confluence of the Peallaidh. Police bur. with ry. sta., L.N.E. On the Tay where it is crossed by one of General Wade's finest bridges. The town has a pop. of 1505. P.O., T.O. Distillery, dyework, sawmills, and woollen factory, hotels, banks. Is in the par. of Dull and lies 32½ m. N.W. of Perth.

Aberfoyle, G. *obar Phuill* confluence of sluggish stream. Vil. and par. S.W. Perthshire. Pop. 1014; 26,810 ac. The vil. stands at S.E. of par., 4 m. S. by W. of the Trossachs between which there is an excellent road. Has P.O., T.O. Ry. sta., L.N.E. Hotels. The vil. or clachan made classic by Scott's novel of *Rob Roy*.

Abergeldie Castle, an estate on His Majesty King Edward's property. The Mansion at an alt. of 840 ft., stands behind Craig-na-ban, a hill (1736 ft.) 6 m. above Ballater and 2 m. below Balmoral.

Aber Isle, small island near S.E. end of Loch Lomond, Dunbartonshire, 2½ m. W. of Buchanan Castle.

Aberlady, par. and vil. with ry. sta., L.N.E. (passenger service withdrawn), E. Lothian. The vil. stands at the mouth of the Peffer Burn on the coast, 5¼ m. N.W. of Haddington; par. 4333 ac. Pop. 1094. P.O., T.O. Vil. pop. 541.

Aberlemno, vil. and par. of central Angus, 5 m. N.E. of Forfar. Pop. 627. P.O., T.O. The par. extends to 8918 ac. S. Esk forms N. boundary. Highest ground (800 ft.) at Tarin Hill Fort. Contains old castle of Flemington and the ruins of Melgund Castle.

Aberlour, par. and police bur. (Charlestown of Aberlour); ry. sta., L.N.E., W. Banffshire, 17 m. S.W. of Keith; par. 14,779 ac. Pop. 2327. Has a distillery. The par., bounded by the r. Spey on N.W., greatest length, 9 m., breadth, 1 to 5 m. Vil. is the seat of Sir Edmund Findlay. Hotel. Ben Rinnes (2755 ft.) is a prominent object. P.O., T.O.

Aberlour, Burn of, Banffshire. Small burn on which the village of that name stands.

Aberluthnet, a rivulet, and ancient name for Marykirk par., Kincardineshire.

Abermilk, or Abermele, Dumfriesshire, ancient par. See ST. MUNGO.

Abernethy, par. and police bur., with ry. sta., L.N.E., S.E. Perthshire, 8½ m. S.E. of Perth. Pop. of par. 1154; of bur. 595. P.O., T.O. Ac. of par. 9232. Vil. stands on the rt. bank of the Nethy, and anciently a seat of the Culdees. Has a round tower similar to those in Ireland. The par. contains the hams. of Aberargie, Glenfoot, and Mugdrum Isl.

Abernethy and Kincardine, par., E. Inverness-shire; 78,088 ac. Pop. 1123. Contains a large deer forest; also vil. of Nethybridge, 4½ m. S.S.W. of Grantown; ry. sta. on L.M.S. P.O., T.O.

Abernyte, par., ham., and seat, 10 m. N.E. of Perth; par. 2533 ac. Pop. 194. P.O., T.O. at Inchture. Contains Dunsinane Hill (1012 ft.) and King's Seat (1235 ft.) near boundary.

Abertarff, formerly a par. in Inverness-shire on the N.W. side of Loch Ness, now united with Boleskine par.

Abertay, sandbank off Buddon Ness, mouth of the Tay, with Lightship.

Abertay Sands, sandbank at S. estuary of the r. Tay.

Aberturret House, a seat 1½ m. W. of Crieff, Perthshire.

Aberuchill, an estate, with mansion house and the remains of an ancient castle, 1¾ m. S.W. of Comrie, Perthshire.

Aberuthven, vil. 2½ m. N. of Auchterarder, Perthshire. P.O.

A' Bhuidheanaich, mt., Inverness-shire, 4 m. N.W. of Kincraig ry. sta.; 2358 ft.

Abington, vil. and ry. sta., L.M.S. Crawfordjohn par. 800 ft. above sea-level, on the left bank of the Clyde, ¾ m. below the entrance of Glengonnar Water. P.O., T.O. Hotel.

Above-the-hill, a ham. in the N.W. corner of the par. of Birsay, Orkney.

Aboyne and Glen Tanar, par., S. Aberdeen-shire, on r. Dee; 25,265 ac. Pop. 1552. Vil. of Aboyne, a charming summer resort with ry. sta., L.N.E., 11 m. E. by N. of Ballater. Hotel. P.O., T.O. Vil. sometimes called Charleston of A. Pop. 902. A. Castle dates from eleventh century; seat of the Marquis of Huntly.

Abriachan, ham. and stream. Latter enters the north side of Loch Ness, 8 m. S.W. of Inverness.

Acha, a dwelling in the S. of the isl. of Coll, Argyllshire.

Achabeg, pl., Morven, Argyll, 1½ m. W. of Lochaline.

Achacharra, refers to standing stones on the isl. of Ulva, Argyllshire.

Achagavel, a habitation at the head of Glen Dubh, Morven, Argyll.

Achaglachgach, shooting - lodge W. of Loch Tarbert, 6½ m. S. of Tarbert, Argyll. State forest here.

Achahoish, Knapdale, Argyll ; ham. at the head of Loch Killisport. P.O., T.O. at Ardrishaig.

A'Chailleach, mt., Inverness-shire, 4 m. N. of Newtonmore. Alt. 3045 ft.

A'Chailleach, mt., Ross and Cromarty, 3276 ft., near the head of Loch Fannich and 8 m. N. of Achnasheen.

Achairn, stream and ham., 4 m. W. of Wick, Caithness.

Achalder, seat, 3 m. W. of Blairgowrie, Perthshire.

Achaleven, Argyll, vil., ¾ m. E. of Connel, on the S. side of Loch Etive ; ry. sta., L.M.S., at Connel. Pop. 147.

Achalick, fishing station on the E. of Loch Fyne, Argyllshire.

Achall, small loch, 70 ft. deep, 1·83 m. long, 1½ m. E. of Ullapool, Ross and Cromarty.

Achallader, a farm and ruined stronghold of the Campbells, Argyll, 1 m. N.E. of the E. end of Loch Tulla. The new road to Ballachulish passes it.

Achamore, pl., on Kyle of Durness, Sutherland.

Achamore House, at the S. end of the island of Gigha, Argyll, opposite Tayinloan, Kintyre.

Achanalt, a ry. sta., L.M.S., with T.O. and ham. with P.O. 21½ m. W. of Dingwall, Ross and Cromarty.

Achan Burn, a tributary of the r. Free, 4 m. S.W. of Kinbrace ry. sta., L.M.S., Sutherland.

Achanduin, a square, roofless building, said to be the residence of the Bishops of Argyll in early times, stands near the S. end of Lismore Isl., Argyll.

Achandunie, a house, 1½ m. N. of Alness, Ross and Cromarty.

Achantiobairt, *field of sacrifice,* a pl. with stone crosses on the E. shore of Loch Fyne, 5½ m. S.S.W. of Inveraray, Argyll.

Achany House, an estate and residence on the rt. bank of the r. Shin, 4 m. N. of Invershin, Sutherland.

Acharacle, see AHARACLE.

Acharacle, ham., Argyll, W. end of Loch Shiel. P.O., T.O. 16 m. S.W. of Glenfinnan ry. sta.

Acharainey, formerly a ham., Caithness, 21 m. S.W. of Wick.

Achareidh, a seat 1 m. W. of Nairn, Morayshire.

Acharn, vil. and stream with falls, 2 m. S.W. of Kenmore, Perthshire. P.O.

Acharnie, ham., 7 m. E. of Huntly, Aberdeenshire.

Acharole Burn, small stream, enters Wick Water, near Loch Watten, Caithness.

Acharosson Burn, head water of Auchalick r., Argyll, 2 m. S. of Kilfinan.

Achateny, ham. and small river, N. coast of Ardnamurchan, Argyll, P.O.

Achavandra, ham., 3 m. N.W. Dornoch, Sutherland. Now obsolete.

Achavanich, ham., 6 m. N. of Latheron, Caithness.

Achavarn, seat, 6 m. S.W. Thurso and E. side of Loch Calder, Caithness.

Achbreck, ham., Banffshire, 6 m. S.E. Ballindalloch ry. sta.

Achdalieu, seat, N. shore of Loch Eil, 5 m. N.W. Fort William ; also the name of an extensive deer forest.

Achduart, ham., 1½ m. N.W. of Ullapool, on the shore of Loch Broom, Ross and Cromarty.

Achenreoch, moorland tract, 4½ m. N.E. Dumbarton.

Achenreoch, loch, Kirkcudbrightshire, 7 m. N.E. of Castle Douglas; alt. 345 ft., depth 34 ft.

Achentorlie, obsolete name, seat, Paisley, Renfrewshire.

Achfarry, pl., west end of Loch More, Sutherland, on the road from Lairg to Scourie.

Achiltibuie, ham. and hotel. P.O. and T.O. ; on the N. coast of Loch Broom, 13 m. N.W. Ullapool, Ross and Cromarty.

Achilty, loch, 4 m. S.W. Strathpeffer, Ross and Cromarty ; alt. 90 ft., depth 119 ft.

Achindrain, ham., 5 m. S.W. of Inveraray, Argyll.

Achingale, ham., 8 m. W. of Wick, Caithness.

Achinner, ham., and lodge in Glen Artney, 5 m. S. of Loch Earn, Perthshire.

Achintee, farm in Glen Nevis, 2 m. E.S.E. of Fort William, Inverness-shire.

Achintoul, lodge, 2 m. N. Kinbrace; ry. sta., L.M.S., Sutherland.

Achintraid, ham., with par. E. side Loch Kishorn, an arm of Loch Carron, Ross and Cromarty.

Achlishie, estate, in Kirriemuir, par. of Angus.

Achluachrach, ham., Inverness-shire, 3 m. W. of Tulloch ry. sta., L.N.E.

Achmelvich, ham., 2 m. W. of Lochinvar, Sutherland.

Achmore, district of Kenmore par. and seat, Perthshire.

Achmore, pl., 4 m. E. of Plockton sta., W. Ross and Cromarty.

Achnaba, seat on W. coast of Loch Fyne, 3 m. E. of Ardrishaig, Argyll.

Achnacarry, ham. and seat, 5 m. N.W. of Spean Bridge, Inverness-shire. P.O., T.O.

Achnacloich, Argyllshire, ham., S. shore of Loch Etive; ry. sta., L.M.S.; pier.

Achnacloich, ham., S. Isle of Skye, Inverness-shire, on Loch Eishort, 6 m. N. of the Pt. of Sleat.

Achnacloich, pl. and small loch, 2¼ m. N. of Alness, Ross and Cromarty.

Achnacone House, 1½ m. N.W. of Creagan ry. sta., L.M.S., and ½ m. E. of Appin Church, Argyll.

Achnacraig, ham., Argyll, Mull Isl., and ferry sta. to Oban. P.O., T.O.

Achnacroish, estate and ham., Argyllshire, Duart Bay, Isl. of Mull; 1½ m. W. of Duart Pt.

Achnacroish, pl., with pier, Lismore Isl., Argyllshire, on the E. side and 5 m. from S. pt.

Achnadrish, shooting lodge, Isl. of Mull, Argyll, on the road from Tobermory to Kilmore, and 6 m. from the former.

Achnagairn, seat 1 m. S.W. of Clunes ry. sta., L.M.S., and 8 m. W. of Inverness.

Achnagarron, ham., 2 m. N.W. of Invergordon, Ross and Cromarty.

Achnagart, pl., on S. side of r. Oykell, and 2 m. N.W. of Invershin; ry. sta., L.M.S., Ross and Cromarty.

Achnahaird House, 4 m. N. of Achiltibuie, Ross and Cromarty.

Achnahannet, pl., S.W. Morayshire, 3½ m. W. of Grantown.

Achnahannet, ham., Kincardine par., Ross and Cromarty, 7 m. N.W. of Bonar Bridge.

Achnairn, on Colabole farm, E. side Loch Shin, and 7 m. N.W. of Lairg, Sutherland, P.O.

Achnamara, seat, at the head of Loch Sween, Argyll. N. Knapdale. P.O., T.O. Pier, ½ m. S.

Achnarrow, ham., S.W. Banffshire, 9 m. S.E. of Ballindalloch sta. P.O. called Auchnarrow.

Achnasheen, ry. sta., L.M.S., Ross and Cromarty, 27¾ m. by ry. from Dingwall. P.O., T.O. It is the starting-place of buses for Lochmaree, Gairloch, and Aultbea.

Achnashellach, a State forest and ry. sta., L.M.S., 40¼ m. by ry. from Dingwall, Ross and Cromarty. P.O., T.O. at sta.

Achnashellach Hostel, on Dingwall-Kyle road in Glen Carron, near A. sta., Ross-shire. A Scottish Youth Hostel.

Achnavarn, ruins of ancient castle on the E. of Loch Calder, par. of Halkirk, Caithness.

Achneigie, pl., 2 m. S.E. of Loch na Sheallag, Ross and Cromarty.

Achorachan, ham. on r. Livet, 6 m. S. of Ballindalloch, Banffshire.

Achosnich, ham., Ardnamurchan par., Argyll, 2 m. E. of Ardnamurchan Pt. P.O.

Achoy, pl., ¾ m. S. of Loch Watten, Caithness.

Achrain, small loch 4 m. S.W. of Reay, Sutherland.

Achrannie, Slugs of, two waterfalls on the r. Isla, 3½ m. N.E. of Alyth, Angus.

Achray, well-known loch in Trossachs, 1 m. E. of Loch Katrine; alt. 276 ft., depth 97 ft.

Achridigill, loch in Sutherland, 3 m. S.W. of Melvich Hotel.

Achriesgill, ham., near the head of Loch Inchord, 16 m. S.W. of Durness by road.

A' Chruach, a mt. in Isl. of Arran (1679 ft.), 3 m. S.W. from Brodick Bay.

Achtercairn, ham., 1½ m. N.W. of Gairloch, Ross and Cromarty.

Achterneed, ham. and ry. sta., L.M.S.; P.O., T.O. at sta.; 4⅓ m. by ry. from Dingwall, Ross and Cromarty.

Achtriochtan Loch, in Glencoe, Argyll, 4 m. S.E. of Ballachulish.

Achuaran, seat, Lismore Isl., Argyll, 1½ m. S.W. from N. pt. of isl.

Achvaich, pl. and strath, Sutherland, 6 m. N.W. of Dornoch.

Achvarasdal Burn, a stream issuing at the foot of Beinn nam Bad Mhor ; it runs for 6 m. into Sandside Bay, N. Caithness.

Achvarasdal Lodge, 1 m. E. of Reay, Caithness.

Ackergill Tower, an ancient stronghold with modern mansion, 1½ m. N. of Wick, Caithness. T.O. at lifeboat sta.

Ackerknowe, pl. and reservoir at Stobs military camp, Roxburghshire, 2½ m. S. of Hawick.

Ackerness, headland on N. of Westray Isl., Orkney.

Acurrach, pl. 2 m. S. of Cladich and 8 m. N. of Inveraray, Argyll.

Adam's Row, ham., 5 m. S.E. of Edinburgh, par. of Newton.

Adamton House, seat in co., and 4 m. N.E. of Ayr.

Add, r., issuing from Loch Sitheanach, it meanders in a S.W. direction into Crinan Loch, Argyll.

Adderlaw, hill summit in the E. par. of Applegarth, Dumfriesshire ; alt. 822 ft.

Addie, hill (893 ft.), 4 m. S.E. of Buckie, Banffshire.

Addiewell, manufacturing vil. in par. of W. Calder, Midlothian. Sprang up in 1866 in connection with paraffin oil works. 1½ m. from W. Calder ; ry. sta., L.M.S. P.O., T.O. ; Pop. 3141.

Aden House, seat, Aberdeenshire, ¾ m. E. of Mintlaw sta., Old Deer.

Adigo, loch in Uig par., Lewis.

Advie, q.s. par., with ry. sta., L.N.E., Morayshire, on r. Spey, 8 m. N.E. of Grantown. P.O., T.O.

Ae Water, r., rises in Queensberry Hill (2285 ft.), 6½ m. W.S.W. of Moffat, and flows for 16 m. to join the Kennel Water 2 m. N. of Lochmaben. State Forest of Ae, near Amisfield ry. sta., 5 m. N.E. of Dumfries.

Affleck, seat and ancient castle in Monikie par., Angus.

Afforsk, picturesque ravine in Gamrie par., Banffshire.

Affric, one of the most picturesque glens, lochs, and waters in Scotland. The loch, 14 m. N.W. of Fort Augustus, stands 744 ft. above sea level and 221 ft. deep. Inverness-shire.

Afton, water flowing 9 m. N. to Nith, in New Cumnock par., S.E. Ayrshire.

Afton-Bridgend, vil. immediately S. of New Cumnock, Ayrshire.

Afton Glen Sanatorium, 3 m. S. of New Cumnock, Ayrshire.

Afton Lodge, 1½ S.W. of Tarbolton, Ayrshire.

Agabatha, ancient military fort in Collessie par., Fife.

A'Ghlas-Bheinn, mt., Ross and Cromarty, 2½ m. N. of Beinn Fhada (Ben Attow) ; alt. 3006 ft.

A-Ghlinne, loch, N. Harris, near the mouth of Loch Resort, Outer Hebrides.

Agie Burn, trib. of r. Roy, Inverness-shire.

Aharacle, eccl. par. in Ardnamurchan and Morven pars., Argyll, and Arisaig and Moidart pars., Inverness-shire.

Aigas, seat, islet, and ferry onr. Beauly, 6 m. S.W. of Beauly, Inverness-shire. P.O.

Aikengall Burn, small stream, enters Braidwood Burn, 4 m. S.E. of Dunbar, E. Lothian.

Aikenhauld, site of an ancient castle in Oathlaw par., Angus.

Aikenhead, seat in Cathcart, S. of Glasgow.

Aikenway, peninsula in a bend of the r. Spey, 1 m. N. of Rothes.

Aikerness, Orkney, a prom. on the N.E. of the mainland, opposite Rousay.

Aiket Castle, Ayrshire, ruined castle, 7 m. N.W. of Kilmarnock.

Aikey Brae, Aberdeenshire, pl. on the border of Old Deer par.

Ailort, Loch Eilt, Inverness-shire, r., issuing from it enters Loch Ailort, Sound of Arasaig.

Ailsa Craig, Ayrshire, prominent island in Firth of Clyde, rising to 1114 ft., 12 m. S. of Arran, and forming part of Dailly par. Its base is elliptical, and measures 3300 ft. in one direction and 2200 ft. in another. Its rock is columnar syenitic trap. "If Ailsa Craig," says Dr. Macculloch, "has not the regularity of Staffa, it exceeds that isl. as much in grandeur and variety as it does in absolute bulk." Pop. 11.

Ailsh, loch, 6½ m. S.S.E. of Assynt, S.W. Sutherland ; alt. 498 ft., depth 24 ft.

Ainort, sea loch, Skye, Inverness-shire, 7 m. W. of Broadford.

Ainshval, mt., Isl. of Rhum ; 2552 ft.

Aird, waterfall, on r. Shinnel, W. Dumfriesshire.

Aird, ham., Wigtonshire, 2½ m. E. of Stranraer.

Aird, ham., Inverness-shire, N.E. extremity of Skye, opposite Trodday Isl.

Aird, ham., Sleat par., Skye, Inverness-shire, 2 m. N.E. of the Pt. of Sleat.

Aird, The, district near Beauly, Inverness-shire.

Aird Castle, ancient castle ruin, Ayrshire, 1 m. N. of Carradale Pt.

Airdeglais, loch, Argyll, 2½ m. N. of the head of Loch Buie.

Airdit, hill (515 ft.), 2½ m. S.W. of Leuchars ry. sta., Fife.

Aird Lamisheadar, lighthouse, visible 15 m. at entrance to Loch Carloway, Lewis, Ross and Cromarty.

Aird Point, see AULTBEA.

Airdrie, mun. bur. and q.s. par., Lanarkshire with ry. stas., L.M.S. and L.N.E. ; 11 m. E. of Glasgow. Pop. 25,954. P.O., T.O. Has extensive iron, boiler, engineering, and railway wagon works, collieries, brick and tile works, paper-making and cotton weaving. A. House, a seat.

Airdrie, seat and estate, Fifeshire, 3 m. W.N.W. from Crail.

Airdriehill, pl. in New Monkland par., Lanarkshire, 1½ m. N.E. of Airdrie.

Airds, seat, on Loch Ken, Kirkcud-brightshire.

Airds, dist., with seat, N. Argyllshire, between Lochs Linnhe and Creran.

Airds Bay, bay on Loch Etive, 1½ m. N. of Taynuilt.

Airds House, Argyll, ¼ m. S.W. of Port Appin.

Airds Moss, Ayrshire, large tract of mossy land, 3 m. N. of Cumnock.

Airgold, one of the summits of the Bengloe mountain range, Perthshire, Blair-Atholl par.

Airhouse, estate, Channelkirk par., Berwickshire, 5¾ m. N.N.W. Lauder.

Airhouse Law, Lammermuir Hills, 1096 ft., Berwickshire.

Airidh Brutach, pl. at head of Loch Sea-forth, Lewis Isl., Hebrides.

Airidh Mhic Ruairidh, sheiling, 2 m. W. of Sollas, North Uist, Ross and Cromarty.

Airidh na Beinne, loch, N.W. Sutherland, 4½ m. W.S.W. of Durness.

Airlequhillart, Flow of, dist. in co., and 6 m. S.W. Wigtown.

Airleywight, seat, in Auchtergaven par., Perthshire, 3½ m., N.N.W. of Dunkeld sta.

Airlie, par., W. Angus, Kirkton of—5¼ m. W.S.W. of Kirriemuir ; 8923 ac. Pop. 687.

Airlie Castle, seat of the, Earl of Airlie, 5 m. S.W. of Kirriemuir Angus.

Airor, ham., Sound of Sleat, Knoydart, W. Inverness-shire.

Airth, vil. and par., E. Stirlingshire, 8¼ m. S.E. of Stirling ; 5561 ac. Pop. 2226. Airth Castle in vicinity. P.O., T.O., and P.O. at sta. Vil pop. 691.

Airthrey Castle, estate, with mineral wells, Logie par., Stirlingshire, 2 m. N. of Stirling.

Airyligg, vil., Wigtownshire, 4½ m. N.E. of Glenluce sta.

Aith, or Skaill, a loch in Sandwich par., Orkney.

Aith, ham., bay, loch, Shetland, 14 m. N.W. Lerwick. P.O., T.O.

Aithernie, estate, E. Fifeshire, 2 m. N.W. of Largo. An ancient tumulus here was opened in 1821 and found to contain twenty stone coffins.

Aithness, headland, W. Shetland, 14 m. N.W. of Lerwick.

Aithova, harbour, Bressay Sound, E. Shetland, near Lerwick.

Aithsting, an ancient par. in Shetland on the W. side of mainland now united with Sandsting, q.v.

Aithsvoe, a creek or bay in S.E. Shetland, N. of Mousa Isl., 9½ m. S.W. of Lerwick.

Aith Wards, southern part of Hoy, Orkney, almost insulated by Long Hope Bay.

Aitkenhead, collier vil., Lanarkshire, 2 m. W. of Bellshill. Pop. of A. and Tan-nochside 3060.

Ait-Suidhe-Fhinn, or Fingal's Sitting Place, mt., Skye, Inverness-shire, at head of Loch Portree, over 1000 ft.

Akermoor Loch, Selkirkshire, 8 m. S.W. of Selkirk.

Albyn Distillery, Campbeltown, Argyll.

Alcaig, ham. and par. of Urquhart and Logie Wester, 1 m. S. of Dingwall, Ross and Cromarty. P.O.

Aldarder, a burn in Knockando par., Morayshire, running 4 m. into the Spey. Originally leaping over an 80-ft. fall, it changed its course in the flood of 1829 and cut a chasm 750 ft. long and 60 to 100 ft. deep.

Aldbar Castle, seat, S.E. Angus, 2½ m. S.W. of Brechin.

Aldcambus, an ancient par. on the coast of Berwickshire, now united to Cockburnspath.

Aldcathie, an ancient par., now in Ecclesmachan and Kirkliston pars., West Lothian.

Aldclune, ham., 2 m. N.E. Pass of Killiecrankie, N. Perthshire.

Alder Bay, on W. shore of Loch Ericht, 2 m. from S.W. end Inverness-shire.

Aldernan or **Allt-Arnan,** rivulet, with course of 3¾ m. into the Falloch, W. Perthshire.

Alderston, convalescent home, 1¼ m. W.N.W. of Haddington, E. Lothian.

Alderstone, seat, 2 m. W.S.W. Mid-Calder, Midlothian.

Aldgirnaig, pl. on r. Garry, near Killiecrankie sta., Perthshire.

Aldie Castle, ham. and ruins of ancient seat of Mercers of Aldie, 5 m. S.W. of Kinross.

Aldivalloch, ham. in Cabrach par., 17 m. S.W. Huntly, Banffshire.

Aldouran Lodge, 3½ m. N.W. of Stranraer, Wigtownshire.

Aldourie Castle, seat, with pier, in par. of Dores, stands on E. side of Loch Ness, 7 m. S.W. of Inverness.

Aldreguie, streamlet falling into r. Livet, at E. side of the Bochle, Banffshire.

Aldroughty, seat, 2 m. W. of Elgin, Morayshire.

Aldyonlie, rivulet rising in hills of par. of Knockando, Morayshire; trib. to r. Spey, into which it falls 3 m. N. of Ballindalloch.

Alemuir, loch, par. of Roberton, Selkirkshire, 6½ m. S.W. of Ashkirk and is in parts 180 ft. deep.

Ale Water, affl. of r. Eye, rises in par. of Coldingham, N.E. Berwickshire, and runs 6 m. S.E. to the Eye, near Eyemouth.

Ale Water, affl. of r. Teviot, Selkirkshire and Roxburghshire, rises in Henwoodie Hill (1189 ft.), S.E. Selkirkshire, and joins Teviot near Ancrum.

Alexandra Parade, ry. sta., L.N.E., Glasgow, 1¾ m. E. of Queen st.

Alexandra Park, dist. in E. of Glasgow. P.O., T.O.

Alexandria, q.s. par. and town on r. Leven, Dunbartonshire, 3¼ m. N. of Dumbarton; ry. sta. on L.N.E. P.O., T.O. Has extensive cotton dyeing, calico-printing, and bleaching works. Pop. 15,565.

Alford, par., vil., and seat on r. Don, in mid Aberdeenshire, 29½ m. from Aberdeen; 9105 ac.; ry. sta. on L.N.E. P.O., T.O. Pop. of par. 1336; pop. of vil. 570. An important agricultural centre. Hotel.

Alford, Howe or **Vale of,** part of Don Valley, Aberdeenshire; 9 m. long by 5 to 7½ m. broad, extends from Forbes to Bridge of Keig.

Aline, sea loch, par. of Morven, Sound of Mull, Argyll. See LOCHALINE. Loch Aline House, a seat.

Allachy, Water of, head stream of the Tanar, Aboyne par., Aberdeenshire.

Alladale, glen, deer forest, and lodge, 10 m. W. of Bonar Bridge, Ross and Cromarty.

Allallaidh, loch in Isl. of Islay, Argyllshire, 7 m. E. of Bowmore.

Allan, seat, 1 m. S. of Fearn sta., Ross and Cromarty.

Allan, affl. of r. Forth, rises in par. of Blackford in the Ochil Hills, Perthshire, and enters r. Forth, 1½ m. above Stirling.

Allan or **Allen,** affl. of r. Tweed, rises in N. Roxburghshire. After a course of 6 m. through the Fairydean, it enters the Tweed 2 m. W. of Melrose.

Allan, affl. of r. Teviot, which it enters 4 m. above Hawick, S.W. Roxburghshire.

Allanbank, mansion, Kirkcaldy, Fifeshire.

Allanbank, pl., Berwickshire, 1½ m. S.W. of Chirnside.

Allanbank, pl., Lanarkshire, 2 m. N.E. of Newmains.

Allanbank, seat, Berwickshire, ½ m. W. of Lauder.

Allanbank House, E. Ross and Cromarty, 1½ m. S.W. of Munlochy.

Allan, Bridge of, see BRIDGE OF ALLAN.

Allander, trib. of the Kelvin, Dunbartonshire, which it joins 2 m. S.E. of Milngavie.

Allanfearn, ry. sta. on L.M.S., in co., and 3¼ m. E. Inverness. T.O.

Allanfield, seat, ½ m. W. side of Dingwall, Ross and Cromarty.

Allangrange, ry. sta. on L.M.S., 8 m. W. of Fortrose, Ross and Cromarty. Allangrange House is 1 m. S.E. of the ry. sta.

Allangrange Muir, pl., par. of Urquhart, Ross and Cromarty.

Allan, Port of, harb., par. of Sorbie, S.E. Wigtownshire, 2 m. E. of Whithorn.

Allanshaw, formerly an important seat in par. of Hamilton, Lanarkshire.

Allanton, ham., N.E. Ayrshire, 5¾ m. E. Newmilns ry. sta.

Allanton, vil., E. Berwickshire, at junction of the Blackadder and Whitadder. P.O., T.O.

Allanton, ham. and seat, Lanarkshire, 3 m. N.E. of Wishaw. Allanton House near. P.O., T.O.

Allanton, mining vil., Lanarkshire, 1¾ m. S.E. of Hamilton.

Allan Water, pl. and stream in par. of Teviothead, Roxburghshire, 8 m. S.W. Hawick.

Allardice, Castle of, former seat, 1 m. N.W. of Bervie, in Arbuthnot par., Kincardineshire.

Allargue, pl., 14 m. N.W. of Ballater, Aberdeenshire.

Allathan, seat, 3 m. N.W. of New Deer, Aberdeenshire.

Allean House, pl. on the lower part of the Tummel, 1½ m. E. of Loch Tummel, Perthshire.

Aller, loch, near Oban, Argyllshire.

Allerly, pl. near left bank r. Tweed, 1 m. N. of Melrose, Roxburghshire.

Allermuir, one of summits in Pentland Hills, alt. 1617 ft.; outskirts of Edinburgh.

Allershaw, pl., 3½ m. from Elvanfoot ry. sta., Lanarkshire.

Alligin, ham., with pier, on N. shore of Loch Torridon, in par. of Applecross, W. Ross and Cromarty. P.O., T.O.

Alloa, par., police bur., and seaport., and chief town of Clackmannanshire, on N. side of Forth, 8 m. N. of Falkirk and 7 m. E. of Stirling; ry. stas., L.M.S. and L.N.E. Pop. 13,322. P.O., T.O. Large trade in beer, whisky, and yarn; exports coal. There is also a small fishing trade. The par. extends to 7161 ac. Pop. 18,244. Hotel. Ferry across r. Forth to S. Alloa. A. House, seat of the Earl of Mar and Kellie.

Alloa Inch, isl., r. Forth, par. of Alloa, Clackmannanshire. Pop. 3.

Alloa, South, small seapt. on r. Forth, opposite Alloa, connected by ferry. P.O., T.O.

Alloway and **Alloway Kirk,** q.s. par. and ham., with ry. sta., L.M.S. (passenger service withdrawn), 2½ m. S. of Ayr.

The birthplace of Robert Burns; old ruin of church made famous by " Tam o' Shanter." The cottage where the poet was born is now a Burns's museum. P.O., T.O.

Alltacoileachan, rivulet, par. of Inveraven, Banffshire. Rises in the Carn a' Bhodaich (2149 ft.) and falls after a N. Wly. course of 4½ m. into the Tervie.

Allt a' Mhuilinn, small stream, S. Jura, Argyll, joining Loch a' Mhuilinn and Loch a' Bhaile Mhargaidh.

Allt an Eoin, trib. Black Water, issuing from Loch an Eoin, Rogart, Sutherland.

Alltcailleach State Forest, 2½ m. S.W. of Ballater, Aberdeenshire.

Allt-Grand, or **Aultgrande,** see GLASS, r. and loch.

Allt Mor, stream, joins Spey at Kingussie, Inverness-shire.

Alltna Caillich, stream in the par. of Durness, Sutherlandshire. In the pl. that was here at one time the famous Gaelic poet, Rob Donn, was born.

Alltshellach, seat, 2 m. E. of Onich, Inverness-shire.

Almagill, hill (720 ft.), with ancient Caledonian camp on its N. slope called Range Castle, in Dalton par., Dumfriesshire.

Almond, r., Lanarkshire, W. Lothian and Midlothian, joining Firth of Forth at Cramond. It has an eastward flow of 24 m. Rises in Shotts par. at an alt. of 700 ft.

Almond, trib. of r. Tay. Flowing 30 m. in an easterly direction, it joins the Tay 2½ m. N. of Perth.

Almond, or **Haining Castle,** a ruin on the left side of the Avon, 5½ m. E.S.E. of Falkirk. It dates from the time of James III., and was often held for the Crown by the Earls of Linlithgow.

Almond Bank and Bridgeton, vil., with ry. sta., L.M.S., par. of Methven, 4 m. N.W. of Perth. P.O., T.O. Pop. 301.

Almondell, W. Lothian, 2½ m. S.E. of Uphall, seat of the Earl of Buchan.

Almond Glen, or **Sma' Glen,** q.v., glen, 6 m. N. of Crieff, Perthshire. See also GLENALMOND.

Alness, r., par., and vil., about 10 m. N.E. of Dingwall, E. Ross and Cromarty. The r. rises in the mts. N.W. of Loch Morie and flows S.E. to Cromarty Firth. The par. measures 44,060 ac. Pop. of par. 849. Ry. sta. L.M.S., P.O.,

T.O. Dalmore and Teaninich distilleries. Vil. pop. (including Bridgend of Alness) 937.

Alnwickhill, pl. on the S. side of the city of Edinburgh, of which it forms a part. Service reservoirs for the city's water supply.

Alnwick Lodge, or **Annick,** q.v., colliery vil. in Irvine par., Ayrshire, 3 m. N.E. of Irvine town.

Alpitty, or **Alpety,** pl. in par. of Arbuthnot, 4 m. N.W. of Bervie, Kincardineshire.

Alsh, Loch, sea inlet, S.W. Ross and Cromarty. Forms part of the channel between the mainland and Isl. of Skye.

Altandhu, ham. in par. of Lochbroom, in Rhu More, Ross and Cromarty, 32 m. N.W. Ullapool.

Altass, ham., Sutherland, 7 m. N.W. Invershin; ry. sta., L.M.S. P.O.

Altavaig, isl., with site of ancient chapel, off N.E. coast of Skye, 2 m. N. of Staffin Bay.

Alt Darrarie, affl. of the Muick, S. Aberdeenshire, which it enters 1 m. N. of Loch Muick.

Altens, ham., N.E. Kincardineshire, 2¼ m. S.E. of Aberdeen. Once an important fishing station.

Altimarlich, stream flowing through Winless Loch, and falling into Wick Water, 4 m. W. of Wick.

Altimeg Hill, 1270 ft., S.E. of Ballantrae, Ayrshire.

Altin, glen in par. of Snizort, Skye, W. Inverness-shire.

Altirlie, headland, 5 m. N.E. of Inverness, Inverness-shire.

Altmore, stream entering r. Isla, 1½ m. E. of Keith, Banffshire.

Altnabreac, pl., with ry. sta., L.M.S., 27½ m. W.S.W. of Wick. P.O. T.O. at sta.

Altnacealgach, hotel, 8 m. S.S.E. of Inchnadamff, S.W. Sutherlandshire.

Altnacraig, mansion on S.W. side of Oban, Argyllshire. It was built by the late Professor Blackie.

Altnaguisach, or **The Hut,** lodge in Balmoral forest near Loch Muick, 9 m. S.E. of Balmoral Castle. Queen Victoria used to stay here.

Altnaharra, ham. on Loch Naver, and 17 m. S. of Tongue, mid Sutherlandshire. Hotel. P.O. Fishing centre.

Altnamain, pl., 7½ m. N. of Alness, E. Ross and Cromarty. Inn.

Altnarie, upland trib. of the Findhorn, with grand cascade, Ardclach par.,Nairnshire.

Alton, vil., Ayrshire, in Loudoun par., 1½ m. N. of Galston.

Alton Burn, Nairn. Boys' preparatory School.

Altonside, State forest, 4 m. S.W. of Fochabers, Morayshire.

Altries House, Kincardineshire, on r. Dee, 1 m. S. Culter ry. sta.

Altrive, stream and farm, Selkirkshire, 1½ m. S.E. of foot of St. Mary's Loch. The farm was the last residence of James Hogg, the "Ettrick Shepherd" 1772-1835.

Altyre, r., Morayshire, flowing past Forres to Findhorn Bay. A. House, 4 m. S. of Forres.

Alva, par. and police bur., with ry. sta., L.N.E., Clackmannanshire, 7 m. E.N.E. of Stirling. In the Ochil Hills ;7632 ac. Pop. 4853; bur. pop. 3820. P.O., T.O. Has spinning mills and woollen mfrs. Close by, Alva Glen with A. House.

Alvah, par. on N.E. border of Banffshire ; 12,537 ac. Pop. 1101. The quaint bridge of A. lies 2 m. S. of Banff. A. Hill 578 ft.

Alves, par. and vil. in Morayshire, 5¼ m. W. of Elgin. L.M.S. sta. ; 9403 ac. Pop. 888. P.O., T.O.

Alvie, par., with loch (70 ft. deep), alt. 685 ft. same name, Inverness-shire, on r. Spey; 86,593 ac. Pop. 541. A. Church, 4 m. S.W. of Aviemore.

Alyth, par. and town (police bur.) on E. border of Perthshire, 4½ m. E.N.E. of Blairgowrie, L.M.S. sta. ; 23,296 ac. Pop. 2629; pop of bur. 1662. P.O., T.O. Has mfrs. of linen and woollen goods, and jute spinning.

Alyth, Burn of, Perthshire, flows S.E. to the Isla, 2 m. E. of Alyth.

Alyth Junction, ham. and L.M.S. ry. sta., on borders of Perthshire and Angus, 24½ m. N.E. of Dundee.

Amat Deer Forest and **Lodge,** Ross and Cromarty, 9 m. W. of Ardgay and 8½ m. W. of Bonar Bridge. Forest 3500 ac.

Am Bodach, mt., Inverness-shire, 7 m. S.E. of Fort William; alt. 3382 ft.

Am Bodach, mt., Inverness-shire, 5 m. N.W. of Kingussie; alt. 2709 ft.

A'Mharconaich, mt., Inverness-shire. See MARCAONACH.

A-Mhorra, loch, N. Jura, Argyllshire.

Amhuinnsuidh, stream and deer forest, N. Harris, Outer Hebrides, Inverness-shire.

Amisfield, vil. and seat; L.M.S. sta. 5 m. N.E. of Dumfries. P.O., T.O.

Amisfield, East Lothian, 1 m. E. of Haddington; a seat of the Earl of Wemyss and March.

Amperlaw, seat, Lanarkshire, 2¾ m. N.N.E. of Carnwath.

Ample, glen and r., Perthshire. See GLEN AMPLE.

Amulree, q.s. par. and vil., in Dull, Fowlis Wester, Kenmore, Little Dunkeld, and Monzievaird and Strowan pars., Perthshire, 12 m. N.N.E. of Crieff. P.O., T.O. Hotel. Here the clans were armed and sworn in, at the rising of 1715.

Anabuich, a place on Loch Moaruig—a branch of Loch Seaforth—on the N.E. of North Harris, Ross and Cromarty. A small island, *Eilean Anabuich*, near the habitation gives the locality its name.

An Cabar, peak of Ben Wyvis, E. Ross and Cromarty, 6½ m. N.W. of Strath-peffer; alt. 3106 ft.

Anchor Paper Works, Denny, Stirlingshire.

An Coileachan, mt., Ross and Cromarty, 8 m. N.E. of Achnasheen; alt. 3015 ft.

Ancrum, par. and vil., N. Roxburghshire; 10,302 ac. Pop. 858. Vil. stands on rising ground on the rt. bank of the r. Ale; pop. 164.

Anderston, q.s. par. and dist. of Glasgow, L.M.S. sta.

Andrewhinney Hill, S. Selkirkshire, 4 m. S.S.W. of Loch of the Lowes; alt. 2220 ft.

Andunty, loch, Petty par., N.E. Inverness-shire.

An Gead Loch, Ross and Cromarty, 12 m. E. of Lochcarron; length 1¼ m., alt. 830 ft., depth 30 ft.

An Gearna, mt., Mull, Argyllshire, 1 m. N.W. of Benmore; 1848 ft.

Angel's Hill, hill topped with stone circle and cairn, Iona, Argyllshire.

Angry, or **Lennox Burn,** Morayshire, flowing 4 m. N. between Birnie and Dallas pars. to the Lossie.

Angus, (formerly Forfarshire). A maritime and agricultural co. on E. coast, bounded by Kincardineshire and Aberdeenshire on the N., North Sea on the E., Firth of Tay on the S., and Perthshire on the W. The fertile valley of Strathmore or the Howe of Angus, 80 m. in length and and a mean breadth from 6 to 8 m., lies between the Braes of Angus, spurs of the Grampians in the N.W., and the lower Sidlaw Range in the S.W. Between the Sidlaw Hills and the Firth of Tay is the Carse of Gowrie, an alluvial tract, 15 m. by 2 to 4 m. Jute, coarse linen, canvas boots, jam and machinery are among the mfrs. Shipbuilding is carried on at Dundee. Its antiquities include vitrified Caledonian and Roman hill forts, Roman camps, cairns and standing stones. Carnoustie, Monifieth, and Montrose are famous for their golf links, and Barry as a military training camp. Principal towns, Dundee, Arbroath, Montrose, and Forfar (co. town). Area 873 sq. m. or 559,034 ac. Pop. 270,190. The co. returns 1 member to Parliament.

Angusfield House, Aberdeen. Boys' preparatory school.

Ankerville, vil., N.E. Ross and Cromarty, 5 m. S. of Tain.

Ann, affl. of Irvine Water, Galston, par., Ayrshire.

Annan, par., royal and mun. bur. seapt. and mkt.-town; L.M.S. sta.; S. Dumfriesshire, on r. Annan, 15½ m. S.E. of Dumfries by rail; par. 10,915 ac. Pop. 6302; bur. pop. 3959. Industries include quarrying, corn milling, and salmon fishing.

Annan, r., Dumfriesshire, rising near Hartfell Mt., on the Peeblesshire border, it flows 49 m. S. to the Solway Firth, 1¾ m. below Annan. Chief trib., Evan, Moffat, Kinnel, Dryfe, and Milk.

Annandale, basin of the r. Annan, and forms the mid of the three divisions of Dumfriesshire in the Annan Valley; includes the vast hollow, the Devil's Beef Tub. Length 30 m., breadth 15 to 18 m.

Annandale's Beef Stand, or **Devil's Beef Tub,** a great abyss 5 m. N.W. of Moffat. Said to have received its name " because the Annandale loons used to put their stolen cattle in there."

Annan Fort, see BARNKIRK.

Annanhill, seat, Ayrshire, 1 m. W. of Kilmarnock.

Annat, pl., Kilchrennan par., W. side Loch Awe, Argyllshire. It means " a mother church."

Annat, ham., Ross and Cromarty, at head of Upper Loch Torridon.

Annat, seat, near Errol, Perthshire.

Annathill, ham., Lanarkshire, 3½ m. N. of Coatbridge. P.O., T.O. Pop. 1063.

Annbank, q.s. par. and mining vil., Ayrshire ; ry. sta., L.M.S., 4½ m. N.E. of Ayr. P.O., T.O., and P.O. and T.O. at sta. Vil. pop. 1190.

Annet Burn, affl. of r. Teith, S. Perthshire.

Annfield, seat, E. Fife, ¾ m. S.E. of Kingskettle sta.

Annick, affl. of r. Irvine, Renfrewshire and Ayrshire.

Annick Lodge, Ayrshire, 3 m. N.E. of Irvine.

Anniesland, dist., N.W. Glasgow.

Anniston, seat, Angus, 5 m. N.N.E. of Arbroath.

Ann's Bridge, dist., Dumfriesshire, on r. Kinnel, 7½ m. N. of Lochmaben.

An Riabhachan, mt., Ross and Cromarty, 4 m. S. of Loch Monar ; 3696 ft.

An Ruadh-Mheallan, mt., W. Ross and Cromarty, 2 m. W. of Ben Alligin ; 2196 ft.

An Sgarsoch, mt., S.W. border of Aberdeenshire, 16 m. W.S.W. of Braemar ; 3300 ft.

An Sligearnach, mt., Perthshire, 9 m. N.E. of Blair-Atholl ; 2577 ft.

An Socach, mt., S.W. Aberdeenshire, 8½ m., S.W. Braemar ; 3059 ft.

An Socach, mt., Ross and Cromarty, 8 m. N.W. of Dingwall ; 3295 ft.

Anstruther (locally **Ainster**), seapt., mkt.-town, coastguard and lifeboat sta.,ry. sta. L.N.E., E. coast of Fife, 18¾ m. E. of Thornton Junction, by rail. Pop. (Anstruther, Easter and Wester, and Kilrenny) 3325. P.O., T.O. In addition to being the chief fishing port on the Fife coast, oilskins and woollen goods are manufactured. Pier lights : E. pier, red, seen 5 m. ; W. pier, white, seen 6 m.

Anstruther Easter, royal and mun. bur. and par., Fife ; L.N.E. sta., Anstruther ; par. 28 ac., bur. 123 ac. (Pop., see ANSTRUTHER.) Connected with A. Wester by bridge over Dreel.

Anstruther Wester, royal and mun. bur. and par., Fife ; L.N.E. sta., Anstruther ;

par. 1034 ac. (Pop., see ANSTRUTHER.) Isle of May opposite.

An Suidhe, mt., Argyllshire, 6 m. W.S.W. of Inveraray ; 1687 ft.

Antermony, seat and loch, Stirlingshire, ¾ m. E. of Milton.

Antoninus' Wall, or **Graham's Dyke,** quondam Roman rampart, from Carriden on Firth of Forth to a point near Old Kilpatrick on the Clyde. Constructed about A.D. 140 as a protection against the Caledonians. Length 36½ m.

Anton's Hill, seat, Berwickshire, 4 m. N.W. of Coldstream.

Anwoth, coast par. and vil. (sub. of Gatehouse), 9 m. N.W. of Kirkcudbright. 11768 ac.

Aonach Beag and **Aonach Mor,** summits near Ben Nevis, Inverness-shire ; alts. 4060 ft. and 3999 ft.

Aonach Beag, mt., S. Inverness-shire, 6 m. W. of Loch Ericht ; 3647 ft.

Aonach Buidhe, mt., Ross and Cromarty, 9½ m. S.E. of Strathcarron ry. sta. ; 2949 ft.

Aonach Dubh, mt., Argyllshire, S. side Glencoe ; 2849 ft.

Aonach Eagach, mt., Argyllshire, N. side of Glencoe ; two peaks, 3168 ft. and 3118 ft.

Aonach Mor, mt., Argyllshire, 7 m. N.W. Bridge of Orchy ry. sta. ; two peaks, 2953 ft. and 2839 ft.

Aonach-Shasuinn, mt., N.W. Inverness-shire, 3 m. S. of Loch Affrick ; 2902 ft.

App, rivulet, S.W. Ayrshire, flowing into Loch Ryan.

Appin, dist., E. side of Loch Linnhe, mid W. Argyllshire.

Appin House, seat, E. Loch Linnhe, W. Argyllshire.

Appin, q.s. par., vil., and port ; L.M.S. sta., 15 m. N.N.E. Oban, Argyllshire. P.O., T.O.

Appin, estate, with colliery, 1½ m. N.E. of Dunfermline, Fifeshire.

Appin, valley, mid Perthshire, near Aberfeldy.

Appin Burn, rivulet, W. Dumfriesshire, flowing 4 m. E. to Shinnel.

Applecross, par. and ham. on bay of same name, 11 m. N.W. of Lochcarron, S.W. coast of Ross and Cromarty. P.O., T.O. ; par. 109,282 ac. Pop. 1033.

Applegarth and Sibbaldbie, par., Dumfriesshire, containing Nethercleugh and Dinwoodie ry. stas., L.M.S., 3 m. N.W. of Lockerbie ; par. 11,871 ac. Pop.807.

Appletree Hall, vil., W. Roxburghshire, 2 m. N.E. of Hawick.

Arabela, seat, E. Ross and Cromarty, 4½ m. S.S.E. of Tain.

Aray, r. in mid Argyllshire, flowing from Loch Awe to Loch Fyne.

Arbigland, coast, estate, and mansion, Kirkcudbrightshire, 1½ m. S.E. of Kirkbean.

Arbiklie, pl. on r. Lunan, Angus, 1 m. N. of Inverkeilor.

Arbirlot, coast. par., and vil., Angus, 2½ m. W. of Arbroath. P.O. ; par. 6748 ac. Pop. 726.

Arbory Hill, remains of an ancient fort, S.E. Lanarkshire, near Lamington ; 1406 ft.

Arbroath, formerly Aberbrothock, par., A. and St. Vigeans, Angus ; parl., royal, and mun. bur., seapt., coastguard, and lifeboat sta. ; ry. stas., L.N.E. and L.M.S., at mouth of Brothock, 17 m. N.E. of Dundee. Opposite Bell Rock lighthouse ; par. 12,763 ac. Pop. 19,437 ; parl. and mun. bur. 17,635. Mfrs. include— flax, jute, linen, sailcloth, and leather. Has also bleachfields, engineering, iron works, and boot factories. The ruined abbey was founded by William the Lion in 1178. The pier shows a red, a white, and two green lights, and a fog gong. United with Brechin, Forfar, Montrose, and Inverbervie in returning one member to Parliament.

Arbuckle, vil., N.E. Lanarkshire, near Airdrie.

Arbuthnott, par., S.E. Kincardineshire, on Bervie Water, 2 m. N.W. of Bervie ; 9584 ac. Pop. 577. P.O. A. House, near Fourdoun, the seat of Viscount Arbuthnott.

Arcan, ham., E. Ross and Cromarty, 6½ m. N.W. of Beauly.

Archasig Haven, small harb., W. Rona Isl., Inverness-shire.

Archerbeck, stream and coalfield,Canonbie par., Dumfriesshire. Flows to r. Liddel.

Archerfield, seat, E. Lothian, 3 m. S.W. of North Berwick.

Archiestown, vil., Morayshire, 6 m. S.W. of Rothes. P.O. Pop. 209.

Ard, loch, S.W. Perthshire, 10 m. S.W. of Callander ; 107 ft. deep, alt. 105 ft.

Ard Achadh, loch, Islay, Argyllshire, 4 m. W. of Port Ellen.

Ardachdail, pl. and headland, Lochbroom par., Ross and Cromarty, 2½ m. N.W. of Ullapool.

Ardachie, seat, Inverness-shire, 1 m. from Fort Augustus.

Ardalanish, Rudh', headland, S.W. Mull, Argyllshire.

Ardali (Ardalum), pl., 13 m. S.W. of Tobermory, Argyllshire, on the isl. of Ulva.

Ardallie, q.s. par. and vil., E. Aberdeenshire, 6 m. S. of Mintlaw sta. P.O. at sta.

Ardanaiseig, seat, N.W. shore of Loch Awe, Argyllshire, 6 m. S.E. of Taynuilt.

Ardardan, seat, Dunbartonshire, 1½ m. N.W. of Cardross.

Ardargie, estate, Perthshire, 6 m. S.W. of Perth.

Ardbeg, vil. and headland, between Rothesay Bay and Kames Bay, Isle of Bute. Has a good harb. and a large distillery.

Ardbeg, vil., with steamboat pier, Islay, Argyllshire, 3 m. E. of Port Ellen.

Ardbennie House, ½ m. S.W. of Madderty ry. sta., Perthshire.

Ardbrecknish, pl. on Loch Awe, Argyll, 1 m. N.E., Port Sonachan.

Ardchattan and Muckairn, mountainous par., Lorn dist., W. Argyll, situated on both sides of Loch Etive, 6½ m. N.E. of Oban. Pop: 2083. 144,941 ac.

Ardcheanochrochan (prom. at the end of the hill), hill, Perthshire, at E. end of Trossachs.

Ardchonnel, ham., Argyllshire, 7 m. N.W. of Inveraray, on E. side of Loch Awe.

Ardchonnel Castle, a ruined stronghold of the Argyll family, on an isl. near E. shore of Loch Awe, 7 m. W. of Inveraray.

Ardchullarie, seat, S.W. Perthshire, on E. side of Loch Lubnaig, 5 m. N.W. of Callander.

Ardchyle, ham., Perthshire, near Killin Junction.

Ardclach, par. and ham., E. Nairnshire. on r. Findhorn, 5½ m. S.W. of Dunphail sta. ; 37,953 ac. Pop. 639.

Arddarroch, seat, Dunbartonshire, 2½ m. N. of Garelochhead.

Ardeer House, Ayrshire, near Stevenston.

Ardeer Works, in par., and 1 m. S. of Stevenston sta., and 3 m. N.W. Irvine, Ayrshire. Ironworks and dynamite factory. P.O., T.O.

Ardelve, vil., 6 m. E. of Kyle of Lochalsh, Ross and Cromarty. P.O., T.O. Pop. 86. Ferry to Dornie across Loch Long.

Arden, vil., Alexandria, Dunbartonshire. P.O., T.O.

Arden, seat, Dunbartonshire, 2½ m. N.W. of Balloch ry. sta.

Arden, vil., New Monkland par., N. Lanarkshire, 3 m. N.E. of Airdrie.

Ardencaple Castle, formerly a seat of the Argyll family, on Gare Loch, 8 m. N.W. of Dumbarton.

Ardenconnel, hostel, Rhu par., Dunbartonshire.

Ardencraig, vil., Bute, 1 m. E. of Rothesay.

Ardendee, seat in co., and 1½ m. N.N.E., of Kirkcudbright.

Ardentallan, seat, on Loch Feochan, Argyll, 4 m. S. of Oban.

Ardentinny, q.s. par. and vil., W. side of Loch Long, at entrance to Glen Finnart, Argyll, 7½ m. N.W. Helensburgh. P.O., T.O.

Ardentraive, seat, S. Argyllshire, near Colintraive, in the Kyles of Bute.

Ardeonaig, ham., Perthshire, on S. side of Loch Tay, 7 m. N.E. of Killin. P.O., T.O.

Ardersier, coast par. and vil. in co., and 10½ m. N.E., of Inverness; 3826 ac. Pop. 1760. P.O., T.O.

Ardessie, ham., and stream with waterfalls, W. Ross and Cromarty, 5½ m. S.W. from Ullapool.

Ardfern, ham. on Loch Craignish, W. Argyll, 4 m. N.W. of Kilmartin. P.O., T.O.

Ardfin, pl., S. Jura, Argyll, 3¼ m. S.W. of Craighouse.

Ardgartan, peninsula between Glencroe and Loch Long, Argyll. A. House, 2 m. from Arrochar, now Scottish Youth Hostel. State forest in neighbourhood.

Ardgay, vil., Ross and Cromarty, on Dornoch Firth, 13 m. N.W. of Tain. P.O., T.O., 1 m. S.E. of Bonar Bridge.

Ardgoil, mountainous estate of 9000 ac., between Loch Goil and Loch Long, Argyll. Gifted to the Corporation of Glasgow by Lord Rowallan.

Ardgour, par. and ham., pier and seat, Argyllshire, 10 m. S.W. of Fort William;

90,600 sc. Pop. 509. Extensive deer forest. P.O., T.O.

Ardgowan Distillery, Greenock, Renfrewshire.

Ardgye, seat, Morayshire, 4 m. N.W. of Elgin.

Ardhallow, pl., Argyll, 2 m. S. of Dunoon.

Ardhasig, pl. on W. Loch Tarbert, Harris, Outer Hebrides.

Ardheslaig, ham. and dist., head of Loch Beag, an arm of Loch Torridon, W. Ross and Cromarty.

Ardhu, pl., Mull, Argyll, 5½ m. S.W. of Tobermory.

Ardilistry, pl., prom., and bay on the E. of Islay, Argyll, 4 m. N.E. of Port Ellen.

Ardincaple, seat, Seil Isl. Argyll, 2 m. from Easdale.

Ardindrean, ham., on the E. side Loch Broom, N.W. Ross and Cromarty, 4 m. S. of Ullapool.

Ardinning, loch, Stirlingshire, 1 m. S. Strathblane.

Ardintoul, seat, on Loch Alsh, Glenshie par., Ross and Cromarty.

Ardivachar Point, cape, W. of S. Uist, Inverness-shire.

Ardkenneth, pl., with R.C. Chapel, S. Uist par., Outer Hebrides.

Ardkinglas Castle, seat, head of Loch Fyne, 6 m. E.N.E. of Inveraray, Argyll.

Ardlair, pl. and pier on Loch Maree, Ross and Cromarty, 4 m. N.W. of Letterewe.

Ardlamont Point, headland, between Kyles of Bute and Loch Fyne, Argyll. A. House.

Ardlaw, ham., Aberdeenshire, 4 m. S.W. Fraserburgh.

Ardle, r. in N.E. Perthshire, flowing to the Shee, uniting to form the Ericht.

Ardler, q.s. par., Kettins par., Angus, and partly in Coupar-Angus and Meigle pars., Perthshire, 3 m. N.E. of Coupar-Angus. P.O., T.O. at sta., L.M.S.

Ardler, pl., Aberdeenshire, Strathdon dist.

Ardlethen, seat, Aberdeenshire, 2 m. W. of Ellon.

Ardlui, a steamboat sta., N. end of Loch Lomond, Dunbartonshire, 7 m. N. of Tarbet. Ry. sta., L.N.E. P.O., T.O. at sta.

Ardlussa, deer forest and mansion, N. Jura, Argyll.

Ardlussa Bay, N. Jura, Argyll, steamboat pier.

Ardmaddy Castle, seat, Argyll, 22 m. S.W. Oban, 3 m. N. of the head of Loch Melfort.

Ardmair, ham., W. Ross and Cromarty, 2½ m. N.W. Ullapool.

Ardmarnock, seat, Argyll, on E. side Loch Fyne, 4¼ m. N.E. of Tarbert.

Ardmay House, Dunbartonshire, 1½ m. S.W. of Arrochar.

Ardmeanach, or **Black Isle,** peninsular dist. between Cromarty and Moray Firths, E. Ross and Cromarty.

Ardmeanach, dist., Mull, Argyll, on N. shore of Loch Scridain.

Ardmellie, seat, N.E. Banffshire, 1 m. N.W. of Marnock.

Ardmichael, rocky prom., with ancient burying-ground, on W. coast of S. Uist Isl.

Ardmiddle, hill and seat, N. Aberdeenshire; hill 557 ft.; seat 2½ m. S.W. of Turriff.

Ardmile, headland, W. side of S. Uist Isl., Outer Hebrides.

Ardmillan House, 3 m. S. of Girvan, Ayrshire.

Ardminish, pl. and bay, with inn, Gigha, W. Argyll. Ferry 3 m. to Monimore, on the opposite shore of Kintyre.

Ardmohr, Whiting Bay, Arran. Co. hospital.

Ardmolich Bridge, pl., head of Loch Moidart, Inverness-shire.

Ardmore, bay, headland, and islets on E. side of Islay Isl., Argyll, 7 m. N.E. of Port Ellen.

Ardmore, wooded prom., with seat, on Firth of Clyde, between Dumbarton and Helensburgh.

Ardmore, headland in Vaternish dist., Isle of Skye, Inverness-shire.

Ardmore, harb., Eddertoun par., Ross and Cromarty, on Dornoch Firth.

Ardmore Distillery, Kennethmont, Aberdeenshire.

Ardmucknish Bay, on the N. side of, and at the mouth of, Loch Etive, Argyll.

Ardnacallioch, prom. at east end of Ulva, Argyll.

Ard-na-Coille, seat, Carrbridge, Inverness-shire.

Ardnacross, bay and estate, 6 m. N.E. of Campbeltown, Argyll, on E. side of the Kintyre.

Ardnadam, vil. on Holy Loch, Argyll, 2½ m. N. of Dunoon, forming part of Sandbank, q.v.

Ardnahien, pl. on Loch Goil, Argyll, 9 m. N.W. Helensburgh.

Ardnamurchan, ham., headland, and par., W. Argyll; 109,402 ac. Pop 1137; q.s. par. P.O., T.O. at Kilchoan. A. Point, the most westerly point of the mainland of Scotland, crowned with a lighthouse showing fixed white light, seen 18 m.

Ardnastang, ham., Argyll, near head of Loch Sunart, close to Strontian.

Ardnave, headland and loch, Argyll, N.W. side of Islay.

Ardniel Bank, seacliff, a S.W. extremity of West Kilbride par., Ayrshire, 6 m. N.W. of Ardrossan; alt. 300 ft.

Ardnish, pl., Inverness-shire, S.W. and at the head of the Sound of Arisaig. P.O.

Ardnoe, headland, Argyll, at mouth of Loch Crinan.

Ardoch, eminence. Kirkcudbrightshire, 2 m. N.E. of Dalry; alt. 700 ft.

Ardoch, pl., Dunbartonshire, 3 m. N.E. of Balloch county hospital.

Ardoch, par. and seat, containing Braco and Greenloaning vils. Perthshire, 7½ m. S.W. of Crieff; 20,715 ac. Pop. 1029. In the grounds of A. House is the celebrated Roman camp, the best preserved in Great Britain.

Ardoch Bridge, pl. with school, Kilmarnock par., Dunbartonshire.

Ardoch House, Kishorn, S.W. Ross and Cromarty, 4½ m. W. of Loch Carron.

Ardochy House, Inverness-shire, on N. shore of Loch Garry.

Ardoe, estate, N.E. Kincardineshire, 1 m. S. of Cults ry. sta., L.N.E., 3¼ m. S.W. Aberdeen.

Ardonald, pl., with limeworks, N.Aberdeenshire, near Cairnie.

Ardovie House, Forfarshire, 2½ m. S. of Brechin.

Ardoyne, hill in Oyne par., Aberdeenshire; alt. 600 ft.

Ardpatrick, headland, seat and ham., S.W. extremity of Knapdale, Argyll, at entrance to W. Loch Tarbert; ham., 10 m. S.W. of Tarbert. Ferry across W. Loch Tarbert to Portachoillan. P.O.

Ardrishaig, q.s. par. and small port at entrance to Crinan Canal, 2 m. S.W of Lochgilphead, Argyll. P.O., T.O. Pop. 956. Hotel. Has an extensive fishing industry. Steamboat communication with Glasgow and western ports. Lighthouse, showing a six seconds flashing light of white, red, and green sectors, visible 4 m.

Ardross, ruined castle, Fifeshire, 1 m. N.E. of Elie.

Ardross, seat and ham., 4½ m. N. of Alness, Ross and Cromarty. P.O. The ham. is 5 m. N. of Alness.

Ardrossan, par., police bur., seapt. and lifeboat sta., and watering-place on N. side of Ayr bay, 31½ m. S.W. of Glasgow; ry. stas., L.M.S.; 6663 ac. Pop. 13,736; town pop. 6888. P.O., T.O. Has a well-equipped harbour, and exports coal and iron. Shipbuilding. Maintains regular steamboat communication with Arran and Ireland. Is greatly frequented in summer as a watering-place. Hotel. Has occulting light, visible 10 m. An Ayrshire bur.

Ardscalpsie Point, headland, Isl. of Bute, W. coast, 4½ m. N.W. of Garroch Head.

Ardsheal, seat, Argyll, S. shore of Loch Linnhe.

Ardstinchar Castle, ruin, near Ballantrae, Ayrshire.

Ardtalnaig, ham., Perthshire, Kenmore par., 9¼ m. N.E. of Killin, on Loch Tay. P.O., T.O.

Ardtalnaig Hostel (Kindrochit), on S. shore of Loch Tay, 11 m. N. W. of Killin and 12 m. S.W. of Aberfeldy; a Scottish Youth Hostel.

Ardtarig, seat, Inverchaolain, Argyll.

Ardtealla, headland and bay, Argyll, E. side of Islay.

Ardtoe, small bay, with pier, on N. coast of Ardnamurchan, Argyll.

Ardtornish, headland and bay, Sound of Mull, Argyll.

Ardtornish Castle, ruined ancient castle on coast of Morven par., Argyll. A stronghold of the Lords of the Isles. A. House close by.

Ardtun, basaltic headland at mouth of Loch Scriden, Mull, Argyll; alt. 130 ft.

Arduaine, pl., with school, Argyll, near Kilmelford, P.O., T.O.

Ardullie, seat, E. Ross and Cromarty, 3 m. S.W. of Evanton.

Ardvare, sea loch, W. Sutherland, 1 m. S.W. of Loch Cairnbawn. A. House on shore.

Ardvasar, bay, headland, and ham., Sleat Sound, Isle of Skye, 12 m. S. of Broadford. P.O., T.O.

Ardveich Castle, N. side of Loch Earn, ½ m. S.E. of Lochearnhead, Perthshire.

Ardven, mt., Arran, Buteshire, 5½ m. W.N.W. Lamlash; 1676 ft.

Ardverikie, seat and deer forest (40,000 ac.), on S.E. side of Loch Laggan, Inverness-shire, 8 m. W. of Dalwhinnie sta.

Ardvoirlich, small bay on W. side, near head of Loch Lomond, Dunbartonshire, 3 m. N. of Tarbet.

Ardvorlich House, seat on S. side of Loch Earn, Perthshire. The " Darlinvarach " of Sir Walter Scott's *Legend of Montrose.*

Ardvourie, deer-forest and seat, on Loch Seaforth, Harris Isl., Inverness-shire, Outer Hebrides.

Ardvreck, Crieff, Perthshire. Boys' preparatory school.

Ardvreck Castle, ruined ancient castle on Loch Assynt, Sutherland. Belonged to the Macleods, and was the place of the Marquis of Montrose's durance after his capture in 1650.

Ardwall, isl. in Borgue par., Fleet Bay, Kirkcudbrightshire, joined to mainland at low water.

Ardwall, seat, in Anwoth par., Kirkcudbrightshire, 2 m. S.W. of Gatehouse-of-Fleet.

Ardwell, q.s. par., ham., seat, headland, and bay, in Stoneykirk par., Wigtownshire, 9 m. S.E. of Portpatrick. P.O., T.O.

Areeming, small loch, Kirkpatrick Durham par., Kirkcudbrightshire, 10 m. N. Castle Douglas.

Argaty House, seat, Perthshire, 1 m. N. of Doune.

Argrennan, seat, Kirkcudbrightshire, 4 m. S.W. Castle-Douglas.

Argyll, dist. in mainland of Argyllshire. Separated from Lorn by Lochs Awe, Avich, and Melford, from Cowal by Loch Fyne, from Knapdale by Loch Gilp and the Crinan Canal. Name signifies *The Land of the Gael.*

Argyll Distillery, Campbeltown, Argyll.

Argyll's Bowling Green, group of mts. overhanging Loch Long, from Glencroe to Loch Goil, Argyll, culminating in a summit 2479 ft. high.

Argyllshire, a maritime and mountainous co. in the W. of Scotland, with an extreme length from N. to S. of 115 m., and a coast-line, owing to the numerous sea-lochs, of 2290 m. The chief mountains are Ben Cruachan, 3689ft.; Ben More, in the Isl. of Mull, 3169 ft.; Ben Starav, 3541 ft.; Stobghabhar, 3565 ft., and Bidean nan Bean, 3766 ft. The rivers are the Orchy and the Awe. The lochs comprise Loch Fyne, famed for its herring; Loch Shiel, for its salmon and sea-trout; and Loch Awe (freshwater), for its salmon, trout, and eels. Glencroe and Glencoe are the most picturesque glens. The principal isls. are, Coll, Mull, Staffa, Iona, Islay, Jura, Lismore, Tyree, and Colonsay. The co. comprises the districts of Lochiel, Ardgour, Sunart, Ardnamurchan, Morven, Lorn, Argyll, Cowal, Knapdale, and Kintyre. It returns one member to Parliament. The chief town is Inveraray. Herring fishing, slate (Ballachulish) and granite quarrying (near Inveraray), and whisky-distilling (Campbeltown and Islay) are the chief industries. Highland cattle are largely reared. The L.N.E. ry. traverses the eastern part of the co. Area, 1,990,472 ac. Pop. 63,014.

Arichlinie, loch, E. Sutherland, 2 m. N.W. of Kinbrace ry. sta.; alt. 451 ft., depth 7 ft.

Arienas, loch, Morven par., Argyll, near head of Loch Aline; 116 ft. deep, alt. 31 ft.

Arinangour, seaport vil., in Coll Isl., Argyll.

Arisaig, dist. and vil., with ry. sta., L.N.E., Inverness-shire, 7½ m. S. of Mallaig. P.O., T.O. Hotel. A. Point, 3½ m. S.W. Regular steamer communication with Glasgow.

Arisaig and Moidart, par., on W. coast of Inverness-shire, includes Shona Isl., 107,552 ac. Pop. 1175. The par., which contains an extensive deer forest, is mountainous and sterile.

Arisdale, stream, Yell Isl., Shetland.

Arity, stream, S. Angus, 1¾ m., N.E. Glamis, enters the r. Dean.

Arkaig, loch, 12 m. long, 359 ft. deep, alt. 94 ft., Inverness-shire, 10 m. N. Fort William.

Arkindeith, ruined tower in Avoch par., Easter Ross and Cromarty; only the dungeon now remains.

Arkle, mt., Sutherland, 4 m. E. head of Loch Raxford, alt. 2580 ft.

Arklet, small loch, Stirlingshire, between Inversnaid and Loch Katrine; now part of the Glasgow Waterworks and its drainage changed from Loch Lomond to Loch Katrine.

Arkleton Hill, seat, E. Dumfriesshire, 5 m. N. of Langholm. A. Hill, 1708 ft.

Arlary, seat, N. Kinross-shire, 1¼ m. N.E. of Milnathort.

Armadale, q.s. par. and town (police bur.), and ry. sta., L.N.E., 1 m. S. West Lothian, 2½ m. W. of Bathgate. Pop. 4854. P.O., T.O. and P.O. at sta. Has important paraffin, chemical, brick and shale works, and is the centre of an extensive mining district.

Armadale, fishing vil., bay, and stream, Farr par., Sutherland, 24¼ m. W. of Thurso. P.O., T.O. A. House, a seat.

Armadale Bay and **Pier,** Sleat par., Isle of Skye, Inverness-shire. Has regular steamboat communication with Glasgow. Has fixed white light, seen 5 m.

Armadale Castle, seat of Lord Macdonald, on S.E. coast of Skye, 7 m. N.E. of Sleat Point.

Armit, stream flowing into Gala Water, 5 m. W. of Stow, Berwickshire and Midlothian.

Arnabost, ham., Coll. Isl., Argyll.

Arnage, seat, and ry. sta., L.N.E., 3½ m. N.W. of Ellon, Aberdeenshire. T.O. at sta.

Arnal, stream, in Barvas par., Lewis, Outer Hebrides, flows 6 m. to the Atlantic.

Arnbarrow, hill, Kincardineshire, Fordoun par., 3½ m. N. Fettercairn; alt. 1060 ft.

Arnbeg, pl., Stirlingshire, 1 m. W. of Kippen. Gathering of Covenanters took place here in 1676.

Arnbrae, ham., Stirlingshire, 1 m. W. of Kilsyth.

Arncroach, vil., E. Fife, Carnbee par., 2¾ m. N.E. of Colinsburgh. P.O., T.O.

Arndean, seat, Kinross-shire, 2 m. N.E. of Dollar.

Arndilly, seat, on the Spey, Banffshire, 2¼ m. N. of Craigellachie.

2

Arneybog, mineral dist., Dunbarton-shire, near Cumbernauld.

Arnfinlay, ancient castle, near Kippen, Stirlingshire.

Arngask, par., Perthshire, 12 m. S. of Perth ; 6446 ac. Pop. 650. A. House a seat.

Arngibbon, seat and glen, 2½ m. E. of Buchlyvie, Stirlingshire.

Arngomery, seat, N. Stirlingshire, ¾ m. W. of Kippen.

Arnhall, Dundee, Angus. Boys' private school.

Arnhall, ham. and seat, Fettercairn par., Kincardineshire, 1 m. E. of Edzell.

Arnhall Castle, Perthshire, 1¼ m. N.W. of Bridge of Allan, ry. sta.

Arnicle, pl., Kintyre, Argyll, near Glen-barr.

Arniefoul, vil., Angus, 5½ m. S.W. of Forfar.

Arnisdale, vil., stream, and seat, W. Inverness-shire, N. side of Loch Hourn. P.O., T.O. A. stream flows 9 m. W. to Loch Hourn.

Arnish, headland and lighthouse, S. side Loch Stornoway, Lewis, Ross and Cromarty. Thirty-seconds white flash-ing light, visible 13 m. A. reef has a six-seconds flashing light.

Arnish, ham. and loch, W. side of Raasay Isl., Inverness-shire.

Arnisort, ham. and loch, N.W. Skye, Inverness-shire, loch branch of Loch Snizort. P.O.

Arniston and Hunterfield, colliery dist. (and vil.), Cockpen par., Midlothian, 3 m. S. of Dalkeith.

Arniston House, late seat of the Dundas family, on the South Esk, Midlothian, 1¾ m. S.W. of Gorebridge.

Arnloss Colliery and Balquhatson Row, vil., S.E. Stirlingshire, near Slamannan.

Arnol, vil. and stream, Ross and Crom-arty, 14 m. N.W. of Stornoway, Lewis.

Arnot Tower, seat, W. Kinross-shire, 2¾ m. W. of Leslie.

Arnprior, vil. and estate, Stirlingshire, 2¼ m. W. of Kippen.

Arns, pl., Dunbartonshire, 2½ m. E. of Cumbernauld.

Arnsbrae House, Clackmannanshire, 1 m. N.W. of Alloa.

Arnsheen, q.s. par. and ham., in Col-monell par., Ayrshire, 12 m. S. of Girvan.

Arnton Fell, hill, Roxburghshire, 2 m. S. Riccarton Junc.; alt. 1464 ft.

Arntully, vil. and estate, E. Perthshire, 8 m. N. of Perth.

Aros, Mull, Argyll. P.O., T.O. See SALEN.

Aros Castle, ruined castle of the Lords of the Isles, E. coast of Mull, Argyll, 7 m. E.S.E. of Tobermory.

Aros House, or **Druimfin Castle,** N.E. coast of Mull, Argyll, 1 m. S.E. of Tobermory.

Aros Water, flows from Loch Frisa to Sound of Mull.

Arpafeelie, pl., Black Isle, Ross and Cromarty, 6 m. E. of Muir of Ord.

Arradoul, pl., Rathven par., Banffshire, near Fochabers.

Arran, isl. in S. of Buteshire, in Firth of Clyde, 6½ m. S.W. of the nearest part of Bute. Its length is 20½ m. and its mean breadth about 9 m. Sheep and cattle rearing is carried on and there is considerable herring fishing. Much frequented in summer by tourists and holiday-makers. The surface in the northern part is very rugged, rising in Goatfell to 2866 ft. Area, 106,149 ac. or 166 s. m. Pop. 4506.

Arran, Cock of, hill, N. coast of Arran, 2 m. E. of Lochranza; alt. 975 ft.

Arrat, ham., Angus, 3 m. E. of Brechin.

Arrinacrinach, ham., Ross and Cromarty, 5½ m. N.W. of Shieldaig. P.O. called Arrina.

Arrochar, par. and vil., with ry. sta., L.N.E., in N.W. Dunbartonshire, on E. side near head of Loch Long, 17½ m. N. of Helensburgh. Hotels. Much frequented in summer for its romantic scenery. Steamer communication with Glasgow. Area of par., 25,872 ac. Pop. 670.

Arrochar Hostel, Arrochar, Dunbarton-shire, 1 m. from vil., 11 m. N. of Gare-lochhead, 12 m. W. of Lochgoilhead. A Scottish Youth Hostel.

Arron, loch, W. Kirkcudbrightshire, 15 m. N. of Minnigaff.

Artafaille, pl., Black Isle, Ross and Cromarty, 7 m. E. Muir of Ord. P.O.

Arthrath, pl., E. Aberdeenshire, 5 m. N. of Ellon. P.O.

Arthur, loch, Kirkcudbrightshire, ¾ m. E. of Killywhan; depth, 48 ft., alt. 239 ft.

Arthurhouse, pl., S.E. Kincardineshire, Garvock par. ; ancient cairn.

Arthurlie, part of Barrhead, Renfrewshire. Pop. 538. Cotton mills, bleachfields, and print works. A. House, a seat.

Arthur Seat, mt., Dumfriesshire, 5 m. N.E. of Moffat; alt. 2398 ft.

Arthur's Seat, conspicuous hill in King's Park, S.E. Edinburgh. Outlined like a lion couchant, commands from its summit a magnificent panoramic view; alt. 822·9 ft.

Arthur's Seat, rock, N. side of Dunbarrow Hill, Dunnichen par., Angus.

Arthurstone, seat, Perthshire, 3 m. N.E. of Coupar Angus.

Artney, stream, trib. of r. Ruchill, in Glen Artney, S. Perthshire.

Artrochie, pl., E. Aberdeenshire, 3 m. N.E. of Ellon.

Ascaig, loch, Sutherland, 12½ m. N.W. of Helmsdale.

Ascog, vil., seat, bay, isl., and loch, E. Bute Isl. 1½ m. S.E. of Rothesay. P.O., T.O. Pop. 178.

Ascreavie, seat, Angus, 4 m. N.W. Kirriemuir.

Ascrib Isles, uninhabited islets in Loch Snizort, N.W. Skye, Inverness-shire.

Ashcraig, seat, N.W. Ayrshire, 4 m. N. of Largs.

Ashdale, stream and glen, Buteshire, S.E. Arran.

Ashdow, a deep ravine, S.W. Stirlingshire, Killearn par., 3 m. S.W. of Killearn.

Ashenyard, or **Ashgrove**, loch, Ayrshire, 3 m. N.E. of Ardrossan.

Ashey, loch, Inverness - shire. See Ashie.

Ashfield, ham., Perthshire, 2 m. N. of Dunblane.

Ashfield, ham., N. Knapdale, S. Argyll.

Ashgill, pl., colliery dist., Lanarkshire, ½ m. S.E. Dalserf. P.O., T.O.

Ashgillhead, par. of Old Monkland, Lanarkshire; special drainage dist. Pop. 1103.

Ashgrove, small loch, Ayrshire, Kilwinning par.

Ashgrove House, Ayrshire, 1¼ m. N.E. of Stevenston.

Ashie, loch, 6½ m. S.S.W. of Inverness; source of the water supply of Inverness.

Ashiesteel, or **Ashiestiel**, former seat on the Tweed, 7½ m. N.W. of Selkirk, Sir Walter Scott's residence prior to his removal to Abbotsford.

Ashiesteel Hill, in co., and 5 m. N., of Selkirk; alt. 1314 ft.

Ashievat, small loch, N. Harris, Outer Hebrides.

Ashintully, seat in Kirkmichael par., Perthshire.

Ashkirk, par., vil., and seat, Selkirkshire, on Ale Water, 6 m. N.N.W. of Hawick ; 13,159 ac. Pop. 308. P.O., T.O.

Ashleigh, Eskbank, Midlothian. Girls' private school.

Ashley, seat, Midlothian, 2 m. S.E. of Ratho sta.

Ashludie Sanatorium, Angus, ¾ m. N.E. of Monifieth.

Ashmore House, Perthshire, 5 m. N. of Blairgowrie.

Ashtie Loch, in co., and 7 m. S., of Inverness.

Ashton, S.W. part of Gourock, Renfrewshire.

Ashval, mt., Rhum Isl., Inverness-shire; alt. 2552 ft.

Askaig Port, seapt. and vil., N.E. coast of Islay, Argyll. Inn. P.O.

Askernish, seat, in W. of S. Uist, Outer Hebrides.

Askival, mt., Isl. of Rhum., Inverness-shire ; alt. 2659 ft.

Asknish, seat on Loch Gair, Argyll.

Asliesk, ruined castle and modern seat, Morayshire, 5 m. E. of Forres.

Assapoll, loch, S.W. Mull, Argyll.

Assel, r., Ayrshire, entering Stinchar R.

Assleed, or **Little Water,** affl. of the Ythan, Aberdeenshire.

Asslich, or **Astlich,** loch, N. Inverness-shire, near Drumnadrochit.

Assloss House, Ayrshire, 1½ m. N.E. of Kilmarnock.

Assynt, coast. par. and ham., S.W. Sutherland; 110,091 ac. Pop. 1342. A mountainous and rocky dist., with numerous lochs. Ham., also called Inchnadamph, 33 m. W.N.W. of Lairg. P.O. Loch Assynt, the chief loch, is 6¼ m. long ; alt. above sea level 215 ft., depth, 283 ft. The chief vil. in the par. is Lochinver, which stands at the head of the sea loch of the same name. It has P.O., T.O., churches, and hotel.

Assynt House, Ross and Cromarty, 1 m. N.W. of Evanton.

Asta, vil. and loch, Shetland, 1 m. N. of Scalloway.

Ath Dubh (Black Ford), stream, S. Harris, Inverness-shire, flows from Loch na Moracha.

Athelstaneford, par. and vil., East Lothian, 3 m. N.E. of Haddington; 5061 ac. Pop. 945. P.O., T.O. Vil. pop. 251.

Atholl, dist., extreme N. of Perthshire, includes a proportion of the Central Grampians. Area, 450 sq. m. Extensive deer forest. Gives the title of duke to the ancient house of Murray.

Atholl Sow, mt., Perthshire, 7 m. S. of Dalwhinnie; alt. 2600 ft.

Attadale, seat, deer forest, and ry. sta., L.M.S., Ross and Cromarty, 5 m. N.E. of Strome Ferry.

Attadale, stream, Ross and Cromarty, flows into Loch Carron.

Attonburn, seat, Roxburghshire, Morebattle par.

Attonburn Hostel, on Bowmont Water, 4 m. S. of Yetholm, Roxburghshire. A Scottish Youth Hostel.

Auch, shooting-lodge, Argyll, 3 m. from Bridge of Orchy sta.

Auchaber, seat, in Forgue par., Aberdeenshire, 7 m. E. of Huntly.

Auchairne, seat, Ayrshire, 2 m. S.E. Ballantrae.

Auchall, loch, Ross and Cromarty, 2¼ m. N.E. of Ullapool.

Auchallader, pass at head of Loch Tulla, N.E. Argyll.

Auchanacie, pl., Banffshire, 3 m. S.W. of Keith.

Auchans House, Dundonald par., Ayrshire, 3 m. N.E. of Troon. Visited by Dr. Johnson, 1773.

Aucharroch, seat, Angus, 3¾ m. N.W. Kirriemuir.

Auchedly, pl., Aberdeenshire, 3 m. N.W. of Ellon.

Auchenbathie, pl., Renfrewshire, 3¼ m. S.E. of Lochwinnoch. Ruined ancient castle, once the property of Sir William Wallace.

Auchenbeatty, stream, Dumfriesshire, flowing 6 m. S.E. to the Nith, near Kirkmahoe.

Auchenbowie, vil. and seat, Stirlingshire, 1¾ m. S.W. of Bannockburn. P.O.

Auchencairn, q.s. par., bay, and vil., Kirkcudbrightshire, 7½ m. S. of Castle-Douglas. A sea-bathing resort. P.O., T.O. Pop. 241. A. House, 1 m. S.E.

Auchen Castle, ruined castle, 2 m. S.W. of Moffat, Dumfriesshire. Modern castle, same name, nearby.

Auchencheyne, seat, Dumfriesshire, 3 m. S.W. of Moniaive.

Auchencorsh Hill, Ayrshire, 3 m. S.E. of Ballantrae; alt. 1067 ft.

Auchencorvie, pl., Argyll, 2½ m. S.W. of Campbeltown.

Auchencrow, vil., Berwickshire, 2 m. S.W. of Reston. P.O., T.O.

Auchendavie, or Auchendowie, vil., Dunbartonshire, 2 m. N.E. of Kirkintilloch. Situated on site of a fort of Antoninus' Wall.

Auchendennan House, Dunbartonshire, on W. bank of Loch Lomond, 1½ m. N.W. of Balloch ry. sta.

Auchendinny, vil., seat, and ry. sta., L.N.E., Midlothian, Glencorse par., on N. Esk, 8¼ m. S. of Edinburgh. Paper mills. Pop. 371.

Auchendolly, seat, Kirkcudbrightshire, 5 m. N. of Castle - Douglas. Mineral well.

Auchendrane, seat in co., and 4 m. S., of Ayr. Old A. is 1 m. S.

Auchendryone, vil., W. Aberdeenshire, forming part of Castleton of Braemar, on opposite side of the Clunie.

Aucheneck House, Stirlingshire, 2 m. S.W. of Killearn ry. sta.

Auchenflower, pl., S.W. Ayrshire, 2½ m. E. of Ballantrae.

Auchengeith, hill, Dumfriesshire, 9 m. N. of Maxwelltown; alt. 984 ft.

Auchengibbert Hill, Dumfriesshire, 2½ m. N.E. of Moniaive; alt. 1221 ft.

Auchengool, estate and mansion, 4½ m. S.E. of Kirkcudbright.

Auchengray, seat, ham., and ry. sta., L.M.S. (762 ft.), Lanarkshire, 6 m. N.E. of Carstairs. P.O., T.O. at ry. sta.

Auchenhalrig, vil., Morayshire, 2½ m. N.E. of Fochabers.

Auchenharvie, castle (in ruins) and ham., N. Ayrshire, 5 m. N.E. of Irvine.

Auchenharvie, seat, Ayrshire, near Stevenston.

Auchenheath, colliery vil., Lanarkshire, ry. sta., L.M.S., 5½ m. W. of Lanark. P.O., T.O.

Auchenlaich, pl., 1¼ m. E. of Callander, Perthshire.

Auchenleck, hill, Dumfriesshire, 3¼ m. N.E. of Thornhill; alt. 1431 ft.

Auchenlochan, pier, with fixed white light, Tighnabruaich, Argyll.

Auchenmade, ry. sta., L.M.S. (passenger service withdrawn), N.W. Ayrshire, 4 m. N.E. of Kilwinning ; alt. 305 ft.

Auchenmalg, pl. and bay, Wigtownshire, 4 m. S.E. of Glenluce. P.O.

Auchenreoch House, Angus, 3½ m. N. of Brechin.

Auchenreoch Loch, Kirkcudbrightshire, ½ m S.W. of Crocketford ; depth 34 ft., alt. 345 ft.

Auchenroddan State Forest, nearest ry. sta., Dinwoodie, Dumfriesshire.

Auchensaugh, hill, 2½ m. S. of Douglas, Lanarkshire ; alt. 1286 ft.

Auchenskeigh, dell, N.W. Ayrshire, 2 m. from Dalry.

Auchenskeoch, estate, with ruined castle, Kirkcudbright, 5 m. S.E. of Dalbeattie.

Auchentibber, vil., Lanarkshire, 1 m. S. High Blantyre.

Auchentorlie, seat, and vestiges of ancient hill-fort, 'Old Kilpatrick par., Dunbartonshire, 1½ m. W. of Bowling.

Auchentoshan, seat, Dunbartonshire, near Old Kilpatrick.

Auchentoshan Distillery, Dunbartonshire, Littlemill, near Bowling.

Auchentroig, pl., Stirlingshire, 1½ m. S.W. of Buchlyvie sta.

Auchernach, seat in Strathdon par., Aberdeenshire, 2½ m. N.W. of Strathdon.

Aucherran, shooting-lodge, 6 m. E. of Crianlarich, Perthshire.

Auchilick Bay, Loch Fyne, Argyll, 3 m. S.W. of Kilfinan.

Auchinairn, vil., Lanarkshire, 3 m. N. of Glasgow. P.O. A. House close by. Pop. of A. and Bishopriggs 5302.

Auchinblae, vil., Kincardineshire, 5½ m. N. of Laurencekirk. P.O., T.O. called Auchenblae. Pop. 372. A. Distillery here.

Auchincarroch, seat, Dunbartonshire, 2 m. N.E. of Alexandria.

Auchincheid, cliff-walled amphitheatre, vale, and cascades on S. coast of Arran, 7 m. S. of Lamlash.

Auchincloich, ruined ancient castle and ham., Ayrshire, 5¼ m. S.W. of Cumnock, Ochiltree par.

Auchincloich, ham., S. Stirlingshire, 3½ m. N.E. Kilsyth.

Auchincloich, ham., E. Ayrshire, 2 m. N.E. of Mauchline.

Auchincruive, seat and ry. sta., L.M.S., in co., and 3 m. N.E., of Ayr. T.O.

Auchindachy, pl., with ry. sta., L.N.E., Banffshire, 3 m. S.W. of Keith. T.O. at sta.

Auchindarroch, seat, Argyll, ½ m. N.W. of Lochgilphead.

Auchindoir and Kearn, par., W. Aberdeenshire, 8 m. S.W. of Gartly sta. ; 15,309 ac. Pop. 834. Containing Lumsden. P.O., T.O.

Auchindoune Castle, ruin on Lismore Isl., Loch Linnhe, Argyll, formerly the residence of the Bishops of Argyll.

Auchindown Castle, ruined old castle in Mortlach par., Banffshire, 2½ m. S.E. of Dufftown.

Auchindrain, ham., Argyll, 6 m. S.W. of Inveraray.

Auchinellan, shooting-lodge, on Loch Awe, Argyll, 1 m. from Ford.

Auchingane, pl., Stirlingshire, 3 m. S.W. of Falkirk.

Auchingill, vil., N.E. Caithness, 10 m. N. of Wick. P.O. called Auckingill. Has fixed lights.

Auchingilloch, a wild tract, notable for conventicles of the Covenanters, in Avondale par., Lanarkshire ; alt. 1514 ft.

Auchingoul, seat, Inverkeithny, Banffshire.

Auchingramont, a sub. of Hamilton, Lanarkshire.

Auchingray House, Lanarkshire, 2 m. E of Caldercruix.

Auchingray Rows, Lanarkshire, 1¼ m. N.E. Forrestfield ry. sta.

Auchingree, ham., N.E. Ayrshire, 2 m. S.W. of Beith.

Auchingree, burn, Kilpatrick Hills, Dunbartonshire ; a large reservoir at the junction with the Auldmennach.

Auchinheglish House, Dunbartonshire, 2 m. N.W. of Balloch ry. sta.

Auchinleck, par., town, and ry. sta., L.M.S., S. Ayrshire, on Lugar Water ; 24,152 ac. Pop. 6624. P.O., T.O. Town 4½ m. S.E. of Mauchline. Pop. 3645. A. House built by the father of James Boswell.

Auchinleish, seat, Angus, 1¼ m. W. of Glenisla.

Auchinlennan, seat, on Loch Lomond, 1½ m. N.W. of Balloch ry. sta., Dunbartonshire.

Auchinloch, vil., Lanarkshire, near Lenzie. Contains the Glasgow Convalescent Home.

Auchinmully, vil., S. Stirlingshire, E. side of Kilsyth par.

Auchinraith, colliery sub. of Blantyre, N.W. Lanarkshire.

Auchinraith House, Lanarkshire, 1 m. S. of Bothwell.

Auchinreoch, seat, Stirlingshire, 1½ m. E. of Milton.

Auchinroath, seat, Morayshire, 1 m. N.W. of Rothes.

Auchinstarry, vil., in detached portion of Dunbartonshire, on Forth and Clyde Canal.

Auchintibber, vil., Ayrshire, 4½ m. N.E. Kilwinning.

Auchintoul, seat, N.E. Banffshire, ½ m. S.W. of Aberchirder.

Auchintroig, ham., Stirlingshire, 1½ m. S.W. of Buchlyvie ry. sta.

Auchlane, ham. and stream, Kirkcudbrightshire, 3 m. S.W. of Castle-Douglas. The stream flows N.W. to the Dee.

Auchlee, estate, with two well-preserved ancient Caledonian stone circles. Kincardine, Banchory-Devenick par.

Auchleeks, ham. and seat in Blair-Atholl par., Perthshire. P.O., T.O.; 8 m. from Blair Atholl.

Auchleven, vil., Aberdeenshire, Premnay par., 2½ m. S. of Insch. P.O.

Auchline House, seat, Perthshire, 4 m. S.W. of Killin.

Auchlochan, seat and ham., near Lesmahagow, Lanarkshire, on r. Nethan.

Auchlossan, quondam loch in Aboyne and Lumphanan pars., Aberdeenshire, 4 m. N.E. of Aboyne.

Auchlunies House, Kincardine, Maryculter par., 2 m. S. of Cults.

Auchlunkart House, seat, Banffshire. Boharm par., 1½ m. S.W. of Mulben ry. sta.

Auchmacoy, ry. sta., L.N.E. (passenger service withdrawn), Aberdeenshire, 3 m. from Ellon. P.O., T.O. at sta. Seat of the Earl of Caithness.

Auchmannoch, seat, Ayrshire, 5 m. N.E. of Mauchline.

Auchmeddan, ham., Aberdeenshire, Aberdour par., 3 m. N.W. of New Aberdour.

Auchmillan, vil., ½ m. N.E. of Mauchline, N. Ayrshire.

Auchmithie, q.s. par. and fishing vil., Angus, 3½ m. N.E. of Arbroath.

Auchmore, pl., Lewis Isl., Ross and Cromarty, 11 m. S.W. of Stornoway.

Auchmore, seat, Ross and Cromarty, 3 m. N.W. of Beauly.

Auchmuty and Balbirnie Mills, vil., Fife, Markinch par. Paper mills.

Auchnacraig, ham., ½ m. S.E. of entrance of Loch Don, Mull, Argyll. P.O., T.O.

Auchnacree, seat, Angus, near Fern.

Auchnafree, shooting-lodge, Perthshire, 13 m. N.N.W. of Crieff.

Auchnafree Hill, Perthshire, Glen Turret; 2565 ft.

Auchnagatt, vil. and ry. sta., L.N.E., Aberdeenshire, 7½ m. N. of Ellon. P.O., T.O.

Auchnagoul, ham., Argyll, 2½ m. S.W. of Inveraray.

Auchnahow, glen, S.E. Sutherland, W. side of Kildonian par.

Auchnaskeoch, stream, Argyll, flowing into Loch Fyne.

Auchopecairn, mt., Roxburghshire; alt. 2382 ft.

Auchriddie, ham. and stream, Aberdeenshire, ½ m. S.W. of New Deer.

Auchries, vil., Aberdeenshire, 7 m. S.W. of Peterhead.

Auchry House, seat, Monquhitter par., Aberdeenshire, 4½ m. E. of Turriff.

Auchter, trib. of r. S. Calder, Lanarkshire, flowing from Bontyhillock to Bridgend.

Auchterarder, par. and police bur., ry sta., L.M.S., Perthshire. 14 m. S.W. of Perth; par. 11,166 ac. Pop. 3098; bur. 2254. P.O., T.O. Hotel. Mfrs. woollen goods, fancy shirtings, etc. Ruins of A. Castle and seat of A. House.

Auchterave Hostel, 2 m. S.W. of Fort Augustus. A Scottish Youth Hostel.

Auchterderran, par. and vil., Fifeshire, 5 m. N.W. of Kirkcaldy; 7872 ac. Pop. 16,664; vil. pop. 6917.

Auchtergaven, vil. and par., in Strathtay dist., Perthshire, 8 m. N.W. of Perth; par. 13,056 ac. Pop. 2193. Has linen works.

Auchterhouse, par. and vil., ry sta., L.M.S., Angus, 8 m. N.W. of Dundee; 6185 ac. Pop. 600. P.O., T.O. Ruins of ancient castle. Contains Sidlaw Sanatorium.

Auchterhouse Hill, Angus, 4 m. E. of Newtyle; alt. 1399 ft.

Auchterless, par. and vil., with ry. sta., L.N.E., N.W. Aberdeenshire, 4 m. S. of Turriff; 16,834 ac. Pop. 1,459. P.O., T.O., and P.O. and T.O. at sta.

Auchtermuchty, par., royal and mun. bur., with ry. sta., L.N.E., Fifeshire, 4¾ m. N.W. of Ladybank Junc.; par. 3531 ac. Pop. 1748; bur. 1253. P.O., T.O. Linen mfrs., malting and distillery; also weighing machine mfrs.

Auchtertool, par. and vil. in S.W. Fife, 4 m. W. of Kirkcaldy. Vil. 2¼ m. E. of Cowdenbeath ry. sta; 2735 ac. Pop. 683; vil. pop. 238. P.O., T.O. A. House is a seat.

Auchtertyre, vil., Newtyle par., Angus.

Auchtertyre, pl., Ross and Cromarty, 5 m. E. of Kyle of Lochalsh. Hotel.

Auchtertyre Hill, Ross and Cromarty, 1 m. N.E. of Lochalsh; alt. 1000 ft.

Auchtoo, ham., S.W. Perthshire, 2 m. E. of Balquhidder.

Auchtyfardle, seat, Lanarkshire, 1 m. N. of Lesmahagow.

Augmund's Row, ruined ancient fort, Elsness prom., Sanday Isl., Orkney.

Augustine, q.s. par., Greenock, Renfrewshire.

Auldbar, ry sta., L.M.S., Angus, 5 m. E.N.E. of Forfar. A. Castle.

Auldclune, ham., Perthshire, 2 m. S.E. of Blair-Atholl.

Auldearn, par. and vil., ry. sta., L.M.S., N.E. Nairnshire, 2½ m. S.E. of Nairn; par. 14,282 ac. Pop. 1193; vil. pop. 331. P.O., T.O.

Auldfield, part of Pollockshaws, Renfrewshire.

Auldgirnaig, ham., Perthshire, 4 m. N.W. of Pitlochry on r. Garry.

Auldgirth, ham., ry. sta., L.M.S., 8 m. N.W. of Dumfries. P.O., T.O.

Auldhill, hill, Ayrshire, W. Kilbride par.

Auldhouse, ham., Lanarkshire, 3 m. S. of Kilbride. P.O.

Auldhouse, burn, entering the White Cart at Pollockshaws, Renfrewshire.

Auldmuir, pl., Ayrshire, 2 m. W. of Dalry. Limeworks. A. Reservoir supplies Dalry.

Auldna, coalfield, Ayrshire, New Cumnock par.

Auldnachuirn, burn entering the Lossie at Dallas par., Moray.

Auld Wick Castle, dismantled ancient baronial stronghold, on coast of Wick par., Caithness.

Auld Wives' Lift, cromlech, 18 ft. long, and 11 ft. broad, Stirlingshire, Baldernock par. 3 m. S.W. of Lennoxtown.

Aulich, ham., Perthshire, N. side of Loch Rannoch.

Auliston Point, N.W. Morven, Argyll.

Aultbea, ham., Ross and Cromarty, E. side of Loch Ewe. Pier, 5 m. N.E. of Poolewe Hotel. P.O., T.O. Regular steamboat communication with Glasgow. Fixed white light on Aird Point in front of pier, white light in rear.

Aultdearg Lodge, Ross and Cromarty, 3 m. N.W. of Lochluichart ry. sta.

Aultgrande, or **Glass,** stream entering Cromarty Firth, 10 m. N.E. of Dingwall, Ross and Cromarty.

Aultguish, stream, flowing through the forest of Rinsky to Loch Ness, Inverness-shire. A continuous cataract, with one leap of 100 ft.

Aultiphurst, pl., Sutherland, 2 m. N.W. of Strathy.

Aultkollie, deep ravine, Loth par., Sutherland, between Brora and Port Gower.

Aultmore, moor, 5 m. S.E. of Buckie, Banffshire.

Aultmore, ham., Banffshire, 2 m. N.W. Keith. Distillery.

Aultnacaillich, stream and waterfall, Sutherland, 11½ m. N.W. of Altnaharra.

Aultnacealgach, pl., Sutherland, 10 m. N.W. of Oykell Bridge.

Aultnaharrie, pl., Lochbroom, Ross and Cromarty. Ferry service with Ullapool.

Aultsigh, stream issuing from Mealfourvounie mt., descending in cataracts and leaps to N.W. side of Loch Ness, Inverness-shire.

Auquharney, seat, Aberdeenshire, 6 m. N.E. of Ellon.

Auquhirie, ham., Kincardineshire, Dunnottar par., 3 m. W. of Stonehaven.

Ausdale, ham., Caithness, 4 m. S.W. of Berriedale and stream leaping into the sea over a lofty cliff.

Auskerry, small isl., 2½ m. S. of Stronsay, Orkney. Lighthouse, white fixed light, visible 16 m.

Aven, trib. of the Feugh in Strachan par., Kincardineshire.

Avendhu (Blackwater), r. entering the Duchray at Aberfoyle, uniting to form the Forth. Stirlingshire.

Aven-nan-geran, stream, Harris, Inverness-shire, falling into Loch Finsbay.

Avernish, ham., on Lochalsh, Ross and Cromarty.

Avich, loch and stream, Argyll. Issuing from the loch, the stream runs 2 m. to Loch Awe. Depth of loch 188 ft., alt. 311 ft.

Aviemore, vil. with ry. sta. and junc., L.M.S., Inverness-shire, 12¼ m. S.W. of Grantown. P.O., T.O. Alt. of sta. 697 ft. The vil. commands a magnificent view of the Cairngorms and is a renowned holiday resort. Hotels.

Aviemore Hostel, on Great North Road, ⅓ m. S. of A. sta., 32 m. S.W. of Inverness, 12 m. N.W. of Kingussie. A Scottish Youth Hostel.

Avoch, par. and fishing vil., ry. sta., L.M.S., in Black Isle, Ross and Cromarty, 1¾ m. S.W. of Fortrose ; par. 8061 ac. Pop. 1408. P.O., T.O. A green fixed light. A. House, 2 m. N.W. Vil. pop. 1002.

Avon, loch and r. in the Cairngorms, S. Banffshire, the latter issuing from the loch at 2377 ft. and flows E.N.E. for 9 m., then 13 m. N. to join the Spey at Ballendalloch. The famous Shelter Stone is at the head of the loch.

Avon, r., running about 18 m. eastward and north-eastward, partly between Stirlingshire and West Lothian to Firth of Forth, 2½ m. W. of Bo'ness.

Avon, r., trib. of the Clyde, rising in Ayrshire, flows mostly in a N.E. direction, joins the Clyde near Hamilton, Lanarkshire.

Avonbridge, vil. on Avon, with ry. sta., L.N.E. (passenger service withdrawn), Stirlingshire, 3½ m. N.E. of Slamannan. P.O., T.O. Pop. 667.

Avondale, par., Lanarkshire ; 37,470 ac. Pop. 5529. Contains ruined famous A. Castle, also Drumclog. A Covenanting centre.

Avondale House, Stirlingshire, 1 m. E. of Polmont.

Avonhead, colliery vil., Lanarkshire, New Monkland par., 4 m. N.E. of Airdrie.

Avonhill House, Stirlingshire, 1 m. W.S.W. Avonbridge ry. sta.

Avonholm, seat in Glassford par., Lanarkshire, ½ m. W. of Stonehouse.

Avonsuidh, or **FinCastle,** seat, West Coast of Harris, Outer Hebrides, Inverness-shire.

Avontoun, seat, West Lothian, 1½ m. S.W. of Linlithgow.

Awe, loch, extending 24 m. N.N.E. to base of Ben Cruachan, Argyll. Mean breadth 1 m., expanding to about 2 m. at foot. Exhibits great variety of scenery. Looks, in its broadest parts, to be remarkably small as compared with the magnitude of the surrounding mountains. Alt. 117 ft., greatest depth 307 ft. Loch Awe ry sta., L.M.S., at head of loch.

Awe, r., Argyll, flowing 4 m. through the Pass of Brander to Loch Etive.

Awe, loch, Assynt, Sutherland, 3½ m. S. of Inchnadamff ; depth 7 ft., alt. 504 ft.

Aylort, sea loch, between Moidart and Arisaig dists., Inverness-shire ; length 9 m., 4 m. wide at mouth.

Ayr, par., parl., mun., and royal bur., and co. town of Ayrshire ; ry. sta., L.M.S., at mouth of Ayr R., 41½ m. S.W. of Glasgow. An attractive town which owes something at least of its prosperity to its connection with Robert Burns. Pleasure steamers run to Arran and other Clyde resorts. Par. 12,493 ac. Pop. of par. 40,412 ; bur. pop. 36,784. Industries—coal, iron, shipbuilding, engineering, carpets, boots and shoes. Lifeboat sta. Two lighthouses with four lights, and an automatic fog-gun. The Ayr Burghs (Ardrossan, Ayr, Irvine, Prestwick, Saltcoats, and Troon) return one member to Parliament.

Ayr, r., Ayrshire, rising in the Muirkirk par., flows 38 m. W. to Firth of Clyde at Ayr.

Ayre Point, S.E. Raasay, Inner Hebrides. Four-seconds flashing light.

Ayrshire, a maritime co. in the S.W., bounded on the W. by the Firth of Clyde and North Channel. Area 1132 sq. m., or 724,251 ac. Pop. 285,217. Its greatest breadth is 28 m. and its length 78 m. The rivers Irvine and Doon cut the entire area of the co. into three sections—Cunninghame in the N., Kyle in the middle, Carrick in the S. ; also drained by the rivers Ayr, Stinchar, and Girvan ; chief loch, Loch Doon. Principal heights, Shalloch on Minnoch

2520 ft. ; Kirriereoch Hill 2562 ft. ;
and Cairnsmuir 2613 ft. Famous for
early potatoes and dairy farming. Has
coalfields, ironworks, fireclay, limestone, and honestone deposits. Chemicals, explosives, leather, tools, and
fisheries. Returns three members to
Parliament.

Ayton, par. and vil., ry. sta., L.N.E.,
Berwickshire. Vil., on r. Eyre, 21 m.
S.E. of Dunbar ; 6700 ac. Pop. 1413.
P.O., T.O. Vil. pop. 472. A. Castle.

Ayton House, Perthshire, 1½ m. S.W. of
Abernethy.

Aytonwood House, Berwickshire, near
Ayton.

Aytoun Hill, seat, Fife, 3½ m. E. of Newburgh.

Aywick, pl., S.E. coast of Yell, Shetland.

B

Ba, loch (2½ m. by 1 m.) and riv. on Moor
of Rannoch, N.E. Argyllshire ; depth
30 ft., alt. 957 ft.

Ba, loch (2 m. by ½ m. and 144 ft. deep,
alt. 41 ft.) and stream in Torosay par.,
isl. of Mull, Argyllshire.

Ba, isl. in par. of Applecross, 4 m. S. of
Applecross pier, W. Ross and Cromarty.

Ba, or **Bahill,** hill in Drumblade par., N.W.
Aberdeenshire ; alt. 700 ft.

Baads, moorland area in the W. of par.
of Cullen, Banffshire.

Babbet Ness, headland, 6 m. E. by S. of
St. Andrews, Fifeshire.

Baberton, golf course and pl., 5 m. S.W. of
Edinburgh G.P.O. Access by bus or
rail (Juniper Green sta., L.M.S.).

Bac Beag and **Bac Mor,** two of the Treshinish Isls., off N.W. Mull, Argyllshire.

Bachnagairn Falls, on rt. S. Esk, above its
junction with the Prosen, N. Angus.

Bachragairn and Glenmuick, deer forest
on borders of Aberdeenshire and Angus.

Bachuil, ham. in N. of Lismore Isl.,
Argyllshire. P.O.

Back, stream rising in Loch Ronach, 5 m.
S. of Forres and flowing down valley of
Pluscardine, Morayshire.

Back, vil. on E. coast of Lewis, and 7 m.
N.N.E. of Stornoway, Outer Hebrides.
P.O., T.O.

Back, small stream flowing round base of
Tower Hill, in Pittencrieff Glen, Dunfermline, Fifeshire.

Backaland, ham., with bay, S. Eday Isl.
Orkney. P.O.

Backaskail, bay in Sanday Isl., Orkney.
Enormous quantities of shell-fish.

Backies, ham. in par. of Golspie, Sutherlandshire. It lies 2 m. N. of Golspie.
It has ruins of an ancient tower, said to
be of Norwegian origin.

Backlass, hill (alt. 300 ft.), in Watten par.,
and 2¾ m. W.S.W. of Watten vil.,
Caithness.

Backmuir, vil., 5½ m. N.W. of Dundee,
Angus.

Backmuir, vil. in Largo par., 2½ m. S.E.
Ceres, Fifeshire.

Back o' Loch Halt, ry. sta., L.N.E.,
between Lenzie Junction and Kirkintilloch, Dunbartonshire.

Back o' Stone, shooting-lodge, 5 m. W.S.W.
of Banchory, Kincardineshire.

Backwater, ham. in par. of Lintrathen
and 10 m. N.W. of Forfar, Angus.

Backwater, ham. and trib. of the Melgam,
W. Angus.

Badachro, ham., 5 m. S.W. of Gairloch,
W. Ross and Cromarty. P.O., T.O.

Bad a' Ghaill, loch (180 ft. deep), 10 m.
N.W. of Ullapool, N.W. Ross and
Cromarty.

Badanden Hill, N.W. Angus, 2¼ m. N.E.
of Inverharity ; alt. 2429 ft.

Badbea Hostel, on Little Loch Broom, 5 m.
N.W. of Dundonnell, Ross-shire. A
Scottish Youth Hostel.

Badcall, ham., rivulet, and bay, N.W.
Sutherland, 2¼ m. S. of Scourie.

Baddanloch, shooting - lodge, and deer
forest, 5 m. W. Kinbrace ry. sta., E.
Sutherlandshire.

Baddoch, trib. of the Clunie, S. Aberdeenshire.

Baden, or **Baddanloch,** loch in upper part
of Kildonan par., Sutherlandshire, 5½ m.
W. by N. of Kinbrace sta. ; depth, 42
ft., alt. 394 ft.

Badenoch, dist. in S.E. Inverness-shire ;
comprises the basin of the Spey from
the source of that river to the vicinity of
Upper Craigellachie ; bounded on the
one side by summits of the Central
Grampians and on the other by the
summits of the Monadhleadh Mts.
It belonged anciently to the Comyns
and passed, in the time of Robert II.
to the " Wolf of Badenoch," the Earl
of Buchan. It is bounded on the

W. by Lochaber and on the E. by Braemar.

Badenscoth, or **Kirkton,** ham. in par. of Auchterless, N.E. Aberdeenshire.

Badensgill, ham. and small stream in par. of, and 2½ m. N.W. of, West Linton, Peeblesshire.

Badentarbet, pl., 12 m. N.W. Ullapool, W. Ross and Cromarty.

Badentoy, pl., in par. of Banchory-Devenick, N.E. Kincardineshire, 6 m. S.W. Aberdeen.

Badenyon, pl., with relics of old castle, in par. of Glenbucket, Aberdeenshire.

Badicaul, ham., 1½ m. N. of Kyle of Lochalsh, W. Ross and Cromarty.

Badlieu, stream, par. of Tweedsmuir, S.W. Peeblesshire; rises on Clyde Law and flows 2½ m. N.E. to the Tweed. Badlieu Rig (1374 ft.) rises here.

Badluachrach, loch, W. Ross and Cromarty, 4 m. N.E. of Poolewe; alt. 309 ft., depth, 143 ft.

Badrallach, pl., 5 m. W.S.W. Ullapool, W. Ross and Cromarty.

Baggage Knowe, hill in par. of Kilsyth, S. Stirlingshire; associated with battle of Kilsyth, 1645.

Baidland Hill, hill (alt. 1099 ft.), in par. of Dalry, N. Ayrshire.

Baikie, seat, and formerly castle, in Airlie par., 5 m. S.W. Kirriemuir, Angus.

Bailford, estate, near Penpont, W. Dumfriesshire.

Bail Hill, hill (alt. 1684 ft.), 4½ m. N.W. Moniaive, Dumfriesshire.

Bailivanich, ham. in par. of S. Uist, Benbecula Isl., Outer Hebrides.

Baillieston, q.s. par. and mining town, Lanarkshire, with ry. stas., L.M.S. and L.N.E., 6¾ m. E. of Glasgow. P.O., T.O. The q.s. par. also bears the name of Crosshill. B. House, ½ m. W. Town pop. 5736.

Baing, loch, 7 m. S.W. Dalmellington, S. Ayrshire.

Bainie, loch, Perthshire, 6½ m. N.W. of Glenisla.

Bainsford, suburb of Falkirk, stands on Forth and Clyde canal, about 1 m. N. of Falkirk. P.O.

Bains Hill (alt. 1042 ft.), N.W. Aberdeenshire.

Bainshole, ham., 5 m. N.W. Insch sta., N.W. Aberdeenshire. P.O., T.O.

Baintown, ham., 1 m. N.E. of Kennoway, Fifeshire.

Balabraes of Aytoun, seat, 1½ m. S.W. Eyemouth, Berwickshire.

Balachladich, ham., 6½ m. N. of Lochinver, W. Sutherlandshire.

Balado, seat (B. House) and ry. sta., L.N.E., 3 m. E. of Crook of Devon, Kinross-shire.

Balagich, or **Ballagioch,** hill (alt. 1084 ft.), S.E. Renfrewshire.

Balaklava, or **Clippens Square,** vil. in par. of Kilbarchan, 6 m. W. of Paisley, Renfrewshire.

Balallan, vil. in Loth par., and 14 m. S.W. of Stornoway, Lewis, Outer Hebrides. P.O., T.O.

Balavil, seat, 3 m. E.N.E. of Kingussie, Inverness-shire.

Balavil, pl. and seat, 1¼ m. S. of Conon, Ross and Cromarty.

Balavreed, seat, 9 m. N.W. of Dunbeath, Caithness.

Balbardie, seat, ½ m. N. of Bathgate, W. Lothian.

Balbeg, seat, 2 m. S.S.W. of Straiton, Ayrshire.

Balbeggie, vil., 4½ m. N.E. of Perth. P.O., T.O. Pop. of B., Guildtown, and Wolfhill, 425.

Balbegno Castle, castle, built in 1509, near Fettercairn, Kincardineshire.

Balbirnie, ham. on borders of Perthshire, in the par. of Ruthven, W. Angus.

Balbirnie House, ½ m. N.W. of Markinch, Fifeshire. The village is called Balbirnie Mills, and has paper mills. See AUCHMUTY and BALBIRNIE MILLS.

Balbithan, old seat on r. Don, in Keithhall par., Aberdeenshire.

Balblair, ham., about 1 m. S.W. of Nairn, was camping place of the royal army on the eve of battle of Culloden.

Balblair, vil., in Resolis par., and 5¼ m. W. of Tain, Ross and Cromarty. P.O., T.O.

Balblair, ham. on Kyle of Sutherland, 2 m. N.W. of Bonar Bridge, Sutherland. State forest nearby.

Balblair, mansion, 1 m. S.E. of Beauly, Inverness-shire.

Balbrogie, vil., in par. of Coupar-Angus, Perthshire.

Balcaithly, estate, 4 m. S.E. of St. Andrews, Fifeshire.

Balcarres House, seat of the Earl of Crawford, in par. of Kilconquhar, Fife, and about 1 m. N.N.W. of Colinsburgh.

Balcary, seat and headland in Rerwick par., and 6½ m. S. of Dalbeattie, Kirkcudbrightshire. Lifeboat sta.

Balcaskie, seat, 2 m. N.W. Pittenweem, Fifeshire.

Balcastle, ham. in Slamannan par., 6 m. S.W. of Falkirk, Stirlingshire.

Balchreck, pl., Sutherland, 4 m. N. of Kinlochbervie. P.O.

Balchrystie, seat on site of ancient Culdee Cell, near Colinsburgh, Fifeshire.

Balcomie Castle, formerly seat of Earls of Kellie, now a farmhouse, near Crail, Fifeshire.

Balconie, seat, 1 m. S.E. of Evanton, Ross and Cromarty.

Balconnel, seat, 5 m. W.N.W. of Brechin, Angus.

Balcraig, remains of old castle, near ruins of Hatton Castle, near Newtyle, S.W. Angus.

Balcruvie, or **Pitcruvie,** ruined ancient castle near Largo, Fifeshire.

Balcurvie, vil., 3 m. E. of Markinch, Fifeshire. Pop. (B. and Windygates) 1744.

Baldarroch, seat, near Murthly, Perthshire.

Baldernock, par. and ham. on south border of Stirlingshire; 4320 ac. Pop. 674. Ham. stands in the W. of the par., 2¼ m. E.N.E. of Milngavie sta.

Baldinnie, ham., 2 m. E. of Ceres, Fifeshire.

Baldoon Castle, ruin, 1½ m. S.S.W. Wigtown. The castle was the scene of the incident which suggested to Sir Walter Scott the tragic end of his *Bride of Lammermoor.*

Baldovan, vil., seat, and ry. sta., L.M.S., 7¾ m. N.W. by rail from Dundee, Angus.

Baldovie, ham., 3 m. N.E. Dundee on road to Arbroath, Angus. P.O.

Baldowrie, seat, 3 m. E. of Coupar-Angus, Perthshire.

Baldragon, pl., ry. sta., L.M.S., 8¾ m. by rail N.of Dundee, Angus.

Baldridge, pl., 1 m. N.W. Dunfermline, Fifeshire.

Balduff Hill, 4 m. N.W. Alyth, Perthshire, alt. 1394 ft.

Baledgarno, ham., 1 m. N.W. Inchture, Perthshire.

Baledmund, seat, 1 m. N. of Pitlochry, Perthshire.

Balemartine, pl., S.E. Tiree Isl., Argyllshire. P.O., T.O.

Balendoch, seat, 2 m. N. of Meigle, Perthshire.

Balenock, mt., (alt. 2092 ft.), 4 m. E. Garelochhead, Dunbartonshire.

Balephetrish Bay, N. coast, Tiree Isl., Argyllshire.

Balephuill Bay, S.W. coast Tiree Isl., Argyllshire.

Balerno, vil., on the Water of Leith, Midlothian, 7 m. S.W. Edinburgh; ry. sta. (alt. 506 ft.), L.M.S. Has paper mills. P.O., T.O. Pop. 1255.

Baleshare, isl. off S.W. coast of N. Uist, Outer Hebrides. P.O. (spelt Balishare). Pop. 136.

Balfield, ham., 3½ m. W. of Edzell, Angus.

Balfluig, old castle, 1 m. S.E. Alford, Aberdeenshire.

Balfour, ruined castle, 4 m. W. Kirriemuir, Angus.

Balfour, ham., in par. of Shapinshay, and 5 m. N.N.E. Kirkwall, Orkney. The castle in the vicinity is a building of mixed architecture. P.O., T.O.

Balfour, seat, in par. of Menmuir, Angus.

Balfour, seat, 2 m. S.E. of Markinch and on r. Leven, Fifeshire.

Balfour House, 2 m. S.E. Aboyne, Aberdeenshire.

Balfron, par. and vil. on the Endrick, W. Stirlingshire; ry. sta. (1½ m. W.), L.N.E.; 7821 ac. Pop. 1200. Hotel. P.O., T.O., and T.O. at sta.; vil. pop. 919.

Balfunning House, seat, 1 m. W. Balfron sta., Stirlingshire.

Balgair, moor in par., Balfron, W. Stirlingshire.

Balgarvie, seat, 1½ m. N.W. Cupar, Fifeshire.

Balgavie, seat and loch, 5 m. E. of Forfar, Angus.

Balgay, eccl. par., and hill on N. side of Dundee, Angus.

Balgedie, Easter and **Wester,** two hams., 3 m. E.S.E. Milnathort, Kinross-shire.

Balgersho House, 1 m. S. Coupar-Angus, Perthshire.

Balgillo House, ¾ m. N. Broughty Ferry, Angus.

Balgonar, pl. and seat, 1 m. N. Saline, Fifeshire.

Balgone, seat, 2½ m. S.E. of N. Berwick, E. Lothian.

Balgonie, Milton of, q.s. par., and vil. on the Leven, Fife, 1½ m. E.S.E. Markinch sta., L.N.E. P.O., T.O. Pop. 517. B. Castle in the vicinity.

Balgowan, pl., 2 m. W.S.W. Methven, Perthshire; ry. sta., L.M.S. T.O. B. House ¾ m. N.

Balgowan, pl., between Cluny Castle and Laggan Church, 5½ m. S.W. of Newtonmore, Spey Valley, Inverness-shire.

Balgown, a small bay on E. side Kirkmaiden par. and 9 m. N.W. of Galloway, Wigtownshire.

Balgownie, Brig o', see BRIG O' DON.

Balgray, ham., 7 m. N. of Dundee, Angus.

Balgray House, 3 m. N. of Lockerbie, Dumfriesshire.

Balgray Reservoir, 1 m. S.E. Barrhead, Renfrewshire.

Balgreggan, seat, 7 m. S. of Stranraer, Wigtownshire; a moat nearby, 460 ft. in circumference and 60 ft. deep.

Balgy, small stream flowing into Loch Torridon, S.W. Ross and Cromarty.

Balhaldie, ham., 3½ m. N.W. of Dunblane, Perthshire.

Balhall, seat, Menmuir par., and 3½ m. N.W. Brechin, Angus.

Balhary, seat, 2 m. S.E. Alyth, Perthshire.

Balhinny, ham., 3 m. W. of Rhynie, W. Aberdeenshire.

Balholmie House, ½ m. S. Cargill, 4½ m. S.W. of Coupar-Angus, Perthshire.

Balhousie, pl., 1½ m. S. of Woodside, E. Fifeshire.

Balhousie Castle, old castellated mansion just N. of Perth, seat of the Earl of Kinnoull.

Baliasta, stream, falling into Loch Cliff, Unst, Shetland.

Baligill, ham., 1 m. N.E. Strathy, N. Sutherland.

Baligill, loch, 2 m. S.W. Melvich, N.E. Sutherland.

Balintore, fishing vil., 7 m. S.E. of Tain, Ross and Cromarty. P.O., T.O. Harb. has fixed white light. Pop. 349.

Balintore Castle, 6¾ m. W.N.W. Kirriemuir, Angus.

Balintraid, pl., with harb., on Cromarty Firth and 3 m. N.E. of Invergordon, Ross and Cromarty.

Balishare, see BALESHARE.

Balkail, seat, ¼ m. S.E. Glenluce, Wigtownshire.

Balkello, ham., 5 m. N. of Dundee, Angus.

Balkerr, ham. and moor, 3 m. N.E. Stranraer, Wigtownshire.

Balkissock, seat, 3 m. E. of Ballantrae, S. Ayrshire.

Ballachantuie, see BELLOCHANTUY.

Ballachulish, vil., and small port with steamboat pier, on S. shore of Loch Leven and 12 m. S. of Fort William, Argyllshire; terminus of Ballachulish branch of Oban and Callander section of L.M.S. Hotel. Extensive slate quarries. P.O., T.O. Regular steamboat communication with Glasgow, Oban, and Fort William. Continuous daily ferry service across Loch Leven. Pop. 727.

Ballaggan, seat and waterfall on Blane R., in Strathblane par., Stirlingshire.

Ballagioch, see BALAGICH.

Ballallan, vil., 14 m. S.W. Stornoway, Lewis, Outer Hebrides.

Ballamenoch, seat, 1¼ m. E. Buchlyvie sta., Stirlingshire.

Ballanbriech, ruined ancient castle on a steep bank overhanging Firth of Tay, 2¾ m. N.E. Newburgh, Fife.

Ballandarg, burn, rising in Kirriemuir par., and running S. to Dean R. in Glamis par., Angus.

Ballanlay, pl., 2½ m. S.W. Rothesay, Isl. of Bute.

Ballantrae, par., vil., and seapt., 34 m. S.W. Ayr, and in that co.; nearest sta., Pinwherry, 8 m. distant; vil. lies at mouth r. Stinchar; 33,581 ac. Pop. 1076; vil. pop. 484. Hotels. P.O., T.O. Summer resort and head of a fishing district.

Ballat, pl., par. of Drymen, and 2 m. N.W. Balfron, Stirlingshire.

Ballater, vil. and police bur. in par. of Glenmuick, on r. Dee, Aberdeenshire; terminal sta. section of L.N.E., and 17 m. E. of Braemar. Hotels. To the N. is a rocky defile, the Pass of Ballater. P.O., T.O. Pop. 1198. Pannanich Wells in vicinity.

Ballaterach, farmhouse, 4½ m. E. of Ballater, S. side r. Dee. Byron spent his boyhood here.

Ballater Youth Hostel, Deebank Road, Ballater, Aberdeenshire. A Scottish Youth Hostel.

Ballathie, seat, 3 m. N.E. Stanley, Perthshire.

Ballavoulin, seat, near Killiecrankie, Perthshire.

Balleave, ham., ½ m. S.W. Kinross, Kinross-shire.

Ballechin, seat in Logierait par. and 3 m. N.W. Ballinluig Junc., Perthshire.

Balledgarno, or **Ballerno,** vil., 7 m. W. of Dundee, Perthshire.

Balleloch, small loch, N. Uist, Outer Hebrides.

Ballenach, ham., 6 m. W.N.W. Lochgilphead, Argyllshire.

Ballencleroch, seat, Stirlingshire, ½ m. N.W. of Campsie Glen ry. sta.

Ballencleuch Law, hill on borders of Lanarkshire and Dumfriesshire; alt. 2267 ft.

Ballencrieff, ruined seat of Lord Elibank, 1¾ m. S.E. Aberlady, E. Lothian.

Ballencrieff, rivulet near Bathgate, W. Lothian. B. House is a ruin.

Ballendrick, seat, 1 m. S.W. Bridge of Earn, Perthshire.

Ballengeich, steep pass behind Stirling Castle.

Ballevullin, ham. in N.W. of Tiree Isl., W. Argyllshire.

Ballewan, ham., 2 m. N.W. Strathblane, Stirlingshire.

Balliasta, pl., Isl. of Unst, Shetlands. An ancient chapelry, with ruins of an old church.

Ballikinrain Castle, seat, 1 m. S.E. of Balfron, Stirlingshire. B. Stream flows into the r. Endrick.

Ballimore, ham. on r. Tummel, 2¾ m. E. of Kinloch Rannoch, Perthshire.

Ballimore, pl. on Loch Fyne, 2 m. S.E. of Lochgilphead, Argyllshire.

Ballinaby, seat in Isl. of Islay, 8 m. N.W. Bridgend, W. Argyllshire.

Ballinakill, seat on coast of Kintyre and 10 m. S.W. Tarbert, Argyllshire.

Ballinbreich, ruined seat of Earls of Rothes, 2½ m. N.E. Newburgh, Fifeshire.

Ballindalloch, seat and ham., with ry. sta., L.N.E., 12 m. N.E. Grantown, Banffshire. P.O., T.O.

Ballindalloch, seat, 1½ m. E. of Balfron sta., Stirlingshire.

Ballindarroch, seat in. co. of, and 5 m. S.W. of Inverness.

Ballindean, ham. and seat, 1½ m. N.W. Inchture, Perthshire.

Ballindown, seat, 2 m. S. of Beauly, Inverness-shire.

Ballingry, par. and ham., 3 m. N.W. Lochgelly, Fifeshire; 3910 ac. Pop. 10,353.

Ballinloan, pl. and stream, 3½ m. W. of Dunkeld, Perthshire.

Ballinluig, vil. and ry. junc., L.M.S., 8 m. N.W. of Dunkeld, Perthshire. P.O., T.O.; alt. of ry. sta., 209 ft.

Ballintomb, trib. of r. Spey, which it joins in the vicinity of Knockando House, Morayshire.

Ballintore, coastal vil., 2½ m. S.E. of Hill of Fearn, N.E. Ross and Cromarty.

Ballintuim, ham., 7¼ m. N.W. of Blairgowrie, Perthshire. P.O., T.O. B. House is a seat.

Balloch, vil., with ry. stas., L.N.E. and L.M.S., Dunbartonshire, at foot of Loch Lomond, 4¾ m. N. of Dumbarton, 19½ m. N.W. of Glasgow by rail. Hotel. Balloch Pier, with ry. sta., is the starting-place of the Loch Lomond steamers. P.O., T.O. B. Castle, early seat of the Earls of Lennox in vicinity.

Balloch, ruined castle on Loch Tay, in vicinity of Kenmore, Perthshire.

Balloch, bay in Great Cumbrae Isl., Buteshire; affords good anchorage, and has oyster beds.

Balloch, loch at base of Torlun Hill, 4 m. S.W. of Crieff, S. Perthshire. It drains into r. Earn.

Balloch, pl. in par. of Kirriemuir, Angus; Area largely peat bog.

Balloch, vil. in co., and 4 m. N.E., of Inverness.

Ballochbuie, deer forest, 4 m. S.W. of Balmoral Castle, S.W. Aberdeenshire.

Balloch Castle, ruins on isl. in Loch Doon, Ayrshire.

Ballochdoan, pl., 2½ m. S.W. Ballantrae, S. Ayrshire.

Balloch Hill, alt. 1199 ft.; close to Keith, Banffshire.

Ballochling, loch, 10 m. from Dalmellington, S. Ayrshire.

Ballochmorie House, 8 m. S. of Girvan, and 2 m. distant from ry. sta. of Pinwherry, S. Ayrshire.

Ballochmyle, seat on the Ayr, and 1½ m. S.E. of Mauchline. It is associated closely with the life of Robert Burns.

Balloch na Houigh, loch, 5 m. S. of Glenelg, W. Inverness-shire.

Ballochroy, ham. in N.W. of Kintyre, Argyllshire.

Ballogie, pl. and seat, 5 m. S.E. of Aboyne, E. Aberdeenshire. P.O., T.O.

Ballo Hill, alt. 1029 ft. ; one of the Sidlaw Hills, par. of Longforgan, Perthshire.

Ballo Reservoir, Lomond Hills, 2½ m. N.W. of Leslie, Fifeshire.

Ballowmill, stream, trib. to Eden, which it joins 2½ m. N.E. of Kettle, Fifeshire.

Ballumbie, seat and ruined castle in par. of Murroes and 4 m. N.E. of Dundee, Angus.

Ballygrant, vil. and loch, 3 m. S.W. of Portaskaig, Isl. of Islay, Argyllshire. P.O., T.O. Pop. 102.

Ballyoukan, seat, 1½ m. S. of Pitlochry, Perthshire.

Balmacaan, a seat of the Countess of Seafield. B. deer- orest, 17 m. S.W. of Inverness, and near Loch Ness.

Balmacara, vil. on N. shore of Loch Alsh and 3 m. N.E. of Kyleakin, Ross and Cromarty. P.O., T.O. B. House is a seat.

Balmaclellan, par. and vil., N. Kirkcudbrightshire, and 2 m. N.E. of New Galloway. P.O., T.O. ; 22,976 ac. Pop. 628.

Balmae, seat, 4 m. S. of Kirkcudbright.

Balmaghie, par. and seat, 6 m. N.E. of Gatehouse-of-Fleet, Kirkcudbrightshire. Threave Castle (in ruins), a Roman camp and a moat are the principal antiquities. Pop. 650. Par. 21,059 ac.

Balmaha, ham., with pier on Loch Lomond and 4 m. N.W. of Drymen, Stirlingshire. P.O. B. Pass close by.

Balmains, seat, 3 m. N.E. of Stanley, Perthshire.

Balmakelly, trib. of N. Esk, which it joins near Brigend, S.W. Kincardineshire.

Balmakewan, seat, 5 m., S.W. Laurencekirk, Kincardineshire.

Balmalcolm, vil., 1½ m. E. of Kettle, Fifeshire.

Balmaleedie, trib. to N. Esk, S. Kincardineshire.

Balmangan, small bay opening off Kirkcudbright Bay, S. Kirkcudbrightshire.

Balmanno, ancient mansion, now a farmhouse, 3 m. S.W. Abernethy, S.W. Perthshire. Nearby is a curious rocking-stone.

Balmanno, seat, 5 m. S.E. of Fettercairn, Kincardineshire.

Balmaqueen, ham., N. Isle of Skye, Inner Hebrides, 5 m. N.E. of Kilmuir.

Balmashanner, hill (alt. 572 ft.), ¾ m. S. of Forfar, Angus.

Balmblae, pl., Falkland par., Fife. Pop. 57.

Balmedie, pl. and seat in co., and 7½ m. N., of Aberdeen. P.O.

Balmenach Distillery, 2½ m. E. of Grantown, Morayshire.

Balmerino, par. and vil. on Firth of Tay, and 3¼ m. W. of Wormit ; 3431 ac. Pop. 599. Remains of an abbey founded by the Queen of William the Lion, 1227.

Balminnoch, seat, 6 m. N.E. of Glenluce, S. Wigtownshire.

Balmoor House, 2¼ m. N.W. of Peterhead, Aberdeenshire.

Balmoral Castle, royal residence on the r. Dee in the par. of Crathie, and 9 m. W. of Ballater, Aberdeenshire. The deer forest covers an area of 30,000 ac. On Craigrowan, a neighbouring mt., is a cairn erected by the late Queen Victoria in memory of the Prince Consort. The ry. sta. is Ballater, L.N.E. P.O., T.O.

Balmore, vil. on the r. Kelvin in Stirlingshire, and 3½ m. E. of Milngavie. The Forth and Clyde Canal passes in the neighbourhood, and there is a ry. sta. on the L.N.E. Pop. B. and Torrance, 1101.

Balmuir, seat, 2½ m. N. of Dundee, Angus.

Balmule, seat, 3 m. N.E. of Dunfermline, Fifeshire.

Balmullo and Lucklawhill, vil., 1¾ m. W.S.W. of Leuchars, Fifeshire. P.O., T.O.

Balmungo, seat, 1½ m. S.E. of St. Andrews, Fifeshire.

Balmuto, seat, 3 m. N.W. of Burntisland, Fifeshire.

Balm Well, St. Catherine's, ¾ m. S. of Liberton, Edinburgh. An old bituminous spring.

Balmyle, seat, 9 m. N.W. of Blairgowrie, Perthshire.

Balnaboth, seat, 9 m. N.W. of Kirriemuir, Angus.

Balnabruach, vil. on peninsula between the Dornoch and Moray Firths, and 8 m. S.E. of Tain, Ross and Cromarty. A little fishing is carried on.

Balnacarn, ham., 8 m. N.W. of Fort Augustus, Inverness-shire.

Balnacoil, seat, ½ m. W. of Aboyne, Aberdeenshire.

Balnacoil Lodge, 8 m. N.W. of Brora, Sutherlandshire.

Balnacoul, ry. sta., L.M.S., 2½ m. W. of Fochabers, Morayshire.

Balnacra, pl., 6½ m. N.E. of Lochcarron, Ross and Cromarty.

Balnageith, vil., 1 m. S.W. of Forres, Morayshire.

Balnagowan, Eilean, small isl. in Loch Linnhe, Argyllshire, S.W. of the mouth of Loch Leven.

Balnagown, seat, 5¼ m. S. of Tain, Ross and Cromarty.

Balnaguard, vil. on banks of r. Tay, 2 m. W. of Ballinluig, Perthshire.

Balnain, ham., 4 m. W. of Drumnadrochit, Inverness-shire. P.O.

Balnakiel Bay, 1 m. W. of Durness P.O., N. Sutherlandshire. B. House, at head of bay, was built about 1744.

Balnakeilly, seat, close to Kirkmichael, Perthshire.

Balnakyle, waterfall on the Black Water, near Brora, Sutherlandshire.

Balnald, seat, 1 m. from Kirkmichael, N.E. Perthshire.

Balnamoon, seat, 4½ m. N.W. Brechin, Angus.

Balnellan, or **Balneilan,** ferry on r. Spey, near the mouth of the Avon ; on borders of Moray and Banffshire.

Balnespick Lodge, 1½ m. E.N.E. Tomatin sta., Inverness-shire.

Balone, ham., 1¾ m. S.W. of St. Andrews, Fifeshire.

Balone, old castellated ruin, 7 m. E.S.E. of Tain, Ross and Cromarty.

Balornock House, 1 m. N.E. of Springburn, Glasgow.

Balquhain, ruined castle, 3 m. W. of Inverurie, Aberdeenshire. It was occupied by Queen Mary immediately before the battle of Corrichie, 1562. Burned by Duke of Cumberland, 1746.

Balquhandy, seat, 2 m. S. of Dunning, Perthshire.

Balquharn, seat, 1½ m. N.W. of Portlethen, N.E. Kincardineshire.

Balquhatstone, seat and collieries adjoining Slamannan, Stirlingshire.

Balquhidder, par.. vil.. and ry. sta., L.M.S., at E. end of Loch Voil, W. Perthshire ;

vil., 2 m. W. of sta. ; 54,675 ac. Pop. 619. P.O., T.O. Rob Roy lies buried here.

Balquholly, ancient castellated building, now forming part of Hatton Castle, and lying 3 m. S.E. of Turriff, Aberdeen shire.

Balranald, seat and small harb. on W. coast of N. Uist, Outer Hebrides.

Balruddery, seat, 6 m. W. of Dundee, Angus. The Den of B. is attractive to geologists and botanists.

Balsarroch, seat, 2 m. W. Kirkcolm, N.W. Wigtownshire.

Balshagray, q.s. par., Govan par., Glasgow.

Balshando, pl. and loch, 9 m. N.W. Dundee, Angus.

Balta, islet and ham. on E. side of Unst, Shetland. Balta Sound is about 2 m. long by 1 m. broad. Pop. 83. Has harb. and lighthouse.

Balthayock, ruin which belonged to Knights Templars, 3 m. E. of Perth. Perthshire.

Baltilly, seat, a little to W. of Ceres, Fifeshire.

Balvag, riv. flowing out of Loch Voil into Loch Lubnaig, S.W. Perthshire.

Balvaird Castle, seat near Strathmiglo, Fifeshire.

Balvarran, seat, 1 m. from Kirkmichael, N.E. Perthshire.

Balvicar, vil. in Isl. of Seil, Argyllshire, 14 m. S.W. of Oban. P.O., T.O. Pop. 180.

Balvie, seat, 1½ m. W. of Milngavie, S. Dunbartonshire.

Balvraid, ham., near Dornoch, Suther landshire.

Balvraid, seat, 2½ m. N.E. of Tomatin sta., Inverness-shire.

Balvraid, ham., 2½ m. S.E. of Glenelg, Inverness-shire.

Balwearie, ruined stronghold, 2 m. W. of Kirkcaldy, Fifeshire. Was the birth place of Michael Scott, the wizard celebrated by Sir Walter Scott.

Balygrundle, pl., Lismore Isl., Argyllshire.

Bamff House, seat, 3½ m. N.W. of Alyth, Perthshire.

Banavie, vil. and ry. sta., L.N.E., at S. end of Caledonian Canal, and 2½ m. N.N.E. of Fort William. P.O., T.O.

Banchory, town (police bur.) on r. Dee, N. Kincardineshire, 17 m. S.W. of Aberdeen ; ry. sta.. L.N.E. P.O.. T.O. Pop. of bur. 1690. Hotels.

Banchory-Devenick, par. on r. Dee in N. Kincardineshire ; 7737 ac. Pop. 1335. P.O.

Banchory House, seat, 2½ m. S.W. Aberdeen, Kincardineshire.

Banchory-Ternan, par. on the r. Dee, N. Kincardineshire. Contains town of Banchory. Pop. 3335 ; 20,080 ac.

Bandirran, ham. and seat, 7 m. N.E. of Perth, Perthshire.

Bandrum, Carnegie Trust Country Home for Children, 4 m. N.W. Dunfermline, par. of Saline, Fifeshire.

Banff, par. and royal mun. bur. ; ry. sta., L.N.E.; mkt.-town, seapt., and capital of Banffshire, at the mouth of the Deveron. Banff proper is on the left bank with Macduff on the rt., a bridge connecting the two. Coastguard and lifeboat sta.; fixed white light seen 8 m. ; par. 6072 ac. Pop. 4136 ; pop. bur. 3489. Hotels. P.O., T.O. Extensive fishing is conducted. Chief industries, boat building and rope and sail making.

Banff Bay, between Banff and Macduff, Banffshire.

Banff Bridge, ry. sta., L.N.E., at mouth of r. Deveron, Banffshire.

Banff Hill, alt. 618 ft., 3 m. W. of Bervie, Kincardineshire.

Banffshire, maritime co., bounded on N. by Moray Firth, E. and S. by Aberdeenshire, W. by Inverness and Moray shires ; area 630 sq. m. or 403,053 ac ; greatest length 59 m., greatest breadth 32 m. Pop. 54,907. Contains some of the highest mts. in Scotland, including Cairngorm, 4060 ft., and part of Ben Macdhui, 4296 ft. The principal rivers are the Spey, Avon, and Deveron. The valleys are fertile and well cultivated, and cattle raising is largely carried on. There are numerous distilleries. The minerals found are marble, limestone, ironstone, granite, marl, slate, topazes, and rock crystal. Co. town, Banff. Returns one member to Parliament.

Bangour, vil., W. Lothian, 1¾ m. S.W. of Uphall, to which it is connected by light railway, L.N.E. It contains the Edinburgh District Board of Control's Asylum for Mental Ailments.

Bank end, vil., 6 m. S.E. Dumfries, Dumfriesshire. P.O.

Bankend, ham., 3½ m. S. of Lesmahagow, W. Lanarkshire.

Bankfoot, vil., 3 m. W.N.W. of Stanley Junc., S.E. Perthshire, ry. sta., L.M.S. P.O., T.O. Pop. 726.

Bankglen, pl., 1½ m. S.W. of New Cumnock, Ayrshire. P.O., T.O. called Bank.

Bankhead, vil., with ry. sta., L.N.E., 4½ m. N.W. of Aberdeen, Aberdeenshire. P.O.

Bankhead, vil., Ayrshire. See SPRINGSIDE.

Bankhead, vil., 2 m. E. Carstairs Junc., Lanarkshire, ry. sta., L.M.S. ; alt. 652 ft.

Bankhead, ham., 7 m. N.W. of Carnoustie, Angus.

Bankhead, ham., 1½ m. S.W. of Denny, S. Stirlingshire.

Bankhead, ham., ¾ m. N. Forteviot sta. Perthshire.

Bankhead, seat, near Rutherglen, N.W. Lanarkshire.

Bankier, ham., 3 m. S.W. Denny, E. Stirlingshire. Has a distillery.

Banknock, ham., 3¾ m. S.W. of Denny, E. Stirlingshire, ry. sta., L.M.S. and L.N.E. P.O., T.O. B. House, E. of sta.

Banks, East and **West,** hams. near Wick, Caithness.

Bankshill, ham., 4 m. E. of Lockerbie, Dumfriesshire. P.O. Pop. 34.

Bankton, seat near Tranent, E. Lothian.

Bankton, seat, 2 m. S.W. Mid-Calder, Midlothian.

Bankton Park, suburb of Kettle, Fifeshire.

Bannachra, ruined old fortalice and ham., in Glenfruin, S.W. Dunbartonshire.

Bannatyne Home of Rest, Newtyle, Angus, 1¼ m. S.E. of Alyth.

Bannatyne House, seat, 7½ m. N.W. of Dundee, Angus.

Banniskirk, pl., with old chapel, 2½ m. S.E. of Halkirk, Caithness-shire.

Bannock, rivulet, running 9 m. E.N.E. to the Forth, 2½ m. below Stirling.

Bannockburn, small town, q.s. par., and famous battlefield, 8 m. N. of Falkirk, Stirlingshire; ry. sta., L.M.S. The scene of King Robert the Bruce's victory over the English, under Edward II. in 1314. Coal mining and mfr. of tweeds, tartans, and carpets. P.O., T.O. Pop. town, 4481.

Banovie, rivulet near Blair-Atholl, N. Perthshire.

Bantaskyne, seat, ¾ m. W. of Falkirk, Stirlingshire.

Banton, q.s. par. and vil., 2½ m. N.E. of Kilsyth, Stirlingshire. P.O. Pop. 374.

Bar, loch near Ballantrae, S.W. Ayrshire.

Bara, pl., 1 m. W. of Garvald, E. Lothian.

Barachan, creek, Loch Scridain, Mull Isl., Argyllshire.

Barachuie, vil. near Shettleston, E. Glasgow.

Baranlongart Burn, flows W. to head of Loch Killisport, Knapdale, Argyllshire. B. House near.

Barassie, coast, vil., 1 m. N.E. of Troon, Ayrshire. Golf course ; ry. sta., L.M.S. Pop. 163.

Barbaraville, vil., 3½ m. N.E. of Invergordon, Ross and Cromarty. Pop. 121.

Barbaswalls, vil., 2¼ m. N. of Meigle, Angus.

Barbauchlaw, stream, Lanarkshire, and W. Lothian, trib. to r. Avon, S.W. of Linlithgow, W. Lothian.

Barbeth House, seat ½ m, S. of New Abbey, Kirkcudbrightshire.

Barbhic, small stream, 3 m. N.W. of Crieff, Perthshire.

Barbieston, seat adjoining Dalrymple, W. Ayrshire.

Barbreck House, 1 m. N.E. of the head of Loch Craignish, Argyll.

Barcaldine House, old castle, seat, and ry. halt, L.M.S., on Loch Creran, 11 m. N.E. of Oban, Argyll. P.O., T.O, State forest in neighbourhood.

Barcaple, seat in co. of, and 6 m. N. of, Kirkcudbright.

Barclay Castle or **Towie Barclay,** ruined keep, 4 m. N. of Fyvie, N. Aberdeenshire.

Barclosh, ruined castle, 1½ m. N.E. of Dalbeattie, E. Kirkcudbrightshire.

Barcloy, ham., 5½ m. S.E. of Dalbeattie, Kirkcudbrightshire.

Bardarroch, small loch near Crinan, Argyllshire.

Bardennock, trib. of r. Nith, Dumfriesshire, into which it falls near Closeburn.

Bardennock Hill, 2 m. S. of Carsphairn, Kirkcudbrightshire ; alt. 1082 ft.

Bardhead, prom. at S. end of isl. of Bressay, and 5 m. S.E. of Lerwick, Shetland ; alt. 265 ft.

Bardister, ham., 2 m. S. of Ollaberry, E. Mainland, Shetland.

Bardister, loch, ¼ m. from vil. of Walls, Shetland.

Bardock, headwater of r. Don, which it joins a little above Towie, W. Aberdeenshire.

Bardowie, loch, 1½ m. S.E. Milngavie, Stirlingshire. B. Castle is a seat.

Bardowie, ry. sta., L.N.E., 2½ m. N. Maryhill, Stirlingshire.

Bardrochat, seat, ½ m. S.E. of Colmonell, N. Ayrshire.

Bardrochwood, seat, near Palmure sta., Kirkcudbrightshire.

Bardykes Rows, pl., 1½ m. N. of High Blantyre, Lanarkshire.

Barean, loch, 3½ m. S.E. of Dalbeattie, Kirkcudbrightshire.

Baremman, estate, 1½ m. N.W. Roseneath, W. Dunbartonshire.

Bargaly, seat, 3 m. E. of Newton Stewart, W. Kirkcudbrightshire.

Bargany, seat, 4¼ m., N.E. of Girvan, Ayrshire.

Bargarran, seat, 2 m. N.E. of Bishopton, Renfrewshire.

Bargattin, loch, 4½ m. S.W. of Castle-Douglas, Kirkcudbrightshire.

Bargeddie, q.s. par., vil., and ry. sta., L.N.E., 2 m. S.W. of Coatbridge, Lanarkshire. P.O.

Bargrennan, q.s. par. and ham. on the Cree, 9 m. N.W. of Newton Stewart, Wigtownshire. P.O., T.O.

Bargrug Sanatorium, 2½ m. N.E. of Dalbeattie, Kirkcudbrightshire.

Barholm, seat, on N. side of Creetown, Kirkcudbrightshire.

Barholm Hill, alt. 1163 ft., 4 m. S.E. of Creetown, Kirkcudbrightshire.

Barjarg, seat, 3½ m. S. of Thornhill, Dumfriesshire.

Barlanark House, 1 m. N.E. of Shettleston, Lanarkshire.

Barlay, seat, 2½ m. S.E. of Balmaclellan, Kirkcudbrightshire.

Barlay Mill, pl., ½ m. N. of Gatehouse-of-Fleet, Kirkcudbrightshire.

Barleith, ry. sta., L.M.S., 2½ m. W. of Galston, Ayrshire.

Barleyknowe, ham., ½ m. N. of Gorebridge, E. Midlothian.

Barleyside, vil. near Falkirk, Stirlingshire.

Barlinnie Prison, in bur. of Glasgow, 2¾ m. E.N.E. of Glasgow G.P.O.

Barlocco, isl. and ham. at S.E. entrance of Fleet Bay, S. Kirkcudbrightshire.

Barlockhart, moor, S.E. of Glenluce, Wigtownshire.

Barmekin Hill, alt. 900 ft., 12½ m. W. of Aberdeen, Aberdeenshire.

Barmore Isl., 2¼ m. N. of Tarbet in N. of Kintyre, Argyllshire.

Barmure, estate, 12 m. N.E. of Ayr, Ayrshire.

Barnagad, loch near Crinan Canal, Argyllshire.

Barnaline, seat, on W. side of Loch Awe, Argyllshire.

Barnbarroch, estate, 3 m. S. of Wigtown, Wigtownshire.

Barnbarroch, pl. and seat, 2¼ m. S. of Dalbeattie, Kirkcudbrightshire.

Barnbougle Castle, seat, on Firth of Forth, in par. of Dalmeny, W. Lothian ; the building is of great antiquity and was restored by Lord Roseberry in 1880 ; it stands in the same ground as Dalmeny House and is 7 m. W. of Edinburgh.

Barncailzie Hall, 1½ m. E.N.E. of Kirkpatrick Durham, Kirkcudbrightshire.

Barncluith, seat adjoining Hamilton, Lanarkshire.

Barndennoch, ham., 4½ m. S. of Thornhill, Dumfriesshire.

Barnhill, N.E. sub. of Glasgow, L.N.E. sta.

Barnhill, ham. and seat, on r. Tay, opposite Perth.

Barnhill, seat, 2 m. E. of Dumbarton, Dunbartonshire.

Barnhill, ry. sta., L.N.E. and L.M.S., 1 m. N.E. of Broughty Ferry, Angus.

Barnhill and Larkfield, vil. near Blantyre, N. Lanarkshire.

Barnhourie, dangerous sandbank and scene of many shipwrecks, off S.E. coast of Kirkcudbrightshire ; extends from a pt. off the mouth of the Nith to a pt. off the mouth of the Urr.

Barnkirk, headland at the mouth of the r. Annan, Dumfriesshire ; has two white lights (fixed) and a fog-bell.

Barns, seat in co. and 3 m. S.W. of Kinross.

Barns, old tower (1498) in co. of, and 4 m. S.W. of, Peebles.

Barns, ruined baronial tower, 1 m. S.W. of Crail, Fifeshire.

Barns, East, vil 3 m., S.E. Dunbar, E. Lothian.

Barnshean, small loch, 3m. E.N.E. of Kirkmichael, Ayrshire. The road

between Dalmellington and Maybole passes near its south shore.

Barnside, hill (865 ft.), par. of Abbey St. Bathans, N. Berwickshire.

Barnslee, seat, ¼ m. E. of Markinch, Fifeshire. Macbeth, thane of Fife is said to have had a castle here.

Barnsmuir, seat, 1 m. S.W. of Crail, Fifeshire.

Barns Ness, prom. 3 m. S.E. of Dunbar, E. Lothian. Lighthouse and white group-flashing light, visible 17 m.

Barns West or **Westbarns,** vil., 2 m. W. of Dunbar, E. Lothian. P.O., T.O.

Barnton, pl. on outskirts of Edinburgh, of which it forms a part. Hotel and golf course, ry. sta., L.M.S. P.O.

Barnyards, vil. near Kilconquhar, E. Fifeshire. Pop. of B. and Kilconquhar 336.

Barochan Mill, ham. and seat, ¾ m. N. of Houston, N. Renfrewshire.

Barone Hill or **Barons Hill,** alt. 532 ft., behind Rothesay, Bute Isl.

Baron's Point, headland at E. entrance to Loch Long, Dunbartonshire.

Barony, eccl. par. in Glasgow.

Bar Point, in Luce Bay, 2 m. N.W. of Port William, Wigtownshire.

Barr, par. and vil. on the Stinchar, and 8 m. E.S.E. of Girvan, Ayrshire ; 54,970 ac. Pop. 494 ; vil pop. 175. P.O., T.O.

Barr, hill overlooking Water of Milk, S. Dumfriesshire.

Barr, seat, with ruined tower in neighbourhood, near Lochwinnoch, S.W. Renfrewshire.

Barr, burn, falling into the Atlantic 2 m. S.E. of Glendarroch Pt., Kintyre, Argyllshire.

Barra, isl. and par. in the Outer Hebrides ; 22,222 ac. Pop. 2250 ; pop. of isl. 2001. The par. includes the inhabited isls. of Barra, Berneray, Mingclay, Gabbay, and Watersay. B. Isl. is 8 m. long by 5 m. broad. The coastline is very irregular, mostly rocky, and pierced by numerous caves. P.O. at Castlebay. Has steamboat communication with Oban.

Barra Castle, seat, 2 m. S.W. of Old Meldrum, Aberdeenshire.

Barracks, The, seat, at W. end of Loch Rannoch, Perthshire, 6 m. E. of Rannoch sta.

Barra Head, prom. at the S. end of

Bernera Isl., the most southerly of the Barra Isls. Lighthouse seen 33 m.; stands 683 ft. above high water.

Barrapoll, ham., in S.W. of isl. of Tiree, Argyllshire.

Barras, sub. of Lochmaben, Mid-Dumfries-shire

Barrhead, q.s. par. and bur. on the r. Leven, E. Renfrewshire, and 7 m. S.W. of Glasgow by rail; ry. sta., L.M.S. P.O., T.O. Pop of bur. 12,308. Industries: engineering, ironfounding, calico printing, and the manufacture of cotton.

Barrhill, dist. and vil., 12½ m. south of Girvan, Ayrshire; ry. sta., L.M.S. Alt. of ry. sta. 444 ft. P.O., T.O. Vil. pop. 277.

Barrhill, hill, alt. 400 ft., 3½ m. S.W. of Cumbernauld, Dunbartonshire; was the site of one of the fortresses built to defend the Wall of Antoninus.

Barriemore, seat, 1 m. N. of Port Appin, W. Argyllshire.

Barrisdale, glen, stream and bay at head of Loch Hourn, Inverness-shire. B. House is the seat.

Barmill, vil., with ry. sta., L.M.S., 2 m. S.E. of Beith, N. Ayrshire. P.O., T.O. Pop. 559.

Barrock, ham., and seat, 11 m. N.W. of Wick, Caithness. P.O.

Barrogil Castle, seat, on Pentland Firth, 14 m. N.E. of Thurso, Caithness.

Barrowfield, q.s. par. in S.W. Glasgow.

Barrs, seat, 7 m. N.E. of Taynuilt, Argyllshire, near Loch Etive.

Barrschol, dist., 4 m. N.W. of Rogart, Sutherlandshire.

Barr Water, river, 10 m. N. of Campbeltown, Kintyre, S.W. Argyllshire.

Barry, coastal par. and vil., with golf course, 9¼ m. N.E. of Dundee, Angus; ry sta. (B. Links) L.M.S. and L.N.E.; par. 5286 ac. Pop. 4508; vil. pop. 245. Large military camp. P.O., T.O. called Barry Camp.

Barry Hill, hill, with remains of a British fort, 1½ m. N.E. of Alyth, Perthshire.

Barsalloch Point, headland in Luce Bay, Wigtownshire.

Barscobe Loch, Hill, and **Castle,** 1 m. N. of Balmaclellan, Kirkcudbrightshire.

Barshaw, hospital and public park, close to Paisley, Renfrewshire.

Barshell, hill, 1¼ m. N.E. of Amisfield, Dumfriesshire, crowned with ruins of Caledonian fort.

Barskimming House, 2 m. S.W. of Mauchline, Ayrshire.

Bar, The, isl. 4 m. long, off coast of Nairn.

Barthol Chapel, ham., 4 m. N.E. of Tarves, N.W. Aberdeenshire. P.O.

Bartonholm, colliery vil., 2 m. N.W. of Irvine, North Ayrshire.

Bartonshill Colliery, in par. and 1½ m. N.E. of Old Monkland, Lanarkshire.

Barvas, par., loch, and two villages (Lower and Upper), 12 m. N.W. of Stornoway, Lewis, Outer Hebrides. P.O., T.O.; par. 97,543 ac. Pop. 5876.

Barvick, stream, falling into the Turret 2 m. N.W. of Crieff, Perthshire. B. Falls have a leap of between 500 and 600 ft.

Barwhapple, loch, 3 m. N.E. of Glenluce, Wigtownshire.

Barwhillanty, seat, 6 m. N.W. of Castle Douglas, Kirkcudbrightshire.

Barwhinnock, seat, in co. of, and 4 m. N.E. of, Kirkcudbright.

Bass, mound (alt. 40 ft.), in town of Inverurie, E. Aberdeenshire, on right bank of r. Urie.

Bassendean House, seat, 5 m. W. of Greenlaw, Berwickshire.

Bass Rock, an isl. rock off E. Lothian, near the mouth of the Firth of Forth, 3¼ m. E.N.E. of North Berwick. It is 1 m. in circumference, 350 ft. high, and is inaccessible on all sides except the S.W.; myriads of Solan geese and other birds inhabit it. It was long used as a prison for Covenanters. Lighthouse with white group-flashing light, visible 18 m.

Bastavoe, bay on E. side of Yell, Shetland.

Bath, small stream flowing W. into Kilbirnie loch, N. Ayrshire.

Bathgate, par. and market town (police burgh) in West Lothian and 18 m. S.W. of Edinburgh; ry. sta. (alt. 470 ft.), L.N.E.; par. area, 10,875 ac. Pop. 18,064. Town pop. 10,097. P.O., T.O. Has extensive coal, iron, and oil industries.

Bathville, vil., ½ m. S. of Armadale, W. Lothian.

Batties Bog, moor, 7 m. W. of Hutton, Mid. Berwickshire.

Battle-Drum, hills, with prehistoric cairns, about 8 m. N. of Arbroath, Angus.

Battle Dykes, pl. with traces of Roman camp, $3\frac{1}{2}$ m. N. of Forfar, Angus.

Battle Hill, eminence, in par. of Annan, Dumfriesshire. Reputed scene of a border battle with the English.

Battle Hill, eminence, near Huntly, Aberdeenshire. Traditional scene of a battle between the Comyns and the Gordons.

Battock Mount, alt. 2555 ft., summit of Grampians at meeting-place of counties of Aberdeen, Kincardine, and Angus.

Bauden, hill, $1\frac{1}{2}$ m. S.E. of Kettle, Fifeshire; crowned by remains of ancient defensive works.

Bavelaw Burn, stream, on which are two reservoirs for Edinburgh water supply. After a course of 6 m. it enters the Water of Leith at Balerno. B. Castle $2\frac{1}{2}$ m. S. of Balerno.

Bawhelps, mt. (alt. 2712 ft.), 6 m. W. of Milton of Clova, Angus.

Bawkie, or **Balgie,** bay, off Firth of Clyde, at S. end of Dunoon; frequently known as West Bay of Dunoon.

Bay, ham. and sea loch, 15 m. N.W. of Portree, Skye, Inner Hebrides.

Bayble, Lower and **Upper,** 2 vils., Lewis Isl., Ross and Cromarty, 8 m. E. of Stornoway, in Eye Peninsula. P.O.

Bayfield, seat, 1 m. N.E. of Nigg, E. Ross and Cromarty.

Bayhead, pl., 12 m. W. of Lochmaddy, N. Uist, Outer Hebrides. P.O., T.O.

Bayhervagh, natural harbour, on east coast of Barra Isl., Outer Hebrides.

Bay of Martyrs, small bay, east side of isl. of Iona, off Mull, Argyllshire.

Bea, small loch in isl. of Sanday, Orkney.

Beach Church, q.s. par. in Broughty Ferry, near Dundee, Angus.

Beacon Hill, alt. 743 ft., in Bressay Isl., Shetland.

Beag a' Mhuilean, loch, in S. of isl. of Jura, Argyllshire.

Bealach nam Bo. pass across N. shoulder of Ben Venue, leads to L. Katrine, Perthshire.

Bealach nam Bo, pass, $4\frac{1}{2}$ m. E.S.E. of Applecross, W. Ross and Cromarty.

Beal an Duine, pl., near entrance to Trossachs, Perthshire. Supposed to be spot described by Scott in "The Lady of the Lake," as the scene of death of Fitz-James's steed.

Bealid, shooting-lodge, 1 m. S.W. of Kingussie, Inverness-shire.

Beallochantuy, ham. and small bay, on W. side of Kintyre, Argyllshire, 10 m. N.W. of Campbeltown.

Beamer, rocky isl. in Firth of Forth, $\frac{3}{4}$ m. S.W. of N. Queensferry, Fifeshire. Marked by white group-flashing light, visible 6 m.

Beancross, ham., $\frac{3}{4}$ m. N.W. of Polmont, Stirlingshire.

Beannach, loch, 6 m. N.E. of Rogart ry. sta., Sutherlandshire.

Beannachan, loch (alt. 465 ft., depth, 176 ft.), in Strath Connon, Ross and Cromarty.

Beannacharan, loch, 14 m. S.W. of Beauly, N. Inverness-shire.

Beannach Mhor and **Beg,** two small lochs near Fionn Loch in W. Ross and Cromarty; depth 27 ft.

Beannoch, loch, 4 m., N.N.E. of Lairg, Sutherlandshire.

Beanoch, loch, $4\frac{1}{2}$ m. N.E. of Lochinver, W. Sutherlandshire; depth 38 ft., alt. about 240 ft.

Beansburn, vil., 1 m. N. of Kilmarnock, Ayrshire.

Bearnock, seat, 6 m. W. of Drumnadrochit, N. Inverness-shire.

Bearsden, or **New Kilpatrick,** town and residential suburb in Dunbartonshire, 5 m. N.W. of Glasgow; ry. sta., L.N.E. P.O., T.O. Pop. 5551.

Beasdale, dist. at head of Loch nan Uamh, S.W. Inverness-shire.

Beast, or **Loch na Bhaisq,** small loch in N. Uist, Outer Hebrides.

Beath, par., 5 m. E.N.E. of Dunfermline, Fifeshire; 6331 ac. Pop. 21,522.

Beaton's Mill, see MILTON, WHINS OF, and SAUCHIEBURN.

Beattock, vil. and seat, 2 m. S.W. of Moffat, N. Dumfriesshire; ry. sta. (alt. 361 ft.), L.M.S., junc. for Moffat. P.O., T.O. B. Hall (alt. 851 ft.).

Beattock Summit, on L.M.S. route between Beattock and Elvanfoot, in S. Lanarkshire; alt. 1014 ft.

Beaufort Castle, seat of Lord Lovat, on the r. Beauly and 4 m. S.W. of Beauly, N. Inverness-shire.

Beauly, town on r. of same name, and 10 m. W. of Inverness; ry. sta., L.M.S. P.O., T.O. Near are the ruins of the priory of St. John, founded 1232. Pop. 799. Hotel.

Beauly, r., formed by junc. of the Farrar with the Glass ; runs for about 16 m. and falls into Beauly Firth near town of Beauly.

Beckford Lodge, Hamilton, Lanarkshire. Town hospital.

Beckton, pl., ½ m. N.W. of Lockerbie, Dumfriesshire. Mineral spring.

Bedlormie, old baronial fortalice, 5 m. N.W. of Whitburn, W. Lothian.

Bedrule, par. and ham. on the Rule, and 4 m. S.W. of Jedburgh ; 3922 ac. Pop. 174.

Bee, sea loch in N.W. of S. Uist, Outer Hebrides.

Beechhill, seat, 2½ m. S.E. of Haddington, E. Lothian.

Beechmount, mansion, Murrayfield, Edinburgh. Gifted in 1926 to Edinburgh Royal Infirmary.

Beechwood, seat near Corstorphine, Edinburgh.

Beechwood, seat, 1½ m. N. of Arbroath, Angus.

Beefstand Hill, on borders of Northumberland, 4 m. S.E. of Hounam, Roxburghshire ; alt. 1844 ft.

Beeslack, seat, 1 m. N. of Penicuik, Midlothian.

Beeswing, ham., E. Kirkcudbrightshire, 6½ m. N.E. of Dalbeattie. P.O.

Beg, headwater of r. Shee, N.E. Perthshire.

Beg, sea loch, branch of Loch Bracadale, Skye, Inner Hebrides.

Beg, loch, 9 m. S. of Halkirk, Caithness.

Beglie, Wicks of, pass in the Ochil Hills, 3½ m. W. of Abernethy, Perthshire.

Beich, small stream flowing S. into Loch Earn, Perthshire.

Beinn a Chait, mt., 6 m. N. of Blair Atholl, Perthshire ; alt. 2942 ft.

Beinn a' Chaolais, mt., one of Paps of Jura, Argyllshire ; alt. 2407 ft.

Beinn a Chaoruinn, mt., S.W. Banffshire, 4 m. N.E. of Ben Macdhui ; alt. 3553 ft.

Beinn a Chaoruinn, mt., 4 m. N.E. of Tulloch ry. sta., Inverness-shire ; three peaks—alts. 3437, 3422, 3394 ft.

Beinn a Chlachair, mt., 3 m. S. of Loch Laggan, Inverness-shire ; alt. 3569 ft.

Beinn a Chruinnach, mt., 5 m. E. of Tomintoul, and on borders of Aberdeenshire and Banffshire ; alt. 2536 ft.

Beinn a Chrulaiste, mt., Argyllshire, 1½ m. N.W. of Kingshouse Inn ; alt. 2805 ft.

Beinn a Mheirlich, mt., 4 m. N. of Tulloch ry. sta., Inverness-shire ; alt. 2984 ft.

Beinn an Dothaidh, mt., 2 m. N.E. of Bridge of Orchy ry. sta., Argyllshire ; alt. 3283 ft.

Beinn-an-Oir, highest of the Paps of Jura, Argyllshire ; alt. 2571 ft.

Beinn an t-Seilich, mt., 4 m. N. of Lochgoilhead, Argyllshire ; alt. 2359 ft.

Beinn an Tuirc, mt., 3 m. S.W. of Saddell church, Kintyre, Argyllshire; alt. 1491 ft.

Beinnavain, loch in Glen Affric, Inverness-shire ; depth 167 ft., alt. 703 ft.

Beinn Bhan, two peaks, 7 m. N.N.E. of Fort William, Inverness-shire ; alts. 2613 ft. and 2522 ft.

Beinn Bhan, mt., Islay Isl., Argyllshire ; alt. 1544 ft.

Beinn Bhan, mt., 2½ m. S.S.W. of Ford Hotel, Argyllshire ; alt. 1049 ft.

Beinn Bharrain, mt., N.W. Arran, Buteshire, alt. 2368 ft. ; another top, 2345 ft.

Beinn Bheag, mt., 1½ m. N.W. of Tyndrum, borders of Perthshire and Argyllshire ; alt. 2149 ft.

Beinn Bhealach, mt. Sutherland, 4 m. S.W. of Helmsdale, alt. 1940 ft.

Beinn Bheigeir, mt., isl. of Islay, Argyllshire ; alt. 1609 ft.

Beinn Bheoil, mt. on W. of Loch Etive, Inverness-shire ; alt. 3333 ft.

Beinn Bhoidheach, mt., 3 m. S.E. of Dalmally, Argyllshire ; alt. 1953 ft.

Beinn Bhreac (Ben Vrack), mt., 4 m. S.E. of Ben Macdhui, W. Aberdeenshire ; alt. 3051 ft.

Beinn Bhreac, mt., 6 m. S.W. Port Appin, Argyllshire ; alt. 2324 ft.

Beinn Bhreac, two mts., isl. of Arran, Buteshire ; alts. 2333 ft. and 1649 ft.

Beinn Bhreac, mt., 3 m. S. of S. end of Loch Treig, Inverness-shire ; alt. 2863 ft.

Beinn Bhreac, mt., 4 m. N. of Kingussie, Inverness-shire ; alt. 2618 ft.

Beinn Bhreac, mt. on E. side of Loch Tay and 4 m. S.W. of Kenmore, Perthshire ; alt. 2341 ft.

Beinn Breacliath, two peaks, 3 m. N.W. of Tyndrum, Argyllshire ; alts. 2544 ft. and 2633 ft.

Beinn Chaluim, mt. at head of Glen Lochay, Perthshire ; alt. 3334 ft.

Beinn Chaolais, see BEINN A' CHAOLAIS.

Beinn Chaorach, mt., 2¼ m. N.E. of Tyndrum, on borders Perthshire and Argyllshire ; alt. 2655 ft.

Beinn Chumhann, mt., 2 m. W.S.W. of Ben Alder, borders Perthshire and Inverness-shire; alt. 2962 ft.

Beinn Dearg, mt., 3 m. N. of Torridon, W. Ross and Cromarty; alt. 2995 ft.

Beinn Donachain, mt., 3 m. N.E. of Dalmally, Argyllshire; alt. 2127 ft.

Beinn dubh Chraige, mt., alt. 3204 ft., 5 m. W. of Crianlarich, Perthshire.

Beinn Each, mt., alt. 2660 ft., 5 m. N.W. of Callander, Perthshire.

Beinn Eagach, mt. Perthshire, 5 m. N. of Aberfeldy; alt. 2259 ft.

Beinn Fhada (Ben Ad'a), mt., alt. 3383 ft., borders of Inverness-shire and Ross and Cromarty.

Beinn Fhada, mt., alt. 3497 ft., 6 m. E.S.E. of Ballachulish, Argyllshire.

Beinn Fhada, mt., isl. of Mull, 1½ m. N.E. of Ben More, alt. 2302 ft.

Beinn Fhionnlaidh, mt., alt. 3294 ft., 2 m. S. of Loch Lungard, Ross and Cromarty.

Beinn Fhuaran, mt., alt. 2632 ft., 5½ m. N.N.E. of Tyndrum, on borders of Perthshire and Argyllshire.

Beinn Ghlas, mt., alt. 3085 ft., 6 m. N.E. of Killin, Perthshire.

Beinn Iutharn Bheag, mt., alt. 3011 ft., N. Perthshire and 10 m. S.W. of Braemar.

Beinn Iutharn Mhor, (Ben Uarn), mt., alt. 3424 ft., N. Perthshire, and 10 m. S.W. of Braemar.

Beinn Liath Mhor, mt., alt. 3034 ft., 7 m. S.W. of Kinlochewe, Ross and Cromarty.

Beinn Maol Chalium, mt. with two peaks, alts. 2967 ft. and 2740 ft., 5½ m. S.E. of Ballachulish, Argyllshire.

Beinn Mhanach, mt. at the head of Glenlyon, Perthshire, alt. 3125 ft.

Beinn Mheadhoim, mt. in Cairngorms, 1 m. S.E. of Loch Avon. alt. 3883 ft.

Beinn Mhic Chasgaig, mt., alt. 2766 ft., 4¾ m. S.W. Kingshouse Inn, Argyllshire.

Beinn Mhor, mt., with two peaks, alts. 3505 ft. and 3370 ft., in Glen Shiel, Ross and Cromarty.

Beinn na Croise, mt., isl. of Mull, Argyllshire; alt. 1649 ft.

Beinn nan Oigreag, mt., alt. 2978 ft., 5½ m. N.W. of Killin, Perthshire.

Beinn Tee, mt., alt. 2956 ft., in Inverness-shire, 3 m. S. of Loch Garry.

Beinn Uidhe, mt., N. of Loch Assynt Sutherland, alt. 2384 ft.

Beith, par., mkt. and mnfrg. town in N. Ayrshire and 18½ m. S.W. of Glasgow by L.M.S. to the ry. sta. P.O., T.O. Par. area, 11,187 ac. Pop. 5977; town pop. 3891. Mfrs. linen thread, fishing nets and furniture.

Belchester House, seat, 4 m. N.W. of Coldstream, Berwickshire.

Beld Craig, dell, transversed by burn and waterfall, 3¼ m. S.E. of Moffat, Dumfriesshire.

Belhaven, coastal vil. and q.s. par., 1 m. W. of Dunbar, E. Lothian.

Belhaven Hill, Dunbar, E. Lothian, boys' preparatory school.

Belhelvie, par. and ham., 8 m. N. of Aberdeen, 12,155 ac. Pop. 1514. P.O. Coastguard sta. B. Lodge, a seat.

Belivat, ham. on a small loch of the same name in co. of, and 7 m. S.E. of, Nairn. P.O.

Bella, small stream flowing 8 m. W. to join the Glenmore, with which it forms the r. Lugar, E. Ayrshire.

Bellabeg, seat, to the N. of Strathdon Church, Aberdeenshire.

Belladrum, seat, 4 m. S. of Beauly, Inverness-shire.

Bellahouston, q.s. par. in the S.W. suburbs of Glasgow; ry. sta. L.M.S. P.O., T.O.

Bellanoch, vil. on Crinan Canal in N. Knapdale, and 6 m. N.W. of Ardrishaig, Argyllshire. P.O., T.O.

Bellart, stream flowing 8 m. into Loch Cuan, isl. of Mull, Argyllshire.

Bellaty Lodge, seat, 6 m. N. of Alyth, Angus.

Bell Craig, mt., alt. 2046 ft., S.W. Selkirkshire, 5 m. S,W. of Ettrick ch.

Belleisle, seat, 2 m. S. of Ayr, Ayrshire, now property of Corporation of Ayr.

Bellfield, sub. of Strathmiglo, Fifeshire.

Bellfield, seat, 1 m. S.E. of Kilmarnock, Ayrshire.

Bellfield, sub. of the town of Stirling, Stirlingshire.

Bellgrove, ry. sta. on L.N.E. in E. of Glasgow.

Bellie, par., on the Moray Firth in N.E. Morayshire. Contains the town of Fochabers; 12,778 ac. Pop. 1842.

Belmont House, Newton Mearns, Renfrewshire, boys' preparatory school.

Bellochantuy, ham. on small bay of same name, 9 m. N.W. of Campbelltown, Argyllshire. P.O.

Bell Rock, or **Inch Cape,** reef off the coast of Angus, at the mouth of the Firth of Tay, and 12 m. E.S.E. of Arbroath; covered at high water but partly uncovered at low water. Lighthouse built by Robert Stevenson and Rennie, 1807–11, flashes red and white lights alternately, and is visible for about 15 m. The rock was originally known as The Inch Cape Rock.

Bellsdyke, seat, 1¼ m. S. of Airth, Stirlingshire.

Bellshill, q.s. par. and town, 11 m. S.E. of Glasgow, Lanarkshire; ry. stas., L.M.S. and L.N.E. P.O., T.O. Collieries and ironworks. Pop. of B. and Mossend, 18,094.

Bells Hill, alt. 1330 ft., one of the Pentland Hills, Edinburgh; lies 3 m. E.S.E. of Balerno.

Bellside, vil. 7 m. S.E. of Coatbridge, Lanarkshire.

Bellsquarry, vil. on road between Kirknewton and West Calder, and 2¾ m. S.W. of Mid Calder, Midlothian. P.O. Pop. 143.

Bellwood, seat, 1½ m. N.W. of Penicuik, Midlothian.

Bellyclone, pl. 5 m. E. of Crieff, Perthshire.

Belmaduthy House, seat in Black Isle, 2 m. N. of Munlochy, Ross and Cromarty.

Belmeanach, bay, E. side of Skye, Inner Hebrides.

Belmont, eccl. par., Govan par., Glasgow.

Belmont, seat, isl. of Unst, Shetland Isls.

Belmont, seat in co. of, and 1 m. S.E. of, Ayr.

Belmont Castle, pl. to the S. of Meigle, E. Perthshire; here is a convalescent home.

Belnagoak, hill, alt. 560 ft., 6 m. N. of Methlick, Aberdeenshire.

Belnahua, isl., 1½ m. W. of Luing Isl., Argyllshire. One of the State Isls.

Belses, vil., 7½ m. N.N.E. of Hawick, Roxburghshire; ry. sta., L.N.E., also serves Ancrum and Lilliesleaf.

Belsgrove, shooting-lodge, 3 m. from Strontian, Sunart, Argyllshire.

Belston, pl., with coal and iron, near Carluke, Lanarkshire.

Belsyde, seat, 2½ m. S.W. of Linlithgow, W. Lothian.

Beltie, see CANNY.

Belton, seat, 2¾ m., S.W. of Dunbar, E. Lothian.

Beltonford, ham., 2 m. W. of Dunbar, E. Lothian.

Beltrees, ham, 1½ m. S.E. of Lochwinnoch, W. Renfrewshire.

Bemersyde, family seat of the Haigs since time of Malcolm IV. Stands on r. Tweed, 2½ m. N.E. of St. Boswells. B. Hill in vicinity.

Ben A'an, mt., alt. 1326 ft., overlooking L. Katrine, Perthshire.

Ben a' Bourd, mt., alt. 3924 ft., 6 m. E. of Ben Macdhui, on the N. border of W. Aberdeenshire.

Ben Achallader, mt., alt. (N. top) 3404 ft., (S. top) 3288 ft.; on borders of Perthshire and Argyllshire, 2½ m., E. of Loch Tulla.

Ben Achally, mt. and loch, alts. 1594 ft. and 1004 ft. respectively; 5 m. N.E. of Dunkeld, Perthshire. A. Loch, 64 ft. deep, alt. 1004 ft.

Bennachie Distillery, 3 m. N. of Insch, Aberdeenshire.

Ben Aden, mt., alt. 2905 ft., N.E. end of Loch Nevis, Inverness-shire.

Ben Aigan, mt., alt. 1544 ft., 2¼ m. E. of Rothes, Banffshire.

Ben Airidh Charr (Arry Car), mt., alt. 2593 ft., 5 m. S.E. of Poolewe, W. Ross and Cromarty.

Ben Alder, mt., alt. 3757 ft., and deer forest (30,000 ac.) on W. of Loch Ericht, S.E. Inverness-shire.

Ben Alisky, mt., alt. 1142 ft., 12 m. N.W. of Berridale, Caithness.

Ben Alligin, mt., alt. 3232 ft., on N.E. shore of Loch Torridon, W. Ross and Cromarty.

Benan, prom. on S. side of isl. of Arran, Buteshire.

Ben-an-Armuinn (Ben Armine), mt., alt. 2338 ft., and deer forest, 13 m. N.E. of Lairg, Sutherlandshire.

Ben Arkle, mt., alt. 2582 ft., 25 m. S.W. of Tongue, N.W. Sutherlandshire.

Ben Arthur (the Cobbler), mt., alt. 2891 ft., 2¾ m. N.W. of Arrocher, Argyllshire.

Benarty, mt., alt. 1167 ft., 1 m. S. of Loch Leven, Kinross-shire.

Benarty House, 1½ m. E.S.E. of Blairadam, ry. sta., Fifeshire.

Ben Auskaird, mt., alt. 1265 ft., 5 m. S. of Laxford, W. Sutherlandshire.

Ben a Vain, or **Beinn Mheadhoin,** mt., alt. 3883 ft., one of Cairngorm group, S.W. Banffshire.

Ben a Vair, or **Beinn a Bheithir,** mt., alt. 3362 ft. and 3284 ft., between Loch Leven and Loch Linnhe, N. Argyllshire.

Benaveallich, mt., at. 1940 ft., 5 m. W. of Helmsdale, E. Sutherlandshire.

Ben Avon, mt., alt. 3843 ft., forming part of Cairngorms, 7 m. N.W. of Castleton of Braemar.

Ben Barvas, mt., alt. 900 ft., N. Lewis Isl., Outer Hebrides.

Benbecula, isl., par. of S. Uist, Outer Hebrides; lies between N. and S. Uist and measures 8 m. in diameter; low and marshy, coast much indented and soil poor. Pop. 961.

Ben Beoch, mt., alt. 1521 ft., 2 m. N. of Dalmellington, Ayrshire.

Ben Bha'ach Ard or **Ben Vacher,** mt., alt. 2826 ft., 12 m. W. of Beauly, Inverness-shire.

Ben Bhalgairean, mt., alt 2085 ft., 4 m. S. of Dalmally, Argyllshire.

Ben Bharrain, mt., alt. 2345 ft., on coast of Kilbrennan Sound, Arran, Buteshire.

Ben Bheag, mt., alt. 2029 ft., rising above Loch Eck, Argyllshire.

Ben Bheula, mt., alt. 2557 ft., at head of Loch Goil, E. Argyllshire.

Ben Bhragie, mt., alt. 1256 ft., surmounted by a colossal statue of the first Duke of Sutherland, 1½ m. N.W. of Golspie, Sutherland.

Ben Bhrotain, mt., alt. 3795 ft., 13 m. W. of Braemar, S.W. Aberdeenshire.

Ben Bhuidhe, mt., alt. 3106 ft., E. Argyllshire, 5 m. N. of Cairn Dow Inn.

Ben Bhuidhe Mor, mt., alt. 1797 ft., N.E. Inverness-shire.

Ben Bhuiridh, mt., alt. 2935 ft., N. end of Loch Awe, Argyllshire.

Ben Bowie, mt., alt. 1028 ft., 3 m. E. Helensburgh, Dunbartonshire.

Benbrack, hill, alt. 1621 ft., 3 m. E. of Dalmellington, Ayrshire.

Benbrack, mt., alt. 1475 ft., 5 m. N.E. of Carsphairn, N. Kirkcudbrightshire.

Ben Bui, mt., alt. 3106 ft., 9½ m. N.E. of Inveraray, W. Argyllshire.

Ben Chabhair, mt., alt. 3053 ft., 3 m. N.E. of head of Loch Lomond, or borders of Perthshire and Stirlingshire.

Ben Chaisteil, mt., alt. 2897 ft., 5 m. N.E. of Tyndrum, borders of Perthshire and Argyllshire.

Ben Chait, mt., alt. 2942 ft., part of Atholl Grampians, North Perthshire.

Ben Chaorach, mt., alt. 2848 ft., 2½ m. E. of head of L. Etive., Argyllshire.

Ben Charorach, mt. Dunbartonshire, 5 m. W. of Luss; alt. 2338 ft.

Benchil, estate, in co. of and 3 m. N. of Perth.

Benchinnan, that part of the Grampians in Co. of Angus.

Ben Chleibh, mt., alt. 3008 ft., 6 m. E. of Dalmally, borders of Perthshire and Argyllshire.

Ben Chochail, mt., alt. 3215 ft., 3½ m. N.E. of Ben Cruachan, Argyllshire.

Ben Chonzie, or **Ben-y-Hone,** mt., alt. 3048 ft., 5½ m. N. of Comrie, Perthshire.

Ben Chroin, mt., alt. 3101 ft., 4 m. N. of L. Katrine, Perthshire.

Ben Chuallaich, mt., alt. 2925 ft., 2¼ m. N.E. of Kinloch Rannoch, Perthshire.

Ben Cleuch, mt., alt. 2363 ft., 2½ m. N. of Tillicoultry, Clackmannanshire, highest summit of the Ochils.

Ben Creachan, mt., alt. 3540 ft., 4 m. E. of Tulla, borders of Perthshire and Argyllshire.

Ben Cruachan, mt., alt. 3689 ft., between L. Etive and L. Awe and 15 m. N. of Inveraray, Argyllshire.

Ben Damh, mt., alt. 2958 ft., W. Ross and Cromarty.

Ben Damph, deer-forest (18,000 ac.) and loch, near L. Torridon, S.W. Ross and Cromarty. B. D. House in vicinity.

Ben Dearg, alt. 3304 ft., in Forest of Atholl, Perthshire.

Ben Dearg, alt. 2974, ft., 11 m. E. of Poolewe, Ross and Cromarty.

Ben Dearg, alt. 3547 ft., 11 m. S.E. of Ullapool, Ross and Cromarty.

Benderloch, peninsular dist. between Loch Etive and Loch Creran, W. Argyllshire; ry. sta., L.M.S.

Bendochy, par., 7435 ac., 1 m. N. of Coupar-Angus, Perthshire. Pop. 504.

Ben Doran or **Doireann,** mt., alt. 3523 ft., 3 m. S.E. of Loch Tulla, E. Argyllshire. Celebrated in Duncan Ban's Gaelic poem.

Ben Eay, or **Ruadh Stac Mor,** mt., alt. 3309 ft., Gairloch dist., W. Ross and Cromarty.

Ben Eibhinn, mt., alt. 3611 ft., 6 m. W. of Loch Ericht, S. Inverness-shire.

Ben Erda, Trotternish, Isle of Skye, Inverness-shire ; alt. 2003 ft.

Ben Eunaich, mt., Argyll, 4 m. N.W. of Dalmally ; alt. 3242 ft.

Benevrich, deer-forest at S. end of Loch Treig, Inverness-shire.

Ben Fhionnlaidh, mt., alt. 3149 ft., N.W. end of Loch Etive, Argyllshire.

Ben Fin or **Fionn,** mt., alt. 3060 ft., 2 m. N.W. of Achnasheen, Ross and Cromarty.

Ben Freiceadain, hill and old fort, 8 m. S.E. of Reay, Caithness ; alt. 781 ft.

Bengall, vil., Dumfriesshire, 3 m. S.W. Lockerbie.

Ben Ghulbhuinn or **Gulabin,** mt., N.E. Perthshire, at head of Glenshee ; alt. 2641 ft.

Ben Glamaig, mt., Isle of Skye, S. of Loch Sligachan ; alt. 2537 ft.

Ben Glass, mt., Dunbartonshire, Luss par. ; alt. 2149 ft.

Ben Gnuis, mt. peak., isl. of Arran, Bute-shire ; alt. 2597 ft.

Ben Griam More and **Ben Griam Beg,** two outstanding mts. in the N. of Sutherland, 5½ and 6½ m. N.W. of Kin-brace ry. sta. ; alts. 1936 and 1903 ft.

Benhar, East, vil., 1½ m. N.W. of Fauld-house, W. Lothian.

Benhar, West, or **Muirhead,** vil., Lanark-shire, 3 m. E.N.E. of Kirk of Shotts. P.O.

Ben Hee, mt. (alt. 2864 ft.) and deer-forest, 5 m. E. of head of Loch More, Sutherland.

Ben Hiaut, mt., Argyll, Ardnamurchan par. ; alt. 1729 ft.

Benholm, par. and seat, on the coast of S.E. Kincardineshire, and 4 m. E. of Laurencekirk ; 4892 ac. Pop. 1092.

Benholm, pl., 2½ m. from Johnshaven, S.E. Kincardineshire. P.O., T.O.

Ben Hope, mt., alt. 3040 ft., 10 m. S.W. of Tongue, N. Sutherland.

Ben Hutig, long hill, N. Sutherland, 3 m. E. of Loch Eriboll ; alt. 1340.

Ben Hynish, highest land on Tiree, Argyll ; alt. 460 ft.

Ben Ime, mt., alt. 3318 ft., at head of Loch Long, Argyllshire.

Ben Killilan, mt., Ross and Cromarty, 7 m. N.E. of head of Loch Duich ; alt. 2466 ft.

Ben Kitlan or **Ceitlein,** mt., Argyll, alt. 2897 ft., 8 m. S.E. of Ballachulish.

Ben Klibreck, one of the highest mts. in Sutherland, 2 m. E. of Altnaharra ; alt. 3154 ft.

Ben Lair or **Larig,** mt., Ross and Cromarty, N. side of Loch Maree ; alt. 2817 ft.

Ben Lawers, alt. 3984 ft., mt. flanking the N.W. shore of Loch Tay, Perthshire. A cairn recently built on the summit (20 ft.) raises the mt. to 4004 ft.

Ben Ledi, alt. 2875 ft., 4½ m. W. by N. of Callander, Perthshire.

Ben Lochain, mt., alt. 2955 ft., 5 m .N. of Lochgoilhead, Argyllshire.

Ben Lomond, alt. 3192 ft., in N.W. Stirlingshire, on E. side of Loch Lomond, and 13½ m. N. of Dumbarton.

Ben Loyal, mt., N. Sutherland, on W. side of Loch Loyal ; alt. 2504 ft.

Ben Lui or **Laoigh,** mt., Perthshire and Argyll, 7 m. E. of Dalmally ; alt. 3708 ft.

Ben Lundie, hill, S.E. Sutherland, 3 m. W.N.W. of Golspie vil. ; alt. 1464 ft.

Ben Macdhui, alt. 4296 ft., in S.W. Aber-deenshire ; one of the Cairngorms, 18 m. W.N.W. of Castletown-of-Braemar. Direction indicator on summit.

Ben Mhanarch, mt., Dunbartonshire, 3 m. N.E. of Garelochhead ; alt. 2328 ft.

Benmore, Perthshire, 10 m. W. of Killin ; alt. 3843 ft.

Benmore, mt. in Assynt par., Sutherland-shire ; alt. 3273 ft.

Benmore, mt. in isl. of Mull, Argyllshire ; alt. 3185 ft.

Benmore, seat, on Holy Loch, Argyllshire, 3 m. from Kilmun ; 10,000 ac. gifted to form a demonstration forest.

Benmore Coigach, mt., Ross and Crom-arty, N. side of Loch Broom ; alt. 2438 ft.

Benmore Distillery, near Campbeltown, Kintyre, Argyllshire.

Ben na Cailich, two mts., S.E. Skye, 2½ m. W. of Broadford ; alt. 2403 ft. ; the other near Kyle Rhea ; alt. 2396 ft.

Bennachie, group of hills, highest point 1733 ft., 6 m. N.E. of Alford, Aber-deenshire.

Ben-na-Lap, mt., alt. 3066 ft., E. of Loch Treig, S. Inverness-shire.

Ben nam Aighean, mt., Argyll, Ard-chattan par. ; alt. 3141 ft.

Bennan Hill, State forest, 1½ m. S.W. of New Galloway ry. sta., Kirkcudbright-shire ; alt. 1750 ft.

Ben Nevis, alt. 4406 ft., lies 7 m. S.E. of Fort William, Inverness-shire, and is the highest mt. in the British Isles. Precipices of great height on the N. side.

Ben Nevis Distillery, N. of Fort William, Inverness-shire.

Ben Odhar, mt. on borders of Perthshire and Argyll, 2½ m. N. of Tyndrum ; alt. 2948 ft.

Ben Oir, one of the Paps of Jura, Argyll ; alt. 2571 ft.

Ben Oss, mt., alt. 3374 ft., in Perthshire, 6 m. W. of Crianlarich.

Ben Pharlagain, mt., N.W. Perthshire, 2½ m. W. of the head of Loch Ericht ; alt. 2836 ft.

Ben Reid, mt., Angus, 1 m. N. of Milton of Clova ; alt. 2577 ft.

Ben Reoch, mt. and seat, Dunbartonshire, 1½ m. S.E. of Arrochar ; alt. 2168 ft.

Ben Resipol, mt., N.W. Argyll, N. side of Loch Sunart.

Benrig, seat, Roxburghshire, 1 m. E. of St. Boswells.

Ben Rinnes, mt., alt. 2756 ft., 5 m. S.W. of Dufftown, W. Banffshire.

Benrinnes Distillery, 2½ m. N. of Ben Rinnes, Banffshire.

Benromach Distillery, near Forres, Moray-shire.

Ben Ruadh, mt., Argyll, Cowal dist. ; alt. 2178 ft.

Ben Screel or **Scriol,** mt., alt. 3196 ft., on Loch Hourn, W. Inverness-shire.

Ben Sguliaird, mt., alt. 3058 ft., 3 m. E.N.E. of Loch Creran, Lorne, Argyll-shire.

Ben Shianta, summit of the Paps of Jura, Argyll ; alt. 2477 ft.

Ben Slioch, mt., alt. 3217 ft., 5 m. N.W. of Kinlochewe, W. Ross and Cromarty.

Ben Smeorale, mt., Sutherland, 5¾ m. N.W. of Brora ; alt. 1592 ft.

Ben Spionnaidh, mt. range near Durness, Sutherland ; alt. 2537 ft.

Ben Stack, conical mt. near Laxford Bridge, N.W. Sutherland ; alt. 2364 ft.

Ben Starav, mt., Argyll, on N.E. shore of Loch Etive ; alt. 3541 ft.

Benston, pl., with limeworks, 3 m. N.W. of New Cumnock, Ayrshire.

Benston, pl. and loch, mainland, Shetland, 7 m. N. of Lerwick.

Bent, a pl., Lanarkshire, 2¼ m. N.E. of Lesmahagow.

Ben Tarsuinn, mt., Arran, Buteshire ; alt. 2706 ft.

Ben Tharsuinn, mt., Ross and Cromarty, N.E. of Loch Maree ; alt. 2750 ft.

Ben Tharsuinn, mt. between Lochs Lomond and Long, Dunbartonshire ; alt. 2149 ft.

Ben Tharsuinn, mt. E. Ross and Cromarty, 7 m. N.W. of Alness ; alt. 2270 ft.

Ben Tigh or **Ben Tee,** mt., S.W. Inverness, N. of Loch Lochy, Lochaber dist. ; alt. 2956 ft.

Ben Tirran, mt., Angus, 13 m. N. Kirrie-muir ; alt. 2860 ft.

Bentpath, ham., 6 m. N.W. of Langholm, Dumfriesshire. P.O., T.O.

Ben Trilleachan, mt., Argyll, near head of Loch Etive ; alt. 2752 ft.

Bents vil., with ry. sta., L.N.E. (passenger service withdrawn), W. Lothian, 5 m. S. of Bathgate.

Bents, stream, mid Aberdeenshire, flow-ing 4½ m. N. to the Don.

Ben Tue, mt., Inverness-shire, 2½ m. W. of Laggan ; alt. 2956 ft.

Ben Tuire, hill, Argyll, Kintyre, 10 m. N.E. of Campbeltown.

Ben Tulachan, mt., S.W. Perthshire, 4½ m. N.E. of head of Loch Katrine ; alt. 3099 ft.

Ben Tulloch, conical mt. (The Sugar Loaf), alt. 2496 ft., Mull, Argyllshire.

Ben Uaig, hill, Argyll, Mull Isl., 3 m. S.E. of Salen ; alt. 1320 ft.

Ben Uary, mt., Sutherland, 6½ m. W. of Helmsdale ; alt. 2046 ft.

Ben Udlaidh, mt., Argyll, N.E. of Dal-mally ; alt. 2529 ft.

Ben Udlaman, mt., alt. 3306 ft., 1 m. W. of Loch Ericht, Perthshire.

Ben Uie, mt., Sutherland, N.E. of Loch Assynt ; alt. 2384 ft.

Benula, deer forest, 32 m. S.W. of Beauly, Inverness-shire.

Ben Ushinish, hill peak, 1000 ft., E. Lewis Isl., Ross and Cromarty.

Ben Vane, mt., alt. 3004 ft., N.W. border of Dunbartonshire.

Ben Vane, mt., Dunbartonshire, 3 m. N.E. of Garelochhead ; alt. 2328 ft.

Ben Venue, alt. 2393 ft., flanks Loch Katrine, on S. side Perthshire.

Ben Vorlich, mt., alt. 3224 ft., E. side of Loch Earn, Perthshire.

Ben Vrackie, mt., alt. 2757 ft., on E. side Pass of Killiecrankie, N. Perthshire.

Benwhat, vil., 2½ m. N.W. of Dalmellington, Ayrshire. P.O.

Ben Wyvis, mt., alt. 3429 ft., and deer forest (24,000 ac.), 10 m. N.W. of Dingwall, Ross and Cromarty.

Ben-y-Gloe, mt., alt. 3671 ft., 8 m. N.E. of Blair Atholl, N. Perthshire.

Beoraid, loch, 2 m. long, Inverness-shire, in Glenmeoble, depth 159 ft.

Beregonium or **Selma,** vitrified fort on N. side of Loch Etive, N.W. Argyllshire.

Bernera, pl., with ruins of barracks, near vil. of Glenelg, W. Inverness-shire. Ferry communication with Kylerhea.

Bernera, Great and **Little,** isls. of N.W. coast of Lewis, Outer Hebrides. Great B. has interesting remains in the shape of standing stones. Pop. 514.

Berneray, isl. and q.s. par. in par. of Harris, and 1 m. N. of N. Uist, Outer Hebrides. P.O., T.O. Pop. 331.

Berneray, isl., 14 m. S.S.W. of Barra, Outer Hebrides. Here is Barra Head lighthouse with occulting light, visibility 33 m. Pop. 6.

Bernisdale, ham., 3 m. E. of Edinbain, isl. of Skye, Inner Hebrides.

Berridale, q.s. par. and vil., 10 m. N.E. of Helmsdale, Caithness. P.O., T.O.

Berriedale, stream, isl. of Hoy, Orkney.

Berriedale Water, stream, rising at foot of Morven, and after a course of 21 m. falls into sea at Berriedale, Caithness.

Berrybank, seat, 1 m. S.E. of Reston, Berwickshire.

Berryhill, seat, 2½ m. W. of Peterhead, Aberdeenshire.

Berryhill, estate, 1¾ m. N.E. of Kilsyth, Stirlingshire.

Berrywell, seat, ½ m. S.E. of Duns, Berwickshire.

Berstane Bay, 1½ m. E. of Kirkwall, Orkney.

Bervie, par. in S.E. Kincardineshire, 13 m. N.E. of Montrose; 2331 ac. Pop. 2116. **Inverbervie,** parl. and royal bur., mkt.-town, and seapt. at mouth of r. Bervie. The bur., combined with Arbroath, Brechin, Forfar, and Montrose, returns one member to Parliament. Pop. of bur. 1032. P.O., T.O.; ry. sta., L.N.E. Industries: flax and tow spinning.

Bervie, riv., falling into B. Bay, at the town of Bervie, Kincardineshire.

Bervie Brow, headland, with cliffs, 451 ft., ¾ m. N.E. of Bervie, Kincardineshire.

Berwick, North, see NORTH BERWICK.

Berwickshire, a border co. in the S.E. of Scotland, bounded by E. Lothian, North Sea, Northumberland, Roxburgh, and Midlothian; area 457 sq. m. or 292,535 ac. Coast rocky and bold, rising at St. Abb's Head and other points to heights of from 177 to 528 ft., and having only two bays, at Eyemouth and Coldingham. The Tweed is the principal r., traces about half of the S. boundary and receives several tributaries, e.g. Leader, Blackadder, and Whitadder; to the N. are the Lammermuir Hills, to the W. the upland region of Lauderdale, while the fertile lower part of the Tweed is called the Merse. Agriculture is practised scientifically and the principal mfrs. are paper, woollens, blanketings, etc. The constituency of Berwick and Haddington returns one member to Parliament. Pop. 26,612.

Bettyhill, ham., near mouth of r. Naver, N. Sutherlandshire; P.O., T.O. Hotel.

Bharanich, loch, near Loch Maree, W. Ross and Cromarty.

Bhasapoll, loch in N.W. of isl. of Tiree, Argyllshire.

Bhrollum, sea loch, S.E. coast of Lewis, Outer Hebrides.

Bhuie, Bein, mt., alt. 2884 ft., 2 m. N. of E. end of Loch Hourn, W. Inverness-shire.

Bickerton Burn, trib. of r. Almond, S.W. of W. Lothian.

Bieldside, vil. and ry. sta., L.N.E., 4½ m. S.W. of Aberdeen. P.O.

Biggar, par. and town (police bur.), with ry. sta., L.M.S., Lanarkshire, 3½ m. N.E. of Symington junc., and 13½ m. S.E. of Lanark by rail; 7278 ac. Pop. 2074; bur. pop. 1323. P.O., T.O. Hotel.

Bighouse, seat, N. Sutherland, 1 m. from Melvich.

Bigton, pl., S.W. Shetland, opposite St. Ninian's Head. P.O., T.O.

Bilbster, ham., seat, and ry. sta., L.M.S., 5¼ m. N.W. of Wick, Caithness.

Billie Castle, ruins, 2 m. S.W. of Reston, Berwickshire.

Bilsdean, ham. and stream, 6¾ m. S.E. of Dunbar, E. Lothian.

Bilston, pl., with inn and burn, 5 m. S. of Edinburgh, Midlothian.

Bilston Lodge, seat near Loanhead, Midlothian.

Binn, hill, alt. 555 ft. adjoining Kinfauns Castle, E. Perthshire.

Binnaness, headland and bay, 7 m. E. of Lerwick, Shetland.

Binnein Beag, mt., alt. 3083 ft., 8 m. S.E. of Fort William, Inverness-shire.

Binnein en Fhidhleir, mt., alt. 2658 ft., 6 m. N. of Lochgoilhead, Argyllshire.

Binnein More, mt., alt. 3700 ft., 8 m. S.E. of Fort William, Inverness-shire.

Binnend, vil., 1 m. N.N.E. of Burntisland, Fifeshire.

Binnie, East and **West,** two hams., 1¾ m. N.W. of Uphall, W. Lothian.

Binniehill and Southfield, vil. on ry. between Manuel and Bathgate, and 10 m. W. of Bathgate, Stirlingshire.

Binning, a wood, 1¼ m. N.W. of Tyninghame, E. Lothian.

Binn of Cullen, hill, alt. 1050 ft., 3 m. S.W. of Cullen, Banffshire.

Binns, seat, 3¼ m. N.E. of Linlithgow, W. Lothian.

Binn, The, hill, alt. 632 ft., ¾ m. N. of Burntisland, Fifeshire.

Binny House, 1 m. N. of Uphall, W. Lothian.

Binram's Cross, mound adjoining St. Mary's Loch, W. Selkirkshire.

Binrock House, Perth road, Dundee; burnt 1925.

Binscarth, seat, 7 m. W.N.W. of Kirkwall, Orkney.

Bin, The, hill (1027 ft.) with State forest, 2 m. N.W. of Huntly, Aberdeenshire.

Birdston, vil., 2½ m. S.E. of Lennoxtown, S. Stirlingshire.

Birgham, vil. on the Tweed, 3½ m. S.W. of Coldstream, Berwickshire. P.O.,T.O.

Birka Water, loch, 45 ft. deep, 4 m. N.W. Collafirth, Mainland, Shetland; alt. about 400 ft.

Birkenbog, seat, 2 m. S.E. of Cullen, Banffshire.

Birkhall, royal seat and estate, 2¼ m. S.W. of Ballater, Aberdeenshire.

Birkhill, pass on borders of Selkirkshire, and Dumfriesshire, 10 m. N.E. of Moffat. Was a favourite haunt of the Covenanters. Alt. 1080 ft.

Birkhill, seat, 2 m. S.W. of Balmerino, N. Fifeshire.

Birkhill, vil., 4½ m. N.W. of Dundee, Angus. B. and Muirhead special lighting dist. Pop. 440.

Birkwood, seat adjoining Lesmahagow, Lanarkshire.

Birnam, vil. on r. Tay and ¾ m. S.E. of Dunkeld, Perthshire; ry sta. (Dunkeld and Birnam; alt. 218 ft.), L.M.S. Pop. 629. P.O., T.O. Duncan's Camp, a circular fortification, is said to have been the place where King Duncan held his court. On B. hill, between the camp and the Tay, is B. Pass.

Birnam Youth Hostel, ½ m. S. of Dunkeld sta., 14 m. N. of Perth. A Scottish Youth Hostel.

Birness, ham., 4 m. N.E. of Ellon, E. Aberdeenshire.

Birnie, par. on the Lossie R. in Morayshire, and 3½ m. S. of Elgin; postal town, Elgin; par. 6788 ac. Pop. 380.

Birnie Road Siding, ry. sta. on L.N.E., 3 m. S. of Bervie, Kincardineshire.

Birns Water, trib. of r. Tyne, E. Lothian.

Birran or **Dundurn,** mt., alt. 2011 ft., 1 m. S. of St. Fillans, Perthshire.

Birrens, pl., with well-preserved Roman Camp, 1 m. N.E. of Ecclefechan, Dumfriesshire.

Birsay, 20 m. N.W. of Kirkwall, N.W. Mainland, Orkney. P.O., T.O.

Birsay and Harray, united pars. on Pomona Isl., N.W. Orkney; 26,749 ac. Pop. 1608. Contains Birsay, Swannay and Harray vils.

Birse, par. and vil. on r. Dee, and 2 m. S.E. of Aboyne, Aberdeenshire; par. 31,592 ac. Pop. 944. Forest of B. is 4½ m. S. of vil.

Birsley, pl. with colliery, ½ m. S. of Tranent, E. Lothian.

Birthwood, seat, 5 m. S. of Biggar, Lanarkshire.

Bishopbriggs, vil., 3¼ m. N.E. of Queen Street ry. sta., Glasgow; ry. sta., L.N.E. P.O., T.O. Pop. B. and Auchinairn, 5302.

Bishopmill, pl. on r. Lossie, and sub. of Elgin, Morayshire. P.O.

Bishop's Burn, stream flowing 6½ m. S.E. and falling into Wigtown Bay, N.E. Wigtownshire.

Bishop's Forest, hill, alt. 1285 ft., N.E. Kirkcudbrightshire.

Bishop's Loch, one of the reservoirs for the Forth and Clyde Canal, 2¾ m. N.W. of Coatbridge, N. Lanarkshire.

Bishop's Loch, in co. of, and 7 m. N.W. of Aberdeen.

Bishop's Seat, mt., alt. 1651 ft., 2½ m. W. of Dunoon, E. Argyllshire.

Bishopton, vil., 5¾ m. N.W. of Paisley, Renfrewshire; ry sta., L.M.S. P.O., T.O. Pop. 820. B. House 1 m. N.W.

Bixter, ham. and bay, 11 m. N.W. of Lerwick, Shetland. P.O., T.O..

Blabhein, or **Blaven,** mt., alt. 3042 ft., 6 m. S. of Loch Sligachan, Skye, Inner Hebrides.

Blackadder, stream, with a course of 20 m. before joining Whitadder above Allanton, Berwickshire.

Black Benwee Hill, alt. 1208 ft., 5 m. NE. of Newton Stewart, Kirkcudbrightshire.

Blackbraes, vil., E. Stirlingshire, 3¼ m. S.E. of Falkirk.

Blackburn, vil. in co. of, and 9 m. N.W. of, Aberdeen.

Blackburn, vil. on r. Almond, and 2½ m. S. of Bathgate, W. Lothian. P.O., T.O. Pop. 1970. Close by, B. Hall and B. House.

Black Burn, trib. of r. Liddel, S.W. Roxburghshire.

Black Burn, trib. of r. Annan, Dumfriesshire.

Black Burn stream, Morayshire, enters the Lossie, near Dallas.

Black Burn, trib. of the N. Esk, which it joins a little below Brigend, Kincardineshire.

Black Cairn, hill, alt. 854 ft., crowned by a cairn, 12 m. W. of Tarves, mid Aberdeenshire.

Blackcastle, ancient camp on the Blackadder, 1¾ m. N.W. of Greenlaw, Berwickshire.

Blackcastle, hill, alt. 917 ft., near Innerwick, E. Lothian.

Blackchester, ancient camp, 2½ m. N.W. of Lauder, W. Berwickshire.

Blackcorries and **Kinlochbeg,** deer forest, 2 m. E. of Ballachulish, N.E. Argyllshire.

Black Country, a term used to describe the coal-mining and iron manufacturing region E. of Glasgow, Lanarkshire.

Black Cove, cavern on S. coast of Kirkcudbrightshire, near Barlocco.

Blackcraig, vil., 2 m. S.E. of Newton Stewart, Kirkcudbrightshire.

Blackcraig, seat, 6 m. N.W. of Blairgowrie, Perthshire. State forest here.

Black Craig, mt., alt. 2298 ft., 5 m. S.E. of New Cumnock, E. Ayrshire.

Black Craig, hill, alt. 665 ft., 6½ m. N.E. of Newburgh, Fifeshire.

Black Craig, mt., alt. 1730 ft., S.E. side of Loch Doon, Ayrshire.

Blackerstone, pl., 7 m. E.N.E. of Longformacus, N. Berwickshire.

Blackford, seat, 1¼ m. W. of Rothie Norman sta., Aberdeenshire.

Blackford, par. and vil., 9 m. N.E. of Dunblane, Perthshire; ry. sta., L.M.S.; alt. 394 ft.; par. 20,991 ac. Pop. 1399; vil. pop. 522. P.O., T.O.

Blackford Hill, ry. sta. on suburban line of L.N.E., on S. side of Edinburgh. The hill, a public park, is 500 ft. high and crowned by a Royal Observatory.

Blackfriars, eccl. par. of Glasgow.

Blackhall, dist. in W. of Edinburgh; ry. sta., Craigleith. P.O., T.O.

Blackhall, estate, S.E. of Paisley, mid Renfrewshire.

Blackhall, seat, 2 m. W. of Banchory, N.W. Kincardineshire.

Blackhall Paper Mills, in Paisley, Renfrewshire.

Blackhead, prom., 2 m. N.N.W. of Port Patrick, Wigtownshire. Here is Killantringan lighthouse.

Blackhill, q.s. par. and ham., 4 m. S.W. of Peterhead, Aberdeenshire. P.O. spelt Blackhills.

Blackhill, pl. near Forth and Clyde Canal, E. of Glasgow.

Black Hill, alt. 1031 ft., 1 m. S.E. of Earlston, S. Berwickshire.

Black Hill, alt. 1628 ft. One of the Pentland Hills, and 2½ m. S.E. of Balerno, Midlothian.

Black Hill, alt. 1182 ft.; Sidlaw Hills, in co. of, and 8 m. N.E. of, Perth.

Blackhills House, 4 m. S.E. of Elgin, Morayshire.

Blackhouse, seat, 1½ m. N.W. of Kippen, W. Stirlingshire.

Blackhouse Heights, mt. ridge, alt. 2214 ft., on borders of Peeblesshire and Selkirkshire, 5 m. N. of St. Mary's Loch.

Blackhouse Tower, ruined keep, 2½ m. N.N.E. of St. Mary's Loch, W. Selkirkshire. Said to have been the scene of " the Douglas Tragedy."

Black Isle, peninsular region between Cromarty Firth and the Firths of Beauly and Moray.

Black Knowe, hill, alt. 1481 ft., $3\frac{1}{2}$ m. S.E. of the Ettrick, S.W. Selkirkshire.

Black Knowe, hill, Selkirkshire, 3 m. S.W. of Ettrick Church ; alt. 1804 ft.

Black Knowe Head, hill, alt. 1806 ft., borders of pars. of Ettrick and Yarrow, Selkirkshire.

Black Knowe Head, mt., alt. 1938 ft., 3 m. N.N.E. of Tushielaw, mid Selkirkshire.

Black Lakes, small lochs, 3 m. E. of Oban, Argyllshire.

Blacklarg, mt., alt. 2231 ft., 7 m. S.E. of New Cumnock, on borders of Ayrshire, Kirkcudbrightshire, and Dumfriesshire. It contains the source of the Afton Water, celebrated by Burns.

Blacklaw, ruined tower, $2\frac{1}{2}$ m. N.W. of Moffat, Dumfriesshire. Was a stronghold of the Douglases of Fingland.

Blacklaw, hill, alt. 929 ft., 8 m. N.W. of Dundee, in Perthshire.

Blacklaw, ham., $1\frac{1}{2}$ m. S.W. of Aberchirder, Banffshire. P.O.

Black Law, mt., alt. 1339 ft., 5 m. S.E. of Muirkirk, E. Ayrshire.

Black Law, mt., alt. 2285 ft., 4 m. N.W. of St. Mary's Loch, S. Peeblesshire.

Black Linn Reservoir, in co. of., and $3\frac{1}{2}$ m. N.E. of, Dumbarton. Water supply for town.

Black Loch, 4 m. S.W. of Newton Mearns, S.E. Renfrewshire.

Black Loch, 2 m. S.E. of Slamannan, on borders of Stirlingshire and Lanarkshire.

Black Loch, 1 m. E. of Dumfries, S. Dumfriesshire.

Black Loch, $\frac{1}{2}$ m. from Loch Glow, on borders of Fifeshire and Kinross-shire.

Black Loch, 10 m. N.W. of Coupar-Angus, Perthshire.

Black Lochs, series of small tarns, 2 m. S.E. of Loch Aven, S. Banffshire.

Black Lochs, Lorne dist., and 3 m. E. of Oban, Argyllshire.

Blacklorg, hill, alt. 2231 ft., 9 m. S.W. of Sanquhar, on borders of Ayrshire and Lanarkshire.

Blacklunans, dist. on the Blackwater, $5\frac{1}{2}$ m. E. of Kirkmichael, E. Perthshire. P.O., T.O.

Blackmile Pass, between Loch Lochy and Loch Arkaig, S.W. Inverness-shire.

Black Mill Bay, bay on W. coast of Luirg Isl., Argyllshire ; fixed red light, visible for 5 m.

Black Mount, alt. 1689 ft., 7 m. E. of Carnwath, E. Lanarkshire.

Black Mount, deer forest (80,000 ac.) and shooting-lodge, 14 m. S. of Loch Creran, in Appin dist., N. Argyllshire.

Blackness, coastal vil. on Firth of Forth, W. Lothian, and $3\frac{1}{2}$ m. N.E. of Bo'ness. P.O., T.O. Pop. 154. B. Castle, formerly a State prison.

Blackridge, vil., $5\frac{1}{2}$ m. W. of Bathgate, W. Lothian. P.O., T.O. Pop. 2019.

Blackridge Colliery, 2 m. E.N.E. of Chapelhall, Lanarkshire.

Black Rock, a cañon, 100 ft. deep, on r. Glass, Ross and Cromarty, $1\frac{1}{4}$ m. from Novar sta.

Blackrock, ham., 2 m. W. of Brigend, Islay Isl., Argyllshire.

Black Rocks, cluster of rocks, off Leith, in Firth of Forth.

Black Ruthven House, 3 m. N.W. of Perth.

Blacksboat, on r. Spey, ry. sta., L.N.E., 23 m. N.E. of Boat of Garten, Morayshire. P.O.

Blackshaw, vil. on Solway Firth, 8 m. S.E. of Dumfries, Dumfriesshire. Pop. 230.

Blackshiels, ham., 15 m. S.E. of Edinburgh, in par. of Humbie, E. Lothian. P.O., T.O.

Blacksmill, ham., hill, and burn, $2\frac{1}{2}$ m. S.E. of Longformacus, mid Berwickshire.

Black Spout, cascade of 120 ft. on the Edradour Burn, and $1\frac{1}{2}$ m. E. of Pitlochry, Perthshire.

Blackston, ham., with ry. sta., L.N.E. (passenger service withdrawn), 4 m. E. of Slamannan, Stirlingshire. T.O.

Blackstone, vil., on the Black Cart, 2 m. N.W. of Paisley, Renfrewshire.

Blackwater, trib. of the Deveron, N. Banffshire.

Blackwater, trib. of r. Ken, Kirckudbrightshire.

Blackwater, trib. of r. Conon, S.W. Ross and Cromarty.

Blackwater, small stream falling into Drimadown Bay, Arran Isl., Buteshire.

Blackwater, small stream, Isl. of Lewis, Outer Hebrides.

Blackwater, a lower reach of the Shee, and unites with the Ardle, 4 m. N.W. of Blairgowrie, to form the Ericht; N.E. Perthshire.

Blackwater, stream with course of 16 m., after which it falls into the Brora 2 m. N.W. of Loch Brora, Sutherland-shire.

Black Water, stream flowing through Glen Dubh, Morven, Argyllshire.

Blackwaterfoot, ham. on W. side of Arran Isl., and 10 m. S.W. of Brodick, Buteshire. P.O., T.O. Hotel.

Blackwater Lochs, series of lochs on borders of Argyllshire and Inverness-shire.

Blackwater Lodge, 7 m. S. of Dufftown, Banffshire.

Blackwood, vil., with ry. sta., L.M.S., 4 m. S.E. of Stonehouse, mid Lanarkshire. Pop. 1823.

Blackwood, seat and hill, 5¾ m. S.E. of Thornhill, Nithsdale, Dumfriesshire.

Blackwood, hill, 1000 ft., 3 m. S.W. of Eaglesham, S.E. Renfrewshire.

Blackwood, seat, 3½ m. E. of Ecclefechan, Dumfriesshire.

Black Wood, forest, on S. shore of Loch Rannoch, Perthshire.

Blackyett, seat, 2½ m. S.E. of Kirtle-bridge ry. sta., Dumfriesshire.

Bladenoch and **Bladnoch Distillery,** vil. on r. Bladenoch, E. Wigtownshire. P.O. called Bladnoch.

Blaich, vil. on Loch Eil and 4 m. N.W. of Fort William, Argyllshire.

Blainslie, vil., 3 m. S.E. of Lauder, in N.W. Roxburghshire.

Blair, vil., 1¼ m. S.E. of Dalry, N.W. Ayrshire.

Blair, mt., alt. 2441 ft., 15 m. N.W. of Kirriemuir, Angus.

Blair, seat, now a convalescent home, 5 m. W.N.W. of Dunfermline, Fifeshire.

Blair, small loch, W. Inverness-shire, drains into Loch Arkaig.

Blair, seat, 4 m. N.E. of Girvan, Ayrshire.

Blairadam, ham, 4 m. S.E. of Kinross, ry. sta., L.N.E. P.O., T.O. B. House is a seat. State forest nearby.

Blair-Atholl, par. and vil. on the Tilt and Garry, and 21 m. N. of Aberfeldy, Perthshire, ry. sta., L.M.S. P.O., T.O.; par. 187,926 ac. Pop. 1557. Hotels. B. Castle. seat of the Duke of Atholl is nearby.

Blair-Atholl Distillery, 1 m. N. of Pitlochry, Perthshire.

Blairbeth House, 1½ m. W. of Cambuslang, N.W. Lanarkshire.

Blairburn, vil. on r. Forth, and 1½ m. W. of Culross, Fifeshire. Blair Castle, a seat, is 1 m. W.

Blairdaff, q.s. par. and ham., 5½ m. S.W. of Inverurie. Aberdeenshire.

Blairdardie, pl. within mun. bur. of Glasgow, 2 m. E. of Clydebank and on Forth and Clyde Canal.

Blairdenon, alt. 2072 ft., a summit of the Ochils, Clackmannanshire, and 6 m. N.W. of Dollar.

Blair Drummond, seat on the Teith and 2 m. S.E. of Doune, Perthshire. P.O.

Blairessan, seat, ¼ m. N. of Killearn, West Stirlingshire.

Blairfindy Lodge, seat, 5 m. S. of Ballin-dalloch sta., Banffshire, nearby are the ruins of B. Castle.

Blairgowrie, par., seat, and town (police bur.) on the Ericht, N.E. Perthshire, and 5 m. N.W. of Coupar-Angus, ry. sta., L.M.S. P.O., T.O. Town pop. 2914. Hotels. Fruit growing and canning centre, agricultural machinery, linen and jute mills. Par. area, 15,870 ac. Pop. 4049; bur. pop. 4676, inclusive of Rattray which adjoins.

Blairhall, vil., par. of Culross, Fifeshire. P.O. Pop. 1172.

Blairhill and **Gartsherrie,** ry. sta., L.N.E., ¾ m. N. W. of Coatbridge, Lanarkshire.

Blairhill, seat, 2½ m. E. of Dollar, E. Perthshire.

Blairhill, seat, 1 m. S.W. of Rumbling Bridge, Kinross-shire.

Blairhoyle, seat, 4½ m. S. of Callander, Perthshire.

Blairhullichan, seat, at W. end of Loch Ard and 5 m. W. of Aberfoyle, Perthshire.

Blairingone, q.s. par. and vil., 2½ m. E.S.E. of Dollar, Clackmannanshire. P.O.

Blair Lodge, Borstal Institution, 1¼ m. S. of Polmont, Stirlingshire.

Blairlogie, vil., 3 m. E.N.E. of Stirling and in that county. P.O.

Blairmore, seat, 6 m. W. of Huntly, Aber-deenshire.

Blairmore and Strone, watering-place, 9 m. N.W. of Greenock, and at entrance to Loch Long, Argyllshire. P.O., T.O.

Blairnathort, seat, 1½ m. N.E. of Milna-thort, Kinross-shire.

Blairour, pl. near Spean Bridge, Inverness-shire.

Blairquhan Castle, seat, on the r. Girvan, 1 m. N.W. of Straiton, Ayrshire.

Blairquhosh, seat, 2½ m. W.N.W. of Strathblane, Stirlingshire.

Blairs, pl. in N. Kincardineshire on r. Dee, and 6 m. S.W. of Aberdeen. P.O. R.C. College.

Blairs Ferry, from N.W. Bute Isl. to Kames on the mainland.

Blairton House, 4½ m. S.W. of Newburgh, E. Aberdeenshire.

Blairvadock, seat, 3 m. N.W. of Helensburgh, Dunbartonshire.

Blane, or **Ballagan,** stream flowing into the Endrick, 1½ m. S.W. of Killearn, Stirlingshire.

Blanefield, vil. in Stirlingshire, 17 m. N.W. of Glasgow by rail, ry. sta., L.N.E. P.O., T.O.

Blantyre, par. and town on the Clyde, 3 m. N.W. of Hamilton, Lanarkshire. Contains fragments of a thirteenth cent. priory, Par. 3953 ac. ; ry. sta. (High Blantyre), L.M.S. Low Blantyre was the birthplace of Dr. Livingstone(1803-1873), the African missionary and explorer. P.O., T.O. Pop. 17,015. The centre of a coal-mining district ; town pop. 15,975.

Blantyre Reservoir (Lanarkshire County Council), 2 m. W. of High Blantyre.

Blarannich, seat, 2 m. E. of Arrochar and on Loch Lomond, Dunbartonshire.

Blarghour, Falls of, 14 m. S.W. of Dalmally, W. Argyllshire.

Blarich, pl., 9 m. N.W. of Golspie, Sutherlandshire.

Blarmachfoldach, ham., 3 m. S. of Fort William, Inverness-shire.

Blasted Heath, a region between Brodie and Forres stas., where Macbeth is supposed to have met the witches who foretold his future.

Blawarthill, Scotstoun, Renfrewshire, Hospital.

Bleachfield Paper Mills, vil. of Ayton, Berwickshire.

Bleaton Hallet, seat, 8 m. N. of Blairgowrie, Perthshire.

Blebo Craigs, vil., 1 m. N.E. of Blebo Mills vil. P.O.

Blebo Mills, vil., close to the vil. of Dura Den, and 5½ m. S.W. of St. Andrews, Fifeshire.

Bleedy Pots, pl. near Gamrie Bay, N. Banffshire.

Blegbie, seat, 2 m. S. of Humbie sta., E. Lothian ; 1 m. S. is B. Hills (1336 ft.).

Blelack, seat, 6 m. N.W. of Aboyne, Aberdeenshire.

Blencathara, see SADDLEBACK.

Blervie, seat, 4 m. S.E. of Forres, Morayshire.

Blind Burn, stream flowing into the Calder, N. Lanarkshire.

Blinkbonny, ham. and hill (654 ft.), 1 m. N.W. of Nenthorn, S. Berwickshire.

Bloak Moss, 3 m. W. of Stewarton, N. Ayrshire.

Blochairn, pl. with cairns, said to be memorials of a battle with the Norse, 2 m. E.N.E. of Milngavie, Stirlingshire.

Bloch Hill (878 ft.), 2½ m. S.W. of Langholm, Dumfriesshire.

Blood, loch in Knapdale, S.W. Argyllshire.

Bloodhope, stream falling into the White Esk above Cassock, N.E. Dumfriesshire.

Bloodstone Hill (1273 ft.), Rhum Isl., Inverness-shire.

Bloody Bay, Mull Isl., Argyllshire, ½ m. S.E. of Ardmore Point.

Bloody Bay, creek, S. end of the isl. of Iona, Argyllshire.

Bloody Burn, stream falling into Fail water, W. Ayrshire.

Bloody Faulds, pl. 1½ m. S.E. of Alford, Aberdeenshire.

Bloody Laws, one of the Cheviots, alt. 809 ft., 4 m. S.W. of Hownam, Roxburghshire.

Bloomhill, seat, 3 m. S.W. of Alexandria, Dunbartonshire.

Blue Cairn Hill, alt. 1715 ft., on N. border of Dumfriesshire and 2½ m. N. from Ettrick Pen.

Blyth Bridge, vil. with P.O., in Peebleshire, and 5 m. S.W. of West Linton.

Blyth Edge, hill, alt. 1522 ft., on S. slopes of Lammermuirs, and 4½ m. N. of Westruther, Berwickshire.

Blythe Burn, trib. to Boon Water, N.W. Berwickshire.

Blythswood, seat in co. of, and 1 m. N.W. of, Renfrew. On S. bank of Clyde is a group-flashing light.

Boardhouse, small loch, mainland, Orkney.

Boarhills, q.s. par. and ry sta., L.N.E., 3¾ m. E.S.E. of St. Andrews by rail, E. Fifeshire. P.O.

Boar of Badenoch, mt., alt. 2422 ft., 5½ m. S. of Dalwhinnie and on W. side of Drumochter Pass, on borders of Perthshire and Inverness-shire.

Boars of Duncansby, whirlpool, off N.E. coast of Caithness.

Boat Cave, cavern, 150 ft. long, 16 ft. high, and 12 ft. broad, in Staffa Isl., W. Argyllshire.

Boath, ham., 6 m. N.W. of Alness, E. Ross and Cromarty. P.O.

Boath, hill, alt. 600 ft., 4 m. N.W. of Arbirlot, Angus.

Boath, seat, near Auldearn, Nairnshire.

Boathouse Bridge, on r. Almond, 1 m. E. of Kirkliston, W. Lothian.

Boat of Bridge, pl., with suspension bridge, across Spey, 7 m. W. of Keith, on borders of Banffshire and Moray.

Boat of Garten, vil. on r. Spey and 5 m. N. of Aviemore, Inverness-shire; ry. stas. (712 ft.), L.M.S. and L.N.E. P.O., T.O. Pop. 247. Hotels.

Boat of Insh, 5¾ m. N. of Kingussie, Inverness-shire. See KINCRAIG.

Boblainy, deer forest (8500 ac.) near Beauly, N. Inverness-shire.

Bocastle, hill, about 750 ft., rising abruptly on all sides, and crowned by ruins of an old defensive work, 1 m. W. of Callander, Perthshire.

Bochel, hill, alt. over 1600 ft., 5 m. N.E. of Tomintoul, Banffshire.

Bochruben, ham. in co. of, and 12 m. S.S.W. of, Inverness.

Boddam, q.s. par. and coastal vil., 3 m. S. of Peterhead, Aberdeenshire; ry. sta., L.N.E. (passenger service withdrawn). Vil. pop. 646. P.O., T.O. Fishing. B. Castle close by (a ruin).

Boddin, coast ham., Angus, 3 m. S. of Montrose.

Boddom, vil., mainland, Shetland.

Bodesbeck Law, hill, alt. 2173 ft., 7 m. N.E. of Moffat, borders of Selkirkshire and Dumfriesshire.

Boes Cave, cavern, near Dunaverty, Kintyre, S. Argyllshire.

Bogany, headland, 1¼ m. N.E. of Rothesay, Buteshire; one red and one white fixed light (Craigmore).

Bogbrae, ham. 4 m. W. of Cruden Bay, Aberdeenshire.

Bogfoot, ham., 5½ m. S.E. of Dalbeattie, Kirkcudbrightshire.

Bogha Cloiche, mt., alt. 2945 ft., 6 m. N.E. of Dalwhinnie, Inverness-shire.

Boghall, ham., 4¼ m. N. of Lauder, Berwickshire.

Boghall, seat, 1¼ m. from Beith, N. Ayrshire.

Boghall, pl., with remains of ancient castle, ½ m. S. of Biggar, Lanarkshire.

Boghead, ham., 3 m. S. of Dalbeattie, Kirkcudbrightshire.

Boghead, vil., 3½ m. S.E. of Stonehouse, Lanarkshire.

Boghead, seat, 1¼ m. S.W. of Bathgate, W. Lothian.

Boghead, vil., 3½ m. N. of Huntly, Aberdeenshire.

Boghead, pl., 4 m. N.E. of New Cumnock, Ayrshire.

Bogheadly, pl., 5 m. N.E. of Stonehaven, Kincardineshire.

Boghole, ham. in co. of, and 6 m. E. of, Nairn.

Bogie, stream, with course of 14 m. before joining Deveron at Huntly, Aberdeenshire.

Bogie House, 2 m. N.W. of Kirkcaldy, Fifeshire.

Bogle Burn, stream, rising in Eildon Hills and flowing to the Tweed at Melrose.

Bogmuchals, ham. and seat, 5 m. N.E. of Fordyce, Banffshire.

Bogmuir, ham., 3 m. N. of Fochabers, Banffshire.

Bogroy, pl. in co. of, and 7 m. W. of, Inverness.

Bogside, ry. sta., L.N.E., 8 m. W.N.W. of Dunfermline, Fife. P.O.

Bogside, pl., with golf course, race course, and ry. stas., L.M.S., 1 m. N. of Irvine, N.W. Ayrshire.

Bogston, ry. sta., L.M.S., 1½ m. E. of Greenock, N. Renfrewshire.

Bogton, loch, S.E. Ayrshire, 1 m. S.W. of Dalmellington.

Bogton, vil., 5½ m. N.E. of Cupar, Fifeshire.

Bogton and Muirend, drainage area, Cathcart and Eastwood pars., Renfrewshire. Pop. 653.

Bogue, ham., 3 m. N.E. of New Galloway, Kirkcudbrightshire.

Bohally, seat, 12 m. N. of Pitlochry, Perthshire.

Boharm, par., 5 m. W. of Keith, Banffshire; 17,751 ac. Pop. 804.

4

Boisdale, sea loch and ham., S. Uist, Outer Hebrides.

Boleskine and Abertarff, par. on both sides of Loch Ness and containing Fort Augustus, Inverness-shire ; par. 110,825 ac. Pop. 1976.

Bolfracks House, seat, on r. Tay, 2 m. S.W. of Aberfeldy, Perthshire.

Bolshan, pl., 2 m. N.E. of Friockheim, Angus.

Bolton, par. and ham., 3 m. W. of Haddington, E. Lothian ; par. 3101 ac. Pop. 296.

Bona Ferry, a landing place at S. end of Loch Dochfour, and 6 m. S.W. of Inverness. Fixed white light.

Bonaly Tower, seat, within the city, and 5 m. S.W. of Edinburgh G.P.O. ; on N. slope of Pentland Hills. Burn and reservoir here.

Bonar Bridge, vil. at head of Dornoch Firth, and 13½ m. N.W. of Tain, Sutherlandshire ; ry. sta., L.M.S. P.O., T.O. Pop. 288. Hotel.

Bonawe, vil. and seat on the r. Awe, and 1½ m. N.E. of Taynuilt, Argyllshire. Has a ferry across Loch Etive. P.O., T.O. called B. Quarries. In vicinity are granite quarries.

Bonchester Bridge, ham. and hill on the Rule, and 6 m. S.E. of Hawick, Roxburghshire. P.O., T.O. B. Hill close by ; alt. 1059 ft.

Bo'ness, see BORROWSTOUNNESS.

Bonhill, par., town, and seat on the r. Leven, S. Dunbartonshire, and connected to Alexandria by a suspension bridge ; par. 8380 ac. Pop. 15,565. Calico bleaching and dyeing. P.O., T.O.

Bonjedward, ham. and seat, 2 m. N. of Jedburgh, Roxburghshire.

Bonkle, vil., 3 m. N.E. of Wishaw, N. Lanarkshire.

Bonnavoulin, vil., Morven, Argyllshire, 2 m. S.E. of Auliston Pt.

Bonnington, ry. sta., L.N.E., on N. side of Edinburgh, and 2¾ m. from Waverley sta. P.O., T.O. Paper mills.

Bonnington, seat, 2¾ m. S. of Lanark. B. Linn, the upper of the four falls on the Clyde.

Bonnington, vil. and seat, Midlothian, 1½ m. W. of Ratho.

Bonnington, or **Bonnyton,** ham., Angus, 2 m. S.W. of Arbirlot.

Bonny, small stream in cos. Dunbarton and Stirling, flowing 7 m. to the r. Carron.

Bonnybank, ham., 1 m. N.E. of Kennoway, Fifeshire.

Bonnybridge, q.s. par. and town, 4 m. W. of Falkirk, Stirlingshire ; ry. stas. on L.M.S. and L.N.E. P.O., T.O. Pop. 6036. Mfrs. cooking and heating appliances, and has large fire-brick works.

Bonnyrigg, a town and police bur. (united with Lasswade in 1929), in the pars. of Cockpen and Lasswade, Midlothian, 6 m. S.E. of Edinburgh and 2 m. S.W. of Dalkeith ; ry. stas., L.N.E. at Bonnyrigg and Broomieknowe. P.O., T.O. Carpet weaving. Pop. 4481.

Bonnyside House, seat, Stirlingshire, 1 m. E. of Bonnybridge.

Bonnyton, sub. of Kilmarnock, N. Ayrshire.

Bonnyton, seat, Ayrshire, 2½ m. Drongan sta.

Bonnyton, S.E. Angus, see BONNINGTON.

Bonnytoun, seat, W. Lothian, 1½ m. N.E. of Linlithgow.

Bonskeid, seat, on r. Tummel, and 3 m. W.N.W. of Pitlochry, Perthshire.

Boon, hill, Berwickshire, 3 m. S.E. Lauder ; alt. 1070 ft.

Boondreigh, stream, W. Berwickshire, flowing 7 m. to the Leader, 2 m. S.E. of Lauder.

Boquhan, seat, glen, and burn, 1½ m. E. of Kippen, W. Stirlingshire.

Boquhanran, eccl. par., in Clydebank, Dunbartonshire.

Bora Holm, uninhabited isl. in the Orkneys, near Milburn.

Boreland, colliery, vil., 1¼ m. N. of Dysart, Fifeshire. Pop. 317.

Boreland, vil., 7 m. N.E. of Lockerbie. Dumfriesshire. P.O., T.O.

Boreland Hill, Kirkcudbrightshire, 7 m. E. of Dalbeattie ; alt. 1632 ft.

Boreray, isl., 3 m. W. of north pt. of N. Uist, Outer Hebrides ; about 1½ m. long and ½ m. wide. Pop. 8.

Boreray, rocky isl., 4 m. N.E. of St. Kilda, Outer Hebrides ; about 2¼ m. in circumference ; alt. 1245 ft. Pasture for 300 to 400 sheep.

Borestone, sub. of St. Ninian's, Stirlingshire. Overlooks battle-grounds of Bannockburn (1314) and Sauchieburn (1488).

Borgie, a State forest and r., N. Sutherlandshire. The r. reaches the sea at Torrisdale Bay after a course of 11 m.

Borgue, par., vil., and seat on the coast, and 5¾ m. S.W., of Kirkcudbrightshire; par. 13,502 ac. P.O., T.O. Pop. 990.

Borodal, pl., 5½ m. W. of Dunvegan in Skye, Inner Hebrides.

Boroholm, ham., Dumfriesshire, in Canonbie par.

Boroughmuir, dist. on W.S.W. of Edinburgh. Formerly a wooded region, but was rapidly cleared of trees after 1504, when citizens were allowed to buy wood at a cheap rate. Here James IV. mustered and reviewed his army previous to marching to Flodden, and near by the Borestone is built into a wall. Now a populous sub. of the city. P.O., T.O. called Boroughmuirhead.

Borreraig, vil. on W. side of Loch Dunvegan, and 3 m. N. of Colbost, Skye, Inner Hebrides. P.O.

Borrobol, seat and ry. sta., L.M.S., 12¼ m. N.W. of Helmsdale, E. Sutherlandshire.

Borrodale, seat, on Loch-na-Nuagh and 35 m. N.W. of Fort William. Prince Charles Edward landed here 25th July 1745.

Borrowstounness and Carriden, (more commonly Bo'ness), par., police bur., and seapt., with an excellent harb. on S. shore of Firth of Forth. W. Lothian, and 24 m. W.N.W. of Edinburgh by rail; ry. sta. (Bo'ness), L.N.E. Two fixed green lights on pier heads. Shipbreaking. Exports coal, bricks, and pottery; imports timber. P.O., T.O.; par. 5939 ac. Pop. 14,098; pop. of bur. 10,095.

Borthwick, par., ham., and stream, S.E. Midlothian, 12¾ m. S.E. of Edinburgh; ry. sta. at Fushiebridge, L.N.E. B. Castle associated with Mary Queen of Scots, as she lived there for three weeks after her marriage with Bothwell. The renovated castle is still occupied. Par. 9376 ac. Pop. 3169.

Borthwick, stream, cos. of Selkirk and Roxburgh, flows 16 m. to Teviot, which it joins 2 m. above Hawick.

Borthwickbrae and **Borthwickshiels,** two seats in W. Roxburghshire and 1½ m. S.W. of Roberton. P.O., T.O. at Borthwickbrae Burnfoot.

Borthwick Hall, seat, S.E. Midlothian, 2 m. S.W. Heriot sta.

Borve, ham., 14 m. N.W. of Stornoway, Lewis Isl., Outer Hebrides. P.O., T.O. B. Lodge, seat of the Earl of Dunmore.

Borve, dist., Inverness-shire, Snizort par., Skye.

Borve, ancient castle, Outer Hebrides, W. side Benbecula Isl.

Borve, seat and stream, Outer Hebrides, W. side of Harris.

Borve, ham., 1½ m. N. of Castlebay, Barra Isl., Outer Hebrides. P.O.

Bothkennar, q.s. par., Falkirk and Grangemouth pars., 4 m. N.E. of Falkirk. Stirlingshire. P.O., T.O.

Bothwell, par. and town on r. Clyde in Lanarkshire, 8 m. S.E. of Glasgow. Bothwell Bridge is famous as the scene of encounter between Monmouth and the Covenanters, 1679. Ry stas., L.M.S. and L.N.E. Coal mines; iron and steel works. P.O., T.O. Par. 13,582 ac. Pop. 60,660; town pop. 3890. B. Castle, a seat of the Earl of Home.

Bothwell Park, vil., 1 m. N. of Bellshill, Lanarkshire. Pop. 3884.

Botriphnie, par. (9461 ac.), 6 m. S.W. of Keith, Banffshire. P.O. Pop. 646.

Bound Skerries Lighthouse, one of the Outer Skerries, E. Shetland; has white flashing light, visible 18 m.

Bourtie, par. (5693 ac.) and vil., 2 m. S.W. of Old Meldrum, E. Aberdeenshire. Pop. 421.

Bourtie House, Aberdeenshire, 1½ m. N.E. of Inverurie.

Bourtree Bush, vil., 7 m. N. of Stonehaven, Kincardine.

Bowden, par. and vil., 1 m. W. of St. Boswells ry. sta., L.N.E., Roxburghshire; 7665 ac. Pop. 561; vil. 136. P.O.

Bower, par. and ham. with sta. L.M.S., in N.E. Caithness; par. 19,684 ac. Pop. 1059. Ham. is 11½ m. N.W. of Wick. P.O., T.O.

Bower, ancient castle on r. Clyde, near Roberton.

Bowershall, vil., 2½ m. N. of Dunfermline, Fife.

Bowhill, colliery vil., 1 m. S. of Auchterderran, S.W. Fife.

Bowhill, seat of the Duke of Buccleuch, 4 m. S.W. of Selkirk.

Bowholm, vil., Dumfriesshire, close to Canonbie.

Bowhouse, vil., 4½ m. N.E. of Wishaw, Lanarkshire.

Bowhouse, ry. sta., L.N.E. (passenger service withdrawn), S.E. Stirlingshire, 2 m. S.W. of Manuel.

Bowhousebog, or **Liquo,** pl., Lanarkshire, 2½ m. N.E. of Newmains.

Bowland, seat and ry., sta. L.N.E., on Gala Water, Midlothian, and 3 m. S.E. of Stow. P.O.

Bowling, vil. on r. Clyde, in S.E. Dunbartonshire, and 3¼ m. E.S.E. of Dumbarton. Occulting red light on pier head; ry. stas., L.M.S. and L.N.E. P.O., T.O. Pop. 1245.

Bowmore, seapt. on Loch Indal, Islay Isl., Argyllshire. P.O., T.O. Pop. 620. Has a distillery. Regularly called at by steamers from Glasgow.

Bow of Fife, pl., 3½ m. W. of Cupar, Fife. P.O.

Boyndie, par., on coast of Banffshire, and 3 m. W. of Banff; 6950 ac. Pop. 2014. P.O. at Whitehills. Par. contains county lunatic asylum. R. Boyndie, short stream flowing into Banff Bay.

Boyndlie, pl. and seat, 6 m. W. by S. of Fraserburgh, N.E. Aberdeenshire. P.O.

Bracadale, par., ham., and sea loch, Skye, Inner Hebrides; par. 93,164 ac. Pop. 1120; ham. 10 m. S.W. of Portree.

Bracara, or **Bracora,** ham. on N. shore of Loch Morar and 5 m. S.E. of Mallaig, S.W. Inverness-shire. P.O.

Brackla, seat and distillery, 3½ m. S.W. of Nairn.

Brackletter, vil., 7½ m. N.E. of Fort William, Inverness-shire.

Bracklinn Falls, on Keltie Water, 1½ m. N.E. of Callander, Perthshire.

Brack, The, mt., alt. 2580 ft.; 3½ m. W.S.W. of Arrochar, Argyllshire.

Braco, vil., 7½ m. S. of Crieff, S. Perthshire. P.O., T.O. Pop. 130. Braco Castle is 1½ m. N.W.

Bradan, loch, the source of Troon water supply, 6 m. S.S.E. of Straiton, Ayrshire.

Brae, ham., 22 m. N.W. of Lerwick, Shetland. P.O., T.O.

Braehead, group-flashing light on S. bank r. Clyde, ½ m. W. of Whiteinch, Glasgow.

Braehead, vil., 1 m. S.S.W. of Coatbridge, Lanarkshire.

Braehead, vil., 3½ m. N.W. of Carnwath, E. Lanarkshire. P.O.

Braeheads, special water dist., Lasswade par., Midlothian. Pop. 156.

Braemar, q.s. par. and vil., 16½ m. W. by S. of Ballater, S.W. Aberdeenshire. P.O. T.O. Capital of Deeside; health resort and tourist centre. Pop. 440. Hotels. Has a meteorological observatory. B. Castle, a seat.

Braemore, seat and deer forest, Ross and Cromarty, 12 m. S.E. of Ullapool.

Braemore, ham. and deer forest, Caithness, 8 m. N.W. Berriedale.

Braendam House, Perthshire, 1¾ m. N.W. of Thornhill.

Braeriach, mt., alt. 4248 ft., one of the Cairngorm group, on borders Aberdeenshire and Inverness-shire.

Braeroy, seat, Inverness-shire, 9 m. N.E. of Roy Bridge.

Braeside Halt, ry sta., L.N.E., between Dumferline and Charlestown, Fife.

Braes of Abernethy, dist. of Grampians, Inverness-shire.

Braes of Balquhidder, see BALQUHIDDER.

Braes of Coul, pl., 7 m. W.N.W. of Kirriemuir, Angus. P.O.

Braes of Doune, Perthshire, 3 m. N.W. of Doune.

Braes of Gleniffer, hills, 3 m. S.W. of Paisley, Renfrewshire; reservoirs for Paisley here.

Braes The, Isle of Skye, the whole dist. between Loch Sligachan and Camastianavaig. See also THE BRAES.

Bragar, North and **South,** vils. in Lewis Isl., Outer Hebrides, and 5½ m. S.W. of Upper Barvas. P.O., T.O.

Brahan Castle, Ross and Cromarty, 4 m. S.W. of Dingwall and 1¾ m. above Conan Bridge. Here General Wade received the submission of the Mackenzies, 1725.

Braid Burn, flows from Pentland Hills to Firth of Forth at Portobello, Edinburgh.

Braid Cairn, hill borders of cos. Aberdeen and Angus, 6 m. S.E. of Ballater; alt. 2807 ft.

Braid Hills, hills, hotel, and golf course within and on S. side of Edinburgh.

Braidon Bay, E. Kincardineshire, 5 m. N.E. of Bervie.

Braidwood, vil., seat, and ry sta., L.M.S., 1¼ m. S.E. of Carluke by rail, mid Lanarkshire. P.O., T.O. at B. sta. Pop. 195.

Braigh Sron Ghorm, mt., Perthshire, 8 m. N. of Blair-Atholl; alt. 2882 ft.

Bran, stream, Ross and Cromarty, flows into Loch Luichart.

Branahuie Bay, Isl. of Lewis, 2½ m. E. of Stornoway.

Branault, ham., Mull Isl., Argyll, 1 m. S. of Kilmory.

Branderburgh, town, now included in Lossiemouth, Morayshire.

Brander, Pass of, famous pass and battle site between Lochs Etive and Awe, Argyll, 7 m. W. of Dalmally.

Brandy, or **Branny,** loch, 16 m. N.W. of Kirriemuir, Angus; alt. 2000 ft.

Brankston Grange, seat, Fife, 3 m. N. of Culross.

Bransly Hill, E. Lothian, 6 m. S. of Dunbar; alt. 1300 ft.

Branxholm, ancient feudal castle, now modernised, standing 3 m. S.W. of Hawick, Roxburghshire; early residence of the Buccleuchs; figures in Scott's "Lay of the Last Minstrel."

Breadalbane, dist. amid the Grampians in W. Perthshire, an elevated and wild tract scored by deep glens with numerous lochs and mountain torrents; extends from Lochaber and Atholl on the N. to Strathearn and Menteith on the S.

Breakish, ham., 5 m. W. of Kyleakin, Skye, Inner Hebrides. P.O.

Breakish, Kyle of Lochalsh, W. Ross and Cromarty. P.O., T.O.

Breascleit, vil., 2 m. N.N.W. of Callernish in S.W. of Lewis Isl., Outer Hebrides; has flashing light (Greinam Isl.), visible 10 m.

Brechin, par., 13,880 ac., parl., royal, and mun. bur., and mkt.-town on r. S. Esk, and 9½ m. by rail W. of Montrose, Angus. Has twelfth century cathedral (now par. church); mfrs. sailcloth and brown linen; spinning, bleaching, and distilling; ry. sta., L.M.S.; P.O., T.O.; the town unites with Montrose, Forfar, Arbroath, and Bervie in returning one member to Parliament. Pop. of par. 8201; pop. of bur., 6838. Hotel. B. Castle, a seat of the Earl of Dalhousie.

Breich, pl., 20¼ m. S.W. of Edinburgh, Midlothian; ry. sta. (alt. 710 ft.), L.M.S. P.O.

Bressay, one of the Shetland Isls.; B. Sound, a harbour between Bressay and mainland. The lighthouse has alternate flashing light, visible 16 m.; fog-siren. P.O., T.O. Pop. 1452.

Brettabister, pl., 10 m. N.N.E. of Lerwick, mainland, Shetland. P.O., T.O.

Brewhead, vil., 4½ m. N.W. of Dundee, Angus.

Briach, ham., N.W. Morayshire, 5½ m. S.E. of Forres.

Brick-Kiln, ham., 1 m. from Canonbie, Dumfriesshire.

Bridesness, prom., N. Ronaldshay, Orkney, 5 m. N. of Start Point.

Bridgefoot, ham., with halt sta., L.N.E., Banffshire, near Banff.

Bridgefoot, ham. on S. border of Angus; post-town Auchterhouse.

Bridgend, vil. at head of Loch Andail and 3 m. N.E. of Bowmore, Islay, Argyllshire; P.O., T.O. Regular steamboat communication with Glasgow.

Bridgend, vil., Ayrshire, 1 m. S. of New Cumnock.

Bridgend, ham., Berwickshire, on r. Leader.

Bridgend, vil., Berwickshire, close to Duns.

Bridgend, vil., Midlothian, near Dalkeith.

Bridgend, vil., on r. Leven and suburbs of Dumbarton. P.O., T.O.

Bridgend, vil., Fife, suburb of Ceres.

Bridgend, vil., in Angus, 3 m. E. of Alyth.

Bridgend, ham., Angus, 6 m. W. of Kirriemuir.

Bridgend, ham., Inverness-shire, near Beauly.

Bridgend, vil., on W. side of Bathgate, West Lothian.

Bridgend, an E. suburb of Perth, on r. Tay. P.O., T.O.

Bridgend, vil., S. Perthshire, on r. Earn, near Crieff.

Bridgend, vil., Perthshire, W. of Kenmore.

Bridgend, ham., Renfrewshire, near Lochwinnoch.

Bridgend, ham., N.W. Roxburghshire, near Melrose.

Bridgend, ham. in the Liberton dist. of Edinburgh, adjoining Cameron Bridge.

Bridgend Hyndford, ham., Lanarkshire, on right bank of the r. Clyde, 2½ m. S. of Lanark.

Bridgend of Alness, vil. in par. of Rosskeen, E. Ross and Cromarty. Pop. 617.

Bridgeness, vil. on Firth of Forth, and 1 m. E. of Bo'ness, W. Lothian; one green and two red lights, visible 5 m. P.O., T.O.

Bridge of Allan, q.s. par. and town (police burgh) on the Allan Water, $2\frac{1}{2}$ m. N.W. of Stirling, Stirlingshire. A residential town; mineral springs attract visitors; ry. sta., L.M.S. P.O., T.O. Pop. of bur., 2897. Hotels.

Bridge of Avon, $1\frac{1}{2}$ m. S.E. of Ballindalloch, Banffshire. P.O.

Bridge of Cally, $4\frac{1}{2}$ m. N.W. of Blairgowrie, Perthshire. P.O., T.O.

Bridge of Canny, 3 m. W. of Banchory, Kincardineshire. P.O.

Bridge of Dee, vil. on r. Dee and 3 m. S.W. of Castle-Douglas, Kirkcudbrightshire; ry. sta. (alt. 158 ft.), L.M.S., P.O., T.O.

Bridge of Dee, ham. in co., 2 m. S., of Aberdeen. P.O.

Bridge of Don, coastguard sta. on the Don, and 2 m. N. of Aberdeen. P.O. T.O. Also called Brig o' Don.

Bridge of Dun, ry. sta. and junction on L.M.S. and 4 m. E. of Brechin, Angus. P.O., T.O.

Bridge of Earn, vil. on the Earn, $4\frac{1}{2}$ m. S.E. Perth, ry. sta., L.N.E.; holiday place. P.O., T.O. Pop. 544. Hotel.

Bridge of Feugh, pl., Kincardineshire, close to Banchory.

Bridge of Fiddich, near Dufftown, Banffshire. P.O.

Bridge of Gairn, 2 m. N. of Ballater, Aberdeenshire. P.O.

Bridge of Garry, 3 m. N.W. of Pitlochry, Perthshire. Here the Garry is spanned by a picturesque bridge just above its junction with the Tummel at the south end of the Pass of Killiecrankie.

Bridge of Gaur, ham. at W. end of Loch Rannoch, W. Perthshire. P.O., T.O.

Bridge of Marnoch, 2 m. S.W. of Aberchirder. P.O., T.O.

Bridge of Oich, pl., Inverness-shire, on Loch Oich, 2 m. from Invergarry.

Bridge of Orchy, pl., 12 m. N.E. of Dalmally, E. Argyllshire, ry. sta. (alt. 618 ft.), L.N.E. P.O., T.O. Hotel.

Bridge of Roy, ham. and hotel, 13 m. N.E. of Fort William, Inverness-shire, ry. sta. (R. Bridge), L.N.E. P.O.

Bridge of Tilt, vil. near Blair-Atholl ry. sta., Perthshire, 8 m. N.W. of Pitlochry. Pop. 130.

Bridge of Turk, ham., 7 m. W. of Callander, near Loch Vennachar, Perthshire. P.O., T.O. Called Brig o' Turk.

Bridge of Urr, vil., 4 m. N. of Castle Douglas, Kirkcudbrightshire. P.O.

Bridge of Walls, pl., 24 m. N.W. of Lerwick, Shetland. P.O., T.O.

Bridge of Weir, q.s. par. and town, 6 m. N.W. of Paisley, Renfrewshire; ry. sta. (alt. 157 ft.), L.M.S. P.O., T.O. Pop. 2740. Quarrier Orphan Homes in vicinity.

Bridge of Westfield, pl., 5 m. S.W. of Thurso, Caithness. P.O.

Bridge Street, ry. sta., Glasgow dist. sub.

Bridgeton, q.s. par. in S.E. Glasgow; ry. stas., L.M.S. and L.N.E. P.O., T.O.

Bridgeton, parl. division, city of Glasgow. Pop. 70,213.

Bridgeton, vil. in Perthshire, near Almondbank. Has remains of a Roman camp. Pop., with Almondbank, 301.

Brig o' Turk, see BRIDGE OF TURK.

Brig o' Turk Scottish Youth Hostel, Creag Dhu Hostel, in the Trossachs, between Loch Achray and Loch Vennachar, Perthshire, 7 m. W. of Callander.

Brindister, vil., prom., and bay, 1 m. E. of Burrafirth Holm, mainland, Shetland. Inner part of the bay known as Unifirth.

Broad Cairn, mt., alt. 3268 ft., 8 m. S.E. of Braemar, on borders of Angus and Aberdeenshire.

Broadford, fishing vil., stream, and bay, 8 m. W. of Kyleakin, Skye, Inner Hebrides. Pop. 157. Hotels. P.O., T.O. White fixed light on pier head, visible 5 m. A centre for tourists exploring the Coolin Mts. Regular steamboat communication with Kyleakin, Portree, Oban, Glasgow, etc.

Broadmeadows, former seat, now an hotel, in co. of, and $4\frac{1}{2}$ m. W. by N. of, Selkirk, on the Yarrow. There is also a Scottish Youth Hostel here.

Broadsea, fishing vil., included in Fraserburgh, N.E. Aberdeenshire. P.O.

Broadstone, Renfrewshire, 2 m. E. of Port Glasgow. Hospital.

Brodick, q.s. par., bay, and vil., with pier on E. side of Arran Isl., Buteshire, and lying 4 m. N.W. of Lamlash. A favourite watering-place. Hotel. P.O., T.O. B. Castle on the N. shore of bay is mainly modern, but incorporates part of an old stronghold. Much of the isl. is given over to deer forests. Pop. 510. Regular steamboat communication with Ardrossan.

Brodie, ry. sta., L.M.S., 3½ m. N.W. of Forres, Moray ; P.O., T.O. Nearby is B. Castle.

Brookfield, vil., 1 m. N.W. of Kilbarchan, Renfrewshire. P.O., T.O. Pop. 286.

Brooklea, Broughty Ferry, Angus. Girls' private school.

Broomhall, the seat of the Earl of Elgin, S.W. Fifeshire, on Firth of Forth, 2½ m. S.W. of Dunfermline.

Broomhall, vil., E. Perthshire, 2 m. S.W. of Longforgan.

Broomhall, pl., Angus, 4 m. N. of Alyth.

Broomieknowe, vil., with ry. sta., L.N.E., near Lasswade, Midlothian, 6 m. S.E. of Edinburgh. Golf course.

Broomielaw, that part on the N. side of Glasgow harb. between Jamaica Bridge and York Street. P.O., T.O.

Broomlands, vil., Inchinnan par., Renfrewshire. Pop. 109.

Broomlee, seat and ry. sta. for West Linton, L.N.E. (passenger service withdrawn), N. Peeblesshire, 6½ m. S.W. of Leadburn.

Brora, vil., loch., riv., and bay, 6¼ m. N.E. of Golspie by rail, Sutherlandshire ; ry. sta., L.M.S. P.O., T.O. Holiday place with boating, fishing, golf (most northerly coal mine in Scotland). Brick and tile works. Pop. 783.

Brough, fishing vil., 11 m. by road, N.E. of Thurso, Caithness. P.O.

Brough, ham. in N.W. Whalsey, Shetland.

Brough Lodge, seat in Fetlar Isl., Shetland.

Brough Ness, headland, S. pt. of Ronaldshay Isl., Orkney.

Brough of Birsay, small isl. off N.W. mainland, Orkney. P.O. T.O. Called B. Light Station.

Broughton, vil. in co. of, and 11 m. S.W. of, Peebles ; ry. sta. (636 ft.), L.M.S. P.O., T.O.

Broughton, Glenholm, and Kilbucho, united pars., Peeblesshire, containing Broughton vil. ; area 19,797 ac. Pop. 632.

Broughty Ferry, q.s. par. and town annexed to Dundee in 1913, 4 m. E. of Taybridge sta. (L.M.S. and L.N.E.), Dundee. P.O., T.O. A much-frequented holiday resort, with good bathing facilities. Railway ferry to Tayport. Lifeboat sta. The fifteenth-century B. Castle stands on the shore.

Broulin Lodge, Inverness-shire, in Glen Strath Farrar.

Brouster Loch, and r., mainland, Shetland. R. falls in B. Voe.

Brow, coast ham., Dumfriesshire, near Ruthwell.

Browhouses, vil. and bay, Dumfriesshire, 5½ m. E. of Annan.

Browndean Laws, two peaks, S. Roxburghshire, 5 m. S.E. of Oxnam ; alts. 1358 and 1309 ft.

Brownhills, ham. near St. Andrews, Fife.

Brownie's Taing, pier, with fixed red light, visibility 5 m., Shetland.

Brown Knowe, mt., Selkirkshire, 4 m. S.E. of Innerleithen ; alt. 1718 ft.

Brownlee, seat, Lanarkshire, near Carluke.

Broxburn, q.s. par. and town in W. Lothian, and 10¼ m. W. of Edinburgh ; ry. sta. is 1 m. S. at Drumshoreland, L.N.E. P.O., T.O. Pop. 7163. Coal and oil-shale mines.

Broxmouth, ham., E. Lothian, at mouth of Broxburn R., 1½ m. S.E. of Dunbar. B. Park, a seat of the Duke of Roxburgh.

Bruach na Frithe, one of the summits in the Cuillin Hills, Isle of Skye ; alt. 3143 ft.

Bruan, ham. and seat, 8 m. S.W. of Wick, Caithness.

Bruar, pl. and stream, with falls, highest 200 ft., flowing into Garry, 3 m. W. of Blair-Atholl, N. Perthshire.

Brubster, vil., N.W. Caithness, 8½ m. S.W. of Thurso.

Brucefield, adjoining Dunfermline, Fife. P.O.

Brucefield, seat in co., and 2 m. E., of Clackmannan.

Brucehaven, har., Firth of Forth, 3 m. S. of Dunfermline, Fife.

Brucehill, seat, Perthshire, near Port of Menteith ry. sta.

Brucklay, vil., seat., and ry. sta., L.N.E., 4 m. S. of Strichen by rail, Aberdeenshire. P.O., T.O. at sta.

Bruichladdich, pl., with distillery, in W. of Isl. of Islay, Argyllshire. Fixed white light, visible at 5 m. P.O., T.O.

Brunton, vil., 6 m. N.E. of Newburgh, Fife. P.O.

Bruntsfield Links, public space in S.W. Edinburgh ; putting greens and short-hole golf.

Bruray Skerries, isl. group, Shetland, E. of Housay. Pop. 34.

Brux, seat of Lord Forbes, 7 m. W. of Alford, Aberdeenshire.

Brydekirk, q.s. par. and vil. $2\frac{1}{2}$ m. N. of Annan, Dumfriesshire. P.O., T.O.

Buachaille Booshala, or Herdsman, basaltic isl. off S. of Staffa, Argyll.

Buachaille Etive, conspicuous mt. at the head of Loch Etive; alt. 3345 ft.

Bualintur, ham., at the head of Loch Brittle, S. Isle of Skye.

Bualnaluib, ham., near Gairloch, W. Ross and Cromarty.

Buccleuch, a cleugh in the glen of Rankle burn, 1 m. S.W. of Selkirk; gives the title of Duke to a branch of the Scott family.

Buccleuch, q.s. par. of Edinburgh.

Buchan, dist. extending for 40 m. from the Ythan to the Deveron in N.E. Aberdeenshire. Chief towns are Peterhead and Fraserburgh. Bullers of Buchan, natural rocky cauldron, 6 m. S. of Peterhead.

Buchanan, par., in W. Stirlingshire, 4 m. N.W. of Drymen; area, 41,644 ac. Pop. 438. Contains Buchanan Castle, former seat of Duke of Montrose, $1\frac{1}{4}$ m. W. of Drymen, and now a hotel.

Buchan, Bullers of, see BULLERS OF BUCHAN.

Buchanhaven, fishing vil., 1 m. E. of Peterhead, Aberdeenshire.

Buchan Ness, most easterly pt. on mainland of Scotland, 3 m. S. of Peterhead. Lighthouse with flashing light, visible 17 m.

Buchanty, vil. on r. Almond, and 10 m. N.E. of Crieff, Perthshire. P.O.

Buchany, vil., Perthshire, $1\frac{1}{4}$ m. N.W. of Doune.

Buchlyvie, q.s. par. and vil, $4\frac{1}{4}$ m. N.N.E. of Balfron, Stirlingshire; ry. sta., L.N.E. P.O., T.O. Vil. pop. 349. Hotel.

Bucholie, ruined castle, 4 m. N.E. of Keiss, Caithness.

Buchromb, seat, Banffshire, N.W. of Dufftown.

Buchtrig, seat, Berwickshire, 4 m. N.W. of Coldstream.

Bucinch, very small wooded islet in Loch Lomond, N. of Inchcruim, Stirlingshire.

Buckhaven, forms with Methil a united police bur., on Firth of Forth, $1\frac{1}{4}$ m. S.W. of Methil, with ry. sta. on L.N.E. Has a good harbour and pier with fixed light, visible at 6 m. Fishing and making of nets and cordage. Lifeboat sta. One of the Kirkcaldy dist. burs. for Parl. P.O., T.O. Bur. pop. 17,643.

Buckie, q.s. par., town, and police bur., 6 m. W. of Cullen, Banffshire. Two harbours with two fixed red lights, visible 13 m. and 5 m.; one white, visible 11 m.; and one green, visible, 5 m. Ry stas., L.N.E. and L.M.S. Fishing and fish curing; net and rope making. P.O., T.O. Pop. of bur. 8688. Hotels.

Bucklerheads, ham., 1 m. N. of Murroes, Angus.

Buck of Cabrach, mt., on borders of Aberdeen and Banff; alt. 2368 ft.

Buckpool, ry. sta., L.N.E., 1 m. W. of Buckie, Banffshire. P.O.

Buckreddan, Kilwinning, Ayrshire. County hospital.

Bucksburn (Auchmull), vil. in co. of, and $4\frac{1}{4}$ m. N.W. of, Aberdeen; ry. sta., L.N.E. P.O., T.O. Has large paper mills. Bucksburn House is 1 m. S.W.

Buddon Ness, prom. on N. side Firth of Tay, Angus; 3 fixed lights.

Bught Hill, S. Lanarkshire, $7\frac{1}{2}$ m. S.W. of Douglas; alt. 1481 ft.

Bught House, 1 m. S.S.W. of Inverness.

Buidhe Mor, loch, N.E. Sutherland, 4 m. S. of Armadale.

Buie, or Buidhe, loch, S.E. Sutherland, $7\frac{1}{2}$ m. N.N.E. of Bonar Bridge ry. sta.

Builg, loch (86 ft deep) and stream, S. Banffshire, 3 m. S. of Inchrory.

Builg, The, hill, Kincardineshire, 7 m. N.W. of Drumlithie, 1291 ft.

Buittle, par. in S.E. Kirkcudbrightshire, 3 m. S.E. of Castle-Douglas; area 11,446 ac. Pop. 778.

Bulg, hill, Angus, 6 m. N.W. of Edzell, 1986 ft.

Bull, loch, Buteshire, 9 m. N.W. of Rothesay.

Bulldoo, ham., N.W. Caithness, $2\frac{3}{4}$ m. N.E. of Reay.

Bullers Buchan, fishing vil., E. Aberdeenshire, near Bullers of Buchan.

Bullers of Buchan, huge rocky cauldron, E. coast of Aberdeenshire, 6 m. S. of Peterhead.

Bullionfield Paper Mills, $3\frac{1}{2}$ m. W. of Dundee, Angus.

Bulvicar, bay, Sell Isl., W. Argyll.

Bunacharan, loch, Ross and Cromarty, 6 m. W. of Struy; alt. 367 ft., depth 113 ft.

Bunachton, loch, 6½ m. S. of Inverness; alt. 701 ft.

Bunaveneader, pl. with school, Harris, Inverness-shire, 3 m. N.W. of Tarbert.

Bunchrew, pl. in co. of, and 3½ m. W. of, Inverness; ry sta., L.M.S. P.O., T.O.

Bundalloch, fishing vil., S.W. Ross and Cromarty, N.E. shore of Loch Long, 10 m. N.E. of Kyle of Lochalsh.

Bu Ness, peninsula, in Fair Isl., Shetland.

Buness, ham., Unst Isl., near head of Balta Sound, Shetland. House of B., a seat.

Bunessan, vil. on S.W. of isl. of Mull, Argyllshire. Hotel. P.O., T.O. White flashing light on Liathanach Isl. on Lathaich Loch. P.O., T.O. Steamers from Glasgow call regularly.

Bunkle and Preston, par. in N.E. Berwickshire, 5 m. N.E. of Duns; 9194 ac. Pop. 507. B. Castle is a ruin.

Bunloit, ham. on W. shore of Loch Ness, and 3 m. S.W. of Urquhart Bay, Inverness-shire.

Bunnahabhain, pl., with distillery, in N. of Islay, Argyllshire, and 2½ m. N. of Port Askaig. P.O., T.O., called Bonahaven.

Buntait Hostel, in Glen Urquhart, W. of Loch Meikle, and near Bearnock, Inverness-shire. A Scottish Youth Hostel.

Burdiehouse, ham. in S. of city of Edinburgh, 4½ m. S. of G.P.O. Limekilns.

Burdsyards, seat, Morayshire, 1¾ m. S.E. of Forres.

Burg, head, Kilfinichen par., S.W. of Mull, Argyll

Burgar, seat, Orkney, 11 m. N.W. of Kirkwall.

Burga Water, loch, 3 m. N.N.W. of Walls, mainland, Shetland.

Burghead, q.s. par., police bur., seapt., and coastguard sta., 9 m. W. of Elgin, Morayshire. Prom. has lighthouse, with fixed white light, visible 10 m., and two red lights, visible 5 m.; ry. sta., L.M.S. P.O., T.O. Pop. of bur. 1255.

Burgie, ham. and seat, N.W. Morayshire, 3¼ m. E. of Forres. Ruins of B. Castle in vicinity.

Burleigh Castle, Kinross-shire, ½ m. E. of Milnathort. Ruined baronial edifice of the family of Balfour.

Burn Anne, stream, Ayrshire, flows into the Irvine near Galston.

Burnbane, vil., E. Perthshire, S.E. of Dunkeld.

Burnbank, q.s. par., vil., and ry. sta., L.N.E., Hamilton par., Lanarkshire. P.O., T.O.

Burnbank, fishing vil., N.E. Kincardineshire, 3 m. S. of Aberdeen.

Burnbrae, pl., Perthshire, 3 m. from Methven ry. sta. P.O.

Burnbrae, vil., 3¼ m. from Tarbolton, Ayrshire.

Burnbrae, vil., 1 m. S.W. of Shotts ry. sta., Lanarkshire.

Burnbrae, seat, 1½ m. E.N.E. of Johnstone, Renfrewshire.

Burnbutts, vil., N.W. Lanarkshire, near Glasgow.

Burness, pl. in N. Sanday, Orkney.

Burness, seat, 2 m. N.E. of Finstown, Orkney.

Burness, loch, in Westray Isl., Orkney.

Burnfoot, ham., with small harb., Kirkcudbrightshire, 5½ m. S.W. of Kirkcudbright.

Burnfoot, small harb., at head of Luce Bay, Wigtownshire.

Burnfoot, seat, 1 m. E. of Ecclefechan, Dumfriesshire.

Burnfoot, pl., W. Lothian, 1 m. E.S.E. of Bridgeness.

Burnfoot, ham., 4 m. N. of Dollar, Perthshire.

Burnfoot, ham., 8 m. S.W. of Paisley, Renfrewshire.

Burnfoot, ham. and seat, 3 m. N.W. of Langholm, Dumfriesshire.

Burnfoot, vil., 5½ m. S.W. of Hawick, Roxburghshire.

Burnfoot, reservoir, Ayrshire, 2 m. E.S.E. of Stewarton.

Burnfoothill, vil., E. Ayrshire, 4 m. N.W. of Dalmellington. Has ironworks. P.O. at Lethanhill.

Burnhaven, vil., Aberdeenshire, 1¾ m. S.W. of Peterhead.

Burnhead, vil., 1¼ m. W. of Thornhill, Dumfriesshire. P.O.

Burnhervie, pl., 3 m. from Kenmay, Aberdeenshire. P.O.

Burnhouse, vil., N. Ayrshire, 3 m. S.E. of Beith.

Burnhouse, seat, Midlothian, 3½ m. N.W. of Stow.

Burnmouth, fishing vil. and coastguard sta. in Berwickshire, and $5\frac{1}{2}$ m. N.W. of Berwick-on-Tweed. Pop. 49. P.O., T.O.

Burnmouth, ham., S.E. Roxburghshire, 3 m. S. of Newcastleton.

Burnoch, affl. of Lugar Water, at Ochiltree, mid Ayrshire.

Burn of Cambus, pl. in Perthshire, 10 m. N.W. of Stirling.

Burn of the Vat, an enormous chasm, resembling a vat, 5 m. E. of Ballater, Aberdeenshire.

Burn Row, vil., E. Stirlingshire, 1 m. S. of Slamannan.

Burnside, pl. near Philpstoun, W. Lothian. P.O., T.O. Pop. 123.

Burnside, vil., $1\frac{1}{2}$ m. N. of Dalry, Ayrshire.

Burnside, vil., $\frac{1}{2}$ m. S.E. of Newburgh, N.W. Fife.

Burnside, vil., sub. of Dumbarton.

Burnside, ham., 6 m. N.W. of Forfar, Angus.

Burnside, ry. sta., L.M.S., Lanarkshire, $5\frac{1}{2}$ m. E. of Glasgow. P.O.

Burnside, seat, E. Perthshire, 2 m. S.W. of Alyth.

Burnside, seat on S. side of Fochabers, Morayshire.

Burnside, ham., Banffshire, 2 m. N.W. of Cornhill sta.

Burnside of Duntrune, ham., 4 m. N.E. of Dundee, Angus.

Burn, The, seat, S.W. Kincardineshire, $1\frac{1}{2}$ m. N. of Edzell.

Burntisland, par., royal and parl. bur., seapt., and holiday centre, on the N. shore of the Firth of Forth, 5 m. by ferry from Granton and $20\frac{1}{4}$ m. by rail from Edinburgh; ry. sta., L.N.E. On the inner and outer heads of the E. pier three fixed red lights, and on outer heads of W. pier, two fixed green lights, all visible at 6 m. Shipbuilding, aluminium works, Exports coal; par. 2606 ac. P.O., T.O. Burgh unites with Kinghorn, Buckhaven, Methil, and Innerleven, Dysart, and Kirkcaldy in sending one member to Parliament. Pop. of par. 5809; pop. of bur. 5389.

Burnt Isles, off N. coast of Bute, shows a white occulting light, visible 5 m.

Burntown, vil. near Gargunnock, N. Stirlingshire.

Burntshiels, pl., Renfrewshire, $1\frac{3}{4}$ m. S.W. of Kilbarchan.

Burnturk, vil., $1\frac{1}{2}$ m. S.E. of Kettle, Fife.

Burnwynd, ham., $1\frac{1}{2}$ m. S. of Ratho, Midlothian.

Burra, East and **West,** two of the Shetland isls., 8 m. S.W. of Lerwick. P.O. Pop. E. Burra Isl. 176; W. Burra Isl. 591.

Burrafrith, ham. and bay, N. Unst Isl., Shetland.

Burraland, pl. and loch, 5 m. S. of Collafirth, mainland, Shetland.

Burraness, prom. on E. side of Yell Isl., Shetland.

Burrastow, seat, on Vaila Sound, mainland, Shetland.

Burra Voe, ham. and bay at S.E. end of Yell Isl., Shetland. P.O., T.O.

Burray, one of the Orkneys, lying between the mainland and S. Ronaldshay. P.O., T.O. Fixed white light, visible 4 m. Pop. 379.

Burrelton and Woodside, vil., $2\frac{1}{2}$ m. S.W. of Coupar-Angus, Perthshire; ry. sta., L.M.S. P.O. Pop. 454.

Burron, hill, near Byeloch, S.W. Dumfriesshire. Has remains of an ancient Caledonian camp.

Burrow Head, prom. at S.E. extremity of Wigtownshire; alt. 150 ft.

Burrow Head, headland, S.E. Stronsay Isl., Orkney.

Burwick, ham. at S. end of S. Ronaldshay Isl., Orkney. P.O.

Busbie Castle, ruin, Ayrshire, 2 m. S.W. of Kilmaurs.

Busby, vil. on the White Cart R., $7\frac{1}{4}$ m. S. of Glasgow, ry. sta., L.M.S. P.O., T.O. Pop. 724.

Busby, moor, N.W. Ayrshire, 5 m. N. of Ardrossan.

Bush, seat, Midlothian, 3 m. N.E. of Penicuik.

Bushy Fell, hill, Roxburghshire, 4 m. S. of Hounam; alt. 1580 ft.

Busta, seat and bay, Delting par., Shetland.

Bute, isl. in Firth of Clyde, separated from Argyllshire by the Kyles of Bute; about 18 m. long by an average of 5 m. broad. North part of the isl. is hilly but southern part is suitable for agricultural and pastoral activities; salubrious climate makes the island suitable as a holiday resort; the royal burgh of Rothesay is the chief town. Pop. 12,112.

Bute, Kyles of, see KYLES OF BUTE.

Bute, North, par., Bute Isl., 18,569 ac. Pop. 1828. Includes Inchmarnock Isl.

Buteshire, composed of the isls. of Bute, Arran, Great and Little Cumbrae, Holy Isl., Pladda, and Inchmarnock, and covering 139,658 ac. of land and 794 of water ; twenty-ninth in size, and thirtieth in pop. of the cos. of Scotland. The rivers are short and the only hill of note is Goatfell, in Arran. Chief industries—agriculture and fishing. The counties of Ayr and Bute return three members to Parliament. Pop. 18,822.

Butlaw Youth Hostel, Port Edgar, 2 m. W. of South Queensferry, West Lothian, 10 m. N.W. of Edinburgh. A Scottish Youth Hostel.

Buttergask, vil, S. Perthshire, 2½ m. E. of Greenloaning.

Butterstone, vil., 2¼ m. N.E. of Dunkeld, Perthshire. P.O., T.O. B. House and Loch (depth 25 ft., alt. 314 ft.).

Butt of Lewis, prom. (142 ft.), forming the N. end of Lewis Isl., Outer Hebrides. Lighthouse (120 ft.) with flashing light, visible 19 m. ; is a Lloyd's signal station.

Bynack, or **Boynag,** stream, S.W. Aberdeenshire, flows 7 m. N.E. to r. Geldie.

Byreburn, a mining locality and stream, between Langholm and Canonbie, S.E. Dumfriesshire.

Byrecleugh, hill ridge, Longformacus par., Berwickshire. Here, a shooting-box of the Duke of Roxburghe.

Byres, estate, 3 m. N.W. of Hamilton, Lanarkshire.

Byth, New, q.s. par. and vil., Aberdeenshire, 6 m. N.E. of Turriff vil. P.O., T.O. B. House, a seat.

C

Caaf, small burn, N.W. Ayrshire ; rises on confines of Kilbride, and after a terminal cascade of 20 ft. it falls into the Garnock, ½ m. below town of Dalry.

Cabrach, ham. and par. in Aberdeenshire. P.O. The ham. is on rt. bank of Deveron, 17 m. S.W. of Huntly, The par. extends 34,103 ac. Pop. 326. Hilly, with the Buck of the Cabrach rising to 2368 ft.

Cadboll, estate and farm, 7 m. S.E. of Tain, Ross and Cromarty.

Caddel, burn, N. of Ardrossan par., Ayrshire. Falls into the Caaf.

Cadden, ancient fortification on coast of Kineff par., Kincardineshire.

Cadder, town and par., N.W. Lanarkshire. Town on site of Roman wall, 3 m. W.S.W. of Kirkintilloch. The par. contains the vils. of Bishopbriggs, Moodiesburn, Garnkirk, Auchenairn, Auchenloch, Chryston, Muirhead, Mollenburn, and part of Lenzie; extends 14,088 ac. Pop. 19,192 ; town pop. 16,033.

Caddonfoot, par. and ham., 3 m. S.W. of Galashiels, N. Selkirkshire. Par. 19,252 ac. Pop. 628.

Caerlaverock, coast par., Dumfriesshire, 5½ m. S.E. of Dumfries ; 7229 ac. Pop. 681. C. Castle, a ruin on N. shore of Solway Firth, was Sir Walter Scott's "Ellangowan." In the churchyard is a monument to Robert Paterson, the "Old Mortality" of Scott, who was buried here in 1801.

Caerwinning, hill, alt. 634 ft., 2¼ m. N. of Dalry, Ayrshire.

Cailam, or **Chaluim,** loch, N.W. Caithness, 9 m. S.S.E. of Reay; alt. 435 ft.

Caillach, see INCH-CAILLACH.

Cailleach, headland, in Lochbroom par., Ross and Cromarty.

Cainail, glen in Torosay par., Isl. of Mull, Argyll, extends for 3 m. near Benmore.

Caiplich, ham. and upland moss on borders of Kiltarlity, Inverness and Urquhart pars., Inverness-shire.

Cairirenan, pl., E. Ross, 5 m. S. of Dingwall; supposed burial pl. of Irenan, a Danish prince.

Cairn, ridge of hills on borders of Midlothian and Peeblesshire.

Cairn, or **Carn,** ham. in Kirkconnel par., Dumfriesshire.

Cairn, small r., cos. of Dumfries and Kirkcudbright ; rises in Glencairn par., lower down changes name to Cluden, and after a total run of 23 m. it joins the Nith, 1 m. N.W. of Dumfries.

Cairnacay, hill-range, alt. 1605 ft., in Inveravon par., Banffshire.

Cairn-a-Crubie, hilltop, resembling human head, near Achnasheen ry. sta., Ross and Cromarty.

Cairnaig, or **Carnach,** small stream, Creich and Dornoch pars., S.E. Sutherland; rises at 800 ft. alt., falls into r. Fleet at Torroboll.

Cairnaire, ancient cairn standing close to the fairway into Inverness harb., surmounted by a beacon.

Cairnaqueen (cairn of remembrance), Aberdeenshire, 1½ m. W. of Crathie Church; stands where ancient clans used to gather for battle.

Cairnavain, stone tumulus on N. border Orwell par., Kinross-shire. Rude stone coffin with urn full of bones and ornaments found near.

Cairnban, or **Cairnbaan**, pl. on Crinan Canal, Argyll, 2½ m. W.N.W. Lochgilphead. Hotel used by anglers. P.O., T.O.

Cairnbanno, ham., 2½ m. S.W. of New Deer, Aberdeenshire.

Cairnbeddie, ham., St. Martin's par., Perthshire, 6½ m. N.N.E. Perth; said to be site of Macbeth Castle.

Cairnbran, large cairn, Loth par., Sutherland, 6 m. N.W. Loth ry. sta.; said to be the pl. where Ossian's dog Bran was buried.

Cairnbroe, or **Carnbroe**, seat, near Bothwell, Lanarkshire.

Cairnbulg, headland, fishing vil., with ry. sta., L.N.E., and estate in Rathen par., Aberdeenshire, 2½ m. S.W. of Fraserburgh. Pop. (C. and Inverallochy) 1308.

Cairnburgbeg and **Cairnburgmore**, two of the five principal isls. of the Treshinish group, 5 m. N.N.W. of Staffa, Argyll.

Cairnchunaig, mt., alt. 2749 ft., on borders of Kincardine and Rosskeen pars., Ross and Cromarty.

Cairncubie, springs on town moor, 1½ m. N.E. Dunfermline, Fife; form part of water supply for Dunfermline.

Cairndow, ham. in Lochgoilhead par., Argyll, E. side of Loch Fyne, 9¾ m. N.E. of Inveraray. Hotel; pier and steamboat communication with Inveraray, 40 m. distant. P.O., T.O.

Cairness, ham. and seat, Lonmay par., Aberdeenshire, 2½ m. N.E. Lonmay ry. sta.

Cairneyhill, vil. on borders of Carnock par., Fife, 1 m. E. of Torryburn. P.O., T.O. Pop. 449.

Cairney, Little, Arbroath, Angus. Town hospital.

Cairnfield, estate and seat, Rathven par., Banffshire, 1½ m. S.E. of Port Gordon.

Cairngall, estate, seat, and granite quarries in Longside par., Aberdeenshire, 5 m. W. of Peterhead.

Cairngorm, mt. on borders of Kirkmichael par., Banffshire, and Abernethy par., Inverness-shire; alt. 4084 ft.; and close to it is Ben Macdhui, 4296 ft.

Cairnharrow, hill, alt. 1497 ft., S.W. Kirkcudbrightshire, 4 m. S.E. of Creetown.

Cairn Hill, seat and estate in Craigie par., Ayrshire, 5 m. S.S.E. of Kilmarnock.

Cairnholy, tumulus in Kirkmabreck par., Kirkcudbrightshire, 6 m. S.E. of Creetown.

Cairnie, or **Cairney**, par. and ham., with ry. sta. (C. Junction), L.N.E., N.W. Aberdeenshire; ham., 4½ m. N.W. of Huntly. P.O., T.O.; par. extends to 15,816 ac. Pop. 1017.

Cairnie, an estate, with seat, in Cupar par., Fife.

Cairniehill, see CAIRNEYHILL.

Cairniemount, mt., alt. 1488 ft., on the borders of Strachan and Fordoun pars., Kincardineshire.

Cairnies, The, seat, estate, and Scottish Episcopal College, Glenalmond, 10 m. W.N.W. of Perth.

Cairnkinna, summit in Penpont par., N.W. Dumfriesshire, 5 m. S.E. of Sanquhar; alt. 1819 ft.

Cairn-o'-Mount, see CAIRNIEMOUNT.

Cairnorrie, see METHLICK.

Cairnpat, or **Cairnpiat**, hill, alt. 593 ft., in Portpatrick par., Wigtownshire, 3¼ m. S.S.W. of Stranraer. Ancient remains on summit.

Cairnryan, seapt. vil. in Inch par., Wigtownshire. P.O., T.O. Pier; hotel; and lighthouse on Cairnryan Pt.

Cairns, ruined ancient castle, Mid-Calder par., Midlothian.

Cairnsmore, mt., alt. 2612 ft., in Carsphairn par., N. Kirkcudbrightshire, 3¾ m. N.E. of Carsphairn vil.

Cairnsmore of Fleet, mt. range, borders of Minnigaff and Kirkmabreck pars., Kirkcudbrightshire, 6 m. E. of Newton-Stewart. It has two summits—Minnigaff, 2331 ft.; Kirkmabreck, 2152 ft.

Cairntable, mt., on borders Muirkirk and Douglas pars., cos. Ayr and Lanark; its summit, 1944 ft., is 2 m. S.E. of Muirkirk vil.

Cairntable Halt, ry. sta., L.M.S., 9½ m. S.E. of Ayr.

Cairntaggart, mt., alt. 3430 ft., 3 m. S.W. of Lochnagar, Aberdeenshire.

Cairnton, estate, 3½ m. W. of Banchory, Kincardineshire.

Cairntoodlie, vil., near town of Peterhead, Aberdeenshire.

Cairntoul, mt., alt. 4241 ft., Crathie par., Aberdeenshire. It forms one of the Cairngorm range, and is 2 m. S.W. of Ben Macdhui.

Cairn Water, stream, cos. Dumfries and Kirkcudbright; flows 11½ m. S.E. to join the Cluden.

Cairn Well, pass. between Perthshire and Aberdeenshire. 10 m. S. of Braemar.

Cairn William, hill, alt. 1469 ft., 4 m. E. of Alford, Aberdeenshire.

Cairnyaran, moorland, with two summits —Cairnerzean (734 ft.), Cairnscarrow (751 ft.), 6 m. N.E. Stranraer, Wigtownshire.

Cairston, seat, 1½ m. N.E. of Stromness, Orkney.

Caitha, ham., Stow par., Midlothian, 4 m. S. by W. of Stow vil.

Caithness, maritime co. in extreme N.E. of the mainland of Scotland; bounded on N. by Pentland Firth, E. by North Sea, S.E. by Moray Firth, S.W. and W. by Sutherland. Measures N.E. to S.W. 43 m., N.W. to S.E. 28 m; extent 438,878 ac., or 712 sq. m.; coast extends 105 m. Co. includes Stroma Isl., chief headland, Dunnet Head (306 ft.), and the Ord (652 ft.). Chief bays are Dunnet Bay, Sinclair's Bay; smaller bays are Thurso, Wick, Sandside, and Cannis. The surface in the N.E. is a plain, but in the W. and S.W. the land is hilly, rising at Morven to 2313 ft. Chief rivers are the Thurso, Wick, Forss, Berriedale, Lingwall, Wester, and Dunbeath The towns are royal bur. of Wick, town of Thurso; vils. Halkirk, Lybster, Castletown, Keiss, Sarclet, and Berriedale; lochs Watten, Calder, More, Hempriggs, Westfield, Stempster, and others. Chief stone, old Red Sandstone. Railway (L.M.S.), enters at Strath Halladale, passing through Halkirk, Thurso and Wick with a light ry. to Lybster. Steamer communication with Leith, Wick, Thurso, and Orkneys. Aeroplane, Inverness to Wick. Chief industry, fishing, mainly at Wick. Flagstones, once an important industry, is now of little account. Pop. 25,656.

Caitnish, pl. in Glenorchy par., Argyll, on r. Orchy, 6 m. N.E. of Dalmally.

Cakemuir, ancient castle, 3 m. E. of Borthwick Castle, Midlothian.

Caladail, shallow loch, S. of P.O., Durness, Sutherland. Supplies water for the ham.

Calair, burn in Balquhidder par., Perthshire, running 7 m. N.E. and N., through Glens Dubh and Buckie.

Calbruar, ham. in Blair-Atholl par., Perthshire, 3 m. from vil. of Blair-Atholl.

Calcots, ry. sta., L.N.E., Morayshire, 3 m. N.E. of Elgin. T.O.

Caldarvan, ry. sta., L.N.E. (passenger service withdrawn), Dunbartonshire. 3¾ m. N.E. of Balloch. C. Loch, 1 m. W; C. House, a seat, 1 m. N.

Calder, large dist. in extreme W. of Midlothian.

Calder, ham. and loch, N.W. Caithness, 6½ m. S.S.W. of Thurso.

Calder, stream, trib. of the Avon Water, at Calderbank, Lanarkshire.

Calder, stream in Kingussie par., Inverness-shire, rising in Monadhliath Mts., it falls into the Spey near Newtonmore.

Calder, stream, rising in Largs par., Ayrshire, passes through Calder dam and falls into Castle-Semple Loch, Renfrewshire.

Calder, stream, N.W. Lanarkshire, rising on Elrig Moor (1000 ft.) and falling into the Clyde at Turnwheel.

Calder, par. in Lanarkshire, see CADDER.

Calder, town, with ry. sta., L.M.S., N. Lanarkshire, 2 m. S.S.W. of Airdrie. T.O. at sta. Numerous ironworks.

Calder, in Nairnshire, see CAWDOR.

Calderbank, town, with ry. sta., L.M.S., in Bothwell and Monkland pars., Lanarkshire, 2 m. S.S.E. of Airdrie. P.O., T.O. Pop. 2254.

Calderbank, estate and seat in Blantyre par., 5 m. N.W. of Hamilton, Lanarkshire.

Calderbank, vil., with bleach works on r. Calder, Lochwinnoch par., Renfrewshire.

Caldercruix, q.s. par. and vil., with ry. sta., L.N.E., in Monkland par., Lanarkshire, 4½ m. E.N.E. of Airdrie. P.O., T.O. Has paper mills and calico printing works. Pop. of C. and Glengowan. 1870.

Calder, East, vil. and ancient par., W. Midlothian ; vil. on rt. bank r. Almond, 1½ m. W.N.W. of Mid-Calder ry. sta. P.O., T.O.

Caldergrove, estate and seat, Cambuslang par., 4½ m. W.N.W. of Hamilton, Lanarkshire.

Calderhall, estate and seat in Kirknewton par., Midlothian, ½ m. S.S.W. of E. Calder vil.

Calder House, Mid-Calder par., Midlothian, S. of town of Mid-Calder. Seat of Lord Torphichen.

Calder, Mid-, par., vil., and ry. sta., L.M.S., Midlothian, near left bank of r. Almond ; 10 m. S,W. of Edinburgh; par. 12,101 ac. Pop. 2793. Has beds of shale,from which paraffin-oil and other products are extracted. P.O.,T.O.at vil. Pop.of vil. 587.

Caldermill, ham., 3 m. S.W. of Strathaven, Lanarkshire. P.O.

Calder, North, stream, rising in W. Lothian, flowing 13 m. S.W., and falling into the Clyde at Daldowie.

Calderpark, seat, Lanarkshire, ½ m. S. of Baillieston.

Calderside, seat, N. Lanarkshire, 1½ m. S.W. Blantyre.

Calder, South, r., rising on moors 2 m. E.N.E. Shotts ironworks. After a run of 11 m. it falls into the Clyde, 1½ m. above Bothwell Bridge, Lanarkshire.

Calder, West, par. and town, with ry. sta., L.M.S., Midlothian, 16 m. S.W. of Edinburgh ; par. 20,979 ac. Pop. 6817. P.O., T.O. Industries—coal, shale, etc. Pop. of town 3404.

Calderwood Castle, ancient seat of the Maxwells, N.W. Lanarkshire, 1½ m. S.W. of H. Blantyre. Now nurseries for Scottish Wholesale Co-operative Society.

Calderwood, Long, ham., 1½ m. N.E. of E. Kilbride, Lanarkshire.

Caldhame Fens, ham., S. Kincardineshire, 3½ m. S.W. Laurencekirk.

Caldons, seat, N.W.Kirkcudbrightshire, on Loch Trool, 9 m. N. of Newton-Stewart.

Caldons Burn, stream, flowing into Luce Bay, Wigtownshire.

Caldra, seat, S.E. Berwickshire, 3 m. S.W. of Duns ry. sta.

Caldron, waterfall, on r. Lednock, Perthshire, 1 m. N. of Comrie.

Caldron Linn, several waterfalls on r. Devon, S.E. Perthshire, 3 m. E. of Dollar.

Caldwell, seat, in Beith par., Ayrshire, 4 m. E. of Beith.

Caldwell, q.s. par., ry. sta., L.M.S., Renfrewshire, 5 m. S.W. of Barrhead.

Caldwell Hall, seat, Neilston, Renfrewshire.

Caledonian Canal, line of inland navigation from Inverness to Loch Eil, near Fort William. It traverses the Great Glen, passing through Lochs Ness, Oich, and Lochy and 22 m. of artificial cut and 38½ m. through lochs. Has a depth of 17 ft. Begun in 1803, opened in 1822, completed 1847.

Caledonian Distillery, in W. Edinburgh.

Caledonian Forest, see CADZOW.

Calf, small isl., N.E. Orkney, off N.E. Eday Isl.

Calf, small isl., at entrance Tobermory, Argyll.

Calf, small isl., near Tiree Isl., Argyll.

Calf Sound, narrow sea strait, N.E. Orkney, between Eday Isl. and Calf Isl. ; lighthouse.

Calgary, seat, on small bay of same name, isl. of Mull, Argyll, 13. m. S.W. of Tobermory. P.O., T.O.

Caliach Point, headland, 9½ m. W. of Tobermory, Mull, Argyll.

Califer, vil., 4½ m. S.E. of Forres, Morayshire.

California, vil., 3 m. S.E. of Falkirk, Stirlingshire. P.O.

Callakille, see KALNAKILL.

Callander, par. and mkt.-town (police bur.), on r. Teith, S.W. Perthshire ; lies 16 m. N.W. of Stirling ; par. 51,186 ac. Pop. 2423. Town is a popular tourists' centre for the Trossachs, etc. Hotels. P.O., T.O. Callander Hydropathic is 1 m. S.W. of town.

Callandrade, seat, Callander, Perthshire.

Callands, seat, on Lyne Water, Peeblesshire.

Callange, Coaltown of, N. and **S.,** adjoining mining hams., 3½ m. S.E. of Cupar, E. Fife.

Callanish, see CALLERNISH.

Callart House, 2 m. E. of N. Ballachulish, in S.W. Inverness-shire.

Callater, glen, S.W. Aberdeenshire, 5½ m., S.E. of Braemar.

Callendar House, seat, ¾ m. S.E. of Falkirk, E. Stirlingshire.

Callernish, district and vil., Isl. of Lewis, Ross and Cromarty, on coast of Loch Roag, 16 m. W. of Stornoway. Important standing stones. P.O., T.O. Called Callanish.

Calligray, or **Killigray,** isl., 2 m. by 1 m., in Sound of Harris, Inverness-shire.

Cally, seat, S. Kirkcudbrightshire, 1 m. S. of Gatehouse-of-Fleet. Nearby is the ruins of House of Cally.

Cally, seat, N.E. Perthshire, on r. Ardle, 1¾ m. N.W. of Bridge of Cally.

Calnadulach, vil., near Connel Ferry, Argyll.

Calrossie, seat, E. Ross and Cromarty, 1 m. N.E. of Nigg.

Calton, district, in E. of Glasgow, near Glasgow Green.

Calton, district, to the E. of Edinburgh, once a burgh.

Calton Hill, prominent hill in centre of Edinburgh; alt. 355 ft.; with public buildings, monuments, and City Observatory; commands an extensive view as far as Ben Lomond to the W.

Calva, islet and harbour, 5½ m. S.E. of Scourie, Sutherland.

Calvay, isl., between S. Uist and Eriskay, Inverness-shire.

Calve, isl., N.E. coast of Mull Isl., at entrance of Tobermory harb., Argyll. Pop. 6.

Calvie, loch, 12 m. from Strathcarron, Ross and Cromarty.

Calvine, ham., 4½ m. W. of Blair-Atholl, Perthshire. P.O., T.O. Hotel.

Calzeat, pl., W. Peebles, 1 m. S. of Broughton.

Cam, or **Cama,** loch (2½ m. long), Assynt par., S.W. Sutherland.

Camas an Duin (bay of the Fort), bay on the E. side of Loch Eriboll, Durness, Sutherland.

Camascross, ham., N.E. Sleat, Isle of Skye.

Camashie, part of Strathconon deer forest, near Achnasheen, Ross and Cromarty.

Camastianavaig, ham., Isle of Skye, 3 m. S.E. of Portree. P.O.

Camas Mòr, bay, N.W. Trotternish, Isle of Skye.

Camasunary, ham., on Loch Scavig, 11 m. N.W. Broadford, Isle of Skye.

Camb, ham., Yell Sound, Shetland, at the head of Mid Yell, Voe. P.O.

Cambie, trib. of r. Leven, mid Fife.

Cambo, seat, 2½ m. N.W. of Crail, Fife.

Cambridge, ham., Berwickshire, 3½ m. E. of Lauder.

Cambus, vil., and ry. sta., L.N.E., 2 m. W. of Alloa, Clackmannanshire. P.O., T.O. Pier and distillery.

Cambusavie, ry. sta., L.M.S., 6 m. N.N.W. of Dornoch, Sutherland.

Cambusbarron, vil., 1½ m. S.W. of Stirling. P.O., T.O. Pop. 1363.

Cambus, Burn of, ham., 2 m. N.W. of Doune, Perthshire. P.O., T.O.

Cambuscurry, hill (600 ft.) and bay, Ross and Cromarty, 7 m. N.W. Tain, Sutherland.

Cambusdoon, on r. Doon, in co. of, and 2½ m. S. of, Ayr. Boys' preparatory school.

Cambuskenneth Abbey, ruins in co. of, and 1 m. E. of, Stirling. Founded by David I. in 1147. Was fine example of early Gothic. Burial-place of James III. and his queen. Cambuskenneth ham. is nearby.

Cambuskethan, pl., with extensive grain mills, W. end of Kilmarnock par., Ayrshire.

Cambuslang, town and par., with ry. sta., L.M.S. P.O., T.O.; 5 m. S.E. of Glasgow by rail; par. extends to 5146 ac. Pop. 27,128. Connected to Glasgow by electric tram. Town pop. 26,478.

Cambusmore, seat, on Keltie Water, 2 m. S.E. of Callander, Perthshire.

Cambusmore, seat, 5 m. S.W. of Golspie, Sutherland.

Cambusnethan, par. and vil., with ry. sta., L.M.S., 1½ m. N.E. of Wishaw, Lanarkshire; par., 16,556 ac. Pop. 36,096. P.O. Par. contains Wishaw, Newmains, Overtown, Coltness Iron Works, and 1¼ m. S.W. Wishaw ry. sta. is C. House; vil. now incorporated with Wishaw.

Cambus-o-May, ry. sta., L.M.S., 4 m. E. of Ballater, S.W. Aberdeenshire.

Cambustane, or **Camustane,** hill, with old stone, 5 m. N.W. Buddon Ness, Angus.

Cambusvrachan, pl., near Innerwick, Glenlyon, Perthshire.

Cambus-Wallace, pl., with Caledonian tombs, 1 m. N.W. of Doune, Perthshire.

Cambus-Wallace, seat, 1 m. N.E. of Biggar, E. Lanarkshire.

Camelon, q.s. par., also ry. sta., L.M.S., forming part of Falkirk, E. Stirlingshire. P.O. C. stands on site of Roman town. It has iron foundries and nail factories.

Camend, ham., 1 m. from Carnwath, Lanarkshire.

Cameron, par., E. Fife, on Cameron Water, 3½ m. S.W. of St. Andrews; par. 9325 ac. Pop. 625.

Cameron, seat on S.W. shore of Loch Lomond, 1½ m. N.W. of Balloch ry. sta., Dunbartonshire.

Cameron Bridge, vil., with ry. sta., L.N.E., on r. Leven, Fife, 3½ m. N.E. Thornton Junction. Distillery.

Cameron's Stone, stone erected to commemorate the Covenanters' battle (1680) on Airdsmoss, Ayrshire.

Camghouran, vil. in Fortingal par., N.W. Perthshire, on the shore of Loch Rannoch.

Camiestane, pl. in Kintore par., Aberdeenshire. Said to have been a battlefield.

Camieston, estate in St. Boswells par., 2½ m. S.W. of Newtown St. Boswells, Roxburghshire.

Camis Eskan, see CAMUS ESKAN.

Camismore, vil. and bay, Kilmuir par., W. coast of Skye.

Camlachie, parl. division, city of Glasgow.

Camlarg, ham. in Dalmellington par., Ayrshire, 1 m. from Dalmellington vil.

Cam Loch, small loch, Knapdale, Argyll, 2 m. N.W. of Ardrishaig.

Cammo, seat in W. Edinburgh, 1 m. S. of Cramond Bridge.

Cammusmore, see CAMISMORE.

Camp, ham., 2½ m. S. by E. of Bannockburn, Stirlingshire.

Camp, ancient fort on summit of Turin Hill, 4½ m. E.N.E. of Forfar.

Camp, ancient fort in Roberton par., Roxburghshire.

Camp, a farm in Old Monkland par., Lanarkshire, near Baillieston.

Camp, hill (1153 ft.), in Yetholm par., Roxburghshire.

Camp, Renfrewshire, see CAMPHILL.

Campbell's Cairn, near Knockando, Morayshire.

Campbelltown, vil. in Ardersier par., 9 m. N.E. of Inverness. Pop. 691.

Campbeltown, mun. and royal bur. and par., 11 m. N.E. of Mull of Kintyre, Argyll. P.O., T.O. Hotels. Lifeboat sta., pier. Steamer communication with Glasgow. Pop. 6309. Par. extends to 44,256 ac. Pop. 7928. Fishing and many distilleries. Lighthouse.

Camperdown House, seat, 1¾ m. N.W. of Lochee, Dundee, Angus.

Campfield, ham. and seat, 6 m. N.W. of Banchory, Aberdeenshire.

Campfield, pl. on r. Don, 16 m. N.W. of Aberdeen.

Camphill, hill and seat in Glasgow. Seat now a museum.

Camphill House, seat, 2 m. N.W. Lumphanan ry. sta., S. Aberdeenshire.

Camphill Reservoir, 3½ m. N.W. Dalry, Ayrshire. Supplies Paisley.

Cample, stream, rises in Wedder Law (2185 ft.), runs 8 m. S. and falls into the Nith, 1 m. S. of Thornhill, Dumfriesshire.

Campmuir, ham. in Kettins par., Angus, 2 m. S. of Coupar-Angus.

Campmuir, pl. and traces of an encampment where troops were stationed in 1715 to overawe the Jacobites, Langton par., Berwickshire.

Campsbank, eminence (400 ft.), Carnock par., S.W. Fife, S.E. of Carnock vil. Supposed Roman camp.

Campsie, par. and ham., S. Stirlingshire. Contains town of Lennoxtown. P.O., T.O.; ry. sta., Campsie Glen; par. 17,860 ac. Pop. 5296.

Campsie Glen, ravine and ry. sta., L.N.E. See CAMPSIE.

Camptown, ham. in Jedburgh par., Roxburghshire, 6 m. S.S.E. of Jedburgh town. P.O., T.O.

Camsericht, seat in Perthshire, 7 m. E. of Rannoch ry. sta.

Camserney, pl. and burn, N.W. Perthshire, 2½ m. W. of Aberfeldy. P.O., T.O.

Cam Sgriob, hill (1591 ft.), 4 m. N.W. of Grantown, Morayshire.

Camster, seat, stream, and loch, E. Caithness, 6 m. N. of Lybster.

Camstraddan, bay and house, W. side Loch Lomond; house ½ m. S. of Luss.

Camus, seat, 1 m. E. of N. Ballachulish, Inverness-shire.

Camusdarroch, ham., W. Inverness-shire, 3½ m. S.W. of Mallaig.

Camusnagaul, ham. opposite Fort William, Inverness-shire.

Camusteel, ham., 2 m. S. Applecross, Ross and Cromarty.

Candacraig, seat, on r. Don, 1½ m. S.W. of Strathdon, W. Aberdeenshire.

Cander, stream, rises in Lesmahagow par. and joins r. Avon, 1 m. N.N.E. of Stonehouse vil., Lanarkshire.

Candida Casa, see WHITHORN.

Candie House, seat, Stirlingshire, 3 m. N.E. of Avonbridge.

Candren, saline spring, 2½ m. E. of Paisley, Renfrewshire.

Candy or **Spittal,** stream, flowing 5 m. S. to Biggar Water.

Canisbay, coast par., N.E. Caithness, on Pentland Firth ; par. 26,247 ac. Pop. 1365. P.O., T.O.

Canisp, conspicuous mt., alt. 2779 ft., on side of Loch Assynt, Sutherland.

Canna, isl. in Inverness-shire, 3 m. N.W. of Rhum Isl., is 4¼ m. long, 1 m. broad, and rises to 690 ft. P.O., T.O. Pier. Soil fertile, sea-fishing good. Pop. 40.

Cannachy Bridge, ham., with inn on r. N. Esk, Angus, 7 m. N. of Brechin.

Cannerton, ham., 2 m. S.E. of Lennoxtown, Stirlingshire.

Cannich, stream, N.W. Inverness-shire, rising in Loch Mullardoch and passing through Glen Cannich to join the Glass at Cannich Bridge.

Cannich Bridge, pl., 7 m. S.W. of Struy Bridge, N.W. Inverness-shire. P.O., T.O.

Cannich Hostel, opposite Cannich Hotel, at the cross-roads leading to Glen Affric, Beauly, Glen Cannich, and Glen Urquhart. A Scottish Youth Hostel.

Canniesburn, ham., E. Dunbartonshire, 1½ m. S.W. Milngavie. Electric tram to Glasgow.

Cannor or **Kinord,** loch, 5 m. E.N.E. of Ballater, S.W. Aberdeenshire.

Canny, stream, cos. Aberdeen and Kincardine ; rises in Hill of Fare and after 8 m. course, falls into r. Dee at Invercanny.

Canonbie, par. and vil., with ry. sta. (Canonbie), L.N.E., S.E. Dumfriesshire, on r. Esk, 6 m. S.E. of Langholm. P.O., T.O. ; par. 24,156 ac. Pop. 1508 ; vil. pop. 109. Contains remains of Roman sta. and ham. of Rowan Burn and Overtown.

Canongate, part of, and q.s. par. of Edinburgh, old town of Edinburgh.

Canonmills, dist. of Edinburgh, on the Water of Leith, where the Canons of Holyrood had their grain mills.

Canter or **Ceannmor,** tarn at alt. 2196 ft. in S.W. Aberdeenshire, 2 m. S.W. of head of Loch Callater.

Canterland, dell, S. Kincardineshire, 5 m. S. of Laurencekirk.

Cant Hills, ridge, 6 m. long, N. of Shotts, Lanarkshire.

Cantick Head, S.E. extremity of South Walls Isl., Orkney. Lighthouse, light visible 16 m.

Cantray, seat, on r. Nairn, in co. of, and 9 m. N.E. of, Inverness.

Cantsdam, vil., 4½ m. N.W. of Auchtertool, S.W. Fifeshire.

Canty Bay, small bay and hotel, 2½ m. E. of North Berwick, Haddingtonshire.

Cantyre, see KINTYRE.

Caol, sea loch in S.W. Mull Isl., Argyll, 13 m. W. of Loch Buir.

Caolas-Uist, channel between Uist and Bernera Isl., Outer Hebrides, Inverness-shire.

Caol Ila Distillery, Port Askaig, Argyll.

Caolisport or **Killisport,** coast district and sea loch, Argyll ; includes the peninsula of Knap. The loch, which runs inland for 6 m., has several small bays and affords good anchorage.

Caolvalloch, ham., near Weem, N.W. Perthshire.

Cape Difficulty, S.W. extremity of Harris, Outer Hebrides.

Capehope, stream, E. Roxburghshire ; rising in Cheviots, it flows to the Kale above Hounam vil.

Cape Law, mt., alt. 2364 ft., N. Peeblesshire, 6 m. N.E. of Moffat.

Capel Fell, mt., alt. 2223 ft., between cos. Selkirk and Dumfries, at sources of Ettrick Water.

Capelrig, seat, Renfrewshire, 2½ m. S.W. Pollokshaws.

Capenoch, seat, 3½ m. S.W. Thornhill ry. sta., Dumfriesshire.

Cape Wrath, extreme N.W. pt. of the mainland of Sutherland. Lighthouse built in 1828 ; light seen for 27 m. The Norsemen called it Hvraf, a turning point.

Caplaw Dam, loch, S. Renfrewshire, 3 m. S. of Johnstone.

5

Cappers, vil., West Lothian 1 m. N. of Whitburn. Pop. 450.

Caprington Castle, seat, Ayrshire, on r. Irvine, 2 m. S.W. of Kilmarnock.

Caputh, par. and vil., Perthshire, on left bank of r. Tay, 5 m. S.E. of Dunkeld; par. 10,083 ac. Pop. 1079. P.O., T.O.

Cara, isl. (1 m. by ½), S. Argyll, 1 m. S. of Gigha, W. of Kintyre.

Carberry Hill, hill (500 ft) 3 m. S.E. of Musselburgh, Midlothian. Carberry Tower (1579), seat of Lord Elphinstone; here Queen Mary surrendered, 15th June 1567.

Carbeth, seat, left bank of Endrick Water, 2 m. S.W. of Balfron, Stirlingshire.

Carbeth-Guthrie, seat, 2 m. W. of Strathblane vil., Stirlingshire.

Carbost, vil., with distillery (Talisker), W. Skye, at head of Loch Harport, 10 m. S.W. of Portree. P.O., T.O. Steamers from Glasgow call regularly.

Carbrook, seat, 3 m. N.W. of Larbert, Stirlingshire.

Cardean House, on borders of Perthshire and Angus, 6½ m. N.E. of Coupar-Angus.

Carden State Forest, 1½ m. S.W. of Auchterderran, 5½ m. S.W. of Thornton Junction, Fife.

Cardenden, glen and vil., with ry. sta. L.N.E., 1 m. S.W. of Auchterderran, Fife. P.O., T.O.

Cardeston, see ALBERBURY.

Cardney House, seat, 2½ m. N. of Dunkeld, Perthshire.

Cardonald, q.s. par., vil. and ry. sta., L.M.S., 4½ m. W. of St. Enoch sta., Glasgow. Pop. 488. P.O., T.O.

Cardon Hill, mt., alt. 2218 ft., W. Peebles, 4 m. S.E. of Biggar.

Cardon Law, hill, alt. 1928 ft., in co. of, and 4 m. N.E. of, Peebles.

Cardow, pl., with distillery, Morayshire, 2 m. N.W. Carron ry. sta. C. House. ½ m. S.E.

Cardowan, ham. and seat, N.W. Lanarkshire, 1 m. S.W. Garnkirk ry. sta. C. House ½ m. S.E.

Cardrona, seat and ry. sta., L.N.E., in co. of, and 3½ m. S.E. of Peebles.

Cardross, seat, on r. Forth, 2½ m. S.E. of Port of Menteith, S.W. Perthshire.

Cardross, par. and vil., with ry. sta., L.N.E., on Firth of Clyde, Dunbartonshire,

4 m. N.W. Dumbarton; par. 8281 ac. Pop. 11,105; vil. pop. 826. P.O., T.O. Lighthouse. C. Park is a seat.

Careston, par., ham, and seat, ry sta., L.M.S., Angus, 5 m. W. of Brechin; par. 2117 ac. Pop. 182. P.O., T.O.

Carewe Hill, alt. 872 ft., N.E. Kincardineshire, 4 m. N.W. of Stonehaven.

Carey, seat, Perthshire, 1 m. W. of Abernethy.

Carfin, mining town, with ry. sta. (Carfin Halt), L.M.S., N.E. Lanarkshire, 2 m. N.E. of Motherwell. P.O., T.O. Celebrated for a Grotto of Our Lady of Lourdes (opened in 1922) with statues of the Virgin and of Bernadette, the sainted shepherdess. Pop. of C., Holytown, and New Stevenson, 12,011.

Carfraemill, ham., N. Berwickshire, 4 m. N.W. Lauder.

Cargen, seat and stream, E. Kirkcudbrightshire; stream flows 7 m. to join Nith, 2½ m. S. of Dumfries. C. House 3¼ m. S.S.W. of Dumfries.

Cargen Bridge, ham. on Cargen Pow, E. Kirkcudbrightshire, 1½ m. S.W. of Dumfries.

Cargenholm, seat, E. Kirkcudbrightshire, on Cargen Water, 2 m. S.W. of Dumfries.

Carghadoun, said to be an ancient fortification on coast, 3 m. S.W. of Whithorn par., Wigtownshire.

Cargilfield, Cramond Bridge, Midlothian, Boys' Preparatory School.

Cargill, par. and vil., with ry. sta., L.M.S., Perthshire; vil. on r. Tay, 4½ m. S.W. of Coupar-Angus; par. 10,216 ac. Pop. 1257. P.O., T.O. at ry. sta.

Carham, ry. sta., L.N.E., Roxburghshire; vil. in Northumberland, 1 m. N.E. of sta.

Carhurlie, coast ham., 2 m. N. of Largo, E. Fifeshire.

Carie, ham., N. side of Loch Tay Perthshire, 6 m. N.E. of, Killin.

Carigil or **La' Cherigal,** loch, Caithness, 9 m. S. of Halkirk.

Carim, seat, S.E. Perthshire, 4 m. S.W. of Blackford.

Carinish, vil. on S. coast of N. Uist, Inverness-shire, 13 m. S.W. of Lochmaddy.

Carity, stream, mid Angus, flowing 9 m. S. and E. to the r. Esk at Inverquharity.

Carleffin Fell, mt., alt. 1385 ft., Ayrshire, 7 m. S.E. of Maybole.

Carlestown House, seat, Stirlingshire, 2 m. W. of Kirkintilloch.

Carleton, bay and fishing pl., 6½ m. S.W. of Girvan, Ayrshire. Balsalloch Hill (611 ft.) to the E.

Carlingheugh Bay, Angus, 2 m. N.E. of Arbroath.

Carlins Cairn, mt., alt. 2650 ft., borders of Carsphairn par., N.W. Kirkcudbrightshire, 3 m. S.S.E. of head of Loch Doon.

Carlin Skerry, rocky islet, often called Barrel of Butter, S. of Orkney mainland.

Carlin Tooth, summit of the Cheviots (1801 ft.), S. of Southdean par., Roxburghshire.

Carlinwark, loch in the N. of Kelton par., Kirkcudbrightshire.

Carlops, vil. in Linton par., N.W. Peeblesshire, on N. Esk R., 14 m. S.S.W. of Edinburgh ; Carlops Hill ¾ m. W. by N. (1490 ft.). P.O.

Carloway, dist. in Lewis, Outer Hebrides, par. of Uig, and close to Loch Roag. Famous for the ancient fort. The vil. is at the head of Loch Caloway. P.O., T.O.

Carlowrie Castle, seat, 1¼ m. E. by N. of the vil. of Kirkliston, Linlithgowshire.

Carlow's Linn, small waterfall on r. Tweed, near Tweedsmuir Church, Peeblesshire.

Carlton Fell, hill, alt. 475 ft., on the coast of Glasserton par., Wigtownshire, 3½ m. S.W. of Whithorn.

Carluke, town and par., Lanarkshire ; par. 15,295 ac. Pop. 10,507. Town stands on Jock's Burn, 2¼ m. E. of the Clyde, 5½ m. N.N.W. of Lanark. Has an elevation of 600 and 700 ft. P.O., T.O. ; ry. sta., L.M.S. Pop. 6027. Chief industries : mining, brick-making, corn-milling, fruit-growing, and jam-making.

Carmacoup, estate and seat, 3½ m. S.W. of the town of Douglas, S. Lanarkshire.

Carmel, burn, rises W. of Kingswell Inn and runs 11 m. S.W. through Fenwick and Kilmaurs pars. to r. Irvine, 1 m. E.S.E. of Dreghorn, Ayrshire.

Carmichael, ham. and par., 2½ m. E. of Sandilands ry. sta., Lanarkshire ; par. 11,374 ac. Pop. 1794.

Carmont, ry. sta., L.M.S., Kincardineshire, 4 m. W.S.W. of Stonehaven.

Carmunnock, par. and vil., N.W. Lanarkshire, 1½ m. N.E. Busby ry. sta. ; par. 3472 ac. Pop. 931. P.O., T.O. Vil. pop. 577.

Carmyle, vil., S.W. border of Old Monkland par., Lanarkshire ; ry. sta., L.M.S. Vil. originated in a muslin manufactory erected about 1741. P.O. Pop. of C. and Mount Vernon 4435.

Carmyllie, vil., with ry. stas., L.N.E. and L.M.S., and par., S.E. Angus ; vil. is 6½ m. W. by N. of Arbroath. P.O., T.O. ; par. 7553 ac. Pop. 753.

Carn, small loch, Argyll, 2 m. from Cairnbaan Inn.

Carna, small isl. in Loch Sunart, Morven par., Argyll.

Carnan, pl. on Loch Carnan, in N.E. of S. Uist, Outer Hebrides. Steamers from Glasgow, etc., call regularly.

Carnabattan (Carn nam Badan), loch in Kiltarlity par., N. Inverness-shire.

Carnach, q.s. par. in Contin, Fodderty, S. Ross and Cromarty.

Carnach, vil. at Bridge of Coe, 1 m. E. of Ballachulish, Argyll. Pop. 201.

Carn a' Mhuilt, mt., alt. 2218 ft., S.E. Ross and Cromarty, 4 m. S. of Carn Bàn.

Carnassary, ruined mansion in Kilmartin par., Argyll.

Carn Bàn, mt., alt. 2410 ft., S.E. Ross and Cromarty, 3 m. S.W. of Carn nam Pollan.

Carn Bàn, mt., alt. 2762 ft., N.W. Ross and Cromarty, 14 m. S.E. of Ullapool.

Càrn Ban., mt., 9 m. S. of Aviemore, Inverness-shire, alt. 3443

Carn Beag, mt., alt. 1805 ft., N.W. Ross and Cromarty, 6 m. S.E. of Kinlochewe.

Carnbee, ham. and par. in E. Neuk of Fife ; ham. 3 m. N.N.W. of Pittenweem ry. sta., L.N.E. ; par. 8396 ac. Pop. 736.

Carnbo, vil. in Fossoway par., Kinrossshire, 4½ m. W. by N. of Kinross. P.O., T.O. 1 m. S.W. is C. House.

Carnbroe, vil., estate, and seat, N. border of Bothwell par., Lanarkshire, on the N. Calder Water, 1¼ m. N.N.E. of Bellshill. P.O. Vil. has extensive ironworks. Pop. 883.

Carnchaivichin, ancient burial cairn, 14 m. N.W. of Forteviot, Perthshire.

Carn Dearg Hostel, on a point of land on the N. shore of Gairloch, Ross-shire, 17 m. from Craig. A Scottish Youth Hostel.

Carneil, eminence, commanding fine view of the Forth basin (400 ft.), $\frac{1}{2}$ m. W.S.W. of Carnock, S.W. Fife.

Carn Eiteige, mt., alt. 2891 ft., S.W. Ross and Cromarty, 10 m. S.E. of Glencarron ry. sta.

Carnethy, second highest summit of the Pentland range, $2\frac{1}{2}$ m. N.W. of Penicuik; alt. 1890 ft.

Carn Geur Aodainn, mt., alt. 1950 ft., S.W. Ross and Cromarty, 3 m. from Strathcarron ry. sta., alt. 1950 ft.

Carn Gorm, mt., S.W. Ross and Cromarty, 5 m. E. of Glencarron ry. sta. alt., 2866 ft.

Carn Liath, summit of Ben-y-Gloe, 5 m. N.E. of Blair-Atholl; alt. 3193 ft.

Carn Mairg, mt., N.W. Perthshire, 5 m. S.E. of Kinloch Rannoch. Alt. 3419 ft.

Carn Mòr, mt., alt. 2122 ft., N.W. Ross and Cromarty, 9 m. S.E. of Ullapool.

Carn na Caillich, ancient mound, S.W. of Morven par., Argyll.

Carn na Cloiche Moire, mt., alt. 1936 ft., S.E. Ross and Cromarty, 5 m. S.W. of Garve ry. sta.

Carn nan Feithean, mt., alt. 1820 ft., N.W. Ross and Cromarty, 10 m. S. of Ullapool.

Carn-nan-tri-tighearnan (*Cairn of the three Lords*), mt., alt. 2013 ft., borders of Nairn and Inverness.

Carn na Pollan, mt., alt. 2773 ft., S.E. Ross and Cromarty, 10 m. S. of Garve ry. sta.

Carnoch, see CARNACH.

Carnock, par. and vil., S.W. Fife, $3\frac{1}{2}$ m. N.W. of Dunfermline; par. 3493 ac. Pop. 1443. P.O., T.O.

Carnon, small stream, Ardchattan, Argyll, $3\frac{1}{2}$ m. S.E. of the Etive.

Carnousie, estate, with ancient seat, Banffshire, on r. Deveron, 4 m. W. of Turriff.

Carnoustie, coast town and police bur., ry. sta., L.N.E., $10\frac{1}{2}$ m. N.E. of Dundee, Angus. Favourite watering place. Hotels. Has linen and chemical works; also boot mfrs., stone-working machinery and crane works. Good golf links. Pop. 4806. P.O., T.O.

Carnsalloch, seat in co. of, and 3 m. N. of, Dumfries.

Carntyne, vil., with ry. sta., L.N.E., Lanarkshire, $\frac{1}{2}$ m. W. of Shettleston. P.O., T.O.

Carnwath, par. and vil., with ry. sta., L.M.S., Lanarkshire, 26 m. S.W. of Edinburgh, and 31 m. S.E. of Glasgow by rail. P.O., T.O.; par. 30,468 ac. Pop. 5258. Is rich in coal and iron-stone. Vil. pop. 787. C. House at end of vil.

Caroline Park, on Firth of Forth, $\frac{3}{4}$ m. W. of Granton; ry. sta., L.N.E. Formerly a seat; now printing-ink and chemical works.

Carolside, seat, Berwickshire, $1\frac{1}{4}$ m. N.W. of Earlston.

Carphin, seat, $6\frac{1}{2}$ m. N.W. of Cupar, Fife.

Carpow, seat, S.E. Perthshire, 3 m. S.W. of Newburgh.

Carr, small stream, S.W. Aberdeenshire, flowing to the r. Dee. Near Braemar.

Carr, reef with beacon at entrance of Firth of Forth; 1 m. E. is the North Carr Lightship, visible 11 m.

Carradale, bay, headland, stream, and vil. with steamboat pier, Kintyre, Argyll, 13 m. N.E. of Campbeltown. P.O., T.O.; also P.O. and T.O. at C. Pier. C. Point (or Ard of C.), 133 ft. high, has remains of old fort. C. Water flows 7 m. S.E. to C. Bay. C. House, a seat. Steamers from Glasgow and Campbeltown call regularly at vil.

Carraig Fadda Point, at W. entrance of harb., Port Ellen, isl. of Islay, Argyll.

Carrbridge, vil. and ry. sta., L.M.S., on Dulnan R., Inverness-shire, 8 m. N.W. of Aviemore; alt. of sta., 914 ft. P.O., T.O. Pop. 233. Hotel.

Carribber, seat in co. of, and $3\frac{1}{2}$ m. S.W. of, Linlithgow.

Carrick, one of the three ancient divisions of the co. of Ayr. Carrick includes all the co. S. of Doon. Prince of Wales, as steward, bears the title of Earl of Carrick. This part of the co. is an assemblage of brae, hill, and mountain.

Carrick, seat, N. extremity of Eday Isl., Orkney.

Carrick, pl. at middle of W. side of Loch Goil, Argyll. P.O., T.O. C. Castle, a ruin. Pier.

Carriden, q.s. par. and vil., at Bridgeness, 1 m. E. of Bo'ness, W. Lothian. Pop. 290.

Carrington, vil. in mid Midlothian, $5\frac{1}{2}$ m., S.W. of Dalkeith; vil. also known as Primrose. Pop. 543.

Carrol Rock, height, close to Loch Brora, Clyne par., S.E. Sutherland.

Carron, vil., with ry. sta., L.N.E., E. Morayshire, 5½ m. S.W. of Craigellachie sta. P.O. called C. Sta.

Carron Hill, alt. 960 ft., and C. House, ¾ m. E. of Carron.

Carron, r. in Stirlingshire, rises C. Bog, flows for 20 m. to join Firth of Forth at Grangemouth.

Carron, vil. near r. Carron in Stirlingshire, and 2 m. N.W. of Falkirk. P.O., T.O. Oldest established ironworks in Scotland (1759); produced small ordnance—naval guns known as Carronades—used in battle of Trafalgar. Carron Iron Company still have lease of the Kinnaird Mineral Field in Stirlingshire, the first lease of which was entered into in 1760 with James Bruce, the Abyssinian traveller. Smeaton and Roebuck are among the famous men associated with the Carron Iron Works. Adjacent collieries supply fuel for the blast furnaces, iron, brass, and aluminium foundries. Makes domestic iron goods as well as large engineering material.

Carron, r. in Nithsdale, Dumfriesshire, rises at 2000 ft. in the Lowther Hills, flows 9 m. S. to the Nith, into which it falls at Carronbridge vil.

Carron, stream, Kincardineshire, rises in the Grampians and flows to the sea at Stonehaven.

Carron, stream, S.W. Ross and Cromarty, rises near Luibgargan Inn and flows S.W. to join head of Loch Carron.

Carronbridge, vil., with ry. sta., L.M.S., Dumfriesshire, on Carron Water, 1¼ m. N.W. of Thornhill. P.O., T.O.

Carronbridge, vil., on r. Carron, near Carron Iron Works, Stirlingshire. P.O.

Carronhall, vil., Stirlingshire, on r. Carron, 2 m. E. of Larbert ry. sta.

Carronshore, vil., Grangemouth par., Stirlingshire, on r. Carron, 1 m. N.E. of Carron Iron Works dry docks. Here are collieries.

Carron Vale, seat, 1 m. S.E. of Larbert, Stirlingshire.

Carrot, wooded hill, alt. 851 ft., S. Angus, 6 m. N.W. of Broughty Ferry.

Carroy, sea loch, part of Loch Bracadale, 12 m. W. of Portree, Skye

Carrubber, see CARRIBER.

Carruchan, seat, Kirkcudbrightshire, 2 m. S.W. of Dumfries.

Carruth, seat, N.W. Renfrewshire, 2 m. N.W. of Bridge of Weir sta.

Carruthers, former par., Annandale, Dumfriesshire, now merged into Middlebie par.

Carrutherstown, ham., S.E. Dumfriesshire, 8 m. S.E. of Lockerbie. P.O.

Carsaig, ham. and bay, N. Knapdale, ½ m. N.W. of Tayvallich, Argyll.

Carsaig, seat, W. of mouth of Loch Buie, isl. of Mull. Argyll, P.O.

Carse, small bay in estuary of the Nith, Kirkcudbrightshire, 1 m. N. of Kirkbean vil. Carse sands here.

Carse, farm, in co. of, and par. of, Kirkcudbright.

Carse, seat, S. Knapdale, Argyll, 2 m. from Ardpatrick.

Carsebreck, curling-pond, in S.E. Perthshire, 2 m. W. of Blackford.

Carsebridge, distillery, N.E. of Alloa, Clackmannanshire.

Carseburn, vil., 1 m. N.E. of Forfar, Angus.

Carsecreugh, ruined castle, 2½ m. N.E. of Glenluce, Wigtownshire.

Carsegowan, hill, alt. 593 ft., 1 m. W. of New Abbey, S.E. Kirkcudbrightshire; Waterloo monument on summit.

Carsegownie, farm, mid Angus; has ancient mansion and Caledonian Cairn with urn.

Carse Gray, seat, 2½ m. N. of Forfar, Angus.

Carse of Clackmannan, section of Carse of Forth, on left bank of the r. Forth.

Carse of Falkirk, section of Carse on right bank of the r. Forth.

Carse of Forth, fertile tract of land extending both sides of the r. Forth in Perth, Stirling, and Clackmannan.

Carse of Gowrie, fertile tract of land in cos. of Perth and Angus, along N. Bank of r. Tay, from Kinnoull Hill to near Dundee, 15 m. long and 1 to 2 m. broad.

Carse of Henryie, tract, E. of Sanday Isl., Orkney.

Carse of Kinneil, part of Carse of Falkirk.

Carse of Stirling, part of Carse of Forth.

Carsethorn, coast vil., 1 m. N.E. of Kirkbean, S.E., Kirkcudbrightshire.

Carsfad, pl., with one of the Galloway Water Power Company's Hydro-electric stas., near New Galloway, Kirkcudbrightshire.

Carsie, vil., 3 m. S. of Blairgowrie, Perthshire.

Carskey, bay and seat, S. coast of Kintyre, 2 m. S.W. of Southend, Argyll.

Carslogie, estate and seat, 1½ m. W. of Cupar, Fife ; for long the seat of the Clephane family.

Carsluith, pl., with old tower, S.W. Kirkcudbrightshire, 3¼ m. S.E. of Creetown. P.O.

Carsphairn, par. and vil., N. Kirkcudbrightshire, on Water of Deugh, 10 m. S.E. of Dalmellington ; par. 5487 ac. Pop. 355 ; vil. has P.O., T.O. Pop. 97. Lead, copper, silver, and zinc are found in the par.

Carstairs, par., seat, and vil. in co. of, and 4 m. E. of, Lanark ; par. 9818 ac. Pop. 2289. P.O., T.O. Vil. 1 m. N.W. of C. Junction. Carstairs House, 1 m. S.W. of C. Junction. C. is supposed to have been a Roman *Coria*. Vil. pop. 500.

Carstairs Junction, important ry. centre, L.M.S., Lanarkshire, 29 m. S.E. of Glasgow. P.O., T.O. Ry sta., alt. 645 ft. Pop. 1191.

Carsthorn, coast ham., S.E. Wigtownshire, 4½ m. S.E. of New Abbey.

Cart, r., trib. of the Clyde, Renfrewshire, formed by the joining of the rs. White and Black Cart, falls into the Clyde 1¼ m. N. of Renfrew.

Carter Bar, road summit, Cheviot Hills, between Roxburghshire and Northumberland ; alt. 1405 ft.

Carter Burn, trib. of the Jed, near Southdean, S.E. Roxburghshire.

Carter Fell, alt. 1815 ft., a summit of the Cheviot Hills, Roxburghshire, on the border, 11 m. S.E. of Jedburgh.

Carterhaugh, wooded peninsula at confluence of rs. Ettrick and Yarrow, Selkirkshire.

Cartland, vil. in co. of, and 2 m. N.W. of, Lanark ; 1 m. N.W. of Lanark are C. Crags, a chasm forming the bed of Mouse Water. Bridge of three arches, designed by Telford, spans the ravine.

Cartsburn, dist. in E. of Greenock, Renfrewshire.

Cartsdyke, E. suburb of Greenock, with ry. sta., L.M.S.

Carty, harb., E. Wigtownshire, on r. Cree, 2 m. S.E. of Newton-Stewart.

Carvie, stream, W. Aberdeenshire, 3 m. N. of the Don.

Carwood House, 2 m. N. of Biggar, E. Lanarkshire.

Cassillis, pl., with ry. sta., L.M.S., 6 m. S. of Ayr. T.O. at sta. C. House, 1 m. S.E. of r. Doon, the seat of the Marquis of Ailsa.

Cassley, r., S. Sutherland, issues from Gorm Loch, and flows into Oykell, 8 m. S.W of Lairg.

Cassock, site of magnetic observatory, Eskdalemuir, Dumfriesshire, 15 m. N.W. of Langholm.

Castle, part of New Cumnock, E. Ayrshire.

Castle, ham., near Milton, Stirlingshire.

Castlebay, fishing ham., S. end of Barra Isl., Outer Hebrides, Inverness-shire. P.O., T.O. Regular steamboat communication with Oban, Glasgow, etc.

Castlecary, pl., with ry. sta., L.N.E., and old castle, 6½ m. S.W. of Falkirk, S.E. Stirlingshire. P.O. Originally a Roman station.

Castle Clanyard, ruined tower, S.W. Wigtownshire, ½ m. S.E. of Clanyard Bay.

Castle Cluggy, ruined fortalice, Perthshire, on peninsula at N. end of Monzievaird Loch.

Castle Coeffin, ruins of ancient stronghold, Argyll, on a peninsula of Loch Linnhe.

Castle Cole, ancient tower, S.E. Sutherland, on Black Water, 2 m. above its junction with the Brora.

Castle Craig, farm, with vestiges of ancient stronghold, E. Ross and Cromarty ; in Nigg par., facing Cromarty.

Castlecraig, seat, W. Peeblesshire, near Tarth Water, 6 m. S.E. of Dolphinton.

Castle Craig Hill, eminence on Hill of Nigg, E. Ross and Cromarty ; alt. 645 ft.

Castle-Douglas, mkt. town with ry. sta., L.N.E., Kirkcudbrightshire, 19½ m. S.W. of Dumfries by rail. P.O. Hotel. Originally an important working of marl. Is the centre of business for E. Galloway. Pop. bur. 3008. Has tanning and agricultural implement works.

Castle Drumin, ruined fortress, Banffshire, at confluence of Avon and Livet, 6 m. S.W. of Ben Rinnes.

Castle Duart, ruined castle of the Macleans of Mull, Mull Isl., Argyll, E. side of Duart Bay.

Castledykes, site of an ancient castle of the Comyns, in a bend of the r. Nith, in co. of, and ¾ m. S. of, Dumfries.

Castledykes, site of Roman sta. in co. of, and 4 m. E. of, Lanark.

Castle Feather, ancient fort, S.E. Wigtownshire, on coast, ½ m. N.W. of Burrow Head.

Castlefern, small stream, Dumfriesshire, flows 7 m. to join Craigdarroch and Dalquhat Waters to form the Cairn.

Castle Forbe,s seat of Lord Forbes, Keig, Aberdeenshire, on r. Don, 3 m. N. of Whithouse sta.

Castle Fraser, old seat of a family of Frasers, Aberdeenshire, 3 m. S.E. of Monymusk.

Castle Girnigoe and **Castle Sinclair,** two ruins, 3¼ m. N.E. of Wick, Caithness.

Castle Gloom, see CASTLE CAMPBELL.

Castle Gower, vitrified fort, S. Kirkcudbrightshire, near Castle-Douglas.

Castle Grant, seat of the Countess of Seafield, in S. of Morayshire, 1½ m. N.E. of Grantown.

Castlehaven, creek, N. E. Ross and Cromarty, at point of Tarbet peninsula.

Castlehill, Dumfriesshire, 2½ m. S.W. of Lockerbie. County hospital.

Castlehill, pl. in co. of, and 3 m. S.E. of, Inverness.

Castlehill, vil., 2 m. N. of Carluke, Lanarkshire.

Castlehill, seat, Castletown, Dunnet Bay, Caithness.

Castle Hill Point, prom., S. Kirkcudbrightshire, 5⅓ m. S. of Dalbeattie.

Castle Huntly, estate and seat, Longforgan par. 7 m. W. of Dundee, Perthshire.

Castle Island, islet off S.E. side of Eigg Isl., Inverness-shire.

Castle Island, islet in Loch Leven in co of, and 2 m. E.S.E. of, Kinross. Its chief interest is that Queen Mary Stuart was imprisoned from 17th June 1567 to 2nd May 1568.

Castle Kennedy, ham. and loch, with ry. sta., L.M.S., and ruined castle, once the seat of the Earls of Cassilis and now the property of the Earl of Stair, 3 m. S.E. Stranraer, W. Wigtownshire. P.O., T.O. at sta.

Castle Kilchurn, ruined stronghold, Argyle, 2½ m. S.W. of Dalmally on Isl. in Loch Awe.

Castle Lachlan, seat of the Maclachlans, Argyll, E. side of Loch Fyne, 10 m. S.W. Inveraray.

Castle Law, hill, alt. 921 ft., N. spur of the Lammermuirs, in par. of, and 3 m. S.W. of, Gifford, E. Lothian.

Castle Law, conical hill, 2 m. S.E. of Forgandenny, S.E. Perthshire.

Castle Law, alt. 1595 ft., a summit of the Pentland Hills, 3 m. N.W. of Penicuik, Midlothian.

Castle Law, hill in par., and close to Linton vil., N.W. Peeblesshire.

Castle Law, alt. 1873 ft., hill in Yetholm par., Roxburghshire.

Castle Law, seat 3 m. N.W. of Coldstream, S.E. Berwickshire.

Castle Leod, seat, S.E. Ross and Cromarty, 1 m. N. of Strathpeffer.

Castle Loch, small loch, near Lochmaben, Dumfriesshire. Depth 18 ft., alt. 135 ft.

Castle Maoil, ruined stronghold on Kyle Akin Sound, S.E. Skye.

Castle Mearnaig or **Castle Glensands,** ruin on summit of rock on W. side of Loch Linnhe, Argyll, opposite Castle Coeffin.

Castle Menzies, seat, 1½ m. N.W. of Aberfeldy, Perthshire.

Castlemilk, ancient seat, once the property of the Bruces, 3 m. S.E. of Lockerbie, S. Dumfriesshire.

Castle Milk, seat, 1½ m. S. of Rutherglen, N.W. Lanarkshire.

Castle Newe, seat, W. Aberdeenshire, on r. Don, 10 m. S.W. of Rhynie.

Castle O'er, Roman camp, N. Dumfriesshire, at confluence of Rae Burn and White Esk.

Castle O'er, name given to a supposed Saxon camp at confluence of White Esk and Black Esk.

Castlepark, vil., 1 m. N. of Auchterarder, S.E. Perthshire.

Castle Rachal, ancient stronghold, Argyllshire, N.W. side of Lismore Isl.

Castle Rankine, stream and ham., S.E. Stirlingshire; the stream flows 4½ m into r. Carron, at Dennybridge. The ham., with a dye-work and a chemical work, stands on its bank, near site of ancient fortalice, 2 m. S.W. Denny.

Castle Roy, ruined fortalice, 3⅓ m. S. of Grantown, N.E. Inverness-shire.

Castles, hill, with camp, E. of Yester, E. Lothian.

Castles, cliff and cave, S. side of Ulva Isl., Argyll.

Castle Semple, loch and estate, with seat, Renfrewshire, close to Lochwinnoch. The loch, 2 m. by ½ m., receives the r. Calder, and the Black Cart flows out of it.

Castle Shuna, ruined fortalice, Argyll, on Shuna Isl., Loch Linnhe.

Castle Spiritual, ancient fort, in co. of, and 6 m. S.W. of, Inverness.

Castle Spynie, ancient and vitrified fort, 3 m. S.E. of Beauly, Inverness-shire.

Castle Stalker, ancient hunting seat of James IV., on insulated rock in Loch Linnhe, Inverness-shire.

Castle Stalker, ruined fort, Eriskay Isl., Inverness-shire.

Castle Stewart, ruined square tower, near rt. bank of r. Cree, Wigtownshire, 3 m. N.W. of Newton-Stewart.

Castle Stuart, seat of the Earl of Moray (1624), near Moray Firth, in co. of, and 6 m. N.E. of, Inverness.

Castle Swin, ruins on E. shore, and 2 m. from mouth, of Loch Swin, Argyll.

Castle Tirrim, ruined castle, Clanranald's stronghold, on isolated rock, S. side of Loch Moidart, S.W. Inverness-shire.

Castleton, par. and vil. (Newcastleton), with ry. sta., L.N.E., 21 m. S. of Hawick, Roxburghshire; par. 67,909 ac. Pop. 1794.

Castleton, ham., S.W. Argyll, 2½ m. S.E. of Lochgilphead. P.O., T.O.

Castleton of Braemar, see BRAEMAR.

Castle Toward, seat in Cowal dist., Argyll, 3 m. N.E. of Rothesay.

Castletown, vil., with harb., on Dunnet Bay, N. Caithness, 5¼ m. E. of Thurso. P.O., T.O. Pop. 600.

Castle Urquhart, ancient stronghold, N. Inverness-shire, on W. shore of Loch Ness, 1½ m. S.E. of Drumnadrochit.

Castle Varrich, ruined castle near the head of Kyle of Tongue, Sutherland.

Castlewalls, hill, 700 ft., 3 m. S.W. of Johnstone, Renfrewshire. Has remains of ancient camp on summit.

Castle Wemyss, seat of Lord Inverclyde, Renfrewshire, on Firth of Clyde, 1¼ m. N.W. Wemyss Bay.

Castlewigg, seat, 2¼ m. N.W. Whithorn, S.E. Wigtownshire.

Castramont, seat, Kirkcudbrightshire, on Water of Fleet, 3 m. N.W. of Gatehouse.

Catacol, fishing ham., N.W. coast of Arran Isl., Buteshire. Pop. 51.

Catcune, ruined castle, Roxburghshire, early seat of the Borthwicks, 4½ m. S.E. of Dalkeith.

Caterthun, White and **Brown,** two hills (alts. 976 and 945 ft.), with Caledonian forts, Angus, 5½ m. N.W. of Brechin.

Catfirth, ham., mainland of Shetland, 9 m. N.E. of Lerwick.

Cathcart, parl. div., city of Glasgow.

Cathcart, town, with ry. sta., L.M.S., and par. in cos. of Lanark and Renfrew, 1 m. S. of Langside. P.O., T.O. Pop. 2623; par. 3132 ac. Pop. 63,318.

Catherinefield, ham. in co. of, and 3 m. N.E. of, Dumfries.

Cathkin, seat N.W. Lanarkshire, 4 m. S.S.E. of Glasgow.

Cathkin Braes, hill ridge, alt. 691 ft., Lanarkshire, 2 m. E. of Carmunnock.

Cathlaw, seat, W. of E. Lothian, 3 m. N.E. of Bathgate.

Cat Law, alt. 2196 ft., spur of the Grampians, Angus, 6 m. N.W. of Kirriemuir.

Cat Lodge, seat, Inverness-shire, 7 m. N. of Dalwhinnie.

Catpair, seat, Midlothian, 1 m. N.E. of Stow.

Catrail, ancient ditch and wall, extending from Galashiels in Selkirkshire to Peel Fell, Northumberland; goes under such names as " The Catrail," " The Deil's Dyke," and " The Picts' Work Ditch." It appears to have been erected as a boundary, not a defence.

Catrine, town and ry. sta., L.M.S., mid Ayrshire, on r. Ayr, 2½ m. S.E. of Mauchline. P.O., T.O. Pop. 2169. Across r. is C. House. Town has cotton mfrs.

Catslack, stream, Selkirkshire, flowing 3 m. S.E. and E. to Yarrow Water, near Yarrow Fens.

Catstane, ancient monumental stone, Midlothian, between r. Almond and Gogar Burn, 3½ m. N.W. of Corstorphine.

Catter, mansion, S. Dunbartonshire, 1½ m. S.W. of Drymen.

Catterline, fishing vil. and ancient par., Kincardineshire, 4 m. S. of Stonehaven. P.O., T.O.

Cattofield, local service reservoir, Aberdeen.

Caulcots, ry. sta., L.N.E., Angus, 5 m. N. of Arbroath.

Cauldchapel, farm and two ancient camps, near Biggar, Lanarkshire.

Cauldcleuch, hill, alt. 1996 ft., Roxburghshire, 9 m. S.W. of Hawick.

Cauldhame, ham., Perthshire, 1½ m. S.W. of Kippen.

Cauldhane, ham., Stirlingshire, 1 m. from Bannockburn.

Cauldmill, ham., Roxburghshire, 2 m. N.E. of Hawick. P.O.

Cauldshields, hill, Roxburghshire, with ancient fort and loch, 1 m. S.E. of Abbotsford. Alt. 1076 ft.

Cauldside, ham., S.E. Dumfriesshire, 3 m. N.E. of Canonbie ry. sta.

Cauldstane Slap, pass, on Pentland Hills, Midlothian, 6½ m. S.E. of Mid Calder.

Caulkerbush, ham. and stream, S.E. Kirkcudbrightshire, 8 m. S.E. of Dalbeattie; burn rises in Clonyard, flows 2½ m. S.E. to Southwick.

Caulside, Dumfries. Town hospital.

Causewayend, pl. and ry. sta., L.N.E. (passenger service withdrawn), S.E. Stirlingshire, 1 m. S. of Manuel sta.

Causewayend, ham., E. Wigtownshire, 3 m. S.E. of Newton-Stewart.

Causewayend, Easter and **Wester,** two hams., Midlothian, 4 m. S.W. of Mid-Calder.

Causewayhead, vil., with ry. sta., L.N.E., in co. of, and 1½ m. N.E. of, Stirling. Paraffin works and quarry. Vil. pop. 998. P.O.

Causewayside, vil., N. Lanarkshire, close to Tollcross.

Causewaystones, vil., Lanarkshire, forming part of Blantyre, q.v.

Cava Island, isl., with ruined chapel, Orkney, 2 m. S. of Houston Head, mainland. Pop. 14.

Cavens, seat, close to Kirkbean, S.E. Kirkcudbrightshire.

Cavers, par., Roxburghshire, 3 m. N.E. of Hawick; par. 24,464 ac. Pop. 1179. C. House an old seat of the Douglases. John Leyden, the poet, was a native of the par.

Cavers Carre, seat, N.W. Roxburghshire, near Ale Water.

Cavers Hill, E. Selkirkshire, 1½ m. S.E. of Kirkhope; alt. 1209 ft.

Caverton Hill, ham., N.E. Roxburghshire, on Kale Water, 4½ m. S.S.E. of Kelso.

Cawdor, par., vil., and stream, Nairnshire; par. 34,744 ac. Pop. 767;

vil. 5½ m. S.W. of Nairn. P.O., T.O. Adjacent is C. Castle (1454), excellent example of the old feudal keep.

Cawpla, or **Caplaw,** hill and loch, 2 m. N.W. of Neilston, Renfrewshire; alt. of hill, 652 ft. Loch formed by a dam across the r. Patrick.

Ceannacroe, seat and deer forest, Inverness-shire, 3 m. S.W. of Dalreichart.

Ceannard, see KENNARD.

Ceannmor or **Kander,** loch, S.W. Aberdeenshire, 3 m. S. of Loch Callater.

Ceathragarbh, dist. of Edderachillis par., Sutherland. Rough, rocky peninsula between lochs Inchord and Latford.

Cellardyke, or **Nether Kilrenny,** fishing vil. and q.s. par., S.E. Fife, an eastward extension of Anstruther Easter, but forming part of the royal bur. of Kilrenny. P.O.

Central Station, terminus of L.M.S. in Glasgow.

Ceres, par. and vil., Fife; par. 10,063 ac. Pop. 1425; vil., 2½ m. S.E. o Cupar. P.O., T.O. Trade in linen.

Cessford, ham., N.E. Roxburghshire, 6 m. N.E. of Jedburgh.

Cessnock, small stream in Kyle dist. of Ayrshire, flowing from Distinkhorn Hill to join the Irvine, 3 m. S.E. of Kilmarnock.

Challoch, ham., N.E. Wigtownshire, 3 m. N.W. of Newton-Stewart.

Challoch, House, 3 m. N.W. of Stranraer, Wigtownshire.

Champfleurie, seat in W. Lothian, 2½ m. S.E. of Linlithgow. Shale-oil works.

Chance Inn, vil., 2 m. S.W. of Ceres, Fife.

Chanlock, rivulet, N.W. Dumfriesshire, flowing 6 m. S.E. to Scar Water, 5 m. N.W. of Penpont Bridge.

Channelkirk, par., N.W. Berwickshire; par. 14,197 ac. Pop. 418. Contains the ham. of Old Channelkirk.

Chapel, vil., Renfrewshire, between Neilston and Barrhead.

Chapel, ham. in W. Angus, 3 m. from Meigle.

Chapel, seat, 1 m. from Kettle, mid Fife.

Chapel, farm, with ruined chapel, close to Moffat, Dumfriesshire.

Chapel, ham. and seat, Fife, 2½ m. W. of Kirkcaldy. P.O.

Chapel and Stirling Bridge, vil., 5½ m. S.E. of Holytown, Lanarkshire, in Cambusnethan par.

Chapelden, pl. in N. Aberdeenshire, 8 m. W. of Fraserburgh, in Aberdour par.

Chapel Donan, pl., with old chapel, Ayrshire, and near Girvan.

Chapelgill, mt. summit, alt. 2282 ft., 5 m. S.W. of Drumelzier, Peeblesshire.

Chapelhall, town, with ry. sta., L.M.S., Lanarkshire, 2½ m. S.E. of Airdrie. P.O., T.O. Pop. 2126.

Chapelhill, ham., Perthshire, near r. Almond, 6 m. N.W. of Methven Junc.

Chapelhill, height, Dunbartonshire, near r. Clyde, and at terminus of Roman wall.

Chapelhope, farm and stream, S.W. Selkirkshire, ½ m. W. of Loch of the Lowes, into which the stream falls. Is the scene of Hogg's tale of the "Brownie of Bodesbeck."

Chapelhope Hotel, near Loch of the Lowes, Selkirkshire, 14 m. W. of Tweedsmuir. A Scottish Youth Hostel.

Chapelknowe, ham., S.E. Dumfriesshire, 6 m. W.S.W. of Canonbie. P.O., T.O.

Chapel Ness, prom., Fife, jutting into Forth, 1 m. S.W. of Elie.

Chapel of Garioch, par. and vil. (Pitcaple ry. sta.) in co. of, and 21 m. N.W. of, Aberdeen. P.O.; par. 13,059 ac. Pop. 1413.

Chapel Point, prom., N.E. Berwickshire, 1½ m. N.W. of Innerwick sta.

Chapelton, ham., Angus, 4½ m. N.W. of Arbroath.

Chapelton, q.s. par., Lanarkshire; par. (E. Kilbride, Glassford, and Hamilton pars.); vil. (Glassford par.). P.O., T.O.

Chapelton Point, headland in Cromarty Firth, opposite Invergordon, Ross and Cromarty.

Chapeltown of Glenlivet, vil., Banffshire, 13 m. S.E. of Ballindalloch ry. sta. P.O., T.O. called Chapeltown. See GLENLIVET.

Charing Cross, ry. sta., L.N.E., in W. of Glasgow.

Charleston, vil. on coast, N.E. Aberdeenshire, near St. Combs.

Charleston, Ross and Cromarty, on Beauly Firth, 2½ m. N.W. of Inverness (Black Isl.). Pop. 80.

Charleston, vil., 2 m. W. of Fraserburgh, Aberdeenshire.

Charleston, vil., Dunfermline par., Fife. Pop. (C. and Limekilns) 1061.

Charleston of Aboyne, see ABOYNE.

Charleston of Glamis, vil., 2½ m. S.E. of Glamis ry. sta., S.W. Angus.

Charleston of Nigg, vil., N.E. Kincardineshire, 1½ m. W. of Cove ry. sta.

Charleston of Paisley, suburb of Paisley, N.E. Renfrewshire.

Charlestown, seapt., vil., and ry. sta., L.N.E., 3½ m. S.W. of Dunfermline, S.W. Fife. P.O., T.O.

Charlestown of Aberlour, police bur., N.W. Banffshire, 4 m. N.W. of Dufftown. Pop. 1175.

Charnac, or **Chuinneag,** small loch (alt. 1500 ft.), Ross and Cromarty, 7½ m. S.W. of Kincardine.

Charnwood, Dumfries. Town hospital.

Charr Water, stream, 4½ m. long, trib. of the Dye Water, W. Kincardineshire.

Charterhall, seat, 3 m. N.E. of Greenlaw, ry. sta., Berwickshire.

Charters, or **Chesters,** vil., Roxburghshire, 7 m. S. of Jedburgh.

Charters-Chest, recess in the slope of Craig Cluny, Aberdeenshire, 1½ m. E. of Castleton of Braemar. Title Deeds of Invercauld were deposited here during the rebellion of 1715, hence the name.

Chatelherault, formerly summer house of the Duke of Hamilton, N.W. Lanarkshire, Hamilton par., now dismantled.

Cheallich, or **Teallach,** picturesque mt., alt. 3483 ft., 4 m. S.W. of Ullapool, Ross and Cromarty.

Cheese Bay, natural harbour, Invernessshire, N.E. of N. Uist, Outer Hebrides.

Cherrybank, vil., in co. of, and 1½ m. S.W. of, Perth. P.O.

Cherrytrees, seat, Roxburghshire, 1¼ m. N.W. of Yetholm.

Chesterhill, vil., 3 m. S.E. of Dalkeith, Midlothian.

Chesters, seat, Roxburghshire, on r. Teviot, 4 m. N.W. of Jedburgh.

Cheviots, broad range of lofty hills extending from Cheviot Hill 25 m. S.W. along the English Border to Peel Fell, in the cos. of Roxburgh and Northumberland. Cheviot itself is 2676 ft., Auchope Cairn, 2422 ft., Peel Fell, 1975 ft., Carter Fell 1815 ft.

Chicken Head, S. point of Eye peninsula, isl. of Lewis, Ross and Cromarty, 5½ m. S.E. of Stornoway.

Chirmory, loch, S.W. Ayrshire, 4 m. S.W. of Barrhill.

Chirnside, par., vil., and ry. sta., L.N.E., Berwickshire, 5 m. N.E. of Duns ; par. 5553 ac. Pop. 1414. P.O., T.O. The vil. stands 1 m. E. of ry. sta. Has mfrs. of woollens and paper. Vil. pop. 872.

Chisholm, seat, near Borthwick Water, W. Roxburghshire, 7¾ m. S.W. of Hawick.

Chisholm, extensive estate on N.W. border of Inverness-shire. Contains romantic gorge named Chisholms' Pass.

Choarie, or **Chorrie,** an isl. in Loch Eriboll, N. Sutherland.

Chon, loch (2½ by ½ m.), Perthshire, 7 m. N.W. of Aberfoyle.

Chon, loch, N. Perthshire, 8 m. W. of Struan.

Chravadail, loch, Harris, Inverness-shire, 20 m. N.W. of Tarbert.

Christ's Kirk, or **Rathmuriel,** ancient par. now forming part of Kennethmount par., W. Aberdeenshire.

Chroadle Bay, S. Uist, 2½ m. N.E. of Ben More, Inverness-shire.

Chroisg, loch, see Rosque.

Chroisg, loch, N.W. Ross and Cromarty, 7¼ m. N.N.E. of Ullapool.

Chryston, q.s. Cadder par., N.W. Lanarkshire.

Chryston and Muirhead, vil., 7½ m. N.E. of Glasgow. P.O. and T.O. named Chryston. Pop. 2933.

Ciaran, loch, Argyllshire, 6 m. N.E of Tayinloan, Kintyre.

Cir Mhòr, high peak, alt. 2618 ft., in N.E. of the Isl. of Arran, Buteshire.

Clachaig, pl., with inn, 5½ m. N.W. of Dunoon, Argyll. P.O., T.O.

Clachaig, one of a cluster of lochs, 3½ m. S.W. of Cairnbaan Inn, Crinan Canal, S.W. Argyllshire.

Clachamish, ham., on Loch Snizort Beag, Skye.

Clachan, vil., N.W. coast of Kintyre, 11½ m. S.W. of Tarbert, Argyll. P.O., T.O.

Clachan, ham. at the head of Loch Fyne, Argyll.

Clachan, loch in co. of, and 8 m. S. of, Inverness.

Clachan, or **Clauchan,** ham., with inn, Arran Isl., Buteshire, 2 m. N.E., of Blackwaterfoot.

Clachan, seat, S.W. Ross and Cromarty, head of Loch Carron.

Clachaneasy, ham., E. Wigtownshire, 8 m. N. of Newton-Stewart.

Clachan Gal, loch, N.E. Sutherlandshire, near Forsinard.

Clachan Heughs, rocky headland on W. side of Loch Ryan, Wigtownshire.

Clachan Inair, pl., with burying-ground, Inverness-shire, Glen Moriston.

Clachan of Glendaruel, see Glendaruel.

Clachan-Seil, pl., W. Argyll, 8 m. S.S.W. of Oban. P.O.

Clachan Sound, channel between isl. of Seil and mainland of Lorne, Argyll.

Clachantiompan, ancient memorial stone near Strathpeffer, Ross and Cromarty.

Clachbhein, hill, alt. 912 ft., isl. of Jura, Argyll, near the N. point.

Clachmore, see Clashmore.

Clachnaben, mt., alt. 1944 ft., W. Kincardineshire, 9 m. S.W. of Banchory.

Clachnaharry, ry. sta., L.M.S., in co. of, and 2 m. N. of, Inverness.

Clachnamben, two huge stones, Inverness-shire, on the moor near Kildermory.

Clach Ossian, large boulder in the Sma' Glen, close to the r. Almond.

Clachshant, or **Clayshant,** ancient par., now included in Stoneykirk par., Wigtownshire.

Clackmannan, par. and co. town, with ry. sta., L.N.E. (C. and Kennet), on r. Devon near its confluence with the Forth, 2 m. S.E. of Alloa. P.O., T.O. ; par. 8416 ac. Pop. 2585 ; pop. of town 1728.

Clackmannan Pow, harb. on N. side of the Forth, Clackmannanshire, 1½ m. S.E. of Alloa.

Clackmannan Road, ry. sta., L.N.E., Clackmannanshire, 2¼ m. E. of Alloa.

Clackmannanshire, smallest co. in Scotland ; bounded on the N. and E. by Perthshire and Fife, S.W. by Firth of Forth, and W. by Stirlingshire ; area, 54·6 sq. m. Pop. 31,948. It has the burghs of Alloa, Alva, Dollar, and Tillicoultry. Its chief antiquities are stone circle in Tillicoultry par., Sauchie Towers, Castle Campbell, and Cambuskenneth Abbey. Surface along the Forth is rich, but the co. also takes in part of the Ochil Hills, including Bencleuch, 2363 ft.

Clackmarras, ham., 4 m. S.E. of Elgin, Morayshire. P.O.

Cladach Point, prom., S.W. of Islay, Argyll.

Claddach, group of small lochs, E. coast N. Uist, Outer Hebrides.

Cladich, pl. with inn, N.E. Argyll, on E. side of Loch Awe. P.O., T.O.

Claggan, pl., Morven par., N.W. Argyll.

Claig Castle, ruined stronghold of the Lords of the Isles, Argyll, on Fraoch Isl., Sound of Islay.

Claigean Bay, small bay in Kildalton par., isl. of Islay, Argyll.

Clair, loch, S.W. Ross and Cromarty, 4 m. S. of Kinlochewe.

Clairdon, ham., N. Caithness, 2 m. E. of Thurso. P.O. named Claredon.

Clairinch, small isl., Stirlingshire, S. end of Loch Lomond.

Clans, Loch of the, 4 m. S.W. of Elgin. Shows remains of lake dwellings.

Clanyard, bay, (2m. by 1 m.), S.W. Wigtownshire, 5½ m. N.W. of the Mull of Galloway.

Claonaig, stream, Kintyre, Argyll, 3 m. S.W. of Skipness Pt. C. House is a seat.

Clarebrand, ham., 2½ m. N. of Castle-Douglas, Kirkcudbrightshire. P.O., T.O.

Claredon, P.O. at CLAIRDON, N. Caithness, which see.

Clarencefield, vil., S.W. Dumfriesshire, 6½ m. E. of Annan P.O.

Clarilaw, ham., Roxburghshire, 3 m. N.E. of Hawick.

Clarkston, q.s. par. and vil., with ry. sta., L.N.E., N. Lanarkshire, 1½ m. E. of Airdrie. P.O., T.O.

Clarkston and **Busby,** town and ry. sta., L.M.S., E. Renfrewshire, 6½ m. S. of Glasgow. P.O., T.O. Pop. 2978.

Clashcarnach, har. for lighthouse boats, 3 m. E. of Cape Wrath, Sutherland.

Clashgulloch, ham. on r. Stinchur, 3 m. N.E. of Barr, S. Ayrshire.

Clashindarroch State Forest, 6 m. S.W. of Huntly, Aberdeenshire.

Clashmack, hill, alt. 1229 ft., N.W. Aberdeenshire, 2 m. S.W. of Huntly.

Clashmore, pl. with inn, 4 m. S.W. of Dornoch, Sutherland. P.O., T.O.

Clash More, loch, S.W. Sutherland, 7 m. N. of Lochinver.

Clash More, loch, Sutherland, 9 m. N.W. of Oykell Bridge.

Clashnessie, ham. and bay, Sutherland, 5½ m. N.W. of Lochinver. P.O., T.O.

Clathick, seat, 4½ m. W. of Crieff, Perthshire.

Clathy, vil., S.E. Perthshire, 2½ m. N.W. of Dunning. Close to are seats of Clathy Park, Clathybeg, and Clathymore.

Clatt, par. and vil., W. Aberdeenshire. 3 m. S.W. of Kennethmont ; par. 5711 ac. Pop. 340. P.O., T.O. in vil.

Clatto, seat and hill (547 ft.). Fife, 5 m. S.W. of St. Andrews.

Clauchandolly, ham. in co. of, and 3½ m. S.W. of, Kirkcudbright.

Clauchlands Point, Arran Isl., Buteshire, 2 m. N.E. of Lamlash.

Clava, plain on borders of Inverness-shire and Nairnshire, 3 m. S.E. of Culloden Moor ry. sta.

Claverhouse, vil., S. Angus, 3½ m. N.E. of Dundee ; monument to Viscount Dundee here.

Clay, sea loch, S.E. Lewis, Outer Hebrides, 2 m. E. of Loch Seaforth.

Claybarns, vil., Midlothian, 1½ m. N.W. of Dalkeith.

Clayburn, stream issuing from Loch-na-Clay, E. Harris, Outer Hebrides.

Claygate, (**Gilnockie**), ry. sta., L.N.E., S.E. Dumfriesshire, 4¼ m. S.E. of Langholm.

Clayhole, sub. of Stranraer, Wigtownshire.

Clayhouses, vil., Midlothian, near Gorebridge ry. sta.

Claypots, old castle, 1 m. N. of Broughty Ferry, Angus.

Claypotts, Easter seat, Caputh par., Perthshire.

Clayquhat, seat, 7 m. N.W. of Blairgowrie, Perthshire.

Clearburn, small loch, Selkirkshire, 9 m. E.S.E. of St. Mary's Loch.

Cleat, seat, Orkney, near Pierowall vil., Westray Isl.

Cleaven Dyke, remains of fort, E. Perthshire, on the peninsula formed by the junction of rs. Isla and Tay ; said to be where Agricola's victory of the Grampians took place.

Cleddin, small burn, S. Dunbartonshire, enters Clyde at Old Kilpatrick.

Cleekhimin, Bothwell par., Lanarkshire. C. and Carfin special drainage dist. Pop. 2785.

Cleghorn, seat and ry. sta., L.M.S., in co. of, and 2 m. N.E. of, Lanark.

Cleikhimin, pl., 4 m. N. of Lauder, Berwickshire.

Cleish, par. and vil. in co. of, and 2 m. S.E. of, Kinross ; par. 6403 ac. **Pop.**

483. P.O., T.O. C. hills, 1241 ft., and C. Castle, a seat.

Cleland, town and ry. sta., L.M.S., Lanarkshire, on S. Calder Water, 3½ m. N.E. of Motherwell. P.O., T.O. Collieries and ironworks here. Pop. of C. and Omoa, 3355.

Clendrie Burn, small burn, trib. of Piltanton Burn, W. Wigtownshire.

Clenoch, Stranraer, Wigtownshire, County hospital.

Clenterty, pl. in co. of, and 7 m. S.E. of, Banff.

Cleongart and **Killocraw,** seat and shooting, S.W. Argyll, 8 m. N.W. of Campbeltown.

Clerkington, seat, 1¼ m. S.W. of Haddington, E. Lothian.

Clermiston, seat, 1 m. N. of Corstorphine, Midlothian.

Clestron, seat, sound, and skerries, Orkney mainland, 15 m. S.W. of Kirkwall.

Clett, rock, 2 m. N.N.W. of Thurso, Caithness.

Cleugh, small stream, Ayrshire, enters r. Ayr.

Cleughearn, seat, Lanarkshire, 6 m. S. of E. Kilbride sta.

Clibberwick Hill, hill, alt. about 500 ft., in isl. of Unst, Shetland.

Clickhimin, loch, Shetland, 1 m. S.W. of Lerwick.

Cliff, Loch of, in Unst Isl., Shetland. Receives the Burn of Baliasta, and sends off Cliff Burn to the Bay of Burrafirth.

Clifford Park, Menstrie, Clackmannanshire. Boys' preparatory school.

Clift Sound, Channel, Shetland, between E. and W. Burra Isls.

Clifton, vil., and Tyndrum Hotel, Perthshire, ½ m. N. of Tyndrum ry. sta.

Clifton, seat, 6 m. S.E. of Kelso, Roxburghshire.

Clifton Hall, boys' preparatory school, Midlothian, on r. Almond, 2 m. S.W. of Ratho Junction.

Clifton Park, seat, N.E. Roxburghshire, 6½ m. S.E. of Kelso.

Clintmains, ham., S.W. Berwickshire, near r. Tweed, 1½ m. N.E. of St. Boswells. P.O.

Clints Dod, mt., alt. 1307 ft., 8 m. S.W. of Dunbar, E. Lothian.

Clints of Dromore, hill, alt. 900 ft., S.W. Kirkcudbrightshire, 1 m. N. of Dromore.

Clippens, or **Clippens Square,** vil., 1 m. N.E. of Johnstone, Renfrewshire.

Cloan, seat, 2 m. S.E. of Auchterarder, Perthshire.

Clober, seat and bleachfield, Stirlingshire, ½ m. N.N.W. of Milngavie.

Clochan, vil. N.W. Banffshire, 2½ m. S.E. of Port Gordon. P.O., T.O.

Clochcan, ham., N.E. Aberdeenshire, 3 m. S.W. of Stuartfield.

Clochnaben, see CLACHNABEN.

Clochoderick, large boulder, 2 m. S.W. of Kilbarchan, Renfrewshire.

Cloch Point, headland with lighthouse, Renfrewshire, 2 m. S.W. of Gourock.

Clocksbriggs, pl., with ry. sta., L.M.S., 2½ m. N.E. of Forfar, Angus.

Cloffin, stream, Dumfriesshire, falls into Evan Water, 2½ m. N.W. of Moffat.

Cloghill, seat in co. of, and 5 m. N.W. of, Aberdeen.

Cloichfoldich, seat, Perthshire, 4 m. N.E. of Aberfeldy.

Clola, ham., E. Aberdeenshire, 3 m. S. of Mintlaw. P.O.

Cloncaird Castle, seat on Girvan Water, Ayrshire, 5 m. S.E. of Maybole.

Clonyard, hill (alt. about 500 ft.) and loch, S.E. Kirkcudbrightshire; loch 3½ m. S.E. of Dalbeattie.

Closeburn, par. and vil., with ry. sta., L.M.S., in co. of, and 11 m. N.W. of, Dumfries; par. 29,158 ac. Pop. 1074. P.O., T.O. Slate and stones quarried, and there are mineral springs. C. Castle in vicinity.

Cloud, loch in Lewis, Ross and Cromarty, 14 m. W.S.W. Stornoway.

Clousta, ham. in N. of Sandsting par., Shetland, on Clousta Voe. Good anchorage here.

Clova, q.s. par., ham., and seat, Angus, on r. Esk, 15 m. N.W. of Kirriemuir. P.O., T.O.

Clova House, 1½ m. W. of Lumsden, Aberdeenshire.

Clova Youth Hostel, 2 m. S. of Milton of Clova, Angus, and 13 m. N. of Kirriemuir. A Scottish Youth Hostel.

Clovenfords, ham., with ry. sta., L.N.E., Selkirkshire, on Caddon Water, 3¼ m. W. of Galashiels. P.O., T.O. Famous for its vineries.

Clovulin, vil., Argyll, 1 m. S.W. of vil. of Ardgour.

Clow, stream, Pettinain par., Lanarkshire, enters Clyde.

Cloy, stream, Buteshire, on E. of Arran Isl., flowing to Brodick Bay.

Clubbiedean, reservoir, Edinburgh Water Works, Midlothian, 1½ m. S.W. Colinton.

Cluden, small vil. on Cluden Water, S.W. Dumfriesshire, 3 m. N.W. of Dumfries.

Cluden Water, stream, S.W. Dumfries-shire, formed by the joining of the Cairn and Old Water of Cluden ; runs for 8 m. to join Nith at Lincluden Abbey.

Clugston, loch, E. Wigtownshire, 2 m. S.S.E. of Kirkcowan.

Clumlee, ham. and headland, Shetland, 17 m. S.W. of Lerwick. Baas of C. is a dangerous reef.

Clunas, pl., 8 m. S. of Nairn. P.O.

Clunes, pl., seat, and ry. sta., L.M.S., in co. of, and 7½ m. W. of, Inverness.

Clunes State Forest, nearest ry. sta., Speanbridge, 24 m. S.E. of Fort Augustus, Inverness-shire.

Clunie, par., N.E. Perthshire, 5 m. S.W. of Blairgowrie ; 11,869 ac. Pop. 479. On an islet in Clunie Loch are the remains of Clunie Castle. P.O.

Clunie, loch, on borders of Ross and Cromarty and Inverness-shire, 606 ft. above sea-level, with depth of 69 ft. and 16 m. W. of Fort Augustus.

Clunie Water, flowing into Loch Clunie. Also Clunie Bridge Inn, an anglers' resort.

Clunie and Ratagan, seat and deer forest, S.W. Ross and Cromarty.

Cluny, par., S.E. Aberdeenshire, 2 m. S.W. of Monymusk ; par. 10,049 ac. Pop. 951. Close by is C. Castle.

Cluny, vil., S.E. Fife, 4 m. N.W. of Kirkcaldy.

Cluny, ham. and castle, N.W. Kincardine-shire, 3 m. N. of Banchory.

Cluny Castle, seat in Strathdon, W. Aberdeenshire.

Cluny Castle, seat of the Cluny Macpher-sons, on r. Spey, 9 m. S.W. of Kingussie, Inverness-shire.

Cluny Hill Hydropathic, on Cluny Hill, S.E. of Forres, Morayshire.

Clunymore, Tips of, mt., alt. 1296 ft., 1½ m. N.E. of Dufftown, N. Banffshire.

Clyde (the *Clota* of Ptolemy), Scotland's most important r. ; rises abou an alt. of 1400 ft. in the extreme S.E. of Lanark-shire and has a total length of 106 m. ; its drainage area is about 1481 sq. m. Is navigable to Glasgow for all ocean-going vessels, but this was only achieved by expensive deepening and kept by constant dredging. The Vale of Clyde is famous for its horses, orchards, and coal and iron mines.

Clyde, vil., Old Monkland par., Lanark-shire, close to Coatbridge.

Clydebank, large industrial town, with ry. stas., L.N.E. and L.M.S., in Old Kilpatrick par., 6 m. N.W. of Glasgow, S. Dunbartonshire, on rt. bank of r. Clyde. Pop. 46,963. P.O., T.O. Has large engineering works and ship-building works, chemical works, and distilleries. Singer's sewing machine works at Kilbowie.

Clyde Ironworks, vil., Lanarkshire, on rt. bank of Clyde, 3 m. S.E. of Glasgow.

Clyde Law, hill, alt. 1789 ft., borders of cos. Lanark and Peebles, 8 m. N.W. of Moffat.

Clydesdale, vil., with ironworks, Lanark-shire, near Holytown Junction.

Clynder, vil., Dunbartonshire, on W. side of Gare Loch, 1 m. N.W. of Roseneath. P.O.

Clyne, par. and vil., S.E. Sutherland, 9 m. S.W. of Helmsdale ; 75,614 ac. Pop. 1723.

Clyneleish, distillery, Clyne par., 1½ m. N.W. of Brora ry. sta., Sutherland.

Cnoc Coirena Fearna, hill, alt. 1434 ft., S.W. Caithness, 12 m. N.W. of Berrie-dale.

Cnoe Cromuilt, hill, alt. 1199 ft., Caith-ness, 5 m. S.W. of Altnabreac.

Coalburn, mining vil. and ry. sta., L.M.S., Lanarkshire, 4 m. S. of Lesmahagow ry. sta. (784 ft.). P.O., T.O. Pop. 1681.

Coalden, vil., Fifeshire, 4½ m. N.W. of Kirk-caldy.

Coalford, vil., 2 m. W. of Ladybank, Fife.

Coalhall, vil. in co. of, and 7 m. E. of Ayr. Has extensive pottery works.

Coalheugh, Chalybeate spring close to Cromarty, Ross and Cromarty.

Coalsnaughton, mining vil., 1 m. S.E. of Tillicoultry, Clackmannanshire. Pop. C. and Devonside and Langour. 1240. P.O.

Coalstoun, ancient seat, on Gifford Water, 2¼ m. S. of Haddington, E. Lothian.

Coalton, vil., Fife, 1½ m. S.E. of Kettle.

Coaltown of Balgonie, vil., Fife, 2 m. S. of Markinch. P.O., T.O. Pop. 1013.

Coaltown of Wemyss, coast vil., Fife, 2 m. N. of Wemyss. P.O. Pop. 1259.

Coam, loch, Argyll, isl. of Islay, 2 m. N. of Bridgend.

Coatbank, see COATS and COATBANK LOW.

Coatbridge, mun. town and ry.stas.,L.N.E. and L.M.S., Lanarkshire, 9 m. E. of Glasgow, 34 m. W. of Edinburgh. P.O., T.O. Pop. 43,056. Its chief industries are ironworks, coal mining, and engineering.

Coatdyke, vil. and ry. sta., L.N.E., N.W. Lanarkshire, 1 m. E. of Coatbridge. P.O., T.O.

Coathill, Coatbridge, Lanarkshire. Town hospital and sanatorium.

Coats, q.s. par., includes part of Coatbridge.

Coats and Burnbank High, vil., Lanarkshire, now part of Coatbridge.

Coats and Coatbank Low, vil., Lanarkshire, part of Whifflet and Rosehall, q.v.

Cobbinshaw, or **Cobinshaw,** vil., ry. sta., L.M.S., Lanarkshire, 18½ m. S.W. of Edinburgh. Shale works. C. Reservoir is close to ry. sta. P.O., T.O. at ry. sta.

Cobbler, mt. and common name for Ben Arthur, q.v.

Cobblers of Lorn, three rocks near Black Mill Bay, isl. of Luing, Argyll.

Cochno, hill, loch, stream, and seat, S.E. Dunbartonshire; hill alt. 1140 ft., 2 m. N.W. of Duntocher; loch, N.E. side of hill; burn issues from loch. C. House to Duntocher Burn.

Cochrage Muir, tract of barren land, N.E. Perthshire, 4 m. N.W. of Blairgowrie.

Cochrane, ancient barony, Abbey par., Renfrewshire.

Cockairney, seat, Fife, 1½ m. W. of Aberdour.

Cockburnlaw, round hill, alt. 1065 ft., in par. of, and 4 m. N.W. of, Duns, Berwickshire.

Cockburnspath, coast par. and vil., with ry. sta., L.N.E., Berwickshire, 7 m. S.E. of Dunbar; par. 12,759 ac. Pop. 876; vil. pop. 236. P.O., T.O. Close by is C. Castle, an ancient tower.

Cock Cairn, mt., alt. 2387 ft., borders Aberdeen and Kincardine, 8. m S.E. of Ballater.

Cockenzie and Port Seton, seapt., police bur., and fishing centre, on Firth of Forth, 1 m. E. of Prestonpans, E. Lothian. Pop. 2526. P.O., T.O. Pier and lighthouse. Export trade in coal and fish. Open-air bathing-pool at Port Seton. C. House is a seat.

Cock Hill, mt., alt. 1960 ft., N.W. Kincardineshire, 11 m. S.W. of Banchory.

Cock Law, hill, alt. 1046 ft., East Lothian, 1½ m. N.W. of Oldhamstocks.

Cocklaw, pl. and remains of ancient camp in co. of, and 4 m. E. of, Lanark.

Cocklaw Castle, remains, 1½ m. S.E. Hawick, Roxburghshire.

Cockle, stream, W. Lothian, flowing to Firth of Forth at Dalmeny Park.

Cocklerue, or **Cuck-le-Roi,** conspicuous hill, alt. 912 ft., 2 m. S.W. of Linlithgow, W. Lothian.

Cock of Arran, alt. 976 ft., N. isl. of Arran, Buteshire. High prom.

Cockpen, par. and vil., 7½ m. S.E. of Edinburgh, Midlothian; par. 2938 ac. Pop. 6161. Has coal mines.

Cockpool, old fort, S. Dumfriesshire, 1¾ m. W. of Ruthwell.

Cockum, stream, cos. of Berwick and Midlothian, flowing into Gala Water at Stow.

Cog Burn, stream, rising N.W. border Dumfriesshire, and flowing 6 m. N.W. into Crawick Water.

Cogrieburn, pl., 7 m. N.W. of Lockerbie, Dumfriesshire.

Cogsmill, ham., W. Roxburghshire, 4½ m. S.E. of Hawick.

Coigach, dist. and ham., N.W. Ross and Cromarty, 8 m. N. of Ullapool.

Coignafearn, deer forest, Inverness-shire, S. of Tomatin.

Coila, r. in Ayrshire, see COYLE.

Coilantogle, once a ford on r. Teith, Perthshire, 2½ S.W. of Callander. Made famous by Scott in "Lady of the Lake." Now part of Glasgow waterworks.

Coileader, r., S.W. isl. of Mull, Argyll. It rises in Glen More, and after a run of 8 m. falls into Loch Scridain.

Coilessan, stream and glen, Ardgoil estate, Argyll, descending into Loch Long.

Coille, or **Castle Coille,** broch, Sutherland, on the Black Water, 7 m. W. of Brora.

Coillebhar, hill, 1747 ft., 3¾ m., W. of Alford, Aberdeenshire.

Coiltie, r., enters Loch Ness at mouth of Glen Urquhart, after a run of 9 m. from the S.W.

Coire, loch, in co. of, and 10½ m. S. of, Inverness.

Coire nam Meann and **Coire na Sith,** two lochs in E. Sutherland, 5 m. from the head of Loch Broom, depth 33 ft., alt. 801 feet.

Coiresalach, waterfall on r. Broom, N.W. Ross and Cromarty, 5 m. from the head of Loch Broom.

Coir-na-Fearn and **Ben Armine,** deer forest, Sutherland. See LOCH CHOIRE.

Coir nan Ceud Creach (*hollow of the hundred spoils*), valley, with cairns, on road between Kinlochewe and Torridon, W. Ross and Cromarty.

Coir nan Uriskin (*the Goblin's cave*), corrie in side of Ben Venue, at the E. end of Loch Katrine, mentioned by Scott in the "Lady of the Lake."

Colaboll, ham., 3 m. N.W. of Lairg, Sutherland.

Colbost, ham., Skye, 2 m. N.W. of Dunvegan.

Coldbackie, pl., 3 m. N.E. of Tongue, Sutherland. P.O., T.O.

Coldingham, par. on E. coast of Berwickshire, and vil., 3 m. W. of Eyemouth. P.O., T.O.; par. 25,379 ac. Pop. 2386. Has remains of a priory. C. Loch, 1½ m. N.E. of vil. Vil. pop. 354. Hotel.

Coldinghamshore, originally the name of St. Abb's, q.v., fishing vil., 1½ m. N.E. of Coldingham vil.

Coldside, ham., 5 m. N.E. Canonbie, Dumfriesshire.

Coldstone, par., Aberdeen, see LOGIE-COLDSTONE.

Coldstream, par. and vil., police bur., with ry. sta., L.N.E., S. Berwickshire, 13½ m. S.W. Berwick-upon-Tweed; par. 8324 ac. Pop. 1922. P.O., T.O. Coldstream Guards were raised here, 1659. Pop. bur. 1233.

Coldwells, pl., 10 m. E. of Ellon, Aberdeenshire.

Coleburn, distillery, N.E. Morayshire, 3 m. N.N.W. of Rothes.

Coleburns Mill, pl., 5 m. S. of Elgin, Morayshire.

Colfin, glen and ry. sta., 3 m. N.E. of Portpatrick, Wigtownshire.

Colgrain, seat, 2½ m. S.E. of Helensburgh, Dunbartonshire.

Coligarth, moorland tract, Isl. of Sanday, Orkney.

Colinsburgh, vil., 3 m. N.W. of Kilconquhar ry. sta., Fife. Pop. 257.

Colinton, vil., now part of Edinburgh, with ry. sta., L.M.S.; par. 5605 ac. Now part of the City of Edinburgh. P.O., T.O.

Colintraive, vil., with pier, Argyll, on Kyles of Bute, 7 m. N.W. of Rothesay. P.O., T.O. Steamers from Glasgow call here.

Coll, isl. and par., Argyll, 19 m. W. of Tobermory; 18,316 ac. Pop. 322. P.O., T.O. Black cattle, sheep, and pigs are exported. Hotel and good loch fishing.

Coll, vil. and burn, 6 m. N.E. of Stornoway, Lewis, Outer Hebrides.

Colla Firth, bay and ham., Shetland, E. side of Northmavine Isl.

Collace, par. and vil. in co. of, and 8 m. N.E. of, Perth; par. 3498 ac. Pop. 362.

Collairney Castle, ruins, 4½ m. S.E. of Newburgh, Fife.

Coll-Earn, seat in Auchterarder, Perthshire.

College, ry. sta., L.N.E., ¼ m. E. of Queen Street sta., Glasgow.

Collessie, par. and vil., with ry. sta., L.M.S., N. Fife; par. 8782 ac. Pop. 1955. Vil. 2 m. N. of Ladybank ry. junction. P.O., T.O.

Collie Law, hill, alt. 1255 ft., Berwickshire, 3⅓ m. N.W. of Lauder.

Collieston, fishing vil., E. Aberdeenshire, 6 m. E. of Ellon. P.O., T.O. Pop. 147. Coastguard sta.

Collin, vil., near Lochar Water, in co. of, and 3½ m. E. of, Dumfries. P.O.

Collin, seat, 8 m. S.E. of Castle-Douglas, Kirkcudbrightshire.

Collinswell, seat, 1 m. from Burntisland, Fife.

Colliston, q.s. par. and vil., with ry. sta., L.M.S., E. Angus, 3½ m. N.W. of Arbroath. C. House in vicinity.

Colliston, seat, 7 m. W. of Auldgirth ry. sta., S.W. Dumfriesshire.

Collsfield, seat, on Water of Faile, Ayrshire, 1¼ m. S.E. of Tarbolton.

Colmonell, coast par. and vil., S. Ayrshire; vil. stands on left bank of the r. Stinchar, 10½ m. S. by W. of Girvan. P.O., T.O.; par. 47,532 ac. Pop. 1713; vil. pop. 207.

Colmslie, ruined tower at the head of Fairy Dean, near Melrose, Roxburghshire.

Colonel's Bed, gorge, S.W. Aberdeenshire, 6 m. S.W. of Braemar.

Colonsay and **Oronsay,** two isls. in W. Argyll, separated only by a narrow channel, which is dry at low water. Colonsay, the most northernmost and larger of the two, is 8 m. in length and a breadth varying from $3\frac{1}{4}$ to $1\frac{1}{2}$ m. The surface rises to 493 ft.; par. 11,076 ac. Pop. 238. The most interesting antiquities are ecclesiastical and almost equal those of Iona. P.O., T.O. A steamer from Glasgow calls regularly.

Colonsay, Little, small isl. off W. coast of Mull, Argyll, and 1 m. S. of Ulva Isl. Pop. 2.

Colpleden, ham., 2 m. E. of Kettle, Fife.

Colpy, ham., mid Aberdeenshire, 4 m. E. of Insch. P.O.

Colquhonny, pl., with ruins of castle and inn, Aberdeenshire, on r. Don, 16 m. S.W. of Alford.

Colsay, small isl., W. side of mainland of Shetland, 8 m. N.W. of Sumburgh Head.

Colstoun, see COALSTOUN.

Coltbridge, dist. of S.W. Edinburgh, with ry. sta., L.M.S.

Coltfield, ham., with ry. sta., L.M.S., Morayshire, 8 m. N.W. of Elgin.

Colt Hill, mt., alt. 1961 ft., W. Dumfriesshire, $8\frac{1}{2}$ m. S.W. of Sanquhar.

Columba's Isle, at mouth of Loch Erisort, E. of Lewis, Ross and Cromarty.

Colvend, coast par. and ham., S.E. Kirkcudbrightshire, 5 m. S.E. of Dalbeattie; par. 18,142 ac. Pop. 1128 (Clovend and Southwick). P.O.

Colvister, ham., North Yell, Shetland, on Basta Voe.

Colzean, or **Culzean Castle,** seat of the Marquis of Ailsa, Ayrshire, 4 m. W. of Maybole.

Colzium, glen and seat and ruined castle, ry. sta., L.N.E., L.M.S., 1 m. N.E. of Kilsyth, S. Stirlingshire.

Comb Head, alt. 1998 ft., on borders of cos. Lanark and Dumfries.

Combs, St., vil., Aberdeenshire, see ST. COMBS.

Comely Bank, sub. of Edinburgh.

Comerton Home, $2\frac{1}{2}$ m. S.E. of Newport, Fife.

Comiston, dist., S. of the Braid Hills, Edinburgh.

Comlongan Castle, ancient remains of castle, 1 m. S.W. of Ruthwell, S.W. Dumfriesshire.; long the seat of the Murrays of Cockpool.

Commondyke, mining vil., Ayrshire, 3 m. N.E. of Cumnock; ry. sta., L.M.S.

Commonhead, pl., with ry. sta., L.N.E. (passenger service withdrawn), adjoining Airdrie, Lanarkshire.

Compass Hill, alt. 458 ft., Canna Isl., Argyll, N.W. of Canna harb.

Compstone, seat, on Tarf Water, in co. of, and $2\frac{1}{2}$ m. N. of, Kirkcudbright.

Comrie, par. and town, Perthshire, $6\frac{1}{2}$ m. W. of Crieff; par. 62,303 ac. Pop. 1775; pop. of town 925; ry. sta., L.M.S. P.O., T.O. Hotels.

Comrie, ham., Fife, 5 m. W. of Dunfermline, close by is C. Castle.

Comrie Ferry, on r. Lyon, 2 m. N. of Kenmore, Perthshire.

Con, loch, N. Perthshire, 6 m. N. of Kinloch Rannoch, depth 9ft.

Cona, r., N.W. Argyll, flowing $9\frac{1}{2}$ m. to Loch Linnhe, 6 m. S.W. of Fort William, at Inverscaddle Bay. Conaglen, seat of the Earl of Morton, $\frac{1}{2}$ m. N. of Inverscaddle Bay. Conaglen deer forest, 10,260 ac.

Conait, burn, N.W. Perthshire, flows $4\frac{1}{2}$ m. S.E. to r. Lyon, 8 m. N.W. of Killin.

Conan House, seat in Ross and Cromarty, $3\frac{1}{2}$ m. S. by W. of Dingwall.

Conchra, ham., 4 m. E. of Glendaruel, Argyll.

Condie, seat, 4 m. S.W. of Bridge of Earn, S.E. Perthshire.

Condorrat, q.s. par. and vil., $2\frac{1}{2}$ m. W. of Cumbernauld, Dunbartonshire. P.O. Pop. of vil. 839.

Coneach (**amhuinn na Connich**), stream, N. Argyll, flows 7 m. from L. Uisge to Loch Linnhe, which it enters at Loch a' Choire.

Conerock, hill, alt. 808 ft., $1\frac{1}{2}$ m. S.W. of Rothes, Morayshire.

Conglass, stream, S.W. Banffshire, flowing 8 m. S.W. to r. Avon, 3 m. N.W. of Tomintoul.

Conheath, seat in co. of, and $4\frac{1}{2}$ m. S.E. of, Dumfries.

Connal Burn, trib. of r. Nith, Ayrshire, at New Cumnock.

Connat, burn, Perthshire, out of Lochs Damh and Girre into r. Lyon at head of Glen Lyon.

6

Connel, shallow loch, W. Wigtownshire, 6 m. N.W. of Stranraer.

Connel Ferry, vil. and ry sta., L.M.S., at entrance to Loch Etive, 6 m. N.E. of Oban, Argyll. Hotel. The falls of Connel, or falls of Lora, are crossed by ry. bridge with a span of 500 ft., and with a track for motor-cars alongside line. P.O., T.O. called Connel. Pop. 178.

Connel Park, vil., 1½ m. S.W. of New Cumnock, E. Ayrshire.

Conningsburgh, ancient par. and vil. and hill range, E. coast of mainland of Shetland, 9 m. S.W. of Lerwick. P.O., T.O. under Cunningsburgh.

Conon, an important salmon r., Ross and Cromarty, formed by the meeting of the Meig and Sheen, after which it flows for 12 m. to enter the sea at the head upper waters of Cromarty Firth.

Conon Bridge, vil. in Urquhart and Logie Wester par., Ross and Cromarty, 2½ m. S.S.W. of Dingwall; ry. sta., L.M.S. (Conon), ¼ m. S. of C. Bridge. P.O., T.O. Pop. 350.

Content, sub. of the town of Ayr.

Contin, par. and ham., E. Ross and Cromarty, 4 m. S.W. of Strathpeffer; par. 197,493 ac. Pop. 1105. P.O., T.O.

Conval Meikle and **Little,** two summits, mid Banffshire, 3 m. S.W. of Dufftown; alts. 1867 and 1810 ft.

Convalmore, distillery, near Dufftown, Banffshire.

Convinth, ancient par., now part of Kiltarlity par., Ross and Cromarty.

Cookney, q.s. par. and ham. Fetteresso par., Kincardineshire, 5½ m. N.W. of Stonehaven.

Cook's Cairn, mt., alt. 2478 ft., mid Banffshire.

Cookston House, seat, ½ m. N. of Brechin, Angus.

Coolin Hills, see CUILLIN HILLS.

Coomb Dod, mt., alt. 2082 ft., S.E. Lanarkshire, 7½ m. E. of Abington.

Coomb Hill, mt., alt. 2096 ft., W. Peeblesshire, 7 m. S.W. of Broughton.

Cooms Fell, mt. ridge between cos. of Dumfries and Roxburgh; alt. 1609 ft.

Copay, isl. in the Sound of Harris, Ross and Cromarty.

Copinshay, isl. (1 m. × ½ m.), Orkney, S.E. coast of mainland. Pop. 25.

Shows a white group-flashing light, visible 22 m. Horse of C., an islet lying ½ m. off N.E.

Copland Road, ry. sta., Glasgow dist.

Coppercleuch, pl., S. Selkirkshire, near W. shore of St. Mary's Loch. P.O., T.O. called Cappercleuch.

Corah, ruins of old castle, 3½ m. E. of Urr, Kirkcudbrightshire.

Coran of Portmark, mt., N. Kirkcudbrightshire, 3 m. W. of Carsphairn; alt. 2042 ft.

Corbelly N. and **S.,** pls., 3 m. S. of New Abbey, Kirkcudbrightshire.

Corbet, old tower, now renovated, 1½ m. S.E. of Morebattle, E. Roxburghshire.

Corbie, loch, E. Aberdeenshire, 2 m. from Parkhill.

Corbie Hall, sub. of Bo'ness, W. Lothian.

Corellan, islet, S. Knapdale, Argyll.

Corennie Hills and **Forest,** 5 m. S.E. of Alford, Aberdeenshire; Benaquhallie, the highest peak, 1621 ft.

Corgarff, q.s. par., Strathdon par., Aberdeenshire. C. Castle on site of ancient stronghold. P.O.

Corgarff Hostel, on the S. side of the Don, 1½ m. below Cockbridge, Aberdeenshire. A Scottish Youth Hostel.

Corkerhill, vil., between Bellahouston and Crookston, Glasgow; ry. sta., L.M.S.

Corkindale Law, hill, Renfrewshire, 2½ m. W.S.W. of Neilston; alt. 848 ft.

Cormorant's Cave, basaltic cavern, 224 ft. long, and 50 ft. high, isl. of Staffa, Argyllshire.

Cornhill, vil., with ry. sta., L.N.E., 8½ m. S.W. of Banff. P.O., T.O.

Cornsilloch, vil. in par. of Dalserf, Lanarkshire.

Cornton, pl. in co. of, and 1¾ m. N. of, Stirling.

Corpach, vil., 1 m. W. of Banavie, Inverness-shire; ry. sta., L.N.E. P.O., T.O. Pop. 209.

Corr, loch, Sutherlandshire, 6 m. S.E. of Altnaharra. Hotel.

Corr, loch, N.E. Ross and Cromarty, 2 m. N.W. of Culrain.

Corra Linn, second of the four falls on r. Clyde, in co. of, and 1½ m. S. of, Lanark.

Corran, headland, at junction of Lochs Goil and Long, Argyll.

Corran Ferry, ferry across west end of W. Loch Tarbert, Argyll.

Corran Narrows, N. and **S.,** strait, between Loch Linnhe and E. end of Loch Eil, Inverness-shire. Ferry between C. Inn, Inverness-shire, 8¾ m. S.W. of Fort William and Ardgour ham., Argyll.

Corrichie, Howe of, boggy marsh, borders of Aberdeenshire and Kincardine-shire, close to Hill of Fare (1562 ft.). C. burn flows 7 m. to Loch Drum. Scene of battle, 1562. Earls of Huntly and Moray.

Corrie, coast vil. and harb., Arran, Buteshire, 5 m. N.E. of Brodick. Hotel. P.O., T.O. Pier for small boats only.

Corrie, ancient par., N.E. Dumfriesshire, now part of Hutton par.

Corrie Common, pl., S. Dumfriesshire, 4 m. N.E. of Lockerbie. P.O., T.O.

Corriecravie, ham., isl. of Arran, Bute-shire, 2 m. S.E. of Blackwaterfoot.

Corrie Funerach, loch, S.W. Ross and Cromarty, 4 m. from Lochcarron vil.

Corriegills, point, E. coast of Arran, Buteshire, directly S. of entrance to Brodick Bay. Bold eruptive rock very conspicuous.

Corriehallie and **Glengowrie,** seat and deer forest, S.E. Ross and Cromarty, 12 m. W. of Muir of Ord ry. sta.

Corrie Kinloch and **Glendhu,** deer forest, W. Sutherland House, 4 m. N.W. of Overscaig Hotel.

Corrieknowes, farm, 1½ m. W. of Annan, Dumfriesshire.

Corriemoillie, seat, S.E. Ross and Cromarty, 3 m. N.W. of Garve ry. sta.

Corriemuchloch, ham., Perthshire, 1½ m. S.W. Amulree.

Corriemulzie, stream, with many cascades, Aberdeenshire, falling through a deep ravine to the Dee, 3 m. S. of Braemar.

Corriemulzie Lodge, on Corriemulzie R., 6 m. S.W. of Oykell Bridge, Ross and Cromarty.

Corrie Varkie, seat, N.W. Perthshire, on Loch Ericht, 10 m. S.W. of Dalwhinnie.

Corrievrechan, strait and whirlpool, be-tween isls. Jura and Scarba, Argyll.

Corrieyairack, pass, corrie, and hill, Inverness-shire. General Wade's road reaches its highest level, 2507 ft., over this pass, 8 m. N.W. of Loch Laggan; alt. of hill, 2922 ft.

Corrigour, seat, S. Inverness-shire, on shore of Loch Lochy, 3 m. S.W. of Laggan Locks.

Corrimony, seat, N.E. Inverness-shire, in Glen Urquhart, 9 m. N.W. of Drumna-drochit.

Corrour, ry. halt and siding, L.N.E., S.E. Inverness-shire, 7 m. N.W. of Rannoch ry. sta. P.O.

Corrour and **Fersit,** deer forest, S. Inver-ness-shire, near Loch Treig, on Rannoch Moor.

Corry, seat, near Broadford, Skye.

Corrybeg, ham., Inverness-shire, on N. shore of Loch Eil, 8 m. N.W. of Fort William.

Corryhabbie Hill, two summits, 7½ m. S.W. of Dufftown, Banffshire; alts. 2563 and 2393 ft.

Corryhallie, deer forest, Ross and Crom-arty, 10 m. W. of Beauly.

Corryhollie, deer forest, Inverness-shire, 5 m. E. of Fort William.

Corry Lair, loch, S.W. Ross and Cromarty, 3 m. N.W. of Auchnashellach ry. sta.

Corsancone, hill, Ayrshire, 3 m. N.E. of New Cumnock; alt. 1547 ft.

Corsbie, ruined tower, 1½ m. N.E. of Legerwood, Berwickshire.

Corsbie, pl., near Newton-Stewart, Wig-townshire.

Corse, hill, stream, seat, and old castle, S. Aberdeenshire. Hill 1383 ft. Burn flows to Leochel Burn. C. House, 3½ m. N.W. of Lumphanan ry. sta. Castle dates 1581. P.O.

Corse, seat, 5 m. E. of Huntly. P.O.

Corse, pl., 6 m. S.W. of Alford, Aberdeen-shire.

Corseglass, pl., 5 m. N.E. of New Gal-loway, N. Kirkcudbrightshire.

Corsehill, seat and remains of Castle Stewarton, N. Ayrshire.

Corsemill, or **Crossmill,** vil., 1 m. N.N.E. of Barrhead, Renfrewshire.

Corserine, mt., Kirkcudbrightshire, 11 m. N. of Newton-Stewart; alt. 2668 ft.

Corsewall, seat and ruined castle, and headland with lighthouse, Kirkcolm par., W. Wigtownshire.

Corsford, pl., Renfrewshire, 1½ m. S.W. of Johnstone.

Corsindae, pl., S. Aberdeenshire, 5 m. S. of Monymusk.

Corskie, hill, 5 m. S. of Huntly, Aberdeen-shire; alt. 1369 ft.

Corsock, q.s. par., Balmaclellan, Kirk-patrick-Durham and Parton, mid Kirk-cudbrightshire. P.O., T. O.

Corstorphine, now part of the city of Edinburgh, q.v.

Cortachy and Clova, par., N.W. Angus; 62,924 ac. Pop. 537. Cortachy vil., 5 m. N. of Kirriemuir. P.O., T.O. C. Castle, seat of the Earl of Airlie, in vicinity.

Cortes, ham., N.E. Aberdeenshire, near Lonmay sta.

Coruisk, dark loch in S. of the Cuillin Hill, S. Skye, near the head of Loch Scavaig.

"Rarely human eye has known
A scene so stern as that dread lake."

Cor Water, one of the head streams of the Tweed. Joins Glencraigie below Tweedhope, Peeblesshire.

Coshieville, small ham. and hotel, Perthshire, 5¼ m. W. of Aberfeldy.

Cossins, pl., with sculptured obelisk, 5 m. S.W. of Forfar, Angus.

Costa, headland and ham., N. end of mainland, Orkney.

Cotburn, hill, 4 m. N.E. of Turiff, Aberdeenshire ; alt. 559 ft.

Cotehill, loch, 1 m. N.W. of Slains, Aberdeenshire.

Cothal, pl., N.W. Aberdeenshire, on r. Don, 7½ m. N.W. of Aberdeen. P.O. Has tweed and woollen mills.

Cothiemuir, upland, with stone circle, Aberdeenshire, N. of Kincardine-O'Neil.

Cottack, vil. in co. of, and 9 m. N.W. of, Dumfries.

Cotton of Lownie, vil., 3 m. S.E. of Forfar, Angus.

Cottown, vil., 1 m. E. of Rhyne, Aberdeenshire.

Couffurach, vil., 2 m. N.E. of Buckie, N. Banffshire.

Coul, seat, S.E. Ross and Cromarty, 7 m. S. of Dingwall.

Coulatt, loch and stream, S.E. Morayshire. Stream enters Spey 1 m. S.E. of Knockando church.

Coulin, deer forest and loch, Ross and Cromarty, 5 m. N. of Auchnashellach ry. sta.

Coull, par., S. Aberdeenshire, 3 m. N.W. of Aboyne ; 9044 ac. Pop. 486. Remains of C. Castle here, and C. House is a seat.

Coull, a mining ham., 2 m. N.W. of Markinch, Fife.

Coulmony, seat, in co. of, and 9 m. S.E. of, Nairn.

Coul More, or **Coulmore,** seat and deer forest, Ross and Cromarty, 1 m. from Balmacarra.

Coulnakyle, see REVACK.

Coulport, ham., with pier, Dunbartonshire, on E. shore of L. Long, 4 m. N.W. of Cove. A ferry between Coulport and Ardentinny.

Coulter, see CULTER ; ry. sta., L.M.S., 1¼ m. S.W. of Biggar, Lanarkshire. P.O., T.O.

Coultra, ham. and hill (584 ft.), Fife, 8½ m N.W. of Cupar.

Coultree (Loch Coal na Doire), loch, S. Inverness-shire, 4 m. N.W. of Dalwhinnie ry. sta.

Coultry, loch, W. Ross and Cromarty, near Loch Torridon.

Countam Hill, mt., N.W. Dumfriesshire, 7½ m. N.W. of Thornhill ; alt. 1640 ft.

Countesswells, pl. and old mansion, 4½ m. W. of Aberdeen. P.O., T.O.

Coupar-Angus, town and police bur. and par., with ry. sta., L.M.S., Perthshire, 15¾ m. N.E. of Perth ; par. 4699 ac. Pop. 2435. P.O., T.O. Remains of abbey and Roman camp. Pop. of town, 1883. Has linen and jute factories, preserve works, etc.

Cour, seat, Kintyre, Argyll, 7 m. N.E. of Carradale.

Cour, r., S. Inverness-shire, falls into the Spean, 3 m. E. of Spean Bridge.

Courance, ham. and seat, Dumfriesshire, 9 m. N.W. of Lockerbie. P.O., T.O.

Courthill, see KISHORN.

Cousland, vil., E. Midlothian, 3½ m. N.E. of Dalkeith. P.O.

Cousland, pl., with shale and oil works, W. Lothian, 1½ m. W. of Livingstone.

Couston, ancient castle, W. Lothian, near Bathgate. Refuge Stone, Boar Stone, and Standing Stones in the N.E.

Coustonholm Paper Mill, Renfrewshire, 1 m. from Pollokshaws.

Couthally Castle, see COWTHALLY.

Couttie, ham., E. Perthshire, 1 m. N.W. of Coupar-Angus.

Cove, fishing vil., with ry. sta., L.M.S., Kincardineshire, 5 m. S. of Aberdeen. P.O., T.O., at sta. (Cove Bay).

Cove, ham., near Cockburnspath, Berwickshire.

Cove, a seat, 1 m. W. of Kirkpatrick ry. sta., Dumfriesshire.

Cove, Dunbartonshire, ham. and watering place, E. entrance to Loch Long. P.O., T.O. C. and Kilcreggan, Pop. 954.

Cove a Chiaran, a cave near Campbeltown, Argyll.

Cove and Kilcreggan, police bur. and watering pl., S.W. Dunbartonshire, 4½ m. N.W. of Greenock. P.O., T.O. called Cove. Pop. 954. Steamboat communication with Glasgow, Greenock, etc.

Covesea, vil., Morayshire, 6 m. N.W. of Elgin. Among the rocks are sculptures. C. skerries, a dangerous reef close by.

Covington and Thankerton, united pars. of Lanarkshire; 5114 ac. Pop. 408. Contains vils. Covington and Thankerton; ry. sta., L.M.S. (Thankerton). C. has ruins of castle.

Cowal, Argyll, large dist., including land between Loch Fyne and the boundary with Perthshire.

Cowdenbeath, police bur., Fife, chiefly a mining town; ry. sta., L.N.E., 6 m. N.E. of Dunfermline. P.O., T.O. Pop. 12,732.

Cowden Castle, seat, S. Perthshire, 2 m. N.E. of Dollar. Built on the site of an ancient fort.

Cowdenhill, ham., near Bo'ness, W. Lothian.

Cowdenknowes, seat, Berwickshire, on Leader Water, 1 m. S. of Earlston.

Cowglen, seat and ham., Renfrewshire, 2 m. W. of Pollokshaws.

Cowhill, tower seat in co. of, and 4½ m. N.W. of, Dumfries.

Cowie, vil., seat, harb., stream (Cowie Water), Kincardineshire. C. Water, after a run of 13 m., falls in Stonehaven Bay; vil. and harb. are 1 m. N. of Stonehaven, and close to it is C. House.

Cowie, vil., St. Ninian's par., Stirlingshire. Coal mining is the chief industry. Ry. sta. (Plean), L.M.S., 1 m. S. of Cowie. P.O., T.O. Pop. 2672.

Cowiefauld, ham., 2 m. W. of Strathmiglo, Fife.

Cowies Linn, waterfall in Eddleston Water, Peeblesshire; leap, 35 ft.

Cowlairs, sub. of Glasgow; ry. sta., L.N.E. Locomotive works.

Cowpits, vil., Midlothian, 1½ m. S. of Musselburgh.

Cowshaven, cavern, Aberdeenshire, near Aberdour, used as a hiding-place by Lord Pitsligo after Culloden (1745).

Cowthally, ruined castle, Lanarkshire, 1½ m. N.W. of Carnwath.

Coylum Bridge, ham., Inverness-shire, 2 m. S.E. of Aviemore.

Coyle, or **Coila,** stream, Ayrshire, rises N. of Dalmellington, and after a run of 14 m. falls into the r. Ayr, 4 m. E. of the town of Ayr.

Coylton, par. and vil. in mid co. of., and 5⅓ m. S.E. of, Ayr; par. 11,589 ac. Pop. 2366. P.O., T.O. Collieries, quarries of sandstone and limestone worked here. Vil. also known as Hillhead of New Coylton.

Cracail, loch, Sutherland, 3 m. S.E. of Lairg.

Cragganmore, pl. and distillery, W. Banffshire, close to Ballindalloch. P.O.

Craggie, or **Creagach,** loch, Sutherland, 4½ m. N.E. of Lairg. Depth 40 ft.

Craggie, loch, 6 m. E. of Tongue, Sutherland.

Craggie, loch between Oykell Inn and Aultnacealgach, Sutherland.

Craibstone House, seat in co. of, and near, Aberdeen.

Craichie, vil., Angus, 3½ m. S.E. of Forfar. P.O.

Craig, ham. and par., Angus; 4324 ac. Pop. 1565. Vil. 1½ m. S.W. of Montrose. Salmon and other fisheries.

Craig, seat, Aberdeenshire, 2 m. W. of Kennethmont ry. sta.

Craig, seat, Ayrshire, 2 m. N.E. of Colmonell.

Craig, seat, in N. Ayrshire, 4 m. S.W. of Kilmarnock.

Craig, seat, N.E. Argyll, 1 m. from Dalmally.

Craigallion, seat and loch, Stirlingshire, 2 m. W. of Strathblane sta.

Craig Alvie, hill, with Waterloo Cairn on summit, Inverness-shire, 4 m. N.E. of Kincraig ry. sta.; alt. 1175 ft.

Craigancoin, or **Craiganoin,** pass, 1 m. S.E. of Moy church, Inverness-shire.

Craig and Millenderdale, seat and shooting, Ayrshire, 3 m. N.E. of Colmonell.

Craig an Fhithich (*raven's rock*), rock close to the ry. between Strathpeffer and Auchternead, Inverness-shire.

Craig-an-Fhithich, W. side of Loch Oich. It was the motto of the Macdonnells of Glengarry.

Craigangower, height, Kincardineshire, 5⅓ m. W. of Banchory; alt. 892 ft.

Craiganour or **Craganour,** shooting-lodge, Perthshire, N. shore of Loch Rannoch.

Craiganroy, small harb., Ross and Cromarty, S. corner of Loch Duich.

Craig an Tarmachain, mt., S. Morayshire, 3 m. E. of Advie ry. sta. ; alt. 2121 ft.

Craigarestie, summit of Kilpatrick hills, Dunbartonshire, 1½ m. N.E. of Bowling ; alt. 1165 ft.

Craigbank, vil., Ayrshire, 2 m. S.W. of New Cumnock. Pop. 2328.

Craigbarnet, seat, Stirlingshire, 2 m. W. of Campsie Glen ry. sta.

Craigbeg, hill, Kincardineshire, 6 m. S.E. of Banchory ; alt. 1054 ft.

Craigben, seat, isl. of Mull, Argyll, ½ m. from Lochbuie.

Craigbhockie and **Craigboddich,** two high cliffs, Sutherland, in Glen Loth, 4½ m. N. of Clyne.

Craigcaffie Castle, old square tower, Wigtownshire, 3½ m. N.E. of Stranraer.

Craig Castle, seat, Aberdeenshire, 2 m. N.W. of Lumsden.

Craigcleuch, seat, Dumfriesshire, 2 m. N.W. of Langholm.

Craig Cluny, huge cliff overhanging the public road, W. Aberdeenshire, 2 m. E. of Castleton.

Craigcrook Castle, seat in S.W. Edinburgh ; once the residence of Lord Jeffrey (1773–1850).

Craigdaimve, sea inlet, W. side of N. Knapdale, Argyll.

Craigdam, ham. on r. Ythan, Aberdeenshire, 1½ m. S.W. of Tarves.

Craigdarroch, seat, Dumfriesshire, 2 m. W. of Moniaive.

Craigderg, granite ridge in co. and par., near Loch Dochfour, Inverness.

Craigdhu, height, Inverness-shire, on left bank of r. Truim, 6 m. S.W. of Kingussie.

Craigdhuloch, cliff, Aberdeenshire, at the head of Glen Muick, overhanging Loch Dubhloch.

Craigdow, loch, W. Ayrshire, 3 m. S.W. of Maybole.

Craigearn, pl., Aberdeenshire, 2½ m. E. of Monymusk.

Craigellachie, vil., distillery, with ry. sta., L.N.E., W. Banffshire, on r. Spey, 13 m. S.E. of Elgin. P.O., T.O. Hotel.

Craigellachie, rock, near Aviemore ry. sta., Invernesshire. The warcry of the Clan Grant is, " Stand fast, Craigellachie."

Craigencat, hill on N. border of Dunfermline par., Fife, ¾ m. E. by S. of Loch Glow ; alt. 921 ft.

Craigend, vil. in co. of, and 2 m. S. of, Perth.

Craigend, ham. and moor, 2 m. S.E. of Strathblane, Stirlingshire.

Craigend Castle, 2 m. S.W. of Strathblane ry. sta.

Craigend, loch, 3 m. N.W. of vil. of New Abbey, Kirkcudbrightshire.

Craigend, seat and vil., 3 m. S.E. of Edinburgh.

Craigendarroch, wooded hill, W. Aberdeenshire, on left bank of r. Dee, opposite to Ballater ; alt. 1250 ft.

Craigendoran, pl., with ry. sta., L.N.E., and steamboat pier, ¾ m. S.E. of Helensburgh. Centre for Clyde steamers. Pop. 82. Hotel.

Craigend Park, Liberton, Edinburgh. Boys' public school.

Craigends, seat, on r. Clyde, Renfrewshire, 3 m. N.W. of Johnstone.

Craigengar, hill near the border of cos. Lanark, Midlothian, and Peebles, and 16 m. S.W. of Edinburgh ; alt. 1700 ft.

Craigengelt, seat, Stirlingshire, 5 m. W. of Denny.

Craigengillan, mt., Kirkcudbrightshire, 4 m. N.E. of Carsphairn ; alt. 1300 ft.

Craigengower, hill, Ayrshire, 1 m. E. of the vil. of Straiton ; alt. 1086 ft.

Craigenputtock, farm, Dumfriesshire, par. of Dunscore. C. Moor (1038 ft.), 15 m. N.W. of Dumfries. Carlyle resided here (1828–1834) and wrote *Sartor Resartus*.

Craigenscore, mt., (Creag an Sgor) Aberdeenshire, 3 m. N.W. of Glenbucket ; alt. 2076 ft.

Craigenterive, seat, Argyll, 9½ m. from Ardrishaig.

Craigentinny, dist. of N.E. Edinburgh.

Craigenveoch, seat, Wigtownshire, 3 m. S.E. of Glenluce.

Craigflower, boys' preparatory school, Torryfarn, Fife.

Craigfoodie, seat, Fife, 2 m. N.W. of Dairsie sta.

Craigford, vil., Stirlingshire, 1 m. from Bannockburn.

Craigforth, seat, in co. of, and 2 m. N.W. of, Stirling.

Craig Gibbon, mt., Perthshire, 3½ m. W. of Dunkeld ; alt. 1263 ft.

Craig Gowan, wooded hill (1437 ft.), Aberdeenshire, 1 m. S.E. of Balmoral. Many cairns erected by members of the Royal family.

Craighall, vil. on r. Ayr, in co. of, and 4 m. N.E. of, Ayr.

Craighall, Perthshire, seat of the Rattrays, 3 m. N. of Blairgowrie.

Craighall, old mansion, Fife, 3 m. S.E. of Cupar.

Craighall, New, q.s. par. and colliery vil., Midlothian, 2 m. S.W. of Musselburgh.

Craighall, Old, ham., Midlothian, 1½ m. S. of Musselburgh.

Craighead, vil., Perthshire, 1 m. N.W. of Almondbank.

Craighead, Stirlingshire, 1½ m. E. of Milton of Campsie.

Craighead, on r. Clyde, Lanarkshire, 1 m. S. of Bothwell.

Craighead, pl., Ayrshire, 7 m. S.W. of Dunlop.

Craigheads, vil., now joined to Barrhead, Renfrewshire.

Craighelder, mt., Kirkcudbrightshire, 6 m. N.E. of Newton-Stewart; alt. 1971 ft.

Craighirst, one of the Kilpatrick hills, Dunbartonshire, 2½ m. N. of Duntocher; alt. 1074 ft.

Craighlaw, seat, Wigtownshire, 1¼ m. N.W. of Kirkcowan.

Craighorn, hill, in co., 7½ m. N.E. of Stirling; alt. 1904 ft.

Craig Hostel, on N. shore of Loch Torridon, between Diabeg and Red Point, Rossshire. A Scottish Youth Hostel.

Craighouse, ham., with inn, Argyll, E. of isl. of Jura. P.O., T.O. Good harb. and quay.

Craig House, mansion, now used as an asylum, in the S. of Edinburgh.

Craigie, par. and vil., Ayrshire, 4 m. S. of Kilmarnock; par. 6576 ac. Pop. 470. Has ruins of old castle. P.O.

Craigie, vil., W. Lothian, 1 m. W. of Cramond Bridge; also Craigie Hall.

Craigie, vil., Perthshire, 4 m. W. of Blairgowrie. P.O.

Craigie, sub. of Perth. P.O.

Craigie, seat in co. of, and on r., Ayr, 1¼ m. S.E. of Ayr.

Craigie, seat, 2 m. N.E. of Dundee, Angus.

Craigie, pl., Aberdeenshire, 1½ m. S.E. of New Machar sta.

Craigiebarns, hill, near Dunkeld, Perthshire; alt. 900 ft.

Craigiebuckler, q.s. par., Aberdeenshire, now Newhills and Peterculter. C. House, 2 m. S.W. of Aberdeen. The Macaulay Institute for Soil Research.

Craigieburn, seat, on Moffat Water, N. Dumfriesshire, 3 m. E. of Moffat. Close by is Craigieburn wood, mentioned in Burns's songs.

Craigielands, vil., Dumfriesshire, 2½ m. S.W. of Moffat. Nearby is C. House.

Craigie Thieves, mt., Angus, 5 m. N.E. Inverharity; alt. 2256 ft.

Craigievar, ham., Aberdeenshire, 4 m. N.W. of Lumphanan ry. sta. C. Castle here. P.O. C. Castle, a seat of Lord Sempill.

Craiglea, hill and slate quarry, Perthshire, 6½ m. N.W. of Methven Junction; height of hill 1737 ft.

Craigleith, sub. of Edinburgh, ry. sta., L.M.S.

Craigleoch, cliff, E. Perthshire, in gorge of r. Ericht.

Craiglockhart, seat and hill, part of Edinburgh, S.W.; ry. sta., L.N.E.

Craiglockhart Castle, ruin near Lanark, on Mouse Water.

Craiglour-achin, hill, close to Lochnagar, Aberdeenshire. On it Queen Victoria had a cairn erected to the memory of the Prince Consort.

Craiglure Lodge, Ayrshire, 6 m. S.E. of Straiton.

Craiglush, loch, E. Perthshire, 1¼ m. N.E. Dunkeld.

Craigmaddie, seat, with ruin of moated tower, Stirlingshire, 2 m. N.E. of Milngavie.

Craigmark, mining vil., 1 m. N.E. of Dalmellington, Ayrshire.

Craigmarloch, vil., Dunbartonshire, 1 m. S.E. of Kilsyth.

Craigmill, vil., Stirlingshire, at S. base of Abbey Craig.

Craigmill, pl., Perthshire, 1½ m. N. of Rattray.

Craigmillar and **Castle,** ham. and ruins of castle, once the home of Mary, Queen of Scots. Now a sub. of S.E. Edinburgh. P.O., T.O.

Craigmore, steep hill, Perthshire, 1 m. N.W. of Aberfoyle; alt. 1271 ft.

Craigmyle, seat, Aberdeenshire, 1 m. E. of Torphins sta.

Craignaban, wooded hill, Aberdeenshire, in Crathie par., 1½ m. S.E. of Abergeldy; alt. 1736 ft.

Craig na Caillich, summit, Perthshire, 3 m. N.W. of Killin; alt. 2990 ft.

Craignafeile, stack, Isle of Skye, near N.E. coast at Loch Staffin.

Craignaiolar, hill, Inverness-shire, 3½ m. N.N.W. of Duthill; alt. 1750 ft.

Craig-nan-Eun State Forest, 2½ m. S.W. of Inverness.

Craigneil Castle, old tower, Ayrshire, 1 m. S. Colmonell vil.

Craignethan Castle, ruins in co. of, and 4½ m. N.W. of, Lanark. Said to be the Tullietudlem of Scott's *Old Mortality.*

Craigneuk, mining vil., N. Lanarkshire, 1¾ m. W.N.W. of Wishaw. P.O., T.O.

Craignish, par., W. Argyll, near Crinan Canal. Has vil. of Ardfern, with P.O., T.O.; par. 928: **ac.** Pop. 280.

Craignure, steamboat pier and ham., E. of isl. of Mull, Argyll, 2 m. N.W of the head of Loch Don. P.O., T.O.

Craigo, vil., with ry. sta., L.M.S., Angus, 5 m. N.W. of Montrose. P.O., T.O. C. House, 1½ m. S. of sta.

Craigoch, stream, W. Wigtownshire, flows 4½ m. to the sea at Dunskey Castle.

Craig of Madderty, estate, Perthshire, 2 m. S.W. of Madderty sta. Has the vil. of St. David's.

Craigowl, summit of Sidlaw Hills, 4 m. S.W. of Glamis, Angus; alt. 1250 ft.

Craig Phadrig, wooded hill, with vitrified fort, in Co. of, and 2½ m. W. of, Inverness; alt. 430 ft. State forest here. St. Columba is said to have met King Brude at this hill in 565.

Craigrath, mt., Kincardineshire, 5 m. N. of Banchory; alt. 1429 ft.

Craigrie, vil. in co. of, and near town of, Clackmannan.

Craigrossie, conspicuous summit of the Ochils, Perthshire, 2 m. S.E. of Auchterarder; alt. 1349 ft.

Craigrothie, vil., Fife, 2 m. W. of Ceres.

Craigrownie, q.s. par., Roseneath par., S.W. Dunbartonshire.

Craigroy, height, Ross and Cromarty, 5 m. E.S.E. of head of Loch Maree.

Craigroyston, or **Rob Roy's Cave,** cave on E. side of Loch Lomond, near Inversnaid, Stirlingshire.

Craigruie, seat, W. Perthshire, on N. side of Loch Voil.

Craigs, ham. near Liberton, part of S.W. Edinburgh.

Craigs, Hightown of, pl., Dumfriesshire, 3 m. S.E. of Dumfries. Pop. (Netherwood, Kelton, and Craigs special water dist.), 340.

Craigs, The, pl., Ross and Cromarty, 1 m. from Ardgay. P.O., T.O.

Craigskean, old baronial fortalice, Ayrshire, in Maybole par.

Craigston, pl. on Barra Isl., Invernessshire.

Craigston, pl., Renfrewshire, 1 m. S. of Johnstone.

Craigston Castle, seat, 4 m. N.E. of Turriff, Aberdeenshire. Founded 1604-7.

Craigthornhill, seat, Lanarkshire, 5 m. S.E. of Hamilton.

Craigton, estate, with old mansion (1635), Dunbartonshire, on C. burn, 2 m. N.W. of Milngavie.

Craigton, vil., S. Angus, 5 m. N.W. of Carnoustie.

Craigton, vil., W. Angus, 4 m. S.W. of Kirriemuir. P.O.

Craigton, seat, W. Lothian, 2 m. N.W. of Winchburgh.

Craigton, ham., Aberdeenshire, N. of Culter.

Craig-y-Barns, or **Craigwood,** wooded hill, Perthshire, near Dunkeld; alt. 558 ft.

Craik Cross Hill, mt., N.E. Dumfriesshire, 6 m. N.E. of Eskdalemuir; alt. 1482 ft.

Crail, parl. and royal bur. and par., Fife, with pier. P.O., T.O., and ry. sta., L.N.E., 11 m. S.E. of St. Andrews; par. 6348 ac. Pop. 1596; town 1058. Industry, fishing, chiefly herring. Watering place, hotels, and golf links.

Crailing, par. and vil., Roxburghshire, 3½ m. N.E. of Jedburgh; par. 5974 ac. Pop. 475. P.O. C. House is a seat.

Cramond, vil. and par., forming part of Edinburgh; ry sta., L.M.S., Barnton. C. House 1 m. S. of vil. C. Bridge. Hotels. C. is a summer resort. P.O. C. Isl. in the Forth 1 m. from the shore, and is accessible at low water.

Cran, loch, in co. of, and 4 m. N.E. of, Nairn.

Crane, loch, Lanarkshire, 3¼ m. N.W. of Dunsyre.

Cranley Moss, bog, Lanarkshire, 1 m. N. of Carstairs.

Cranloch, pl. in Morayshire, 2 m. E. of Lhanbryde.

Crann, loch, W. Ross and Cromarty, 4½ m. W. of Achnasheen; depth 17 ft.

Crannoch, Grange par., Banffshire, special water dist. Pop. 65.

Cranshaws, par. and ham., N. Berwickshire, 9 m. N.W. of Duns; par. 6045 ac. Pop. 119. P.O., T.O.

Cranston, par., Midlothian, 4½ m. S.E. of Dalkeith. Extent, 4324 ac. Pop. 761. Coal, limestone, and sandstone.

Crarae, pl., with granite quarries, Argyll, W. side of Loch Fyne, 9 m. S.W. of Inveraray.

Craskie, seat and deer forest, Invernessshire, forms part of Glencannich.

Crathes, pl., with ry. sta., L.N.E., N.W. Kincardineshire, 2½ m. E. of Banchory. P.O., T.O. C. Castle 1½ m. N.W.

Crathie, vil. near Balmoral, Aberdeenshire.

Crathie and Braemar, united pars.; 182,257 ac. Pop. 1245. Contains vils. of Crathie and Braemar. H.M. residence at Balmoral. Many high mountains including Ben Macdhui, 4296 ft. P.O., T.O.

Crathie Point, headland, N. Banffshire, 2½ m. E. of Cullen.

Craufurdland Castle, on Craufurdland Water, 3 m. N.E. of Kilmarnock, Ayrshire.

Crawford, par., vil., and ry. sta., L.M.S., Lanarkshire, on r. Clyde, 3 m. S.E. of Abington. P.O., T.O. Hotels. Par., 65,485 ac. Pop. 1572. Clyde crossed here by a chain bridge; vil. pop. 337.

Crawford Priory, seat, Fife, near r. Eden, 3 m. S.W. of Cupar.

Crawfordjohn, par. and vil., Lanarkshire, 4 m. W. of Abington; 26,369 ac. Pop. 572.

Crawfordton, seat, W. Dumfriesshire, 2 m. S.E. of Moniaive.

Crawick, burn, Dumfriesshire, after a run of 8 m. falls into the Nith near Sanquhar.

Crawick Bridge, ham., Dumfriesshire, 7 m. W. of Leadhills. Pop. (Crawick special water dist.) 270.

Crawick Mill, suburb of Sanquhar, Dumfriesshire.

Crawley, reservoir near Milton Bridge, 6½ m. S. of Edinburgh.

Crawton, fishing vil. on coast, Kincardineshire 4 m. S. of Stonehaven.

Cray, loch and pl., with church, Perthshire, on Skee Water, 15 m. N.W. of Blairgowrie; close to it is C. House.

Creach, hill in Renfrewshire, 5 m. S.W. of Greenock; alt. 1446 ft.

Creachben, mt., Argyll, Isle of Mull; alt. 2289 ft.

Creagan, ferry, inn, and ry. sta., L.M.S.; side of Loch Creran, N. Argyll.

Creag Dhu Hostel, Brig o' Turk, between Loch Achray and Loch Vennachar, Perthshire, 7 m. W. of Callander. A Scottish Youth Hostel.

Creag Meaghaidh, mt., Inverness-shire, 4 m. E. of Loch Laggan; alt. 3700 ft.

Creagorry, pl., 4 m. S.E. of Nunton, Lochmaddy, Inverness-shire, P.O., T.O.

Creagvain, mt., S.W. Banffshire, 5 m. S. of Tomintoul.

Creak, vil., Aberdeenshire, 3½ m. S.W. of Rhyne.

Creanwall Isles, 2 islets, Barra par., Hebrides.

Creca, vil., Dumfriesshire, 5 m. N.E. of Annan.

Cree, r. in Galloway, issuing from Loch Moan; alt. 675 ft.; on boundary of cos. Ayr and Kirkcudbright; has a run of 25 m. to join the sea in Wigtown Bay. It is navigable for small vessels to Carty. Salmon, sea and brown trout fished.

Creebridge, vil., Kirkcudbrightshire, on r. Cree, opposite Newton Stewart. Pop. (C. and Minnigaff) 405.

Creed, or **Amhuinn Ghride** (*stony river*), important salmon r., Lewis, Ross and Cromarty; enters sea at Stornoway after a run of 9½ m., during which it traverses two lochs.

Creetown, small seaport, with ry. sta., L.M.S., W. Kirkcudbrightshire, at the head of Wigtown Bay. P.O., T.O. Hotel. Granite quarries worked here. Pop. 759.

Creggans, ferry across Loch Fyne, Argyll, at Strachur.

Creich, pl. in S.W. Isl. of Mull, Argyll.

Creich, par., S. Sutherland, contains Invershin ry. sta., L.M.S., and vil. of Bonar Bridge; 110,737 ac. Pop. 1515.

Creich, par., N.E. Fife, 4½ m. N.W. of Cupar; 2338 ac. Pop. 272.

Creran, sea loch, glen, and stream, N. Argyll, between Appin and Ardchattan. Loch spanned by ry. bridge, L.M.S.

Crianlarich, vil., with ry. stas., L.M.S. and L.N.E., W. Perthshire, 5½ m. S.E. of Tyndrum. P.O., T.O. Hotel. Fine bridge of 300 ft. span across the Fillan Water.

Crianlarich Hostel, Perthshire; at L.N.E. sta., 13 m. W. of Killin, and 41 m. E. of Oban. A Scottish Youth Hostel.

Crib Law, hill (1389 ft.), Selkirkshire, 11 m. S. of Hawick.

Crichhope Linn, dell, with waterfalls on C. Burn, near Closeburn ry. sta., Dumfriesshire.

Crichie, hill (500 ft.), Aberdeenshire, 1¾ m. S. by W. of Inverurie.

Crichie House, seat, N.E. Aberdeenshire, Old Deer par., ¾ m. S.E. of Stuartfield.

Crichton, par. and vil., Midlothian, contains vil. of Pathhead; 5498 ac. Pop. 797. Rocks mostly carboniferous limestone. C. moss is 900 ft. above sealevel. Near vil., ruins of C. Castle.

Crichton, pl., 16 m. N. of Dunscore, Dumfriesshire.

Crieff, town (police bur.) and par., Perthshire, with ry. sta., L.M.S. P.O., T.O. Town stands in a picturesque spot above the banks of the r. Earn, 100 to 400 ft. above sea-level; has fame as a health resort. 3 m. N. is the Sma' Glen, with Ossian stone. Par. 11,965 ac. Pop. 6058, of which the town is 5544. Hydro and hotel.

Crieff Junction, ry. sta., L.M.S., Perthshire, 2 m. S.W. of Auchterarder.

Criffel, hill ridge (1867 ft.), Kirkcudbrightshire, 2 m. S. by W. of New Abbey vil.

Crige Duibhe (Loch na Creige Duibhe— *Loch of the Black Rock*), loch, Inverness-shire, 3 m. S. of Loch Morar; alt. 360 ft., depth 93 ft.

Crimond, coast par. and ham., E. Aberdeenshire; ham., 3 m. E.S.E. of Lonmay ry. sta.; par. 6281 ac. Pop. 535. Has the fishing ham. of Rattray and Rattray Head, with lighthouse. P.O.

Crimonmogate, par., 2 m. E. of Lonmay, N.E. Aberdeenshire; a seat of the Earl of Southesk.

Crinan, vil., canal, loch, and pier. Canal between Loch Fyne and Sound of Jura, 9 m. long; 15 locks, 10 ft. deep. Vil. with P.O., T.O., and hotel near W. end. Steamers from Glasgow and Greenock call.

Cringie Law, mt., W. Roxburghshire, 4 m. N.W. of Hawick; alt. 1155 ft.

Cringletie, seat, in co. of, and 3 m. N.W. of, Peebles.

Crocach or **Crokach,** loch, 3 m. N. of Lochinver, N.W. Sutherland.

Crocketford, vil., N.E. Kirkcudbright, 10 m. N.E. of Castle Douglas. P.O., T.O.

Croftfoot, ry. sta., L.M.S., S.E. Glasgow.

Crofthead, vil. and ry. sta., L.N.E. (passenger service withdrawn), W. Lothian; forms part of Fauldhouse, 6 m. S.W. of Bathgate.

Crofthead, near Ayr. County hospital.

Croft Head, mt., Dumfriesshire, 4 m. E. of Moffat; alt. 2085 ft.

Croftmartaig, ham., Perthshire, close to Acharn, q.v.

Croftness, ham., 6 m. S.E. of Ballindalloch, Banffshire.

Croftouterly, part of Leslie, Fife.

Crofts, camp, 4 m. N.W. of Castle-Douglas, Kirkcudbrightshire.

Croggan, pl., Torsay par., isl. of Mull, at entrance to Loch Spelve. P.O., T.O. On pier, fixed white light, visible 5 m.

Crogo, ham., N.E. Kirkcudbrightshire, 1 m. N.W. of Corsock.

Croick, q.s. par. in Kincardine par., Ross and Cromarty, 10 m. W. of Ardgay.

Crom (*crooked*), loch, Ross and Cromarty, 7½ m. N.W. of Loch Glass.

Cromalt Hills, range on the boundaries of Sutherland and Ross, 9 m. N.E. of Ullapool; highest pt., Meall Coir an Lochan, 1692 ft.

Cromar, dist. of Aberdeenshire, extending from Aboyne to Badenoch.

Cromarty, mun. bur., seapt., and par., Ross and Cromarty, 5 m. E. of Invergordon. P.O., T.O.; par. 8017 ac. Pop. 1232; town 837. Industries: fishing, making sail-cloth. Lighthouse has a wireless sta. Hugh Miller, the geologist, born here. Ferry service with Nigg.

Cromarty Bay and **Firth,** bay an expansion of the Firth, the latter an arm of the sea. 18 m. inland to near Dingwall entrance are two bold headlands called the " Sutors of Cromarty."

Cromartyshire, now entirely joined with Ross-shire.

Crombie, vil. and ancient par., latter now part of Torryburn par., Fife. P.O.

Crombie, pl., Fife, 12 m. E. of Dundee.

Crombie Castle, ancient seat near Aberchirder, Banffshire; also C. Burn.

Cromdale, Inverallan, and **Advie,** par. and ry. sta., Morayshire, in Strathspey, 3 m. N.E. of Grantown ; par. 63,354 ac. Pop. 2948, of which Grantown has 1577. P.O., T.O. Pop. of C. Bridge, 71.

Cromore, ham., with small harb., E. of Lewis, Ross and Cromarty. P.O.

Cromwell Park, vil., with bleachworks, Perthshire, 1½ m. N.W. of Almondbank.

Cromwell's Fort, remains of forts at Leith, Inverness, and Ayr.

Cromwell's Mount, small mound near Dunbar, where Cromwell is said to have stood while directing the battle of Dunbar.

Crona, two small islets, Assynt par., Sutherland, 1 m. S.W. of Oldany Isl.

Cronberry, vil., with ry. sta., Ayrshire, 2 m. N.E. of Lugar. P.O.

Crongart, loch, 3 m. N.E. of Barrhill, S.W. Ayrshire.

Crook, inn in Tweedsmuir par., 7 m. S. of Broughton, Peeblesshire. P.O.

Crookedholm, vil., Ayrshire, 1½ m. E. of Kilmarnock. Pop. 648.

Crooked Loch, very small loch on the boundary between cos. Roxburgh and Selkirk, 9 m. S.W. of Hawick.

Crook of Alves, vil., ½ m. N. of Alves, Morayshire. Pop. 220.

Crook of Devon, vil., with ry. sta., L.N.E., on r. Devon, in co. of, and 6 m. S. of, Kinross.

Crookston, seat, S.E. Midlothian, 2 m. N. of Fountainhall ry. sta.

Crookston, pl. and ry. sta., 5½ m. W. of Glasgow, Renfrewshire. P.O., T.O.

Croot, small loch, 3 m. N.E. of Kirkmichael, Ayrshire.

Crospol, loch, ½ m. W. of Durness church, Sutherland, and close to the manse.

Cross, q.s. par., now part of Barvas par., Isle of Lewis, Ross and Cromarty, Outer Hebrides.

Crossaig, ham., Argyll, E. coast of Kintyre, 6 m. S.W. of Skipness.

Crossall, height, W. Lothian, surmounted by a stone cross, 1¾ m. E.S.E. of Queensferry.

Cross and Burness, united par., Orkney, isl. of Ronaldshay and part of Sanday; 8265 ac. Pop. 971. P.O., T.O.

Crossbasket, seat, Lanarkshire, 1 m. S.W. of High Blantyre sta.

Crossbost, vil., isl. of Lewis, Ross and Cromarty, 9 m. S.W. of Stornoway. P.O., T.O.

Crossford, ham., 1½ m. W. of Dunfermline, Fife. P.O. Pop. 365.

Crossford, vil., Lesmahagow par., Lanarkshire. P.O., T.O. Pop. of C. and Hazelbank, 618.

Crossford, ham. and ry. sta., Dumfriesshire, 3 m. N.W. of Dunscore.

Crossgatehall, ham., 2 m. S.E. of Inveresk, Midlothian.

Crossgates, mining vil., ry. sta., L.N.E., Fife, 3½ m. N.E. of Dunfermline. P.O., T.O. Pop. 2354.

Crossgills, ham. near Ruthwell, Dumfriesshire.

Crosshall, mining vil., Stirlingshire, 2 m. S.E. of Falkirk.

Crosshands, vil., Ayrshire, 2 m. N.W. of Mauchline.

Crosshill, sub. of Glasgow, with ry. sta., L.M.S. P.O., T.O.

Crosshill, q.s. par. and vil., 3 m. S.E. of Maybole, Ayrshire. P.O., T.O. Vil. pop. 545.

Crosshill, q.s. par., Old Monkland par., N. Lanarkshire.

Crosshill, ham., 4 m. N.W. of Chapelton, Lanarkshire.

Crosshill, 6 m. S.W. of Strathaven, S.W. Lanarkshire.

Crosshouse, q.s. par. and vil., with ry. sta., Ayrshire, 2 m. W. of Kilmarnock. P.O., T.O. Vil. pop. 824.

Crosshouses, ham., 2 m. S.E. by E. of Kettle vil., Fife.

Cross Isle, isl., Shetland, at entrance of Quendal Bay, 3½ m. W.N.W. of Sumburgh Head.

Crosslaw, dist. in co. and par. of Lanark. Pop. 459.

Crosslee, vil., with ry. sta. (Houston), L.M.S., N. Renfrewshire, 5 m. W.N.W. of Paisley. Pop. 355.

Crosslee, ham., Midlothian, 3 m. S. of Stow.

Crossmichael, par. and vil., with ry. sta., Kirkcudbrightshire, 3½ m. N.W. of Castle-Douglas : par. 9931 ac. Pop. 1160. P.O., T.O.

Crossmount House, seat, near r. Tummel, 3 m. S.E. of Kinloch Rannoch, N.W. Perthshire.

Crossmyloof, sub. and ry. sta., S.W. Glasgow.

Crosspol, bay in S. of Coll Isl., Argyll.

Crossraguel, ruined abbey (1240), Ayrshire, 2 m. S.W. of Maybole.

Cross Roads, or **Grange Cross Roads,** ham., Banffshire, 4 m. N.E. of Keith. P.O. named C. Roads.

Crossroads, ham., 5 m. S. of Falkirk, Stirlingshire.

Crossroads, pl., 6 m. W. of Kilmarnock, Ayrshire.

Crossroads, pl., 9 m. E. of Thurso, Caithness.

Crossroads, pl., 5 m. N.E. of Blackshiels, S.W. East Lothian.

Crossroads, pl., N. Kincardineshire, 2 m. S. of Crathes.

Crossroads, in Wemyss par., Fife. Pop. 313.

Cross Street, sub. of Stornoway, Lewis, Ross and Cromarty.

Crosston, ham., Angus, 6 m. N.E. of Forfar.

Cross Wood, reservoir, Edinburgh Waterworks, 9 m. S.W. of Balerno.

Crottinloan, seat, Perthshire, 2 m. S.E. of Pitlochry.

Croulin Island, Applecross par., S.W. Ross and Cromarty, off entrance to Loch Carron.

Crovie, fishing vil., 9 m. E.N.E. of Macduff, Banffshire. Pop. 170.

Crowe Water, stream, S.W. Ross and Cromarty, flowing to head of Loch Duich.

Crowlista, pl., Uig par., Lewis. Ross and Cromarty.

Crow Row, pl., Lanarkshire, forming part of Garnkirk.

Croy, vil., ry. sta. (L.N.E.), and quarries, Dunbartonshire, 1¾ m. S.S.E. of Kilsyth. P.O. Pop. 98.

Croy and **Dalcross,** ham. and par. in cos. of Nairn and Inverness ; ham., 8 m. S.W. of Nairn ; par. 17,954 ac. Pop. 911. P.O., T.O.

Cruach, mt., Perthshire, 1½ m. N.W. of Loch Lydoch ; alt. 2420 ft.

Cruachan, Falls of, cascade in the Pass of Brander, near head of Loch Awe, Argyll.

Cruachlussa, height, Argyll, 3½ m. S.W. of Ardrishaig ; alt. 1530 ft.

Crubenmore, seat, 8½ m. S.W. of Kingussie, S.E. Inverness-shire.

Cruden, coast par., vil., and ry. sta. (C. Bay passenger service withdrawn), Aberdeenshire, 8 m. S.E. of Peterhead ; par. 18,236 ac. Pop. 2554, C. Bay. P.O., T.O. Pier at Port Erroll. Pop. of vil. 446. Hotel.

Cruden Water, flows 11 m. S.E. to Cruden Bay, at Port Erroll.

Cruffell, mt., N.W. Dumfriesshire, 6 m. S.W. of Sanquhar ; alt. 1825 ft.

Cruggelton, ancient par., now part of Sorbie par., Wigtownshire. Has remains of castle.

Cruicksfield, seat, 4 m. N.E. of Duns, Berwickshire.

Cruick Water, stream, N.E. Angus, flows 16 m. E. to North Esk, at Stracathro.

Cruline House, seat, isl. of Mull, Argyll, 3½ m. S.W. of Salen.

Crutherland, seat, Lanarkshire, on Calder Water, 2½ m. S.E. of East Kilbride.

Cruys, mt., N. Angus, 12 m. N.W. of Edzell ; alt. 2424 ft.

Crynock, stream, N.E. Kincardineshire, flows into r. Dee at Inch of Culter.

Cuaich, loch, Inverness-shire, 8½ m. S.W. of Kingussie.

Cuan Sound, narrow channel between isls. of Teil and Luing, Argyll.

Cuen, loch in Kildonan par., Sutherland.

Cuffabouts, ham., W. Lothian, 1½ m. E.S.E. of Bo'ness.

Cuil, bay, in Loch Linnhe, Argyll.

Cuil, loch, 2 m. E. of Altnabreac, Caithness.

Cuilabhacaidh (Coldbackie), vil., 3 m. N.E. of Tongue, Sutherland. P.O.

Cuilhill, vil., with ry. sta., L.N.E., Lanarkshire, 2 m. W. of Coatbridge.

Cuillie, or **Culaidh,** small loch, Sutherland, 2½ m. from Forsinard ry. sta.

Cuillin Hills, group of the most romantic hills in Britain. Chiefly composed of Gabbro, they represent the wildest and most picturesque scenery in Scotland. Situated at the head of Loch Scovaig in Skye, they can be approached from the sea or from Glen Brittle or Sligachan.

Cuiltrannich, ham., 9 m. N.E. of Killin, Perthshire.

Cuilunum Moss, ham., 2 m. N.W. of Port of Menteith ry. sta., Perthshire.

Cuithir, or **Quire,** remains of a loch, Kilmuir par., Skye, from where diatom deposits were found and worked, 11 m. N.E. of Portree.

Culachy, deer forest and seat, Inverness-shire, 2 m. S. of Fort Augustus.

Culag, loch and r., the latter flowing into Loch Inver, ½ m. S. of the vil., and close to where the C. Hotel stood.

Culardoch, mt., W. Aberdeenshire, 5 m. N.E. of Braemar; alt. 2955. ft. The highest public road in Scotland passes 3 m. E. of it, where it rises to 2224 ft.

Culbin, formerly a sandy tract forming a desert along the shore of the Moray Firth, extending westward from Findhorn to Auldearn, par. of Nairn. Taken over by the Forestry Commission and now an extensive State forest.

Culblean, hill range, 5 m. long, Glenmuick par., S.W. Aberdeenshire, extending from Morven (2862 ft.) towards Culblean (1567 ft.) and to the r. Dee, N. of Ballater.

Culbokie, vil., E. Ross and Cromarty, on E. side of Cromarty Firth, 6 m. N.E. of Conon Bridge. P.O., T.O.

Culburnie, pl., 2 m. W. of Kiltarlity, Inverness-shire.

Culcabock, vil. and golf course, E. side of Inverness. Pop. 284.

Culchary, ham. in co. of, and 5 m. S.W. of, Nairn.

Culcreuch, seat, Stirlingshire, 7 m. N. of Campsie.

Culdees Castle, seat, 4 m. S. of Crieff, Perthshire.

Culdrain, seat, 4 m. S. of Huntly, N. Aberdeenshire.

Culduthel, ham. and seat in co. of, and 3 m. S. of, Inverness. Inverness town hospital.

Culen, loch, 1¼ m. E. of Achanalt, mid Ross and Cromarty.

Culeryin, loch, 4 m. N.E. of Walls, Shetland.

Culhorn House, seat, Wigtownshire, 2 m. S.E. of Stranraer.

Culkein, coast ham., W. Sutherland, 8 m. N.W. of Lochinver.

Cullalo Hills, 6 m. E. of Dunfermline, Fife.

Cullavoe, bay and ham., Yell par., Shetland. P.O., T.O.

Cullen, parl. and royal mun. bur., Banff-shire; small seapt. and par., 5½ m. W.N.W. of Portsoy. Chief industry, fishing. C. House close by. C. water, with a run of 8 m., falls into C. Bay. P.O., T.O. Par. 881 ac. Pop., par. 1814; bur. 1688. C. House a seat of the Countess of Seafield.

Cullenoch, old name of Laurieston vil., Kirkcudbrightshire, par. of Balmaghie.

Cullen Park, seat, near Strathaven, Lanarkshire.

Cullerley, ham., Aberdeenshire, 3 m. S.E. of Echt.

Culliecudden, ham., Ross and Cromarty, 4¼ m. S.E. of Invergordon.

Cullipool, pl., W. Argyll, 13 m. S.W. of Oban. P.O., T.O.

Cullivoe, see CULLAVOE.

Culloch, burn, S.E. Kirkcudbrightshire, flows 3 m. S.W. of Kirkgunzeon Water.

Cullochy Locks, on Caledonian Canal, N. end of Loch Oich, Inverness-shire.

Culloden Moor, on border of Nairnshire and Inverness-shire, with ry. sta., L.M.S. P.O., T.O. Also P.O. called Culloden, 6 m. E.N.E. of Inverness. Was the scene of the defeat of Prince Charles Edward by the Duke of Cumberland, and the cause of the Stuarts finally crushed, 16th April 1746. A cairn, 20 ft. high, marks the battle-field. C. House ¾ m. from ry. sta. An extensive State forest in vicinity.

Cullow, pl., 5 m. N. of Kirriemuir, Angus.

Cullykhan, ravine, Gamrie par., Banffshire.

Culmaile, old name of Golspie, Sutherland.

Culnacnock, ham., Isl. of Skye, 12 m. N. of Portree. P.O., T.O.

Culpleasant, seat, Ross and Cromarty, 3 m. S.W. of Tain.

Cul Point, headland, W. of Islay, Argyll.

Culrain, ham., ry. sta., L.M.S., 3 m. N.W. of Bonar Bridge, Ross and Cromarty. Near is C. Lodge. P.O.

Culross, par., parl. and royal and mun. bur., and port, S.W. Fife, on N. shore of the Firth of Forth. P.O., T.O.; par. 7591 ac. Pop., parish 4088; bur. 495.

Culroy, ham., 3 m. N. of Maybole, Ayrshire.

Culsalmond, par. and ham., Aberdeenshire, 4½ m. N.E. of Insch; 6994 ac. Pop. 508.

Culsh, burn and hill, E. Aberdeenshire. Burn rises in New Deer and flows 3½ m. to Auchreddie Burn. C. hill, alt. 503 ft.

Culshabbin, pl., 4½ m. N.W. of Mochrum, Wigtownshire.

Culter, par. and vil., with ry. sta., L.M.S. (Coulter), E. Lanarkshire; par. 10,187 ac. Pop. 335; vil. 1½ m. S.E. of ry sta., 2¼ m. S.W. of Biggar. C. Water, 6½ m. to the Clyde. C. Fell, alt. 2454 ft., 6½ m. S. of Biggar. Here also C. Allers House and C. Mains.

Culter, vil., burn, and ry sta. in co. of, and 7½ m. S.W. of, Aberdeen. Pop. 2004. Close by are Culter Paper Mills. C. House, an ancient house.

Cultercullen, vil., E. Aberdeenshire, 1 m. E. of Udny.

Cultoquhey, 2½m. N.E. of Crieff,Perthshire.

Cults, q.s. par. and vil. with ry. sta., L.N.E., 4 m. S.W. of Aberdeen; vil. pop. 1869. P.O., T.O. C. House in vicinity.

Cults, par. and ham., Fife; 2811 ac. Pop. 589. P.O., T.O.

Culvain, summit, 2½ m. S.E. of head of Loch Arkaig, Inverness; alt. 3224 ft.

Culverman Fell, hill, Wigtownshire, 3 m. N.W. of Kirkcowan; alt. 702 ft.

Culzean Castle, Maybole, Ayrshire, seat of the Marquess of Ailsa (Earl Cassillis).

Cumberland Stone, huge boulder on Culloden Moor, Inverness-shire, on which it is said the Duke of Cumberland stood after the victory over Prince Charlie's forces in 1746.

Cumbernauld, par. and vil., with ry. sta., L.M.S., Dunbartonshire (detached), 13 m. N.E. of Glasgow; par., 11,675 ac. Pop. 4829; vil. pop. 1201. P.O., T.O., and P.O., T.O. at C. ry. sta. Industries: weaving, mining, quarrying, and fire-clay works. C. House close by.

Cumbrae, par., 3511 ac. The par. comprises the two islands of Great and Little Cumbrae. Great Cumbrae contains the vil. of Millport, a seaside resort. These islands are in the co. of Bute, and halfway between Bute Island and the coast of Ayr, and there is a lighthouse on the W. side of Little Cumbrae, light seen 19 m. Pop. Great C., 2144; Little C. 21.

Cumbria, *land of the Cymry* (*Welsh*), extended from the Clyde to the Dee, and from Dunbartonshire to Chester; the name still appears in Cumberland.

Cuminestone, Cummestone, or **Cumminestown,** vil. and burn, Aberdeenshire, 6 m. E. of Turriff. P.O., T.O. Pop. 412.

Cumledge, ham. and seat, 2 m. N. of Duns, Berwickshire. Ham. has mill for making blankets.

Cumlodden, q.s. par. and seat, Glassary and Inveraray par., Argyll, on Loch Fyne, 8 m. S.W. of Inveraray.

Cumloden, seat, of the Earl of Galloway, Kirkcudbrightshire, 2 m. N.E. of Newton-Stewart.

Cumlongan, castle, Dumfries, see COMLONGAN.

Cummertrees, par. and vil., with ry. sta., L.M.S., Dumfriesshire, 3½ m. W. of Annan; par. 9473 ac. Pop. 970. P.O., T.O.

Cummingston, vil., 1½ m. E. of Burghead, Moray. Pop. 146.

Cumnock, town, with P.O., T.O., ry sta., L.M.S. (Old Cumnock), Ayrshire, on r. Lugar, 16 m. E. of Ayr. Pop. (C. and Holmhead), 3653. Hotel. Mfrs. tweeds, pottery, agricultural and dairy implements; anthracite coal; good limestone and sandstone quarried here.

Cumnock, New. par. and vil., E. Ayrshire, at the meeting of Afton Water and r. Nith, 5 m. S.E. of Cumnock; par. 48,175 ac. Pop. 6419. P.O., T.O. Is rich in minerals.

Cumnock, Old, par., E. Ayrshire, 14,168 ac. Pop. 5637.

Cumrue, loch, 8 m. W. of Closeburn, mid Dumfriesshire.

Cunnigar, mound, near Mid-Calder, Midlothian.

Cunninghame, one of the old divisions of Ayrshire, in the N. part of the co.

Cunninghamhead, seat and ry. sta., L.M.S., Ayrshire, 4 m. N.W. of Kilmarnock. P.O., T.O. at sta.

Cunninghar, hill, with stone circle, Clackmannanshire, 2½ m. W. of Dollar.

Cunningsburgh, ham, E. coast of mainland, Shetland, 9 m. S.W. of Lerwick. P.O., T.O. See CONNINGSBURGH.

Cunnoquhie, seat, 4 m. W. of Cupar, Fife.

Cunzierton, hill, 6 m. E.S.E. of Jedburgh, Roxburghshire. Has ancient camp. Alt. 1100 ft.

Cupar, par., co. town, and royal and mun. burgh, 5½ m. N.E. of Ladybank, Fife; par., 5713 ac. Pop. 7110; pop. of town, 4596. Ry. sta., L.N.E. P.O., T.O. Hotel. Trade in corn; beet sugar factory.

Cuparmuir, vil., 2 m. W. of Cupar, Fife.

Cur, stream, Cowal dist., Argyll. Flows 10 m. to Loch Eck.

Curate's Steps, pass, beside r. Ayr, Ayrshire, near Sorn Castle.

Curate's Wells, spring, in par. and on Glebe of Dunsyre, 6 m. N.E. of Carnwath, E. Lanarkshire.

Curgie, small bay, Wigtownshire, 3 m. N. of Mull of Galloway.

Curling Hall, seat, near Largs, Ayrshire. Estate includes site of the battle of Largs.

Curr, hill, Roxburghshire, 5 m. S.E. Morebattle, near English border; alt. 1849 ft.

Curreath, seat, 3 m. N.E. of Troon, Ayrshire.

Currie, vil., par., and ry. sta., L.M.S., Midlothian, on the Water of Leith, 6 m. S.W. of Edinburgh; par., 13,031 ac. Pop. 3261; vil. pop. 1045. P.O., T.O. C. Castle nearby. Vil. has paper and snuff mills.

Currie Hill, ry. sta., L.M.S., on Edinburgh and Glasgow ry., and 5½ m. S.W. of Edinburgh.

Cushnie, ancient par., Alford dist., Aberdeenshire. Now part of Leochel. P.O., T.O. It contains Soiach Mt.; alt. 2032 ft.

Cuthill, or **Cuttle,** sub. of Prestonpans, E. Lothian. P.O.

Cuthlie, ry. sta., L.N.E., Dundee and Arbroath joint ry., 2 m. N.W. of Elliot Junction.

D

Daal, ham., S.W. coast of Bressay Isl., Shetland, 3 m. S.E. of Lerwick.

Dabton, seat, Dumfriesshire, 1½ m. N. of Thornhill.

Daer Water, longest head-stream of the Clyde, rises on Queensberry Hill, and runs 10 m. northward to confluence of Powtrail Water, near Elvanfoot.

Daff, stream, Renfrewshire, flows 1½ m. from Leap Muir to the r. Kip, at Inverkip.

Daiglen, small stream, Tillicoultry par., Clackmannanshire.

Dail, or **Dall,** stream, Sutherland, Durness par., Cape Wrath side.

Dail Dariach Lodge, near Arrochar, Dunbartonshire.

Dailly, par., vil., ry. sta., L.M.S., S. Ayrshire, 5½ m. N.E. of Girvan; 17,958 ac. Pop. 1726. P.O., T.O. Contains vil. of Dailly or New Dailly and the ham., of Old Dailly, also islet of Ailsa Craig. Pop. of vil of D., 472.

Dailuaine, Carron, Knockando, Morayshire. Distillery; ry. halt, L.N.E.

Daimh, loch, Perthshire, 10 m. N.W. Killin; 93 ft. deep; alt. 1369 ft.

Dairsie, par., ry. sta., L.N.E., Fifeshire, 3 m. N.E. of Cupar; 2550 ac. Pop. 537. P.O., T.O. D. Castle (ruins).

Dalarossie, ancient par., N.E. Inverness-shire. See MOY.

Dalaruan Distillery, Campbeltown, Argyll.

Dalavich, ancient par., Lorn, Argyll. See KILCHRENAN.

Dalbarber, vil., Perthshire, 2 m. S.W. of Methven.

Dalbeathie, seat, Perthshire, 2½ m. S.E. of Dunkeld.

Dalbeattie, town (police bur.) and q.s. par., ry. sta., L.M.S., Urr par., Kirkcudbrightshire, 5 m. E.S.E. of Castle-Douglas. Pop. 3011. Hotel. P.O., T.O. Stands on D. burn. D. loch 1½ m. N.E. Town has granite quarries, polishing works, paper, bobbin, corn, and saw mills. State forest of D. adjoins.

Dalblair, pl., in glen between Wardlaw Hill and Cairntable, Ayrshire.

Dalbreck Lodge, Strathconon, Ross and Cromarty, 17 m. W. of Muir of Ord.

Dalcairnie Linn, waterfall, on D. burn, Ayrshire, one leap over 60 ft. D. burn enters Loch Bogton, ¾ m. S.W. Dalmellington.

Dalcapon Mains, on the r. Tummel, 1½ m. N. of Ballinluig Junction, Perthshire.

Dalcataig, pier and pl., W. side of Loch Ness, Inverness-shire.

Dalchally, glen, Angus, 8 m. W. of Milton of Clova. See GLENCALLY.

Dalchenna House, Argyll, on Loch Fyne, 2 m. S.W. of Inveraray.

Dalchonzie, seat, pl., and ry. sta., L.M.S., on r. Earn, Comrie par., 1½ m. W. of Comrie.

Dalchork, seat, Sutherland, 3½ m. N.W. Lairg.

Dalchosnie, seat, and scene of a victory of King Robert the Bruce, N.W. Perthshire, Fortingall par., Perthshire.

Dalchreichart, ham., Inverness-shire, Urquhart par. P.O., T.O.

Dalchroy House, seat, Morayshire, near Advie.

Dalchully House, Inverness-shire, or r. Spey, 9 m. S.W. Newtonmore.

Dalclathick, shooting-lodge, Perthshire, 6 m. S.W. of Crieff.

Dalcrombie, seat, Inverness-shire, 11 m. S.E. of Inverness.

Dalcross, old par., now united to Croy ; ry. sta., L.M.S., 6¾ m. N.E. of Inverness. P.O., T.O. at sta. D. Castle, a restored baronial castle.

Dalcrue, or **Dalcruvie,** ham., 1¾ m. N.E. of Methven, Perthshire. Handsome bridge crosses the Almond here.

Daldorch House, Ayrshire, ½ m. N.E. of Katrine ry. sta.

Daldowie, seat, Lanarkshire, 1 m. N.W. of Uddingston.

Dalduff, ruined baronial fort, Ayrshire, 3 m. S.E. of Maybole.

Dale, ham., Shetland, 3 m. N.W. of Lerwick.

Dale, pl., Caithness, 5 m. S. of Halkirk.

Dalelia, pl., with pier, N. side of Loch Shiel, 4 m. E. of Shiel Bridge, S.W. Inverness-shire.

Dales Voe, inlet, E. coast of mainland, Shetland, Tingwall par.

Dalfaber, pl., Inverness-shire, 1 m. N.E. of Aviemore.

Dalgain, vil., Ayrshire, 4 m. S.E. of Mauchline, Sorn par.

Dalgairn, seat, near Cupar, Fife.

Dalganachan, pl., Caithness, 14½ m. S.W. of Caithness.

Dalgarnock, old par., now united to Closeburn, Dumfriesshire.

Dalgarven, vil. on Garnock R., Kilwinning par., Ayrshire.

Dalgety, par., Fife coast, 5 m. S.E. of Dunfermline ; 3341 ac. Pop. 1481.

Dalginross, sub. of Comrie, Perthshire. Remains of Roman camp.

Dalgonar, seat, Dumfriesshire, 4 m. N.W. of Auldgirth.

Dalguise, vil., with ry. sta., L.M.S., Perthshire, 4½ m. N.N.W. of Dunkeld. Ry. viaduct crossed r. Tay here. P.O.

Dalhalvaig, pl., Sutherland, in par., and 7½ m. from, Reay. P.O., T.O.

Dalhousie Castle, Midlothian, 1½ m. S.E. of Bonnyrigg, on S. Esk. Formerly seat of Earl of Dalhousie, now a boys' preparatory school.

Daliburgh, hospital, pl., S.W. of South Uist Isl., Outer Hebrides.

Dalintober, vil., distillery, and pier, Campbeltown, Argyll. White light, seen 5 m.

Daljarrock, seat, on Stinchar R., Ayrshire, 4 m. N.E. of Colmonell. P.O.

Dalkeith, mkt.-town, police bur., and par. ; ry. sta., L.N.E., Midlothian, 6½ m. S.E. of Edinburgh ; 2320 ac. Pop. 7854 ; bur. pop. 7502. P.O., T.O. Town has brush works, breweries, and corn mills. D. Palace, chief seat of the Duke of Buccleuch.

Dalk's Law, hill on the border of Coldingham Moor, Berwickshire ; 634 ft.

Dallachulish, seat, Argyll, 3 m. S.E. of Appin.

Dallachy, Upper and **Nether,** two hams., Morayshire, Bellie par., N.E. of Fochabers.

Dallas, par. and vil., Morayshire, 9 m. S.E. of Forres, or r. Lossie ; 22,928 ac. Pop. 541. Loch D., 4 m. S.W. P.O., T.O. Vil. pop. 142. Ruin of Tor Castle.

Dallasdhu Distillery, N.W. Morayshire, 2 m. S. of Forres.

Dallawoodie, seat, S.W. Dumfriesshire, 2 m. N.W. of Maxwelltown.

Dalleagles, ham., Ayrshire, 2½ m. S.W. of New Cumnock.

Dall House, Perthshire, 5 m. S.W. of Kinloch Rannoch.

Dalmacoutter, pl., Lanarkshire, near New Monkland.

Dalmahoy, former seat, now golf club, 2½ m. S.E. of Ratho Junction, Midlothian. D. Hills over 800 ft.

Dalmakerran, seat, Dumfriesshire, 5 m. S.W. of Thornhill.

Dalmally, vil., with ry. sta., L.M.S., on r. Orchy, 3 m. from Loch Awe, 16 m. N.N.E. of Inveraray, Argyll. P.O., T.O. Hotel.

Dalmarnock, q.s. par., sub. of Glasgow, on the Clyde ; ry. sta., L.M.S.

Dalmarnock, vil., Perthshire, 3½ m. N.W. of Dunkeld.

Dalmarnock Paper Mills, Bridgeton, Glasgow.

Dalmelling, former priory and convent, 1½ m. E. of Ayr.

Dalmellington, par. and vil. ; ry. term., L.M.S., in co. of, and 15 m. S.W. of Ayr ; 17,814 ac. Pop. 6151. P.O., T.O. Vil. pop. 2228. Associated with scenes in the persecutions of the Covenanters. Town has coal and ironworks.

Dalmennoch, bay, Loch Ryan, Wigtownshire, 3½ m. N.E. of Stranraer.

Dalmeny, par., vil., ry. sta., L.N.E., N.E. West Lothian, 9½ m. N.W. of Edinburgh; 5864 ac. Pop. 3237; pop. of D. sta. 274; par. contains part of S. Queensferry. P.O., T.O. D. Park, seat of the Earl of Rosebery.

Dalmigavie, seat and romantic dell, Moy par., Inverness-shire, on r. Findhorn.

Dalmoak Castle, Dunbartonshire, 1 m. N.W. Dumbarton.

Dalmore, coast vil., seat, and pier (at Belleport, ½ m. E.) Ross and Cromarty, 2½ m. W. of Invergordon. Has a distillery.

Dalmore, seat, Ayrshire, 1¼ m. S.W. of Tarbolton.

Dalmore Paper Mills, Milton Bridge, Midlothian, near Auchendinny ry. sta.

Dalmuir, q.s .par., with ry. sta., L.N.E., Dunbartonshire, 10 m. N.W. of Glasgow. P.O., T.O. Shipbuilding and engineering works.

Dalmunzie Lodge, seat, Perthshire, 3 m. N.W. of Spittal of Glenshee.

Dalnabreck, seat, Perthshire, 10 m. N.W. of Blairgowrie.

Dalnacardoch, deer forest, Perthshire, 11 m. W.N.W. of Blair Atholl.

Dalnagairn, seat, Perthshire, near Kirkmichael.

Dalnaglar Castle, Perthshire, 15 m. N. of Blairgowrie.

Dalnaglaton, pl., Caithness, 11½ m. S.W. of Halkirk.

Dalnair House, Stirlingshire, 2 m. S. of Drymen.

Dalnamein, seat, Perthshire, 4½ m. N.W. of Struan. D. and Glaschorie, a shooting of 15,000 ac.

Dalnashaugh Inn, Banffshire, 1½ m. S.E. Ballindalloch ry. sta.

Dalnaspidal, ry. sta., L.M.S., Perthshire, 2 m. S. of watershed of Central Grampians, 5 m. N.W. of Dalnacardoch. P.O., T.O. at sta. This is the highest ry. sta. in Scotland.

Dalnavert, seat and deer forest, Inverness-shire, Alvie par., 1½ m. N.E. of Kincraig.

Dalnawillan, shooting lodge, Caithness, 4 m. S.E. of Altnabreac.

Dalness, or **Royal,** deer forest Argyll-shire, Glen Etive, 18 m. N.E. Taynuilt.

Dalness, ham. and romantic cascade on Etive R., Argyll, 5 m. from Loch Etive.

Dalnotter House, Dunbartonshire, ½ m. S.E. of Old Kilpatrick. D. Hill, 753 ft.

Dalnygap, seat, Wigtownshire, 5 m. N.W. of New Luce.

Dalpowie, shooting lodge, Perthshire, 2¼ m. S.E. of Dunkeld.

Dalqueich, ham., 3 m. N.W. of Kinross.

Dalquhairn, seat and burn, Ayrshire, 3 m. N.E. of Barr.

Dalquharran Castle, seat in Dailly par., Ayrshire, on the right bank of Girvan Water.

Dalquhurn, or **Bonhill House,** near Renton, Vale of Leven, Dunbartonshire. Tobias Smollett was born here (1721–71).

Dalrachney, seat, Inverness-shire, 1 m. W. of Carrbridge.

Dalreavoch Lodge, Sutherland, 8 m. N.W. of Golspie.

Dalreoch, q.s. par. and ry. stas., L.M.S. and L.N.E., Dunbartonshire. Sta. at N.W. end of Dumbarton.

Dalreoch, ham., Perthshire, ½ m. N. of Dunning.

Dalreoch, seat, Perthshire, 1 m. W. of Tummel Bridge.

Dalreoch, shooting-lodge, Perthshire, 8 m. N.E. Pitlochry.

Dalreoch, seat, Ayrshire, 2 m. E. of Colmonell.

Dalriada, ancient principality in Western Highlands, or pristine Kingdom of the Scots. It originated with immigrants of the same name in N. of Ireland. They came to Argyll in 503, acquired ascendency over the Western Caledonians, and established a monarchy, with seat at Dunstaffnage. Their king, Kenneth MacAlpine, succeeded to the crown of Northern Pictland, and by uniting it to that of Dalriada, founded the Kingdom of Scotland and removed his court in 843 to Forteviot.

Dalroy Burn, trib. of r. Nairn, Inverness-shire, ½ m. E. of Culloden ry. sta.

Dalruadhain, ancient seat of the Dalriads, on the ground now occupied by Campbeltown, Argyll.

Dalrulzian, shooting - lodge, Perthshire, 10 m. N.W. of Blairgowrie.

Dalry, town and par., with ry. sta., L.M.S. Ayrshire, on Garnock R., 11¼ m. N.W. of Kilmarnock; par. 19,164 ac. Pop. 6827; q.s. par.; town pop., 4602. P.O., T.O. Town has woollen and hosiery mfrs.

Dalry, par. (34,503 ac. Pop. 843; vil. pop., 535) and vil., Kirkcudbrightshire, on r. Ken, 8 m. N.W. of New Galloway. P.O., T.O.

Dalry, dist. and ry. sta. (D. Road), L.M.S., W. Edinburgh.

Dalry, Dalrigh, or **Dalree,** pl. near head of Strathfillan, on W. border of Perthshire. It was the scene in 1306 of a sharp skirmish between King Robert the Bruce and Macdougall, Lord of Lorn, in which Bruce lost the Brooch of Lorn, described in Scott's "Lord of the Isles."

Dalrymple, par. and vil., with ry. sta., L.M.S., in co., and 4 m. S.E. of Ayr, on r. Doon; par. 7832 ac. Pop. 1366; vil. pop. 219. P.O., T.O.

Dalserf, par., vil., and ry. sta., L.M.S., Lanarkshire, on r. Clyde, 7 m. S.E. of Hamilton; 7444 ac. Pop. 17,312. D. House, a seat. Contains part of the town of Larkhall.

Dalsetter, ham. in Yell Isl., Shetland, on Basta Voe. P.O. Called Sellafirth.

Dalshangan, seat, Kirkcudbrightshire, 5½ m. N.W. of Dalry.

Dalskairth, seat, Kirkcudbrightshire, 3 m. S.W. of Dumfries.

Dalswinton, vil. and estate in Kilmahoe par., Dumfriesshire, 7 m. N.W. of Dumfries. P.O.

Dalton, vil. and par., Dumfriesshire, 6 m. N.W. of Annan; 6891 ac. Pop. 570. P.O., T.O. Vil. D. and Kirkwood. Pop. 99.

Dalton, Cambuslang sub., Lanarkshire.

Dalveen, pass over the Lowther Hills, on border of Durisdeer par., Dumfriesshire; alt. 1200 ft.

Dalveich, ham. on Loch Earn, Perthshire, 2 m. E. of Lochearnhead.

Dalvey, pl., Morayshire, 3½ m. N.E. Cromdale ry. sta. Dyke and Moy par.

Dalvey House, Morayshire, 2 m. W. of Forres.

Dalwhat Water, head stream of the Cairn, in Glencairn par., Dumfriesshire.

Dalwhinnie, pl. with ry. sta., L.M.S., Inverness-shire, in Glenstruim, 13 m. S.W. of Kingussie. Alt. of ry. sta., 1174 ft. P.O., T.O. Has a distillery. Hotel.

Dalziel, par., N. Lanarkshire; 3640 ac. Pop. 45,114; q.s. par., S. Dalziel; vils. of D. and Netherton. Pop. 1333. Several collieries and iron and steel works.

Damh, or **Daimh,** loch, Perthshire, 3 m. W. of Killin; depth 95 ft.

Damph, loch, Ross and Cromarty, 12 m. E. of Ullapool.

Damph, or **Damh,** loch, W. Ross and Cromarty, 2 m. E. of Shieldaig; depth 206 ft., alt. 129 ft.

Dams, ham., Fife, 1½ m. S. of Kettle.

Damsay, isl., Firth Bay, Orkney.

Damside, pl., Perthshire, 2 m. N.E. of Auchterarder.

Dams, The, Glasgow Corporation Waterworks, Renfrewshire, 1½ m. S. of Barrhead.

Dandaleith, haugh and ry. sta., L.N.E., Morayshire, on r. Spey, Rothes par.

Danderhall, ham., Midlothian, 2 m. N.W. of Dalkeith.

Danes Dyke, ancient stonework, E. Fife, between Balcomie Castle and Fife Ness.

Danestone House, Aberdeenshire, ½ m. N. of Woodside ry. sta.

Danevale Park, seat, Kirkcudbrightshire, 2½ m. N.W. of Castle-Douglas.

Dankeith, seat, Ayrshire, 4½ m. S.W. of Kilmarnock.

Danna, isl. and seat, Argyll, Knapdale par. Pop. 25.

Danskine, pl., E. Lothian, 5½ m. S.E. of Haddington.

Dara, stream, N. Aberdeenshire, joining the Deveron near Turriff.

Dardar, dell, with cascade, Aberdeenshire, on coast of Aberdour par.

Dardenne, Kilmacolm, Renfrewshire. Boys' private school.

Dargavel, seat, Renfrewshire, Erskine par., 1 m. S. of Bishopton.

Dargie, ham., Angus, 3 m. W. of Dundee.

Dark Mile (Mil Dubh), glen between Loch Arkaig and Loch Lochy, Inverness-shire.

Darleith, estate, Dunbartonshire, Bonhill par., 6 m. N.W. of Dumbarton.

Darmead Linn, headstream of the Breich, Cambuskenneth par., Lanarkshire.

Darnaconner, loch, S. Ayrshire, 2½ m. N.E. of Barrhill.

Darnaway Castle, seat of the Earl of Moray, 3½ m. S.W. of Forres, Morayshire. D. Forest, S. of the castle.

Darnconner, mining vil., Ayrshire, 2½ m. N.E. of Auchinleck. P.O.

Darngaber, vil., Lanarkshire, 3 m. S. of Hamilton. Vestige of ancient castle of D.

Darngavel, vil., Lanarkshire, New Monkland, 2½ m. N.E. of Airdrie.

Darn Hall, Peeblesshire, ½ m. W. of Eddleston. A seat of Lord Elibank.

Darnick, vil., Roxburghshire, 1 m. W. of Melrose. Contains a massive tower of fifteenth century, now furnished as a museum of Border antiquities. P.O., T.O.

Darnley, seat, Renfrewshire, 2 m. S.W. of Pollokshaws. It belonged for ages to a branch of the Stewarts; gave them the peerage title of lord, so prominent in history, as borne by the husband of Queen Mary. D. Hospital, ½ m. N.E.

Darnow, ham., Wigtownshire, 4 m. N.W. of Kirkcowan.

Darra, hill, Aberdeenshire, 2 m. S.E. of Turriff; 283 ft.

Darrach, hill, on W. verge of Denny par., Stirlingshire; alt. 1170 ft.

Darvel, q.s. par. and police bur., with ry. sta., L.M.S., Ayrshire, 9 m. E. of Kilmarnock. Bur. pop. 3232. P.O., T.O. Mfrs. lace and carpets.

Dava, pl., with ry. sta., L.M.S., Morayshire, 8½ m. N.W. Grantown. P.O., T.O. at sta. D. Moor, 1 m. S.

Dava, summit, on L.M.S., Morayshire, 3½ m. S. of Dava sta.; alt. 1052 ft.

Davaar, Davar, or **Devaar,** isl., in mouth of Campbeltown Loch, Kintyre, Argyll. Lighthouse, with revolving light, 120 ft., seen 17 m. Davaar House here.

Davan, loch, Aberdeenshire, 1½ m. W. of Dinnet sta.

Davidshill, vil., Ayrshire, 1 m. S. of Glengarnock. County hospital.

Davidson's Mains, vil. and ry. sta., L.M.S., in city, and 3½ m. N.W. of Edinburgh. P.O., T.O.

Davidston, seat, S.E. Angus, 1½ m. S.E. of Newtyle.

Davie, loch, N. Arran, Buteshire, 3 m. S. of Lochranza.

Davington, ham. and burn, Dumfriesshire, 16½ m. N.W. of Langholm. D. Burn, trib. of the Esk.

Daviot, ham. and ry. sta., L.M.S., 6½ m. S.E. of Inverness. P.O., T.O. D. House, 1 m. N.E. of Nairn. Remains of a castle of the Earls of Crawford, and several ancient Caledonian stone circles.

Daviot, par., eccl. par., and ham., E. Aberdeenshire; par. 4453 ac. Pop. 557; ham., 5 m. N.W. of Inverurie. P.O.

Daviot and **Dunlichty,** par., N.E. Inverness-shire; 59,223 ac. Pop. 789.

Davo, picturesque ravine, Garvock par.. Kincardineshire, 3 m. N.E. of Laurencekirk.

Dawsholm, pl. and paper mill, Maryhill, Glasgow. Glasgow Corporation Gasworks.

Dawyck House, Peeblesshire, ¾ m. S.W. of Stobo sta.

Dead Burn, affl. to the Lyne, Newlands par., Peeblesshire.

Dead Loch, marshy pool, an expansion of Yarrow Water, Selkirkshire.

Deadmangill, burn and ravine, Dumfriesshire, 8 m. W. of Ecclefechan.

Deadmen's Holm, Ayrshire, opposite the mouth of Bloody Burn.

Deadriggs, pl., Berwickshire, 1 m. N. of Eccles.

Deadwater, stream and ry. sta., L.N.E., Roxburghshire, 5¾ m. E. of Riccarton Junction, named from a morass in Castleton par.

Deaf Heights, mt., Selkirkshire, 5 m. W. of Stow; alt. 1844 ft.

Dean, dist. and q.s. par., in W. Edinburgh. Includes the vil. of D., on the Water of Leith.

Dean, ruined castellated mansion, Ayrshire, 1 m. N.E. of Kilmarnock. Seat of the Boyds, Earls of Kilmarnock.

Dean Burn, W. Lothian, flowing into the r. Forth, 1 m. W. of Bo'ness.

Deanburnhaugh, vil., Roxburghshire, Roberton par., 8 m. S.W. of Hawick.

Deanich, deer forest and seat, Ross and Cromarty, 18 m. S.W. of Bonar Bridge.

Deannie Lodge, Ross and Cromarty, 5 m. W. of Struy.

Dean Park, q.s. par. in S.W. subs. of Glasgow.

Deans, W. Lothian, 2 m. N.W. of Livingstone Oilworks.

Deanshaugh, vil., Perthshire, 7 m. S. of Dunkeld.

Deanston, vil. and seat, Perthshire, 1 m. W. of Doune. P.O., T.O. Vil. has cotton mills.

Deantown, colliery vil., Midlothian, 1¾ m. S.E. of Musselburgh.

Dean Water, stream, Angus, joins the Isla, 1 m. N. of Meigle.

Dearg Sgeir (*Green Island*), Sound of Mull, Argyll. White flashing light, seen 9 m.

Deasthack, affl., r. Beauly, Inverness-shire.

Debatable Land, tract of country on the Border, between r. Esk and Sark. Long the subject of contention between Scotland and England.

Dechmont, ham. and ry. sta., L.N.E., West Lothian. P.O., T.O. Pop. 351. D. Hill, 688 ft. D. House, 1 m. S.W.

Dechmont, colliery, Lanarkshire, 2 m. N.W. of High Blantyre. D. Hill, alt. 602 ft.

Dechmont Road, pl., West Lothian, 1 m. W.S.W. of Dechmont.

Dee, r., rising among the Cairngorm Mts., running E. by northward to the sea at Aberdeen. Length about 96 miles; forms above Castleton a series of falls called the Linn of Dee. The main source of the Aberdeen water supply.

Dee, r., Kirkcudbrightshire, formed by conflux of the Ken and Black Dee, flows 50 m. S.E. and S. to Solway Firth.

Deechoid, or **Deadh Choimhead,** mt., Argyll, Muckairn par., 5½ m. E. of Oban; alt. 1255 ft.

Deer, dist., Aberdeenshire, containing the pars. of Old and New Deer. Pop. 24,585.

Deer, or **South Ugie Water,** stream, N. Aberdeenshire, flowing to the N. Ugie Water, 5 m. N.W. of Peterhead.

Deer, pl., with ancient entrenchment, Dumfriesshire, 2½ m. N.W. of Thornhill.

Deer Dyke, remains of earthen division, border of Garvock and Laurencekirk pars., Kincardineshire.

Deer Law, hill, border of Selkirkshire and Peeblesshire, 2 m. N.W. of St. Mary's Loch; alt. 2065 ft.

Deerness, q.s. par., St. Andrews and D. par., Orkney, 8½ m. E. by S. of Kirkwall. P.O., T.O.

Deer, New, par. and vil., E. Aberdeenshire; vil. 2½ m. W. of Maud Junction; par. 26,754 ac. Pop. 3666; vil. pop. 686. P.O., T.O. See NEW DEER.

Deer, Old, par. and vil., Aberdeenshire; vil., 1 m. S.W. of Mintlaw sta.; par. 27,372 ac. Pop. 3380. P.O. Vil. pop. 171. Ruin of a Cistercian Abbey of the thirteenth century.

Deershaw, or **Newtondeershaw,** ham. in co., and 5½ m. S.W. of Banff.

Deer Sound, large harbour, Orkney, mainland, 5 m. S.E. of Kirkwall.

Deer State Forest, nearest ry. sta., Mintlaw, 9 m. W. of Peterhead, Aberdeenshire.

Deeside, valley of the r. Dee, partly in Kincardineshire, but chiefly in Aberdeenshire; also County District of Aberdeenshire. Pop. 10,974.

Deil's Brig, natural arch, Holborn Head, Caithness, 1½ m. N. of Scrabster.

Deil's Dyke, line of denuded trap rock, S.E. coast of Great Cumbrae Isl., Buteshire.

Deil's Dyke, ancient line of fortifications from Loch Ryan in Wigtownshire to upper part of Solway Firth in Dumfriesshire. In some parts still very distinct, but in most parts quite extinct.

Delfour, pl., with remarkable assemblage of ancient Caledonian monuments, in Alvie par., Inverness-shire.

Delfur Lodge, Banffshire, on r. Spey, 1½ m. S.E. of Orton ry. sta.

Delgaty Castle, seat, Aberdeenshire, 2 m. N.E. of Turriff.

Delhaig, pl., Gorgie dist., S.W., city of Edinburgh.

Dell, shooting-lodge, 23 m. S.W. of Inverness.

Dell, South, vil., Ross and Cromarty, Barvas par.

Delnabo, seat, Banffshire, 1 m. S.W. of Tomintoul.

Delnadamph, shooting-lodge, on r. Don, Aberdeenshire, 15 m. N.W. of Ballater.

Delnies, Easter and **Wester,** 2 hams., 2½ and 3½ m. W. of Nairn.

Delny, vil. and ry. sta., L.M.S., E. Ross and Cromarty, 3¼ m. N.E. of Invergordon. P.O., T.O.

Deloraine, tract on Ettrick r., in co., and 13 m. S.W. of Selkirk.

Delting, par., Shetland. Includes part of the mainland between Yell Sound and St. Magnus Bay, also Muckle Roe and other islands; par. 36,966 ac. Pop. 958.

Delvilla State Forest, 1 m. N. of Bogside ry. sta., which is 8 m. W.N.W. of Dunfermline, Fife.

Delvine, seat, Caputh par., Perthshire, 7 m. S.E. of Dunkeld.

Denbeath, sub. of Methil, Fife. P.O.

Denbrae, seat, Fife, 2 m. S.W. of St. Andrews.

Denend, pl., Fife, 4½ m. N.W. of Kirkcaldy.

Den Fenella, romantic ravine, with cascade of 65 ft., Kincardineshire, St. Cyrus par.

Denfield, seat, Angus, 1 m. N.W. of Arbroath.

Denfind, steep ravine, S.E. Angus, 10 m. W.S.W. of Arbroath.

Denhead, ry. stas., L.N.E. and L.M.S., Angus, 1½ m. E.S.E. of Carmylie sta.

Denhead, vil., Angus, 2 m. W. of Lochee.

Denhead, ham., Fife, 3½ m. S.W. of St. Andrews.

Denhead of Auchmacoy, ham., Aberdeenshire, 3 m. E. of Ellon.

Denholm, vil., Roxburghshire, 5 m. N.E. of Hawick. Pop. 365. P.O., T.O. D. Dean, a deep ravine, lies to the W.

Denmill Castle, ruined ancient castellated seat, Fifeshire, 1½ m. S.E. of Newburgh.

Denmore, shooting-lodge in co., and 3 m. N. of, Aberdeen.

Dennis Ness, headland, in N.E. of North Ronaldsay, Orkney. Lighthouse, with flashing light, 140 ft. above high water, seen 18 m.

Dennistoun, eccl. par. and dist. in E. of Glasgow. P.O.

Denny, town and par., with ry. sta., L.M.S., in co., and 12 m. S. of Stirling; par. 8276 ac. Pop. 9488; town pop. (including Dunipace) 5512. P.O., T.O. Town has coal, iron, and paper-making industries. D. Paper Mill, 1 m. from sta.

Dennyloanhead, town and ry. stas., L.M.S. and L.N.E., Stirlingshire, 1¾ m. S. of Denny. P.O., T.O. Pop. 3136.

Dennystown, sub. of Dumbarton, on r. Leven.

Denoon Burn, stream, Angus, 2¾ m. N.W. of Glamis.

Denoon Law, hill, 689 ft., Angus, Glamis par. Site of ancient D. Castle.

Denovan, vil. and seat, Stirlingshire, ½ m. N. of Denny.

Den, The, r., Ross and Cromarty, flows N.E. into Cromarty Firth.

Den, The, vil., N. Ayrshire, 2½ m. N.E. of Dalry.

Derclach, loch, Straiton par., Ayrshire.

Derculich, seat and loch, Perthshire, 3½ m. N.E. of Abernethy; loch 70 ft. deep. D. Burn flowing 2½ m. to r. Tay.

Dergan, stream, Argyll, running N. to Loch Creran, Ardchattan par.

Dernaglar, loch, Wigtownshire, 5 m. E. of Glenluce.

Derry, or **Loch an Dithreibh,** loch, Sutherland, 6 m. S.W. of Tongue; depth 157 ft., alt. 268 ft.

Derry, loch, Wigtownshire, 11 m. N.W. of Newton Stewart.

Derry Lodge, shooting-lodge, Aberdeenshire, 3 m. N.W. of Linn of Dee.

Derry Water, stream, S.W. Aberdeenshire, flowing to Lui Water, at D. Lodge, 9 m. N.W. of Braemar.

Dervaig, ham., Mull Isl., Argyll, 8½ m. S.W. of Tobermory. P.O., T.O.

Deskford, par. and vil., Banffshire; 8156 ac. Pop. 541; vil. 4 m. S. of Cullen. P.O. D. Burn flows 7½ m. N. to Moray Firth.

Deskryshiel, seat, S.W. Aberdeenshire, 7½ m. N.W. Dinnet ry. sta.

Deskry Water, Aberdeenshire, running about 9 m. N. to the Don, at upper boundary of Towie par.

Dess, ry. sta., L.N.E., Aberdeenshire, 3 m. N.E. of Aboyne. P.O., T.O. at sta. D. House, 1 m. E.

Dessary, r. and glen, Inverness-shire, W. end of Loch Arkaig.

Deucaledonian Sea, ancient name for the part of the Atlantic among and around the Hebrides.

Deuchar, seat, Angus, 8 m. W. of Brechin. D. Hill, 970 ft.

Deuchar-Swire, morass, Selkirkshire, on Yarrow Water.

Deuchar Tower, on Yarrow Water, Selkirkshire, near Yarrow church.

Deuchrie Dod, hill, E. Lothian, 4 m. S.E. of E. Linton; alt. 968 ft.

Deugh, stream, Kirkcudbrightshire, running 20 m. to riverside of Ken, 4½ m. N.W. of Dalry.

Deveron, r., Aberdeenshire and Banffshire. Rising near the Buck of Cabrach, flowing mainly N.E., falls into the sea between Banff and Macduff; length 61 m.

Devilly, pl., Kincardineshire, 2 m. N.E. of Fettercairn.

Devil's Beef Tub, vast hollow, N. Dumfriesshire, 5 m. N.W. of Moffat.

Devil's Cauldron, ancient circular structure, Bute Isl., 7 m. S. of Rothesay.

Devil's Cauldron, cavern, with cascade, Lednock R., Perthshire, 1¼ m. N. of Comrie.

Devil's Cave, in Kincraig Hill, Kilconquhar par., Fife, 3 m. S.W. of Colinsburgh.

Devil's Elbow, on road over the Grampians from Alyth to Braemar, 9½ m. S. of Braemar ; alt. 2000 ft.

Devil's Mill, cataract of Devon r., border of Perthshire and Kinross-shire, 1½ m. from Crook of Devon.

Devil's Point, steep mt., Cairngorm Mts., S.W. Aberdeenshire, rising on the E. of the r. Dee and 2 m. S.E. of Cairn Toul ; alt. 3303 ft.

Devil's Staircase, difficult descending track from a point near head of Glencoe to a point near the head of Loch Leven, N. Argyll ; alt. 2206 ft.

Devol's Glen, rocky wooded glen, with stream making two waterfalls of 20 ft. and 100 ft., near Port Glasgow, Renfrewshire.

Devon, r., Perthshire, Kinross-shire, and Clackmannanshire. Rising in the Southern Ochils, enters the Forth at Cambus, 2½m. N. of Alloa ; length 33½ m.

Devon, Black or **South,** r., Fife and Clackmannan, flows 15 m. S.W. to the Forth, S.E. of Alloa.

Devonshaw, hill, with ancient circular camp, Wandell par., Lanarkshire ; alt. 1275 ft.

Devonshaw, seat, Kinross-shire, 2½ m. N.E. of Dollar.

Devonside, vil., Clackmannanshire, Tillicoultry par., 2¾ m. N.E. of Alva. Pop. D., Coalsnaughton, and Langour, 1240.

Dewar, ham. and lofty hill, Heriot par., Midlothian. Hill topped with a stone called "Lot's Wife" ; alt. over 1550 ft.

Dewar Town, vil., Borthwick par., Midlothian, 3 m. N.E. of Gorebridge.

Dews, loch, Fetteresso par., Kincardineshire.

Dewshill Colliery, N.E. Lanarkshire, 1 m. N.E. of Kirk of Shotts.

Dheirrig, or **Eilean Dearg,**isl.,Loch Riddon, Argyll, 2½ m. N.W. of Colintraive.

Dhomhnuil, loch, isl. of Islay, Argyll, 4 m. S.E. of Bowmore.

Dhu, or **Dubh Loch** (**Black Loch**), Aberdeenshire, near head of Loch Muick. Overhung by precipices over 1000 ft. high.

Dhuacrigh, loch, Isle of Skye, 3 m. N.E. of Bracadale.

Dhuisk, or **Dusk,** stream, Ayrshire, flowing 6 m. N.W. to the Stinchar.

Dhullochan, loch, Inverness-shire, 3 m. S.E. of Arnisdale.

Dhu Loch, Perthshire, 10 m. W. of Blair-Atholl.

Dhu Loch, Buteshire, 2 m. S.W. of Rothesay.

Dhu Loch, Wick par., Caithness, 2½ m. of Altnabreac sta.

Dhu Loch, or **Dubh Loch,** Inverness-shire, 5 m. W. Newtonmore.

Dhu Loch, or **Dubh Loch,** Perthshire, 1½ m. S.W. of Loch Chon.

Dhu Loch, or **Dubh Loch,** Sutherland, 7 m. E. of Lochinver.

Dhu Loch or **Dubh Lochan,** S.W. Banffshire, 14 m. S. of Nethybridge.

Diabaig, or **Dibaig,** ham., with harb., W. Ross and Cromarty, on Loch Torridon. P.O., T.O.

Dibiedale, deer forest (17,000 ac.), N.E. Ross and Cromarty, 5 m. N. of Alness.

Dickmont Law, hill, Angus, 1 m. N.E. of St. Vigeans ; alt. 323 ft.

Dieraclet, loch, Harris, 1 m. S. of Tarbert.

Digg, ham., Isle of Skye, 18½ m. N. of Portree.

Dighty Water, stream, Angus, running 15 m. eastward to Firth of Tay, at 2 m. E. of Broughty Ferry.

Digmore, small harb. in N. Uist, Outer Hebrides.

Dikehead, ham., N. Lanarkshire, ¾ m. W. of Shotts.

Dildawn, seat, Kirkcudbrightshire, 3 m. S.W. of Castle-Douglas.

Dillarburn, vil., Lanarkshire, 1¼ m. N.E. of Abbeygreen.

Dilty Moss, morass, Angus, Carmylie and Guthrie pars.

Dindinnie, pl., Wigtownshire, 2½ m. W. of Stranraer.

Din Fell, Roxburghshire, 5 m. S.W. Riccarton Junction ; alt. 1735 ft.

Dingleton, ham., Roxburghshire, 1 m. S. of Melrose.

Dingwall, capital of co., royal, mun., and parl. bur., with ry. sta., Ross and Cromarty, 18½ m. N.W. of Inverness ; par. 6465 ac. Pop. 2763 ; bur. pop. 2554. P.O., T.O. Hotel. Ry. junc., diverging lines to Kyle of Lochalsh and Caithness and Sutherland.

Dingy's How, ancient tumulus, 36 ft. high, St. Andrew's par., mainland, Orkney.

Dinlabyre, pl., on Liddel Water, Roxburghshire, Castleton par. D. Burn with cascade.

Din Law, mt., Peeblesshire, 3 m. S. of Talla reservoir; alt. 2182 ft.

Dinnet, q.s. par. and vil., with ry. sta., L.N.E., Aberdeenshire, 4½ m. W. of Aboyne. P.O., T.O. Pop. 93. D. Burn entering r. Dee in vicinity; Muir of D. to the S.; and D. House, a seat.

Dinnil, stream, flowing into r. Dee, near Aboyne, Aberdeenshire.

Dinvin House, Wigtownshire, 1 m. N. of Portpatrick.

Dinwoodie, ry. sta., L.M.S., Dumfriesshire, 6 m. N.W. of Lockerbie. D. Hill 871 ft. D. Lodge.

Diobaig Harbour, inlet of Loch Torridon, Ross and Cromarty.

Dionaird, r., N.W. Sutherland, flowing 14 m. N. to Kyle of Durness. Good salmon r.

Dippen, seat, Argyll, near Carradale, coast of Kintyre.

Dippen Head, headland, S.E. coast of Arran, Buteshire; alt. 300 ft. D. Lodge, 2 m. S.E. of Whiting Bay.

Dipple, ancient par., now included in Speymouth, Morayshire; also pl., 1½ m. W. of Fochabers.

Dippool Water, rivulet, Lanarkshire, running 7½ m. S.W. to the Mouse.

Diranean, seat, Perthshire, 10 m. E. of Pitlochry.

Dirk Hatteraick's Cave, S.W. coast of Kirkcudbrightshire, between Creetown and Gatehouse.

Dirleton, coast par., vil., with ry. sta., L.N.E., E. Lothian, 2½ m. S.W. of North Berwick; par. 9209 ac. Pop. 2824; vil. pop. 560. P.O., T.O. D. Castle, twelfth-century stronghold, dismantled by Cromwell.

Dirlot Castle, ruins of ancient fort, Caithness, 15 m. S. of Thurso.

Dirrans, vil., Ayrshire, 1¼ m. S. of Kilwinning. Pop. 287.

Dirrie More, pass on road from Dingwall to Ullapool (909 ft.) Ross and Cromarty, near Loch Droma.

Dirrington, Great and **Little,** two conical hills, Lammermuir Hills, Berwickshire; alt. 1309 and 1191 ft.

Dirrochan Fall, on Ardessie Burn, N.W. Ross and Cromarty, 8 m. S.E. of Ullapool.

Disblair, seat, Aberdeenshire, Fintray par., 2 m. W. of New Machar.

Distinkhorn, mt., E. Ayrshire, 5½ m. E.S.E. of Galston; alt. 1283 ft.

Ditch Hall, Fortingall par., Perthshire. Remains of ancient structure of earth and turf on Inverchadain Farm.

Dithreibh, Loch an, Sutherland, 6 m. S.W. of Tongue; depth 157 ft. max., alt. 267 ft.

Divach, affl. of the Coiltie, Invernessshire, Urquhart par. Has a picturesque waterfall.

Divach Lodge, Glen Urquhart, Invernessshire, 2½ m. S.W. of Drumnadrochit.

Divie, stream, rising on Brae Moray, and flows 12 m. N.W. to the Findhorn, Morayshire.

Dobson's Well, chalybeate spring, E. Lothian, ½ m. W. of Haddington.

Dochard, loch, Argyll, 5 m. W.N.W. of Bridge of Orchy ry. sta.; 42 ft. deep, alt. 735 ft.

Dochart, r. Perthshire, flows from Loch Dochart, at the head of Glen Dochart, to Loch Tay, at Killin; length of r. 13 m. Loch Dochart 1 m. E. of Crianlarich ry. sta.; depth 11 ft., alt. 513 ft.

Docherty, glen, Ross and Cromarty, S.E. of Kinlochewe.

Dochfour or **Little Loch Ness,** Invernessshire, near foot of Loch Ness. D. Burn. D. House on W. bank.

Dochgarroch, ham. and lock on Caledonian Canal, Inverness-shire, 4½ m. S.W. of Inverness. P.O., T.O.

Docra Limeworks, Ayrshire, 1½ m. from Beith.

Dod Burn, affl. of Allan Water, Roxburghshire, 8 m. S.W. of Denholm.

Dodd Hills, two mts., N. Kirkcudbrightshire, near Carsphairn; alts. 1500 and 1632 ft.

Dod Fell, mt., N. Dumfriesshire, 8 m. N.E. of Langholm; alt. 1519 ft.

Dod Hill, in Roxburghshire, Teviothead par.; 1620 ft.

Dod Law, mt., Midlothian, 3½ m. S.W. of Heriot sta.; alt. 1550 ft.

Dodridge Law, E. Lothian, 3½ m. S. of Ormiston sta.

Dods-Corse Stone, ancient cross, Berwickshire, 4 m. S.E. of Lauder.

Doe, r., Inverness-shire, affl. of the Moriston, 16 m. S.W. of Invermoriston.

Dogden, extensive moss in Greenlaw and Westruther pars., Berwickshire.

Dog Hillock, mt., Angus, 10 m. N.E. of Kirriemuir ; alt. 2369 ft.

Dog Hillock, mt., S.W. Aberdeenshire, 4½ m. N.W. of Milton of Clova ; alt. 2400 ft.

Dog Island, Lake of Menteith, Perthshire.

Dog's Ferry, on Loch Leven, Argyll.

Dogs, Isle of, in Loch Laggan, Inverness-shire.

Dog's Stone, isolated rock, Oban Bay, Argyll. Subject of curious legends.

Dogton, pl., Fife, 4½ m. N.W. of Kirk-caldy. Contains ancient standing stone.

Doilate, loch, N.Argyll, in Sunart, 3½ m. N. of Strontian ; depth 55 ft., alt. 22 ft.

Doine, loch, Perthshire, adjoining Loch Voil, 4 m. S.W. of Balquhidder ; 65 ft. deep, alt. 414 ft.

Doll, ham., Sutherland, 2 m. W. of Brora.

Doll, or Dole, Glen of the, Angus, Clova par. Overhanging rock called Scorrie of the Dell. Glen D. Lodge, seat.

Dollar, police bur. and par., with ry. sta., L.N.E., Clackmannanshire. Town stands at the foot of the Ochil Hills, 6¼ m. N.E. of Alloa ; par. 4772 ac. Pop. 1804 ; bur. pop. 1485. P.O., T.O. Dollar Academy, established 1818.

Dollarbeg, seat, Clackmannanshire, 1 m. S.E. of Dollar.

Dollar Burn, Clackmannanshire, flows 1½ m. S.E. to the Devon.

Dollarfield, seat, Clackmannanshire, ½ m. S. of Dollar.

Dollar Law, mt., 9½ m. S.W. of Peebles ; alt. 2680 ft.

Dollars, seat, Ayrshire, 4 m. S.E. of Kilmarnock.

Dollerie, seat, Perthshire, Madderty par. 2⅔ m. S.E. of Crieff.

Dolphington, seat and stream, West Lothian, 1 m. S.E. Dalmeny.

Dolphinton, par., ham., seat, and ry. stas., L.M.S., L.N.E., S.E. Lanarkshire, 11¼ m. N.E. of Carstairs ; 3578 ac. Pop. 210. P.O., T.O. ; alt. of ry sta. 721 ft.

Don, sea loch, Mull Isl., Argyll, E. coast.

Don, r., S. Aberdeenshire ; rises on the W. border of Aberdeenshire and flows E. to the sea, 1 m. N.E. of Old Aberdeen ; length, 82 m.

Donald's Cleuch, ravine, Peeblesshire, Tweedsmuir par. Retreat of Donald Cargill, a famous Covenanter. D. C. Head 2616 ft.

Donald's Quay, N. side of Clyde, Dun-bartonshire, E. of Bowling. White occulting light.

Donan, small isl. in Loch Alsh, S.W. Ross and Cromarty.

Donavourd, seat, Perthshire, 1½ m. S.E. of Pitlochry.

Donibristle, seat of the Earl of Moray, Fife, W. of Aberdour.

Donibristle Colliery, vil., Fife, 2 m. E. of Crossgates. P.O. called D.

Donside Paper Mills, in Aberdeen city.

Don Street, ry. sta., L.N.E., 2 m. N.N.W. of Aberdeen.

Doon, r., Ayrshire, dividing Carrick from Kyle, issues from Loch Doon (100 ft. deep, 5¼ m. long) and flows to the Firth of Clyde ; length 26¼ m. Ruin of ancient tower on Castle Isl. in Loch Doon, called Balliol's Castle.

Doon, hill, spur of the Lammermuirs, E. Lothian, 1 m. E. of Spott ; alt. 582 ft.

Doon Castle, ruin, Ardwell Point, Wig-townshire, 7 m. S.E. of Portpatrick.

Doonholm, seat, on r. Doon, 3 m. S. of Ayr.

Doonside, seat, with ruined castle, Ayr-shire, Maybole par., 3 m. S. of Ayr.

Doon Tynron, hill, Dumfriesshire, 4 m. S.W. of Thornhill ; 945 ft.

Door, Dore, or **Tangwick Holm,** islet, St. Magnus Bay, Shetland.

Dorback, pl. Inverness-shire, on D. Burn, 1½ m. from Nethybridge.

Dorbock Burn, stream, Morayshire, joins the Divie, 1 m. S.W. of Dunplail ry. sta.

Dores, par. and vil., Inverness-shire, E. side of Loch Ness ; 25,143 ac. Pop. 572. P.O., T.O. Steamboat pier and inn.

Dorlin House, seat, Inverness-shire, on Loch Moidart, 6 m. N.W. of Salen.

Dormont, seat, Dumfriesshire, 6 m. S.W. of Lockerbie.

Dormont, vale, Roxburghshire, 12 m. S.E. of Ancrum.

Dornadilla, Scandinavian tower, Suther-land, 11 m. S.W. of Tongue. One of the many brochs in the co.

Dornal, loch, Wigtownshire, 10 m. N.N.W. of Newton-Stewart ; alt. 386 ft., depth 10 ft.

Dornie, fishing vil., hotel and ferry, Ross and Cromarty, Kintail par., at mouth of Loch Long. P.O., T.O. Pop. 80. Ferry to Ardelve Hotel across Loch Long, also to Totaig across Loch Duich.

Dornoch, cap. of co., royal and mun. bur., seapt., par., ry. sta., L.M.S., Sutherland, on Dornoch Firth, 6 m. N.E. of Tain, 57½ m. N.E. of Inverness ; par. 33,012 ac. Pop. 2086 ; bur. pop. 725. Retains ancient cathedral (1222) ; rebuilt 1837. Hotels. P.O., T.O.

Dornoch Firth, inlet of the North Sea, between Ross and Sutherland, 22¼ m. long and 11 m. wide at entrance.

Dornoch State Forest, Dornoch, Sutherland.

Dornock, par., vil., and seat, ry. sta., L.M.S., Dumfriesshire, on Solway Firth, 3 m. E. of Annan ; 4788 ac. Pop. 1688. P.O., T.O. called Eastriggs ; vil. pop. 1308.

Dorrery Lodge, Caithness, 1½ m. S.W. of Scotcalder.

Dorusduan, shooting-lodge and deer forest ; 10,000 ac. Ross and Cromarty, 10 m. E.S.E. of Lochalsh.

Dorusmore, strait, Argyll, between Craignish Pt. and Garbhreisa.

Dothan, ham., Fife, 3½ m. N.W. of Kirkcaldy.

Double-Dykes, remains of extensive ancient fortification, Lanarkshire, Stonehouse par.

Dougal Loch, inlet, Sutherland, N. of Laxford.

Dougalston, seat and loch, Stirlingshire, 1 m. S.E. of Milngavie.

Douglas, par., vil., ry. stas. (D. and D. West), L.M.S., Lanarkshire ; 34,162 ac. Pop. 2948 ; q.s. par., D. Water, vil. on D. Water. 10¾m. S.W. of Lanark ; vil. pop. 1441. P.O., T.O. D. Castle, a seat of the Earl of Home.

Douglas Burn, stream, Selkirkshire, flows 6 m. S.E. to Yarrow Water.

Douglasdale, basin of the D. Water, Lanarkshire.

Douglas Hall, seaside resort, Kirkcudbrightshire, 2 m. E. of Colvend.

Douglas Park, colliery, Lanarkshire, 1 m. N.E. of Bothwell.

Douglas Park, a seat, 2 m. E. of Bothwell.

Douglas Pier, on Loch Goil, Argyll, 3 m. S. Lochgoilhead. P.O.

Douglastown, ham. in co. and 3½ m. S.W. of, Forfar. P.O.

Douglas Water, stream, Dunbartonshire, flows through Glen D., to Loch Lomond.

Douglas Water, stream, Argyll, flows E. to Loch Fyne, which it joins 3 m. S.W. of Inveraray.

Douglas Water, r., Lanarkshire, flows 20 m., joining the Clyde 2½ m. S.E. of Lanark.

Douglas Water Rows, vil., Lanarkshire, Carmichael par. P.O., T.O.

Dougrie Lodge, Arran Isl., W. Coast, 8 m. W. of Brodick.

Doule, or **Dhughaill,** loch, Ross and Cromarty, 6 m. N.E. of head of Loch Carron ; alt. 93 ft., depth 179 ft.

Douloch, ham., Wigtownshire, 6 m. N.W. of Stranraer.

Dounby, ham., Orkney, 8 m. N. of Stromness. P.O., T.O.

Doune, police bur., with ry. sta., L.M.S., Perthshire, 3 m. W. of Dunblane, 9 m. N.W. of Stirling. Pop. 822. Hotel. P.O., T.O. D. Castle in ruins. D. Lodge, seat of Earl of Moray, 1½ m. N.W.

Doune, mt., Dunbartonshire, 5 m. N.W. of Luss ; alt. 2409 ft.

Doune, pl., Ross and Cromarty, 3 m. E. of Oykell Bridge.

Doune, The, of Rothiemurchus, seat of the Grants of Rothiemurchus, Inverness-shire, 2½ m. S.W. of Aviemore.

Dounreay, ham., Caithness, 2 m. N.E. of Reay. D. Castle, ruins, ancient seat of the Reays.

Doura, colliery, vil., Ayrshire, 3 m. E. of Kilwinning

Dour Burn, Fife, Aberdour par. Supplies a reservoir for Burntisland.

Dourie Voe, inlet, Shetland, E. mainland.

Dovecotwood, ruins of ancient castle, Stirlingshire, Kilsyth par.

Dowally, vil., Perthshire, 4½ m. N.W. of Dunkeld. D. Loch, 1 m. S.E.

Dowanhill, subs. of Glasgow, N.W.

Dowie Dens, formerly moor, now cultivated, Selkirkshire, near Yarrow Church.

Downfield, q.s. par., with ry. sta., L.M.S. (Baldovan and D.), Angus, near Dundee.

Downie Hills, Angus, Monikie par., Panmure monument on top, 105 ft. high.

Downie Park, seat, Angus, Tannadice par.

Downies, fishing vil., Kincardineshire, 6 m. N.W. Stonehaven.

Downreay, ham., Caithness, 3 m. E. of Reay. P.O., T.O.

Drainie, par., Morayshire, containing Lossiemouth, 6948 ac. Pop. 4371.

Drakemyre, vil., Ayrshire, near Dalry.

Dreghorn, par. and vil. with ry. sta., L.M.S., Ayrshire ; par. 5621 ac. Pop. 4338 ; 2 m. E. of Irvine. P.O., T.O. Vil. pop. 3032.

Dreghorn Castle, former seat, now a boys' preparatory school, Colinton, S.W.Edinburgh. Grounds property of War Office.

Drem, vil. and ry. junction, L.N.E., E. Lothian, 17½ m. E. of Edinburgh by rail. P.O., T.O.

Dressetland, seat, Dumfriesshire, 2½ m. from Thornhill.

Drewsteignton, Bearsden, Dunbartonshire. Girls' private school.

Driesh, mt., Angus, 14 m. N.W. Kirriemuir ; alt. 3105 ft.

Drimlabarra, pl., S.E. Arran, Buteshire.

Drimneil House, Argyll, 3½ m. from Appin.

Drimnin, vil., pier, and seat, Argyll, E. Sound of Mull, opposite Tobermory. P.O., T.O.

Drimsynie, seat, Argyll, ½ m. W. of Lochgoilhead.

Drimycon Bay, Argyll, Gigha Isl.

Drinesheader, ham. and loch, W. coast of Harris, Outer Hebrides.

Drinnishader, ham., Harris, S. Loch Tarbert.

Drip Bridge, on the Forth, 2¼ m. W.N.W. of Stirling.

Drissaig, seat, Argyll, 13 m. S.W. of Taynuilt.

Drochil Castle, unfinished, well-preserved baronial fort. Built by Regent Morton (1530–78), 7 m. N.W. Peebles. D. Hills, 1234 ft.

Dron, par., 5½ m. S. of Perth ; 4342 ac. Pop. 204. D. Hill, 800 ft.

Drongan, colliery vil., and seat, with ry. sta., L.M.S., Ayrshire, 10 m. S.E. of Ayr. P.O., T.O.

Drongs, The, rocks rising to over 100 ft. in St. Magnus Bay, Shetland.

Dronley, vil., with ry. sta., L.M.S., Angus, 11½ m. N.W. of Dundee.

Druidaig, shooting - lodge, Ross and Cromarty, on Loch Duich.

Druidibeg, loch, S. Uist, Inverness-shire.

Druidsmere, seat, Perthshire, 2 m. S. Blairgowrie.

Druimarbin, vil., Inverness-shire, 2¼ m. S.W. of Fort William.

Druimbeg, seat, S.W. Inverness-shire, on Loch Shiel.

Druimdrishaig, seat, Argyll, 11 m. S.W. of Ardrishaig.

Druim Fada, mt., Argyllshire, 6 m. N.W. Fort William, 2385 ft.

Druimneil House, Argyll, 1 m. S. of Port Appin.

Drum, pl., with ry. sta., L.N.E., 10 m. S.W. Aberdeen.

Drum, pl., Edinburgh, Liberton par.

Drum, pl., Kinross-shire, near Crook of Devon.

Drumadoon, bay and prom., S.W. Arran, Buteshire.

Drumalbin, ancient name of Central Grampians.

Drumbag, ham., Sutherland, Assynt par., P.O., T.O. Called Drumbeg.

Drumbain House, Ayrshire, 4 m. S.W. of Ballantrae.

Drumblade, par. and vil., Aberdeenshire, 4 m. E. of Huntly ; 9319 ac. Pop. 828. P.O., T.O.

Drumblair, seat, Aberdeenshire, Forgue par., 10 m. N.E. of Huntly.

Drumbowie, pl., Stirlingshire, Muiravonside par.

Drumbowie Colliery, Lanarkshire, 3 m. E.S.E. of Airdrie.

Drumboy Hill, Ayrshire and Renfrewshire border, 876 ft.

Drumbreck, vil., Lanarkshire, 5½ m. N.E. Airdrie.

Drumbuie, pl., Wigtownshire, 5 m. S.W. Newton-Stewart.

Drumbuie, vil., Ross and Cromarty, 2½ m. N.W. of Lochalsh.

Drumburn, vil., Kirkcudbrightshire, Newabbey par.

Drumcarrow, seat, Fife, 4 m. S.W. of St. Andrews.

Drum Castle, seat, Aberdeenshire, 1 m. N.W. of Drum ry. sta., 10 m. S.W. of Aberdeen. Close to it is Tower of D., Hill of D. (414 ft.), and Loch of D.

Drumchapel, vil. and ry. sta., L.N.E., Dunbartonshire, 6½ m. N.W. of Glasgow. P.O., T.O.

Drumcharry, ham., Perthshire, 7½ m. W. of Aberfeldy.

Drumchork, pl., Ross and Cromarty, ½ m. S.E. of Aultbea.

Drumclair Colliery, vil., Stirlingshire, ½ m. S.E. of Slamannan.

Drumclog, moorland locality, ham., ry. sta., L.M.S., Lanarkshire, 6 m. W. of Strathaven. Scene of a famous skirmish in 1679, in which Claverhouse was defeated by the Covenanters. P.O., T.O.

Drumclyer, seat, N. Kirkcudbrightshire, 6 m. W. of Dumfries.

Drumcoltran, pl., Kirkcudbrightshire, 10 m. S.W. of Dumfries. Old square tower.

Drumduan, seat, Kincardineshire, 2½ m. S.W. of Aberdeen.

Drumduan House, Morayshire, near Forres.

Drumeldrie, ham., Fife, 1½ m. E. of Largo.

Drumellan, seat, Ayrshire, 1 m. N.E. of Maybole.

Drumellie, or **Marlee Loch,** Perthshire, 2 m. S.W. of Blairgowrie; depth 58 ft., alt. 146 ft.

Drumelzier, par. and vil. in co., and 9½ m. S.W. of, Peebles; 17,970 ac. Pop. 200. P.O. Ruins of D. Castle.

Drumfearn, pl., Isle of Skye, Sleat Peninsula, N.

Drumforber House, Kincardineshire, 2 m. N. of Laurencekirk.

Drumfork, lodge, Angus, 9 m. N. Blairgowrie.

Drumgask, ham., Inverness-shire, ½ m. S. of Laggan.

Drumgelloch, Lanarkshire, sub. of Airdrie.

Drumglay, vil., Angus, 2 m. W. of Forfar.

Drumin, seat, Banffshire, Inveraven par. Ruins of D. Castle.

Druminnor House, Aberdeenshire, 4 m. S.W. of Kinnethmont.

Drumintoul Lodge, Inverness-shire, 2½ m. from Aviemore sta.

Drumkilbo, seat, Perthshire, 1 m. N.E. of Meigle.

Drumlamford, seat and loch (26 ft. deep), Ayrshire, Colmonell par., 4½ m. S.E. of Barrhill sta.

Drumlanrig Castle, Dumfriesshire, 3½ m. N.N.W. of Thornhill, on r. Nith. A seat of the Duke of Buccleuch.

Drumlean, ham., Perthshire, 3 m. N.W. of Aberfoyle.

Drumlemble, colliery, 4 m. W. of Campbeltown. P.O.

Drumley, seat, Ayrshire, 2 m. S.W. of Tarbolton.

Drumlithie, vil., ry. sta., L.M.S., Kincardineshire, 7 m. S.W. of Stonehaven. P.O., T.O. Pop. 166.

Drummilling, seat, Ayrshire, W. Kilbride.

Drummoddie, ham., Wigtownshire, 4½ m. N.E. of Port William.

Drummond Castle, Perthshire, 3½ m. S.W. of Crieff. Seat of the Earl of Ancaster.

Drummond Hill, deer forest, Perthshire, Kenmore par. State forest at head of Loch Tay, between Kenmore and Fortingall.

Drummonie House, seat, Perthshire, Dunbarney par.

Drummore, seapt., vil., coastguard sta., and bay, Wigtownshire, 5 m. N.W. of Mull of Galloway. P.O., T.O. Pop. 382.

Drummore, school, E. Lothian, 1½ m. N.E. of Musselburgh. D. Colliery in vicinity.

Drummossie Muir, Inverness-shire. Includes battlefield of Culloden.

Drummuie, pl. and seat, Sutherland, 1 m. W. of Golspie.

Drummuir, ry. sta., L.N.E., Banffshire, 4½ m. N.E. of Dufftown. D. Castle, on r. Isla, in vicinity.

Drumnadrochit, ham. at mouth of Glenurquhart, Inverness-shire. Hotel. P.O., T.O. 14 m. S.W. of Inverness.

Drum-na-Gesk, seat, Aberdeenshire, 2½ m. S.E. of Aboyne.

Drumoak, par., 10 m. S.W. of Aberdeen; 7239 ac. Pop. 798. P.O., T.O. D. House, a seat.

Drumochter, or **Drumouchter,** mt. pass and ry. summit (1484 ft.), border of Perthshire and Inverness-shire, 2 m. N.W. of Dalnaspidal.

Drumore, shooting-lodge and loch, Perthshire, 9½ m. N. of Blairgowrie.

Drumore House, Argyll, near Campbeltown.

Drumour, ham. and seat, Perthshire, 4 m. S.W. Dunkeld.

Drumpark, seat, Kirkcudbrightshire, 3½ m. S. of Dunscore.

Drumpellier, dist. and seat, Lanarkshire, near Coatbridge.

Drumquhassle, Park of, seat, Stirlingshire, Drymen par.

Drumrossie House, Aberdeenshire, near Insch ry. sta.

Drumruinie Lodge, Ross and Cromarty, 7 m. N.E. of Ullapool.

Drumry, estate, Dunbartonshire, 2 m. S.E. of Duntocher.

Drums, seat, Renfrewshire, 2 m. W. of Bishopton sta.

Drums, pl., Perthshire, Errol par.

Drumshangie Colliery, N. Lanarkshire, 2 m. N.E. of Airdrie.

Drumsheugh, dist., W. Edinburgh.

Drumshoreland, ry. sta., L.N.E., W. Lothian, 11½ m. S.W. of Edinburgh.

Drums of Ardgaith, pl., 6 m. E. of Perth.

Drumsturdy, vil., Angus, 6 m. N.E. of Dundee.

Drumtassie, burn and pl., W. Lothian, 6 m. S.W. Linlithgow.

Drumtochty Castle, Kincardineshire, 4½ m. N.W. of Fordoun ry. sta. D. Glen here; State forest here.

Drumvaich, ham., Perthshire, 5 m. N.W. of Doune.

Drumwhirn House, seat, Kirkcudbrightshire, on Urr Water.

Drumyocher, ham., Kincardineshire, 3 m. N.E. of Fordoun sta.

Drunkie, loch, Perthshire, 3 m. N.E. of Aberfoyle; depth 97 ft.

Drunzie, ham., Kinross-shire, 3 m. N.E. of Milnathort.

Dry, isl. in par. of Gairloch, Ross and Cromarty. Pop. 6.

Drybridge, vil., ry. sta., L.M.S., Ayrshire, 5 m. S.W. of Kilmarnock. P.O., T.O. at sta. Pop. 215.

Drybridge, ham., Banffshire, 2 m. S.E. of Buckie. P.O.

Dryburgh Abbey, ruined abbey, national property, Berwickshire, on left bank of Tweed, 3½ m. S.E. of Melrose. Contains the tomb of Sir Walter Scott. D. Abbey and D. House in vicinity.

Dryden, estate, Midlothian, near Roslin.

Dryfeholm, seat, Dumfriesshire, 2¼ m. N.E. Lockerbie.

Dryfesdale (Drysdale), par., Dumfriesshire; 10,256 ac. Pop. 3343.

Dryfe Water, Dumfriesshire, joins the Annan 2 m. W. of Lockerbie.

Drygrange, seat, bridge, and imposing ry. viaduct, Roxburghshire, 2¼ m. N.E. of Melrose.

Dryhope Tower, ancient border peel, Selkirkshire, near foot of St. Mary's Loch, stands on D. Burn, and near it D. Rig; alt. 1712 ft.

Drylaw House, city of Edinburgh, ¾ m. N.W. Craigleith ry. sta.

Drymen, par., vil., with ry. sta., L.N.E. (passenger service withdrawn), Stirling, 5 m. W.S.W. of Balfron; 30,854 ac. Pop. 1128. P.O., T.O. D. sta., 1¼ m. S. P.O., T.O. Pop. of vil. 304.

Drynie, seat, Ross and Cromarty, Kilmuir Wester par., 4 m. N. of Inverness.

Drynoch Lodge, seat and ham., Skye, near head of Loch Harport.

Duagrich, loch, Isle of Skye, 3 m. N. of Bracadale.

Dualt, trib. of the Carnock, Stirlingshire, with waterfall of 60 ft.

Duard, or **Rudha Dubh Ard,** headland, Ross and Cromarty, entrance to Loch Broom.

Duart Bay, N.E. Mull Isl., Argyll. D. Castle, formerly the seat of the Macleans. D. Point, a group-flashing light, white and red, visible 12 m.

Dubford, ham., in co., and 7½ m. E. of Banff. P.O.

Dubh Artach Rocks, off the coast of Argyll, 15½ m. S.W. of Iona Lighthouse, showing white group-flashing light, seen 18 m.

Dubh Loch, N.W. Ross and Cromarty, 8 m. E.S.E. of Poolewe; depth 86 ft.

Dubh Sgeir, Barra, Hebrides. Fixed white light, seen 7 m.

Dubh Sgeir, islet, Argyll, Jura par. Shows a flashing light, visible 9 m.

Dubton, seat and ry. sta., L.M.S., Angus, 3 m. W.N.W. of Montrose. T.O. at sta.

Duchal Law, Renfrewshire, E. summit of the Braes of Gleniffer; alt. 725 ft.

Duchall House, seat, Renfrewshire, 1 m. S. of Kilmacolm sta. D. Castle, in ruins, formerly belonging to the Lords Lyle, 1 m. N.W.

Duchally, seat, Perthshire, 2 m. E. of Blackford.

Duchally, deer forest, Sutherland, 21 m. N.W. of Invershin.

Duchar Law, mt., Peeblesshire, 6½ m. S. of Innerleithen; alt. 1779 ft.

Duchray Water, S. headstream of the Forth, rising on Ben Lomond, joining the Forth, 1 m. W. of Aberfoyle. D. Castle, formerly a stronghold of the Grahams, on rt. bank of D. Water.

Du Craig, rocky islet in Firth of Forth, between Limekilns and N. Queensferry, Fife.

Duddingston, vil., with ry. sta. (D. and Craigmillar), L.N.E., in city, and 2½ m. S.E. of Edinburgh G.P.O. P.O. Breweries and distilleries. D. House, formerly a seat of the Duke of Abercorn, now a golf club and course. D. Loch, at base of Arthur Seat.

Duddy Hill, East Lothian, S.E.; alt. 1120 ft.

Dudhope Castle, Dundee, Angus, at the foot of D. Law.

Dudwick, estate, Aberdeenshire, 4 m. N.E. of Ellon. D. Hill, 572 ft.

Duff House, former seat of the Earl of Fife, near Banff; unoccupied.

Duff Kinnel, stream, Dumfriesshire, affl. of Kinnel Water.

Dufftown, police bur.; ry. sta. (1 m. N.), L.N.E., Banffshire, 11 m. S.W. of Keith; 317 ac. Pop. 1454. P.O., T.O. Limeworks and distilleries.

Duffus, par. and vil., Morayshire; 9565 ac. Pop. 3337; vil. New Duffus, 5 m. N.W. of Elgin. P.O., T.O. Ruins of D. Castle. D. House a seat.

Duich, sea loch, S.W. Ross and Cromarty, extending 5 m. S.E. from head of Lochalsh.

Duich, stream, isl. of Islay, 3 m. S. of Bowmore.

Duin, Loch an, borders of Perth and Inverness cos., 6 m. S.E. of Dalwhinnie; depth 102 ft., alt. about 1590 ft.

Duirinish, par., W. Skye, Inverness-shire; 80,067 ac. Pop. 2146.

Duirinish, ham. and ry. sta., L.M.S., S.W. Ross and Cromarty, 4 m. N.W. of Kyle of Lochalsh.

Duirland, pl., Dunbartonshire, 5 m. S.W. of Luss.

Duisdale, pl. and seat, Isle of Skye, Sleat par.

Duisk Burn, stream, S. Ayrshire, trib. of Stinchar.

Duisky, vil., Argyll, 7 m. N.W. of Fort William.

Duke Street, ry. sta., L.N.E., E. Glasgow.

Dulaich, loch, Sutherland, 2½ m. N.E. of Lairg.

Dull, par. and vil., Perthshire, 3½ m. W. of Aberfeldy; 56,149 ac. Pop. 2354.

Dullan Water, Banffshire, flowing to the Fiddich.

Dullatur, dist., traversed by Forth and Clyde Canal; ry. sta., L.N.E., Dunbartonshire; vil., 2 m. E. of Kilsyth. P.O., T.O. Vil. pop. 134.

Dulnain, r., Inverness-shire, flowing 28 m. N.E. to the Spey, 3 m. S.W. of Grantown.

Dulnain Bridge, ham., Morayshire, 3 m. S.W. of Grantown, on Dulnain R. P.O., T.O.

Dulochan, loch, W. Inverness-shire, Knoidart peninsula.

Dulsie Bridge, ham. and romantic bridge, Nairnshire, on Findhorn R.; ham. 12 m. S.E. of Nairn.

Dumbarton, capital of co., parl., royal, and mun. bur., par., mkt.-town, and seapt., with ry. stas., L.M.S. and L.N.E., Dunbartonshire, 16 m. by rail N.W. of Glasgow. The shipbuilding yards are among the most important on the Clyde. Other industries include engineering and engine and boiler making. Par. 8324 ac. Pop. 16,076; parl. and mun. bur. pop. 21,546. P.O., T.O. D. Castle, a fortress at least a thousand years old, situated on a double-peaked basaltic rock rising 240 ft. above the r. Clyde.

Dumbarton, East, ry. sta., L.M.S., adjoining Dumbarton.

Dumbartonshire, see DUNBARTONSHIRE.

Dumbreck, hill, Stirlingshire, 1½ m. N.E. Strathblane; alt. 1664 ft.

Dumbrock, loch, Stirlingshire, 1 m. S.W. of Strathblane.

Dumbuck, vil., seat, and hill, 1 m. E. of Dumbarton Castle, adjacent to the Clyde. Shows group-flashing light.

Dumbuils, hill, Perthshire, Forgandenny par. Traces of ancient fort; alt. 300 ft.

Dumfries, capital of co., police and royal bur., par. and r. port, with ry. sta., L.M.S., Dumfriesshire, on r. Nith, 33 m. W.N.W. of Carlisle by rail and 90 m. S. by W. of Edinburgh. The mfr. of tweeds is the leading industry and hosiery ranks next. There is a large trade in pork and live stock. Par. 10,084 ac. Pop. 19,359. Police bur. pop. (including Maxwelltown in Kirkcudbrightshire), 22,975. P.O., T.O. Robert Burns died here in 1796.

Dumfries House, a seat of the Marquess of Bute, 2 m. W. of Cumnock, Ayrshire; also a ry. sta., L.M.S.

Dumfriesshire, a border co., bounded on the S. by Cumberland and the Solway Firth. The surface is mostly undulating and, in the N. and N.E., mountainous. The principal heights are Hart Fell, 2651 ft., and White Coomb, 2695 ft. The chief rivers, the Nith, Annan, and Esk, traversing their respective dales. The industries include the mfr. of hosiery, tweeds, leather, and machinery. Coal and lead are mined, but the majority are engaged in agriculture. Area, 10,723 sq. m.

or 686,302 ac. Pop. 81,047. Returns one member to Parliament.

Dumglow, hill, Kinross-shire, 4½ m. S.W. of Kinross ; alt. 1241 ft.

Dumgoyne, ry. sta., L.N.E., Stirling-shire, 1½ m. S.E. of Killearn. P.O.

Dumyat, a summit of the Ochil Hills, S. Perthshire, 3½ m. N.E. of Stirling ; alt. 1375 ft.

Dun, par. and ry. sta. (Bridge of Dun), L.M.S., Angus, 4 m. E. of Brechin ; 4391 ac. Pop. 419. D. House, seat. Dun's Dish, loch.

Dunach, seat, Argyll, 3½ m. S. of Oban.

Dunachton, seat, Inverness-shire, Alvie par., near Kincraig ry. sta.

Dunagoil, bay and headland, S.W. coast of Bute, 1½ m. N.W. of Garroch Head.

Dunain and **Lochend,** shooting-lodge, 6 m. S.W. of Inverness.

Dunairds House, seat, Perthshire, near Birnam.

Dunalastair, pl., Perthshire, 3 m. W. of Tummel Bridge. P.O., T.O. D. seat, formerly the home of the Robertsons of Struan.

Dunan, bold prom., Lochbroom par., Ross and Cromarty, 10½ m. N.W. of Ullapool.

Dunan, ham., Isl. of Skye, Inverness-shire, 4 m. N.W. of Broadford. P.O.

Dunan, seat, Perthshire, 3 m. E. of Rannoch sta.

Dunans, seat, Argyll, in Glendaruel, 5 m. N.W. head of Loch Striven.

Dunardarigh, North Berwick, E. Lothian. Girls' private school.

Dunaskin, pl., Ayrshire, near Waterside sta. P.O., T.O.

Dùn Athad, a high rock on the coast-line at the extreme south end of Kildalton par., Islay, Argyll. The summit had a fort or castle of which there is only a mound now.

Dunaverty, vil., bay, and headland, Argyll, 5 m. E.N.E. of Mull of Kintyre. Ruins of castle of the Lords of the Isles. Lifeboat sta.

Dunbar, royal and mun. bur., seapt.-town, and par., with ry. sta., L.N.E., East Lothian, 29 m. E. of Edinburgh by rail. par. 7486 ac. Pop. 5062 ; town and bur. pop. 3751. Hotels. P.O., T.O. Coastguard and lifeboat stas. Shows on pier head and middle quay, red fixed lights, seen 6 m., and on Bayswell Hill a fixed red and a fixed green light, visible 4 m. and 3 m. respectively. A holiday resort. Exports potatoes. Agricultural implements, rope-making, and brewing. D. Castle, famous for its defence by "Black Agnes," Countess of Dunbar, 1337-38.

Dunbarney, par., Perthshire, 4 m. S.E. of Perth ; area, 3906 ac. Pop. 1085. D. House, a seat.

Dunbarrow, seat and hill, Angus, 7 m. S.E. of Forfar ; alt. of hill, 500 ft.

Dunbartonshire, a western co., bounded on the S. by the Clyde, and E. by Stirlingshire and Loch Lomond. It comprises a main body and a detached portion. The dist., between Lochs Lomond and Long, is unequalled for its scenery, owing to its numerous lochs, glens, and mountains. Ben Vorlich, 3055 ft., and Ben Vane, 3004 ft. are the highest mountains, and the largest lochs are Loch Lomond, Loch Long, and the Gareloch. The Clyde seaboard, Vale of Leven and main portion of co., is rich and pleasantly varied. Industries include cotton works, calico printing, bleaching, dye-ing, and paper-making. There are also extensive shipbuilding yards along the Clyde. Area, 156,927 ac. Pop. 147,744. Returns one member to Parliament.

Dunbeath, coast vil., Caithness, 20 m. S.W. of Wick. P.O., T.O. Pop. 575. Fixed light on pier, red and green sectors, visible 5 m.

Dun-Bhail-an-Righ, or **Balanree,** ex-tensive cliff, W. coast Argyll, 5½ m. N.E. Oban.

Dun Bhoraraic, hill, Islay, Argyll, Kill-arrow par. Ruins of ancient fort ; alt. 619 ft.

Dunblane, police bur., par., and mkt.-town, with ry. sta., L.M.S., Perthshire, on Allan Water, 5 m. N. of Stirling, 41½ m. N.W. of Edinburgh ; par. 22,873 ac. Pop. 4421. Police bur. pop. 2692. Hotels. P.O., T.O. Dun-blane Cathedral was founded in the seventh century and became a cathedral in the reign of David I. The hydro-pathic establishment attracts many visitors. Mineral spring and worsted spinning industry. The Queen Victoria Military School for the sons of Scottish soldiers and sailors is 1 m. N.

Dunblane Scottish Youth Hostel. The Whitehead Hostel, 12 m. S.W. of Callander, 6 m. N. of Stirling, about ½ m. from D. ry. sta.

Dunbog, par., N.W. Fife, 3¾ m. E. of Newburgh; 3456 ac. Pop. 215. D. Hill, 706 ft. D. House. Ruin of Collairnie Castle.

Duncaan, hill, Raasay, Inverness-shire, N. end of Inverarish Burn; alt. 1456 ft.

Duncansbay, ham., N.E. Caithness, 2. m. E. of Huna.

Duncansbay Head, prom. at N.E. extremity of the Scottish mainland, Caithness, 18½ m. N.E. of Wick. A white flashing light, visible 20 m.; also a siren. D. Stacks, 1 m. S.

Duncansburgh, q.s. par., N. Inverness-shire, Kilmallie par. Was one of the early names for Fort William.

Duncan's Height, tumulus, Orkney, on isthmus connecting Deerness peninsula with mainland isl.

Duncan's Hill, mound and an ancient fort, Perthshire, S. of Dunkeld.

Duncanston, ham., Ross and Cromarty, 3½ m. N.E. of Conon ry. sta. P.O. called Duncanstown.

Duncanstone, ham., Aberdeenshire, 3 m. S.W. of Insch ry. sta. P.O.

Duncharloway, pl., with ancient circular fortification on Loch Carloway, Lewis, Outer Hebrides.

Dun Chibhich, ancient strong fortification, on Gigha Isl., Argyll.

Dun Coillich, mt., Perthshire, 5 m. N.W. of Kenmore; alt. 1866 ft.

Duncomb, conical hill (Kilpatrick Hills), with magnificent view, Dunbartonshire, Old Kilpatrick par., 3¼ m. N.W. of Duntocher; alt. 1313 ft.

Dun Connell, isl., near Pladda Isl., Firth of Clyde.

Duncow, vil., seat, and burn, Dumfriesshire; vil., 5 m. N. of Dumfries. P.O., T.O.

Duncraggan, now **Brig o' Turk,** pl., Perthshire, 6 m. W. of Callander. Figures in Scott's "Lady of the Lake."

Duncraig Castle, seat, S.W. Ross and Cromarty, on Loch Carron, 4 m. S.W. of Strome Ferry.

Duncrievie, vil. and seat, Perthshire, 1 m. S. of Glenfarg.

Duncrub Park, seat of Lord Rollo, in Dunning par., Perthshire.

Duncryne, ham. and hill, Dunbartonshire, 4 m. N.E. of Balloch Pier. Hill, 462 ft.

Dundaff Hills, eastern part of the Lennox Hills, Stirlingshire, 6½ m., S.S.W. of Stirling; alt. 1157 ft.

Dundaff Linn, waterfall on the r. Clyde, Lanarkshire, a little above New Lanark.

Dun-da-gu, or **Dun-da-Ghaiothe,** mt., N.E. isl. of Mull; alt. 2512 ft.

Dundalav (Dù-da-laimh, *fort of the two hands*), hill, S. Inverness-shire, 12 m. W. of Kingussie. Crowned with remarkable rude ancient fortress; alt. over 1500 ft.

Dundarave, or **Dunderawe,** on Loch Fyne, Argyll, 4½ m. N.E. of Inveraray. Ruins of old turreted house of the Macnaughtons.

Dundardil, a bold prom. close to the E. side of Loch Ness and 2½ m. N.E. of Foyers. The ridge rises to 929 ft.

Dundargue Castle, ruin, N. Aberdeenshire, 3 m. S.W. of Rosehearty.

Dundas Castle, seat, West Lothian, 1½ m. S.W. of Queensferry. D. Hill, 380 ft.

Dundavid, ancient castle, Isl. of Skye, 19 m. N.W. of Portree.

Dundee, third Scottish city. Parl., royal, and co. bur., sea pt., and par. Angus on the left bank of the estuary of the Tay, 10 m. from the sea and 50 m. N.N.E. of Edinburgh; 21 m. E. of Perth. Stands mostly on the slope between Dundee Law and Balgay Hill and the Tay. In parishes, Dundee, Mains, and Strathmartine. Ry. stas., D. West, L.M.S.; Tay Bridge, L.N.E.; and D. East, L.N.E. and L.M.S. Par. 5991 ac. Pop. 175,933; parl. and mun. bur. pop. 175,583. Staple industry, jute mfrs; other mfrs., hemp and flax goods, linoleum, machinery, preserves and confectionery, shipbuilding, and fruit and fish canning. Extensive dockyards. There is a college incorported with the University of St. Andrews. Returns two members to Parliament.

Dundelchak, loch, 8½ m. S. of Inverness.

Dundonachie, estate, Perthshire, 1½ m. S.W. of Dunkeld.

Dundonald, coast par. and vil., Ayrshire; par. 12,358 ac. Pop. 16,459; vil. 4½ m. S.E. of Irvine. P.O., T.O. Pop. 465. D. Castle, in ruins, belonged to the Cochranes, Earls of Dundonald. Convalescent Home.

Dundonald, ruins of old castle, Argyll, in W. Kintyre.

Dundonnell, pl., N.W. Ross and Cromarty, 5 m. S. of Ullapool, Lochbroom par. Hotel. P.O., T.O. D. River from 12 m. N.W. to Little Lochbroom. D. deer forest, 25,000 ac. D. House and D. Lodge, seats.

Dundreggan, shooting lodge, Glen Moriston, Inverness-shire, 8 m. N.W. of Fort Augustus.

Dundreich, hill, 5 m. N.E. of Peebles; alt. 1954 ft.

Dundrennan, vil., ruined abbey, and seat, Kirkcudbrightshire, 6 m. from Kirkcudbright. P.O., T.O. Pop. 79.

Dunduff, pl., 6 m. S.W. of Ayr. Traces of Roman camp and old baronial fortalice. D. Glen spanned by a ry. viaduct, 90 ft. high.

Dundurcus, ancient par., now annexed mainly to Boharm in Banffshire and partly to Rothes in Morayshire. D. Vale along r. Spey.

Dundurn, q.s. par., Perthshire, Comrie par.

Dundyvan, q.s. par., and sub. of Coatbridge, Lanarkshire.

Dunean, seat and hill, 3 m. S.W. of Inverness; hill, 940 ft.

Dunearn, hill, Fife, 2 m. N.W. of Burntisland, 734 ft.

Dunearn, shooting lodge, Nairnshire, 10½ m. N.W. of Grantown.

Duneaton Water, Lanarkshire, flows 19 m. N.E. to the r. Clyde, at about 2 m. N. of Abington.

Duneaves, seat, N.W. Perthshire, 1¾ m. N.W. of Kenmore.

Dunecht, seat, 12 m. W. of Aberdeen. P.O., T.O. Seat of Viscount Cowdray.

Dunevan, ancient hill fort, Nairnshire, near Cawdor.

Dunfallandy, seat and ancient monumental stone, Perthshire, 1½ m. S. of Pitlochry.

Dunfermline, parl., royal, and mun. bur., town, and par., with ry. stas., D. Upper and D. Lower, L.N.E., Fife, 3 m. N. of the Forth, 16¾ m. N.W. of Edinburgh by rail; par. 22,499 ac. Pop. 40,918;

parl. and mun. bur., pop. 34,954. Hotels. P.O., T.O. Prior to the union it was a favourite residence of Scottish kings, several of whom including Bruce, are buried in its Abbey. Birthplace of Andrew Carnegie, who gave £500,000 for town's benefit. Centre of the damask linen trade of the United Kingdom; bleaching, rubber work, coalfields, and ironfounding. Bur. includes Rosyth on Firth of Forth. D. unites with Cowdenbeath, Inverkeithing, and Lochgelly in returning one member of Parliament.

Dunfillan, or **St. Fillan's Chair,** conical hill (600 ft.) near St. Fillans, Perthshire.

Dun Fionn, vitrified fort, Inverness-shire, 4½ m. S.W. of Beauly.

Dungallan, ham., Argyll, W. side of Colonsay.

Dungarthill, seat, Perthshire, 1½ m. S.E. of Dunkeld.

Dungavel, seat of the Duke of Hamilton, 6 m. S. of Strathaven, Lanarkshire. D. Hill, 1502 ft.

Dungavel, hill, 9 m. S.E. of Lanark, 1675 ft.

Dungeon, one of a series of small lochs, Kirkcudbrightshire, 8½ m. W. of Earlston.

Dunglass, rocky prom. on the Clyde, 2½ m. E.S.E. of Dumbarton.

Dunglass, seat, E. Lothian, 1½ m. N.W. of Cockburnspath, occupies the site of the castle of the Earls of Home. D. Dean, a ravine. D. Burn flows 4½ m. N.E. to the sea.

Dungrain Law, mt., Lowther Hills, Lanarkshire, 2 m. S.E. of Leadhills; alt. 2186 ft.

Dungyle, hill, Kirkcudbrightshire, 3 m. S. of Castle-Douglas; 600 ft. Remains of Caledonian fort.

Dunhead, vestige of ancient triangular camp, Angus, 5 m. W.S.W. of St. Vigeans.

Dunian, hill, Roxburgh, 1¾ m. S.W. of Jedburgh. Alt. 1095 ft.

Dunie, seat, Ross and Cromarty, 2 m. W. of Ardgay.

Dunimarle, seat, Fife, near Culross. A former castle belonged to the Thanes of Fife, and was the place of the murder of Lady Macduff by order of Macbeth.

Dunino, par. and ham., E. Fife, 4½ m. S.E. of St. Andrews; 2966 ac. Pop. 197. P.O. Ham. pop. 188.

Dunipace, town and par., E. Stirlingshire; par. 5593 ac. Pop. 1768. Town forms part of the police bur., of Denny and Dunipace. Hills of D. and D. House 2 m. S.E.

Duniquoich, steep wooded hill, Argyll, adjacent to Inveraray; alt. 800 ft.

Dunira, seat, Perthshire, 2½ m. N.W. of Comrie.

Dunjop, seat, 6 m. N. of Kirkcudbright.

Dunkeld, mkt.-town, with ry. sta. (D. and Birnam), L.M.S., on r. Tay, 15¼ m. N.N.W. of Perth. P.O., T.O. Pop. 465. Hotels. Ruins of twelfth - century cathedral. The choir now used as parish church. D. House, a seat of the Duke of Atholl.

Dunkeld and Dowally, par., Perthshire, on r. Tay; 15,293 ac. Pop. 946.

Dunkeld, Little, par., E. Perthshire, on rt. bank of Tay; 44,607 ac. Pop. 2182. Ry. sta., Dunkeld and Birnam.

Dunkenny, seat, Angus, 2 m. N.W of Glamis.

Dunlappie, old par., Angus, near Brechin. D. House, 1 m. S. of Edzell.

Dun Law, mt., S. Lanarkshire, 2 m. E.S.E. of Leadhills; alt. 2216 ft.

Dun Law, mt. on border of Peebles and Selkirk; alt. 2584 ft.

Dun Leacainn, mt., Argyll, bordering Loch Fyne, 6 m. S.W. of Inveraray. Quarry of very fine granite here; alt. 1173.

Dunliath, Scandinavian fort, Isle of Skye, Kilmuir par., 5 m. S.W. of the extreme N. of the par.

Dunlichty, old par., united to Daviot, Inverness-shire.

Dunlop, par. and vil., with ry. sta., L.M.S., Ayrshire; vil. 2½ m. N. of Stewarton; par. 7180 ac. Pop. 1224; vil. pop. 553. P.O., T.O. Has long been famous for its cheese. D. House 1¼ m. E. of sta.

Dunlop Place, vil., Lanarkshire, Dalserf par., 5½ m. S.E. of Hamilton.

Dunloskin, seat, Argyll, near Dunoon.

Dunlossit, seat, ¼ m. S. of Portaskaig, Islay, Argyll.

Dunlugas, ham. and seat, Banffshire, Alvah par., 4½ m. N.W. of Turriff.

Dunmacsniochan, vitrified fort, Argyll, 5½ m. N.E. of Oban.

Dunmaglass, estate and shooting-lodge, E. Inverness-shire, Daviot par.

Dunman, rocky hill, with vestiges of ancient Caledonian fort, Wigtownshire, Kirkmaiden par.; alt. 522 ft.

Dunmore, vil., with small harb., on r. Forth, Stirlingshire, Airth par. P.O., T.O. D. Park in vicinity.

Dunmore, hill, Perthshire, 1½ m. N.W. of Comrie.

Dunmore, seat and pl., Argyll, on W. Loch Tarbert, Kintyre.

Dunmore, hill, Perthshire, 5 m. S. of Amulree; 1520 ft. Remains of ancient fort.

Dunmurchie, loch, S. Ayrshire, 1½ m. S.E. of Barr.

Dunn, ham., Caithness, Watten par., 9 m. S.E. of Thurso.

Dunnabie, pl., Dumfriesshire, 6 m. N.E. of Kirtlebridge.

Dunnet, coast par. and vil., Caithness, 9 m. N.E. of Thurso; par. 16,842 ac. Pop. 928. P.O., T.O. D. Head, the most northerly point of the Scottish mainland. Lighthouse with group-flashing light, 346 ft. above high-water, and seen 25 m. Royal Naval signal station.

Dunnichen, par. and vil., Angus, 3¼ m. S.E. of Forfar; par. 4091 ac. Pop. 978. D. Hill, 764 ft. D. House a seat.

Dunnideer, or **Dun o' Deer,** hill, Aberdeenshire, 1½ m. W. of Insch. Remains of vitrified forts; alt. 876 ft.

Dunnikier, hill, Fife, 3 m. N. of Colinsburgh; 750 ft.

Dunnikier Colliery, Fife, 2 m. N. of Kirkcaldy. D. House, a seat.

Dunninald, seat, Angus, 2½ m. S.W. of Montrose.

Dunning, par., vil., and ry. sta., L.M.S., 10 m. S. of Perth; par. 14,878 ac. Pop. 1056; vil. 1¾ m. S.E. of sta. P.O., T.O. Pop. 623.

Dunnottar, coast town and par., Kincardineshire; par. 7781 ac. Pop. 1987; pop. of town (forming part of Stonehaven) 1425. 1 m. S.W. of Stonehaven is D. House. D. Castle crowns an insulated salient rock rising from the sea in cliffs 160 ft. high; was erected and occupied by the Keiths, Earls Marischal. Became the hiding-place of the Scottish Regalia in the time of Cromwell, and a State prison in the times of Charles II. and James VII., where many Covenanters were immured and tortured. In the adjoining churchyard, Sir Walter Scott met Robert Paterson, the original of "Old Mortality."

8

Dun of Creich, hill, Sutherland, near church of Creich. Vitrified fort and standing-stone.

Dunollie Castle, seat and ruined ancient castle, Argyll, 1 m. N. of Oban. The castle (twelfth century) was the chief seat of the Macdougals, Lords of Lorne. A fixed light, white and red sectors, seen 8 m.

Dunolly Burn Reservoir, E. Lothian, 1½ m. S.W. of Garvald.

Dunool, mt., N. Kirkcudbrightshire; alt. 1758 ft.

Dunoon, town (police bur.) and eccl. par.; watering-place, Cowall dist., Argyll, extending, with Hunter's Quay and Kirn, 3 m. along W. shore of the Firth of Clyde. 7 m. W. of Greenock. Hotels. Contains the West of Scotland Convalescent Home. Ruin of ancient castle of the Stewarts. P.O., T.O. Pier shows fixed white light, seen 6 m., and on jetty a green light, seen 4 m. Frequent steamboat communication with Glasgow, Greenock, Rothesay; regular ferry service with Gourock.

Dunoon and **Kilmun,** coast par., Cowal, Argyll; 44,606 ac.

Dunphail, seat, ancient castle, and ry. sta., L.M.S., Morayshire, 8½ m. S. of Forres. P.O., T.O.

Dun Point, Islay, Argyll. Lighthouse with fixed light, white and red sectors, visible 12 m.

Dunragit, vil., with ry. sta., L.M.S., Wigtownshire, 5 m. E.S.E. of Stranraer. P.O., T.O. In vicinity, Mote of D., hill, and D. House, seat.

Dunreggan, vil., W. Dumfriesshire, forming part of Moniaive.

Dun-Richnan, ancient fort, Inverness-shire, 1½ m. S.E. of Dores.

Dun Rig, mt., borders of Peeblesshire and Selkirkshire, 5½ m. S. of Peebles; alt. 2433 ft.

Dunrobin, seat of the Duke of Sutherland, with private ry. sta., L.M.S., S.E. Sutherland, 2 m. N.E. of Golspie. D. Glen in vicinity. D. deer forest, 12, 180 ac.

Dunrod, ancient par., now embraced by par. of Kirkcudbright. P.O., T.O.

Dunrod, old barony, Inverkip par., Renfrewshire. Traversed by D. Burn, which supplies the reservoirs of Greenock Waterworks. D. Hill 2 m. N.E. of Inverkip; alt. 936 ft.

Dunrossness, par., S. Shetland; 32,301 ac. Pop. 2704. P.O., T.O. Includes Fair Isl.

Duns, police bur., mkt.-town, and par., with ry. sta., L.N.E., Berwickshire, 7½ m. N.N.E. of Greenlaw, 55 m. S.E. of Edinburgh by rail; par. 11,404 ac. Pop. 2677; bur. pop. 1788. P.O., T.O. Hotel. The capital and largest town in the co. In vicinity, D. Law, 713 ft., and D. Castle, a seat.

Dunsappie, artificial loch, King's Park, Edinburgh; 371 ft. above sea level.

Dunscaith, ruined large ancient castle of the Barons of Sleat, on Loch Eishart, Isle of Skye, Inverness-shire.

Dunscore, par. and vil., with ry. sta., L.M.S., 11 m. N.W. of Dumfries; 14,825 ac. Pop. 962. P.O., T.O.

Dunscriben, vitrified fort, Inverness-shire, on Loch Ness, in Urquhart and Glenmoriston par.

Dunscuddeburgh, ruins of fort, Isle of Skye, Kilmuir par.

Duns Dish, loch, Angus, 5 m. N.W. of Montrose.

Dunshelt, vil., Fife, 1 m. S.E. of Auchtermuchty. P.O.

Dunside Rig, hill, Selkirkshire, 4 m. E. of Ettrick church; alt. 1206 ft.

Dunsinane Hill, one of the Sidlaws summits, 8½ m. N.E. of Perth; alt. 1012 ft. Formerly crowned with a strong castle, said to have been built by Macbeth, and locally called Macbeth's Castle.

Dunsinnan, seat, in co., and 7 m. N.N.E. of Perth.

Dunsiston Row, pl., N. Lanarkshire, 1½ m. N.E. of Chapelhall.

Dunskeig, hill, with two ancient forts, one vitrified at S. side W. Loch Tarbert, Argyll; alt. 469 ft. D. Bay, ¾ m. S.W.

Dunslair Heights, hill range, 3 m. N.E. of Peebles; alt. 1975 ft.

Duns Spa, mineral spring, Berwickshire, 1½ m. S.E. of Duns.

Dunstaffnage Castle, ruined famous castle, 3½ m. N. of Oban, Argyll. Erected in the latter part of the twelfth century; occupies the site of a chief seat of the Dalriadan kings. Belonged to, successively, the Macdougals and the Campbells; was maintained as a fortress till the rebellion of 1745. D. House 1½ m. S.E.

Dunsyre, par. and vil., with ry. sta., L.M.S., Lanarkshire, 6½ m. E. of Carnwarth; 10,777 ac. Pop. 152. P.O., T.O. at sta. In vicinity, D. Hill, 1313 ft., and the ruins of D. Castle.

Duntanlich, seat, Perthshire, 5½ m. W. of Pitlochry.

Duntarvie, estate, W. Lothian, Abercorn par., 3 m. S.W. of Queensferry.

Duntelchaig, loch, 8¼ m. S.S.W. of Inverness, 3½ m. long; depth 205 ft., alt. 702 ft.

Dunterlie, sub. of Barrhead, Renfrewshire.

Duntillan Hill, N.E. Lanarkshire, 1 m. N.W. Kirk of Shotts; alt. 951 ft.

Duntocher, q.s. par. and small town, Dunbartonshire, 9 m. N.W. of Glasgow. Town pop. 3814. P.O., T.O.

Dunton, reservoir, Kilmarnock water supply, Ayrshire, 8 m. N.E. of Kilmarnock.

Duntreath Castle, on Blane R., Stirlingshire, 2 m. N.W. of Strathblane.

Duntroon Castle, modernised ancient castellated mansion, Argyll, on N. shore of Loch Crinan.

Duntrune, seat, Angus, 4 m. N.E. of Dundee; also the ham. of Burnside of D.

Duntulm, ham. and restored castle, N.W. Isle of Skye, Inverness-shire, on Loch Scour, 19 m. N.W. of Portree, and 1½ m. S. of the N. coast of Kilmuir par.

Dunure, or **Fisherton,** fishing vil. and ruined old castle, ry. sta., L.M.S. (passenger service withdrawn), Ayrshire. 5¼ m. N.W. of Maybole. P.O., T.O.

Dunvegan, sea loch, vil., and castellated mansion in N.W. Isle of Skye, Inverness-shire, 22¼ m. from Portree. Inn. P.O., T.O. Fixed light, white and red sectors on N. corner of pier and a fixed white light on the S. corner. D. Castle, seat of the Macleods. D. Loch 7½ m. long, 2¼ m. wide. Regular calling pl. for steamers from Oban and Glasgow.

Dunvegan Hostel, on the Bracadale Road, ½ m. S. of Dunvegan Hotel, Skye. A Scottish Youth Hostel.

Duppin Burn, W. Argyllshire, flows S.W. to the sea at Lochgilphead.

Dupplin Castle, seat of Lord Forteviot, near r. Earn, 5 m. S.W. of Perth. D. Loch 1¼ m. N.W.

Dura Den, ravine and vil., Fife, 2¾ m. E.S.E. of Cupar D. House in vicinity.

Durdie, Nether and **Over,** in co., and 6 m. E. of Perth.

Durhamtown, vil., W. Lothian, 1 m. S.W. of Bathgate.

Durie, seat and colliery, Fife, 1½ m. N.W. of Leven.

Durinemast, loch, Morven dist., Argyll, connected with Loch Arienas.

Durisdeer, par., vil., and hill, Dumfriesshire; par. 19,735 ac. Pop. 745; vil., 6 m. N. of Thornhill. P.O. D. Hill, on Lanarkshire border, 1861 ft.

Durn, hill and stream, Banffshire, Fordyce par.; hill, 2 m. S.W. of Portsoy, 651 ft.

Durness, par. and vil., N.W. Sutherland, 20 m. N.W. of Tongue; par., 140,770 ac. Pop. 529. P.O., T.O. Hotel. D. Ferry, 2¾ m. S.W. to Cape Wrath lighthouse.

Durno, vil., Aberdeenshire, 2 m. N. Pitcaple ry. sta. P.O.

Duror, q.s. par. and vil., with ry. sta., L.M.S., in N. of Appin dist., Argyll. P.O., T.O. Vil. 5 m. S. of Ballachulish. State forest in Glen D.

Durran, ham., Caithness, 6 m. S.E. of Thurso.

Durris, par and vil., Kincardineshire; par. 15,315 ac. Pop. 731; vil. (Kirton of Durris), 4½ m. E. of Banchory. P.O. In vicinity, D. House and D. Tower. State forest nearby.

Durrockstock Dam, Renfrewshire, 2 m. S.W. of Paisley.

Dury Voe, bay, Shetland, E. coast.

Dusk River, stream, Ayrshire, flowing 10 m. into Stinchar R., near Penwherry sta.

Dusk Water, stream, N.W. Ayrshire, flowing 8 m. into Garnock R., 2 m. S. of Dalry.

Dutchman's Cap, hill, Treshnish Isles, Argyll, 5½ m. N.W. of Staffa; alt 284 ft.

Duthiestone House, Perthshire, 1 m. N. of Dunblane.

Duthill and **Rothiemurchus,** par. N.E. Inverness-shire, 7 m. S.W. of Grantown; 78,684 ac. Pop. 2126.

Dwarfie Hammers, rocks, Hoy Isl., Orkney, 2 m. S.E of Ward Hill.

Dwarfie Stone, a huge sandstone boulder, with three apartments hewn, on the S. side of the island of Hoy, and opposite the island of Graemsay, Orkney.

Dwarwick Head, Caithness, E. end of Dunnet Bay.

Dyce, par. and vil., with ry. junction, L.N.E., Aberdeenshire; par. 5231 ac. Pop. 1256. Vil., Gordon Place and Dyce, 6¼ m. N.W. of Aberdeen. P.O., T.O. Vil. pop. 809.

Dye Water, stream, Berwickshire, flowing 14 m. E. to the Whitadder, near Ellem.

Dye Water, stream, Kincardineshire, flows 15 m. through Glen Dye to the Feugh, near Strachan church. Bridge of D., seat, 5 m. from mouth.

Dye, or **West Water,** r., Angus, flowing 20 m. S.E. to the North Esk, 4 m. N.E. of Brechin.

Dyke and Moy, par. and vil. (Dyke), Morayshire, 6 m. E. of Nairn. 13,719 ac. Pop. 987; vil. pop. 96.

Dykebar House, Renfrewshire, 1¼ m. from Paisley.

Dykehead, vil., Lanarkshire, 2 m. S.W. of Coatbridge. Pop. of D. and Shotts. 10,665.

Dykehead, ham., Ayrshire, 2 m. W. of Kilmaurs.

Dykehead, ham., S. Perthshire, 3 m. N.E. of Buchlyvie sta.

Dykends, ham., Angus, 5½ m. N. of Alyth, P.O., T.O.

Dykeneuk, sub. of Bo'ness, West Lothian.

Dyke Water, Sutherland, flows 10 m. to Halladale R.

Dyrock, stream, Ayrshire, flows 4 m. S.W. to Girvan Water, 1 m. N.E. of Crosshill.

Dysart, parl., royal, and mun. bur., q.s. par., and seapt.-town, with ry. sta., L.N.E., Fife, 2 m. N.E. of Kirkcaldy. Pop., see KIRKCALDY. P.O., T.O. D. House, once a seat of the Earl of Rossyln. Unites with Kirkcaldy, Burntisland, Kinghorn, Buckhaven, Methil, and Innerleven, in returning one member to Parliament. White fixed light on E. pier, visible 5 m.; on middle pier a green and red fixed light, seen 3 m. Exports large quantities of coal.

E

Eabost House, W. coast of Isle of Skye, Inner Hebrides, on Loch Bracadale and 2 m. N.W. of Struan.

Eagerness, or **Eggerness,** headland, on N. side Garliestown Bay, in co. of, and 6½ m. S.S.E. of, Wigtown.

Eaglescarnie, seat, 3 m. S. of Haddington, E. Lothian.

Eaglesfield, vil., 1 m. N.E. of Kirtlebride, Dumfriesshire. P.O., T.O. Pop. 569.

Eaglesham, par. and vil. in S.E. Renfrewshire, and 8½ m. S. of Glasgow; par. lies between the White Cart and the Earn; area, 15,660 ac. Pop. 1671. Nearest ry. sta., L.M.S., Clarkston, 3½ m. N. P.O., T.O. Vil. pop. 1218.

Ealag, or **Ellag,** small loch, 6½ m. N.W. of Oykell Bridge, N. Ross and Cromarty.

Ealannach, loch, 8½ m. S.W. of Halkirk, Caithness.

Earlish, ham., Isle of Skye, 2.m. S. of Uig. P.O.

Earl Patrick's Castle, ruin, near Scalloway, Shetland. Erected in 1600.

Earls, or **Mote Hill,** site of open air courts, Buchan, at Ellon, E. Aberdeenshire.

Earls Burn, Stirlingshire, flows 6 m. to Carron Water, 3 m. N.W. of Denny.

Earls Cross, stone pillar, near Dornoch, Sutherland.

Earlsferry, royal bur. in par. of Elie, which adjoins Earlsferry and is the nearest ry. sta., L.N.E.; 13 m. S.W. of St. Andrews, Fife. Holiday resort and golfing place. Mfrs. golf clubs.

Earlshall, seat, ½ m. E. of Leuchars, Fife.

Earlsmill, seat, 4½ m. S.W. of Forres, Morayshire.

Earlston (anciently Ercildoune), par. and vil., with ry. sta., L.N.E., 18 m. S.W. of Duns, Berwickshire; par. 9808 ac. Pop. 1689; vil. pop. 1071. P.O., T.O. Mfrs. woollens and ginghams. Thomas the Rhymer's Cottage is still standing (about 1270).

Earlston, seat in co. of, and 4 m. W.S.W. of, Kirkcudbright.

Earlston, ruined seat of the Gordons, 2 m. N.W. of Dalry, N. Kirkcudbright. E. burn flows 4 m. to Water of Ken. Nearby one of the hydro-electric stations of the Galloway Water Power Co. under construction.

Earn, r. and loch in S. Perthshire; the loch, 12 m. W. of Crieff, is $6\frac{1}{2}$ m. long by an average breadth of half a mile; the river flows out of the loch and runs into the Tay's estuary, 46 m. eastward. Alt. of loch, 317 ft., depth 287 ft.

Earn Bank, seat, Bridge of Earn, Perthshire.

Earn, Bridge of, vil., Perthshire. See BRIDGE OF EARN.

Earnock, Meikle, see MEIKLE EARNOCK.

Earraid, isl., Kilfinichen and Kilvickeon pars., off S.W. coast of Mull Isl., Argyllshire. Pop. 33.

Easdale, isl., 16 m. S.W. of Oban, on the Firth of Lorn, Argyllshire. Large slate quarries. Vil. of same name has a pier. P.O., T.O. Pop. 78. Glasgow steamers pass almost daily.

Eassie, ry. sta., L.M.S., in W. Angus, and $4\frac{1}{4}$ m. N. of Alyth Junction. P.O., T.O.

Eassie and Nevay, united par., Angus; 5052 ac. Pop. 523. E. Burn flows 6 m. N. to Dean Water.

Eastabist, seat, mainland Orkney, 6 m. N.W. of Finstown.

East and **West Laroch,** part of Ballachulish.

East Barns, vil., 3 m. S.E. of Dunbar, E. Lothian.

East Cairn Hill, alt. 1839 ft., one of the Pentland Hills, and 5 m. S.S.W. of Balerno.

East Calder, vil. and former par., Midlothian, $11\frac{3}{4}$ m. S.W. of Edinburgh. P.O. Pop. 1075.

East Church, q.s. par. in Brechin, Angus.

East Church, q.s. par. in Peterhead, Aberdeenshire.

Eastend, seat, 2 m. W. of Thankerton, Lanarkshire.

Easter Denhead House, $1\frac{1}{2}$ m. N.E. of Coupar-Angus, Perthshire.

Easterfield, ham. in co., and 13 m. S. of Banff.

Easterhill House, Lanarkshire, $\frac{1}{2}$ m. S.W. of Tollcross.

Easterhouse and **Swinton,** mining vil., 3 m. W. of Coatbridge, Lanarkshire; ry. sta., L.N.E. P.O., T.O.

Easter Moffat, seat, $2\frac{1}{2}$ m. E. of Airdrie, Lanarkshire.

Easter Moncrieffe, seat in co. of, and 3 m. S.E. of, Perth.

Easter Skene, seat near Loch Skene, in co. of, and 9 m. N.W. of, Aberdeen.

Eastertyre, seat, $2\frac{1}{2}$ m. W.N.W. of Ballinluig, Perthshire.

Eastfield, vil. and paper-mill, 1 m. E. of Rutherglen, N.W. Lanarkshire. P.O.

East Fortune, vil., 3 m. E. of Drem, E. Lothian; ry. sta., L.N.E. Sanatorium. W. Fortune, pl., $1\frac{1}{2}$ m. N.W. of Athelstaneford.

East Girt Hill, mt., W. border of Renfrewshire, $5\frac{1}{2}$ m. N. of Lochwinnoch.

East Grange, pl., with ry. sta., L.N.E., 6 m. W. of Dunfermline, Fife.

East Haugh, seat, 2 m. from Pitlochry, Perthshire.

Easthaven, fishing vil., $4\frac{1}{2}$ m. S.W. of Arbroath, Angus; ry. sta., L.N.E.

East Head, headland, $1\frac{1}{2}$ m. N.E. of Portsoy, Banffshire.

Easthouses, vil., Newbattle par., Midlothian, $1\frac{1}{2}$ m. S.E. of Dalkeith. P.O. See NEWTONGRANGE.

East Kilbride, par. and small town, 12 m. S. of Glasgow by rail; ry. sta. (alt. 504 ft.), L.M.S.; par. 23,680 ac. Pop. 5300. P.O., T.O. Vil. pop. 2329.

East Linga, isl. off E. coast of Whalsey, Shetland.

East Linton, police bur., on r. Tyne, and 6 m. W. of Dunbar, E. Lothian; ry. sta., L.N.E. P.O., T.O. called Prestonkirk. Pop. 882.

East Loch Tarbert (or **Tarbert Bay**), on W. side of Loch Fyne, Argyllshire.

East Longrigg, vil., 1 m. N.E. of Longriggend, Lanarkshire.

East Lothian, formerly Haddingtonshire, co. in S.E. Scotland, bounded on N. by Firth of Forth, E. by North Sea, S. by Berwickshire, and W. by Midlothian; area 267 sq. m., or 170,971 ac. of land and 360 ac. of water. Pop. 47,338. In the S. are the Lammermuir Hills (highest pt., Lammer Law 1750 ft.), the centre and N. are mainly plain, broken by the Garleton Hills, on which is a conspicuous column erected in 1824 to the fourth Earl of Hopetoun, and by the isolated heights of Traprain Law and N. Berwick Law. The Tyne is the only r. of note. Agriculture reaches a high stage of efficiency, and coal mining is carried on in the western part of the co. Mfrs. are of little importance. The cos. of Berwick and E. Lothian return one member to Parliament.

East Neuk o' Fife, no civil or well-defined boundaries ; the area surrounding Fife Ness including the whole of the peninsula between St. Andrews Bay and Largo Bay.

East Pilton, dist. and ry. halt, L.M.S., city of Edinburgh, near Granton. Public health hospital.

East Raffles, pl., Dumfriesshire, ¾ m. S.E. of Mouswald. Pop. Ruthwell and E. Raffles special water dist., 615.

Eastriggs, vil., 3 m. E. of Annan, Dumfriesshire ; ry. sta., L.M.S. P.O., T.O. Pop. 925.

Eastshield, seat, 3 m. N.W. of Carnwath, Lanarkshire.

Eastside Heights, hill on E. border of Peeblesshire, 7 m. N.E. of Peebles ; alt. 1944 ft.

East Wemyss, coastal vil., 3 m. E. of Thornton Junction, Fife. P.O., T.O.

East Whitburn, colliery vil., ¼ m. E. of Whitburn, W. Lothian.

Eastwood, par., chiefly lying in Glasgow but spreading to Renfrewshire. Contains part of Shawlands, Pollokshields, and Thornliebank ; area 5630 ac. Pop. 26,497.

Eastwood House, seat, near Dunkeld, Perthshire.

Eastwood Park, Lord Weir's seat, 1 m. S.E. of Thornliebank, E. Renfrewshire.

Ecclaw, pl., Berwickshire, 2 m. S. of Cockburnspath.

Ecclefechan, vil. in co. of, and 12 m. E. of, Dumfries. The house in which Carlyle was born still stands ; ry. sta., L.M.S. P.O., T.O. Pop. 658. Hotel. Weaving is carried on.

Eccles, par., ham., and seat, on the Tweed, and 5¼ m. S.E. of Greenlaw, Berwickshire ; par. 12,424 ac. P.O., T.O. Pop. 1182.

Eccles, seat, 2 m. W. of Thornhill, Dumfriesshire.

Ecclesmachan, par. and vil., 3 m. S.W. of Winchburgh, W. Lothian ; par. 4066 ac. Pop. 1793 ; vil. pop. 59.

Echt, par., vil., and seat, in co. of, and 12 m. W. of, Aberdeen ; par. 11,952 ac. Pop. 1111. P.O., T.O.

Eck, loch, 6½ m. long, 8 m. W. of Garelochend, Argyllshire ; depth 139 ft., alt. 67 ft.

Eckford, par. (10,000 ac.) and vil., on the Teviot, and 6 m. S. of Kelso, Roxburghshire. P.O., T.O. Pop. 653.

Eday, isl. and par. (includes Pharay Isl.), with two good harbs., and 13½ m. N.N.E. of Kirkwall, Orkney ; 7798 ac. Pop. 470. E. Sound lies between E. and Sanday.

Eday Island, N. Orkney group, lies between Westray and Sanday ; length 7½ m. by 2½ m. broad. P.O., T.O. Pop. 430.

Edderton, par. on S. shore of Dornoch Firth, N.E. Ross and Cromarty ; ry. sta., L.M.S., 5¼ m. N.W. of Tain. P.O., T.O. Pop. 485 ; ac. 22,129.

Eddlestone, par. (18,474 ac.) and vil. on the Eddlestone Water and in co. of, and 4 m. N. of, Peebles ; ry. sta. P.O., T.O. E. Water enters the Tweed at Peebles. Pop. 452.

Eddlewood, colliery vil. and seat, 1½ m. S.S.W. of Hamilton, Lanarkshire.

Eddrachalda, ruined seat of the Mackenzies of Assynt, on N. border of Loch Assynt, Sutherland.

Eddrachillis (Eadar - da - chaolais, *between two fiords*), coastal par. in N.W. Sutherlandshire ; 135,631 ac. Pop. 967.

Eden, r. in Fife, flowing for 30 m. eastward through Stratheden to reach the sea at St. Andrews.

Edenbank, seat, 1½ m. N. of Kelso, Roxburghshire.

Edendon Water, Perthshire, joins the Garry near Dalnacardoch.

Edenhall, neurological hospital, Pinkieburn, Musselburgh, Midlothian.

Eden Hall, seat, 3 m. N.E. of Kelso, Roxburghshire.

Eden House, seat in N. Aberdeenshire ; stands on r. Deveron, and 4 m. S.E. of Banff.

Eden Park House, ½ m. N.E. of Cupar, Fife.

Edenshead House, Strathmiglo, Fife.

Edensmuir, State forest, ½ m. W. of Ladybank, Fife.

Edenstown, vil., 1¼ m. W. of Ladybank, Fife.

Eden Water, stream, Berwick and Roxburgh shires, rises near Lauder, and joins the Tweed 3½ m. N.E. of Kelso.

Edenwood, seat, on the Eden, 2 m. S.S.W. of Cupar, Fife.

Ederline, or **Aligan,** loch, 1 m. S. of Loch Awe, Argyllshire ; alt. 122 ft., depth 58 ft.

Edgarhope State Forest, 1¾ m. N.E. of Lauder, Berwickshire.

Edgefield, special water dist., Lasswade par., Midlothian. Pop. 68.

Edgehead, ham., Midlothian, 3 m. S.E. of Dalkeith.

Edgelaw, reservoir, ¾ m. W.S.W. of Temple, Midlothian.

Edgerston, q.s. par., in pars. of Jedburgh and Southdean, Roxburghshire. E. House 7½ m. S.E. of Jedburgh.

Edinample Castle, seat, on S. side of Loch Earn, and 2 m. N.E. of Lochearnhead sta., L.M.S.

Edinbain, vil., with inn, 10 m. E. of Dunvegan, Skye, Inner Hebrides. P.O., T.O. (spelt Edinbane).

Edinbarnet, seat, 1 m. N.E. of Duntocher, Dunbartonshire.

Edinbellie, estate and old house, 1½ m. E. of Balfron, N. Stirlingshire.

Edinburgh, lat. 55° 57, N., long. 3° 11, W.; co. of a city, par., parl., royal, and mun. bur., co. town of Midlothian, and capital of Scotland. Stands on Firth of Forth and including Cramond, Granton, Newhaven, Leith, and Portobello, has a sea-front of 9 m., with Leith as its port, 47½ m. E. of Glasgow and 392¼ m. N. of London by E. coast route. The town rises by a series of ridges and is overlooked by Arthur's Seat and other hills now within the city boundary. The town grew up around the Castle, which was a stronghold of Edwin of Northumbria, and from whom the town is said to have derived its name. In 1128, David I. founded the Abbey of Holyrood, about 1 m. S. of the Castle, and round it grew the bur. of the Canongate, which remained a separate municipality till 1856. In 1329, Edinburgh was made a bur. by a charter given by Robert the Bruce, who also gave the town the right of establishing a port at Leith, the vassalage of the port to the capital lasting till 1833, when Leith was made a bur. It was during the fifteenth century that Edinburgh began to be recognised as the Capital of Scotland, and James IV. and James V. confirmed this by building a palace within the Abbey of Holyrood, and by establishing the Court of Session as the supreme court of justice for Scotland. In 1450 the first protecting wall was built round the town with an extension after Flodden (1513) to include the sub. of the Canongate to the S. As population increased, high tenements were built up to fourteen stories. In 1582 the University was founded. Shortly after the middle of the eighteenth century the enclosing wall was broken down and town began to expand. The Nor' Loch on the N. side of the Castle was drained, and bridges were thrown across the numerous valleys in the immediate neighbourhood. The modern city covers a large area for its population, and there are many open spaces. Famed for its natural beauty of situation, romantic and literary associations, architecture and educational institutions, the city has earned the title of " the modern Athens." Attractions for visitors are the Castle, with the minute Norman Chapel of St. Margaret and the wonderful Scottish National War Memorial, the historical associations of the Royal Mile from the Castle to the Palace of Holyroodhouse, Parliament House, St. Giles Cathedral, John Knox's House, etc. The principal ry. stas. are Waverley (L.N.E.) and Princes Street (L.M.S.), and there are numerous suburban stas. on both lines. While the city is mainly a residential one, it has some rather important industries, e.g. printing, publishing, bookbinding, mapmaking, insurance and banking, coalmining, brewing, distilling, papermaking (in the outskirts of Juniper Green), rubber works, scientific instruments, and fisheries; while its medical school is world-famous. Area 32,402 ac., or 50·6 sq. m. Pop. 438,998.

Edinburghshire, former name of Midlothian, q.v.

Edinchip House, 1 m. N. of Balquhidder sta., Perthshire, seat of Clan Gregor Chief.

Edingight, seat, 7 m. N.E. of Keith, N. Banffshire.

Edinglassie, estate, with old mansion-house, 28 m. S. of Rothes, S.W. Aberdeenshire.

Edinglassie Lodge, 6½ m. W. of Huntly, N.W. Aberdeenshire.

Edington, ham., 2½ m. E. of Chirnside, Berwickshire.

Edinkillie, par. and ham., on the r. Divie and 8¼ m. S. of Forres, Morayshire. Pop. 731 ; par. 34,846 ac.

Edinkiln, Strathblane par., Stirlingshire. Special water dist. Pop. 147.

Edintore House, 3 m. S. of Keith, Banffshire.

Edinville, ham., on Burn of Aberdour, W. Banffshire, 2 m. S. of Charlestown of Aberlour.

Edmonston, seat, 3¼ m. N.E. of Biggar, E. Lanarkshire.

Edmonston, vil., with colliery, 3½ m. S.E. of Edinburgh G.P.O. E. House in neighbourhood.

Ednam, par. and vil., 3 m. N.E. of Kelso sta. on L.N.E. P.O. Birthplace of Robert Thomson, author of " The Seasons " (1700–1748); 3846 ac. Pop. 394 ; vil. 239.

Edradour Distillery, near Pitlochry, Perthshire. E. Burn flows 4½ m. S.W. to join the Tummel 1½ m. below Pitlochry. Milton of E., a ham, 1 m. E. of Pitlochry.

Edradynate, seat, 3 m. E. of Aberfeldy, Perthshire.

Edramucky, seat, on Loch Tay and 4½ m. N.E. of Killin, Perthshire.

Edrington Castle, ruined stronghold, now a farm place, on the Whitadder, Berwickshire, and 4½ m. W. of Berwick-on-Tweed. Modern castle nearby ; 1 m. N. is E. House.

Edrom, par. and vil., with ry. sta., L.N.E., and 3½ m. N.E. of Duns, Berwickshire ; par. 9558 ac. Agriculture. P.O., T.O. Pop. 1015.

Edzell, par. and vil., with terminal sta. on branch line from Brechin (Angus), L.M.S. ; 6½ m. N. of Brechin, Angus ; par. 18,967 ac. Pop. 891 ; vil. pop. 618. P.O., T.O. Hotels.

Egilsay, isl., 11 m. N. of Kirkwall, Orkney. Red fixed light on pier-head. Has remains of the ancient St. Magnus Church. P.O. Pop. 85.

Eglinton Castle, seat of Earl of Eglinton, 2 m. N. of Irvine, N. Ayrshire.

Eglinton Ironworks, vil., 2½ m. N.W. of Irvine, Ayrshire.

Eglinton Street, ry. sta., L.M.S., in Glasgow.

Eididh nan Clach Geala, mt., 10 m. S.E. of Ullapool, Ross and Cromarty ; alt. 3039 ft.

Eigg, isl. (6½ m. by 4 m.), 7½ m. W. of mainland of Inverness-shire, and 5 m. S.W. of Skye, Inner Hebrides. P.O., T.O. Shows a white flashing light, visible 14 m., on E. point Eilean Chasguidh. Pop. 138. The geological interest in this island is the Scour of Eigg, rising to 1289 ft.

Eigheach, loch, on r. Ghaoir, 15 m. S.W. of Dalnaspidal ; alt. 818 ft., depth. 28 ft.

Eil, sea loch of Argyllshire and Inverness-shire. Consists of Upper Loch E. (striking eastward) and Lower Loch E. (striking south-westward). The latter is rather part of Loch Linnhe.

Eildons, triple-crested height in Roxburghshire, S. of Melrose ; highest peak 1385 ft. On E. side and 1½ m. from Melrose is the ham. of Eildon. E. Hall, the adjacent seat, is the property of the Duke of Buccleuch. The highest has a geographical indicator on its summit.

Eileach-an-Naoimh, *Isles of the Sea,* W. Argyllshire. White flashing light, visible 12 m.

Eileanach, loch, 5 m. E. of Altnabreac sta., Caithness.

Eilean a' Chalmain, isl. off extreme S.W. coast of Mull, Argyll.

Eilean Aigas, islet, with seat. See AIGAS.

Eilean-anabuich, pl., with school, on Loch Maaruig, E. Harris, Inverness-shire.

Eileanan Glasa, in Sound of Mull, Argyll, has a white flashing light, visible 9 m.

Eilean an Tighe, one of the three Shiant isls., Outer Hebrides.

Eilean-Bhroin, islet in Loch Vennachar, Perthshire.

Eilean Buidhe, in Burnt Isl., Kyles of Bute, Argyll, showing white occulting light, visible 5 m.

Eilean Donan, rocky islet at head of Loch Alsh, W. Ross and Cromarty. Ancient castle now renovated and inhabited.

Eilean Dubh, isl. at entrance to Loch Broom, Ross and Cromarty.

Eilean Eiga, isl. off W. coast of S. Morar, 3 m. N.W. of Arisaig, Inverness-shire.

Eilean Fada, islet on W. side of Knapdale, Argyll, near head of Loch Killisport.

Eilean Finnan, islet, with ruins of ancient church, in Loch Shiel, Argyll.

Eilean Fraoch, islet, with old fort, in Sound of Islay, Argyll.

Eilean Jushard, isl. at mouth of Loch Sheil, E. coast of Lewis, Outer Hebrides.

Eilean Liathanaich, Gray Isls., N. coast of Mull, Argyllshire. Flashing light (Bunessan), white with red sector, visible 11 m.

Eilean Maree, islet, with ancient burying-ground in Loch Maree, Ross and Cromarty.

Eilean Mhuire, one of the Shiant Isls., and 4 m. S.W. of Lewis, Outer Hebrides.

Eilean More, Flannan Isls., Outer Hebrides. Has lighthouse, visible 24 m.

Eilean More, islet off N.E. of Coll Is., Argyll.

Eilean More, islet, with ruins of chapel, at S. end of Sound of Jura, Argyll.

Eilean Munde, in Loch Leven, Argyll, has ruins of tenth-century chapel.

Eilean Musdile, largest of a group of islets, S. of Lismore, Argyll.

Eilean-nan-Caorach, in Firth of Lorne, Argyll.

Eilean-nan-Gillean, isl. off N. coast of Skye, Inner Hebrides. Fixed light, red and white sectors, visible 12 m.

Eilean-nan-Gobhar, islet having two vitrified forts, in Loch Aylort, Inverness-shire.

Eilean-nan-Roan, islet, 5½ m. N.N.E. of Tongue, in Kyle of Tongue, N. Sutherlandshire. Precipitous cliffs and fine natural arch.

Eilean Naomh (or **Coomb Island**), off Kyle of Tongue, N. Sutherland, was an ancient burying-ground.

Eilean Ruairidh (or **Rory**), islet in Loch Maree, and 17 m. S.W. of Ullapool, W. Ross and Cromarty.

Eilean Shona, isl., with seat, in Loch Moidart, Inverness-shire.

Eilean Soa, 3 m. by 2½ m. due S. of Cuillin Hills, Skye, Inverness-shire.

Eilt, loch (119 ft. deep), in Moidart dist., W. Inverness-shire and 16 m. W.N.W. of Fort William ; alt. 96·4 ft.

Eishort, sea loch, 4½ m. S. of Broadford, Skye, Inner Hebrides ; penetrates inland for about 6½ m.

Eitachan, loch in the Cairngorms, Aberdeenshire, 1 m. S. of Loch Avon.

Elchies, eccl. par., with two seats, Easter and Wester, near Craigellachie Junction, Morayshire.

Elcho Castle, on the Tay, in co. of, and 5¼ m. S.E. of, Perth ; a seat of the Earl of Wemyss and March.

Elder Park, q.s. par. in Govan par., Glasgow.

Elderslie, q.s. par. and town, 2½ m. W. of Paisley, Renfrewshire ; ry. sta., L.M.S. ; mfrs., carpets. P.O. Town pop. 3113. Said to be the birthplace of Sir William Wallace (1270–1305).

Elderslie, seat, on the Clyde, in co. of, and ½ m. E. of, Renfrew.

Elderslie, vil., 1¼ m. W. of Kilmarnock, Ayrshire.

Eldin House, seat, Lasswade, Midlothian.

Elfhill, pl., Kincardineshire, 4½ m. W. of Stonehaven.

Elf House, stalactite cave, near Dalry, Ayrshire.

Elgin, city, royal bur., and capital of Morayshire, on rt. bank of r. Lossie and 6 m. above its mouth ; ry. stas. L.M.S. and L.N.E. Hotels. Contains many antiquities, including ruins of cathedral founded 1224. Woollen factories, distilleries, and sawmills. P.O., T.O. Pop. 10,192 ; pop. of bur. 8810.

Elgin, New, vil., just to the S. of Elgin. P.O.

Elgoll, ham., 15 m. N.W. of Broadford, Skye, Inner Hebrides. P.O., T.O.

Elibank, seat of Lord Elibank, 5 m. E. of Innerleithen, Selkirkshire.

Elie, vil., par., police bur., and seapt., 14 m. S.E. of St. Andrews, Fife ; ry. sta., L.N.E. ; holiday pl. with golf links. Hotels. P.O., T.O. Coastguard sta. On Elie Ness is a flashing light, visible 12 m. E. House, a sixteenth-century building, is nearby. Par. 1999 ac. Pop. 1251 ; bur. pop. E. and Earlsferry, 1098.

Eliock House, 5 m. S.W. of Kirkconnel, N.W. Dumfriesshire.

Ellag, loch, 4 m. E. of Aultnacealgach, Sutherland.

Ellangowan Paper Mills, at Milngavie, Stirlingshire, and 9½ m. by rail from from Glasgow.

Ellanreach State Forest, nearest ry. sta. Kyle of Lochalsh, W. Ross and Cromarty.

Ellen's Isle, or **Eilean Molach,** islet in E. end of Loch Katrine, Perthshire ; scene of most of the incidents in Scott's " Lady of the Lake."

Elliot Junction, ry. sta., L.N.E. and L.M.S. 1½ m. S.W. of Arbroath, Angus.

Ellishill, seat, 2½ m. N.W. of Peterhead, Aberdeenshire.

Ellisland, farm, on r. Nith, and 2½ m. S.S.E. of Auldgirth, W. Dumfriesshire. Tenanted for three years by Robert Burns.

Elliston House, 2½ m. S. of St. Boswells, Roxburghshire.

Elliston Tower, ancient castle on Castle Semple Loch, 3½ m. S.W. of Johnstone.

Ellon, par. (23,137 ac.), police bur., and ry. sta., L.N.E.; 19 m. N.E. of Aberdeen. P.O., T.O. Pop. 3446; pop. of bur. 1300. Hotel.

Ellrig, loch, 2 m. N.E. of Slamannan, S.E. Stirlingshire.

Ellson Fell, alt. 1761 ft., 9 m. N.E. of Langholm, Dumfriesshire.

Ellwick, vil. and bay, Shapinsay Isl., Orkney.

Elm Park, Inverness. Girls' private school.

Elphin, ham., Assynt, S.W. Sutherland-shire, 23 m. S.E. of Lochinver by road. P.O.

Elphinstone, coal-mining vil., 2 m. S.W. of Tranent, E. Lothian. E. Tower, nearby, is a well-preserved ruin. P.O. Pop. 674.

Elrick, vil. in co. of, and 7½ m. W. of, Aberdeen.

Elrick, seat, 1½ m. S.S.W. of New Macher, Aberdeenshire.

Elrick, ham., 6½ m. S.W. of Rhynie, E. Banffshire.

Elrick More, mt., Perthshire, 4 m. W. of Dunkeld; alt. 1693 ft.

Elrig, vil. and loch (10 ft. deep), 3 m. N.W. of Port-William, Wigtownshire. P.O., T.O. E. House is a seat.

Elshieshields, seat, 2 m. N.W. of Loch-maben, Dumfriesshire.

Elsick, a seat of the Earl of Southesk, 2½ m. S.W. of Portlethen ry. sta., Kincardineshire.

Els Ness, prom., Sanday Isl., Orkney; projects 1½ m. to sea.

Elsrickle, or **Elsridge Hill,** vil., 4 m. N.E. of Biggar, Lanarkshire. P.O.

Elvanfoot, ham., at confluence of E. Water and r. Clyde, 5 m. S.E. of Abington, Lanarkshire; ry. sta. (alt. 894 ft.), L.M.S. P.O., T.O.

Elvan Water, Lanarkshire, rises on Low-ther Hills to join the Clyde at Elvanfoot.

Elvingston, seat, 1½ m. S.E. of Longniddry, E. Lothian.

Emanuel, ruined priory, on r. Avon, and 6 m. S.E. of Grahamstown, E. Stirling-shire.

Embo, coastal vil. and seat, 2½ m. N.E. of Dornoch, S.E. Sutherland. Fishing. P.O., T.O. at sta., L.M.S. Pop. 404.

Enard Bay, inlet, 4 m. S.W. of Lochinver, N.W. Ross and Cromarty.

Eneclate, pl., isl. of Lewis, Outer Hebrides, 5 m. from Miavaig, W. Loch Tarbet. P.O.

Ennoch, ham., 12½ m. N.W. of Blairgowrie, Perthshire.

Enoch, hill, alt. 1865 ft., 5 m. E. of Dalmellington, Ayrshire.

Enoch, loch, Kirkcudbrightshire, 6 m. S.W. of Loch Doon.

Enoch, ham., 1½ m. N.E. Portpatrick, Wigtownshire.

Enochdhu, vil., in Strath Ardle, and 10 m. N.E. of Pitlochry, Perthshire. P.O., T.O.

Enrick, stream, flowing N.N.E. into Loch Ness, near Drumnadrochit, Inverness-shire.

Ensay, isl. in Sound of Harris, and 2 m. S.W. of Lewis, Outer Hebrides. Pop. 8.

Enterkin, stream in N.W. Dumfriesshire; rises on Lowther Hill, flows 5½ m.S.S.W., and falls into r. Nith after going through E. Pass.

Enterkine, seat, 2 m. S.W. of Tarbolton, Ayrshire.

Enterkinfoot, ham., on the Enterkin, and 6 m. N.W. of Thornhill, Dumfries-shire.

Enzie, q.s. par. and ham. in Bellie and Ruthven pars., Moray and Banffshire, and 7¾ m. N. of Keith. P.O.

Eochar, or **Iochdor,** ferry landing pl., Lochmaddy, N. Uist, Outer Hebrides. P.O.

Eoligary, seat, in N. of Barra Isl., Outer Hebrides. P.O.

Eport, sea loch, 3 m. S. of Lochmaddy, N. Uist, Outer Hebrides.

Erchless, q.s. par. and deer forest (12,670 ac.), in Kintail par., Ross and Cromarty and Inverness-shire, 10 m. W.S.W. of Beauly. E. Castle, seat of the Chisholm.

Eredine, seat, on Loch Awe, and 1½ m. S. of Port-in-Sherrich, Argyllshire.

Ericht, loch, 1 m. from Dalwhinnie sta., and 60 m. N.W. of Perth, between Perthshire and Inverness-shire; 1153 ft. above sea-level, 14¾ m. long by about ¾ m. wide (average). A stream of the same name issues from the foot of Loch Ericht and flows 5½ m. S.E. to Loch Rannoch; depth 512 ft., but the level is now being raised for water-power purposes.

Ericht, stream in Perthshire, formed by the junction of the Black Water and the Ardle, and flowing 10 m. S.E. to join the Isla, 2 m. N.E. of Coupar-Angus.

Erisay, islet in Sound of Harris, Outer Hebrides.

Eriska, isl. and seat, in Loch Creran, N. Argyllshire. Pop. 11.

Eriskay, isl., 3 m. by 1½ m., 5 m. E. of Barra, Outer Hebrides. P.O., T.O. First landing-place in Scotland of Prince Charles Edward Stuart. Pop. 420.

Erisort, sea loch, penetrating 10 m. inland, 7 m. S. of Stornoway, Lewis, Outer Hebrides.

Erradale, North and **South,** hams., 4½ m. S.W. of Gairloch, W. Ross and Cromarty. P.O., T.O. at North E.

Erraid, isl., off S.W. of Mull Isl., Argyllshire. Pop. 33.

Erray House, in N. of Mull Isl., Argyllshire.

Eriboll, ham. and sea loch, 11 m. S.W. of Tongue. The ham. is on the E. side of the loch, and the loch is 10½ m. long. On White Head is a fixed light (white, with red sectors).

Errichel, ham., near Aberfeldy, Perthshire.

Errickstanebrae, hill, near meeting point of Lanark, Peebles, and Dumfries, 5 m. N.W. of Moffat; alt. 1527 ft.

Errochty Water, stream, flowing E. through Glen Erochy to join the Garry at Houan, 4 m. W. of Blair-Atholl, Perthshire.

Errodail, Dun, prom. in Lewis Isl., 2½ m. S.E. of Butt of Lewis, Outer Hebrides.

Errogie, ham., 2½ m. E. of Inverfarigaig, Inverness-shire. P.O.

Errol, par. and vil. on the Firth of Tay, Carse of Gowrie, and 10 m. E. of Perth, Perthshire; ry. sta. (1 m. N. of vil.), L.M.S. P.O., T.O. called E. Village. P.O., T.O. at sta; par. 9533 ac. Pop. 1891; vil. pop. 693. E. Park near.

Erskine, par. (7547 ac., pop. 2129) on the Clyde, N. Renfrewshire, and 5 m. N.W. of Paisley; contains Bishopstown ry. sta., L.M.S. E. House is near the Erskine, Old Kilpatrick Ferry, 1 m. N.E. of sta.

Ervie, pl., near Kirkcolm, W. Wigtownshire. P.O.

Esha Ness, pl. and prom. in W. Northmavine par., mainland, Shetland. P.O. Lighthouse.

Esk, two short streams, combining to flow into the Firth of Forth at Musselburgh, Midlothian.

Esk, r., E. Dumfriesshire, formed by confluence of Black and White Esks; flows 40 m. S. and S.E. to head of Solway Firth.

Esk, loch in N.W. Angus, and 11 m. S. of Balmoral Castle.

Eskadale, vil. in Strath Glass, 7 m. S.W. of Beauly, Inverness-shire.

Eskbank, vil., ¾ m. S.W. of Dalkeith, Midlothian; ry. sta., L.N.E. P.O., T.O.

Eskdale, Langholm, Dumfriesshire. Co. hospital.

Eskdalemuir, par., 14 m. N. of Langholm, N.E. Dumfriesshire, and terminal sta. on branch of L.N.E. Government magnetic observatory. P.O., T.O. Pop. 378.

Esk, North and **South,** rs. in Angus; the North E. flows 29 m. S.E. to the sea, 3 m. N.E. of Montrose. Important net fisheries at mouth of r. The South E. rises in N.W. of co. and flows 48½ m. S. and E. past Brechin to the sea at Montrose.

Esk Paper Mills, at Penicuik, on r. N. Esk, and 10 m. S. of Edinburgh.

Esragan, stream, Argyllshire, flowing 4½ m. S. to Loch Etive at Inveresregan, 2 m. N.W. of Bonawe.

Essendy House, Lethendy par., Blairgowrie, Perthshire.

Essenside, loch, 6 m. S.W. of Lilliesleaf, Roxburghshire.

Esslemont, ham., 1¾ m. S.W. of Ellon, E. Aberdeenshire; ry. sta., L.N.E. T.O. at sta. 1 m. N. of the sta., on the r. Ythan, is E. House.

Eswick, headland, 8¼ m. N. of Lerwick, mainland, Shetland. Flashing light, with white, red, and green sectors, visible 15 m.

Ethie, fishing sta. and stream in Black Isl., Ross and Cromarty; the stream flows into the Moray Firth.

Ethie Castle, seat of Earl of Northesk, 5 m. N.N.E. of Arbroath, Angus.

Ethie Haven, fishing vil. on S. side of Lunan Bay, and 6 m. N.E. of Arbroath, Angus.

Etive, salmon r. and sea loch, Argyllshire; the r. rises on the Moor of Rannoch and has a S.W. course of 15 m. to the loch; the loch then extends 10½ m. S.W., changing direction to westward for 8¾ m.; at Dunstaffnage Castle, 3½ m. N.N.E. of Oban, it merges in the Firth of Lorne.

Etterick Bay, on W. side of Bute Isl., 3 m. W.S.W. of Rothesay, Buteshire.

Etteridge, small loch and seat, 7½ m. S.W. of Kingussie, Inverness-shire.

Ettrick, pastoral par. on the r. Ettrick, S. Selkirkshire. James Hogg (The Ettrick Shepherd) was a native of the par; area 42,456 ac. P.O., T.O. Pop. 308.

Ettrick Bank, seat in co. of, and 1½ m. N. of, Selkirk.

Ettrick Bridge, vil., with inn, on the Ettrick, and 7 m. S.W. of Selkirk. Fishing on lower reaches. P.O., T.O.

Ettrick Forest, was a remnant of the great Caledonian Forest which formerly covered most of Selkirkshire and portions of Peeblesshire and Midlothian. During English invasions the timber frequently suffered severely. Robert I. granted the forest to the Douglases, who forfeited it in the fifteenth century when it became a royal hunting-ground. In 1528 James V. began to turn the land into sheep-walks, and most of it is now under sheep.

Ettrickhaugh, special scavenging dist., Selkirk. Pop. 565.

Ettrick Pen, border of Selkirk and Dumfries, alt. 2269 ft.

Ettrickshaws, seat, Kirkhope par., Selkirkshire.

Ettrick Water, stream of Selkirkshire, flowing 32 m. N.E. to join the Tweed 2½ m. below Selkirk. In Ettrick churchyard, Hogg (The Ettrick Shepherd) was buried.

Evanton, vil., with hotel, 6½ m. N.E. of Dingwall, E. Ross and Cromarty. P.O., T.O. Pop. 372.

Evan Water, stream, rising near Crawford, Upper Lanarkshire, and flowing 12 m. S.E. to join the Annan 2 m. S.E. of Moffat, Dumfriesshire.

Evelix, stream, S.E. Sutherland, issuing from Loch-an-Lagain and flowing 14 m. S.E. to Dornoch Firth at Meikle Ferry.

Everland, seat, Fetlar Is., Shetland.

Evertown, pl., S. Dumfriesshire, 3 m. W. of Canonbie. P.O.

Evie and Rendall, united par., mainland, Orkney; includes isl. of Gairsay; par. area, 14,741 ac. P.O., T.O. Pop. 756.

Ewe, isl. in Loch Ewe, N.W. Ross and Cromarty; shows a red and a white light, each visible 15 m. Pop. 37.

Ewe, river, N.W. Ross and Cromarty, issuing from Loch Maree and flowing 3 m. N.W. into Loch Ewe at Poolewe.

Ewe, sea loch opening off the Minch, N.W. Ross and Cromarty, and 18 m. S.W. of Ullapool; penetrates into the land for 10 m.

Ewes, pastoral par. in E. Dumfriesshire; area, 24,978 ac. P.O., T.O. Pop. 241.

Exnaboe, vil., 3 m. N. of Sumburgh Head, mainland, Shetland.

Eye, loch, N.E. Ross and Cromarty, 3½ m. S.E. of Tain; depth 7 ft, alt. 50 ft.

Eye, or **Knock,** or **Uie,** q.s. par., Stornoway par., Lewis, Outer Hebrides.

Eyebroughty, or **Ibris,** islet in Firth of Forth, 3¾ m. N.W. of North Berwick, E. Lothian.

Eyemouth, police bur., seapt., bur. of barony, and par. (1087 ac.), with term. sta. on branch of L.N.E., on estuary of r. Eye, and 21 m. N.E. of Greenlaw, Berwickshire. Fishing centre. P.O., T.O. Headquarters of a large fishery district, and is a lifeboat and coastguard sta. Has a fixed red light, visible 6 m., and on W. pier-head a fixed white light, also visible 6 m. Pop. 2321; pop. of bur. 2231.

Eye Peninsula, on N.E. of Lewis, 4 m. E. of Stornoway, Outer Hebrides; measures 7 m. by 2 m., and has several villages.

Eye Water, stream, rising in S.E. of East Lothian, and flows for 20 m. S.E. and N.E. to the sea at Eyemouth.

Eynhallow, isl., between Ronsay Isl. and mainland, Orkney.

Eynort, sea loch in E. of South Uist Isl., Outer Hebrides.

Eynort, sea loch on W. Coast of Skye, Inner Hebrides.

F

Fad, loch, ½ m. S. of Rothesay, Bute Isl., Buteshire.

Fada, loch, Colonsay Isl., ½ m. S. of Kiloran, Argyllshire.

Fada, loch, Skye Isl., 3½ m. N. of Portree, Outer Hebrides.

Fada, Islay Isl., 4 m. S.W. of Portaskaig, Argyllshire.

Fada, loch, near Erchless Castle, 9 m. S.W. of Beauly, Inverness-shire.

Fada, loch, Ross and Cromarty, 5 m. N.E. of Poolewe ; alt. 498 ft., depth 56 ft.

Fada, loch, N. Uist, Argyll, 3 m. N.W. of Lochmaddy ; depth 45 ft., alt. 30 ft.

Fafernie, mt., alt. 3274 ft., 18 m. N.W. of Kirriemuir, Angus.

Faichfield, seat, 4 m. W. of Peterhead, Aberdeenshire.

Faifley, vil., 3¾ m. E. of Bowling, Dunbartonshire.

Fail Water, stream, mid Ayrshire, flowing 7½ m. S.E. to join the Ayr at Failford. On the right bank of the stream, just N. of Tarbolton, are the ruins of a monastery.

Fair-a-Far, pl. on S. side of Cramond, in N.W. of city of Edinburgh.

Fair Aird Head, or **Farrid Head,** prom. (315 ft.), 8¼ m. S.E. of Cape Wrath, in N. Sutherland ; projects into sea for about 2 m.

Fairburn, seat, 2½ m. S.E. of Contin, S. Ross and Cromarty. A ruined tower formerly a stronghold of the Mackenzies.

Fairfield, shipbuilding yard on R. Clyde, at Govan. P.O., T.O.

Fairfield, seat, 2½ m. S.E. of Troon, Ayrshire.

Fairgirth, ham., 5 m. S.E. of Dalbeattie, Kirkcudbrightshire.

Fairholm, seat, ½ m. W. of Larkhall, Lanarkshire.

Fairies Dykes, trap reefs off the Great Cumbrae, Firth of Clyde, and 14 m. N.W. of Irvine.

Fair Isle (Sheep Island), Shetland isl. between Shetland and Orkney, and 25 m. S.S.W. of Sumburgh Head ; 3 m.

by 2 m., and so rugged that landing can be made only on N.E. side. Crofting, fishing, and sheep-raising. P.O., T.O. Pop. 108. Two lighthouses, one at S.W. end with group-flashing light, visible 16 m., and one at N.E. end with group-flashing light, visible 23 m. Has a surf lifeboat and cliff ladders. The men are engaged in fishing and the women hosiery.

Fairlaw, seat, 1½ m. S.W. of Reston, Berwickshire.

Fairley, seat, in co. of, and 5¼ m. W. of, Aberdeen.

Fairlie, q.s. par. and holiday pl., on Firth of Clyde and 2 m. S. by E. of Largs, Ayrshire ; ry. stas. (Fairlie and Fairlie Pier), L.M.S. P.O., T.O. Pop. 975. F. House is a seat, and the ruins of F. Castle stand on F. Burn.

Fairlie House, 3½ m. S.W. of Kilmarnock, Ayrshire.

Fairmilehead, pl., on S. side of city of Edinburgh and close to Pentland Hills. P.O.

Fairmuir, q.s. par. in Dundee   P.O.

Fairnilee, or **Fernilee,** ham. on the Tweed, in co. of, and 4 m. N.W. of, Selkirk.

Fairnington, seat, 2 m. S. of Rutherford sta., N. Roxburghshire.

Fairybank Colliery and **Fairybank Quarry,** near Holytown, Lanarkshire.

Fairy Bridge, pl., 3 m. N.E. of Dunvegan, Skye, Inner Hebrides.

Fairy Dean, dell, on the Allan Water, N. Roxburghshire, and about 1½ m. W. of Galashiels.

Fala and Soutra, united par. (6577 ac.) in E. Midlothian. Contains vil. of Fala (which is conjoint with Blackshiels), 3½ m. S.E. of Pathhead. Within the par. are F. Hill, 1¼ m. N.W. of Heriot. F. Flow, partly reclaimed moss. F. Muir and F. Burn. Par. pop. 229 ; vil. pop. 70.

Faldonside, seat and loch, 3 m. S.W. of Melrose ; borders N.W. Roxburghshire and N.E. Selkirkshire.

Falfield, seat, 3½ m. S.E. of Ceres, Fifeshire.

Falkirk, par., parl. and mun. bur., and mkt.-town, on a declivity overlooking Carse of Falkirk, in co. of, and 11 m. S.S.E. of, Stirling ; stands 25½ m. W.N.W. of Edinburgh and 22 m. N.E. of Glasgow ; the town includes Falkirk proper, Camelon, Grahamstown, Bainsford, and Laurieston ; par. 15,111 ac. Pop. 45,443 ; pop. of bur. 36,565 ; rys., L.M.S. and L.N.E. P.O., T.O. Falkirk stands in midst of foundries and collieries, and is connected with Grangemouth by rail. Other industries: distilling, brewing, chemical works, brick and tile works. 1 m. N. are the Carron Iron Works. Falkirk, Grangemouth, and Stirling, for parliamentary purposes, return one member.

Falkland, par., town, royal bur., and ancient royal palace and hunting-seat in mid Fifeshire. The remains of the palace were renovated by the Third Marquess of Bute ; ry. sta. (F. Road), L.N.E. ; par. area 8170 ac. P.O., T.O. F. House is a seat, and 1 m. E. of the town is the vil. of Newton of Falkland. Pop. 2181 ; pop. of bur. 791. Has linen and floorcloth mfrs.

Falla, seat, 2½ m. from Auchengray sta., Lanarkshire.

Fallbogue Bay, in co. of, and 5 m. S.E. of, Kirkcudbright.

Fallen Rocks, immense mass of fallen old red sandstone. boulders, 2 m. N.N.W. of Sannox Bay, Arran Isl., Buteshire.

Fallin, mining vil. in co. of, and 3 m. S.E. of, Stirling. P.O., T.O. Pop. 1279.

Falloch, Falls of, on r. Falloch, at mouth of Glen F., S.W. Perthshire. The r. flows S. into the head of Loch Lomond at Ardlui.

Fallside, pl., with ry. sta., L.M.S., 9½ m. S.E. of Glasgow, Lanarkshire.

Falls of Cruachan, ry. sta., L.M.S., on Loch Awe, Argyllshire. The falls are on Allt Cruachan, near the sta.

Falside Castle, ruin, 2½ m. S.E. of Musselburgh, in W. of East Lothian.

Fanna Hill, alt. 1643 ft., 8½ m. S.E. of Hawick, Roxburghshire.

Fankerton and **Stoneywood,** Denny and Dunipace dist., Stirlingshire. Special drainage dist. Pop. 457.

Fannich Mountains, popularly known as " the Ross-shire Alps," and presenting wild and magnificent scenery, the system lies in mid Ross and Cromarty, and 4 m. N. of Loch F. ; depth 282 ft., alt. 822 ft. ; includes a deer forest of 20,000 ac. ; the highest peak in the system is 3637 ft. On the N. bank of the loch is F. Lodge.

Fannyside, loch and moor, 2½ m. S.E. of Cumbernauld, Dunbartonshire.

Fardlehill, vil., Ayrshire, 1½ m. S. of Kilmaurs.

Fare, Hill of, alt. 1545 ft., 5 m. N. of Banchory, on border Kincardineshire and Aberdeenshire.

Farg, r., Perthshire, Fife, and Kinross-shire, rising amid Ochils and flowing for 10 m. N.E. through Glen F. to join the Earn 1½ m. N.W. of Abernethy.

Farigaig, r., Moray and Inverness-shire, flows 8 m. N.W. and S.W. to Loch Ness at Inverfarigaig pier.

Farigaig, r., flows for about 12 m., and traversing a deep ravine to the S. side of Loch Ness, which it joins at Inverfarigaig, Inverness-shire.

Farlairie, loch, 5 m. N. of Golspie, Sutherland.

Farley and **Urchana,** deer forest (5000 ac.), 3½ m. W. of Beauly, Inverness-shire.

Farme, pl. and seat on r. Clyde, 3 m. N.E. of Cathcart, Lanarkshire. P.O., T.O.

Farnachty, ham., 2¼ m. S. of Buckie, Banffshire.

Farnell, par. and ry. sta., L.M.S. ; par. 5662 ac. Is bounded on the N. by the South Esk. P.O., T.O. at sta. Pop. 478.

Farr, mountainous par. on the coast, in N.E. Sutherland ; par. 258,750 ac. P.O., T.O. called Bettyhill. Pop. 1769.

Farr, ham. and seat on r. Nairn, in co. of, and 8 m. S.E. of, Inverness. P.O., T.O.

Farragon Hill, alt. 2559 ft., 4 m. N.W. of Aberfeldy, Perthshire.

Farrar, r., Ross-shire and Inverness-shire, flowing about 25 m. E. through Glen F. to join the Glass near the old castle of Erchless. Half-way down the glen, the F. plunges wildly, later flowing smoothly till it joins the Beauly. The region is one of great beauty and attracts tourists.

Farrmheall, or **Farveall,** hill, 1709 ft., 8 m. S.W. of Durness, Sutherland.

Farthingbank, ham., near r. Nith, and 4½ m. N.W. of Thornhill, Dumfriesshire.

Faseney Water, trib. of Whitadder, above Priestlaw, E. Lothian.

Fasheilach, mt., alt. 2362 ft., 6 m. S.W. of Ballater, borders of Angus and Aberdeenshire.

Faskadale, landing-pl. on coast, 6 m. N.E. of Ardnamurchan Pt., Argyllshire.

Faskally, seat, at junction of Tummel and Garry, and 2 m. N.W. of Pitlochry, Perthshire.

Faskine and **Palace Craig,** vil., on the N. Calder, and 1½ m. S. of Airdrie, Lanarkshire.

Fasnacloich, pl. and seat, 1½ m. N.E. of the head of Loch Creran and 7 m. E. of Appin, Argyllshire. P.O., T.O.

Fasnakyle, seat and deer forest (25,000 ac.), 20 m. S.W. of Beauly, Invernessshire.

Fasque, seat, 4¾ m. N.W. of Laurencekirk, S.W. Kincardineshire.

Fassifern, seat on N. side of Upper Loch Eil, and 8 m. W.N.W. of Fort William, Argyllshire.

Fast Castle, ruined stronghold on the coast of Berwickshire, 3½ m. N.W. of St. Abb's Head. Is the "Wolf's Crag" in Scott's *Bride of Lammermoor.*

Fatlips Castle, ruined keep on Minto Crags, near the Teviot, and ¾ m. N.E. of Minto House, Roxburghshire.

Fatlips Castle, ruined keep on Tinto Hill, 2 m. N.N.E. of Wiston, Lanarkshire.

Fauldhouse, q.s. par. and town, 6½ m. W.S.W. of West Calder, W. Lothian; ry. stas., L.M.S. and L.N.E. P.O., T.O. Coal mines and oil works. Pop. 4243.

Fauldshope Hill, alt. 1448 ft., in co. of, and 4¾ m. S.W. of, Selkirk.

Fealar Deer Forest, 14,500 ac., in N. Perthshire. F. Lodge is 15 m. N.E. of Pitlochry.

Fearn, par. and vil. in N.E. Ross and Cromarty. Vil. is Hill of Fearn and is 1 m. E. of sta. and 4½ m. S.E. of Tain; par. 7868 ac. Pop. 1492. P.O., T.O., and T.O. at sta. F. Abbey is the church. Vil. pop. 145.

Fearnan, ham. and landing-pl., on N. side of Loch Tay, Perthshire, and 3 m. W. of Kenmore. P.O., T.O. Hotel.

Fearn, Easter and **Wester,** two dists. and two streams on the Dornoch Firth, Ross and Cromarty. F. Lodge on Dornoch Firth.

Fearnoch State Forest, 2 m. W. of Taynuilt, Argyllshire.

Feddal Castle, 1 m. N.W. of Greenloaning, Perthshire.

Fedderat Castle, ruined stronghold, 1½ m. N.E. of New Deer, N. Aberdeenshire.

Feddinch, seat, 2½ m. S.W. of St. Andrews, Fife.

Feinn, loch, 1¾ m. N.E. of Kilmelfort loch, W. Argyllshire.

Fender, loch, 1888 ft. above sea level and 78 ft. deep, N.W. end of Glen F., Perthshire. Near the confluence of the Fender and the Tilt are the Falls of F. 4 m. N.W. of Killiecrankie is the ham. of F. Bridge.

Fendoch, stream flowing N.E. to the Almond at Buchanty, Perthshire. On left bank, at foot of Sma' Glen, are remains of a Roman camp.

Fenella Castle, ruined keep, 1 m. W. of Fettercairn, Kincardineshire.

Fenton Barns, 1½ m. N. of Drem Junction, E. Lothian. Fenbars Sun-Ray Coy., producers of certified and sun-ray milk; one of the largest of its kind in Scotland.

Fenwick, par., vil., and burn, Ayrshire; vil. (High F.), 3½ m. N.E. of Kilmarnock; par. 18,030 ac. P.O., T.O. Pop. 1209; vil. pop. 314.

Feochan, sea loch, 4 m. S. of Oban, Argyllshire.

Feolin Ferry, Jura, Argyll, on Sound of Islay and opposite Port Askaig, Islay.

Feor, loch, 3½ m. S.E. of Gairloch, W. Ross and Cromarty.

Fergus, loch, in co. of, and 4 m. S.E. of, Ayr.

Fergushill, q.s. par. and mining vil., 1¾ m. E. of Kilwinning, N. Ayrshire.

Ferguslie, sub. and seat in the W. of Paisley, Renfrewshire.

Ferintosh, ham., 3 m. N.E. of Conon sta., S.E. Ross and Cromarty.

Ferintosh Distillery, Dingwall, Ross and Cromarty.

Fern, par. 8797 ac; 7 m. W. of Brechin, Angus. P.O., T.O. Pop. 406.

Fernie, estate, 4 m. W. of Cupar, Fife. Contains ancient and modern castles. Ham., F. Easter, 3 m. W. of Cupar.

Ferniegair, mining vil., 2 m. S.E. of Hamilton, Lanarkshire. P.O.

Fernie Hill, pl., ½ m. N. of Gilmerton, in S. of city of Edinburgh.

Ferniehirst Castle, on the Jed Water, 2 m. S. of Jedburgh, Roxburghshire. A Scottish Youth Hostel.

Ferniehirst Hill, 3½ m. S.W. of Stow, Midlothian; alt. 1643 ft.

Fern Tower, seat, 1½ m. N.E. of Crieff, Perthshire.

Fernwoodlee, seat, 1 m. S.W. of Oakley ry. sta., Fife.

Ferny Bank, seat, 8 m. N.W. of Edzell, Angus.

Ferrar, pl., 2 m. W. of Aboyne, Aberdeenshire.

Ferrindonald, pl. in Skye, Outer Hebrides, near Church of Sleat.

Ferrindonald, ancient territory of Clan Munro ; between Invergordon and Dingwall, E. Ross and Cromarty.

Ferrybank, seat, 1 m. S.W. of Cupar, Fife.

Ferryden, fishing vil. at mouth of S. Esk, Angus, and ¾ m. S. of Montrose. P.O. Pop. 898.

Ferry Hill, peninsula jutting into Firth of Forth, 6 m. S.W. of Aberdour, Fife ; this is the extreme southern part of the co. and contains vil. of N. Queensferry. The firth is here spanned by the Forth Bridge.

Ferryhill, ry. junction of L.M.S. and L.N.E., in q.s. par. of same name in S.W. Aberdeen. P.O.

Ferry Little, across the outlet of Loch Fleet, 4 m. N. of Dornoch, Sutherland.

Ferry Meikle, or **Muckle,** ferry between Sutherland and Ross and Cromarty, over Dornoch Firth, 4 m. N.W. of Tain.

Ferry-Port-on-Craig, Fife, see TAYPORT.

Fersit Camp, on r. Treig, Inverness-shire, 2 m. S. of Tulloch, 8 m. E. of Roy Bridge. P.O. Temporary camp in connection with a dam being built on r. Spean for increased water supply for the aluminium works at Fort William.

Fersness Bay, on W. of isl. of Eday, Orkney.

Feshie, r. in S.E. Inverness-shire ; rises amid Grampians on borders of Perthshire, and flows 23 m. in a northerly direction to the r. Spey a little above Kincraig.

Feshie Bridge, ham. on the Feshie, 2 m. S.E. of Kincraig. P.O.

Fethaland, Point of, extreme N. pt. of mainland, Shetland.

Fetlar, par. (9906 ac.) and isl. in the Shetlands, 33 m. N.E. of Lerwick. P.O., T.O. Pop. 217.

Fetterangus, vil. on the Ugie, 2 m. N.W. of Mintlaw, N. Aberdeenshire. P.O., T.O. Pop. 276.

Fettercairn, par. and vil. on the Crichie, and 4½ m. N.W. of Laurencekirk, S.W. Kincardineshire ; par. 14,846 ac. Pop. 1087 ; vil. pop. 232. Hotel. P.O., T.O. Has a distillery. F. House, a seat.

Fetteresso, par. and coastal vil., 1½ m. W. of Stonehaven, Kincardineshire ; par. 27,247 ac. Pop. 4909.

Fetternear House, seat (burnt down in 1919) on r. Don, and 1 m. N.W. of Kemnay sta.

Fettes College, boys' school in the N. of city of Edinburgh. Conducted on lines of great public schools of England.

Fettykill Paper Mill, Leslie, Fifeshire.

Feughside, pl., on Water of Feugh, 2 m. W. Strachan, Kincardineshire.

Feugh, Water of, r., Aberdeenshire and Kincardineshire, flows for about 20 m. E. and N.E. before joining Dee opposite Banchory.

Fhiarlaid, Loch an, W. Ross and Cromarty, 6 m. S.E. of Kinlochewe.

Fhuaran, loch, 1 m. W. of Ben Buie, S. of Mull Isl., Argyllshire.

Fiag, loch, r., and glen ; the loch is 6 m. N. of Loch Shin, and the r. is 5 m. and enters Loch Shin 3½ m. S.E. of Overscaig ; depth of loch 71 ft., alt. 700 ft.

Fiaray, islet at N. end of Barra, Outer Hebrides.

Fichlie, Peel of, mound, with vitrified remains of a tower, 1½ m. N.E. of Towie church, Aberdeenshire.

Fiddes, pl., Kincardineshire, 1½ m. E. of Drumlithie. F. Castle near.

Fiddich, stream, in mid Banffshire, flowing nearly 20 m. N.E. and N.W. through Glen F. to join the Spey at Craigellachie.

Fiddler Burn, Lanarkshire, trib. of the Clyde into which it falls, near Mauldslie Castle.

Fidra, isl. in Firth of Forth, E. Lothian, and 2½ m. W.N.W. of N. Berwick. Soil barren. F. Isl. lighthouse, 113 ft. above high tide, has a quarter-minute group-flashing light, visible 16 m.

Fierth, isl. in S. of Lismore Isl., W. Argyllshire.

Fife, or Fifeshire, a peninsular co. in the E. of Scotland, bounded on the N. by the Firth of Tay, E. by the North Sea, S. by the Firth of Forth, and W. by the cos. of Clackmannan, Kinross, and Perth ; area 323,012 ac. of land and 1624 ac. of water. Pop. 276,368. The coastal line is varied but seldom high, and the undulating interior offers a succession of cultivated vales and hills, the most prominent eminences being the E. and W. Lomonds (1471 ft. and

1713 ft.), Largo Law (965 ft.). The Eden and the Leven, though not very important, are the principal rs.; the valley of the Eden, or Howe o' Fife, and the belt of loam along the coast on the Firth of Forth are the most fertile tracts. Agriculture is very highly developed and Fife is a great coal-producing co. The principal mfrs. are linen and artificial silk (especially Dunfermline), wax-cloth and linoleum (Kirkcaldy). Black-band ironstone is mined at Lochgelly and Cowdenbeath, and the co. engages in ironfounding, machine making, shipbuilding (Burntisland), distilling, brick- and tile-making, beet sugar manufacture, paper-making, oil-cake milling, aluminium production; shipping and fishing are largely carried on, the chief ports being Kirkcaldy, Burntisland, and Methil. Historic centres, Dunfermline and St. Andrews. Many popular holiday resorts and golf links. For parliamentary purposes, the co. is divided into E. and W., each of which returns one member.

Fife and Kinross Mental Hospital, 1 m. from Springfield sta., Cupar, Fife.

Fife Keith, town (part of Keith), 4½ m. E. of Mulben, N. Banffshire. P.O., T.O.

Fife Ness, low headland at the E. end of Fife, and 2 m. N.E. of Crail, Fife. T.O. at coastguard sta.

Figgate Burn, continuation of Braid Burn, which rises among the Pentland Hills and flows N.E. to the sea at Portobello, Edinburgh. F. Whins, tract (formerly a forest) between Arthur's Seat and Portobello, 3¼ m. E. of Edinburgh G.P.O.

Fillan Water, r. in W. Perthshire, rising on border of Argyllshire and flowing 11 m. N.E. and S.E. through Strath Fillan into Loch Dochart, opposite Crianlarich.

Finalty, mt., alt. 2954 ft., 8 m. W. of Milton of Clova, N.W. Angus.

Finart, stream in Argyllshire, flowing through Glen F. to F. Bay on Loch Long, near Ardentinny.

Finavon, Angus, 5 m. N.E. of Forfar. P.O. See FINHAVEN.

Finbracks, mt., alt. 2478 ft., 10 m. N. of Kirriemuir, Angus.

Fincastle, dist., glen, and burn, 6 m. N.W. of Pitlochry, and between Garry and Tummel, Perthshire. P.O., T.O. F. House is a seat.

Fincharn Castle, ruined stronghold of the Macdonalds at S. end of Loch Awe, Argyllshire. Fincharn R. flows N.W. from Loch F. to Loch Awe.

Finderlie, seat, 2 m. W. of Milnathort, Kinross-shire.

Findhorn, fishing vil., 5 m. N. of Forres, mouth of r. Findhorn, on bay of same name. P.O., T.O. Pop. 386.

Findhorn, r., rising in the Monadhliath Mts., and flowing 62 m. in a N.E. direction through the cos. Inverness, Nairn, and Moray to the Moray Firth, 2 m. N. of Forres.

Findhorn Bridge, pl. in co. of, and 13½ m. S.E. of, Inverness.

Findlater Castle, ruined castle, 1¾ m. E. of Cullen, N. Banffshire.

Findlay's Seat, hill, Morayshire, 6½ m. S.E. of Elgin; alt. 861 ft.

Findochty, seapt. bur., 3½ m. N.E. of Buckie, Banffshire; ry. sta., L.N.E. P.O., T.O. On W. pier head is a white fixed light and on beach a red fixed light. The ruins of F. Castle is ½ m. S.W. Pop. 1675.

Findo-Gask, par. and ham. in co. of, and 8 m. S.W. of, Perth; ry. sta., L.M.S., Balgowan, 1½ m. N.N.W.; par. 5188 ac. Pop. 329.

Findon, fishing vil., 1½ m. N.E. of Portlethen, N.E. Kincardineshire. Gives name to Findon or Finnan haddocks.

Findon, farm, 1 m. S.W. of Gardenstown, N.E. Banffshire.

Findon House, on S.E. shore of Cromarty Firth, Ross and Cromarty, and 5 m. N.E. of Conon Bridge.

Findon State Forest, near Munlochy, 5½ m. S.W. of Fortrose, E. Ross and Cromarty.

Findouran Lodge, 5 m. N.E. Cairngorm, on r. Avon, S.W. Banffshire.

Findrassie, seat, 2½ m. N.W. of Elgin, Moray.

Findynate, seat, 6 m. W. of Ballinluig Junction, and on r. Tay, Perthshire.

Finella Hill, Kincardineshire, 5 m. N.W. of Laurencekirk; alt. 1358 ft.

Finfan, strong mineral spring, 1½ m. S.W. of Garmouth, N.E. Moray.

Fingal's Cave, see Isl. of Staffa.

Fingal's Grave, hill near Killin, Perthshire.

Fingal's Griddle, ancient Caledonian remains in par. of Ardnamurchan, 10 m. N.W. of Loch Sunart, Argyllshire.

9

Fingal's Oak, old tree, near Barcaldine House, on S. side of Loch Creran, Argyllshire.

Fingal's Seat, mt., alt. over 1000 ft., at head of Portree Loch, Skye, Inner Hebrides.

Fingal's Stair, flight of rough steps in the rock on the side of Ben Eadden, 9 m. N.W. of Port Appin, N. Argyllshire.

Fin Glen, 8½ m. N.W. of Kilsyth, Stirlingshire.

Fin Glen, ravine, between Ben Bhan and Loch Earn, and 2 m. S.W. of St. Fillans, Perthshire.

Fingray, Hill of, alt. 1560 ft., 7½ m. N.W. of Fettercairn, on W. border of Kincardineshire.

Finhaven Castle, ruined castle on the S. Esk and 5 m. N.E. of Forfar, Angus. Associated with the Earls of Crawford. Finhaven House is a modern seat in the vicinity. P.O. Finhaven Hill is crowned by a vitrified fort and commands a fine view.

Finlagan Loch, Islay, 3 m. W. of Port Askaig, Argyllshire. On an islet are the ruins of a castle, a former stronghold of the Macdonalds.

Finlarig Castle, ruined stronghold on Loch Tay and 1½ m. N.E. of Killin, W. Perthshire. Was a seat of the Campbells of Glenorchy. The keep is mentioned in Scott's *Fair Maid of Perth*.

Finlas Loch, 5 m. S.W. of Dalmellington, S.E. Ayrshire. Drains into Loch Doon. A source of Ayr water supply.

Finlay's Castle, ruin, 4 m. S.E. of Moss Side, N.E. Nairnshire.

Finlay's Mire, moss, 3 m. E. of New Deer, N. Aberdeenshire.

Finlayston House, on r. Clyde, 2½ m. E. of of Port Glasgow, Renfrewshire.

Finnan, stream, flowing 5 m. through Glen Finnan to head of Loch Shiel, S.W. Inverness-shire.

Finnan, stream, S.W. Inverness-shire, through Glen F. to Loch Shiel.

Finnart, seat, on Loch Long, 2½ m. N. of Garelochhead, Dunbartonshire.

Finnart, shooting-lodge, 9 m. W. of Kinloch Rannoch and near S.W. corner of Loch Rannoch, Perthshire.

Finnart House, seat, 6 m. S.S.W. of Ballantrae, S.W. Ayrshire.

Finnart Point, headland in S.W. Ayrshire, on E. side of the mouth of Loch Ryan.

Finnich, glen and burn in W. Stirlingshire. The burn rises on the border of Dunbartonshire, and flows for about 5 m. N.E. through the glen to the r. Blane.

Finnich Malaise, seat, 1 m. S. of Drymen, Stirlingshire.

Finnieston, dist. on r. Clyde, near Queen's Dock, in city of Glasgow.

Finsbay, ham. and loch in S.E. of Harris, Outer Hebrides. P.O.

Finstown, coastal vil. and port, with pier, on Bay of Firth, 6½ m. W.N.W. of Kirkwall, Orkney. P.O., T.O.

Fintray, par. and vil. on r. Don, S.E. Aberdeenshire ; par. 7323 ac. Pop. 736; vil. 1½ m. N.E. of Kinaldie ry. sta., L.N.E., which is 10½ m. N.W. of Aberdeen. P.O. ; 1 m. E. of vil. is F. House, seat of Lord Sempill.

Fintray Bay, inlet off Firth of Clyde, Great Cumbrae Isl., Buteshire.

Fintry, par. and vil. on the Endrick, mid Stirlingshire ; par. 13,788 ac. Pop. 281 ; vil. is 16 m. S.W. of Stirling. P.O., T.O. In N. of par. are Fintry Hills (1676 ft.). E. of the vil. are the ruins of Fintry Castle, a stronghold of the Grahams ; vil. pop. 147.

Fintry, estate, 3 m. N.E. of Dundee, Angus.

Fintry Hostel, Fintry, Stirlingshire, ½ m. E. of Kippen Fork, 17 m. N.E. of Glasgow. 14 m. S.W. of Stirling. A Scottish Youth Hostel.

Finzean, q.s. par., pl., and seat, 7 m. E.S.E. of Aboyne, Aberdeenshire. P.O., T.O.

Fionn Bheinn, mt., alt. 3060 ft., 7 m. E. of Kinlochewe, Ross and Cromarty.

Fionn Loch (144 ft. deep and 559 ft. above sea-level), 5½ m. E. of Poolewe, N.W. Ross and Cromarty.

Fionn Loch, 4 m. S.E. of Lochinver, S.W. Sutherland ; depth 90 ft., alt. 370 ft.

Fionn Loch Mor, 5½ m. E. of Inchnadamff, S.W.S. Sutherland.

Fionphort, pl., on Sound of Iona, in S.W. of Mull Isl., Argyllshire. P.O., T.O.

Firkin, bay and prom. on W. side of Loch Lomond, in N. Dunbartonshire, and 2¾ m. S.E. of Tarbet.

Firth, par. on Firth Bay, Mainland, Orkney. In the bay are oyster beds ; area, 8126 ac. Pop. 609.

Firths Voe, inlet, about 16 m. N.N.W. of Lerwick, Mainland, Shetland. It is ½ m. wide at entrance, and on N. shore is an occulting light (white, red and green sectors), visible 11 m.

Fishcross, Clackmannanshire, Special lighting dist. Pop. 330.

Fisherfield and Letterewe Deer Forest, 45,000 ac., 16 m. N.W. of Achnasheen, N.W. Ross and Cromarty.

Fisherie, vil., N. Aberdeenshire, 9 m. N.E. of Turriff. P.O., T.O.

Fisherrow, W. part of town of Mussel-burgh, Midlothian. P.O., T.O. On E. pier-head is a red fixed light, visible 6 m. Fishing.

Fisher's Tryst, inn, on road to Penicuik and 8 m. S. of Edinburgh. Nearest sta., Glencorse.

Fisherton, q.s. par. and fishing ham., in co. of, and 6 m. S.W. of, Ayr.

Fishertown, or **Seatown,** lower part of town of Cullen, Banffshire. Mostly inhabited by fisher-folk.

Fish Holm, isl., 2½ m. S. of S. end of Yell, Shetland.

Fishnish Bay, inlet on N. side of Mull Isl., Argyllshire.

Fishwick, ancient par., now in Hutton par., S.E. Berwickshire.

Fishwives' Causeway, road leading part of way between Portobello and Edinburgh. Considered to be part of Roman road between Inveresk and Cramond and was much used by fisher-women.

Fitch, vil., 3 m. W. of Lerwick, Mainland, Shetland.

Fitful Head, prom. (929 ft.), 6 m. N. of Sumburgh Head, Mainland, Shetland.

Fiunary, pl., in Morven, Argyllshire, 5½ m. W.N.W. of Lochaline. State forest.

Five-Mile-House, ham., 5 m. N.W. of Dundee, Angus. P.O., T.O.

Five Sisters, group of mts., highest peak 3505 ft., in S. corner of Ross and Cromarty.

Fladda, isl., at mouth of Loch Reasort, Lewis Isl., Outer Hebrides.

Fladda or **Flodday,** isl., par., S. Uist, 2 m. S. of Vatersay, Outer Hebrides. Pop. 26.

Fladda, isl., 2½ m. S.E. of N. Uist Isl., Outer Hebrides.

Fladda, one of the Treshnish Isls., 3 m. S.W. of Treshnish Pt., Mull, W. Argyllshire.

Fladdabister, ham. and bay, 8 m. S.W. of Lerwick, Mainland, Shetland.

Fladdachuain, isl., Inverness-shire, 5 m. N.W. of Pt. of Aird, Skye.

Fladda Lighthouse, on Fladda Isl., on N. end of Scarba Sound, Argyllshire. Fixed light (white and red sectors), seen 11 m.

Fladday, isl., Portree par., Inverness-shire, in Sound of Raasay. Pop. 23.

Flanders Moss, now largely reclaimed, 4 m. S.E. of Aberfoyle, Stirlingshire, and on S. side of r. Forth.

Flannan Isles, or **Seven Hunters,** group of seven small isls., 14 m. N.W. of Gallon Head, Lewis, Outer Hebrides; a light-house on one (Eilean More), with half-minute white group-flashing ligh visible 24 m. Pop. 6.

Flashader, dist. in par. of Duirinish, Isle of Skye, Inner Hebrides, 15 m. W. of Raasay.

Flaughton Hill, 328 ft., highest pt. in Eday, Orkney.

Fleet, r. in S.E. Sutherland, rises 2 m. S.E. of Lairg and flows 17 m. S.S.E. to Moray Firth, 3 m. S. of Golspie. Loch F., a land-locked tidal lagoon, 9 m. S.W. of Brora; has two white fixed beacon lights, visible 3 and 4 m. The Mound of F. is an artificial one to carry the road over the estuary.

Fleet, Fell of, mt., alt. 1544 ft., 6 m. W. of New Galloway ry. sta., Kirkcud-brightshire.

Fleet, Islands of, group of isls. at entrance to F. Bay, Kirkcudbrightshire.

Fleetside, pl., with school, Anwoth par., Kirkcudbrightshire.

Fleet State Forest, near Gatehouse-of-Fleet, Kirkcudbrightshire.

Fleet, Water of, r. in S.W. Kirkcudbright-shire, formed by junction of Big and Little Waters of F., and flows for 5 m. S. and S.E. to F. Bay, 7 m. S.E. of Creetown.

Flemington, estate and ancient strong-hold in mid Angus, 5 m. N.E. of Forfar.

Flemington, seat, 1½ m. S.E. of Fort George ry. sta., N.E. Inverness-shire.

Flemington, vil., ½ m. N.E. of Strathaven. Lanarkshire.

Flemington, mining vil., 1½ m. S.E. of Cambuslang, N.W. Lanarkshire. Pop. of F. and Newton 4176.

Flemington, ry. sta., L.M.S., 1 m. S.E. of Motherwell sta., Lanarkshire. P.O., T.O.

Flichity House, seat, in co. of, and 10½ m. S.E. of, Inverness.

Flisk, par. on Firth of Tay, 6 m. N.E. of Newburgh, Fife; par. 2585 ac. Pop. 156.

Float Bay or **Port Float,** small inlet, 6 m. S.E. of Portpatrick, W. Wigtownshire.

Float Moss, low-lying ground on the Clyde, 8 m. S.E. of Carstairs Junction, Lanarkshire.

Flodabay, sea loch on E. coast of Harris, Outer Hebrides.

Flodda, or **Fladday,** isl. in Sound of Raasay, 4 m. E. of the Isle of Skye, Outer Hebrides. Pop. 23.

Flodday, isl., S. Uist, Outer Hebrides, 2 m. S. of Vatersay. Pop. 26.

Flodigarry House, 3½ m. S.S.E. of Aird Pt., Skye, Inner Hebrides.

Floors Castle, seat of Duke of Roxburghe, on r. Tweed, 1 m. W. of Kelso, Roxburghshire.

Flossy, loch, 3½ m. S.W. of Lerwick, Mainland, Shetland.

Flotta, isl. and q.s. par., in par. of Walls and Flotta, 15 m. S.S.W. of Kirkwall, Orkney. P.O., T.O. Pop. 282. Calf of F., isl. off N. coast of F.

Flotterstone Bridge, bridge over Glencorse Burn, Midlothian, and 7 m. S. of G.P.O., Edinburgh.

Flowerburn House, 2½ m. N.E. of Fortrose, E. Ross and Cromarty.

Flowerdale, valley and mansion at head of Gair Loch, and 4 m. S. of Poolewe, W. Ross and Cromarty. F. deer forest, 50,000 ac.

Flowerhill, q.s. par., forming part of Airdrie, N. Lanarkshire.

Fluchary, ham., 2 m. W. of Strathy, Sutherland.

Flugarth, loch, 5 m. N. of Collafirth, Mainland, Shetland.

Fochabers, vil. and holiday resort on the Spey, 7 m. E.S.E. of Elgin, Moray; ry. sta., L.M.S. P.O., T.O. Pop. 880.

Fodderlee, Easter and **Wester,** pls., 5 m. S.W. of Jedburgh, Roxburghshire.

Fodderty, par. in E. Ross and Cromarty, and contains Strathpeffer sta., 4½ m. W. of Dingwall; par. 46,587 ac. Pop. 1487.

Fogo, par. and ham. on the Blackadder, mid Berwickshire; ham., 3½ m. S.W. of Duns. Fogorig is another ham. close to the above. Pop. 369.

Foinaven, or **Fionne Bheinn,** mt., alt. 2980 ft., 6 m. S.W. of the head of Loch Eriboll, N.W. Sutherland.

Folda, ham., 13 m. N.W. of Alyth, N.W. Angus. P.O., T.O.

Foley House, on S. side of Rothesay, in co. and isl. of Bute.

Fonab, seat, on r. Tummel, and 1 m. W. of Pitlochry, Perthshire.

Foodie, ham., 1¾ m. N. of Cupar, Fife. Near it is F. Hills (about 450 ft.).

Foord, pl., 1½ m. N.E. of Drumlithie, Kincardineshire.

Footdee, S.E. dist. of town of Aberdeen, at the foot or mouth of the r. Dee. P.O.

Fopachy (Phopachy), landing pl., near Bunchrew, on S. side of Beauly Firth, Inverness-shire.

Forbes, ancient par., now united with Tullynessie, Aberdeenshire, on r. Don, 2 m. N. of Alford. P.O., T.O.

Forbescourt, seat, Broughty Ferry, Dundee, Angus.

Forbestown, ham. on r. Don, ½ m. E. of par. church of Strathdon, E. Aberdeenshire.

Forbidden Cave, 3 m. N.E. of Arbroath, Angus.

Ford, or **Ford-Loch-Awe,** ham., with hotel and pier on Loch Awe, 9 m. E. of Taynuilt, N.E. Argyllshire. P.O., T.O.

Ford, or **Pathhead Ford,** practically forms a vil. with Pathhead on r. Tyne, Midlothian, and 10¼ m. S.E. of Edinburgh. P.O., T.O.

Fordell, mining vil., ¾ m. S.E. of Crossgates, Fife. F. Castle, a seat.

Fordie, shooting-lodge, 4 m. E. of Dunkeld, Perthshire.

Ford of Frew, ford on r. Forth, 1 m. N.E. of Kippen, on border of Perthshire and Stirlingshire.

Fordoun, par. and seat on the Bervie, 3¼ m. N.E. of Laurencekirk, Kincardineshire; ry. sta. (210 ft.), L.M.S.; par. 26,859 ac. Pop. 1560. P.O., T.O.

Fordyce, par. and vil. on Moray Firth, and 4 m. E.S.E. of Cullen, Banffshire; par. 17,211 ac. Pop. 3462. P.O. Vil. pop. 212. F. Hill in the S. neighbourhood, 580 ft.

Fore Ness, peninsula opposite F. Holm, Mainland, Shetland.

Foresthill Hostel, on the slope of the Meldons, $\frac{1}{4}$ m. W. of the main Edinburgh-Peebles road, near Eddleston, Peeblesshire, 17 m. S. of Edinburgh. A Scottish Youth Hostel.

Forest Lodge, near Tyndrum, W. Perthshire.

Forest Lodge, in Glen Tilt, 8 m. N.E. of Blair-Atholl, N. Perthshire.

Forest Lodge, seat, 5 m. E.S.E. of Boat of Garten, Inverness-shire.

Forest Mill, ham. on Black Devon, $3\frac{1}{4}$ m. E. of Alloa, Clackmannanshire ; ry. sta., L.N.E. Pop. 47.

Forfar, par., co. and mkt.-town, royal and mun. bur., $21\frac{1}{2}$ m. N.E. of Dundee, Angus ; lies in fertile valley of Strathmore ; ry. sta. (alt. 227 ft.), L.M.S. ; par. 8430 ac. Pop. 11,062 ; mun. bur. pop. 9660. Hotels. P.O., T.O., and T.O. at sta. Forfar unites with Arbroath, Bervie, Brechin, and Montrose in returning one member to Parliament. F. Loch ($1\frac{1}{4}$ m. by $\frac{1}{4}$ m.) lies to the W. of the town. The town has linen factories and bleach works.

Forfarshire, former name of Angus (q.v.).

Forgan, par. on Firth of Tay, N. Fife. Contains town of Newport ; par. 5172 ac. Pop. 3809.

Forgandenny, par. and vil. on the Earn, S.E. Perthshire ; par. 6821 ac. Pop. 507. Vil. 4 m. S.W. of Perth ; ry. sta., L.M.S. P.O., T.O. Pop. 59.

Forgebraehead, ham. in the par. of Canonbie, Dumfriesshire.

Forgieside, pl., 4 m. S.E. of Fochabers, Moray.

Forgie, Upper and **Little,** two hams., 2 m. N.W. of Keith, W. Banffshire.

Forglen, par. on the Deveron, and 2 m. W. of Turriff, N.E. Banffshire ; par. 6251 ac. Pop. 574. P.O., T.O. F. House is a seat.

Forgue, par. on the Deveron, N. Aberdeenshire, and 6 m. N.E. of Huntly ; 17,360 ac. Pop. 1574. P.O., T.O. F. Burn flows 10 m. to the Deveron.

Formal, Knock of, wooded hill, alt. 1158 ft., 4 m. N. of Alyth, in Angus.

Formartine, or **Fermartyn,** ancient dist. in mid and N. Aberdeenshire. Occupies the entire coast between mouths of Don and Ythan, and bounded on W.

by Garioch. Ancient thanage and now gives the title of Viscount to Marquess of Aberdeen and Temair.

Forneth, ham. and seat, on Loch of Clunie, $4\frac{1}{2}$ m. W. of Blairgowrie, Perthshire. P.O.

Forres, par., parl., royal, and mun. bur., and mkt.-town on the r. Findhorn, 2 m. from its mouth, and $12\frac{1}{2}$ m. W. of Elgin, Moray ; ry. sta., L.M.S. ; par. 5469 ac. Pop. 4698 ; pop. of bur. 4169. Hotels. P.O., T.O. On a sheltered part of the nearby Clunie Hills is Cluny Hill House, a hydropathic. Forres and the surrounding region is the scene of part of Shakespeare's *Macbeth.* F. House is a seat. The town mfrs. woollen goods and artificial manures.

Forrestburn Reservoir, 2 m. W. of Harthill, Lanarkshire. Supplies Bathgate, W. Lothian.

Forrestburn Water, pl. and burn, Shotts, N.E. Lanarkshire. Burn connects two reservoirs.

Forrestfield, ry. sta., L.N.E., on N.E. border of Lanarkshire, and $6\frac{1}{2}$ m. E. of Airdrie.

Forse, fishing ham. and burn, 2 m. S.W. of Lybster, Caithness ; $2\frac{1}{2}$ m. W. of Lybster is F. House.

Forsinard, pl. and seat, $24\frac{1}{4}$ m. N.W. of Helmsdale ; ry. sta. (505 ft.), L.M.S. P.O., T.O. at sta.

Forss, seat, on F. Water, 1 m. from its mouth, and $5\frac{1}{2}$ m. W. of Thurso, Caithness. P.O., T.O. Flagstone quarries on Hill of Forss.

Fort Augustus, q.s. par. and vil. on Caledonian Canal, at head of Loch Ness and $33\frac{1}{4}$ m. N.W. of Fort William. Fort erected here for holding the Highlanders in check after the rising in 1715 ; enlarged by Wade in 1730, and taken by Highlanders in 1746 ; is now a Benedictine monastery ; ry. sta., L.N.E. (closed). P.O., T.O. Pop. 600. Hotels. Shows a six-seconds white flashing light.

Fort Charlotte, fort adjoining Lerwick, Shetland. Built in the time of Cromwell and now used as coastguard sta., and serves as the northern headquarters of the Royal Naval Reserve.

Forter Castle, ruined stronghold of the Ogilvies, 4 m. N.W. of Kirkton of Glenisla, N.W. Angus.

Forteviot, par. and vil. on the May, in co. of, and 7 m. S.W. of, Perth; ry. sta., L.M.S.; par. 9886 ac. Pop. 467. P.O., and T.O. at sta.

Fort George, fort on Moray Firth, in co. of, and 11¼ m. N.E. of, Inverness; ry. sta., L.M.S. P.O., T.O.

Forth, q.s. par. and mining vil. in N.E. Lanarkshire, and 1 m. S.W. of Wilsontown. P.O., T.O. Pop. of F. and Wilsontown, 2039.

Forth and Clyde Canal (constructed 1768–1790), connects Firths of Forth and Clyde, cos. of Stirling, Dunbarton, and Lanark. Entered from Firth of Forth by the r. Carron, 1 m. above Grangemouth, and follows a S.W. direction for 38 m. to the Clyde at Bowling. Steam navigation was first seriously attempted here.

Forth Bridge, a ry. viaduct from N. to S. Queensferry over the Firth of Forth connecting Fife and W. Lothian. Constructed (1883–1890) on the cantilever principle, the bridge measures 1½ m. long, of which 1 m. is spanned by the cantilevers, the main spans being each of 1710 ft., and the pinnacle of the structure 360 ft. above the water. It shows four white and eight red fixed lights 156 ft. above high water. Cost 2¾ million £. Carries L.N.E. ry.

Forth, Firth of, estuary of r. Forth. From Alloa (where the r. may be said to end and the firth to begin) to its mouth, between Fife Ness and St. Baldred's Cradle, it measures 51 m. in length, with a varying width, being 1 m. broad at Queensferry, 5 m. at Granton, and 17 m. at Elie. The chief islands are Inchkeith, Inchcolm, Inchgarvie, Cramond Island, Isle of May, Fidra, and the Bass Rock; chief ports, Leith, Granton, Burntisland, Bo'ness, Grangemouth; important fisheries.

Forthie Water, falls into Bervie Water, 1 m. S. of Drumlithie, Kincardineshire.

Forth River, cos. of Perth, Stirling, and Clackmannan, formed by two headstreams rising to the N. of Ben Lomond which meet 1 m. W. of Aberfoyle, S.W. Perthshire; flows E. to Stirling and Alloa, where it gradually expands into the Firth of Forth. The r. is sluggish, and its course, especially in the neighbourhood of Stirling, very tortuous (Links of F).

Fortingall, par. and vil. on the Lyon, N.W. Perthshire, and 8 m. W.S.W. of Aberfeldy; par. 231,034 ac. Pop. 1716. Hotel. P.O., T.O. (spelt Fortingal). Traces of prehistoric forts and a stone circle.

Fort Matilda, ry. sta., L.M.S., 1½ m. E. of Gourock, Renfrewshire; headquarters of Clyde submarine defence.

Fortrie, ham., 5½ m. S.W. of Turriff, S. Banffshire. P.O., T.O.

Fortrose, q.s. par., royal and mun. bur., on W. side of inner Moray Firth, Black Isle, Ross and Cromarty; town lies 10½ m. N.E. of Inverness, and nearly opposite Fort George to which there is a ferry; ry. sta., L.M.S. P.O., T.O. Pop. 875.

Fort William, small town (police bur.), on the Caledonian Canal and Loch Linnhe and at the foot of Ben Nevis, in co. of, and 65½ m. S.W. of, Inverness; ry. sta., L.N.E. Hotels. P.O., T.O. Pier with fixed light, visible 5 m. Has distilleries and important aluminium works. Pop. 2527. Regular steamer communication with Glasgow, Oban, Inverness, etc.

Foss, q.s. par. and vil. on the Tummel, 12 m. W. of Pitlochry and 1½ m. from head of Loch Tummel, Perthshire. P.O. Nearby is F. House.

Fossoway, par. (15,792 ac.), in co. of Kinross. Contains the hams. of East F. and West F., 5½ m. W. of Kinross. P.O., T.O. Nearby is F. House. Pop. 1092.

Foswell, seat, 1 m. S.E. of Auchterarder ry. sta., Perthshire.

Fotheringham House, seat in co. of Angus, 4 m. S. of Forfar. F. Hill, 1 m. N., alt. 800 ft.

Foula, isl. (3 m. by 1½ m.), 16 m. S.W. of the nearest pt. of Mainland, and 27 m. W. of Scalloway, Shetland; highest pt. Kaim of Foula, 1372 ft. Frequented by innumerable sea-birds. Very difficult of access. P.O. Pop. 118.

Foulden, par. and vil. on the Whitadder and 4 m. S.E. of Chirnside, E. Berwickshire; par. 3276 ac. Pop. 306. P.O., T.O.

Foulford Place and Road, vil., Beath par., Fife.

Foulis Castle, 4½ m. N.N.E. of Dingwall, S.E. Ross and Cromarty; ry. sta. (Foulis), L.M.S.

Foulis Ferry, pl., 2 m. W. of Evanton, Ross and Cromarty.

Foulshiels, farm, on the Yarrow, in co. of, and 3 m. W. of, Selkirk. Birthplace of Mungo Park (1771–1805), the African explorer.

Foulshiels Colliery, 2 m. S.E. of Whitburn, W. Lothian.

Fountainbleau, farm, in co. of, and close to, Dumfries. Chalybeate spring.

Fountainbridge, dist. in W. of city of Edinburgh. P.O., T.O.

Fountainhall, ham. and ry. junction, L.N.E., on the GalaWater, Midlothian, 22½ m. S.E. of Edinburgh by rail. P.O., T.O.

Fountainhall, seat, now a farmhouse, of the Dick-Lauder family, 1½ m. S.E. of Ormiston sta., E. Lothian.

Fourman Hill, alt. 1127 ft., 4½ m. N.E. of Huntly, on border of Aberdeenshire and Banffshire.

Fourmerkland, seat, 3½ m. N.W. of Lockerbie, Dumfriesshire.

Fourmerkland, ancient tower (1590), in co. of, and 5 m. N.W. of, Dumfries.

Foveran, coastal par. and seat, on the Ythan, 5 m. S.E. of Ellon sta.; par. 10,524 ac. Pop. 1520.

Fowlis Easter, par. and vil., 6 m. N.W. of Dundee, S.W. Angus; par. 2823 ac. Pop. 251. To the S. is F. Castle.

Fowlis Wester, par. (28,736 ac.), and vil., on the r. Almond, 4½ m. N.E. of Crieff, Perthshire. P.O. called Fowlis. Pop. 837.

Foxbar, seat, 2 m. S.E. of Paisley, Renfrewshire.

Fox Hall, seat, on the Almond, about ½ m. S.E. of Linlithgow, W. Lothian.

Foyers, r., mid Inverness-shire, flowing 9 m. N.N.E. and N. to fall into Loch Ness, 10½ m. N.E. of Fort Augustus. Falls of F. near the mouth of the above r., but now of little volume since the Aluminium Co. directed the bulk of it through pipes to develop power. Near the mouth of the r. is a hotel and P.O., T.O.

Frackafield, ham., 3 m. N.W. of Lerwick, Shetland.

Frandy Reservoir, in Ochil Hills, 4½ m. N.N.W. of Dollar, in Perthshire; water supply for Rosyth in Fife.

Frankfield, loch, 3 m. N.E. of Glasgow, Lanarkshire. F. House is a seat.

Fraoch Eilean, islet in Loch Awe, 2¼ m. S.W. of Kilchurn Castle, N.E. Argyllshire. On it are the ruins of a Macnaughton stronghold.

Fraoch Eilean, islet in Loch Hourn, W. Inverness-shire.

Fraoch Eilean, islet in Loch Lomond, opposite Luss.

Fraochaidh, mt., alt. 2883 ft., 5 m. S.W. of Ballachulish, Argyllshire.

Fraserburgh, par., police bur., fishing and seapt. town, on a bay of the same name, 17½ m. N.W. of Peterhead, N.E. Aberdeenshire; ry. sta., L.N.E.; par. 5635 ac. Pop. 10,203; pop. of bur. 9720. P.O., T.O. Coastguard, lifeboat sta., and important seat of the herring fishery. On outer end of S. breakwater is a red six-seconds flashing light, visible 5 m.; on the outer end of Balaclava breakwater a green fixed light, visible 8 m., and a fog bell; and on the middle and N. piers are red fixed lights, visible 5 m.

Freasdail, loch in Kintyre, 6¼ m. S.W. of Tarbert, Argyllshire.

Freeburn, ham. on r. Findhorn, and 1¼ m. N.W. of Findhorn Bridge, N.E. Inverness-shire.

Freefield, seat, 4 m. N.E. of Insch, N. Aberdeenshire.

Freeland, seat, 2¼ m W. of Bridge of Earn, Perthshire.

Freeland, ham. and seat in N. Renfrewshire, 1¼ m. N.N.E. of Bishopton sta.

Freester, loch and ham., 7 m. N. of Lerwick, Shetland.

Freevater Forest, tract of moorland, N.W. of Gleann Mòr, Ross and Cromarty.

Frendraught House, 6 m. N.E. of Huntly, N.W. Aberdeenshire.

Freswick, vil. on bay of same name, 12 m. N. of Wick, Caithness. P.O. Fixed white light, visible 5 m., on pier. On S.W. shore of bay is F. House.

Freuchie, q.s. par. and vil., Fife, Falkland par. Vil. 1 m. N. of Falkland Road sta. P.O., T.O.; vil. pop. 850.

Freuchie, loch, 1¾ m. W. of Amulree, N. Perthshire; depth 62 ft., alt. 867 ft.

Freugh, East and **West,** 2 hams., 2 m. and 2½ m. N.E. of Stoneykirk, Wigtownshire. The shooting of Freugh and Balgreggan covers 6000 ac.

Friardykes, site of a cell of Melrose Abbey. 4¾ m. S.E. of Stenton, E. Lothian.

Friars Brae, hill adjoining Linlithgow. Was the site of a Carmelite monastery, founded 1290.

Friars Carse, seat, on the r. Nith, in co. of, and 6 m. N.W. of, Dumfries. Originally the site of a cell of Melrose Abbey.

Friars Glen, 4¼ m. N.E. of Fettercairn, Kincardineshire. Vestiges of a Carmelite friary.

Friarton, or **Moncrieffe Island,** isl. in r. Tay, in co. of, and 1 m. S. of, Perth.

Friarton, 1 m. S. of Perth. Town hospital.

Friockheim, q.s. par. and vil. on the Lunan, E. Angus, and 6½ m. N.W. of Arbroath; ry. stas., L.M.S. and L.N.E. P.O., T.O. Special lighting dist. Pop. 585.

Frisa or **Freisa, Loch,** 5 m. N.W. of Aros, in isl. of Mull, W. Argyllshire; depth 205 ft., alt. 245 ft.

Frodlaw Height, alt. 1714 ft., 8 m. N. of Langholm, Dumfriesshire.

Frogden, farm, 3½ m. W. of Yetholm, Roxburghshire.

Frostley Water, trib. of the Teviot at Teviothead, Roxburghshire.

Fruid Water, trib. of Tweed, 1¼ m. S.S.W. of Tweedsmuir, Peeblesshire.

Fruin Water, r., rising near W. border of Dunbartonshire and flowing 12½ m., S.E. and E.N.E. into Loch Lomond, 2½ m. N.W. of Balloch.

Fuar Bheinn, mt., alt. 2511 ft., 2½ m. S. of E. end of Loch Sunart, N. Argyllshire.

Fuar Bheinne, loch, 3 m. N.E. of Loch Caolisport, Knapdale, Argyllshire.

Fuda, or **Fuday,** isl., with rocky coasts, ½ m. N.E. of Barra, Outer Hebrides.

Fuinafort, pl., 6 m. W. of Bunessan, Mull Isl., Argyllshire.

Fullarton, q.s. par. and vil. on the Irvine, and close to the town of that name, N.W. Ayrshire.

Fullarton, part of Tollcross dist., N.W. Lanarkshire.

Fullarton House, seat, 1¼ m. S.E. of Troon sta., Ayrshire.

Fullarton Water, stream feeding Edgelaw Reservoir, S. Midlothian.

Fullwood Moss, reclaimed bog, 3 m. N.W. of Paisley, Renfrewshire.

Fulton, ruined peel tower on the Rule, 4 m. S.W. of Jedburgh, Roxburghshire.

Fungal, Glen of, ravine adjoining Aboyne, Aberdeenshire.

Funniach, ham., ¾ m. S.W. of Durris, Kincardineshire.

Funtack Burn, issues from Loch Moy and flows through Strathdearn, nearly 3 m. S.E. and E.N.E. to join the Findhorn, Inverness-shire.

Funzie Bay, inlet, Fetlar Isl. and 9 m. N.E. of Otterswick, Shetland.

Furnace, vil. on Loch Fyne and 8 m. S.W. of Inveraray, Argyllshire. P.O., T.O. Pop. F. and Goatfield, 286.

Fushiebridge, vil. on the Gore Water, Midlothian, and 12¾ m. S.E. of Edinburgh by rail; ry. sta., L.N.E.

Fyn, loch, in co. of, and 4 m. N.E. of, Dumbarton. Water supply for that town.

Fyne, sea loch in S. and mid Argyllshire. Opens off the Firth of Clyde and Kilbrennan Sound and runs 40½ m. N.E., varying in width from a little over a furlong to 5 m. Noted for herrings of highest quality. Haddocks, whiting, and codling also taken. R. Fyne rises on border of Perthshire and flows 6½ m. S.W. to the head of the Loch Fyne.

Fyntalloch, loch, 8½ m. N.W. of Newton-Stewart, Wigtownshire; depth 15 ft., alt. 341 ft.

Fyrish Hill, or **Cnoc Fyrish,** alt. 1478 ft., 1¼ m. N. of Novar House, S.E. Ross and Cromarty. Crowned by a massive artificial erection, representing the Gates of Seringapatam, built by Sir Hector Munro of Novar in 1785 on his return from his Indian Campaign.

Fyvie, par. (29,585 ac.), on the r. Ythan, N. Aberdeenshire, and 7 m. S.E. of Turriff; ry. sta. (alt. 245 ft.), L.N.E. PO., T.O. Pop. 3180. F. Castle is a seat.

G

Gabbert, pl., 6 m. E. of Kyle of Lochalsh, Ross and Cromarty. Ferry to Dornie across Loch Long.

Gabhsun, pl., 7 m. S.W. of the Butt of Lewis, Ross and Cromarty.

Gadgirth, seat, on r. Ayr, in co. of, and 5 m. E. of, Ayr.

Gady Burn, Aberdeenshire. See GAUDY.

Gagie House, seat, 5 m. N. of Broughty Ferry, Angus.

Gaic Deer Forest, Gaic Lodge, Inverness-shire, 9½ m. S. of Kingussie. Forest, 13,000 ac.

Gaich Wood, S. Morayshire, on left bank r. Spey between Grantown and Dulnan Bridge.

Gailes, ry. sta., L.M.S., Ayrshire, 2 m. S. of Irvine. Golf course and military training ground here.

Gaineimh Loch, 6 m. W. of Forsinard sta., Sutherland.

Gainmheich, loch, Ross and Cromarty 10½ m. N. of Ullapool; depth 150 ft., alt., 251 ft.

Gair Loch, W. arm of Loch Fyne, Argyll.

Gairloch, par. and vil., W. Ross and Cromarty ; par. 200,646 ac. Pop. 2380. Steamboat pier, hotel, and golf course. P.O., T.O. at head of Gair Loch, 29 m. N.W. of Achnasheen, where steamer from Glasgow calls.

Gairlochy, ham. and ry. sta. (G. Road End), L.N.E., Inverness-shire, at foot of Loch Lochy, 3 m. N.W. of Spean Bridge. During summer months regular steamer service from Inverness and Banavie (Caledonian Canal).

Gairn, r., S.W. Aberdeenshire, rises on Ben Avon, Banffshire border, and after a run of 20 m. E. and S.E. enters the Dee 1½ m. N.W. of Ballater.

Gairney Water, trib. of Taner Water, Aberdeenshire, 5 m. S.W. of Aboyne.

Gairney Water, Perth and Kinross, flows 8 m. E. to Loch Leven. Gairneybridge is 1¼ m. from its mouth, and 4 m. from Kinross is G. House.

Gairnshiel Shooting-Lodge, on r. Gairn, Aberdeenshire, 5 m. N.W. of Ballater.

Gairnside, pl., with Roman Catholic church, on r. Gairn, 4 m. N.W. of Ballater.

Gairowan, r., W. Inverness-shire, flows 5 m. from Sgor-nan-Coireachan to Loch Quoich.

Gairsay, isl., Orkney, 1 m. E. of Mainland and 1½ m. N.W. of Shapinsay. Pop. 5. G. House is a seat.

Gaitnip, range of coast cliffs, Orkney, on E. side of Scapa Bay, 3 m. S. of Kirkwall.

Gala House, near Galashiels, Selkirkshire.

Gala Lane, stream, cos. Kirkcudbright and Ayr ; issues from Dungeon and falls into Loch Doon.

Galashiels, parl., mun. bur., and manufacturing town on the Gala Water, 1 m. above its confluence with the Tweed, Selkirkshire. Modern town dates from eighteenth century. Dorothy Wordsworth refers to it in 1803 as " the village of Galashiels pleasantly situated on the banks of the stream." Chief centre for tweed manufacture in Scotland, and has dyeworks and tanneries. Ry. sta., L.N.E. P.O., T.O. Free library, hotels. Pop. 13,102.

Gala Water, dist. of co. of Midlothian. Pop. 7002.

Gala Water, r., cos. of Midlothian and Selkirk, rises in the Moorfoot Hills and flows 21 m. to the r. Tweed, which it enters 2½ m. W. of Melrose. The sites of ancient castles and camps and the vil. of Stow are on its banks.

Gallanach, pl. on Loch Gair (arm of Loch Fyne), Argyll, 4½ m. S.W. of Minard.

Gallanach House, seat, on Sound of Kerrera, Argyll, 4 m. S.W. of Oban.

Gallangad Burn, rises near Doughnot Hill and flows first N., then N.E. (as Catter Burn), to Endrick Water, near Drymen sta., Dunbartonshire.

Gallatown (originally Gallowstown), vil., forming N. sub. of Kirkcaldy, Fife. P.O., T.O.

Gallery House, seat in Angus, on r. Esk, 5 m. N.W. of Dubton ry. sta. 2 m. N.W. of Dubton sta. is the vil. of Upper Gallery.

Gallo Hill, in W. of Westry Isl., Orkney, alt. 355 ft.

Gallon Head, prom., W. of Lewis, Ross and Cromarty, 10 m. S.W. of Carloway.

Galloway, a large dist. in S.W. Scotland, originally included parts of the cos. of Ayr and Dumfries, but in recent times its extent has been confined to the shires of Wigtown and the Stewartry of Kirkcudbright. Its greatest length from E. to W. is 63½ m., and breadth 43 m.

Galloway Burn, stream on border of cos. of Wigtown and Ayr, flowing S. to Loch Ryan.

Galloway House, (1740), seat belonging to the Earl of Galloway, Wigtownshire, 1 m. S.E. of Garlieston.

Galloway, Mull of, is an extension of the co. of Wigtown into the North Channel ; for 14 m. this prom. juts into the sea, with a varying width of 2 to 4 m. ; It rises to over 200 ft., and the extreme end is crowned with a lighthouse, the light of which can be seen at a distance of 23 m.

Galloway, New, police and royal bur., Kirkcudbrightshire, on Water of Ken, 5 m. N.W. of its sta. L.M.S. and 25 m. W. of Dumfries. P.O., and T.O., and P.O., T.O. at New G. sta.

Galloway, Rhinns of, lowland W. of Loch Ryan and Luce Bay, Wigtownshire.

Gallowgate, dist. and ry. sta. (G. Central), L.N.E., in E. of Glasgow.

Gallowhill, ham., 1½ m. S.W. of Alford, Aberdeenshire.

Galston, par. and town (police bur.), with ry. sta., L.M.S., Ayrshire, on r. Irvine, 5½ m. E.S.E. of Kilmarnock; par. 15,154 ac. Pop. 6345; of town 4601. P.O., T.O. Mfrs. of lace, muslin, and blankets; also an extensive coalfield.

Galt, The, extreme N.W. of Shapinsay Isl., Orkney.

Galtrigill, ham., Duirinish par., Skye, Inverness-shire, on Loch Dunvegan, 2 m. S.E. of Dunvegan Head.

Galtway, ancient par. in co. and par. of Kirkcudbright.

Gamescleuch Castle, sixteenth-century ruin, Ettrick par., Selkirkshire.

Gameshope Loch, Tweedsmuir par., Peeblesshire. G. Burn flows 5 m. N. to Talla Water.

Gamhna Loch, in Rothiemurchus Forest, Inverness-shire, near Loch an Eilean.

Gamrie, coast par., Banffshire; 17,041 ac. Pop. 5629. Contains the town of Macduff. There are the ruins of a church, founded 1004.

Gana Hill, height, borders of Lanarkshire and Dumfriesshire; alt. 2190 ft.

Gannel Burn, stream, Clackmannanshire; together with the Daiglen Burn forms Burn of Tillicoultry.

Gannoch, mt., Aberdeenshire, 4 m. N.W. of Mt. Battock; alt. 2396 ft.

Gannochy Bridge, bridge (1732) on the N. Esk, border of Angus and Kincardine, 1 m. N. of Edzell.

Gantocks, cluster of rocks with beacon, Firth of Clyde, between Dunoon and Cloch Pt.

Garadhban State Forest, near Drymen ry. sta., 23¼ m. S.W. of Stirling.

Garallan, vil., Ayrshire. See GARRALLAN.

Garan, or **Garanhill,** original name of the town of Muirkirk, Ayrshire.

Garavaig Bridge, near mouth of stream, same name, flowing N. to Loch Maree, Ross and Cromarty, 1½ m. N.W. of Talladale.

Garbat, seat, 4½ m. N. of Garve, Ross and Cromarty.

Garbet Beg, loch, 2 m. S.E. of Rhiconich, W. Sutherland.

Garbet More, loch, 4 m. S.E. of Rhiconich, W. Sutherland.

Garbh Breac, loch, N. Inverness-shire, 3½ m. S.S.W. of Erchless Castle.

Garbhaig, loch, W. Ross and Cromarty, 5¼ m. N.N.W. of Kinlochewe hotel; depth 93 ft.

Garbh Allt, trib. of Glenrosie Water, Buteshire, 2 m. W.N.W. of Brodick, Arran Isl.

Garbh Allt Falls, Aberdeenshire, see GARRAWALT.

Garbh Eilean, one of the Shiant Isls., Outer Hebrides.

Garbh Eilean, isl. in Loch Maree.

Garbh Eilean, Sutherland. See GARVE ISLAND.

Garbhreisa (Garbh Réis), isl., Argyll, off Craignish Point.

Garden, seat, Stirlingshire, 1½ m. N.E. of Buchlyvie.

Gardens, ham., Mainland Shetland, 1 m. S.E. of Mossbank.

Gardensquare, vil., 1 m. from Airdrie, Lanarkshire.

Gardenstown, fishing vil. and q.s. par. in co. of, and 8 m. E. of, Banff. P.O., T.O. Was founded in 1720 by Alex. Garden of Troup. Harbour and pier. Pop. 945.

Garderhouse, ham. and valley, Shetland, 15 m. N.W. of Lerwick. P.O.

Gardie House (seventeenth century), seat, Bressay Isl., Shetland. P.O.

Gardnerside, vil., Lanarkshire, near Bellshill.

Gardyne Castle, old baronial mansion, Kirkden par., Angus, 1 m. S. of Guthrie Junction.

Gare Loch, right arm of the Firth of Clyde, Dunbartonshire. Sheltered, with very good anchorage.

Garelochhead, q.s. par., vil., with pier, and ry. sta., L.N.E. Dunbartonshire, 7½ m. N.W. of Helensburgh. P.O., T.O. Pop. 566. Favourite summer resort. Motor service to Helensburgh.

Garfarran Peel, Stirlingshire. See GARTFARRAN.

Garf Water, trib. of the Clyde, 1¼ m. N.W. of Lamington, Lanarkshire.

Gargunnock, par. and vil., with ry. sta., L.N.E. (passenger service withdrawn), in co. of, and 6 m. W. of, Stirling; par. 9862 ac. Pop. 570. P.O., T.O. Near is G. House, and 2 m. W. the G. Hills, 1591 ft; 1 m. N.E. stood G. Peel. Vil. pop. 255.

Garioch, dist. in Aberdeenshire, bounded by Formartine, Mar, and Strathbogie.

Garioch, Chapel of, see CHAPEL OF GARIOCH.

Garleffin Fell, mt., S.E. Ayrshire, 4 m. S.W. of Straiton; alt. 1385 ft.

Garleton Hills, small group of hills, 1½ m. N. of Haddington, E. Lothian; alt. 590 ft. Hopetoun monument on highest point and at the base the remains of G. Castle.

Garlies Castle, a thirteenth-century building (ruin), Kirkcudbrightshire, 2½ m. W. of Newton-Stewart.

Garlieston, small seapt.; ry. sta. (Millisle), 1 m. W. on G. Bay, in co. of, and 9¼ m. S.E. of, Wigtown. Pop. 385.

Garlogie, vil. in co. of, and 10 m. W. of, Aberdeen.

Garlot Hill, Kincardineshire, 7 m. N.W. of Fordoun; alt. 1116 ft.

Garmond, vil., Aberdeenshire, 5½ m. N.E. of Turriff. Pop. 130.

Garmouth, seapt. vil., with ry. sta., L.N.E, near mouth of Spey, Morayshire, 9 m. E. of Elgin. P.O., T.O. Pop. 338. Coastguard station and salmon fisheries.

Garnethill, dist. in Glasgow, 1½ m. N.W. of the cathedral.

Garngad, dist. and ry. sta., L.N.E., Glasgow, S.E. of Springburn.

Garnkirk, vil., with ry. sta., L.M.S., 5¾ m. N.E. of Glasgow. Has fireclay works and distillery; ¾ m. N. is G. House.

Garnock, r., Ayrshire, rises on border of co., flows for 21 m. into Firth of Clyde, at Irvine Harbour.

Garpel Water, trib. of the r. Ayr, 1 m. S.W. of Muirkirk, Ayrshire.

Garphar House, seat in par., and 2 m. E. of Ballantrae, Ayrshire.

Garrabost, vil., Eye Peninsula, Lewis, Ross and Cromarty, 7 m. E. of Stornoway. P.O., T.O.

Garrachorry Burn, one of the headstreams of the r. Dee, Aberdeenshire.

Garraghuism Cave, coast cave near Coll, 6 m. N.E. of Stornoway, Lewis, Ross and Cromarty.

Garral Hill, Banffshire, 4 m. N. of Keith; alt. 767 ft.

Garrallan, mining vil., 2 m. S.W. of Cumnock, Ayrshire. G. House in vicinity.

Garrawalt Falls, in Ballochbuie Forest, Aberdeenshire, 4 m. S.E. of Braemar.

Garrel, stream, Dumfriesshire, 2 m. N.N.W. of Lochmaben, trib. of the Ae Water.

Garrel Hill, part of Kilsyth range, Stirlingshire, 2 m. N. of Kilsyth; alt. 1503 ft.

Garresdale, pl. on Isl. of Canna, Argyll.

Garrier Burn, trib. of Carmel Water, 1 m. S.E. of Dreghorn, N. Ayrshire.

Garrion Gill, trib. of the r. Clyde, 3 m. S.W. of Wishaw; also Garrion Grain Mills, Garrion Tower (seat), and Garrion Bridge, Lanarkshire.

Garrison, The, seat at Millport, Great Cumbrae Isl., Firth of Clyde.

Garroch, seat, Kirkcudbrightshire, 5 m. N.W. of New Galloway.

Garroch Head, S. point of Bute Isl., Firth of Clyde.

Garrogie, seat, 6¼ m. S.E. of Foyers pier, Loch Ness, Inverness-shire.

Garron Point, headland, on N. side of Stonehaven Bay, Kincardineshire.

Garrows, Easter, seat, 5¼ m. N.W. of Amulree, Perthshire.

Garry Burn, trib. of the Ordie, at Loak, Perthshire.

Garrynahine, ham., with hotel, at head of Loch Roag, Lewis, Ross and Cromarty, 13 m. W. of Stornoway. Good salmon fishing.

Garry River, S.W. of Inverness-shire. Issues from Loch Quoich, flows through Glen Garry to Loch Oich at Invergarry. G. Loch is an extension of the r. Garry.

Garry River, Perthshire, issues from Loch G., flows through Glen G. to join the Tummel below the Pass of Killiecrankie.

Garryside, vil. near Blair-Atholl, Perthshire.

Gars-bheinn, mt., Bracadale par., Skye, Inverness-shire; alt. 2934 ft.

Garscadden, vil., with ironworks, 1¾ m. N.W. of Bearsden, Dunbartonshire; near is G. House with castellated Gothic gateway.

Garscube, seat and vil., Dunbartonshire, 1 m. W. of Maryhill sta. Has colleries and quarries.

Gart, The, seat, on r. Teith, in par. of, and 1¼ m. S.E. of, Callander, Perthshire.

Gartchonzie, tract between Callander and Loch Vennachar. S.W. Perthshire.

Gartcosh, vil., with ry. sta., L.M.S., Lanarkshire, 3 m. N.W. of Coatbridge. P.O., T.O. Pop. 1874 Fireclay, iron, and steel works.

Gartcraig House, seat, 1½ m. N.W. of Shettleston, Lanarkshire.

Garten, Boat of, see BOAT OF GARTEN.

Gartfarran Peel, remains of Roman fort, Stirlingshire, 2 m. N.W. of Buchlyvie.

Gartferry, seat, Lanarkshire, 2 m. N.E. of Garnkirk.

Gartgill, vil., Old Monkland par., 1¼ m. from Coatbridge, Lanarkshire.

Garth, ham., Shetland, on G. Valley, 15 m. N.E. of Papa Stour.

Garthamlock House, 2 m. S.E. of Shettleston, N. Lanarkshire.

Garth Castle, (fourteenth century) ruined stronghold of the Wolf of Badenoch, on Keltney Burn, 2 m. N.E. of Fortingall, Perthshire. Garth House near.

Garthland, seat, near Lochwinnoch, Renfrewshire, 3½ m. N.E. of Kilbirnie.

Garthland Mains, farm, 3¾ m. S.E. of Stranraer, Wigtownshire.

Garth Loch, on E. side of Loch Ness, Inverness-shire, 2½ m. S.E. of r. Foyers; depth 91 ft., alt. 618 ft.

Gartincaber, seat, 2½ m. S.W. of Doune, Perthshire.

Gartlea, vil. in the par. of New Monkland, Lanarkshire. Pop. 306.

Gartloch, Asylum, Lanarkshire, on Bishop Loch, 1 m. S.E. of Garnkirk.

Gartlove, pl., E. Clackmannanshire, 1 m. N. of Kincardine ry. sta.

Gartly, par. and ry. sta., L.N.E., Aberdeenshire; par. 10,088 ac. Pop. 659. P.O., T.O. Contains ruins of old castle of the Barclays. Vil. pop. 49.

Gartmore, vil., with ry. sta., L.N.E., Perthshire, 2½ m. S. of Aberfoyle. P.O., T.O. Also P.O. at sta.; ½ m. N.E. of vil. is G. House. Vil. pop. 292.

Gartmorn Dam, reservoir, 2 m. N.E. of Alloa, Clackmannanshire.

Gartnagrenach, shooting, 7 m. S.W. of Tarbert, Kintyre, Argyll.

Gartnatra, Islay, Argyll, 1 m. N.W. of Bemore. County hospital.

Gartnavel, low eminence on which Glasgow Lunatic Asylum stands, Glasgow.

Gartness, vil., Lanarkshire, on N. Calder Water, 2 m. S.E. of Airdrie.

Gartness, vil. and ry. sta., L.N.E. (passenger service withdrawn), W. border of Stirlingshire, 1½ m. N.W. of Killearn.

Gartney Strath, low ground along the N. side of Loch Katrine, Perthshire.

Gartocharn, ham., Dunbartonshire, 3 m. W. of Drymen. P.O., T.O.

Gartsherrie, town (part of Coatbridge) and q.s. par., ry. sta. (Blairhill and G.), L.N.E., Old Monkland par., Lanarkshire. G. House and G. Ironworks here.

Gartshore, seat, 2½ m. E. of Kirkintilloch, Dunbartonshire.

Gartymore, part of W. Helmsdale, Sutherland.

Garty, West, pl. and seat. Sutherland, S.W. of Helmsdale.

Garva Bridge, ham., Inverness-shire, 6 m. W. of Laggan.

Garvaig, loch, 6½ m. S.E. of Gairloch, W. Ross and Cromarty.

Garvald, par. and vil., 5¾ m. S.E. of Haddington, E. Lothian ; par. 13,412 ac. Pop. 554 (Garvald and Bara). P.O., T.O. Vil. pop. 117. Ancient camp, 480 ft. in diameter. White Castle and Yester Castle are antiquities.

Garvald House, seat, N.W. Peebleshire, 1 m. N.W. of Dolphinton.

Garvald Point, Greenock. See GARVEL POINT.

Garvan, ham. and glen, on S. side and near head of Loch Eil, Argyll.

Garve, ham., with ry. sta., L.M.S., and hotel, Ross and Cromarty, 11¾ m. W. of Dingwall. Hotel. P.O., T.O. To the S.E. is Loch G., alt. 220 ft., depth 105 ft. G. Burn rises near Dirrie More and flows 18 m. S.E. to the r. Conon.

Garve Island, N. coast of Sutherland, 5 m. S.E. of Cape Wrath.

Garvelloch Islands, a group of isls. in the Firth of Lorn ; occupied in early times by monks, who built beehive stone buildings, some of which are still standing.

Garvock, par., S. Kincardineshire, 2 m. S.E. of Laurencekirk ; 7966 ac. Pop. 320 Hill of G., alt. 813 ft.

Garvock, seat, 1 m. E. of Dunning, Perthshire.

Garwald Water, trib. of the White Esk, 2 m. N. of Eskdalemuir, Dumfriesshire.

Garwall Hill, Kirkcudbrightshire, 14 m., N.N.W. of Newton-Stewart, alt. 1125 ft.

Garynahine, Lewis Isl. See GARRYNAHINE.

Gask, par. and ham. in co. of, and 8 m. S. of, Perth. Contains G. House, see FINDO-GASK. Par. 5185 ac. Pop. 329.

Gask, pl., with ancient stone circle, in co. of, and 4½ m. S.E. of, Inverness.

Gask Hill, tumulus, 1½ m. S.E. of Collessie, Fife.

Gask Hill, extreme S.W. of Angus; alt. 1141 ft.

Gask House, old mansion, Aberdeenshire, 1½ m. S. of Turriff.

Gask House, Perthshire. See GASK.

Gasstown, vil. in co. of, and 1½ m. S.S.E. of, Dumfries. P.O., T.O. (Dumfries).

Gass Water, trib. of Lugar Water, E. Ayrshire, 15 m. long.

Gatehead, ry. sta., L.M.S., Ayrshire, 2¾ m. from Kilmarnock.

Gatehope, trib. of the Tweed, rising in Cardon Law (1928 ft.); at 1750 ft. it enters the Tweed ¾ m. E.S.E. of Peebles.

Gatehouse-of-Fleet, r. port (police bur.), with ry. sta., L.M.S., Kirkcudbrightshire, 1½ m. from head of Fleet Bay. P.O., T.O. called Gatehouse. Pop. 888. Hotel.

Gateleybridge, ham., 2 m. N.E. of Thornhill, Dumfriesshire.

Gateside, Greenock, Renfrewshire. Town hospital.

Gateside, vil., 1 m. S.E. of Beith, Ayrshire. P.O.

Gateside, ham. on r. Annan, Dumfriesshire, ½ m. S. of Wamphray sta.

Gateside, vil., 1½ m. N.W. of Markinch, Fife.

Gateside, vil., with ry. sta., L.N.E., 2¼ m. S.W. of Strathmiglo, Fife. P.O. Near is G. House. See EDENSHEAD.

Gateside, vil., close to Glamis sta., Angus.

Gateside, vil., Kirkcudbrightshire, 4½ m. N.E. of Dalbeattie.

Gateside, vil. near Whitburn, W. Lothian.

Gateside, vil., Renfrewshire, S.W. of Barrhead.

Gateside, pl., 2 m. from Prestonkirk, E. Lothian. P.O.

Gattonside, vil., Roxburghshire, on the Tweed, ¾ m. N.W. of Melrose, to which it is connected by footbridge over the Tweed. P.O. Pop. 285.

Gaudy Valley, Aberdeenshire, through which the Gaudy Burn passes and falls into the Ury, 1 m. E. of Oyne church.

Gauer, r., 7 m. long, connecting Lochs Lydoch and Rannoch, Perthshire.

Gauldry, vil., Fife, 4 m. S.W. of Newport. P.O., T.O.

Gauldswell, pl., 3 m. N.W. of Alyth, Perthshire.

Gauldwell Castle, ancient castle, 5 m. W. of Kebble, Banffshire.

Gaviside, vil., 1 m. N.E. of West Calder, Midlothian.

Gavinton, vil., Berwickshire, 2 m. S.W. of Duns. P.O.

Gaylet Pot, cavity with great sea cavern, Angus, 3 m. N.E. of Arbroath.

Geal Charn, mt., E. Inverness-shire, 7 m. S.E. of Nethybridge; alt. 2692 ft.

Geal Charn Mòr, mt., Inverness-shire, 4 m. W. of Aviemore; alt. 2702 ft.

Geanies, pl. and seat, Ross and Cromarty, on Moray Firth, 8 m. S.E. of Tain. P.O., T.O.

Gear Abhainn, river, Inveraray, Argyll. Leaves Dubh Loch and enters Loch Fyne.

Geddes, vil. in co. of, and 2½ m. S. of, Nairn. G. House, a seat.

Gedd Loch, or **Loch an Gead,** Ross and Cromarty, Kintail par.

Gedintailler, vil., 7 m. S.E. of Portree, Skye, Inverness-shire.

Geilston, seat, 1 m. N.W. of Cardross, Dunbartonshire.

Geir House, seat, 9 m. S.W. of Kirkwall, Orkney.

Gelder Burn, Aberdeenshire, see GLEN GELDER.

Geldie Burn, S.W. Aberdeenshire. Flows 8½ m. N.W. to the r. Dee, which it enters 3 m. above the Linn of Dee. G. Lodge is 12 m. from Braemar.

Gellyburn, vil. in co. of, and 11 m. N. of, Perth.

Gelly Loch, ½ m. S.E. of Lochgelly, Fife.

Gelston, vil., Kirkcudbrightshire, 2½ m. S. of Castle-Douglas. P.O., T.O. ½ m. S.E. is G. Castle, a seat.

Gelt Water, Ayrshire. See GUELT WATER.

General's Bridge, across Yarrow Water in co. of and 4 m. S.W. of, Selkirk.

General's Hut, the hut in which General Wade lived (1727) when engineering the road along Loch Ness; now replaced by the Foyers Hotel, at mouth of the r. Foyers, Inverness-shire.

Genoch, seat, Old Luce par., Wigtownshire, 1 m. S.W. of Dunragit.

Gentlemen's Cave, on W. coast of Westray Isl., Orkney. Was the retreat of fugitive Jacobites in 1746.

Geordie's Hill, S.W. Roxburghshire, 7 m. W. of Riccarton Junction; alt. 1521 ft.

Georgemas, pl. and ry. sta., L.M.S. (G. Junction), Caithness, 14½ m. N.W. of Wick.

Georgetown, vil. in co. of, and 2¼ m. E. of, Dumfries.

Georgetown, Perthshire. See Tighnalinn.

Georgetown, vil. and ry. sta., L.M.S., 2¼ m. from Paisley, Renfrewshire. P.O.

Gerryvard, vil. on S. side of Loch Erisort, Lewis Isl., 13 m. S. of Stornoway, Ross and Cromarty.

Gerston, ham. in Halkirk par., Caithness.

Gesto Hospital, Skye, near Edinbain, 13½ m. N.W. of Portree.

Geusachan, seat and burn near the head of Strathglass, Inverness-shire. See Guisachan.

Geylet Pot, see Gaylet Pot.

Ghaoir, stream, N.W. Perthshire, flows 7 m. E. to W. end of Loch Rannoch.

Ghiuragarstidh, loch, W. Ross and Cromarty, 7 m. N.E. of Gairloch; alt. 117 ft., depth 37 ft.

Ghreosabhaigh Loch, sea loch, in E. Harris, Outer Hebrides.

Ghriama, a loch, N.W. Sutherland, 11 m. E. of Ullapool, and at N. end of Loch Shin; alt. 304 ft., depth 64 ft.

Giant's Chair, natural seat in the rocks on Dullan Water, Banffshire, 9 m. E. of Ballindalloch.

Giant's Grave, hollow space on Tail Burn, 10 m. N.E. of Moffat, Dumfries-shire.

Giant's Leg, buttress of natural archway at S. extremity of Bressay Isl., Shetland.

Giant's Steps, hill, 2½ m. N.W. of Pitlochry, Perthshire, at Falls of Tummel.

Giant's Stone, standing stone, ½ m. S.W. of Tweedsmuir church, Peeblesshire.

Gibbiestone, vil., Perthshire, 3 m. N.W. of Bankfoot.

Gibbleston Lodge, at Scalloway, Shetland.

Gibb's Cross, 3 m. S.E. of Westruther, Berwickshire.

Gibliston House, seat, 2 m. E. of Colinsburgh, Fife.

Gicht House of Aberdeenshire, see Gight.

Giffen, ry. sta., L.M.S. (passenger service withdrawn), 2½ m. S.E. of Beith, Ayrshire. G. Castle (ruin) and G. House here.

Gifferton, vil., Fife. See Giffordtown.

Giffnock, vil., with ry. sta., L.M.S., Renfrewshire, 5¼ m. S. of Glasgow; near are sandstone quarries. P.O., T.O.

Gifford, vil., 4 m. S.E. of Haddington, E. Lothian, ry. sta., L.N.E. (passenger service withdrawn). P.O., T.O. Pop. 338. Hotel.

Giffordgate, sub. of the town of Haddington, E. Lothian.

Giffordtown, vil., 1½ m. N.W. of Ladybank, Fife.

Gigalum, small isl. close to the S.E. end of Gigha Isl., Argyll.

Gigha, large isl. off the W. coast of Kintyre, Argyll, 3 m. W. of Tayinloan. P.O., T.O. Ferry to Kintyre. Pop. 240.

Gigha and Cara, united par., Argyll; 3622 ac. Pop. 243.

Gighay, isl. in Sound of Barra, Outer Hebrides.

Gight Castle, or **House of Gight,** ruin, on r. Ythan, Aberdeenshire, 5 m. E. of Fyvie. P.O. at Gight.

Gilbertfield, vil., 1 m. S.E. of Cambuslang, Lanarkshire. Old mansion (1607). P.O.

Gilderumple Head, prom., Mainland of Shetland, 4 m. S.E. of Reawick.

Gilfillan, site of ancient church in co. of, and 6 m. S. of, Wigtown.

Gilfoot, seat, 2½ m. from Carluke, Lanarkshire.

Gillander's Cave, 9 m. S.W. of Loth, Sutherland.

Gill Bay, or **Port Gill,** 8½ m. S.E. of Portpatrick, Wigtownshire.

Gilburn, stream flowing into Freswick Bay, Caithness.

Gillean, isl., Ross and Cromarty, at W. entrance to Loch Alsh; Lighthouse, 1 m. N. of Kyleakin. Pop. 3.

Gillean, coast ham., at S. entrance to Loch Eishort, Isle of Skye.

Gillen House, seat, Isle Oronsay, 8 m. S.E. of Broadford, Isle of Skye.

Gillesbie House, seat, 7 m. N.E. of Lockerbie, Dumfriesshire.

Gillies Hill, close to site of the Battle of Bannockburn, Stirlingshire ; alt. 500 ft.

Gillock, pl., Watten, Caithness. P.O. Near Bower.

Gill of Cree, wooded gorge on r. Cree, 1 m. above Minnigaff Church, border of Wigtown and Kirkcudbrightshire.

Gills, vil. at head of Gills Bay, N.E. Caithness, 1½ m. S. of St. John's Head.

Gillyburn, vil., Perthshire. See GELLY-BURN.

Gilmanscleugh, ravine, Selkirkshire, Black Knowe to Ettrick Water.

Gilmerton, vil., part of Greater Edinburgh, 4 m. S.E. of Edinburgh G.P.O. ; ry. sta., L.N.E. (passenger service withdrawn). P.O., T.O. Coal and iron are mined. G. cave was hewn (1719–1724) of the sandstone rock and used by George Paterson as a smithy and dwelling-place ; near is G. House.

Gilmerton, seat, 3½ m. N.E. of Haddington, E. Lothian.

Gilmerton, vil., 2 m. N.E. of Crieff, Perthshire. P.O., T.O.

Gilminscroft, seat, Ayrshire, 2 m. E. of Catrine ; close by is Gilminscroft Colliery.

Gilmour's Linn, waterfall, on Touch Burn, Stirlingshire, 4 m. W. of Bannockburn.

Gilmourton, vil., Lanarkshire, 4½ m. from Strathaven.

Gilnockie, ry. sta., L.N.E., S.E. Dumfriesshire, 4½ m. S.E. of Langholm. Peel-tower connected with Johnnie Armstrong the freebooter stood here (1529).

Gilp Loch, left arm of Loch Fyne, 3 m. N.W. of Lochgilphead, Argyll.

Gilsay, isl. in the Sound of Harris, Outer Hebrides.

Gilston Hill, S.W. of E. Lothian ; alt. 1362 ft.

Gilston Lodge, seat, Dalkeith, Midlothian.

Gilston, New (Backmuir of), vil., Fife, 2½ m. S.E. of Ceres.

Giorra, loch (Tay Basin), W. Perthshire, 10 m. N.W. of Killin.

Gipsey Point, headland in co. of, and 5 m. S. of, Kirkcudbright.

Girdleness Lighthouse, on Girdle Ness, Kincardineshire ; is 120 ft. high ; headland 65 ft. ; can be seen 19 m.

Girdle Toll, pl., near Irvine, Ayrshire. P.O.

Girgenti, pl., N.W. Ayrshire, 4½ m. E. of Kilwinning.

Girista, ham., Shetland, on Wadbister Voe, 6½ m. N.W. of Lerwick. Loch of G. is N. of ham.

Girnigoe, Caithness, see CASTLE GIRNIGOE.

Girnock Burn, S.W. Aberdeenshire ; flows 6½ m. N.E. to the Dee, 3 m. W. of Ballater.

Girthgate, old bridle-road along the vale of Allen Water from Old Melrose to the hospitium of Soutra, E. Lothian.

Girth-head, seat, on r. Annan, 1½ m. S. of Wamphray sta., Dumfriesshire.

Girthon, par., Kirkcudbrightshire ; 33,364 ac. Pop. 1015. Ruins of old church 2 m. S.E. of G. house.

Girvan, seapt. town (police bur.) and par., Ayrshire, at mouth of Water of Girvan, 21½ m. S.W. of Ayr ; par. 14,580 ac. Pop. 6056. G. town, 5292. P.O., T.O. Favourite watering-place ; sea-bathing ; golf links. Hotels.

Girvan, Water of, Ayrshire ; rises 5 m. S.W. of head of Loch Doon and flows N.W. and S.W. to the sea at Girvan.

Giurigartie, loch, 2 m. E. of Poolewe, Ross and Cromarty.

Gizzen Briggs, sandbanks, E. of Dornoch, Sutherland.

Glack, mental hospital, Aberdeenshire, 5 m. N.W. of Inverurie.

Glack, hill pass, Angus, between Newtyle Hill and Hatton Hill.

Glackharnis, mt. pass, 6 m. E. of Ballindalloch, Banffshire, between Ben Rinnes and Meikle Conval.

Glackingdaline, bay, with steep rocks, and ancient remains on summit, Ulva Isle, Argyll.

Gladhouse Water, headstream of r. S. Esk, Midlothian ; flows N. to Gladhouse Reservoir (Edinburgh Waterworks).

Gladsmuir, par. and vil., 4 m. W. of Haddington, E. Lothian ; par. 7035 ac. Pop. 1779. P.O.

Gladstone Park, seat, Renfrewshire, near Bishopton sta.

Gladstone Park, see DOLLIS HILL.

Gladswood, seat, near Melrose, N.W. Roxburghshire.

Glaidney, part of Ceres vil., Fife.

Glaissean Loch, Argyll, see GLASHAN.

Glaitness House, seat, Kirkwall, Orkney.

Glamaig, mt., Isle of Skye, alt. 2670 ft. See BEN GLAMAIG.

Glamis, par. and vil. with ry. sta., L.M.S. (1½ m. N.W. of vil.), on Glamis Burn, 5½ m. S.W.of Forfar, Angus; par. 14,265ac. Pop. 985. P.O., T.O. ; T.O. at sta. G. Burn flows through Glen Ogilvie to Dean Water. G. Castle, seat of the Earl of Strathmore and Kinghorne. Malcolm II. said to have been murdered here (1034). Castle is considered to be one of the finest examples of the Scottish baronial style of architecture.

Glas Bheinn (Glasven), mt., Sutherland, on N. of Loch Assynt ; alt. 2541 ft.

Glas Bheinn (Glashven), Morvern, Argyll ; alt. 2034 ft.

Glas-bheinn, Sutherland, at head of the Kyle of Durness ; alt. 1085 ft.

Glas Burn, stream, in N. of Glenisla par., Angus. Flows from Monega Hill to r. Isla, 4 m. S.W. of Loch Esk.

Glas Choire, very small loch, 3½ m. N. of Dalwhinnie sta., Inverness-shire.

Glasclune, seat of Lord Kinross, North Berwick, E. Lothian.

Glasclune Castle, ruin, 2 m. N.W. of Blairgowrie, Perthshire.

Glasdale, vil. near Comrie, Perthshire.

Glas Eilean, isl., in Loch Alsh, Ross and Cromarty.

Glasgoego, seat, S.E. Aberdeenshire, 3½m. S.E. of Kintore.

Glasgow, a parl., royal, and mun. bur., is the second city in point of population in the British Isles and largest city in Scotland. It stands on the r. Clyde, about 22 m. from its estuary, in the lower ward of the county of Lanark, and is distant from Edinburgh 44 m., from London 397 m., from Ayr 34 m., and from Aberdeen 137 m. The city covers an area of 29,511 ac. and has a population of 1,088,461 (census 1931). The earliest records refer to St. Kentigern or Mungo who built a church on the site now occupied by the Cathedral, the only building of great historical interest in the city (1179–1446). Glasgow did not fully receive the position of a royal burgh till 1636. The University was founded in 1451. As a commercial centre the city dates from the Union of Parliaments, 1707, when Scotland obtained trading privileges with the English colonies, and the mercantile activity then begun, has been the leading feature ever since. The city owes part of its development and prosperity to the r. Clyde, which after a course of 106 m., enters the Firth of Clyde at Gourock. Originally fordable at Glasgow, the river has been deepened on many occasions, and is now capable of accommodating the largest vessels. The great utility of the river has been almost created by the enterprise of the people and the Clyde Navigation Trust. There is a large American trade, but commercial contracts have been made with all parts of the world. Beyond the harbour are the great shipbuilding yards and marine engineering works. The first trading steamship to be built was the Comet, which plied between Glasgow and Greenock. Another factor in the development of the city is the Lanarkshire coalfields in the immediate hinderland, where formerly were large deposits of iron. Here the first experiments with Neilson's hot-blast in iron furnaces were carried on (1828) and the economies thus effected placed Glasgow well in the forefront of the iron industry, and boiler-making, steam-tube making, locomotive building, and general engineering, are still among the most important industrial features of the city. Bleaching and calico printing were established in 1738, much earlier than in England, and turkey-red dyes were made very early. Towards the end of the eighteenth century, Tennant discovered the bleaching power of chloride of lime, which gave rise to the establishment of large chemical works. Other industries are : brewing, dyeing, cotton and muslin spinning and weaving, the making of paper, china, glass, etc. The Clyde is crossed by numerous bridges, ferries, and subways, and the harbour, which has ten miles of quays, extends from Glasgow Bridge to the junction of the Kelvin and the Clyde. The city is well served by both L.M.S. and L.N.E. railways. Among the noted public buildings are : the City Chambers, Art Galleries, Museum, Libraries, etc. There are also a large number of Public Parks. The city returns fifteen members to Parliament.

Glasgow Green, ry. sta., L.M.S., in the E. of city of Glasgow.

Glasgow Port, Renfrew, see PORT GLASGOW.

Glashen, loch, 3½ m. E. of Kilmichael, W. Argyll.

Glashven, hill, Argyll, see GLAS BHEINN.

Glashven, ham., Skye, on Staffin Bay, 18 m. N. of Portree.

Glas Loch, Ross and Cromarty, on the borders of Kiltearn and Alness pars., at N.E. base of Ben Wyvis; depth 365 ft., alt. 713 ft.

Glas-loch-Mòr, loch, Sutherland, near the source of the r. Brora.

Glas Moal, mt., N.W. of Glenisla par., Angus ; alt. 3502 ft.

Glasnick, High and **Low,** pls., Wigtownshire, 2 m. N.E. of Kirkcowan.

Glass, par., Aberdeenshire, 12,594 ac. Pop. 537. P.O., T.O. The church is on the r. Deveron, 6 m. W. of Huntly.

Glassalt Shiel, lodge, built by Queen Victoria in 1868 at the head of Loch Muick, S.W. Aberdeenshire, 10 m. S. of Balmoral.

Glassary, par., containing Lochgilphead, Argyll ; par. 60,229 ac. Pop. 2970, vil. Kilmichael-Glassary, 4 m. N. of Lochgilphead. P.O., T.O.

Glassaugh, ry. sta., L.N.E., Banffshire, 2 m. S.W. of Portsoy. T.O. G. House, a seat.

Glassburn House, seat, 15 m. S.W. of Beauly, Inverness-shire.

Glassel, pl. and ry. sta., L.N.E., Kincardineshire, 4½ m. N.W. of Banchory. P.O., T.O. Near is G. House.

Glassert, The, seat on N. side of Loch Ard, Perthshire, 2½ m. W. of Aberfoyle.

Glasserton, coast par. and ham., S.E. Wigtownshire ; par. 13,889 ac. Pop. 771. P.O. The rocks are various but chiefly silurian. G. House, ⅓ m. S.W. of the church.

Glassford, par. and ry. sta., L.M.S., Lanarkshire ; par. 6442 ac. Pop. 1177. P.O., T.O.

Glassie Loch, Perthshire, 2½ m. N. of Aberfeldy ; alt. 1260 ft.

Glass Island, or **Eilean Glas,** an isl. at point of Scalpay Isl., at the mouth of Loch Tarbert, Harris. Has a lighthouse 100 ft. high and 130 ft. above high water with a range of 17 m.

Glassmount, an estate, with modern mansion, in Kinghorn par., Fife, 3 m. N.N.E. of Burntisland.

Glass River and **Loch,** r. issues from Loch Glass and flows in S.E. direction, and falls into the Cromarty Firth 6 m. N.E. of Dingwall. The loch, 713 ft. above sea-level. Older name for r. Allt-Grande—*ugly burn,* referring to the deep chasm near the foot of the r. and known as Black Rock.

Glass River, a r. of Kiltarlity and Kilmorack pars., Inverness-shire ; formed, 2¾ m. S.W. of Glen Affric Hotel, by the joining of the rs. Affric and Amhainn Deabhaidh, thence it winds 12 m. northward until it unites with the Farrar to form the r. Beauly.

Glass, or **River Allt-Grande,** Ross and Cromarty, flows from Loch Glass and enters Cromarty Firth, 1 m. E.S.E. of Novar ry. sta.

Glasterlaw, vil. and ry. sta., L.M.S., Angus, 3 m. N.E. of Friockheim.

Glas Thulachan, mt., N.E. Perthshire, 13 m. N.E. of Blair-Atholl; alt. 3445 ft.

Glasvein, ham., Skye, Inverness-shire. See GLASHVEN.

Glasven or **A' Ghlas-Bheinn,** mt., S.W. Ross and Cromarty, 5 m. N.E. of Invershiel ; alt. 3006 ft.

Glasven, see GLAS BHEINN.

Glaudhall, seat, Lanarkshire, 1 m. N.E. of Garnkirk.

Glazert Burn, trib. of Annick Water, Ayrshire, 4 m. N.W. of Kilmarnock.

Glazert Water, Stirlingshire, is formed by three head streams near Campsie Glen, and flows to the Kelvin at Kirkintilloch.

Gledfield House, seat, Ross and Cromarty, on r. Carron, 1¼ m. W. of Ardgay.

Glen, vil., 1¼ m. S. of Falkirk, Stirlingshire.

Glen, vil., ¼ m. N.E. of Castlebay, Barra Isl., Outer Hebrides.

Glen, burn, E. Kirkcudbrightshire, rising on Criffel, it flows to New Abbey Pow.

Glen, pl. and reservoir, 2 m. S. of Paisley, Renfrewshire.

Glen, The, seat in Peebleshire, on Quair Water, 5 m. S.W. of Innerleithen ; seat of the Tennant family.

Glenacardoch Point, W. coast of Kintyre, Argyll.

Glen Achall, glen, N.W. Ross and Cromarty, on r. Rhidorroch, near Ullapool.

Glenaden Distillery, Old Deer par., Aberdeenshire.

Glenae, seat of a branch of the Dalzell family, 5½ m. N. of Dumfries.

Glen Affric, valley of the r. Affric, Inverness-shire. Famous beauty spot.

Glenaffric Hotel, Inverness-shire, 2½ m. N.E. of Fasnakyle. See GLEN AFFRIC.

Glen Afton, glen on r. Afton, 10 m. E. of Dalmellington, Ayrshire; celebrated in Burns's poems. Sanatorium.

Glen Aheurich, glen on r. Aheurich, Argyll.

Glenairney Lodge, 1 m. S. of Dunphail ry. sta., Morayshire.

Glenaladale, stream, Moidart dist., S.W. Inverness-shire.

Glenalbert, pl., with waterfall, 1 m. N.W. of Dalguise sta., Perthshire.

Glenalbyn, distillery, in co. of, and 1½ m. S.W. of, Inverness.

Glenalla Fell, hill, Ayrshire, 3 m. S.W. of Straiton; alt. 1406 ft.

Glenalmond, 4 m. N.W. of Methven ry. sta., Perthshire. P.O., T.O. On the rt. bank of the Almond is G. Trinity College.

Glen Ample, valley of Ample Water, S. side of Loch Earn, Perthshire, in Balquhidder par.

Glenapp, or **Butters Church,** q.s. par., Ballantrae, Ayrshire, 6¼ m. S. of Ballantrae.

Glen App, or **Glenapp,** picturesque glen on the r. App, in Ballantrae par., S.W. Ayrshire.

Glenaray, q.s. par., Argyll, part of Inveraray par.

Glen Aray, glen on the r. Aray, Argyll, close to Inveraray.

Glen Arbuck, ravine, Dunbartonshire, in Old Kilpatrick par.; resembles a deep rent in the hill formed by the vertical stroke of an earthquake.

Glenarbuck House, seat, on the Clyde, in dist. of Bowling, Dunbartonshire.

Glenardoch, seat, near Doune, Perthshire.

Glen Arklet, glen, Dunbartonshire, on Arklet Water between Inversnaid and Loch Arklet. The water of this loch now falls into Loch Katrine, Glasgow Waterworks.

Glenaros House, seat, isl. of Mull, Argyll, at the mouth of the r. Aras, 1½ m. N.W. of Salen.

Glen Artney, a beautiful sylvan glen in Comrie par., Perthshire, and a large deer forest referred to by Scott in the "Lady of the Lake," as "In lone Glenartney's hazel shade."

Glenartney Deer Forest, Perthshire; lodge, 8 m. S.W. of Comrie.

Glen Ashdale, glen on r. Ashdale, Arran Isl., Argyll.

Glen Avon, the high land in the upper part of the r. Avon or Aven, Banffshire, between Loch Avon and Inchrory; forms part of Glenavon deer forest.

Glenavon, seat, on Avon Water, Lanarkshire, 1½ m. N.E. of Stonehouse sta.

Glenavon Deer Forest, S. Banffshire, 37,150 ac; post town, Tomintoul.

Glenays, old castle in co. of, and 4½ m. S.W. of, Ayr.

Glen Bannisdail, deep ravine between Lochan Fada and N.E. side of Loch Maree, Ross and Cromarty.

Glenbar, ham., near mouth of Barr Water, W. coast of Kintyre, Argyll, 6½ m. S. of Tayinloan. P.O., T.O.

Glenbarry, ry. sta., L.N.E., 5 m. N.E. of Grange Junction, Banffshire. T.O.

Glen Beg, glen, 1½ m. S.W. of Glenely, W. Inverness-shire, extending 5 m. W.N.W. to the head of Sleat Sound.

Glenbeg and Inverlael, deer forest (21,060 ac.), with seat, Ross and Cromarty, 8 m. S.E. of Ullapool.

Glen Beich, glen, Perthshire, in the W. of Comrie par.; the glen is traversed by the Beich Burn which falls into Loch Earn.

Glenbennan Hill, N.E. Kirkcudbrightshire, 4 m. S.W. of Drumpark; alt. 1305 ft.

Glenbervie, par., mid Kincardineshire, 15,041 ac. Pop. 644. G. House, 1½ m. W. of Drumlithie sta.

Glenbervie, seat, 1½ m. N W. of Larbert sta., Stirlingshire.

Glenboig, vil., with ry. sta., L.M.S., Lanarkshire, 2½ m. N. of Coatbridge. P.O., T.O. Has fireclay and brickworks. Pop. 2295.

Glen Boltachan, glen in Comrie par., Perthshire, 3 m. S.E. from Loch Boltachan.

Glenborrodale Castle, seat, with deer forest, Argyll, on N. shore of Loch Sunart. P.O., T.O. (Glenborrodale).

Glenbranter House, at head of Loch Eck, 3 m. S.E. of Strachur, Argyllshire. State forest.

Glen Breakerie, or **Glen Breckrie,** glen in S. of Kintyre, Argyll, 9 m. S.W. of Campbeltown.

Glenbreck, seat, S. Peeblesshire, 10½ m. N. of Moffat.

Glen Brerachan, glen, 6 m. N.E. of Pitlochry, Perthshire, on headstream of Ardle Water.

Glen Brighty, glen, 7 m. S. of Loch Callater, N.W. Angus.

Glen Brittle, pl. and wild glen to the W. of the Cuillin Hills, Isl. of Skye, Inverness-shire. The glen is traversed by the r. Brittle which falls into Loch Brittle, S.W. Skye. State forest here.

Glenbruach House, seat, in the Trossachs, on N. side of Loch Achray, S.W. Perthshire.

Glen Bruar Deer Forest and **Glen Bruar Lodge,** 6 m. N. of Struan, Perthshire.

Glenbuck, q.s. par. and vil., with ry. sta., L.M.S., on E. border of Ayrshire, 3½ m. N.E. of Muirkirk. P.O., T.O. Vil. pop. 482. Close by is House of G.

Glenbucket, par. on W. border of Aberdeenshire; 11,084 ac. Pop. 222. P.O., T.O., called Glenbuchat. At confluence of Bucket Water and r. Don there are the ruins of Glenbucket Castle, and 7 m. N.W. is G. Lodge.

Glen Buckie, glen in S.W. Perthshire, in Balquhidder par., extending 5 m. N. to the foot of Loch Voil.

Glenburgie Distillery, 4 m. E. of Forres, Morayshire.

Glenburn Hydropathic, Rothesay, Buteshire.

Glenburnie, vil., with sawmills, 1½ m. S.E. of Newburgh, Fife.

Glencadam Distillery, Brechin, Angus.

Glencaird, seat and shooting, W. Kirkcudbrightshire, 9 m. N.N.W. of Newton Stewart. G. Hill (1025 ft.).

Glencairn, par., Dumfriesshire, 30,078 ac. Pop. 1352. Contains vil. of Moniaive.

Glencaladh House, seat, at W. side of entrance to Loch Riddon, Argyll.

Glencallater, seat, S.W. Aberdeenshire, 5 m. S.E. of Braemar.

Glencally, glen and deer forest (2920 ac.), N.W. Angus; r. Cally goes through the glen.

Glencalvie, seat and shooting, Ross and Cromarty, 7 m. S.W. of Bonar Bridge.

Glencanisp Deer Forest, about 30,000 ac., 1½ m. S.E. of Lochinver, Sutherland. On Loch Suardalain is G. House.

Glen Cannich, valley of the r. Cannich, Inverness-shire; Kilmorack par.

Glencannich Deer Forest, see CRASKIE.

Glencaple, vil. in co. of, m. S. of, Dumfries, on r. Nith. P.O., T.O. Pop. 201. Convalescent Home.

Glen Carradale, glen, on Carradale Water, Kintyre, Argyll.

Glencarradale House, seat, Kintyre, Argyll, 13 m. N.E. of Campbeltown.

Glen Carron, valley of r. Carron, extending 15 m. N.E. from head of Loch Carron, S.W. Ross and Cromarty.

Glencarron Platform, ry. sta., L.M.S., Ross and Cromarty, 16½ m. N.E. of Strome Ferry. Near is G. Lodge.

Glencarse, ham., with ry. sta., L.M.S., in co. of, and 6 m. E. of, Perth. P.O., T.O. N. of sta. is G. Hill (596 ft,), and G. House, a seat.

Glen Cassley, on r. Cassley, S. Sutherland, S.W. of Loch Shin.

Glencassley, seat, S.E. Sutherland, 11 m. N.W. of Invershin.

Glen Catacol, glen, 10 m. N.W. of Brodick in N.W. Arran, Buteshire.

Glen Ceitlein, glen, Argyll, in Ardchattan par., 3 m. N.E. of the head of Loch Etive.

Glen Chalmodale, glen in N. of Arran Isl., descends N.W. to Loch Ranza, Buteshire.

Glen Clova, upper part of the basin of the South Esk; in Cortachy and Clova par., Angus.

Glen Cloy, glen on r. Cloy, isl. of Arran, Buteshire.

Glen Clunie, valley of Clunie Water, S. Ross and Cromarty.

Glen Coe, a wild and desolate defile in N. Argyll, guarded on the south by the Three Sisters and on the N. by the Chancellor. The scenery as viewed from the summit is among the wildest in Scotland. The lower end was inhabited by the Clan M'Donald, and in 1690 an act of treachery took place by which a large number of the inhabitants were massacred in cold blood. The glen is also claimed as the birthplace of Ossian.

Glencoe, vil. and seat, 1¾ m. E. of Ballachulish ry. sta. Hotel. P.O., T.O. See GLEN COE.

Glen Coire-Lair, glen, S.W. Ross and Cromarty, on r. Lair, near Achnashellach ry. sta.

Glen Conrie, glen on r. Conrie, Aberdeenshire, 2 m. E. of Glenbucket.

Glenconvinth, Inverness-shire. See GLEN CONVINTH.

Glen Convinth, glen, Inverness-shire, on Belladrum Burn, a trib. of r. Beauly, 4 m. S. of Beauly vil.

Glencorse, par. and ry. sta. L.N.E. (passenger service withdrawn), on Glencorse Burn, 14 m. S. of Edinburgh by ry., and 8¼ m. by road; par. 4246 ac. Pop. 1278. T.O. at Milton Bridge; par. contains G. Barracks, G. Reservoir (a source of Edinburgh water supply), and G. House.

Glen Coul, glen at head of Loch Glencoul, 7 m. S. of Loch More, Sutherland.

Glencoul Distillery, in co. of Angus, 7 m. W. of Brechin.

Glencoul Loch, arm of the sea, 6 m. N. of. Loch Assynt. For G. Deer Forest, see GLENDHU and GLENCOUL.

Glencraig, vil., Auchterderran par., Fife, near Lochgelly. P.O., T.O. Pop. (G. and Lochore) 6907.

Glencreggan, seat, 5 m. from Tayinloan, W. Kintyre, Argyll.

Glencripsdale House, seat, Argyll, at mouth of Glen Cripsdale, on S. side of Loch Sunart, 4 m. S.W. from Salen.

Glen Croe, glen, E. Argyll, Lochgilphead par. Beginning at the Col (800 ft.), between the heads of Lochs Fyne and Long, it descends 4½ m. to join Loch Long at Ardgartan, 2½ m. S.W. of Arrochar.

Glencross, par. Midlothian. See GLENCORSE.

Glen Dale, glen, 4 m. S.W. of Duvegan Castle, Isle of Skye. Glendale is a seat. P.O., T.O.

Glendale, seat, Isle of Skye, in Glen Dale, 8 m. N.W. of Dunvegan.

Glen Daruel, valley of r. Ruel, Argyll. See GLENDARUEL.

Glendaruel, pl., with hotel, at head of Loch Riddon, Argyll. P.O., T.O. G. House is a seat.

Glendean's Banks, chasm, Traquair par., Peeblesshire. Near the house of The Glen.

Glen Dearg, or **Glen Diridh,** glen, 11 m. N.E. of Dalnacardoch, Perthshire, between Ben Dearg and Glen Tilt.

Glendebadel Bay, on N.W. coast of Jura.

Glen Dee, valley of the r. Dee, Aberdeenshire, 9 m. N.W. of Braemar.

Glendelvine, seat, 3 m. N. of Murthly sta., Perthshire.

Glen Derry, glen, on Derry Water, Aberdeenshire, 9 m. N.W. of Braemar.

Glen Devon, valley of the r. Devon, S. Perthshire. See DEVON and GLENDEVON.

Glendevon, par., S. Perthshire, forming part of the Ochil Hills; 11,097 ac. Pop. 119. Contains an old castle and G. House, 7 m. N. of Dollar.

Glen Dhu, glen at head of Loch Glendhu, S. of Eddrachillis par., W. Sutherland.

Glendhu, arm of the sea, entering at Kylestrome, when it divides into two arms, viz. Glendhu and Glencoul; also the name of a deer forest, Sutherland, N. of Inchnadamph.

Glendinning, estate, with remains of castle, N.E. Dumfriesshire, on Megget Water, 4 m. S.W. of Wisp Hill.

Glen Diridh, Perthshire. See GLEN DEARG.

Glendochart, hill farm in the N.E. of Penninghame par., N.E. Wigtownshire. It is traversed by a rampart that is known as Deil's Dyke and a large circular hill-fort.

Glen Docherty, glen, Ross and Cromarty, 9 m. N.W. of Achnasheen; descends to Loch Maree.

Glendoe Deer Forest, Inverness-shire; 17,190 ac.; 3 m. E. of Fort Augustus.

Glendoick, ham., 1¼ m. N.E. of Glencarse, S.E. Perthshire. P.O. ¾ m. N. is G. House.

Glen Doin, S.W. Lorn, Argyll, at head of Loch Craignish.

Glen Doll or **Glen of the Doll,** see DOLL and GLENDOLL.

Glendoll Deer Forest, N.W. Angus, 18 m. N.W. of Kirriemuir, about 10,000 ac. This forest is noted for rare plants.

Glen Dorcha, glen between Loch-na-Gainimh and Lochan Fada, 7 m. S.E. of Lochinver, S.W. Sutherland.

Glendorch Burn, trib. of Snar Water, Lanarkshire. Near it stood G. Castle.

Glen Douglas, Dunbartonshire, passing pl. for trains on the West Highland Ry.

Glendoune, seat, 1 m. S. of Girvan, Ayrshire.

Glendowran Burn, trib. of the Snar Water, 1½ m. S.S.W. of Crawfordjohn, S. Lanarkshire.

Glen Dubh, glen, N. Argyll, descends to Loch Creran near Barcaldine House.

Glen Dubh, 9 m. W. of Loch Linnhe, Argyll.

Glen Dubh, upper part of the Duchray Water, Stirlingshire, on E. side of Ben Lomond.

Glen Dubh, W. Sutherland.

Glenduckie, vil., 3 m. E. of Newburgh, Fife. Near is G. Hill; alt. 714 ft.

Glendullan, distillery, Banffshire, 2½ m. N.E. of Dufftown.

Glen Duror, glen on r. Duror, 4 m. S.W. of Ballachulish, Appin, Argyllshire. State forest.

Glen Dye, glen on Dye Water, Kincardineshire. Contains the seat of Glendye Lodge.

Glendye Lodge, seat, Strachan par., Kincardineshire, on Dye Water, 7 m. S.W. of Banchory.

Glen Eagles, Perthshire, runs S. over the Ochil Hills, connecting with Glen Devon. See GLENEAGLES.

Gleneagles, ry. sta., L.M.S., famous golfing centre and palatial hotel, Perthshire, 2 m. S.W. of Auchterarder and 9 m. S.E. of Crieff; alt. 395 ft. T.O., and P.O., T.O. at G. Hotel. G. Castle, a ruin, the ancient home of the Haldanes.

Glenearn, seat, 2 m. S.W. of Bridge of Earn, Perthshire.

Glen Einig, glen along the r. Einig, near Oykell Bridge, Ross and Cromarty.

Glenelg, par., estate, and vil., with pier, W. Inverness-shire; par. 134,778 ac. Pop. 1690. P.O., T.O. The par. is divided into three portions by Loch Hourn and Loch Nevis, known as Glenelg, Knoidart and Morar.

Glenelg Hostel, on the Sound of Sleat, at the extreme S. end of G. vil., W. Inverness-shire. A Scottish Youth Hostel.

Glen Elgin, distillery, 3 m. S. of Elgin, Morayshire.

Glenennich, or **Gleneinich,** an Alpine glen in Rothiemurchus par., Inverness-shire, 10 m. S. of Aviemore.

Glenericht House, seat, E. Perthshire, on r. Ericht, 4¼ m. N. of Blairgowrie.

Glen Esk, upper part of the r. N. Esk, Angus, 4 m. S. of Mount Battock.

Glenessland Burn, trib. of Cairn Water, Dumfriesshire.

Glen Etive, glen on r. Etive, Argyll. Glenetive House, a seat.

Glenetive House, seat in Glen Etive, Argyll, 14 m. N.E. of Taynuilt. P.O., T.O.

Glenfalloch House, seat in Glen Falloch, S.W. Perthshire, 2 m. N. of Ardlui pier, near head of Loch Lomond.

Glenfarclas Distillery, Banffshire, 4 m. S.W. of Aberlour.

Glenfarg, glen and vil., with ry. sta., L.N.E., in S.E. Perthshire, 10 m. S. of Perth on borders of Kinross. P.O., T.O. Pop. 295. Hotel.

Glen Farquhar, glen, 7 m. N.E. of Fettercairn, Kincardineshire.

Glen Fearnach, Perthshire, See GLENFEARNACH.

Glenfearnach, glen in E. of Moulin par., Perthshire, on the Allt Fearnach.

Glen Fender, glen on Glenfender Burn, Perthshire, 1 m. N.E. of Blair-Atholl.

Glenfeochan, seat, at head of Loch Feochan, Argyll, 4 m. S.E. of Oban.

Glenfernate (Upper) Lodge, 9½ m. E. of Pitlochry, Perthshire.

Glenferness, seat of the Earl of Leven and Melville, on r. Findhorn, 8 m. S. of Dunphail, P.O., T.O.

Glen Fernisdale, glen, S.E. Inverness-shire, in Kingussie and Insch par.

Glenfeshie Deer Forest (13,260 ac.), Inverness-shire. Glenfeshie Lodge is 10 m. S.E. of Kingussie.

Glenfiddich Deer Forest, (25,000 ac.), Banffshire. G. Lodge, on r. Fiddich, 5 m. S. of Dufftown.

Glenfield, ham., 1½ m. S.W. of Paisley, Renfrewshire.

Glen Finart, glen, 8 m. N. of Helensburgh, Argyll; descends from Loch Eck to Loch Long. State forest here.

Glenfinart House, seat, at mouth of Glen Finart, Loch Long, Argyll, 1 m. N. of Ardentinny.

Glenfincastle, pl., 9 m. N. of Aberfeldy, Perthshire.

Glen Finglas, glen, S.W. Perthshire, in Callander par., traversed by Turk Water, near Brig o' Turk.

Glenfinnan, ham., seat, and ry. sta. L.N.E., at mouth of Glen Finnan and head of Loch Shiel, Inverness-shire. P.O., T.O.

Glenfintaig, seat, at foot of Glen Gloy, Inverness-shire, 9 m. N.E. of Fort William.

Glenforsa House, seat, isl. of Mull, Argyll, 1½ m. E. of Salen.

Glenfoyle Distillery, 5½ m. S.W. of Gargunnock, ry. sta., Stirlingshire.

Glenfyne Distillery, Ardrishaig, Knapdale, W. Argyll.

Glenfyne Lodge, in Glen Fyne, Argyll, 6 m. N.E. of Cairndow at head of Loch Fyne.

Glengaber Burn, trib. of Megget Water, Peeblesshire.

Glengaffock Hill, N.W. Dumfriesshire, 4 m. N.E. of Kirkconnel sta.; alt. 1583 ft.

Glengairn, q.s. par. in Glenmuick, Crathie, and Braemar pars., Aberdeenshire.

Glen Gallie, glen, in the S. of the par. of Durness, Sutherland, 1 m. N. of Gobernuisgach. Praised by the native poet, Rob Donn, as a hunting-ground.

Glengap Burn, trib. of Tarf Water, 3 m. N.E. of Ringford, Kirkcudbrightshire.

Glengarioch Distillery, Old Meldrum, Aberdeenshire.

Glengarnock, town, with ry. sta., L.M.S., and old castle, N. Ayrshire, 1 m. S.E. of Kilbirnie. P.O., T.O. Important ironworks. Pop. (G. and Kilbirnie) 8332.

Glen Garr, glen at head of Garry Burn, 7 m. W. of Stanley, Perthshire.

Glen Garrel, glen, 7 m. E. of Moffat, Dumfriesshire, on Garwald Water.

Glengarrisdale Bay, on N.W. of Jura Isl., Argyll, 4 m. S.W. from the extreme N. of the isl.

Glengarry, q.s. par., Kilmonivaig par., Inverness-shire.

Glengarry Deer Forest, Inverness-shire, 18,000 ac. Post town, Invergarry. Lodge on shore of Loch Oich. State forest here.

Glengavel, Lanarkshire. See GLENGIVEL.

Glengaw Burn, stream in co. of, and par. of, Ayr.

Glen Geallaidh, S. Morayshire, descends from Carn Kitty to Spey Valley, near Ballindalloch Castle.

Glen Gelder, glen in Balmoral Forest, Aberdeenshire, on Gelder Burn, which

flows from Lochnagar to the Dee, near Balmoral Castle.

Glengirnaig, pl., Aberdeenshire, 4 m. from Ballater. P.O.

Glengivel Water, trib. of Avon Water, W. Lanarkshire.

Glenglassaugh Distillery, near Portsoy, Banffshire.

Glen Gloy, glen on E. side of Loch Lochy, Inverness-shire, on which tracts of so-called parallel roads—similar to Glen Roy—can be seen.

Glengloy House, seat, in Glen Gloy, S.W. Inverness-shire, 6 m. W. of Loch Arkaig.

Glengollie, a glen in S. of Durness par., Sutherland. Favourite hunting-ground of Rob Donn, the native Gaelic poet.

Glengolly, ham., N. Caithness, 2 m. S.W. of Thurso.

Glengonner Water, trib. of the r. Clyde, Lanarkshire, which it enters near Abington.

Glengorm (formerly Sorn), castle and seat in N.W. of isl. of Mull, Argyll, 8 m. W. of Tobermory.

Glengowan, vil., N. Monkland par., 5 m. E. of Airdrie, Lanarkshire. Pop. of G. and Caldercruix, 1870.

Glengraecoham, 2½ m. W. of Portree, Isle of Skye.

Glengrant Distillery, Rothes, E. Morayshire.

Glen Grivie, glen, 10 m. S.W. of Fasnakyle, Inverness-shire. Descends E. to Glen Affric.

Glen Grudie, W. Ross and Cromarty, falls S. to Loch Maree.

Glenguin Distillery, Killearn, Stirlingshire.

Glen Gyle, glen at the W. head of Loch Katrine, Perthshire.

Glengyle, seat, Perthshire, at mouth of Glen Gyle, near W. end of Loch Katrine.

Glen Halmadale, in N. Arran, Buteshire. See GLEN CHALMADALE.

Glen Harich, Argyll, glen and loch, 4 m. N. of Strontian. State forest in glen.

Glen Haultin, glen in isl. of Skye, on r. Haultin, near the head of Loch Snizort.

Glenhead Lodge, 11 m. N.W. of Kirriemuir, Angus.

Glen Hinnisdal, glen, isl. of Skye, on r. Hinnisdal, which flows into Loch Snizort.

Glenhinnisdal, vil. in Glen Hinnisdal, Skye Isl., 12 m. N.W. of Portree.

Glenhopeknowe Burn, trib. of the r. Tweed, which it enters along the boundary of Peeblesshire.

Glenhowam, vil. in co. of, and 5 m. S.E. of, Dumfries.

Gleniffer Distillery, near Paisley, Renfrewshire.

Glen Innich, Inverness-shire. See GLEN EUNACH.

Glen Isla, valley of the upper part of the r. Isla, 9 m. N.W. of the Milton of Clova, Angus.

Glenisla, par. and ham., W. Angus, on r. Isla, 9 m. N.W. of Alyth; par. 41,242 ac. Pop. 477. P.O., T.O. 13 m. N.W. of Alyth is G. House, a seat.

Glenkens, N. dist. of Kirkcudbrightshire. Comprises the pars. of Carsphairn, Dalry, Kells, and Balmaclellan.

Glen Ketland, Argyll. See GLEN CEITLEIN.

Glenkill Burn, trib. of the Water of Ae, Annandale, Dumfriesshire.

Glen Killoch, wooded ravine, Fereneze Hills, Renfrewshire, near Neilston.

Glenkilry Lodge, on the Black Water, E. Perthshire, 12 m. N. of Blairgowrie.

Glenkinchie Distillery, E. Lothian, 2½ m. S.E. of Ormiston sta.

Glenkindie House, seat, near the confluence of the Kindie and the Don, 11 m. S.W. of Rhyne, Aberdeenshire. P.O., T.O.

Glenkingie Deer Forest, Inverness-shire, 16,130 ac; post town, Fort William.

Glen Kinglas, glen, Argyll, on Kinglas Water, flowing W. to Loch Fyne at Cairndow.

Glen Kinglass, glen, Argyll, on r. Kinglass, flowing S.W. to Loch Etive, 5 m. N.E. of Bonawe.

Glenkirk, shooting, 2 m. S. of Tomatin sta., Inverness-shire.

Glen Kyllachy, seat, on r. Findhorn, Inverness-shire, 5 m. S.W. of Tomatin ry. sta.

Glenkyllachy Lodge, 5 m. S. of Tomatin, Inverness-shire, on r. Findhorn.

Glenlaggan, seat, Kirkcudbrightshire, near Loch Ken.

Glenlair, seat, Kirkcudbrightshire, on Urr Water, 7 m. N. of Castle-Douglas.

Glen Latterach, glen, 6 m. N.W. of Rothes, Morayshire.

Glen Lean, glen, Cowall, Argyll, in Dunoon par., traversed by the road from Sandbank and Kilmun to the head of Loch Striven.

Glenlednock, valley of the Lednock, 4 m. N.E. of St. Fillans, Perthshire.

Glenlee, Hamilton, Lanarkshire. Town hospital.

Glenlee, seat, Kirkcudbrightshire, on Water of Ken, 3 m. N.W. of New Galloway. Near by (under construction) a hydro-electric sta. of the Galloway Water Power Co.

Glen Lichet, glen, S. Ross and Cromarty, on r. Croe, flowing to Loch Duich.

Glenlivet, or **Glenlivat,** q.s. par., Inveravon par., Banffshire, 6 m. S.E. of Ballindalloch. P.O., T.O. Here is G. Distillery. Battle of G. was fought (1594).

Glen Livet, glen, 6 m. E. of Kirkmichael, Perthshire.

Glenlochar, ham., 3 m. N.W. of Castle Douglas, Kirkcudbrightshire. P.O., T.O. G. Lodge near.

Glenlochay House, seat, 2½ m. N.W. of Killin, Perthshire.

Glenlochsie, or **Dalmunzie, Lodge,** 3 m. from Spittal of Glenshee, Perthshire, N.E. Perthshire.

Glenlochy, valley of the Lochy, W. Perthshire, 9 m. N.E. of Tyndrum.

Glenlochy Distillery, Fort William, Inverness-shire.

Glen Logan, Ross and Cromarty, descends 6 m. to Kinlochewe.

Glenlogan, vil. and ironworks, Ayrshire, on r. Ayr, 3 m. E. of Catrine; close by is G. House.

Glen Logie, glen, 9 m. W. of Kirriemuir, Angus, on Prosen Water.

Glenloin, glen and seat, Argyll, at head of Loch Long.

Glenlomond Sanatorium, Fife, 3½ m. W. of Milnathort.

Glenlora, seat, 2 m. W. of Lochwinnoch, Renfrewshire.

Glenlossie Distillery, 3 m. S.W. of Elgin, Morayshire.

Glen Loth, in Loth par., S.E. Sutherland.

Glenloy State Forest, 6 m. N.E. of Banavie, which is 2½ m. N.N.E. of Fort William, Argyllshire.

Glenluce, vil., with ry. sta., L.M.S., 8¾ m. S.E. of Stranraer. Pop. 750. P.O., T.O. G. Abbey (1190) 1½ m. N.W. on Luce Water.

Glen Lussa, S.E. Kintyre, Argyll, 3 m. N.E. of Campbeltown.

Glen Lyon, the longest glen in Scotland, Perthshire, 32 m. from Fortingall, to the W. border of the co.

Glenlyon, Innerwick in, q.s. par., Fortingall par., N.W. Perthshire. P.O. and T.O. (at Innerwick, 10 m. W. of Fortingall) called Glenlyon ; close by is G. House, the seat of the Campbells of Glenlyon.

Glenmanno Burn, trib. of Scar Water, 7 m. N.W. of Penpont, N.W. Dumfriesshire.

Glen Mark, glen, of Lochlee par., N. Angus, traversed by the Water of Mark.

Glenmarkie Lodge, 9½ m. S.S.W. of Keith, Aberdeenshire.

Glenmarkie Lodge, Angus, 11 m. N. of Alyth.

Glenmarlin, ravine, with waterfall, on Scar Water, Dumfriesshire, 1 m. W. of Penpont.

Glen Massan, glen, in Kilmun portion of Dunoon par., Argyll.

Glenmavis, vil., Lanarkshire, forming part of New Monkland. P.O. Pop. 802.

Glenmavis Distillery, Bathgate, W. Lothian.

Glenmazeran Lodge, 5½ m. S.W. of Tomatin sta., Inverness-shire on r. Findhorn. P.O., T.O.

Glenmillan House, seat, 1 m. N.E. of Lumphanan sta., Aberdeenshire.

Glenmoidart House, seat, in Moidart dist., S.W. Inverness-shire.

Glenmorangie Distillery, Tain, Ross and Cromarty.

Glen More, glen, in Torsay par., isl. of Mull, Argyll, between Loch Don and Loch Scridain.

Glen More, Bute Isl., on Glenmore Burn, which flows to Etterick Bay.

Glen More, glen, E. Inverness-shire, wide glen from Aviemore to Loch Morlich.

Glen More, glen, 9 m. N.W. of Aberfeldy, Perthshire.

Glenmore, r. and glen, Inverness-shire, 4 m. S. of Lochalsh.

Glenmore, vil., and burn, 5 m. S.W. of Portree, Isle of Skye.

Glenmore, S. Ardnamurchan, Argyll, 8 m. S.W. of Salen hotel.

Glenmore, seat, 4¼ m. S.E. of Maybole, Ayrshire.

Glenmore Albin, (**Gleann mòr na-h Albin**), the *great glen of Scotland* ; applies to the great ravine of 60 m. between Inverness and Fort William now traversed by the Caledonian Canal and many lochs, the largest being Loch Ness.

Glenmore Forest, a State forest and deer forest, Inverness-shire, surrounding Loch Morlich, 6 m. S.W. of Aviemore. G. Lodge.

Glenmore House, seat, on Loch Melford, Argyll, near Kilmelford.

Glenmore Water, Ayrshire, See GLENMUIR.

Glenmoriston, ancient par., now joined to Urquhart par., Inverness-shire. P.O., T.O. Nearby is Invermoriston House at the foot of G., 6 m. N.E. of Fort Augustus.

Glenmorven Lodge, Argyll, on Sound of Mull, 4 m. from Tobermory.

Glenmount Hill, S. Ayrshire, 3½ m. S.W. of Dalmellington ; alt. 1252 ft.

Glenmoy Burn, trib. of S. Esk. 6 m. long, 2 m. N. of Cortachy Castle, mid Angus.

Glenmuick, par., Aberdeenshire, on r. Dee ; 8734 ac. Pop. 2157. G. House, 2 m. S.W. of Ballater.

Glenmuick and Bachnagairn Deer Forest, 19,030 ac. Post town, Ballater, Aberdeenshire.

Glenmuir Water, head stream of Lagar Water, E. Ayrshire, on border of co. Glenmuirshaw, a moorland.

Glen Navidale, small glen near the Ord of Caithness, about 2 m. E. of Helmsdale.

Glen Nevis, glen, Inverness-shire, on W. and S. of Ben Nevis.

Glen Nevis Hostel, on S. side of glen, near Glen Nevis House, 3 m. from Fort William. A Scottish Youth Hostel.

Glennevis House, seat, on Water of Nevis, S.W. Inverness-shire, 3 m. S.E. of Fort William.

Glen Noe, glen on r. Noe, Argyll, which falls to Loch Etive, 4 m. N.E. of Taynuilt.

Glenny, streamlet flowing through a ravine to Lake of Menteith, Perthshire, 9 m. S.W. of Doune.

Glenny Law, hill near Abernyte, S.E. Perthshire.

Glenochil, pl., with ry. sta., L.N.E. (Menstrie and Glenochil), 4 m. N.W. of Alloa, Clackmannanshire ; here is Glenochil Distillery.

Glen Ogil, glen in Angus, on Noran Water, a trib. of the S. Esk.

Glenogil House, or **Easter Ogil,** seat in Angus, 9 m. N.E. of Kirriemuir.

Glen Ogilvie, pl. and glen, 6½ m. S.W. of Kingsmuir, Angus, on Glamis Burn. P.O.

Glen Ogle, wild rocky defile at the W. end of Loch Earn, Perthshire; traversed by road and the L.M.S. ry.

Glenorchy and **Inishail,** united par., N.E. Argyll, 12 m. S. of Tyndrum; 146,628 ac. Pop. 1341. G. Lodge is ½ m. from Dalmally.

Glenormiston House, seat, near the Tweed, 1½ m. N.W. of Innerleithen, Peeblesshire.

Glen Prosen, valley of Prosen Water, 10 m. N.N.E. of Kirriemuir, Angus.

Glenprosen, q.s. par., Cortachy and Clova par., Angus. P.O., T.O. G. Lodge is 12 m. N. of Kirriemuir.

Glen Quaich, glen, 6 m. S. of Aberfeldy, Perthshire, on r. Quaich, flowing to head of Loch Freuchie.

Glenquaich Lodge, on S. side of Loch Freuchie, Perthshire, 13 m. S.W. of Dunkeld.

Glenquicken Moor, S.W. Kirkcudbrightshire, 3 m. E. of Creetown.

Glenquiech, seat, Angus, 7 m. N.E. of Kirriemuir.

Glen Quoich, on Quoich Water, Aberdeenshire, a trib. of the r. Dee.

Glen Quoich, Inverness-shire, on r. Quoich, flowing S.E. to Loch Quoich. P.O., T.O.

Glenquoich Deer Forest, Inverness-shire, about 50,000 ac.; G. Lodge is on N. side of Loch Quoich.

Glen Ranza, glen in N. of Arran Isl., on r. Ranza, flowing into Loch Ranza.

Glen Righ, glen, 2 m. N.W. of N. Ballachulish, Argyllshire. State forest.

Glen Rinnes, vale of Dullan Water, Banffshire, S.W. of Dufftown.

Glenrinnes, q.s. par., Aberlour and Mortlach pass., Banffshire. G. Lodge is a seat.

Glenrisdell, seat, 6½ m. S. of Tarbert, Kintyre, W. Argyll.

Glen Rosa, or **Glenrosie,** glen in E. of Arran, Buteshire; commencing at an alt. of 1750 ft., it descends to Brodick Bay.

Glenrossal Lodge, 9 m. from Invershin sta., Sutherland.

Glenrothes Distillery, Rothes, E. Morayshire.

Glen Roy, a narrow, precipitous glen in Kilmonivaig par., Inverness-shire, traversed by the r. Roy which falls into the Spean. The glen is famous for the level lines along its sides designated " parallel roads " but now known to be the result of a dammed-up lake during the " ice age " in Scotland. The glen is entered from Roybridge.

Glen Salach, glen in Argyll, Ardchattan par., 5½ m. N.N.W. of Loch Etive.

Glensanda Castle, on W. coast of Loch Linnhe, Argyll.

Glen Sannox, glen in N.E. Arran, Buteshire; begins on slope of Goatfell at 1680 ft. and winds round the mountain for about 4 m. It is the wildest glen in Arran.

Glen Sassunn, small glen, E. Fortingall par., 3 m. S.W. of Kinloch Rannoch, Perthshire.

Glensax Burn, stream, Selkirk and Peebles; flows 7 m. to the Tweed, 1½ m. S.E. of Peebles.

Glen Shant, applied to the lower part of Glenrosie q.v.

Glenshee, q.s. par. and ham., N.E. Perthshire, on Shee Water; ham. at the Spittal of Glenshee; also G. Lodge. P.O., T.O.

Glen Shee, upper part of Shochie Burn, E. Central Perthshire, 4½ m. E. by S. of Amulree.

Glen Shee or **Glenshee,** in the upper valley of the Shee Water, Perthshire; and a ham. 30 m. N. by W. of Blairgowrie. The road from Blairgowrie to Braemar passes through this glen.

Glenshellish, seat, 3½ m. S.E. of Strachur, Argyll.

Glen Shiel, long, narrow glen, S. Ross and Cromarty, on r. Shiel.

Glenshiel, par., S. Ross and Cromarty, 57,320 ac. Pop. 317. The church is on Loch Duich, 8 m. S.E. of Kyle of Lochalsh. P.O., T.O.

Glenshiel State Forest, 8 m. S.E. of Kyle of Lochalsh, W. Ross and Cromarty.

Glenshieldaig Deer Forest, Ross and Cromarty, 10,860 ac. Post town, Lochcarron.

Glenshirra Lodge, Inverness-shire, on E. side of Loch Crunachan.

Glenside, ry. sta., L.M.S. (passenger service withdrawn), 13 m. S.W. of Ayr.

Glenskiach Distillery, 3 m. S.W. of Evanton, N.E. Ross and Cromarty.

Glenspean Lodge, on r. Spean, Inverness-shire, 16 m. N.E. of Fort William.

Glenspey Distillery, Rothes, Morayshire.

Glenstewart, seat, S. Dumfriesshire, 5 m. N.W. of Annan.

Glenstiven House, seat, on E. side of Loch Striven, Argyll, 9 m. N.W. of Toward Pt. pier.

Glen Strathfarrar, Ross and Cromarty. See FARRAR.

Glen Stuart, seat, S. Dumfriesshire, 4 m. N.W. of Annan.

Glentaggart Burn, trib. of Glespin Burn, 3 m. S.W. of Douglas, Lanarkshire.

Glentana Deer Forest, Aberdeenshire, 4 m. S.W. of Aboyne, 28,000 ac. G. House is a seat.

Glen Tarbert, glen, N. Argyll, between Lochs Sunart and Linnhe.

Glen Tarken, glen, 1¾ m. N.W. of St. Fillans, Perthshire.

Glentarras Distillery, near Keith, Banffshire.

Glentauchers Distillery, near Keith, Banffshire.

Glen Tilt, narrow glen, extending from Blair-Atholl, Perthshire, for 9 m. N.E. with Ben-y-Gloe guarding it on the E. (3671 ft.).

Glentirrow, or **Glenterra,** tract of moorland, with standing stones, 2½ m. S.W. of New Luce, Wigtownshire.

Glentoo Loch, Kirkcudbrightshire, 4½ m. W. of Castle-Douglas.

Glentress, State forest, 2 m. E. of Peebles.

Glentromie, seat in Glen Tromie, 3 m. S.E. of Kingussie, Inverness-shire.

Glen Trool, glen, 10 m. N. of Newton-Stewart, Kirkcudbrightshire; has Glentrool Lodge.

Glentrool Lodge, see GLEN TROOL.

Glentruim, Inverness-shire, seat at foot of the glen, 7 m. S.W. of Kingussie.

Glentulchan, see GLEN TULCHAN.

Glen Tulchan, seat, Perthshire, on r. Almond, 5 m. N.W. of Methven sta.

Glenturrenmuir, vil., ½ m. S.E. of Kippen, Stirlingshire.

Glen Turret, glen, Perthshire, on Turret Water. Contains Glenturret Lodge at N. end of Loch Turret, 7 m. N.W. of Crieff.

Glenturret (Hosh) Distillery, near Crieff, Perthshire.

Glenturret Lodge, Perthshire. See GLEN TURRET.

Glentyan, water dist. in par. of Kilbarchan, Renfrewshire. Pop. 22. G. House adjacent.

Glen Udale, glen, 9 m. N.E. of Strome Ferry, S.W. Ross and Cromarty; descends to head of Loch Carron.

Glenugie Distillery, Peterhead, Aberdeenshire.

Glenuig, glen, hill, seat, and ham. on G. Bay, Moidart dist., S.W. Inverness-shire. P.O. Alt. of hill 943 ft.

Glen Ure, glen, Argyll, on r. Ure, a trib. of the r. Creran. Contains G. House, 12 m. N.E. of Appin.

Glen Urquhart, famous glen in Inverness-shire; extends from Loch Ness for 7 m. W. along the r. Enrick. P.O. at Drumnadrochit. State forest in the glen.

Glenurquhart, pl., 7 m. W. of Drumna-drochit, Inverness-shire. P.O., T.O. called Glen Urquhart. Pop. of special water supply dist., 415.

Glenury Distillery, Stonehaven, Kincardineshire.

Glen Vale, ravine between W. Lomond Hill and Bishop Hill, on border of Fife and Kinross.

Glen Vorlich, glen, Perthshire, descends N. from Ben Vorlich to Loch Earn.

Glen Wharry, glen, 4 m. N.W. of Sanquhar, N.W. Dumfriesshire.

Glenwharry Burn, trib. of the r. Nith at Kirkconnel, N.W. Dumfriesshire.

Glenwhilly, ham., with ry. sta., Wigtownshire. P.O. (Glenwhillie), 10 m. N.E. of Stranraer (direct line).

Glenzier, burn, S.E. Dumfriesshire, rising in Langholm, it flows S.E. to Cumberland and falls into the Esk.

Glespin Burn, trib. of Douglas Water, 1 m. S.W. of Douglas, Lanarkshire. P.O.

Glesterlaw, Angus. See GLASTERLAW.

Gletness North Isle and **South Isle,** two isls., Shetlands, 6 m. N. of Lerwick.

Glimps Holm, isl., 2½ m. W. of Rose Ness, Orkney, in Holm Sound.

Glomach, Falls of, waterfall, S.W. Ross and Cromarty ; a headstream of the Elchaig ; difficult access to these falls which have a drop of over 350 ft.

Glomach Deer Forest, Ross and Cromarty ; see KILLILAN and GLOMACH.

Gloomingside Burn, Clackmannan. See GANNEL.

Glorat, seat, Stirlingshire, 1 m. E. of Lennoxtown.

Gloup, The, cave, Mainland of Orkney, 11 m. S.W. of Stronsay.

Gloup Hall, seat in the N. of Yell Isl. Shetland, on Gloup Voe, 3 m. N.W. of Cullavoe.

Gloup Holm, isl. near mouth of Gloup Voe, Shetland.

Glower-o'er-em Hill, W. Lothian, 4 m. E. of Polmont ; commands an extensive view.

Glow Loch, among the Cleish Hills, on border of Fife and Kinross, 5 m. N.E. of Saline.

Gloy, small stream, Inverness-shire, falls into Loch Lochy at Glen Gloy.

Glunimore, isl. near Sanda Isl., Kintyre, Argyll.

Glupe, The, vast cavern at Duncansbay Head, Caithness.

Gluss Isle, peninsula, in E. of Mainland, Shetland.

Glutt, burn and grouse moor, Caithness ; burn falls into the Thurso. G. Lodge is a seat.

Goales, ravine, Kilmay par., Fife.

Goatfell, one of the most conspicuous mountains in the isl. of Arran, Buteshire, 3 m. N.W. of Brodick ; alt. 2866 ft.

Goatfield, dist. and seat on Loch Fyne, Argyll, 8½ m. S.W. of Inveraray. Pop. of Goatfield and Furnace, 286.

Goatmilk Hill, Fife, 1 m. S. of Leslie ; alt. 561 ft.

Goauch Hill, N.W. Kincardineshire, 2 m. W. of Banchory ; alt. 1104 ft.

Gobernuisgach, seat, 28 m. N. of Lairg, Sutherland, at the head of Glen Golli.

Goblins Cave, cave near the E. end of Loch Katrine, made famous by Scott in the " Lady of the Lake."

Goblin's Den, wild ravine cutting Ardtun headland in S.W. of Mull Isl., Argyll.

Gockston, Paisley, Renfrewshire. Town hospital.

Godman's Walk, a walk along the high terrace overlooking the r. Ness, in the town of Inverness.

Gogar, ancient par., now partly in Corstorphine, Midlothian, 5½ m. N.W. of Edinburgh ; ry. sta., L.N.E. Near the ham., is G. House, G. Burn, G. Mount, and G. Park.

Gogar Burn, flows for over 13 m. N.E. and joins the r. Almond, 1¾ m. N.W. of Gogar, Midlothian. G. House, now Mental Hospital.

Gogar House, seat, Stirlingshire, 3 m. S.E. of Bridge of Allan.

Gogo Water, trib. of the r. Clyde at Largs, Ayrshire.

Goil Loch, sea loch, Argyll, W. arm of Loch Long.

Goin Loch (or **Blackwoodhill Dam**), loch, 3½ m. S.W. of Eaglesham, Renfrewshire, on Ayrshire border.

Goldenacre, a district of Edinburgh, near Ferry Road.

Goldenberry Hill, Ayrshire, ½ m. N.E. of Farland Head ; alt. 244 ft.

Goldielands, 2 m. S.W. of Hawick, Roxburghshire, on r. Teviot; peel-house.

Goldielea, seat, Kirkcudbrightshire, 3 m. S.W. of Dumfries.

Golf Club House Halt, ry. sta., L.N.E., near Banff.

Gollachy Burn, stream, Banffshire, flowing 4 m. N. to the sea between Buckie and Port Gordon.

Gollanfield, ham., with ry. sta. (Gollanfield Junction), L.M.S., 1½ m. S.E. of Campbeltown, Inverness-shire. P.O., T.O. ; nearby is G. House.

Golspie, par. and vil., and Dunrobin Castle ; par. 19,690 ac. Pop. 1391. Vil. on G. Burn ; ry. sta., L.M.S. Hotel. 17¼ m. by rail from Helmsdale. Vil. pop. 883. There are numerous antiquities in the neighbourhood.

Golspie Tower, vil., 1 m. N. of Golspie, Sutherland. P.O., T.O.

Gometra, isl., Argyll, on W. side of Ulva Isl., used as a fishing sta. Pop. 37.

Gonachan, ham. at the mouth of Gonachan Burn (trib. of the Endrick Water), ½ m. E. of Fintry, Stirlingshire.

Gonar Burn, or **Strichen Water,** N. Aberdeenshire, headstream of N. Ugie, into which it falls 2¼ m. N.E. of New Pitsligo.

Gonfirth, vil., Mainland, Shetland, on Gon Firth, 3 m. S.W. of Voe.

Goodbush Hill, on border of Ayr and Lanark, 5 m. N. of Muirkirk ; alt. 1556 ft.

Goodie Water, stream, S. Perthshire, issues from Lake of Menteith and flows 8½ m. S.E. to Forth R ; 1½ m. N.W. of Gargunnock.

Goosey Loch, 3½ m. W. of Barrhill, S.E. Perthshire.

Gorbals, dist., parl. div., and ry. sta., L.M.S., Glasgow, on S. side of the r. Clyde. P.O., T.O. called Gorbals' Cross.

Gordon, par. and vil., with ry. sta., L.N.E., Berwickshire ; par. 9719 ac. Pop. 702 ; vil. (W. Gordon), on Eden Water, 4 m. S.W. of Greenlaw. P.O., T.O. Vil. pop. 286.

Gordon Arms Inn, Selkirkshire, on Yarrow Water, 3 m. N.E. of St. Mary's Loch ; visited by tourists and anglers.

Gordon Castle, seat of the Duke of Richmond and Gordon, on border of Banff and Morayshire, on the rt. bank of the Spey near Fochabers ; part dates (1449–1684).

Gordon Memorial or **Barthol Chapel,** q.s. par. Fyvie, Methlick, Meldrum, and Tarves, pars., Aberdeenshire.

Gordon Place and Dyce, vil., Dyce par., Aberdeenshire.

Gordon, Port, see PORT GORDON, Banff-shire.

Gordon's Mills, vil., Ross and Cromarty, on shore of Cromarty Firth, 2 m. S. of Invergordon.

Gordonstown, vil., Aberdeenshire, 3 m. W. of Fyvie. P.O.

Gordonstown, vil., 3 m. S.W. of Cornhill, Banffshire.

Gordonstown or **Gordonstone,** seat of the Gordon Cummings, 5½ m. N.W. of Elgin, Morayshire.

Gorebridge, large vil., with ry. sta., L.N.E., on Gore Water, 12 m. S.E. of Edinburgh, Midlothian ; important mining dist. P.O., T.O. Pop. 1687.

Gore Water, stream, Midlothian, formed by the junction of the Middleton North Burn and Middleton South Burn ; flows to S. Esk, which it joins 1 m. N.W. of Gorebridge.

Gorgie, dist. of Edinburgh and within the city, with ry. sta., L.N.E. P.O.

Gormach, seat, 3 m. N.W. of Blairgowrie, Perthshire.

Gorm Loch, 9 m. S.W. of Kirkhill, Inver-ness-shire.

Gorm Loch Beg, loch, 16 m. W. of Loch-inver, S.W. Sutherland.

Gorm Loch Mòr, loch, 5 m. E. of Loch Assynt, Sutherland.

Gortan, N.E. Argyll, 7 m. N.E. of the Bridge of Orchy sta., W. Highland ry., L.N.E. ; place for trains to pass on the single line ; alt. 1109 ft.

Gortencorn, pl., N.E. Ardnamurchan, Argyll, 4 m. N.W. of Salen Hotel.

Gortenfern, pl., N.E. Ardnamurchan, Argyll, 5½ m. N.W. of Salen Hotel.

Gortinane, seat, 16 m. S.W. of Tarbert, Kintyre, W. Argyll.

Gortleech or **Gorthlick,** ham., Inverness-shire, 15 m. N.E. of Fort Augustus. P.O., T.O.

Gosford House, seat of the Earl of Wemyss and March, East Lothian, 2 m. N. of Longniddry sta.

Goshen, vil., 4 m. W. of Grangemouth, Stirlingshire.

Gossabrough, coast ham., in S.E. of Yell Isl., Shetland.

Gothie Hill, 6½ m. N.N.E. of Fettercairn, Kincardineshire ; alt. 1468 ft.

Gouen, loch, W. Ross and Cromarty, near Gairloch Hotel.

Goukstane Burn, rt. trib. of Water of Ae, ½ m. S.S.W. of Kirkmichael, Dumfries-shire.

Gour, loch, 7 m. W. of Carsphairn, S.E. Ayrshire.

Gourdie, seat, 4½ m. N.E. of Murthly, Perthshire.

Gourdiehill, seat, 1½ m. E. of Errol sta., Perthshire.

Gourdon, seapt. and fishing vil., with harb., 1 m. S. of Bervie, Kincardineshire. Pop. 994. P.O., T.O. ; ry. sta., L.N.E. Exports grain. G. Hill, a landmark, is 436 ft. high.

Gourock, town (police bur.), in the par. of Inverkip, Renfrewshire. Pop. 8844. The older part of the town lies along the shore on the bay of the same name and extends for two miles. By water, the town is 1½ m. S. of Kilcreggan, 4⅜ m. S. of Helensburgh, 2½ m. E.N.E. of Cloch Lighthouse, 2¾ m. W.N.W. of Greenock, and 26 m. W.N.W. of Glas-gow. The bay affords good anchorage for yachts, and a hill (Barnhill, 490 ft.) rising at the back of the town commands a magnificent view of the surrounding district. P.O., T.O. Hotels. Tramway from Greenock and railway from Glasgow. Steamer communication with all places on the Clyde.

Gourock Burn, stream, Ayrshire, flowing to the Firth of Clyde at Ann's Lodge.

Govan, par. and formerly a mun. town, now part of, and a parl. div. of, Glasgow (q.v.) P.O., T.O. Par. pop. 364,780. At Shieldhall there are the extensive factories of the Scottish Co-operative Wholesale Society. Ferry communication with Partick.

Govanhill, dist. and ward in S.E. Glasgow. P.O., T.O.

Gowanbank, vil., Angus, 3½ m. N.W. of Arbroath. P.O.

Gowkhall, vil., Fife, 3 m. W. of Dunfermline.

Gown, loch, Ross and Cromarty, 1½ m. S.W. of Achnasheen sta.; alt. 522 ft., depth 52 ft.

Gowrie, ancient dist. of Perthshire between Stormount and the Tay.

Goyle, hill, 7 m. N. of Laurencekirk, Kincardineshire; alt. 1527 ft.

Grabhir, or **Gruver,** vil. in Lewis, Ross and Cromarty. See GRUVER.

Graden Burn, Berwickshire, rises 2 m. N.N.W. of Coldstream town, falls into the Tweed at Miln Graden.

Graden Place, farm, in Linton par., Roxburghshire, 4 m. W.N.W. of Yetholm; traces of ancient fortalice here.

Graemsay, isl. in Hoy Sound, Orkney; has P.O. and two lighthouses, one (108 ft. high) at the S. end and a low one at the N. end. Pop. 114.

Graemshall, seat and loch, Mainland of Orkney, on Holm Sound, 7 m. S.E. of Kirkwall.

Graham's Castle, Stirlingshire. See FINTRY.

Graham's Dyke, vil. near Bo'ness, E. Lothian. See ANTONINUS WALL.

Graham's Knowe, eminence, 5 m. S.E. of Alyth, Angus.

Grahamslaw or **Grahamslaw Haugh Head,** vil., Roxburghshire, 5 m. S. of Kelso, on Kale Water; there are a number of artificial caves which were used as hiding-places of the Covenanters.

Grahamston, town and q.s. par., with ry. sta., L.N.E. The town now forms part of the burgh of Falkirk, with a pop. of 36,565. P.O., T.O.

Grahamston, sub. of Barrhead, Renfrewshire. P.O., T.O.

Graitney, Dumfriesshire. See GRETNA.

Gramesall, pl., N. coast of Benbecula, Outer Hebrides.

Grampians, The. The name is generally applied to the range of mts. beginning at Dumbarton in the S.W. of Scotland, forming a high ridge to the N. and N.E. as far as Stonehaven on the E. of Scotland. They vary in altitude from 2000 ft. to 3984 ft. at Ben Lawers; 4241 ft at Braereach, and 4296 ft. at Ben Macdhui. To the early Celtic inhabitants they were known as Druim Ablin—*the ridge* or *backbone of Scotland*—and that form is still retained in Tyndrum (G. Tigh an Druim, *the house on the ridge*.)

Gramsdale Hotel, on N. coast of Benbecula, Outer Hebrides.

Grandholm, vil., with woollen works, in co. of, and 2 m. N.N.W. of, Aberdeen, on r. Don; close by is G. Cottage, a seat, and 2 m. N. of Auchmill is G. House, a seat.

Grandiscole, vil. on W. coast of Bressay isl., Shetland.

Grandtully, or **Grantully,** vil., with ry. sta., L.M.S., and hotel, Perthshire, on r. Tay, 2½ m. N.E. of Aberfeldy. G. Castle is a fine baronial mansion of the sixteenth century; see Scott's "Tully-Veolan" in *Waverley.*

Grandtully Distillery, 2½ m. N.E. of Aberfeldy, Perthshire, nearby sta.

Grange, par. and ry. sta., L.N.E., Banffshire; par. 15,093 ac. Pop. 1154; ry. sta. 4½ m. E. of Keith. P.O., T.O. at Grange sta.

Grange Distillery, 1 m. N.W. of Burntisland, Fife.

Grange Fell, hill, Dumfriesshire, 6½ m. E. of Lockerbie; alt. 1045 ft.

Grange Hall, seat, Kinloss par., 2 m. N.E. of Forres, Morayshire.

Grange House, seat, 3¼ m. N.E. of Maybole, Ayrshire.

Grange House, 1¼ m. E. of Bo'ness, W. Lothian.

Grange of Lindores, vil., Fife. See LINDORES.

Grangemouth, par., seapt. town, and police bur., on the S. of the Firth of Forth, E. Stirlingshire, where the r. Carron enters it. It is a busy shipping centre, doing business in timber and coal with all the Scandinavian countries. The town was founded in 1777 in connection with the building of the Forth and Clyde Canal. The docks have been extended several times and most ocean-going steamers can enter them. The L.M.S. and L.N.E. rys. send branches to the town. Bur. pop. 11,798. P.O., T.O. Par. 8046 ac.; pop. 20,682.

Grangemuir House, seat, Fife, 1¼ m. N.W. of Pittenweem.

Grangepans, part of the town of Bo'ness, W. Lothian; formerly a separate vil.

Grange Vale, seat, near Beith, N. Ayrshire.

Grant Castle, Morayshire. See CASTLE GRANT.

Grant Lodge, near Elgin, Morayshire.

Granton, Edinburgh, seapt. on the Firth of Forth, with commodious harbour. Principal esparto port in the British Isles; links with L.M.S. and L.N.E. Ry. ferry to Burntisland on opposite shore, 5 m. across. The harbour is largely used by trawlers. P.O. Here are the Edinburgh and Leith Gasworks.

Grantown-on-Spey, small mkt.-town and police bur., 12 m. N.E. of Aviemore; was founded in 1776. Pop. 1577; ry. sta., L.M.S. P.O., T.O. Hotels.

Grantshouse, ham., with ry. sta., L.N.E. (Grant's House), Berwickshire, on Eye Water, 16 m. N.W. of Berwick-upon-Tweed. P.O., T.O. Pop. 98.

Grantully, Perthshire. See GRANDTULLY.

Gravir, Grabhir, or **Gruver,** vil., Lewis Isl., Outer Hebrides, 11 m. S. of Stornoway. P.O., T.O.

Gray Cairn, The, at Balnabroich, 3 m. S. of Kirkmichael, N.E. Perthshire.

Gray House, seat (1715), near S.W. border of Angus, 5 m. W. of Dundee.

Graystone, vil., Angus, 7½ m. N.W. of Arbroath; P.O. called Greystone.

Grealin, ham., Skye, Inverness-shire, 12 m. N. of Portree.

Great Cave, on W. coast of Gigha Isl., Argyll.

Great Cumbrae Island, in the Firth of Clyde. See CUMBRAE, Great and Little.

Great Door, The, strait, Argyll. See DORUSMORE.

Great Glen, see GLEANN MOR NA-H ALBIN, *the great glen of Scotland,* generally referred to as Glen More.

Great Hill, Dumfriesshire, 2½ m. E. of Closeburn; alt. 1150 ft.

Great Western Road, ry. sta., L.N.E., Kelvinside dist., in W. of city of Glasgow.

Greenburn, pl., with fairs, Aberdeenshire, 1¼ m. N.W. of Auchmill.

Greenburn, part of the town of Fauldhouse, q.v.

Greenend, ham. originally, but now part of the city of Edinburgh. P.O., T.O.

Greenend, ham., Lanarkshire, 1½ m. S.E. of Coatbridge.

Greenfield, mining vil., Lanarkshire, 4 m. W. of Motherwell.

Greenford, vil., Angus, 8 m. N.E. of Dundee.

Greengairs, mining vil., New Monkland par., 4 m. N.E. of Airdrie. P.O., T.O.

Green Hill, 5 m. W. of Leuchars, Fife; alt. 608 ft.

Green Hill, 6 m. S.E. of Linton, Roxburghshire; alt. 1244 ft.

Greenhill, vil., Dumfriesshire, 3½ m. S.W. of Lockerbie.

Greenhill, seat, Hownam par., E. Roxburghshire.

Greenhill, ry. sta., L.M.S., W. of Falkirk, 15 m. N.E. of Glasgow. P.O., T.O.

Greenhill Dod, mt., N. Dumfriesshire, 4½ m. N.N.W. of Moffat.

Green Holm, isl., Shetland, 4 m. N.E. of Lerwick.

Green Holm, isl., Shetland, ½ m. E. of Fethaland Pt.

Green Holm, Muckle and **Little,** two isls., Orkney, 1½ m. S.W. of Eday Isl.

Greenland, vil., Caithness, 3½ m. E. of Castletown. P.O.

Greenland, vil. in S. of Walls par., Shetland, on Vaila Sound.

Greenlaw, par. and vil., with ry. sta., L.N.E., Berwickshire, on Blackadder Water, 7½ m. S.W. of Duns; par. 12,159 ac. Pop. 889. P.O., T.O. Vil. pop. 528.

Greenlea, ham. in co. of, and 4½ m. E. of, Dumfries.

Greenloaning, vil., with ry. sta., L.M.S., Perthshire, 5½ m. N.E. of Dunblane. P.O., T.O.

Greenmill, vil. in co. of, and 5¼ m. S.E. of, Dumfries.

Greenock, par., parl. and mun. bur., seapt., and town, Renfrewshire, on S. side of r. Clyde, 22½ m. N.W. of Glasgow. Has a spacious harb. and extensive docks. The Prince's Pier is the starting-point for the Clyde tourist steamers. The chief seat of the sugar-refining industry, and has also extensive ship-building yards, engineering works, distilling, and oil refining. Birthplace of James Watt, inventor of the steam-engine. Stas., L.M.S. (Prince's Pier, Lynedoch, Central, West, and Upper). Par. 5866 ac. Pop. 77,928 ; bur. pop. 78,948. Hotels. Regular ferry service with Helensburgh. Returns one member to Parliament.

Greenock Water, trib. of the r. Ayr, E. Ayrshire.

Greenside, small dist. of Edinburgh, beside the Calton Hill.

Greenside, seat, 4 m. N.W. of Largo, Fife.

Greenyards, vil., 1 m. E. of Bannockburn, Stirlingshire.

Greenyards, pl. in Strath Carron, 6 m. W. of Bonar Bridge, Ross and Cromarty.

Greeshop House, seat, 1¼ m. from Forres, Morayshire.

Greeto Water, trib. of Gogo Water, 3 m. W. of Misty Law, Ayrshire.

Greg Ness, headland on S. side of Nigg Bay, N.E. Kincardineshire.

Greinord Bay, Ross and Cromarty. See GRUINARD BAY.

Gremista, seat, 2 m. N.W. of Lerwick, Shetland.

Greshornish Loch, sea loch, Skye, Ross and Cromarty, at the head of Loch Snizort. On W. shore is G. House, 3 m. N.W. of Edinbane.

Greskine State Forest, nearest ry. sta., Beattock, Dumfriesshire.

Gress House, seat, near mouth of r. Gress, Lewis Isl., Ross and Cromarty, 9 m. N.E. of Stornoway. G. River flows 7 m. to Broad Bay.

Gretna, vil. and par., Dumfriesshire, on the border of the Solway Firth ; ry. sta., L.M.S. P.O., T.O. Par. 8510 ac. Pop. 2857. Gretna was long, and is still, famous for its runaway marriages. Vil. pop. 1762. Hotel.

Gret Ness, headland, S.E. Mainland of Shetland Isles.

Greyfriars, Old and **New.** These two churches within the City of Edinburgh were under one roof, but divided into two separate churches, until quite recently ; the two are now joined into one church.

Greyhope Bay, N.E. Kincardineshire, between Girdle Ness and mouth of r. Dee.

Grey Mare's Tail, a famous waterfall (about 200 ft.), being the stream issuing from Loch Skeen just before it enters Moffat Water, 10 m. N.E. of Moffat.

Grey Mare's Tail, waterfall (100 ft. leap), Crichope Burn, 1½ m. N. of Crichope Linn, Dumfriesshire.

Greystone, vil., 7½ m. N.W. of Arbroath, Angus. P.O. See GRAYSTONE.

Griam Loch, see LOCH A' GHRIAMA.

Gribdae, seat in co. of, and 4 m. from, Kirkcudbright.

Gribton, seat, on Cairn Water, in co. of, and 5 m. W. of, Dumfries.

Gribun, peninsula and ham., Mull Isl., Argyll, 8 m. S.W. of Gruline. P.O.

Grif, or **Grief Skerry,** 3½ m. E. of Whalsay Isl., Shetland.

Grimbister, Holm of, in Bay of Firth, 1½ m. E. of Finstown, Orkney.

Grimersta, one of the best salmon rivers in Scotland, in the W. of Lewis, near Garrynahine. Hotel. G. Lodge is 17 m. W. of Stornoway.

Grimes Dyke, see ANTONINUS WALL.

Griminish Point, N.W. extremity of N. Uist Isl.

Grimness, ham. in N.E. of S. Ronaldsay, Orkney, near Grim Ness.

Grimsay, isl., between N. Uist and Benbecula, Outer Hebrides, Inverness-shire. Pop. 259. P.O., T.O.

Grimsay, isl., ¾ m. N. of S. Uist Isl., Inverness-shire. Pop. 19.

Grimsetter House, seat, on E. side of Bressay Isl., and 2¾ m. S.E. of Lerwick, Shetland Isls.

Grimshadar Loch, sea loch on E. side of Lewis, 4½ m. S. of Stornoway, Outer Hebrides.

Grista, loch, 4½ m. N.W. of Lerwick, Shetland Isls.

Groay, isl. in Sound of Harris, Outer Hebrides.

Groban, loch, 5 m. W. of Sligachan Hotel, Isle of Skye.

Grobesness, vil., 3 m. from Voe, Shetland.

Grogarry, ham., 3 m. N.N.E. of Howmore, S. Uist.

Grogport, vil. on E. coast of Kintyre, Argyll, 5 m. N. of Carradale.

Grougar, ham., near Kilmarnock, Ayrshire.

Groundwater, loch, 5 m. S.W. of Kirkwall, Orkney.

Grove, The, seat, on E. border of Kirkcudbrightshire, 3½ m. N.W. of Dumfries.

Grubbit Law, hill, 1½ m. S.E. of Morebattle, Roxburghshire ; alt. 1071 ft.

Grudie, r., Ross and Cromarty, flowing from Ben Eay to Loch Maree. The Bridge of Grudie is 5½ m. N.W. of Kinlochewe.

Grudie, a r. entering the Kyle of Durness, Sutherland, at the N.W. end of the Kyle.

Gruinard Bay, N.W. Ross and Cromarty, between Loch Ewe and Little Loch Broom. Contains G. Isl. Receives G. River flowing from Loch na Sheallag, and Little G. River flowing from Fionn Loch. At mouth of G. River, 15 m. N.W. of Poolewe, is G. House. Shooting (G. and Monkcastle) extends to 16,000 ac.

Gruinards, seat, Strathcarron, N.W. Ross and Cromarty, 6 m. W. of Ardgay.

Gruinart or **Gruinnard,** ham. and sea loch in N. W. of Islay, Argyll ; ham., 6 m. N.W. of Bridgend. P.O., T.O. Sea loch good anchorage for small vessels.

Gruinlochan, loch, 3½ m. S.W. of Croggan, isl. of Mull, Argyll.

Grula, ham., E. Skye, head of Loch Eynort.

Gruline, ham. and seat, 3 m. S.W. of Salen, isl. of Mull, Argyll. P.O., T.O.

Grumach Hill, N.W. Aberdeenshire, 5 m. N.W. of Rhynie ; alt. 1724 ft.

Grunay Skerries, isls. in the outer Skerries group, Nesting par., Shetland. Pop. 6.

Gruting, ham., Sandsting par., Mainland, Shetland, on Gruting Voe.

Grutness Voe, small sea loch in S. of Mainland of Shetland, near Sumburgh Head.

Gruver, or **Grabhir,** vil., Lewis, 10 m. S. of Stornoway, Ross and Cromarty. P.O., T.O. called Gravir.

Gualann, mt., Stirlingshire, 3½ m. N.E. of Balmaha pier ; alt 1514 ft.

Gualinn House, originally an inn and rest for travellers, now a shooting - lodge, on the watershed between Rhiconich

and Durness. 5 m. N. of the former. the road crosses the Cal at an alt. of 596 ft.

Guardbridge, vil., with ry. sta., L.N.E. (Guard Bridge), Fife, at the mouth of the r. Eden, 4 m. N.W. of St. Andrews. P.O., T.O. Pop. 977 ; has paper mills.

Guay, ry. sta., L.N.E., 6 m. N.W. of Dunkeld, Perthshire.

Guildtown, vil. in co., and 5½ m. N. of Perth. P.O., T.O. Pop. of G., Balbeggie, and Wolfhill, 425.

Guildy, vil., Angus, 5 m. N.W. of Carnoustie.

Guinach Loch, 1¼ m. N.W. of Kingussie, Inverness-shire.

Guirshadir, sub. of Stornoway, Lewis, Ross and Cromarty.

Guisachan, seat, Inverness-shire, at the head of r. Glass and 22 m. S.W. of Beauly.

Guisachan and Glasha Deer Forest, N.W. Inverness-shire, 15,580 ac. Post town, Beauly.

Gulberwick, vil., 3 m. S. of Lerwick, Shetland.

Gulbin, r., Inverness-shire, issues from Loch Gulbin, and flows to r. Spean, near foot of Loch Laggan.

Gullane, favourite watering - place and golfing centre on the Firth of Forth, 19½ m. E.N.E. of Edinburgh, E. Lothian ; ry. sta., L.N.E. (passenger service withdrawn). P.O., T.O. Hotels. Pop. 1414.

Gull Rocks, Argyll. See DUN-NA-FEULAN.

Gulvain Deer Forest, Inverness-shire and Argyll, 6 m. N.W. of Fort William.

Gunna Island, in G. Sound, W. Argyll, 4 m. S.W. of Ardnish.

Gunna Sound, sound between the isls. of Coll and Tiree, Inner Hebrides, Argyll.

Gunnister Island, at mouth of G. Voe, Mainland, Shetland.

Gunsgreen House, seat, E. Berwickshire, 1 m. E. of Eyemouth.

Gutcher, ham., on E. coast of Yell, Shetland. P.O., T.O.

Guthrie, par. and ham., with ry. sta. (G. Junction), L.M.S., Angus, on Lunan Water, 7 m. E. of Forfar ; 2810 ac. Pop. 225. P.O., T.O. ; 1 m. from sta. is G. Castle, built 1468, enlarged 1848.

Guynd, The, seat, Angus, on Elliot Water, 5 m. W. of Arbroath ; ancient camp in the Den of G.

Gyle Loch, Mull Isl. See LOCH-NA-KEAL.

Gylen Castle, ruined stronghold of the Macdougalls of Lorne, on S. coast of Kerrera Isl., Argyll, 4 m. S.W. of Oban.

Gyratsmyre, Hill of, ridge, 1 m. long and 1 m. S. of Drumlithie, Kincardineshire.

H

Haaf Grunie, isl., Shetland, 1 m. S.E. of Unst Isl.

Habbie's Howe, scene of Allan Ramsay's " Gentle Shepherd," Midlothian, 4½ m. S.W. of Penicuik. The name of a waterfall on Glencorse Burn, in the Pentland Hills.

Habchester, hill, with vestiges of ancient camp, Berwickshire, 1½ m. S.E. of Ayton ; alt. 712 ft.

Habrahelia, great cavern, resembling an amphitheatre, Orkney, N. end of Papa Westray Isl.

Haco's Ness, headland, Orkney, S.E. end of Shapinsay, 6 m. N.E. of Kirkwall.

Hadden Rig, hill ridge, Roxburghshire ; alt. 541 ft.

Haddington, royal and mun. bur., par., and co. town, East Lothian, on r. Tyne ; ry. sta., L.N.E., 18½ m. E. of Edinburgh. Par. 12,254 ac. Pop. 5682 ; bur. and town pop. 4405. P.O., T.O. Hotel. Has grain mkt. and corn mills, and hosiery and brewing industries. Its abbey church, the " Lamp of Lothian," dates from the twelfth or thirteenth century. Alexander II. was born here in 1198, and John Knox is said to have been born nearby.

Haddingtonshire, see EAST LOTHIAN.

Haddo House, seat of the Marquis of Aberdeen and Temair, Aberdeenshire, 6½ m. N.W. of Ellon.

Haddo, House of, seat, Aberdeenshire, 8 m. N.E. of Huntly.

Hadyard Hill, Ayrshire, 6½ m. S.W. of Maybole ; alt. 1059 ft. Remains of ancient camp and fort.

Haer, dist., S.E. Perthshire, 2 m. S. of Blairgowrie.

Hafton House, seat, near Hunter's Quay, on Holy Loch, Argyll.

Haggie Knowe, hill, on S. border of Roxburghshire, 1 m. N.E. of Peel Fell ; alt. 1611 ft.

II

Haggs, q.s. par. and vil., Stirlingshire. Vil., Hollandbush and H., 5 m. N.E. of Kilsyth.

Haggs Castle, ruined ancient baronial fortalice of the Maxwells of Pollock, Renfrewshire, in S.W. of Glasgow.

Haghill, seat, E. of Glasgow, ¾ m. N.W. of Parkhead. P.O.

Hagshaw Hill, Lanarkshire, 3 m. W. of Douglas ; alt. 1540 ft.

Hagsthorn, vil., N. Ayrshire, 2 m. N.E. of Dalry.

Ha' Hill, mound (about 50 ft. high), 5 m. S. of Banff, Alvah par.

Ha' Hill, mound, Lanarkshire, 14 m. W. of Dunsyre.

Hailes Castle, ruin, E. Lothian, 4 m. N.E. of Haddington. Temporary residence of Mary, Queen of Scots, after her abduction by Bothwell.

Hailes, New, see NEW HAILES, Midlothian.

Hailie, seat, Ayrshire, Largs.

Haining Castle, Stirlingshire, 4¾ m. S.E. of Falkirk. Original name of Almond Castle.

Haining, The, seat, adjacent to Selkirk.

Hairmyres, ry. sta., L.M.S., Lanarkshire, 1¾ m. W. of East Kilbride, 9¾ m. S. of Glasgow ; alt. 501 ft. Colony for consumptives.

Halbeath, mining vil., with ry. sta., L.N.E., Fife, 2½ m. N.E. of Dunfermline. P.O. Pop. 410.

Halberry Head, prom., Caithness, 4 m. N.E. of Lybster. H. Castle, a ruin.

Halcro Head, headland, Orkney, S.E. South Ronaldsay.

Haldane's Mill, vil., Dunbartonshire, ½ m. N. of Balloch.

Halfdavoch, ham., Morayshire, 5 m. S. of Forres.

Halfmerk Hill, on border of Dumfriesshire and Ayrshire, 3 m. N.W. of Kirkconnel ; alt. 1478 ft.

Halfmorton, par., Dumfriesshire, 6081 ac. Pop. 274. ; 6½ m. S.W. of Langholm.

Halfway, mining vil., Lanarkshire, 1 m. S.E. of Cambuslang.

Halfway House, drainage area, Renfrewshire, Paisley par. P.O., T.O.

Halin in Waternish, q.s. par., Isle of Skye, Duirinish par., Inverness-shire.

Halkirk, par. and vil., with ry. sta., L.M.S., Caithness ; par. 102,616 ac.

Pop. 1725. Vil., 8½ m. S. of Thurso. P.O., T.O., and P.O. at H. Roadside. Vil. pop. 363.

Halkshill, seat, Ayrshire, Largs.

Hall, seat, Roxburghshire, Yetholm.

Halladale, r., Sutherland, flows 22 m. N. to the sea at Portskerry.

Hallbar Tower, ancient fortalice of eleventh century, Lanarkshire, Carluke par.

Hall Bay, Kincardineshire, 3 m. N.E. of Stonehaven.

Halleaths, seat, Dumfriesshire, 1½ m. E. of Lochmaben. H. Loch in vicinity.

Hallforest Castle, imposing ruined castle of the Earls Marischal, Aberdeenshire, 1¼ m. S.W. of Kintore.

Hall Gill, ravine, Lanarkshire, 7 m. S.W. of Blackhall.

Hallgreen, pl., with vestiges of ancient priory, Dumfriesshire, 7 m. S.E. of Langholm.

Hallgreen Castle, (1376), strong old castle, on eminence near the sea, Kincardineshire, Bervie par.

Hallguards, site of the original Hoddam Castle, Dumfriesshire, 2 m. S.W. of Ecclefechan.

Hallhead, old mansion (1688), Aberdeenshire, 5½ m. N.W. of Lumphanan.

Hallhill, seat, Lanarkshire, 2 m. N.E. of Strathaven.

Hallhill, Wester, colliery, near Baillieston, Lanarkshire, 6½ m. E. of Glasgow.

Halliday Hill, Dumfriesshire, 6 m. S.W. of Parkgate ; alt. 600 ft.

Halligarth, seat, Unst Isl., Shetland, near Balta Sound.

Hallival, mt., in Isl. of Rhum, Inverness-shire, 2365 ft.

Hallmanor, seat, 4½ m. S.W. of Peebles.

Hallpark, Clackmannanshire, see KEILARS-BRAE.

Hallrule, seat, Roxburghshire, 8 m. E. of Hawick.

Hallside, q.s. par., vil., and seat, Lanarkshire, 2 m. E. of Cambuslang. Steelworks here. P.O., T.O.

Halltree, ancient chapelry, Midlothian, on Gala Water, 5½ m. N.W. of Stow.

Hallyards, ruined tower, Midlothian, 1 m. N. of Ratho.

Hallyards, ruined old mansion, Fife, 2 m. S.E. of Lochgelly.

Hallyards, seat in co. of, and 3 m. S.W. of, Peebles.

Hallyburton House, Angus, 2½ m. S.E. of Coupar-Angus.

Hallydown, seat, Berwickshire, 1 m. N.W. of Eyemouth.

Hallyness, headland, Mainland, Shetland, 8 m. S. of Lerwick.

Halmyre House, seat, N. Peeblesshire, 3 m. S.W. of Lamancha ry. sta.

Halsary, dist., Caithness, 9 m. S.E. of Halkirk.

Halton Burn Reservoir, Fife, 2½ m. N.E. of Leven.

Halyhill, Perthshire, adjacent to Fortevoit vil., site of ancient capital of Pictavia.

Ham, harbour, Caithness, Dunnet par., 6 m. N.E. of Castleton.

Ham, vil., Shetland, Bressay, 1½ m. S.E. of Lerwick.

Ham, vil. and voe, Shetland, E. coast of Foula Isl.

Hamar, ham., Shetland, Northmarine par. P.O.

Hamar, headland, Shetland, S.E. Bressay Isl.

Hamars Ness, Shetland, N.W. Fetlar Isl.

Hamar Voe, sea loch, Shetland, Northmavine par. P.O. at Hamar.

Hametoun, vil., Shetland, Foula Isl.

Hamilton, mun. bur., mkt.-town, and par., with ry. stas. L.N.E. (Hamilton and Peacock Cross), L.M.S. (Central and West), Lanarkshire, near the junc. of the Avon and the Clyde, 10¾ m. S.E. of Glasgow ; par. 13,107 ac. Pop. 44,224 ; bur. pop. 37,863. Hotel. P.O., T.O. The centre of a very rich coal and iron field. H. Palace, until recently the seat of the Dukes of Hamilton, is now dismantled ; town connected with Glasgow by electric tram.

Hamilton Division, parl. division of Lanarkshire. Pop. 65,136.

Hamilton Farm Colliery, Lanarkshire, 1½ m. from Cambuslang.

Hamilton Palace Colliery Rows, vil., Lanarkshire, 1 m. S. of Bothwell.

Hamilton Road, lighting dist., Lanarkshire, Dalziel par. Pop. 594.

Hamna Voe, ham. and sea loch, N.W. Mainland, Shetland, 2 m. S.W. of Lunna Ness.

Hamnavoe, ham. and bay, Shetland, N.W. of West Barra. P.O., T.O.

Hamnaway, or **Thamanabhaidh,** sea loch, W. coast of Lewis Isl., 3 m. N. of Loch Reasort.

Hamrigarth, ham., Mainland, Shetland, 14 m. S.E. of Walls.

Handa Island, W. Sutherland, 2½ m. N.W. of Scourie.

Handwick, Nether, pl., Angus, 3½ m. S.W. of Glamis.

Hangingshaw, pl., S.E. Midlothian, 1 m. S.E. Heriot sta.

Hangingshaw, The, seat and ruined castle, 6 m. N.W. of Selkirk, on r. Yarrow.

Hanley, seat, Midlothian, Corstorphine par.

Harburn, ry. sta., L.M.S., Midlothian, 15 m. S.W. of Edinburgh. H. House here.

Harden, old border tower, and seat of Lord Polwarth, Roxburghshire, 4 m. W. of Hawick. Has interesting historical associations and relics of feudal art.

Hardengreen, pl., with goods sta., L.N.E., Midlothian, 1 m. from Dalkeith.

Hardgate, Dunbartonshire, part of Faifley.

Hardgate, vil., Kirkcudbrightshire, 4 m. N.W. of Dalbeattie.

Hardington House, Lanarkshire, on r. Clyde, near Lamington.

Hardiston, seat, Kinross-shire, 2¼ m. S.W. of Cleish.

Hardler Knowe, hill, Roxburghshire; alt. 1060 ft.

Hardmuir Wood, Morayshire, near Brodie sta., Shakespeare's scene of Macbeth's meeting with the weird sisters.

Harefaulds, pl., Angus, 5 m. S.E. of Forfar. Traces of Roman camp.

Hare Hill, Lanarkshire, 4½ m. N.E. of Carluke; alt. 1029 ft.

Hare Hill, a summit of the Pentland Hills, Midlothian, 3 m. S.S.E. of Balerno; alt. 1470 ft.

Harelaw, ancient border tower, Dumfries-shire, 2¼ m. N.E. of Canonbie.

Hare Law, mt., Peeblesshire, 3 m. S.E. of Walkerburn; alt. 1670 ft.

Harelaw Dam, reservoir, Renfrewshire, 2¼ m. S. of Neilston.

Hare Myre, loch and moor, Perthshire, 5 m. S. of Blairgowrie.

Hare Ness, headland, N.E. Kincardine-shire, 1 m. S. of Cove.

Hare's Den, ravine, Kincardineshire, 3½ m. W. of Kinneff.

Hareshaw, vil., Lanarkshire, 1 m. S.E. of Newhouse ry. sta.

Hareshawmuir, seat, Ayrshire, 5 m. N.E. of Kilmarnock. H. Water flows to Craufurdland Water.

Harewood Glen, seat, 3½ m. W. of Selkirk.

Harland, Hill of (201 ft.), Caithness, 3 m. N.W. of Wick.

Harlash, isl., in Loch Bracadale, Isle of Skye, Inverness-shire.

Harlaw, battlefield of 1411, in Chapel of Garioch par., Aberdeenshire.

Harlaw Reservoir, Midlothian, 1½ m. S.E. of Balerno, depth 54 ft.; Edinburgh water supply.

Harleburn Head, mt., S. Lanarkshire, 4 m. S.E. of Crawford; alt. 1776 ft.

Harleswynd, kennels of Fife Foxhounds, 3½ m. S.E. of Cupar.

Harlosh, ham., Isle of Skye, 4½ m. S.E. of Dunvegan. P.O., T.O.

Harmeny, seat, Midlothian, Balerno.

Harmetray, isl., in Sound of Harris, Outer Hebrides

Harold's Tower, Caithness, 1½ m. N.E. of Thurso. Stands on the grave of Earl Harold, who was slain in battle in 1190.

Haroldswick, vil. and bay, Shetland, E. side of Unst. P.O., T.O.

Harperfield, seat, 4 m. S. of Lanark, adjacent to confluence of the Douglas and the Clyde.

Harper Rigg Reservoir, Midlothian, 6 m. S.W. of Balerno; maximum depth 30 ft. Edinburgh water supply.

Harpleas Reservoir, Fifeshire, 3½ m. N.W. of Leslie.

Harport, sea loch, Isle of Skye; 6 m. long, deflecting south-eastwards from Loch Bracadale. H. House, near Carbost.

Harrabol, ham., Isle of Skye, 1½ m. E. of Broadford.

Harray, q.s. par., Birsay and H. par., Orkney. Harray vil., 4 m. N.W. of Finstown. P.O., T.O.

Harray, Loch of, Mainland, Orkney, 7½ m. W. of Kirkwall.

Harrietfield, vil., S.E. Perthshire, 3½ m. N.E. of Methven.

Harris, par., and S. part of Lewis Isl., Outer Hebrides, Inverness-shire; par. 123,752 ac. Pop. 4467. P.O., T.O. In addition to the S. part of Lewis, the par. comprises the isls. of Bernera, Ensay, Killigray, Pabbay, St. Kilda, Scalpa, Scarp, Scotasay, Tarrensay and Soay. Celebrated Harris tweeds manufactured here.

Harris, Sound of, between Lewis and N. Uist, Outer Hebrides, 8 m. in length and 8 to 12 m. in breadth. Forms the only navigable channel for large vessels through the Outer Hebrides.

Harrow, loch, Kirkcudbrightshire, Kells par. ; depth 29 ft., alt. 811 ft.

Harrow, ham., Caithness, 5 m. E.N.E. of Dunnet.

Harryburn Brae, mt., S. Lanarkshire, 3 m. S. of Abington ; alt. 1629 ft.

Harta Corrie, in Cuillin Hills, Isle of Skye, near Sgurr nan Gillean.

Hart Fell, mt., Dumfriesshire, 5¾ m. N.N.E of Moffat ; alt. 2651 ft.

Hart Fell, hill, Dumfriesshire, 7 m. N.E. of Lockerbie ; alt. 1085 ft.

Hartfell Rig, mt., Peeblesshire, 1 m. E. of Hart Fell ; alt. 2422 ft.

Hartfell Spa, mineral spring, Dumfriesshire, 5 m. N.E. of Moffat.

Hartfield, shooting - lodge, Ross - shire, 2 m. N. of Applecross.

Hartfield House, seat, Dunbartonshire, Cove.

Harthill, ruined castle, Aberdeenshire, Oyne par.

Hart Hill, Morayshire, 5 m. S. of Elgin ; alt. 902 ft.

Hart Hill, Stirlingshire, 4 m. S. of Gargunnock ; alt. 1428 ft.

Harthill, mining vil., Lanarkshire and W. Lothian, 2¼ m. W. of Whitburn, and 1½ m. S. of Westercraigs sta., L.N.E. Pop. 4018.

Harthill and Benhar, q.s. par., Lanarkshire, Shotts par. P.O., T.O.

Hart Law, Berwickshire, 7 m. N.N.E. of Lauder ; alt. 1578 ft.

Hartree House, Peeblesshire, 1 m. S.E. of Biggar. Boys' private school.

Hartrigge House, seat, Roxburghshire, near Jedburgh.

Hartsgarth Fell, Dumfriesshire, 8 m. N.E. of Langholm ; alt. 1806 ft.

Hartshorn Pike, mt., S.E. Roxburghshire, alt. 1789 ft ; near Peel Fell.

Hartside Burn, trib. of r. Clyde, Lanarkshire, Lamington.

Harts Leap, hill pass, Selkirkshire, 2¼ m. S.E. of St. Mary's Loch.

Hartwood, seat, Midlothian, 1½ m. S. of West Calder.

Hartwood, pl. and ry. sta., L.M.S., N. Lanarkshire, 2 m. S.W. of Shotts. P.O., T.O. Lanark county asylum in vicinity.

Hartwoodhill, seat, Lanarkshire, 1 m. N.E. of Hartwood ry. sta. H. Colliery in vicinity.

Harvieston, seat, Midlothian, 1 m. S.E. of Gorebridge. Remains of Catcune Castle.

Harviestoun Castle, Clackmannanshire, 2 m. W. of Dollar.

Harwood, burn, Midlothian, flows 8 m. to the Calder, 1 m. W. of Bellsquarry.

Harwood, seat, Roxburghshire, 5½ m. S.E. of Hawick.

Hascosay, isl., Shetland, between Fetlar and Yell.

Hassendean, stream, dell, ry. sta., L.N.E. and ancient par., Roxburghshire, 4½ m. N.E. of Hawick. H. Burn falls into the Teviot.

Hassockrigg Colliery, Lanarkshire, 2 m. S.W. of Harthill.

Hatchednize, pl., Berwickshire, 3 m. N.W. Coldstream.

Hatson House, Orkney, 1 m. N.W. of Kirkwall.

Hatton, vil. and ry. sta., L.N.E. (passenger service withdrawn), Aberdeenshire, 8¼ m. N.E. of Ellon. P.O., T.O. Pop. 327.

Hattonburn House, Kinross-shire, 1 m. N.E. of Milnathort.

Hatton Castle, seat, N. Aberdeenshire, 3 m. S.E. of Turriff. Remains of ancient castle of Balquholly.

Hatton Castle (1575), ruined fortalice, Angus, vicinity of Newtyle.

Hatton House, Midlothian, 1½ m. S.W. of Ratho.

Hatton House, Kincardineshire, 3¼ m. S.W. of Laurencekirk.

Hatton Law, vil., Fife, 1½ m. N.W. of Largo.

Hatton, Lower and **Upper,** adjacent hams., Perthshire, 1½ m. N. of Dunkeld.

Hatton of Fintray, vil., Aberdeenshire. See FINTRAY.

Hatton of Rattray, pl., Perthshire, 1½ m. N.E. of Blairgowrie.

Haugh, vil., Ayrshire, 1½ m. S. of Mauchline.

Haugh and **H. Bridge,** pl., Glasgow, 2 m. W. of Pollokshaws.

Haugh Burn, stream, W. Lothian, flowing 7 m. into Firth of Forth at Abercorn.

Haugh of Urr, vil., Kirkcudbrightshire, 4 m. N.W. of Dalbeattie. P.O., T.O. Pop. 161.

Haughs Bay, Kincardineshire, 2 m. N.E. of Johnshaven.

Haughs of Cromdale, battlefield of 1690, Morayshire, 3½ m. E. of Grantown-on-Spey.

Haughton, seat, Aberdeenshire, 1 m. N.E. of Alford sta.

Haulkerton, estate, Kincardineshire, Laurencekirk par.

Haultin, stream, Isle of Skye, flowing W. to Loch Eyre.

Hauster, or **Thrumster Burn,** stream, Caithness, flows 8 m. N.E. to Wick Water.

Havra, Little and **South,** isls., Shetland, W. coast of Mainland.

Hawes Brae, steep descent into S. Queensferry, West Lothian ; H. Pier for ferry to N. Queensferry. A fixed white light, seen 6 m.

Hawick, mun. bur., town, and par., with ry. sta., L.N.E., in S.W. Roxburghshire, on the Teviot, 53¾ m. S.E. of Edinburgh, and 45½ m. N.E. of Carlisle ; par. 14,872 ac. Pop. 18,214 ; bur. pop. 17,059. P.O., T.O. Hotels. The town figured often and strongly in the events of Border warfare and forays, and was several times burnt by English armies. The town is noted for hosiery, tweeds, and woollen mfrs. Branxholme Castle, in vicinity, was the scene of Scott's " Lay of the Last Minstrel."

Hawkcraig, cliff, Fife, E. side of Aberdour vil.

Hawkhead, pl., seat, and ry. sta., L.M.S., E. Renfrewshire, 2 m. S.E. of Paisley. H. Mental Hospital is here.

Hawkhill, dist., W. Dundee, Angus. P.O., T.O.

Hawkhill, pl. and racecourse, ½ m. E. of Ayr. P.O.

Hawkiesfield, pl., Fife, 1 m. N. of Dunfermline.

Hawkshaw, vestiges of ancient castle, Peeblesshire ; also stream, affl. of the Tweed, which it joins 2 m. S.W. of Tweedsmuir.

Hawkwood Hill, Lanarkshire, 4 m. S.W. of Strathaven ; alt. 1251 ft.

Hawthornbrae Convalescent Home, Duddingston, S.E. Edinburgh.

Hawthornden, romantic seat and ry. sta., L.N.E., Midlothian, on a cliff overhanging the N. Esk. Picturesque grounds, including remains of very ancient baronial fortalice, two tiers of remarkable caves, and a striking reach of the N. Esk's ravine, 8 m. from Edinburgh. T.O. at sta. Was the property of the Drummonds of Hawthornden.

Haxalgate, ancient road up the Eildon Hills, Melrose, Roxburghshire.

Hayfield House, seat, Argyll, N.W. side of Loch Awe, 7 m. S.E. of Taynuilt.

Haymarket, dist. and ry. sta., L.N.E., in W. Edinburgh.

Hayston Hill, with ancient circular moat, Angus, Glamis par. ; alt. 900 ft.

Haystoun, seat, 2 m. S.E. of Peebles.

Hayton, vil., Perthshire, 2 m. from Glencarse.

Haywood, q.s. par. and mining vil., with ry. sta., L.M.S., Lanarkshire. Carnwath par. P.O. ; T.O. at sta. Vil. pop. 223.

Hazelbank, vil., 3½ m. N.W. of Lanark, on r. Clyde. P.O. Pop of H. and Crossford, 618.

Hazelburn Distillery, Campbeltown, Kintyre, Argyll.

Hazeldean House, Renfrewshire, 4½ m. S.W. of Busby.

Hazlehead, seat, 3 m. W. of Aberdeen.

Heading Hill, Aberdeenshire, 4 m. N.W. of Peterhead.

Headlesscross, dist., Lanarkshire, 2½ m. E.S.E. of Shotts.

Heads, ham., Lanarkshire, ½ m. N.W.of West Quarter, Glassford par.

Headshaw Loch, Roxburghshire, 4½ m. S.W. of Lilliesleaf.

Heads of Ayr, ry. sta., L.M.S. (passenger service withdrawn), 6¼ m. S.W. of Ayr.

Healaval More and **Healaval Beg,** adjacent summits, Isle of Skye, 3 m. and 4 m. S.W. of Dunvegan ; alts. 1538 ft. and 1601 ft. Also called Macleod's Tables, N. and S.

Heardmanston, a seat of Lord Sinclair, Haddington, E. Lothian.

Hearthstane Burn, affl. of r. Tweed, Peeblesshire, ½ m. S. of Crook Inn.

Heart Law, mt., E. Lothian, 3 m. S.W. of Oldhamstocks, 8 m. S.E. of Dunbar ; alt. 1283 ft.

Heast, ham., Isle of Skye, Inverness-shire, 5 m. S.E. of Broadford.

Heathcot, seat, Kincardineshire, 1 m. S. of Cults ry. sta.

Heatherley, Inverness. Girls' private school.

Heatherlie, q.s. par., N.E. Selkirkshire.

Heavyside, seat, Lanarkshire, 1 m. E. of Biggar.

Hebrides, The, or Western Isls., off the W. coast of Scotland, extending 200 m. from the Butt of Lewis to the Kintyre peninsula, and including St. Kilda, 50 m. N.W. The Outer Hebrides are separated from the mainland by the Minch and from the N. isls. of the Inner Hebrides by the Little Minch. The chief isls. in the outer group are, Lewis, Harris, N. Uist, Benbecula, S. Uist, Barra etc., and the St. Kilda group. The Inner Hebrides comprise the three groups of Islay, Mull, and Skye. Area about 3184 sq. m. The isls. in general are mountainous, sandy shores, and melancholy moorland broken by numerous lochs. The raising of cattle and sheep and distilling are the principal industries. Stornoway and Barra are fishing centres, and Harris tweeds are a characteristic production, especially of Harris. The inhabitants mostly use the Gaelic language.

Heck, vil., Dumfriesshire, 3 m. S.W. of Lockerbie.

Hecla, mt., S. Uist Isl., Outer Hebrides; alt. 1988 ft.

Hedderwick Hill, seat, E. Lothian, 3 m. W. of Dunbar.

Heilen Loch, Caithness, 4 m. E. of Castletown; depth 5 ft., alt. 113 ft.

Heilim Ferry, Sutherland, on Loch Eriboll.

Heilinabretta, headland, Shetland, E. coast of Mainland.

Heillisay, islet, 2 m. N.E. of Barra, Outer Hebrides.

Heisker, isl., Outer Hebrides, 7½ m. W. of N. Uist. Pop. 33. Lady Grange was imprisoned on this isl. in 1733 and afterwards taken to St. Kilda.

Heiton, vil., Roxburghshire, 2½ m. S.W. of Kelso. P.O., T.O. Pop. 122.

Heldale Water, loch, Orkney, in S. Hoy Isl.

Heleness, seat, Shetland, 2 m. S.E. of Cunningsburgh.

Helensburgh, q.s. par., coast town (police bur.), with ry. sta., L.N.E., quay, and pier, Dunbartonshire, Rhu par.; entrance to Gare Loch, N. shore of Firth of Clyde, opposite Greenock (4 m. distant); 8 m. W.N.W. of Dumbarton, 23¾ m. N.W. of Glasgow. Pop. police bur. 8893. P.O., T.O. Hotel. A watering and residential place. Shows a red and green fixed light. See CRAIGENDORAN. Frequent steamer communication with Greenock. etc.

Helensdale, seat, Shetland, 1½ m. S.W. of Lerwick.

Helenslee, seat, near Dumbarton.

Hellia Spur, precipice, N.W. of Rousay, Orkney; alt. 500 ft.

Hellier Holm, isl., Orkney, off S. coast of Shapinsay.

Hellisay, isl. in Sound of Barra, Outer Hebrides.

Hellister, vil., and loch, Shetland, 6 m. N. of Scalloway.

Hellmuir Loch, 9 m. S.W. of Selkirk.

Hell's Glen, Argyll, between Lochs Fyne and Goil; traversed by road from Lochgoilhead to Inveraray.

Hell's Hole Valley, glen, Morayshire, near Forres.

Hell's Lum, curious large cavern, Banffshire, 2½ m. S.W. of Kinloss.

Hell's Skerries, cluster of islets, 10 m. W. of Rhum, Inner Hebrides.

Hellyar Holm, isl. Orkney, off S. coast of Shapinsay. Lighthouse with fixed and 10-seconds flashing light, visible 13 m.

Helman Head, prom., Caithness, 3 m. S. of Wick.

Helmsdale, small seapt., with ry. sta., L.M.S., Sutherland, at mouth of r. Helmsdale, 17¼ m. N.E. of Golspie. P.O., T.O. Pop. 732. Hotels. A coastguard and lifeboat sta., and the headquarters of a fishery dist., extending from Embo to Dunbeath.

Helmsdale, r., Sutherland, flows 20 m. S.E. to the sea at Helmsdale; excellent salmon river.

Helshetter, pl., with mineral spring, Caithness, 1¼ m. S. of Reay.

Hempriggs, vil., Caithness, 1½ m. S. of Wick. H. Lodge, a seat, H. Loch, and H. Stacks, lofty insulated rocks in vicinity; depth of loch, 8 ft., alt. 134 ft.

Hen and Chickens, rocks, off E. coast of Lewis Isl., Outer Hebrides, 6 m. S.E. of Stornoway.

Henderland, pl. and ruined fortalice, Peeblesshire, near St. Mary's Loch.

Hendersyde Park, seat, Roxburghshire, Ednam par., 1½ m. N.E. of Kelso.

Hendrey's Course, hill, N.E. Lanarkshire, 2½ m. N. of Cobbinshaw ry. sta.; alt. 1177 ft.

Hensol, or **Duchrae,** seat, Kirkcudbrightshire, 1 m. S.E. of New Galloway sta.

Henwood, strong natural fastness, Roxburghshire, Oxnam par. A great retreat in the Border forays.

Herbertshire Castle, destroyed by fire (1914), adjoining Denny, Stirlingshire. H., sub. of Denny.

Herdmanston, estate, with remains of ancient castle, E. Lothian, Saltoun par., 4 m. S.W. of Haddington.

Herdsman, basaltic islet, adjacent to Staffa, Argyll.

Herd, The, or **Shepherd of Sandwood,** headland, Sutherland, 7 m. S. by W. of Cape Wrath.

Heriot, par. and ry. sta., L.N.E., Midlothian; 16,501 ac. Pop. 417. P.O., T.O. H. sta., 19¼ m. S.E. of Edinburgh by rail.

Hermand New House, ham., Midlothian, 1 m. E. of West Calder. H. House, a seat.

Hermaness, hill, Shetland, N.W. of Unst; alt. 657 ft.

Herma Ness, prom., Shetland, N.W. of Unst Isl.

Herman Law, mt., border of Selkirk and Dumfries, 3 m. W.N.W. of Ettrick Church; alt. 2014 ft.

Hermiston, vil., Midlothian, 1¾ m. N. of Currie. P.O. H. House, a seat.

Hermitage Castle, a ruin, Roxburghshire, 5½ m. N. of Castleton. Figures notably in connection with a visit to it by Queen Mary, to the Earl of Bothwell. It is the most impressing and best preserved of the Border Peels.

Hermitage, The, of Braid, mansion on Braid Burn, Edinburgh, 3¼ m. S. of G.P.O.

Herriot's Dyke, an ancient earthen rampart, subtended by a ditch, through the centre of Berwickshire, still traceable 1 m. N. of Greenlaw.

Hesselhead, pl., with castle ruins, N. Ayrshire, 2 m. E.S.E. of Beith.

Heston Island, at the mouth of Auchencairn Bay, Kirkcudbrightshire, Rerrick par. Pop. 3. Has a group flashing light, visible 17 m.

Heughhead, W. Aberdeenshire, 16 m. S.S.W. of Rhynie.

Hewk, seat, Dumfriesshire, 4½ m. N. of Lockerbie.

Heylipoll, pl., Isl. of Tiree, Argyll, 4 m. S.W. of Scarinish.

Heylor, ham., Shetland, N. Mainland, S.W. coast of Ronas Voe. P.O., T.O.

Higgins' Neuck, pl. in Stirlingshire, on the r. Forth, opposite Kincardine, Fife. Here is a ferry, and bridge in course of construction.

High Bridge, picturesque three-arch bridge, built by General Wade, on the Spean, S.W. Inverness-shire, 1½ m. W. of Spean Bridge.

High Coats and Burnbank, part of Coatbridge, Lanarkshire.

Highfield House, mansion, Ross and Cromarty, 1¼ m. N.N.W. of Muir of Ord sta.

Highlandman, ry. sta., L.M.S., Perthshire, 1½ m. S.E. of Crieff.

Highland Park Distillery, Orkney, 1 m. S. of Kirkwall.

Highlands, in one sense, all the mountainous country N.W. and N. of the Firth of Clyde; in another sense, all the country of the old feudal Celtic clans; in a third sense, all the country inhabited by Gaelic-speaking population. But, in each of the senses, the word requires to be qualified. The mountainous country embraces a large aggregate of valley and plain, and adjoins a still larger aggregate of territory strictly akin in surface, history, and usages, to the Southern Lowlands. Generally speaking, it is applied to that portion of Scotland on and beyond the Grampians.

Highlaws House, N.E. Berwickshire, ½ m. S.W. of Eyemouth ry. sta.

Highlee, boys' home, E. Lothian, 1 m. S.W. of Humbie sta.

High Mark, seat, Wigtownshire, 4½ m. N.N.W. of New Luce.

High Sunnyside, sub. of Coatbridge, Lanarkshire.

Hightae, vil. and loch, Dumfriesshire, 2¾ m. S.S.E. of Lochmaben, and 4 m. S.W. of Lockerbie. P.O., T.O. H. Mill Loch is 1 m. N.W.; depth 18 ft., alt. 187 ft.

Hildesay, isl., Shetland, 3 m. W. of Scalloway.

Hill End, ham., Fife, 2 m. N.N.E. of Saline.

Hillend, public park, 4½ m. S.S.W. of Edinburgh G.P.O.

Hillend, vil., Fife, 1¼ m. N.E. of Inverkeithing. P.O.

Hillend, vil., N.E. Lanarkshire, Shotts par., 5 m. E.N.E. of Airdrie. In vicinity, H. Reservoir.

Hillfoot, holiday home and ry. sta., L.N.E., Dunbartonshire, 1½ m. S. of Milngavie.

Hillfoots, dist., Clackmannanshire. Pop. 2860.

Hillhead, q.s. par. and dist., N.W. Glasgow, with ry. sta., Glasgow Dist. Subway. P.O., T.O.

Hillhead Division, parl. division of Glasgow. Pop. 51,884.

Hillhead of Pitfodels, near Cults, Aberdeenshire. Convalescent hospital.

Hillhouse, seat, Ayrshire, Dundonald par., 3 m. N.N.E. of Troon.

Hilliclay, ham., N. Caithness, 4½ m. S.E. of Thurso.

Hill of Alyth, E. Perthshire, 1½ m. N. of Alyth, alt. 966 ft.

Hill of Beath, mining vil., Fife, Beath par., ½ m. N.W. of Crossgates sta. P.O. Alt. of hill 745 ft.

Hill of Blair, E. Perthshire, 5 m. W.S.W. of Alyth : alt. 690 ft.

Hill of Dale, Mainland, Shetland ; alt. 797 ft.

Hill of Dores, one of the Sidlaw hills, S.E. Angus, 3 m. S.E. of Coupar-Angus. It was crowned with an old castle, traditionally said to have been for some time the residence of Macbeth.

Hill of Drip, seat, Perthshire, Kincardine par.

Hill of Forss, Caithness, 3 m. W. of Thurso.

Hill of Fyrish, E. Ross and Cromarty, 3 m. W. of Alness. Crowned with a monument in the form of a temple : alt. 1483 ft.

Hill of Keillor, vil., Angus, Newtyle par., 4 m. E. of Coupar-Angus.

Hill of Nigg, N.E. Ross and Cromarty, in Nigg par. ; alt. 600 ft.

Hill of Stake, Ayrshire, 5 m. N.E. of Largs ; alt. 1711 ft.

Hill of Strone, Angus, 2½ m. W. of Milton of Clova ; alt. 2778 ft.

Hill of Trusta, Kincardineshire, 5½ m W. of Stonehaven ; alt. 1051 ft.

Hill of Wirren, Angus, 5½ m. W.N.W. of Edzell ; alt. 2220 ft.

Hillrigs, pl., near Biggar, Lanarkshire. Birthplace of George M. Kemp, architect of the Scott Monument, Edinburgh.

Hillshawhead, mt., S.E. Lanarkshire, 7½ m. N.N.E. of Abington sta. ; alt. 2141 ft.

Hillside, seat, Fife, a little N. of the vil. of Aberdour.

Hillside, 1 m. W. of Perth. Town hospital and sanatorium.

Hillside, seat, S.W. Fife, Saline par., 8 m. N.N.W. of Dunfermline.

Hillside, q.s. par. and vil., Angus, 2 m. N.N.W. of Montrose. P.O., T.O. Special drainage dist. Pop. 317.

Hillside, vil., Kincardineshire, Banchory-Devenick par., 1 m. N. of Portlethen sta. P.O.

Hillslap, ruined tower, N.W. Roxburghshire, on Allan Water, 3 m. N.E. of Galashiels.

Hills Tower, ancient tower, Kirkcudbrightshire, Lochrutton par., 5¾ m. W.S.W. of Dumfries.

Hillswick, seapt. vil., and a voe or bay, Shetland, Northmaven par., 36 m. N.N.W. of Lerwick. P.O., T.O. A tourist centre. Hotel. H. Ness has a fixed light, white with red sectors, visible 15 m.

Hill, The, seat, Kirkcudbright, Creetown.

Hilltown, dist. in N. Dundee, Angus. P.O., T.O.

Hillwell, ham., Shetland, 1½ m. S.W. of Dunrossness.

Hillyland, vil., 1½ m. N.W. of Perth.

Hilton, pl., 4 m. S.W. of Banff.

Hilton, ancient par., S.E. Berwickshire, united to Whitsome par.

Hilton, estate, with cairn, Ross and Cromarty, 6 m. S. of Dingwall. P.O.

Hilton, pl., with pottery, Clackmannanshire, 1 m. E. of Alloa.

Hilton, vil., 1 m. S. of Inverness.

Hilton, seat, Fife, 1 m. N. of Cupar.

Hilton, seat, Perthshire, Forgandenny par.

Hilton, seat, Kinross-shire, 1 m. W. of Milnathort.

Hilton of Cadboll, fishing vil., N.E. Ross and Cromarty, Fearn par., on the Moray Firth, 7 m. S.E. of Tain. P.O. called Hilton.

Hinnisdal, a stream, Isle of Skye, Trotternish dist., running 5½ m. W.S.W. to Loch Snizort.

Hirbesta, vil., Isle of Skye, W. Trotternish dist. Its post-town is Kilmuir.

Hirn, vil., N. Kincardineshire, 4 m. N.E. of Banchory.

Hirsel Law, pl., Berwickshire, 2 m. N.W. of Coldstream.

Hirsel, The, seat of the Earl of Home, Coldstream par., Berwickshire, on the rt. bank of the Leet Water, 1¾ m. N.N.W. of Coldstream. H. Loch is here.

Hirst Hill, hill, N.E. Lanarkshire, Shotts par., on the watershed between the Clyde and Forth ; alt. 959 ft.

Hirta, see ST. KILDA.

Hoan, isl., in Durness par., N.W. Sutherland, off the W. side of the mouth of Loch Eriboll.

Hobgoblin Hall, see YESTER CASTLE.

Hobister, seat, Orkney, 4½ m. W.S.W. of Kirkwall.

Hobkirk, par., Roxburghshire, on Rule Water, 7 m. E. by S. of Hawick ; 16,209 ac. Pop. 524.

Hobseat, shooting-lodge, Kincardineshire, on Cowie Water, 9 m. W. of Stonehaven.

Hoddom, par., S. Dumfriesshire ; 7515 ac. Pop. 1242. Contains Ecclefechan.

Hoddom Castle, seat, Cummertrees par., S. Dumfriesshire. Succeeded a strong Border fortalice of its own name on left bank of r. Annan, at Hallguards, and demolished in the terms of a Border treaty. Said to have been a seat of the Royal Bruces.

Hodges, pl., E. Lothian, Gladsmuir par., 4½ m. S.W. of Haddington.

Hogan, ham., Bressay, Shetland, 1½ m. N. of Lerwick.

Hogganfield, q.s. par., vil., and loch, Lanarkshire, and partly in Glasgow.

Hogh Bay, inlet, W. coast Coll Isl., Argyll.

Hoil, loch, Perthshire, 3½ m. S. of Aberfeldy ; alt. 1600 ft. ; depth 46 ft.

Holburn, q.s. par., in S. Aberdeen.

Holburn Head, a magnificent headland, Caithness, 2 m. N.W. of Thurso. On Little Head, adjacent, is a lighthouse showing one light, flashing once every 10 seconds, seen 14 m.

Holburn Street, ry. sta., L.N.E., in city of Aberdeen.

Holehouse Hill, Dumfriesshire, Kirkmichael par. ; alt. 1307 ft.

Holekettle, or **Kettle Bridge,** vil., central Fife, ½ m. S. of Kettle vil.

Hole of Row, natural arch, Orkney, on S. side of the Bay of Skaill.

Holl, shooting-lodge, Angus, 10 m. N. of Alyth.

Holl, vil., N.W. Isle of Skye, Inverness-shire.

Holland, Bay of, in S. of Stronsay, Orkney.

Hollandbush, vil., Stirlingshire, 3 m. S.S.W. of Denny. Stands contiguous to Haggs vil. P.O., T.O.

Holland House, Orkney, in Papa Westray, 20 m. N. of Kirkwall.

Holland House, N. Ronaldsay, Orkney.

Hollee, vil., Dumfriesshire, 1 m. S.W. of Kirkpatrick.

Hollows, a ruined Border tower, S.E. Dumfriesshire, on rt. bank of the Esk, 2 m. N.N.W. of Canonbie vil. Was the stronghold of the notorious freebooter, Johnnie Armstrong of Gilnockie.

Hollow-wood, or **Howwood,** vil., ry. sta., L.M.S., Renfrewshire, 3 m. S.W. of Johnstone. P.O., T.O.

Holl Reservoir, Fife, on Lomond Hills, 2 m. N.W. of Leslie ; depth 38 ft.

Hollybush, seat and ry. sta., L.M.S., Ayrshire, near the rt. bank of the Doon, 6 m. S.E. of Ayr. P.O., T.O. at sta.

Holm, par., S.E. Orkney, Mainland ; 8457 ac. Pop. 710. P.O., T.O. Contains Lambholm Isl. and St. Mary's Holm, vil., on Holm Sound, 6 m. S.E. of Kirkwall.

Holm, uninhabited isl., at the mouth of Stromness Harbour, Orkney.

Holm, ham., Lewis Isl., Outer Hebrides, 2 m. S.E. of Stornoway. H. Point, at entrance to Stornoway Harbour. H. Isl., at mouth of H. Bay.

Holm, burn, Inverness-shire, flows out of Loch Ashie.

Holmains, old baronial tower and a range of hills, Dumfriesshire, Dalton par., 4 m. S. of Lochmaben ; alt. of hills 800 ft.

Holme Rose House, seat, N.W. Nairnshire, 4 m. S.E. of Fort George sta.

Holmes, seat, Ayrshire, 3 m. E.S.E. of Kilmarnock.

Holmes, The, seat, N.W. Roxburghshire, near Newtown St. Boswells.

Holme, The, seat, Kirkcudbrightshire, 2 m. N.E. of New Galloway.

Holm House, Kirkcudbrightshire, ¾ m. N.W. of Balmaclellan.

Holm of Noss, rocky islet, Shetland. Once connected with Noss Isl. by a rope cradle bridge, See Noss.

Holm of Papa, isl., off E. coast of Papa Westray, Orkney.

Holm of Papal, isl., Lerwick par., Shetland.

Holm of Scockness, islet, between Rousay and Egilsay, Orkney.

Holms Colliery, Ayrshire, 1½ m. from Hurlford. H. House.

Holms Water, stream, W. Peeblesshire, flows 8 m. N.N.E. to Biggar Water, ¾ m. above that stream's confluence with the Tweed.

Holoman, pl., on W. coast of Raasay Isl., Inverness-shire.

Holydean Castle, seat, 1 m. W. of Bowden, N.W. Roxburghshire. The site of the castle of the Duke of Roxburghe's ancestors.

Holy Island, at mouth of Lamlash Bay, Arran Isl., Buteshire. It rises in tiers of cliffs from base to summit to an alt. of 1030 ft ; it is 2½ m. in length. Pop. 19. A lighthouse, with a half-minute alternating flashing white and red light, 124 ft. above high water and seen 17 m.

Holy Isle, small isl., with ruined ancient church and large ancient cemetery, off N.W. coast of Islay Isl., Argyll. Traditionally associated with St. Columba.

Holylee, seat, Peeblesshire, or r. Tweed, 2 m. N.W. of Thornilee sta.

Holy Linn, waterfall, Kirkcudbrightshire, on Garpel Burn, Balmaclellan par.

Holy Loch, Firth of Clyde, between the Dunoon and Kilmun sections of D. and K. par., 2 m. in length and 1 m. at greatest breadth.

Holyrood, royal palace and ruined abbey, Edinburgh, adjacent to the King's Park. For centuries the residence of the Scottish sovereigns. The abbey, in ruins, was founded by David I. in 1128. The palace was commenced in 1501, and is especially famous as the residence of Mary Queen of Scots and the scene of the murder of her secretary, David Rizzio, in 1566. Partially destroyed on two occasions, the present edifice was rebuilt 1671–1679.

Holytown, q.s. par., town, and ry. sta., L.M.S., Lanarkshire, 11 m. S.E. of Glasgow. Pop. town (H., New Stevenson, and Carfin) 12,011. P.O., T.O. Centre of a rich mining district.

Holywell Haugh, Ladykirk par., Berwickshire. Place where Edward I. of England and the Scottish nobles met to determine the succession of the Scottish Crown.

Holywood, par. and vil., with ry. sta., L.M.S., W. Dumfriesshire ; par. 8807 ac. Pop. 968. Vil. 1 m. S. of sta., 3½ m. N.W. of Dumfries. P.O.

Home, Berwickshire. See HUME.

Honeynook, pl., Aberdeenshire, near Maud Junction.

Hoove, vil., Shetland, 8 m. N.W. of Lerwick.

Hope, r. and loch, N.W. Sutherland. The r. flows 14 m. N. to Loch Eriboll. H. Lodge, near H. Bridge, 8 m. W. of Tongue. Depth of loch 187 ft., alt. 12 ft.

Hopefield, vil., Dumfriesshire, 5 m. N.W. of Auldgirth.

Hope Head, mt., S.W. Selkirkshire, 1½ m. S.W. of Ettrick church ; alt. 1697 ft.

Hopeman, fishing vil. and summer resort, with ry. sta., L.M.S., Morayshire, 6½ m. N.W. of Elgin. P.O., T.O. Pop. 1200. H. Lodge in vicinity.

Hope Park, seat, S. Berwickshire, Coldstream.

Hopertootie, ham., Dumfriesshire, in Kirkpatrick Juxta par.

Hopes, seat, E. Lothian, Garvald and Bara par., 9 m. S.E. of Haddington. The property of the Marquess of Tweeddale. H. Water, the upper reaches of Gifford Water.

Hopetoun House, seat of the Marquess of Linlithgow, W. Lothian, 3 m. W. of South Queensferry.

Hopetoun Monument, E. Lothian, on the Garleton Hills, 1½ m. N. of Haddington.

Hopewell, seat, Aberdeenshire, 1½ m. W.N.W. of Tarland.

Hop-pringle, ruined ancient castle, Midlothian, Stow par., 1 m. N.W. of Fountainhall.

Horisdale, loch, W. Ross and Cromarty, 3¼ m. S. of Gairloch.

Horisfield, isl., Ross and Cromarty, Gairloch par.

Horn, loch, Sutherlandshire, 4 m. N.W. of Golspie ; alt. 1155 ft.

Horndean, vil. and stream, S.E. Berwickshire ; vil. 7½ m. N.E. of Coldstream.

Horrisdale, isl., W. Ross and Cromarty, in Gair Loch. Pop. 11.

Horsburgh Castle, shattered peel tower of the Horsburghs, 3 m. E. of Peebles.

Horse Island, Inner Hebrides, Argyll, 4 m. S.W. of Eigg Isl.

Horse Island, islet, off coast of N. Ayrshire, 1 m. W. of Ardrossan.

Horse Island, N.W. Ross and Cromarty, 6 m. S.W. of Loch Enard.

Horse of Copinsay, rocky islet, Orkney, off E. coast of Mainland.

Horsley Hall, seat, Argyll, Strontian.

Hoscote, seat, Roxburghshire, 8¼ m. S.W. of Hawick.

Hoselaw Loch, Roxburghshire, 4 m. S.E. of Sprouston.

Hosh Distillery, Crieff, Perthshire.

Hospitalfield, seat, Angus, 1¼ m. W. of Arbroath. The original of "Monkbarns" in Sir Walter Scott's *Antiquary*.

Hospital Mill, ham., Fife, near Pitlessie.

Hostigates, loch, Shetland, Mainland, 7½ m. N.E. of Walls ; 41 ft. deep.

Hoswick, ham., Shetland, 1½ m. W. of Sandwick.

Houbie, pl., Shetland, in S. Fetlar Isl.

Houff, ruins of ancient fort, Aberdeenshire, 10 m. E. of Tarland.

Hough Bay, W. coast of Tiree Isl., W. Argyll.

Houland, ham., Shetland, near Scalloway.

Houllma Water, loch, Shetland, Mainland, 4 m. N.E. of Walls.

Hound Hillock, mt., Kincardineshire, 4 m. N.W. of Fettercairn ; alt. 1698 ft.

Houndland, loch, Orkney, in N. of Pomona.

Hound Point, prom., W. Lothian, 1½ m. N.E. of S. Queensferry.

Houndslow, ham., Berwickshire, 7 m. E. of Lauder.

Hourn Loch, sea loch, opening from Sleat Sound, W. Inverness-shire; length about 14 m.

Housabister, ham., Shetland, 10 m. N. of Lerwick.

Housay, isl., Shetland, 10 m. E. of Mainland, Nesting par. Pop. 68.

House, bay, Orkney, in S. of Stronsay.

Housedale, seat, S.E. Aberdeenshire, 3 m. W. of Skene.

Househillmuir and Cottages, pl. in Glasgow, 2¼ m. N.E. of Barrhead.

House of Auchires, N.E. Aberdeenshire, 2½ m. N.W. of Lonmay ry. sta.

House of Muir, pl. where markets were held, Midlothian, 3 m. N. of Penicuik.

House of Muir, vil., E. Lothian, 9 m. S.W of Haddington.

House of Tongue, a seat of the Duke of Sutherland, Sutherland, 1 m. N.W. of Tongue.

Houster, ham., Shetland, on Lax Firth, 6 m. N.W. of Lerwick.

Houston, vil., with ry. sta., L.M.S., Renfrewshire, 1½ m. E.N.E. of Bridge of Weir, and 6 m. N.W. of Paisley. P.O., T.O. Pop. 460. H. House, a seat.

Houston and Killellan, par., Renfrewshire, 3 m. W. of Paisley ; 7549 ac. Pop. 2591.

Houstoun, or **Houston,** vil., W. Lothian, 1 m. N.W. of Uphall sta. H. House, an old seat.

Houton, ham., Orkney, 9 m. S.W. of Kirkwall. H. Bay with Holm of H., islet.

How, pl., Shapinsay, Orkney.

Howatstone, seat, Midlothian, 2½ m. W.S.W. of Mid-Calder.

Howbeg, ham., in S. Uist, Outer Hebrides.

Howdean, pl., with traces of ancient camp, Roxburghshire, 2¼ m. N.W. of Oxnam.

Howden, seat, Midlothian, 1½ m. W. of Mid-Calder.

Howden's Hall, pl., 3¼ m. S. of Edinburgh G.P.O.

Howe, ham., Caithness, 4 m. W. of Keiss.

Howe of Angus, The, see ANGUS.

Howe of Mearns, The, Kincardineshire.

Howequoy, pl. and prom., Orkney, Mainland, 5 m. S. of Kirkwall.

Howford, vil. in co., and 2 m. S. of Nairn.

Howgate, vil., Midlothian, 1½ m. S.E. of Penicuik. Pop. 79.

Howie, loch, Kirkcudbrightshire, 5¼ m. N.E. of New Galloway ; 39 ft. deep, alt. 757 ft.

Howietoun Fishery, fish hatchery and breeding ponds, Loch Coulter, 5 m. S.W. of Stirling.

Howmore, vil. on W. coast of S. Uist, Outer Hebrides.

Howmuir, vil., Kirkcudbrightshire, 10 m. N. of Castle-Douglas. See CORSOCK.

Hownam, par. and vil., E. Roxburghshire ; par. 15,086 ac. Pop. 174. Vil. 11 m. S.E. of Kelso. P.O. H. Law (1472 ft.), 2 m. N.E. of vil., a Roman camp.

Howpasley, pl., S.E. Selkirkshire, on Borthwick Water, 12 m. S.W. of Hawick.

Howtown, vil., W. Aberdeenshire, 1½ m. N.W. of Rhynie.

Howwood, q.s. par., vil., and ry. sta., L.M.S., Renfrewshire, 3 m. S.W. of Johnstone. P.O., T.O. Vil. pop. 782.

Hoxa Head, peninsula, N.W. of South Ronaldsay, Orkney. A fixed light, white with red sectors, visible 12 m. H. Sound between H. Head and Flotta Isl.

Hoy, sound and isl., in S.W. Orkney. The sound separates the isl. from Stromness par., in Pomona; measures 7 m. in length. The isl. extends southward from the sound to within 6¾ m. of Caithness; measures 13 m. in length. Pop. 955. The Old Man of H. is a detached pillar-shaped rock, 450 ft. high. H. Lodge is 4 m. S.W. of Stromness. Steamer connection with Stromness.

Hoy, ry. sta., L.M.S., Caithness, 5 m. S.E. of Caithness.

Hoy and Graemsay, par., Orkney; 15,183 ac. Pop. 264. Comprises Graemsay Isl. and the N. part of Holy Isl. P.O., T.O.

Hoy Scund Lighthouses (High and **Low),** on Graemsay Isl., Hoy Sound, Orkney. Have fixed lights, seen 16 m. and 12 m.

Hughton, pl., Inverness-shire, 7 m. S.W. of Beauly. P.O., T.O.

Huip, Holm of, isl., Orkney, off N. of Stronsay.

Huip Ness, prom., in N. of Stronsay, Orkney.

Hullion, ham., Orkney, S. Rousay, 11 m. N.W. of Kirkwall.

Hulmawater, loch, Shetland, 4 m. N.E. of Walls.

Humbie, par. and q.s. par., and ry. sta., L.N.E. (passenger service withdrawn), E. Lothian, 17¼ m. E.S.E. of Edinburgh by rail; 8772 ac. Pop. 587. P.O., T.O. H. House, seat of Lord Polwarth, 3½ m. N.E. of Blackshiels.

Humbie, pl., W. Lothian, 1 m. N.W. of Kirkliston. Holiday home.

Humbie Water, E. Lothian, ½ m. S.E. of Salton ry. sta.

Hume, par. and ham., S. Berwickshire, 3 m. S. of Greenlaw; 4108 ac. Pop. 259. P.O., T.O. H. Castle, ruin on H. Craigs.

Humphrey Loch, Dunbartonshire, 1½ m. N.E. of Bowling; Dumbarton water supply.

Huna, ham., Caithness, 3 m. W. of Duncansbay Head. P.O., T.O. Lifeboat sta.

Huna, loch, N. Uist, Outer Hebrides, 6½ m. W.N.W. of Lochmaddy; alt. 22 ft., depth 25 ft.

Hunda Island, Orkney, off W. coast of Burray Isl., 10 m. S. of Kirkwall. H. Sound, between H. and Burray. Uninhabited.

Hundalee, estate, Roxburghshire, 1½ m. S. of Jedburgh.

Hunder, loch, N. Uist, Outer Hebrides, 2 m. S. of Lochmaddy; depth 60 ft., alt. 22 ft.

Hundland, pl., Orkney, on Loch of H., 3 m. E. of Birsay church; alt. of loch 90 ft., depth 7 ft.

Hundleshope Heights, Peeblesshire, 4 m. S. of Peebles; alt. 2249 ft.

Huney Island, Shetland, off E. coast Unst Isl., 3 m. S.E. of Baltasound.

Hunglader, ham., Isle of Skye, near Kilmuir church.

Hungry Bridge, Lanarkshire, crosses Forth and Clyde Canal 1½ m. N. of Bishopbriggs.

Hungry Law, mt., Cheviot Hills, 5 m. E. by N. Carter Fell; alt. 1643 ft.

Hunish Rudha, prom., Isle of Skye, Kilmuir par., 2 m. W. of The Aird. Loch H. on S.W.

Hunterfield and Arniston Engine, mining vil., Midlothian, 3 m. S.E. of Dalkeith.

Hunter's Bog, valley between Arthur's Seat and Salisbury Crags, King's Park, Edinburgh.

Hunter's Hill, Angus, 5 m. S.W. of Forfar; alt. 664 ft.

Huntershill Quarry, Lanarkshire, near Bishopbriggs.

Hunter's Quay, Argyll, adjoining Dunoon. Pier. P.O., T.O. The Royal Clyde Yacht Clubhouse is here.

Hunterston House, seat, Ayrshire, 2¼ m. N.W. of W. Kilbride.

Hunter's Tryst, pl., in city of Edinburgh, 1½ m. E.S.E. of Colinton.

Huntfield, seat, Lanarkshire, 3 m. N.W. of Biggar.

Hunt Hill, Morayshire, 3 m. S.W. of Rothes; alt. 1210 ft.

Hunt Hill, Angus, 4 m. W.S.W. of Milton of Clova; alt. 2384 ft.

Hunthill House, seat, Roxburghshire, 1½ m. S.E. of Jedburgh.

Hunthill Lodge, seat, Angus, 9 m. W.N.W. of Edzell.

Huntingdon, pl., Berwickshire, 2 m. N.W. of Lauder.

Huntington House, seat, E. Lothian, 2 m. N.W. of Haddington.

Huntingtower and **Ruthvenfield,** vil. and ancient castle, Perthshire, Tibbermore par., 2½ m. N.W. of Perth. P.O. H. Castle, once known as Ruthven Castle, now in ruins, was the seat of the Earls of Gowrie, and in 1582 was the scene of the Raid of Ruthven, when the young king, James VI., was kidnapped by the Earl of Gowrie.

Hunt Law, mt., Peeblesshire, 7 m. S.E. of Broughton ; alt. 2094 ft.

Huntly, town (police bur.) and par., with ry. sta., L.N.E., N.W. Aberdeenshire, 41 m. N.W. of Aberdeen, in Strathbogie dist. ; par. 12,475 ac. Pop. 4597 ; bur. pop. 3778. P.O., T.O. Hotels. H. or Strathbogie Castle (in ruins) was granted to the Gordons by Robert the Bruce. H. Lodge, 1 m. N. The town has a large trade in agricultural produce.

Huntly Burn, stream, N.W. Roxburghshire, flows N.E. to the Tweed, 3 m. W. of Melrose. Huntlyburn House, seat, 1 m. S.W. of Melrose.

Huntly Burn, stream, E. Perthshire, flows S.E. to the Firth of Tay, near Longforgan sta.

Huntly Cot Hills, Midlothian, 6 m. S.E. of Leadburn. Part of Moorfoot Hills ; alt. 1694 ft.

Huntly's Cave, Morayshire, 2 m. N. of Grantown-on-Spey.

Hurcars, The, rocks at the entrance to Eyemouth Bay, Berwickshire.

Hurlet, vil., Renfrewshire, 3 m. S.E. of Paisley.

Hurlford, q.s. par. and town, with ry. sta., L.M.S., Ayrshire ; q.s. par. partly in Galston and Kilmarnock pars ; town, on r. Irvine, 2 m. S.E. of Kilmarnock. Pop. 3326. P.O., T.O. Has iron and fireclay works.

Hurly Hawkin, site of ancient castle, Angus, at Liff, 4½ m. N.W. of Dundee.

Hurst Grange, Stirling, boy's preparatory school.

Husabost, pl., with pier, Isle of Skye, W. side of Loch Dunvegan.

Husbandtown, vil., Angus, 7 m. N.E. of Dundee.

Hutcheon Street, ry. sta., L.N.E., city of Aberdeen.

Hutchesontown, q.s. par., Govan par., Glasgow.

Hutchison, vil., Lanarkshire, 1½ m. from Cambuslang.

Hutton, par. and vil., S.E. Berwickshire ; par. 5521 ac. Pop. 614. Vil. 6 m. W. of Berwick-on-Tweed. P.O. H. Castle 1½ m. N.W.

Hutton and Corrie, par., Dumfriesshire, 7 m. N.N.E. of Lockerbie ; 23,941 ac. Pop. 608.

Hutton Lodge, 6 m. S.E. of Dumfries.

Hylipol, q.s. par., Tyree par., Argyll.

Hyndal Hill, Renfrewshire, 7 m. S. of Greenock ; alt. 1244 ft.

Hynd Castle, ruin, Angus, 9 m. N.E. of Dundee.

Hyndford, estate, Lanarkshire, 2 m. S.E. of Lanark.

Hyndfordbridge End, ham., on r. Clyde, 2½ m. S.E. of Lanark.

Hyndhope, seat, on Ettrick Water, 9½ m. S.W. of Selkirk.

Hyndland, ry. sta., L.N.E., W. of Glasgow. P.O., T.O.

Hynish, vil., Tiree Isl., Argyll, on S.W. shore. Ben H., over 460 ft., 1 m. W.

Hyskier, islet, W. of Rhum, Invernessshire. A lighthouse with group-flashing light, seen 17 m. Pop. 3.

I

I, one of the names of Iona. See IONA.

Iasgair (An t-Iasgair), small isl., off the extreme N. of Isle of Skye, 3 m. N.W. of Kilmuir.

Ibris, isl., Firth of Forth, E. Lothian, Dirleton par. See EYEBROUGHTY.

Ibrox, S.W. dist. of Glasgow ; ry. sta., L.M.S. P.O.

Icolmkill, or **I-Columb-Kill,** see IONA.

Idrigill, ham., headland, and cave, Isle of Skye, Duirinish coast; alt. of headland, 400 ft. The cave was one of the prisons of Lady Grange.

Idrigill, ham. and prom., Isle of Skye, Uig Bay, 8 m. N.W. of Snizort.

Idvies, seat, Angus, 3 m. S.W. of Guthrie Junction.

Illeray, isl., contiguous to S.W. coast of N. Uist, Outer Hebrides.

Illiston, old baronial edifice, on r. Almond, W. Lothian, 1 m. S.E. of Drumshoreland sta.

Imachar, pl., W. coast of Arran, Buteshire, 9½ m. S.W. of Lochranza church.

I Mor, isl., with eccl. ruins, in Sound of Jura, Argyll, at mouth of Loch Swin. See EILEAN MOR.

Imperial Distillery, Morayshire, Carron sta.

Inalterie, pl., with remains of very old, strong, massive building, Banffshire, 2½ m. S. of Cullen.

Inch, pl., Kincardineshire, 2½ m. S.W. of Fettercairn.

Inch, par., N.W. Wigtownshire ; 30,875 ac. Pop. 3965.

Inch, isl., Clackmannanshire, Alloa par.

Inchaffray Abbey, vestige of ancient abbey, Perthshire, 6 m. E. of Crieff.

Inchard, loch and r., N.W. Sutherland, flows into Loch Inchard near Rhiconich.

Inchbae Deer Forest, Ross and Cromarty ; 21,070 ac. I. Lodge, 6 m. N. of Garve sta.

Inchbare, ham., Angus, on N. Esk, 4½ m. N. of Brechin. P.O.

Inchbelly, ham., with bridge of the Kelvin, Dunbartonshire, 1 m. N.E. of Kirkintilloch.

Inchberry, ham., Morayshire, 2½ m. S.W. of Fochabers.

Inchbrakie, seat, Perthshire, 2½ m. E. of Crieff. Ruins of I. Castle, destroyed by Cromwell in 1651.

Inchbrayock Island, or **Rossie Island,** in mouth of r. South Esk, Angus. Included in the parl. bur. of Montrose, with which it is connected by a suspension bridge.

Inchcailliach, wooded isl., in Loch Lomond, Stirlingshire, 3 m. S.E. of Luss. Has sites of ancient nunnery and ancient parochial church, the adjoining churchyard being a burying-place of the Macgregors.

Inchcape Rock, the old name of the Bell Rock, q.v.

Inchcolm, isl. (now national property), in the Firth of Forth, Fife, 1½ m. S. of Aberdour, of which par. it is a part. The Monastery of I. was founded by Alexander I., in 1123 ; the monastic buildings are still in a good state of preservation. Pop. 9.

Inchconnachan (Colquhoun's Island), Dunbartonshire, in Loch Lomond, 1 m. S.E. of Luss.

Inchcoonans, pl., with tileworks, Perthshire, 1 m. N.W. of Errol.

Inchcormac, isl. with remains of ancient chapel, Argyll, in mouth of Loch Swin, N.W. Knapdale.

Inchcroe Deer Forest, see KINTAIL.

Inchcruin, isl. in Loch Lomond, Stirlingshire, 1½ m. S.E. of Luss.

Inchdairnie, seat, Fife, 2 m. S. of Leslie.

Inchdrewer Castle, ruins of castellated seat, 3 m. S.W. of Banff.

Inch, East and **West,** on r. Carron, Stirlingshire, 4 m. N.E. of Bonnybridge. Two flat tracts.

Inches Station, ry. sta. (alt. 770 ft.), L.M.S., Lanarkshire, 6 m. S.W. of Douglas. P.O., T.O.

Inchfad, isl. in Loch Lomond, Stirlingshire, near Balmaha pier.

Inchgalbraith, isl., with remains of ancient castle, in Loch Lomond, Dunbartonshire, 6½ m. W.N.W. of Drymen.

Inchgarth, seat, 3 m. S.W. of Aberdeen, near Cults sta.

Inchgarvie, isl., Firth of Forth, Dalmeny par., W. Lothian. Fragment of old fort or castle ; a support of the Forth Bridge. On N.W. pt. a flashing light, seen 10 m.

Inchgarvie, seat, W. Lothian, on W. side of Queensferry.

Inchgower Distillery, Buckie, Banffshire.

Inchinnan, par., N. Renfrewshire, on r. Clyde ; 3334 ac. Pop. 1287. P.O., T.O. Pop. of I. and Erskine 1927.

Inchkeith, isl. in Firth of Forth, Kinghorn par., Fife, 4 m. N.E. of Leith ; alt. 180 ft. Crowned by a lighthouse with revolving light, visible 21 m., and a fog-signalling sta. Fortified. Pop. 16.

Inchkeith Hill, Berwickshire, 3 m. W.N.W. of Lauder ; alt. 1197 ft.

Inchkenneth, isl., W. coast of Mull, Argyll, in Loch-na-Keal. Belonged to the Culdees of Iona. Pop. 10.

Inchlaw, hill, Fife, 4 m. S. of Newport ; alt. 626 ft.

Inchlonaig, isl. in Loch Lomond, Dunbartonshire, 1 m. E. of Luss.

Inchmahome, isl. in Lake of Menteith, S.W. Perthshire, 8 m. S.S.W. of Callander. Has remains of an ancient priory, and was the residence of Mary Queen of Scots, in her girlhood.

Inchmarlo House, seat, N.W. Kincardineshire, 2 m. N.W. of Banchory.

Inchmarnock, isl., on W. side of Bute Isl., Firth of Clyde. Has remains of a chapel. Pop. 14.

Inchmartine House, seat, and ruined small Gothic church, Perthshire, 1 m. W. of Inchture.

Inchmickery, small isl. in Firth of Forth, 2½ m. N. of Cramond shore.

Inchmoan, flat mossy islet in Loch Lomond, Dunbartonshire, 1¼ m. S.E. of Luss.

Inchmore, pl., 4 m. W. of Inverness, Kirkhill par.

Inchmurrin, isl. at S. end of Loch Lomond, Dunbartonshire, 2¼ m. N. of Balloch pier. Pop. 15. Preserved as a deer park. Ruins of a fortalice of the Earls of Lennox.

Inchnacardoch, state forest and shooting-lodge, Inverness-shire, near Fort Augustus.

Inchnadamph, ham., S.W. Sutherland, at S.E. end of Loch Assynt. Hotel.

Inch, North and **South,** see PERTH.

Inchock, seat, Angus, 5 m. N.E. of Arbroath.

Inchrory, pl., Banffshire, 7 m. S. of Tomintoul.

Inchrye Abbey, seat, Fife, on Lindores Loch, 2½ m. S.E. of Newburgh.

Inchtalla, islet in Lake Menteith, Perthshire.

Inchtavannach (Monk's Island), in Loch Lomond, Dunbartonshire, 1 m. S.E. of Luss. Pop. 4.

Inchterf, ham., Stirlingshire, 2 m. E.S.E. of Milton of Campsie.

Inch, The, seat, Edinburgh, Liberton par., 2½ m. S. of G.P.O.

Inchture, par. and vil., with ry. sta., L.M.S., Perthshire, Carse of Gowrie, 8 m. S.W. of Dundee ; 4122 ac. Pop. 507. P.O., T.O. Motor bus service to Dundee.

Inchtuthil, site of Roman station, Perthshire, on left bank of r. Tay, 7 m. E.S.E. of Dunkeld.

Inchyra, vil., with harbour, on r. Tay, 5 m. E. of Perth. I. seat, 1 m. N.E.

Indail, sea loch, S.W. Islay, Argyll. Penetrates 12 m. N.N.E. ; 11 m. wide at entrance.

Inganess, prom., Westray, Orkney, 6 m. N. of Rousay.

Inganess Bay, Orkney, Mainland, 2 m. E. of Kirkwall.

Inglewood, seat, Clackmannanshire, ¾ m. N.W. of Alloa.

Inglisgreen, see SLATEFORD AND INGLIS-GREEN.

Inglismaldie, old castellated seat of the Earl of Kintore, Kincardineshire, 6 m. S.W. of Laurencekirk. State forest nearby.

Ingliston, pl., Angus, 2½ m. S.W. of Forfar.

Ingliston, golf club and course, Midlothian, 1½ m. N. of Ratho.

Inishail, isl. in Loch Awe, Argyll, 2½ m. S.W. of Kilchurn Castle. Contains an interesting ancient cemetery and the remains of a small convent and chapel.

Inistrynich House, seat, Argyll, on N.E. shore of Loch Awe, 5 m. S.W. of Dalmally.

Inkerman, vil., Renfrewshire, 1½ m. N.W. of Paisley. P.O.

Inkstack, pl., N. Caithness, 3 m. S.E. of Dunnet.

Innellan, q.s. par. and vil., with steamboat pier, Argyll, on Firth of Clyde, 3¾ m. S. of Dunoon ; 30 m. from Glasgow by river. P.O., T.O. I. Hill, 935 ft. Vil. pop. 736. Hotel.

Innergellie House, seat, Fife, 1½ m. N.E. of Anstruther.

Innerhadden House, seat, Perthshire, 1 m. S.E. of Kinloch Rannoch.

Inner Hebrides, see HEBRIDES.

Innerleithen, par. and police bur., with ry. sta., L.N.E., Peeblesshire, on Leithen Water, ½ m. from r. Tweed, 6¼ m. S.E. of Peebles and 33½ m. S. of Edinburgh by rail ; par. 23,994 ac. Pop. 3747 ; bur. pop. 2359. P.O., T.O. A seat of Scottish woollen manufacture. Famous for its medicinal spring, believed to be the St. Ronan's Well of Sir Walter Scott's novel.

Innerleven, Fife. See BUCKHAVEN.

Innermessan, site of ancient town and extant moat, Inch par., Wigtownshire, on Loch Ryan, 2½ m. N.E. of Stranraer. An ancient Caledonian settlement, with a moat 336 ft. in circuit and 78 ft. in vertical height.

Innerpeffray, ry. sta., L.M.S., Perthshire, 2¼ m. E.S.E. of Crieff. T.O. I. Castle, ruins, 1½ m. S.E. of sta.

Inner Sound, between Applecross, W. Ross and Cromarty, and Raasay Isl., Inverness-shire, Inner Hebrides, 3 m. to 6 m. broad.

Innertig, see KIRKCUDBRIGHT-INNERTIG.

Innervar Cottage, in Glen Lyon, 13 m. W. of Aberfeldy.

Innerwell House, Wigtownshire, on I. Bay, 3½ m. N.E. of Garlieston.

Innerwick, par., vil., and ry. sta., L.N.E., E. Lothian ; 13,142 ac. Pop. 684. P.O., T.O. Vil. pop. 138. Sta., 1½ m. E. of vil. and 4½ m. S.E. of Dunbar. I. Castle, ruins, 1 m. E.

Innerwick House, seat, Perthshire, Glen Lyon, 9 m. W. of Fortingall.

Innes House, old seat (seventeenth century) of the Earl of Fife, Morayshire, 5 m. N.E. of Elgin.

Innimore Lodge, Mull Isl., Argyll, 4 m. from Pennyghael.

Inninmore Bay, Sound of Mull, Argyll, on the Morven coast, 3½ m. E.S.E. of Ardtornish Pt.

Innischonan, isl. and seat, Argyll, N.E. end of Loch Awe, 1¾ m. S.W. of Kilchurn Castle. Pop. 2.

Innischonell, isl., in Loch Awe, Argyll, 8 m. from S. end of loch. Ruin of ancient chief seat of the Argyll family.

Inniserrech, or **Innis-shearraich,** isl. in Loch Awe, Argyll. Ruined ancient chapel and burying-ground, near Portinisherrich.

Insch, par. and vil., with ry. sta., L.N.E., Aberdeenshire ; par. 8371 ac. Pop. 1249. Vil. 13½ m. S.E. of Huntly and 27½ m. N.W. of Aberdeen. P.O., T.O. Pop. (I. and Rothney) 917.

Insh, q.s. par. and vil., Inverness-shire, partly in Alvie, but chiefly in Kingussie and Insh par. ; vil. 5 m. N.E. of Kingussie. P.O., T.O. Loch I. 100 ft. deep, alt. 721 ft., 1½ m. N. of vil.

Inshes House, seat, 2½ m. S.E. of Inverness.

Inshewan, seat, Angus, on S. Esk, 4½ m. N.E. of Kirriemuir.

Inshlaggan, pl., Glen Garry, Inverness-shire, 2 m. W. of W. end of Loch Garry.

Inshoch Castle, ancient baronial fortalice, 3½ m. E. of Nairn.

Inver, vil., Perthshire, on rt. side of Tay, opposite Dunkeld.

Inver, fishing vil., N.E. Ross and Cromarty, on Dornoch Firth, 6 m. E. of Tain. P.O.

Inver, r., S.W. Sutherland, flows 6 m. S.W. to Loch Inver.

Inver, pl., S. Aberdeenshire, 3 m. from Crathie. P.O.

Inver, seat, Caithness, ⅓ m. E. of Dunbeath.

Inverailort, seat, S.W. Inverness-shire, 28 m. W. of Fort William.

Inverallan, q.s. par., Morayshire and Inverness-shire, Cromdale par. I. House, 1⅓ m. S. of Grantown.

Inver Alligin Hostel, on N. shore of Upper Loch Torridon, Ross-shire. A Scottish Youth Hostel.

Inverallochy, q.s. par. and fishing vil., N.E. Aberdeenshire ; vil. 3½ m. S.E. of Fraserburgh. P.O., T.O. Vil. pop. (I. and Cairnbulg) 1308. I. Castle, an old stronghold of the Comyns, 1½ m. S. of vil.

Inveralmond House, Midlothian, Cramond par.

Inveramsay, ry. sta., L.N.E., 3¾ m. N.W. of Inverurie. T.O.

Inveran, ham., S. Sutherland, 1½ m. N.W. of Invershin, on r. Shin. Inn.

Inveran Lodge, N.W. Ross and Cromarty, on Loch Maree, 2 m. S.E. of Poolewe.

Inverannoty, Scone, Perthshire. Convalescent home.

Inveraray, royal and mun. bur., par., and co. town of Argyllshire ; par. 46,887 ac. Pop. 989. The town (pop. 455) stands on the W. side of Loch Fyne, 30 m. by shortest route N.W. of Greenock. P.O., T.O. Hotels. The ducal mansion, I. Castle, is the chief seat of the Duke of Argyll. Three white lights on pier, and on pier-head two red lights, visible 5 m. Regular steamer communication with Glasgow, Greenock, etc. Ferry communication with St. Catherines on S. side of Loch Fyne.

Inverardoch, seat, S. Perthshire, near Doune.

Inverardran, seat, W. Perthshire, near Crianlarich sta.

Inverarish, pl., Raasay Isl., W. Inverness-shire.

Inverarity, par., Angus, 4 m. S. of Forfar ; 11,070 ac. Pop. 773.

Inverarnan, pl., at foot of Glenfalloch, near head of Loch Lomond, 10 m. N. of Tarbet.

Inverasdale, ham., N.W. Ross and Cromarty, on W. side of Loch Ewe, 4½ m. N.W. of Poolewe. P.O., T.O.

Inveravon, par. and ham., Banffshire; par. 48,967 ac. Pop. 1598. Ham. 1½ m. N.E. of Ballindalloch sta.

Inveravon, old tower on site of Roman sta., W. Lothian, 3 m. N.W. of Linlithgow.

Inverawe House, seat, Argyllshire, on r. Awe, 1½ m. E. of Taynuilt sta.

Inverbeg, pl., Dunbartonshire, on Loch Lomond, 3 m. N. of Luss. There is a ferry to Rowardennan Pier.

Inverbeg Hostel, on Loch Lomond, at entrance to Glen Douglas, Dunbartonshire, 3½ m. N. of Luss and 12 m. N. of Balloch. A Scottish Youth Hostel.

Inverbervie, parl., royal, and mun. bur., with ry. sta. (Bervie), L.N.E., Bervie par., Kincardineshire. Parl. and mun. bur. pop. 1032. P.O., T.O. It unites with Montrose and other four co. burs. in returning one member to Parliament. See BERVIE.

Inverbraan House, Perthshire, near Dunkeld.

Inverbreakie, seat, Ross and Cromarty, 1 m. N. of Invergordon.

Inverbroom Lodge, N.W. Ross and Cromarty, near head of Loch Broom, 8 m. S.E. of Ullapool.

Inverbrothock, q.s. par., Angus, Arbroath and St. Vigeans par. Comprises part of Arbroath.

Invercannich, ham., N.W. Inverness-shire, 17 m. S.W. of Beauly.

Invercanny, estate, N.W. Kincardineshire, 2 m. W. of Banchory. Contains a reservoir for the water supply of Aberdeen.

Invercarron House, see INVERCHARRON.

Invercassley, bridge and seat, S. Sutherland, at mouth of r. Cassley, Creich par.

Invercauld House, seat of the Farquharsons, S.W. Aberdeenshire, in the Dee Valley, 1½ m. N.E. of Braemar.

Inverchaolain, par., in S. of Cowal dist., Argyll; 29,321 ac. Pop. 318. I. church on Loch Striven, 6 m. N.W. of Toward.

Inverchapel, landing-pl. at the foot of Loch Eck, Cowal, Argyll, 5½ m. N.W. of Strone.

Invercharron House, N. Ross and Cromarty, 1 m. N. of Ardgay.

Inverchroskie House, N.E. Perthshire, 2 m. N.W. of Kirkmichael.

Invercloy, adjoining Brodick, Arran, Buteshire. Special scavenging dist. Pop. 391.

Invercreran, shooting-lodge, Argyll, 8 m. S.W. of Ballachulish.

Inverdruie, vil., N.E. Inverness-shire, near Aviemore sta. I. House in vicinity.

Inverdunning House, Perthshire, 1½ m. N. of Dunning.

Inverebrie, pl., Aberdeenshire, Methlick par., 3 m. N.W. of Ellon.

Invereighty, seat, Angus, 3½ m. S.W. of Forfar.

Invererne House, seat, Morayshire, 1½ m. N.W. of Forres.

Inverernie House, seat, 7 m. S.E. of Inverness, on r. Nairn.

Invereshie House, seat, E. Inverness-shire, Alvie par., 1 m. S.E. of Kincraig sta. I. Deer Forest, 9270 ac.

Inveresk, par. and vil., with ry. sta., L.N.E., Midlothian; par. 5132 ac. Pop. 20,686; vil. pop. 688. Vil. ½ m. N.N.W. of sta., on r. Esk, in S. vicinity of Musselburgh, 6½ m. E. of Edinburgh. T.O. at sta. In vicinity, I. House, I. Lodge, and I. Paper Mills. Of historical interest, Pinkie battlefield and Carberry Hill.

Inveresragan, pl., Argyll, on Loch Etive, 2 m. N.W. of Bonawe.

Inverewe House, seat and deer forest, adjacent to Poolewe, N.W. Ross and Cromarty. Deer forest, 13,000 ac.

Inverey, ham., S.W. Aberdeenshire, 4 m. W. of Braemar. Remains of ancient stronghold of the Farquharsons.

Inverfarigaig, landing-pl., E. side of Loch Ness, 17 m. S.W. of Inverness. P.O.

Inverfolla, pl., with ancient obelisk, Argyll, 2 m. S.E. of Appin, on Loch Creran.

Invergarry, ham., at mouth of Glengarry, with ry. sta. (closed for passengers) (2 m. S.W.), Inverness-shire, 7½ m. S.W. of Fort Augustus. P.O., T.O. Hotel. I. House and ruins of I. Castle, the ancient seat of the chieftain of Clan Macdonnell, in vicinity.

Invergelder, pl., S.W. Aberdeenshire, 1¼ m. S.W. of Balmoral.

Invergeldie Lodge, Perthshire, 5 m. N.W. of Comrie.

12

Inverglen House, Argyll, near Strachur pier, on E. side of Loch Fyne.

Invergloy, seat, also steamboat pier for Loch Lochy, and ry. sta., L.N.E., Inverness-shire, 5 m. N. of Spean Bridge. P.O., T.O.

Invergordon, police bur. and seapt. town, with ry. sta., L.M.S., and ferry, E. Ross and Cromarty, on Cromarty Firth, 13 m. N.E. of Dingwall and 12¾ m. S.W. of Tain. Bur. pop. 1417. P.O., T.O. Hotel. On dockyard pier is a white group-flashing, ten-seconds light. Ferry communication with Balblair, on Black Isle, and Cromarty. I. Castle, seat, 1 m. N.W.

Invergowrie, q.s. par. and vil., with ry. sta., L.M.S., Angus ; vil. (in Perthshire) 3½ m. W. of Dundee. P.O., T.O. Pop. 906. I. House, 1½ m. N.E. of sta.

Inverguseran House, W. Inverness-shire, on Sound of Sleat, 5 m. S.E. of Ornsay Isl.

Inverhaggernie, pl., Perthshire, 1 m. N. of Crianlarich.

Inverharity, ham., Angus, 2 m. N. of Kirton of Glenisla.

Inverichnie, pl., N.E. Banffshire, 2½ m. S.W. of Macduff.

Inverie, ham., stream, and seat, W. Inverness-shire, 10 m. S.E. of Ornsay, on Loch Nevis.

Inverinan Lodge, Argyll, 13 m. S. Taynuilt, on W. side of Loch Awe. State forest nearby.

Inverinate, ham. and seat, S.W. Ross and Cromarty, 10 m. S.E. of Lochalsh, on Loch Duich. Hotel. P.O., T.O. State forest nearby.

Inverkeilor, par. and vil., with ry. sta., L.N.E., Angus ; 10,252 ac. Pop. 1206. 6 m. N. of Arbroath, on Lunan Water. P.O., T.O.

Inverkeithing, parl., royal, and mun. bur. and par., with ry. sta., L.N.E., Fife ; par. 2830 ac. Pop. 4969 ; parl. and mun. bur. pop. 3185. The town stands on a bay of its own name, 3¾ m. S.E. of Dunfermline. P.O., T.O. Has two red fixed lights. Papermaking mills, quarrying, etc.

Inverkeithney, par. and vil., Banffshire ; par. 7641 ac. Pop. 640. Vil. 7 m. S.W. of Turriff, on r. Deveron. P.O.

Inverkindie, ham., Aberdeenshire, 10 m. S.W. of Rhynie.

Inverkip, par. and vil., with ry. sta., L.M.S., W. Renfrewshire ; par. (containing town of Gourock) 12,495 ac. Pop. 11,179. Vil. 6 m. S.W. of Greenock. P.O., T.O. Vil. pop. 467.

Inverkirkaig, ham., S.W. Sutherland, near Lochinver.

Inverlael Lodge, N.W. Ross and Cromarty, 6 m. S.E. of Ullapool.

Inverlaidnan, ham., E. Inverness-shire, 3 m. W.S.W. of Carr Bridge.

Inverlair, seat, Inverness-shire, 8 m. E. of Spean Bridge.

Inverleith, northern dist. of Edinburgh. Contains Royal Botanic Gardens and Arboretum and I. Public Park.

Inverliver Forest, a State forest, Argyllshire, on Loch Awe, 9 m. S.W. of Kilchrenan.

Inverliever House, Argyll, 1½ m. N.E. of Ford Pier, Loch Awe.

Inver, Loch, N.W. Sutherland, Assynt par., sea loch.

Inverlochy Castle, two castles, ruin and modern, in S.W. end of Great Glen, Inverness-shire, on Lochy R. Ruin 1½ m., modern castle 3 m., N.E. of Fort William. P.O. called Innerlochy.

Invermallie, pl., S.W. Inverness-shire, 2 m. W. of east end of Loch Arkaig.

Invermark Castle, ruins of castle, Angus, 17 m. N.W. of Edzell. In vicinity, I. Lodge. I. Deer Forest, 20,000 ac.

Invermarkie, seat, W. Aberdeenshire, 7 m. W. of Huntly.

Invermay, seat, Perthshire, 1½ m. S.E. of Forteviot sta. Ruined ancient tower.

Invermoriston, ham. and pier, Inverness-shire, on W. side of Loch Ness, 6 m. N.E. of Fort Augustus. P.O. and T.O. called Glenmoriston. Hotel. I. House, seat. I. Deer Forest, 17,220 ac.

Invermossat, ham., Aberdeenshire, 6 m. W. of Alford.

Invernahavon, ancient battlefield, Inverness-shire, 6 m. S.W. of Kingussie.

Invernahyle, seat, W. Argyll, 2½ m. E. of Appin sta.

Invernan, seat, S.W. Aberdeenshire, Strathdon par., 18 m. S.W. of Rhynie.

Invernauld Lodge, S. Sutherland, on r. Oykell, 7 m. N.W. of Invershin sta.

Invernaver, ham., N. Sutherland, near mouth of r. Naver.

Inverneil, seat, Argyll, S. Knapdale, on Loch Fyne, 2½ m. S. of Ardrishaig.

Inverness, par., parl. and royal bur., mkt. and co. town, seapt., Inverness-shire, 108 m. W.N.W. of Aberdeen, 192½ m. N. of Edinburgh, situated near the mouth of the r. Ness and at the junction of the Beauly and Moray Firths; ry. sta., L.M.S. Frequently called the capital of the Highlands. The Caledonian Canal passes about ½ m. from the town; par. (I. and Bona) 28,438 ac. Pop. 26,584; royal and mun. bur. pop. 22,582. Hotels. Chief industries: brewing, distilling, ironfounding, and the manufacture of woollen goods. In addition, there are railway repair shops. Fixed lights shown: breakwater, one green; Thornbush pier, two red; and on foundry, one green. The I. Division of I. and Ross and Cromarty returns one member to Parliament. Coasting steamers sail to Aberdeen, Cromarty, etc.

Inverness-shire, a maritime Highland co., stretching from the Moray Firth to the Atlantic Ocean, and embracing all the Skye group of the Inner Hebrides and the Outer Hebrides southward from Lewis proper and the tiny distant isle-group of St. Kilda. It is the largest co. in Scotland, with an area of 4211 sq. m., or 2,695,094 ac. Pop. 82,108. The mainland is divided into two sections by Glenmore, through which runs the Caledonian Canal, connecting Loch Linnhe and Beauly Firth. The surface is extremely mountainous. Among the loftiest summits are Ben Nevis, 4406 ft., the highest peak in Great Britain; Breariach, 4248 ft.; and Cairngorm, 4084 ft. The chief rivers are the Spey, the Ness, and the Beauly. The most important lochs, Ness, Oich, and Lochy in the course of the Caledonian Canal; Ericht and Laggan in the S., Shiel in the W., Eil and Arkaig to the W. of the Canal. The chief sea lochs are Hourn, Nevis, Aylort, Eil, and Leven. Old red sandstone prevails in the N.E. tract of low country, and is mostly covered with fertile soil, but granite, trap, and metamorphic rocks prevail, the co. generally being wood and forest, heath and stony waste. Striking traces of glacial action are found in the terraces of Glen Roy. Sheep-farming is extensively carried on. On the W. coast

the most important industry is herring-fishing, while electrical power—chiefly for the manufacture of aluminium—is carried on at Foyers, on Loch Ness, Fort William on the W., and a number of the Highland lochs are now being connected to form another large electrical power scheme. I. and Ross and Cromarty form three parliamentary divisions, each returning one member.

Invernettie, vil., Aberdeenshire, 1¼ m. S. of Peterhead.

Invernoaden, pl., Argyll, 3 m. S.E. of Strachur, on Loch Fyne.

Invernochty, Doune of, Aberdeenshire. See DOUNE.

Inveronich House, Argyll, near Lochgilphead, Cowal.

Inveroran Hotel, Argyll, 2¾ m. N.W. of Bridge of Orchy, on Loch Tulla.

Inveroykell Lodge, N. Ross and Cromarty, 10 m. N.W. of Culrain sta.

Inverpattack Lodge, Inverness-shire, 1½ m. E. Loch Laggan. Hotel.

Inverpefferan, G. name for Dingwall, Ross and Cromarty.

Inverpolly Lodge, N.W. Ross and Cromarty, 7 m. S. of Lochinver, on Loch Enard.

Inverquharity Castle, old baronial castle, Angus, 3 m. N.E. of Kirriemuir.

Inverquhomery, seat, 2 m. S.W. Longside, Aberdeenshire.

Inverquiech, site of ancient castle, E. Perthshire, 2 m. N.E. of Alyth.

Inverroy, vil., Inverness-shire, on r. Spean, 11 m. N.E. of Fort William.

Inversanda, pl., Argyll, 6 m. S.W. of Ardgour, on W. side of Loch Linnhe.

Inverscadle, bay, S.W. Inverness-shire, Loch Linnhe, 9 m. S.W. of Fort William.

Invershiel House, S. Ross and Cromarty, 1½ m. S. of Kintail, on Loch Duich.

Invershin, ham., with ry. sta., L.M.S., S. Sutherland, 5½ m. S. of Lairg. P.O., T.O.

Inversnaid, ham., with hotel and pier, Stirlingshire, on E. side of Loch Lomond, 3 m. N.E. of Tarbet, 4¾ m. S.E. of Ardlui Pier. P.O., T.O. In vicinity, I. Lodge and I. Waterfall. Ferry for Inveruglas, opposite.

Invertiel, q.s. par., Abbotshall and Kinghorn pars., Fife.

Invertrossachs, seat, S.W. Perthshire, 5 m. S.W. of Callander, on Loch Vennachar.

Inverugie, seat, Morayshire, 3 m. E. of Burghead.

Inverugie, vil., with ry. sta., L.N.E. (¾ m. S.W.), Aberdeenshire, 2½ m. N.W. of Peterhead. P.O., T.O. at sta. I. Castle, ruined castle of the Earls Marischal.

Inveruglas, ham., Dunbartonshire, 3¼ m. N.W. of Luss, on W. side of Loch Lomond. Inn.

Inveruglas, Upper, ferry station (for Inversnaid), Dunbartonshire, on I. Water, W. side of Loch Lomond. Adjacent is I. Isle, with ruins of stronghold of the Macfarlanes.

Inverurie, royal and police bur. and par., with ry. sta., L.N.E., N.W. Aberdeenshire, 16 m. N. of Aberdeen, at confluence of the Ury with the Don ; par. 4951 ac. Pop. 4415 ; police bur. pop. 4524. P.O., T.O. Hotel. Large railway repairing works, and is centre of an important agricultural dist. The Bass of I., a natural mound, is close to the town.

Inverurie Paper Mill, Port Elphinstone, Inverurie, Aberdeenshire.

Invervar, ham. and seat, Perthshire, 4½ m. W. of Fortingall.

Invervegain, pl., Argyll, 2½ m. N. of Inverchaolin, on Loch Striven.

Inverwick, pl., Inverness-shire, 6 m. S.W. of Invermoriston Hotel, on Dundreggan Loch.

Invery House, seat, N.W. Kincardineshire, 1½ m. S. of Banchory.

Inzie Head, headland, N.E. Aberdeenshire, 4 m. N.W. of Rattray Head.

Inzievar, seat, Fife, 4 m. W. of Dunfermline.

Iochdar, Outer Hebrides. P.O., T.O. See EOCHAR.

Iona, isl., q.s. par. and vil., with pier, Kilfinichen and Kilvickeon par., Argyll, Inner Hebrides. P.O., T.O. 1½ m. W. of the Ross of Mull and 47¾ m. W. by S. of Oban ; it is about 3½ m. long by 1½ m. broad. I. is also called Icolmkill, or simply I. It was famous from 563 till the beginning of eighth century for the evangelising establishment of Columba and his Culdee successors ; it afterwards suffered seriously by invasions of Norsemen and Picts ; it became notable again, in 1203 and following years, by erection on it of Roman Catholic establishments, and the remains are chiefly of that period. The chapel of St. Oran is supposed to date from the eleventh century, and the attached burying-ground is said to contain the remains of Scottish, Irish, and Norwegian kings. The inhabitants are engaged in crofting and fishing. Pop. 141.

Iona, Sound of, between Mull and Iona, 1 m. wide.

Iorsa Water, Arran Isl., flows 8 m. S.W. to Machrie Bay, through Glen I.

Ireland, Bay of, Orkney, Stenness par.

Ireland, Ness of, prom., mainland, Shetland, Dunrossness par.

Irongath Hills, W. Lothian, 1½ m. N. of Linlithgow ; alt. 559 ft.

Irongray, ry. sta., L.M.S., Dumfriesshire. See KIRKPATRICK-IRONGRAY.

Iron Isle, sunken rock, visible at low water, S.W. coast of Arran Isl., off Brown Head.

Ironlosh, loch, S.E. Kirkcudbrightshire, 1½ m. E. of Kippford.

Ironmacannie, pl., Kirkcudbrightshire, 2½ m. S. of Balmaclellan.

Iron Rock, or **Sgeir Maoile,** in Sound of Jura, Argyll. Lighthouse, with half-minute flashing light, seen 14 m.

Irvine, parl., royal, and mun. bur., mkt., and seapt. town, and par., with ry. stas., L.M.S., Ayrshire ; par. 3953 ac. Pop. 8318 ; bur. pop. 12,032. The r. Irvine rises at the watershed with Lanarkshire and flows 21 m. W. to the Firth of Clyde, and there has a joint mouth with the Garnock. The town stands on the N. side of the mouth, 7 m. W. of Kilmarnock and 11 m. N. of Ayr. P.O., T.O., and P.O., T.O. at I. Harbour. Coal and chemicals are exported. The industries are shipbuilding, steam sawmills, engineering and foundry yards, and chemical works. A fixed light at inner side of the entrance to the harb., visible 6 m., and at N. side of entrance a white flashing light. Returns one member to Parliament.

Irvine House, seat, S.E. Dumfriesshire, 3 m. S.S.E. of Langholm.

Irving, ancient par., now in Kilpatrick-Flemington par., S.E. Dumfriesshire.

Irvingtown, S.E. Dumfriesshire, 3½ m. N.W. of Gretna.

Isauld, ham., N.W. Caithness, 10½ m. W. of Thurso.

Isay Island, Isle of Skye, 18 m. N.W. of Portree, in Loch Dunvegan.

Isbister, seat, Orkney, mainland, 6 m. N.W. Kirkwall.

Isbister, ham. and loch, Shetland, in E. of Whalsay.

Isbister, loch, Orkney, mainland, 9 m. N.W. of Finstown; alt. 40 ft., depth 3 ft.

Isla, r. of Angus and Perthshire, flows 46 m. through Glenisla and Strathmore to the Tay, in the vicinity of Cargill sta., 4½ m. S.W. of Coupar-Angus. In the Den of Airlie, Angus, forms the Reekie Linn and the Slugs of Auchrannie.

Isla, r., Banffshire, rises near Dufftown, and flows 18 m. N.E. to the Deveron, near Rothiemay.

Islabank, seat, E. Perthshire, 1½ m. N.W. of Coupar-Angus.

Isla Distillery, Perth.

Island Vow, Dunbartonshire, 2 m. S.E. of Ardlui Pier.

Isla Park, seat, E. Perthshire, 1 m. N.E. of Coupar-Angus.

Islay, chief isl. of southernmost group of Inner Hebrides. Lies 17 m. W. of Kintyre; is divided by I. Sound (1 m. wide) from Jura; length 25 m., breadth 19 m.; area of co. dist. 255,455 ac., or 235 sq. m. Pop. 4970. Loch Indail penetrates its S.W. side, and almost meets Loch Gruinart, which opens on its W., forming a peninsula known as the Rhinns of Islay. The chief industries are dairy-farming, stock-raising, and whisky distilling. Slate is plentiful and marble is obtained. Chief town, Bowmore. Pop. 620. On Rudha Mhail, the N. pt., is a fixed light (white and red sectors), seen 18 m. On M'Arthur's Head, at S. end of sound, is a fixed light, seen 17 m. There is daily communication by steamer from W. Loch Tarbert to Port Ellen.

Islay House, ancient seat of the Campbells of Islay, at head of Loch Indail, Islay, Argyll.

Islay, Mull of, or **Mull of Oa,** prom., S. end of I. Isl., Argyll.

Islay, Sound of, between Islay and Jura Isls., Argyll, 14 m. long, 1 m. wide.

Isle Maree, islet, with ancient burying-ground, in Loch Maree, Ross and Cromarty.

Isle Martin, isl., in Loch Broom, N.W. Ross and Cromarty, 4½ m. N.W. of Ullapool. Pop. 3.

Isle of May, see MAY, ISLAND OF.

Isle of Whithorn, seapt. vil. with pier, S.E. Wigtownshire, 3½ m. S.E. of Whithorn. P.O., T.O. Pop. 233.

Isle Ornsay, isl., off N. coast of N. Uist Isl., Outer Hebrides, Inverness-shire.

Isle Ornsay, isl. and vil., Isle of Skye, Inverness-shire. Vil., Isleornsay, with pier and good harbour, 9 m. S.E. of Broadford. P.O., T.O. Off S.E. coast, lighthouse with a fixed light, seen 12 m. Off N. end a white flashing six-seconds light, seen 9 m. Pop. 2.

Isle Toll, pl., Dumfriesshire, 2 m. S.E. of Auldgirth. I. Tower, seat, in vicinity.

Islivig, pl., Ross and Cromarty, W. coast of Lewis Isl., Uig par. P.O.

Itlaw, ham., 5 m. S. of Banffshire. P.O., T.O.

Iubhair, loch, S.W. Perthshire, 10 m. S.W. of Killin; depth 65 ft., alt. 512 ft.

J

Jackton, vil., Lanarkshire, 3 m. S.W. of E. Kilbride.

Jamestown, town and q.s. par., with ry. sta., L.N.E. (passenger service withdrawn), Dunbartonshire, 1 m. S.E. of Balloch Pier, Loch Lomond. P.O., T.O. Calico print works.

Jamestown, vil., Fife, near Inverkeithing.

Jamestown, vil., Ross and Cromarty, 1 m. S. of Strathpeffer.

Janefield, seat, 1¼ m. N.N.E. of Kirkcudbright.

Janetown, fishing vil., S.W. Ross and Cromarty, on W. side of Loch Carron, 4 m. S.W. of Strathcarron sta., 5 m. N.E. of Strome Ferry. P.O., T.O., called Lochcarron. Pop. 259.

Janetstown, vil., Caithness, 2½ m. S.W. of Thurso. P.O.

Jardine Hall, seat, Dumfriesshire, on r. Annan, 5 m. N.W. of Lockerbie.

Jarlshof, ruin, S. Shetland, mainland, on W. Voe, 1¼ m. N.W. of Sumburgh Head.

Jeantown, see JANETOWN and LOCH-CARRON.

Jedbank, seat, Jedburgh, Roxburghshire.

Jedburgh, royal and mun. bur., par., and co. town of Roxburghshire, with ry. sta., L.N.E.; par. 23,204 ac. Pop. 4110; bur. pop. 3057. P.O., T.O. Hotel. The town stands on Jed Water, 10¾ m. S.W. of Kelso and 56 m. S.E. of Edinburgh by rail. It has an Abbey founded by David I. in 1118 and partially restored by the ninth Marquess of Lothian, 1875. The manufacture of woollen goods is the chief industry.

Jedfoot Bridge, ry. sta., L.N.E., Roxburghshire, 1¾ m. N. of Jedburgh.

Jed Forest, now extinct, formerly covered the valley of the Jed, Roxburghshire.

Jedforest, shooting-lodge, Roxburghshire, 4½ m. from Jedburgh.

Jed Water, Roxburghshire, flows 21 m. N. past Jedburgh to the Teviot, near Jedfoot Bridge sta.

Jedworth, an ancient name of Jedburgh.

Jeffries Corse, hill, 6 m. N. of Peebles; alt. 2004 ft.

Jeminaville, vil. on S. side of Cromarty Firth, E. Ross and Cromarty, 2 m. S.S.E. of Invergordon by ferry. P.O., T.O. called Poyntzfield.

Jennyswell Convalescent Home, Arbroath, Angus.

Jerviston House, Lanarkshire, on S. Calder Water, ¾ m. N.E. Motherwell.

Jerviston Square (including **Coalhall**), Lanarkshire, Bothwell par., 1½ m. N.E. of Motherwell.

Jerviswood, estate, 1½ m. N.E. of Lanark.

Jethart, see JEDBURGH.

Jewel Cottages, mining vil., city of Edinburgh, 3½ m. S.E. of G.P.O.

Jock's Lodge, dist., Edinburgh, 1¾ m. E. of G.P.O. P.O., T.O. See PIERSHILL.

Jock's Shoulder, mt., N. Dumfriesshire, 6 m. E.S.E. of Moffat; alt. 1754 ft.

Jock's Thorn, pl., Ayrshire, ½ m. E. of Kilmaurs.

John Dippie's Well, on Watch Water, Berwickshire, near Longformacus.

John Knox's Church, q.s. par., Aberdeen.

John o' Groats House, Caithness, 1½ m. W.S.W. of Duncansbay Head, formerly curious habitation, built in early part of sixteenth century, now represented by grassy mound. P.O., T.O., and hotel. John o' Groats denotes the N. extremity of the mainland.

Johnsburn House, seat, Midlothian, Balerno.

Johnshaven, fishing vil., Kincardineshire, 8½ m. N.E. of Montrose. Coastguard and lifeboat sta., with ry. sta., L.N.E. P.O., T.O. Pop. 719.

John's Hill, a summit of the Ochils, Perthshire, 4 m. S. of Dunning; alt. 1500 ft.

Johnston, seat, Kincardineshire, 1 m. S.E. of Laurencekirk.

Johnstone, town (police bur.) and q.s. par., with ry. stas. (J. and J. North), L.M.S., Renfrewshire, on the Black Cart, 3 m. W. of Paisley. Pop. 12,837. P.O., T.O. Principal industries: cotton and flax spinning, manufacture of machinery, brass, and iron. J. Castle, 1 m. S.E., a seat.

Johnstone, par., Dumfriesshire, on r. Annan; 13,486 ac. Pop. 661. P.O., T.O. at J. Bridge, 1 m. N.W. of Dinwoodie sta.

Johnstone, seat, Dumfriesshire, 15 m. N.W. of Langholm.

Johnston Loch, N. Lanarkshire, ½ m. N. of Gartcosh sta.

Johnstounburn, seat, E. Lothian, 2¼ m. N.E. of Blackshiels.

Johnstoun's Mill, ham., Fife, 1 m. E. of Largo.

Joppa, vil., 5½ m. E. of Ayr.

Joppa, ry. sta., L.N.E., and E. sub. of Portobello, within the city of Edinburgh, 3½ m. E. of G.P.O. P.O., T.O. Salt works.

Jordanhill, dist., N.W. Glasgow, seat and ry. sta., L.N.E. P.O. Teachers' training college.

Jordanstone, seat and ry. sta., L.M.S., Perthshire, 2 m. E. of Alyth.

Junction Bridge, ry. sta., L.N.E., Leith, Edinburgh.

Juniper Bank, seat, Peeblesshire, 1 m. N.E. of Walkerburn.

Juniper Green, q.s. par., with ry. sta., L.M.S., on Water of Leith, in city of Edinburgh, 5 m. S.W. of G.P.O. P.O., T.O.

Jura, isl. and par., Argyll, Inner Hebrides, extends 18 m. S.S.W.; widens from 3½ m. near its N. end to 8½ m. at its S. end, and sends off from its E. side Lochs Crinan, Swin, and Killisport. Is nearly bisected in the middle by Loch Tarbert in the W. Consists chiefly of one bleak, rugged mountain ridge; rises in its S. section into three conical peaks, called Paps of Jura, the loftiest with an alt. of 2571 ft. Par. (including the isls. of Balnahua, Garvelloch, Lunza, Pladda, Scarba, and Skervuile), 93,794 ac. Pop. 382; pop. of isl. 364. P.O. Hotel.

Jura Deer Forest, Jura Isl., Argyll. Shooting-lodge near Craighouse Hotel, 6 m. E. of Port Askaig, Islay.

Jura House, seat of Campbell of Jura, S. coast of Jura, 3½ m. S.W. of Craighouse Hotel.

Jura, Sound of, separates the isl. of J. from S. part of Knapdale and Lorn, Argyll. Extends about 20 m. S.S.W., and is from 3 to 7 m. broad.

Justinhaugh, ry. sta., L.M.S., Angus, 4¼ m. N. of Forfar. T.O.

K

Kailzie, seat, in co. of, and 2½ m. S. of, Peebles. Ruins of church in old church-yard.

Kaimes Hill, alt. 800 ft., 3 m. S. of Ratho, Midlothian. Vestiges of a Danish camp.

Kaim Park, seat, 1 m. S.E. of Bathgate, W. Lothian.

Kaims, ham., par. of Liberton, and 3½ m. S. of Edinburgh.

Kaims, vil., 1 m. S.W. of Muirkirk, Ayrshire.

Kaims, seat, Newlands par., Peeblesshire.

Kaims Castle, a Roman camp, 2¼ m. N.N.E. of Ardoch Camp, S.E. Perthshire.

Kaimeshill, seat, ½ m. S.W. of Muirkirk, Ayrshire.

Kaimshill Sanatorium, Ayrshire, 1 m. E. of Kilmarnock.

Kair House, seat, near Bervie Water, and 1½ m. N.E. of Fordoun sta., Kincardine-shire.

Kale Water, stream, in E. Roxburghshire, rising among Cheviots, flows N.E. for about 20 m., and falls into Teviot 4½ m. S. of Kelso.

Kalloch (An Teallach), mt., alt. 3483 ft., N.W. Ross and Cromarty, see CHALLICH.

Kalnakill, ham., on Raasay Sound, 6½ m. N. of Applecross, N.W. Ross and Cromarty.

Kame of Hoy, prom. on Hoy Sound, Hoy Isl., Orkney, and 15 m. S.W. of Kirkwall. A landmark for mariners.

Kames, vil., on Kyles of Bute, and 1½ m. S.W. of Tighnabruaich, Argyllshire. Has inn and pier. P.O., T.O. Nearby are K. Powder Mills.

Kames, vil., on W. side of Loch Fyne and 3¾ m. E. by N. of Lochgilphead, Argyll-shire. Has saltpetre works. Pop. 297.

Kames, seat, 5 m. E. of Greenlaw, S. Berwickshire.

Kames Bay, opens off Firth of Clyde, 2½ m. N.W. of Rothesay, Buteshire.

Kames Castle, on Kames Bay, at base of Kames Hill (875 ft.); in N.W. vicinity is Wester Kames Castle, a ruined tower, once a seat of the Spences.

Kames, Mid, and Kames, West, adjacent hams. on Loch Fyne, 3½ m. N.E. of Lochgilphead, Argyllshire.

Kanaird, stream, N.W. Ross and Cromarty, with a course of 5 m., and flowing into Loch Broom opposite Isle Martin.

Katewell, seat, Evanton, Ross and Cromarty.

Katrine, Loch, 8½ m. W. of Callander, S.W. Perthshire and N.W. Stirling-shire; lies E.S.E. and W.N.W. and is 8 m. long, with an average breadth of ¾ m.; 364 ft. above sea-level and 495 ft. deep; engirt by mts., the highest being Ben Vane on the N. and Ben Lomond and Ben Venue on S. Celebrity partly due to natural beauty and partly to association with Scott's "Lady of the Lake." Source of Glasgow water supply.

Kearon, loch, 11 m. S.W. of Tarbert, Kintyre, W. Argyllshire.

Kearvaig, pl. and stream in N.W. Suther-landshire; rising in Loch na Gainmhich and reaching the sea in a bay of the same name, 2½ m. S.E. of Cape Wrath.

Keavil, seat, 2¼ m. S.W. of Dunfermline, Fife.

Kebbaty House, 4 m. S. of Monymusk, Aberdeenshire.

Kebock (or Cabag) Head, prom. on E. side of Lewis, 13 m. S. of Stornoway, Outer Hebrides.

Kedslie, farm, 3 m. N.W. of Earlston, Berwickshire. Site of a pre-Reformation chapel.

Keen, Mount, cone-shaped peak (3077 ft.) of Grampians, 7 m. S.S.E. of Ballater, border of Aberdeenshire and Angus.

Keglish, or Geo-glas, loch, 4 m. S. of Dores, N.E. Inverness-shire; alt. 763 ft., depth 32 ft.

Keig, par. and vil., on the r. Don and 4 m. N.E. of Alford, Aberdeenshire; par. 8062 ac. Pop. 492. P.O.

Keilarsbrae and Hallpark, water dist. in Clackmannanshire, in vicinity of Alloa. K. House is a seat. Pop. 1372.

Keil House, seat, 10½ m. S.W. of Campbeltown, on S. coast of Kintyre, Argyllshire.

Keillor, Hill of, see HILL OF KEILLOR.

Keillour Castle, seat, 2½ m. N.W. of Balgowan, mid Perthshire.

Keills, ham., 13¾ m. S.W. of Crinan, W. Argyllshire. P.O. K. Head, prom. between Loch Sween and the Sound of Jura ; narrow, and over 8 m. long.

Keils, vil. on S.E. coast of Jura Isl., Argyllshire.

Keil School, Dumbarton. Boy's private school.

Keir, par. in Nithsdale, W. Dumfriesshire ; 7813 ac. Pop. 463. On the r. Scar and 4 m. S.W. of Thornhill is K. Mill (alt. 1171 ft.), with P.O.

Keir, seat, 1¾ m. N.W. of Bridge of Allan and 1¾ m. S.W. of Dunblane, S. Perthshire.

Keirs Hill, alt. 1005 ft., 4 m. W.N.W. of Dalmellington, Ayrshire.

Keiss, q.s. par. and fishing vil., on Sinclairs Bay, and 8 m. N.E. of Wick, Caithness. P.O., T.O. Hotel. Shows a fixed light, visible 5 m. Nearby the ruins of K. Castle, and also K. House, a modern structure.

Keith, par. and town (police bur.), on the Isla, N. Banffshire ; par. 19,395 ac. Pop. 6082 ; pop. of bur. 4424. Town is 12½ m. N.W. of Huntly. P.O., T.O. Hotel ; ry. stas. (K. and K. Town), L.N.E. Has distilleries, woollen mfrs., and meal mills.

Keith, ancient par., now incorporated in par. of Humbie, E. Lothian. P.O., T.O. called Humbie. K. Marischal, on K. Water, 3 m. N.E. of Blackshiels, is a seat ; 1½ m. S.E. is Upper K.

Keith-hall and Kinkell, par., on the Urie, mid Aberdeenshire ; 7600 ac. Pop. 661. Contains Keith Hall, seat of Earl of Kintore.

Keith-Hundeby, ancient name of Humbie, E. Lothian.

Keithick, seat, 1½ m. S.W. of Coupar-Angus, Perthshire.

Keith-Inch, Aberdeenshire. See PETERHEAD, of which this fishing vil. now forms a part.

Keithock, seat, 2 m. N. of Brechin, Angus.

Keith Paper Mills, Inverkeithing, Fife.

Keithtown, ham., 1¾ m. W. of Maryburgh, E. Ross and Cromarty.

Keith Water, E. Lothian stream, flowing into Humbie Water, 1¼ m. S.E. of Salton ry. sta.

Kelburne Castle, 2 m. S.S.E. of Largs, Ayrshire. Since thirteenth century has been in possession of the Boyle family, now Earls of Glasgow.

Kelhead, vil., 3¼ m. N.W. of Annan, Dumfriesshire.

Kelk Place, vil., Kilmaurs par., Ayrshire.

Kellas, vil. on r. Lossie, and 6½ m. S.S.W. of Elgin, Moray. P.O.

Kellas, or **Hole of Murroes,** ham. in par. of Murroes, 5 m. N.E. of Dundee, Angus.

Kellerstain, home for mental defectives, 5½ m. W. of Edinburgh, near Gogar sta.

Kellie Castle, now decayed, 3½ m. N.W. of Pittenweem, Fife ; property of Earl of Mar and Kellie.

Kellingtrongan, loch, 2¼ m. S. of Ballantrae, S.W. Ayrshire.

Kelloe, seat, 1½ m. S.E. of Edrom sta., Berwickshire. Nearby is Killoe Bastile, a ham.

Kello Water, rises on Blacklarg Hill, flows N.E. for about 9 m., and falls into the Nith, 2½ m. W.N.W. of Sanquhar, Dumfriesshire.

Kells, par., on the Ken, mid Kirkcudbrightshire ; 48,521 ac. Pop. 848. Contains New Galloway.

Kells Range, or **Rhynns of Kells,** mts. (2668 ft.), in pars. of Kells and Carsphairn, N. Kirkcudbrightshire.

Kelly Burn, Aberdeenshire, flows 7 m. N. to the r. Ythan.

Kelly Castle, ancient tower and seat, on Elliot Water, and 2½ m. W. of Arbroath, Angus.

Kelly House, on the Kelly Burn, ¾ m. N.N.W. of Wemyss Bay, W. Renfrewshire.

Kellysyde, seat, ¼ m. N.E. of Dollar, E. Perthshire.

Kelphope Burn, headstream of the r. Leader, 6 m. long, meeting Soonhope Burn 3 m. N. of Lauder, Berwickshire.

Kelso, par., police bur., and mkt. town in Roxburghshire, 52 m. S.E. of Edinburgh ; par. 5466 ac. Pop. 4279. Occupies an extensive plain on the N. bank of the Tweed and opposite the confluence of that r. with the Teviot ;

ry. sta., L.N.E. P.O., T.O. Pop. of bur. 3854. Hotel. Has important corn mkt., agricultural trade, and other industries. Contains the ruins of K. Abbey.

Keltie Castle, seat, 4 m. E. of Auchterarder, Perthshire.

Keltie Water, stream, S.W. Perthshire; joins the Teith, 3 m. S.E. of Callander. The famous Bracklinn Falls are in the vicinity of Callander.

Keltney Burn, stream in Breadalbane, N.W. Perthshire; rises on Carn Mairg, flows N.E. and then S.E., and falls into the Lyon 1½ m. above its mouth at Coshieville Inn. Above Coshieville are the Falls of K.

Keltneyburn, ham., 3 m. N.E. of Fortingall, mid Perthshire. P.O.

Kelton, small river-port on the Nith, in co. of, and 3½ m. S. of, Dumfries.

Kelton, par., on the Dee, S. Kirkcudbrightshire; 11,208 ac. Pop. 3621. Contains part of Castle-Douglas; also K. Hill, or Rhonehouse; vil. 2½ m. S.W. of Castle-Douglas. P.O. called K. Hill. Pop. Netherwood, K., and Craigs special water dist. 340.

Kelty, q.s. par. and coal-mining town in Fifeshire; town 5 m. S.E. of Kinross; ry. sta. (1 m. E.), L.N.E. P.O., T.O. Pop. 6439.

Kelty Bridge, adjoining vil., in par. of Cleish, Kinross-shire.

Kelty Water, trib. of Forth, forming part of boundary between cos. of Perth and Stirling, 2¼ m. N. of Buchlyvie.

Kelvin, stream, Stirlingshire, Lanarkshire, and Dunbartonshire, rising in Kilsyth Hills and flowing S.W. for 21 m., when it joins the Clyde at Partick, in W. of Glasgow; 4 m. N.W. of Glasgow it enters the wooded dell of Kelvingrove.

Kelvin Bridge, ry. sta. on L.M.S. and Glasgow District Subway, in the N.W. of the city. A bridge across the Kelvin. P.O., Kelvindale.

Kelvindale Paper Mills, Kirklee, Maryhill, Glasgow.

Kelvingrove, parl. division of Glasgow. Pop. 67,870.

Kelvinhaugh, q.s. par. and dist. on N. side of Clyde, Glasgow, and 3½ m. S. of Maryhill. Contains Queen's Docks, and has important shipbuilding yards.

Kelvinside, q.s. par. and residential dist., on r. Kelvin, in the N.W. of Glasgow; ry. sta., L.M.S.

Kemacoil, seat in par. of Little Dunkeld, Perthshire.

Kemback, par. on r. Eden, Fife. Contains K. House, a seat, 3 m. E. of Cupar par. 2585 ac. Pop. 519.

Kemnay, par. and vil. on r. Don, E Aberdeenshire; par. 5113 ac. Pop. 1468. Vil. is 18 m. W.N.W. of Aberdeen; ry. sta., L.N.E. P.O., T.O. Vil. pop. 870. K. House is a seat in the neighbourhood. State forest in neighbourhood.

Kemp, loch (51 ft. deep, alt. 578 ft.), near E. shore of Loch Ness, and 4½ m. S.W. of Foyers Hotel, mid Inverness-shire.

Kempock, prom. on Gourock Bay, Firth of Clyde, Renfrewshire. Two fixed green lights.

Kendoon, pl., with one of the Galloway Water Power Co.'s hydro-electric stas., near New Galloway, Kirkcudbrightshire.

Kenley Green House, ½ m. N.E. of Boarhills sta., Fife.

Ken, Loch, depth 62 ft., alt. 142 ft. See KEN, WATER OF.

Kenlum, hill, 2½ m. N.W. of Gatehouse-of-Fleet, S.W. Kirkcudbrightshire; alt. over 950 ft.

Kenmore, fishing vil. on W. side of Loch Fyne, and 4½ m. S.W. of Inveraray, Argyllshire.

Kenmore, par. and vil., on Loch Tay, Breadalbane, mid Perthshire; par. 57,273 ac. Pop. 961. Hotel. Vil. at efflux of r. Tay from Loch Tay. P.O., T.O.

Kenmuir, q.s. par., in par. of Old Monkland, Lanarkshire. Pop.

Kenmure Castle, ¾ m. S.E. of New Galloway, mid Kirkcudbrightshire. Seat of the Gordons of Kenmure.

Kenmure House, 1 m. N.W. of Bishopbriggs sta., and 4 m. N. of Glasgow G.P.O., N.W. Lanarkshire.

Kennagary, pl., 3 m. E. of Tarbert, Lewis, Outer Hebrides.

Kennageall, Sutherland. See WHITEN HEAD.

Kennard, or **Ceannard,** small lake, 4 m. S.E. of Aberfeldy, Perthshire; alt. 1342 ft., depth 72 ft.

Kennart, rivulet in Ross and Cromarty. See KANAIRD.

Kennavarah Point, prom., forming S.W. extremity of Tiree Is., W. Argyllshire.

Kennedy Castle, see CASTLE KENNEDY.

Kennet, mining vil. in co. of, and 1 m. E.S.E. of, Clackmannan; ry. sta. (Clackmannan and K., 1 m. N.W.), L.N.E. K. House, in the neighbourhood, is the seat of Lord Balfour of Burleigh.

Kennethmont, par. and ham. in N.W. Aberdeenshire; par. 8468 ac. Pop. 663. Ham. (Kirkhill of K.) is 8 m. S. of Huntly; ry. sta. (563 ft.), L.N.E. P.O., T.O., and T.O. at sta.

Kennetpans, vil., with harb., on Firth of Forth, 3½ m. S.E. of Alloa, Clackmannanshire.

Kennishead, ry. sta., L.M.S., 4½ m. S.W. of St. Enoch sta., Glasgow.

Kennoway, par. and vil. in mid Fife; par. 3943 ac. Pop. 1709. Vil. 3 m. N.W. of Leven. Pop. 946. P.O., T.O.

Kennox House, seat, 2¼ m. W. of Stewarton, N. Ayrshire.

Kenny, Loups of, series of falls on the Melgam, W. Angus, 2 m. N.N.E. of Airlie Castle.

Kensaleyre, pl. at head of Loch Eyre, Skye, Inner Hebrides.

Kensalroag, vil., at head of Loch Bracadale, W. Skye, Inner Hebrides.

Kentallen, vil., with pier, on bay of same name, E. side of Loch Linnhe and 3 m. S.W. of Ballachulish, Argyllshire; ry. sta., L.M.S. P.O., T.O. Two fixed red lights on pier.

Kentangval, vil. in isl. of Barra, Outer Hebrides.

Kentra, ham. on bay of same name, N.E. Ardnamurchan, W. Argyllshire.

Ken, Water of, stream in Kirkcudbrightshire; rises on the N. border of the co., flows for 28 m. S. to the Dee; during the last 4½ m. of its course it expands into Loch Ken (depth 60 ft., alt. 142 ft.)

Keoldale, ham., hotel, and ferry over the Kyle of Durness to Cape Wrath, 76 m. W. of Thurso, Caithness.

Keose, ham. on N. shore of Loch Erisort, E. of Lewis, Outer Hebrides. P.O., T.O.

Keppel Pier, on S.E. coast of the Great Cumbrae, Firth of Clyde, Buteshire.

Kepplehill Colliery, 1 m. S.W. of Stane, Lanarkshire.

Kepplestone, pl., 4 m. W.N.W. of Aberdeen.

Keppoch, pl. in par. of Kintail, on N. shore of Loch Duich, Ross and Cromarty.

Keppoch, seat, 2 m. N.W. of Cardross, Dunbartonshire.

Keppoch, ham., near Arisaig, S.W. Inverness-shire.

Keppoch, seat, on the Spean, S.W. Inverness-shire, 14 m. N.E. of Fort William.

Kerbet Water, 3½ m. S.W. of Forfar, Angus.

Kerelaw, seat, ¾ m. N.E. of Stevenson, N. Ayrshire; nearby are ruins of K. Castle.

Kerfield, seat, on the Tweed, in co. of, and ¾ m. S.E. of, Peebles.

Kerloch, cone-shaped hill (1747 ft.), 5 m. S. of Banchory, Kincardineshire.

Kernsary, loch, 1½ m. W. of Poolewe, W. Ross and Cromarty; alt. 68 ft., depth 93 ft.

Kerrera, isl. in Firth of Lorne, 4 m. E. of isl. of Mull, Argyllshire; length 4¼ m. by average breadth 1½ m.; separated from mainland by Sound of K. On Sgeirean Dubha is a ten-seconds group-flashing light, visible 9 m. P.O. Pop. 79.

Kerriff, or **Kerry,** S. division of par. of Kilfinan, Cowal, Argyllshire.

Kerrow House, seat, 13¼ m. N.W. of Fort Augustus, Inverness-shire.

Kerry, r. in W. Ross and Cromarty; flows for 16 m. N.W. before it falls into Gairloch at Flowerdale. There are fine falls on the r. 4 m. S.E. of Gairloch.

Kerrycroy, vil., on a bay of the same name in isl. of Bute, and 2½ m. S.E. of Rothesay, Bute.

Kerrylamont, vil., 4 m. S.S.E. of Rothesay, isl. of Bute.

Kerrysdale, Pass of, ravine, traversed by the Kerry, 2 m. S.E. of Gairloch, W. Ross and Cromarty.

Kerse, small loch in co. of, and 6 m. S.E. of, Ayr.

Kerse, q.s. par. and seat of Earl of Zetland, 2 m. N.E. of Grahamston, Stirlingshire.

Kersewell, seat, 2¼ m. E.N.E. of Carnwath, E. Lanarkshire.

Kershader, pl. on N. side of Loch Erisort, Lewis, Outer Hebrides. P.O., T.O.

Kershope Burn, S. Roxburghshire, rises on English border, which it follows in a S.W. direction for 9 m., falling into the Liddel 2 m. from Newcastleton.

Kershopefoot, pl., with ry. sta., L.N.E., at junc. of Kershope Burn and Liddel Water, on N.W. border of Cumberland, and 11¾ m. N.E. of Longtown, Roxburghshire. P.O., T.O. at sta.

Kersknowe, seat, 4 m. S.E. of Kelso, Roxburghshire.

Kersland Barony, seat, 3 m. S. of Beith, N. Ayrshire. Vestiges of ancient castle.

Kessock, ham. in E. Ross and Cromarty, 2 m. N.W. of Inverness. P.O., T.O. Ferry over the narrow strait between the Moray and Beauly Firths.

Kessock State Forest, 3 m. S.E. of Allangrange ry. sta., which is 8 m. W. of Fortrose, E. Ross and Cromarty.

Ket, The, stream in S.E. Wigtownshire, falling into the sea, after a course of 5½ m., at Portyerrock.

Kettins, par. and vil. in S.W. Angus; par. 7461 ac. Pop. 660. Vil. is 1 m. E.S.E. of Coupar-Angus. P.O.

Kettle, par. and vil. on the Eden, mid Fife; par. 7587 ac. Pop. 1518. Vil. 1 m. S. of Ladybank Junction. Pop. 844. Ry. sta., L.N.E. P.O., T.O. (called Kingskettle); 1 m. S.E. is a ham., Kettlehill, and 1 m. S. is Kettlebridge.

Kettleholm, ham. on the Water of Milk, and 3 m. S.S.W. of Lockerbie, S. Dumfriesshire. P.O., T.O.

Kettlester, loch, ¾ m. N. of Burravoe, Yell, Shetland.

Kettletoft Bay, in S. of Sanday Isl., Orkney. A fixed light, red and green sectors, on pier-head.

Kevock, ham. adjoining Polton, Lasswade, Midlothian.

Kiachnish Water, stream, 8 m. long, flowing into Loch Linnhe, 3½ m. S.W. of Fort William, Inverness-shire.

Kiavan, loch, 1½ m. S.E. of Clachan of Kintyre, Argyll.

Kidshielhaugh, seat, 5 m. N.W. of Duns, Berwickshire.

Kiel, pl., with ruins of an ancient church, 3 m. N.W. of Connel Ferry, N. Argyllshire.

Kiells, ham., 1 m. W.S.W. of Portaskaig, Islay, Argyllshire.

Kierfold House, 7 m. N. of Stromness, mainland, Orkney.

Kilanailean, ham., 6 m. N.N.W. of Brigend, isl. of Islay, Argyllshire.

Kilantringan Loch, 2 m. S.E. of Ballantrae, S.W. Ayrshire.

Kilarrow and Kilmeny, par. in N.E. of isl. of Islay, Argyllshire; par. 65,956 ac. Pop. 2104.

Kilbady, seat, 3½ m. N.E. of Keith, Banffshire.

Kilbagie, pl., 1½ m. N. of Kincardine, Clackmannanshire. Forth Paper Mills are here.

Kilbarchan, par. and town in mid Renfrewshire; par. 9037 ac. Pop. 7510. Town 5 m. W. by S. of Paisley; ry. sta., L.M.S. P.O., T.O. Town pop. 2747.

Kilberry, prom., seat, and ham., 16 m. S.W. of Tarbert, Knapdale, S. Argyllshire. P.O., T.O.

Kilbirnie, par. and town on the Garnock, N. Ayrshire; par. 10,377 ac. Pop. 8193. Town 10 m. N. of Irvine and 4¼ m. N.E. of Dalry; ry. sta., L.M.S. P.O., T.O. Pop. (K. and Glengarroch) 8332. K. Castle was an ancient seat of the Earls of Crawford. Various mfrs., including linen thread, fishing nets, rope, wincey and flax spinning mills.

Kilblane, ancient chapelry in par. of Kirkmahoe, Dumfriesshire.

Kilblane, ancient par., now part of Southend par., Kintyre, Argyll.

Kilbowie, pl., forming part of Clydebank, Dunbartonshire, with large works of Singer Sewing Machine Co.; q.s. par. ry. sta., L.M.S. P.O.

Kilbrandon and Kilchattan, united par. (partly inland, partly insular) on Firth of Lorne, Argyllshire, and containing the isls. of Easdale, Luing, Seil, Shuna, and Torsay; par. 13,384 ac. Pop. 823.

Kilbrennan Sound, strait between Arran Isl. and Kintyre, Buteshire and Argyllshire; 27 m. long by 3 to 15 m. wide. Herring fisheries. Takes its name from St. Brendon, an Irish monk who visited the Hebrides in the sixth century.

Kilbride, pl., 4 m. S.W. of Broadford, Skye, Inner Hebrides.

Kilbride, ham. in S. Uist, 9 m. N. of Lochboisdale, Outer Hebrides. P.O., T.O.

Kilbride, par. in Arran, Buteshire, on Firth of Clyde; contains Lamlash; par. 39,013 ac. Pop. 2658.

Kilbride, ancient chapelry, 2¾ m. S.S.E. of Creetown, S.W. Kirkcudbright.

Kilbride, ham., 4 m. S. of Tighnabruaich, Argyllshire.

Kilbride, East and **West,** see EAST KILBRIDE and WEST KILBRIDE.

Kilbridemore, seat, Glendaruel, Argyllshire, and 7 m. N. of Loch Riddon (N. of Kyies of Bute).

Kilbryde Castle, seat, 3 m. N.W. of Dunblane, S. Perthshire.

Kilbucho, ham. on the Kilbucho Burn, Peeblesshire, and 4 m. S.E. of Biggar.

Kilcalmonell, par., Argyll ; 28,403 ac. Pop. 1360. Contains a portion of Tarbert on Loch Fyne.

Kilcalmonell and Kilberry, eccl. par. in pars. of Kilcalmonell and S. Knapdale, S. Argyllshire.

Kilchamaig, seat, on bay of same name in W. Loch Tarbert, 5½ m. S.W. of Tarbert, Argyllshire.

Kilchattan, ham., Colonsay Isl., 4½ m. N. of Oronsay Isl., Inner Hebrides, Argyllshire.

Kilchattan, see KILBRANDON AND KILCHATTAN.

Kilchattan, pl., with ruins of ancient church, Gigha Isl., W. of Kintyre, Argyllshire.

Kilchattan Bay, vil. and pier on bay of same name, 8 m. S. of Rothesay, Bute. P.O., T.O. Pop. 156.

Kilcheran House, pl. and loch (depth 60 ft.) in Lismore Isl., Loch Linnhe, and 6 m. N.W. of Oban, Argyllshire.

Kilchiaran, pl., bay, and house, in S.W. of Islay Isl., Argyllshire, 2 m. S.S.W. of Kilchoman church.

Kilchoan, ham., with harbour and inn, par. of Ardnamurchan, 6 m. N.W. of Tobermory, Argyllshire. P.O., T.O.

Kilchoan (locally **Coan** or **Cuan**), ancient par., now in par. of Kilbrandon and Kilchattan, Lorne, W. Argyllshire.

Kilchoman, par. in S.W.of isl. of Islay, 4 m. W. of Bowmore, W. Argyllshire ; 37,428 ac. Pop. 1500. Contains Port Charlotte.

Kilchousland, ancient par., now in Campbeltown par., Kintyre, S. Argyllshire.

Kilchrenan and Dalavich, united par. on Loch Awe, mid Argyllshire ; 45,953 ac. Pop. 322. Contains vil. of K., 7 m. S.E. of Taynuilt. P.O., T.O.

Kilchrist, loch and ancient church, par. of Strath, Skye, Inner Hebrides ; lies 2½ m. S.W. of Broadford, and there is a monument to the Mackinnons.

Kilchrist Castle, 2 m. S.W. of Campbeltown, Kintyre, Argyllshire.

Kilchurn Castle, at N.E. end of Loch Awe and 2½ m. W. of Dalmally, Argyllshire.

Kilcolmkill House, 5 m. n. of Brora, Sutherland.

Kilconquhar (locally pronounced **Kinneuchar**), par. and vil. in Fife ; vil. stands on N. side of loch of same name and is 4 m. E. of Largo ; ry. sta., L.N.E. P.O. Par. 6631 ac. Pop. 1211 ; vil. pop. (K. and Barnyards) 336. Nearby is K. Castle, a seat of the Earl of Lindsay.

Kilcoy, ham., par. of Killearnan, S.E. Ross and Cromarty, and 8 m. N.W. of Inverness. K. Castle is a seat.

Kilcreggan, holiday pl., with pier, at S.W. corner of the peninsula separating Loch Long and Gare Loch, Dunbartonshire, and 6 m. N.W. of Greenock. P.O., T.O. Continuous with Cove on its N.W., with which it forms a police bur.

Kildalloig, seat, 3 m. S.E. of Campbeltown, Kintyre, Argyllshire.

Kildalton, par. and seat, 3 m. S.E. of Bowmore, S.E. Islay, Argyllshire ; par. 47,201 ac. Pop. 1380. P.O., T.O.

Kildary, ham. on r. Balnagowan, par. of Kilmuir Easter, and 5¼ m. N.E. of Invergordon, E. Ross and Cromarty ; ry. sta., L.M.S. P.O., T.O. Nearby is the seat of K. House.

Kildavanan, pl. on N. side of Ettrick Bay, 5 m. N.W. of Rothesay, Bute Isl., Firth of Clyde.

Kildean, 1½ m. W. of Stirling. Town hospital.

Kildermorie Deer Forest, 25,000 ac., in E. Ross and Cromarty. K. Lodge is 10 m. N.W. of Alness.

Kildonan, ham. and seat, 8 m. S. of Lamlash, in S.E. of Arran Isl., Bute. P.O., T.O. K. Castle, originally a Dalriadan fortalice ; behind its cliff is the Plain of K., with a stone circle.

Kildonan, vil. on Loch Greshornish and 2 m. N. of Edinbain, Skye, Inner Hebrides.

Kildonan, par. on r. Helmsdale, Sutherland ; par. 134,532 ac. Pop. 1454 ; ry. sta. (192 ft.), L.M.S. P.O., T.O. Contains Helmsdale, 9½ m. S.E. of sta. Near sta. is K. Lodge. The burns in the par. were the scene of a "gold rush" in 1868–1869.

Kildonan House, seat, 1 m. N.W. of Barrhill sta., S. Ayrshire.

Kildrochet House, 3 m. S. of Stranraer, Wigtownshire.

Kildrummy, par. and vil. in mid Aberdeenshire; par. 10,349 ac. Pop. 364. Vil., Milltown of K., stands on r. Don, 7 m. W. of Alford. Nearby is the imposing ruin of K. Castle.

Kilduff, seat, in co. of, and 4 m. W. of, Kinross.

Kilduff, seat, 2 m. S. of Drem ry. junc. E. Lothian.

Kilduncan, seat, 5 m. S.E. of St. Andrews, Fife.

Kilfinan, par., vil., and bay in the Cowal dist. of Argyllshire; bay on E. side of the S. end of Loch Fyne; par. 32,306 ac. Pop. 1489. Vil. on the bay and 5¾ m. N.W. of Tighnabruaich. P.O., T.O.

Kilfinichen and Kilvickeon, par. in S.W. of Mull Isl., Argyllshire; includes also isls. of Errait, Inch Kenneth, Iona, and Dhuheartach; par. 60,937 ac. Pop. 1020.

Kilfinnan, pl. on Caledonian Canal, and 4 m. S.S.W. of Invergarry, Invernessshire. Occulting light, red and green sectors.

Kilgarth, ham. in par. of Old Monkland, 2 m. N.W. of Coatbridge, N. Lanarkshire.

Kilgour, ancient par., now in Falkland par., Fife.

Kilgrammie, mining vil., 1 m. W. of Dailly, W. Ayrshire.

Kilgraston, seat, the property of the Government, now a military hospital for wounded soldiers, 1 m. S.W. of Bridge of Earn, Perthshire.

Kilkatrine, Duke of Argyll's house on E. shore of Loch Fyne, opposite Inveraray, Argyllshire.

Kilkenneth, ham. in W. of isl. of Tiree, Inner Hebrides, Argyllshire.

Kilkenzie, vil., 4¼ m. N.W. of Campbeltown, Kintyre, Argyllshire. P.O., T.O.

Kilkerran, ry. sta., L.M.S., in par. of Dailly, 4½ m. S. of Maybole, Ayrshire. T.O. at sta. Nearby is K. House.

Kilkivan, ancient par., now in par. of Campbeltown, Kintyre, Argyllshire.

Killantrae Burn, 7 m. long, and falls into Luce Bay at Port William, S. Wigtownshire.

Killantringan, lighthouse on prom. of Black Head, 2 m. N. of Portpatrick, W. Wigtownshire. Has group-flashing light, visible 19 m. T.O.; telephonic eommunication with Portpatrick.

Killarrow and Kilmeny, par., Islay, Argyll; area 65,956 ac. Pop. 2014. Contains Bowmore and Bridgend.

Killean and Kilchenzie, par. in W. of Kintyre, Argyllshire; par. 42,138 ac. Pop. 870. Contains ham. of Killean and Killean House, a seat, 1 m. S. of Tayinloan.

Killearn, par. and vil. in W. Stirlingshire; par. 15,375 ac. Pop. 908. Vil. on the Endrick, 9½ m. N.W. of Lennoxtown. Pop. 328. Ry. sta., L.N.E. P.O., T.O. P.O., T.O. at K. sta. K. House is 1½ m. S.W. of vil.

Killearnan, par. on Beauly Firth, Black Isle, S.E. Ross and Cromarty; par. 7735 ac. Pop. 705. P.O., T.O. Par. church, 3 m. E. of Muir of Ord sta.

Killegray, small isl., 2½ m. S. of Harris, Outer Hebrides. Pop. 5.

Killelan, seat, 5 m. S.W. of Campbeltown, Kintyre, Argyllshire.

Killen, pl., 3 m. W. of Fortrose, S.E. Ross and Cromarty. P.O.

Killermont House, seat, 1 m. N. of Maryhill, Dunbartonshire.

Killeter, hill (978 ft.), 3 m. E. of Helensburgh, Dunbartonshire.

Killichonan, vil. on N. side of Loch Rannoch and 8 m. W. of Kinloch Rannoch, N.W. Perthshire.

Killichronan, seat at head of Loch-na-Keal and 3 m. S. of Aros, Mull Isl., Argyllshire.

Killiechassie, seat, 1½ m. N.E. of Aberfeldy, Perthshire.

Killiecrankie, ry. sta. on L.M.S., 3 m. S.E. of Blair-Atholl, and 32¼ m. N.W. of Perth by rail. P.O., T.O. The Pass of K. is on r. Garry, and extends from the sta. for 1½ m. S.E. to Garry Bridge, 2½ m. N.W. of Pitlochry; near the sta. are K. House and K. Cottage. The celebrated battle of K. was fought at the N. end of the Pass, 27th July 1689.

Killiemore House, on N. shore of Loch Scridain, in isl. of Mull, Argyllshire.

Killiemore Lodge, 2½ m. S.E. of Kirkcowan, Wigtownshire.

Killigray, see CALLIGRAY.

Killilan, pl. and seat, 9 m. N.E. of Loch-alsh, S. Ross and Cromarty. P.O. Nearby is K. Deer Forest which, with Glomach, covers 25,000 ac.

Killimster, ham., Caithness, 5 m. N.W. of Wick. K. Loch and K. Moss in vicinity.

Killin, par. and vil. in W. Perthshire; par. 116,176 ac. Pop. 1414. Vil. at confluence of Lochay and Dochart, near W. end of Loch Tay and 23¼ m. S.W. of Aberfeldy; ry. sta. (360 ft.), L.M.S. Pop. 423. P.O., T.O. Summer resort; hotels.

Killiness Point, prom. on Luce Bay, par. of Kirkmaiden, S. Wigtownshire, and 3 m. N. of Mull of Galloway.

Killin Junction, ry. sta. (765 ft.), L.M.S., 4 m. S. of Killin, Perthshire.

Killin Loch, (alt. 1044 ft., depth 67 ft.), 9 m. E.N.E. of Fort Augustus, Inverness-shire. K. Lodge is nearby.

Killin Pier, on Loch Tay, 1 m. N. of vil. of K.

Killisport (Caolisport), sea loch and dist. in W. of Knapdale, Argyllshire. The dist. includes the Point of Knap, while the loch, which penetrates 6½ m. N.E., has many safe anchorages in the small bays.

Killochan, ry. sta. (151 ft.) on L.M.S., 2¼ m. N.E. of Girvan, Ayrshire. T.O. Nearby is the seat of K. Castle (sixteenth century).

Killoch Estate, in pars. of Galston, Mauchline, and Sorn, Ayrshire. Pop. 107.

Killochyett, ham., adjoining Stow, Midlothian.

Killoran Bay, N.W. of Colonsay Isl., Inner Hebrides, Argylishire.

Killundine, ancient castle and seat in Morven, Argyllshire, 7 m. S.E. of Tobermory, on Sound of Mull.

Killywhan, ry. sta. (236 ft.), L.M.S., 8¼ m. S.W. of Dumfries, in Kirkgunzeon par. of Kirkcudbright.

Kilmacolm, par. and town in N. Renfrewshire; par. 19,477 ac. Pop. 5402; town (pop. 2756) on Gryfe Water, and 7¾ m. S.E. of Greenock; ry. sta. (291 ft.), L.M.S. P.O., T.O. A holiday resort with hydropathic.

Kilmadock, par. on r. Teith and 7 m. N.W. of Dunblane, S. Perthshire; par. 24,814 ac. Pop. 2282.

Kilmahew Castle, seat, 1 m. N. of Cardross sta., S.W. Dunbartonshire. Nearby is a ruined castle of the Napiers.

Kilmahog, vil. on the r. Leny and 1 m. N.W. of Callander, S.W. Perthshire.

Kilmailing, part of Old Cathcart, W. Renfrewshire.

Kilmalieu, seat on W. side of Loch Linnhe, Argyllshire, and 10 m. S.W. of Ardgour.

Kilmallie, par., containing Fort William, Lochaber dist., Argyllshire; par. 205,640 ac. Pop. 5049.

Kilmaluag, pl., with hotel (temp.), in extreme N.E. of Trotternish, Skye, Inner Hebrides.

Kilmany, par., vil., and seat in N. Fife; par. 5340 ac. Pop. 521. Vil. 5 m. N.E. of Cupar; ry. sta., L.N.E. P.O., T.O.

Kilmardinny, seat in par. of New Kilpatrick, and 1½ m. S. of Milngavie, S.E. Dunbartonshire.

Kilmarie Lodge, on W. side of Loch Slapin, and 12 m. S.W. of Broadford, Skye, Inner Hebrides.

Kilmarnock, par., mun. bur., and manufacturing town in Cunningham dist., N. Ayrshire, 12 m. N. by E. of Ayr and 24¼ m. S.W. of Glasgow by rail; ry. sta. (154 ft.), L.M.S.; par. 9437 ac. Pop. 36,393; bur. pop. 38,099. P.O., T.O. This town, together with Dumbarton, Port Glasgow, Renfrew, and Rutherglen, forms the K. parl. dist., returning one member to Parliament. Surrounding dist. is rich in coal, supplying the mfrs. of the town, namely, carpets, woollens, boot and shoe, winceys, whisky, lace, fireclay goods; the iron trade now the staple trade, and there are foundries and machine-making shops. K. House an ancient seat.

Kilmarnock Division, parl. division of Ayrshire. Pop. 73,911.

Kilmarnock Water, trib. of r. Irvine, N. Ayrshire.

Kilmaronaig, seat and shooting, 5 m. N.E. of Oban, Argyllshire.

Kilmaron Castle, seat, 1¾ m. N.W. of Cupar, Fife.

Kilmaronock, par. on S.E. side of Loch Lomond, E. Dunbartonshire; par. 10,339 ac. Pop. 863. K. Castle is a ruined stronghold. The par. includes several isls. in Loch Lomond.

Kilmartin, par. and vil. on Loch Craignish, mid Argyllshire; par. 25,104 ac. Pop. 455. Includes MacCaskin and Ree Isls., in Loch Craignish. Vil. 8 m. N.W. of Lochgilphead. P.O., T.O.

Kilmartin, pl. and seat on Loch Meikle, Glen Urquhart, and 5 m. W. of Drumnadrochit, Inverness-shire.

Kilmartin River, flows N. into Staffin Bay, N.E. Skye, Inner Hebrides.

Kilmaurs, par. and small town near Kilmarnock, N. Ayrshire ; par. 5909 ac. Pop. 4396. Vil. 2¼ m. N.N.W. of Kilmarnock ; ry. sta. (154 ft.), L.M.S. P.O., T.O. Pop. 1820.

Kilmaveonaig, ancient par., now in par. of Blair-Atholl, Perthshire.

Kilmein Hill, alt. 1406 ft., 2½ m. E.N.E. of Patna, Ayrshire.

Kilmelfort, vil. on Loch Melfort, 16 m. S. of Oban, Argyllshire. P.O., T.O. called Kilmelford. Hotel.

Kilmeny, q.s. par. in par. of Killarrow and K., 3 m. S.W. of Portaskaig, Islay Isl., Argyllshire.

Kilmichael, ancient par. and ham. ; par. now in par. of Campbeltown, and ham. 2 m. N.W. of that town, Kintyre, Argyllshire.

Kilmichael, seat, par. of Kilbride, 1½ m. S. of Brodick, Arran Isl., Buteshire.

Kilmichael-Glassary, vil. on r. Add, and 3½ m. N. of Lochgilphead, Argyllshire. P.O., T.O.

Kilmichael-Inverlussa, vil. at head of Loch Swin and 7 m. S. of Crinan, Knapdale, Argyllshire.

Kilmodan, par. on the Ruel, Cowal, mid Argyllshire ; par. 25,447 ac. Pop. 213. Church at Clachan of Glendaruel, 8 m. W. of Tighnabruaich.

Kilmonivaig, par. of Loch Lochy, S. Inverness-shire, and 15 m. W. of Dalnaspidal ; par. 267,223 ac. Pop. 2894. Church is near Spean Bridge, 9¾ m. N.E. of Fort William.

Kilmorack, par. and ham. in N.W. Inverness-shire ; par. 139,782 ac. Pop. 1741. Ham. on the Glass, 2½ m. S.W. of Beauly.

Kilmore, Argyllshire. P.O. See KILMORE AND KILBRIDE.

Kilmore, ham., 2 m. N.E. of Armadale, S. of Skye, Inner Hebrides ; Sleat parish church here.

Kilmore and Kilbride, united pars. on Firth of Lorne, mid Argyllshire ; 29,503 ac. Pop. 6577. P.O., T.O. called Kilmore. Contains the parl. bur. of Oban and Kerrera Isl. ;

Kilmore church (fifteenth century) 4½ m. S.E. of Oban, and Kilbride church 3 m. S. of Oban.

Kilmorich, see LOCHGOILHEAD AND KILMORICH.

Kilmorie, ancient and ruined chapel, between Lochs Swin and Killisport, S. Knapdale, Argyllshire.

Kilmorie, Arran Isl., Buteshire. P.O., T.O. See KILMORY.

Kilmorie Castle, ancient keep in par. of Rothesay, Bute Isl., Buteshire ; 10 m. W. of Largs ; was a stronghold of the Jamiesons.

Kilmory, ancient seat and graveyard, 1 m. S.E. of Lochgilphead, Argyllshire.

Kilmory, par. and ham. in S.W. of Arran, Buteshire ; par. 67,136 ac. Pop. 1874. Par. includes isl. of Pladda ; ham. 10 m. S.W. of Lamlash. P.O., T.O.

Kilmory, pl. on N. coast of isl. of Rhum, Inner Hebrides.

Kilmuir, par. and ham. in N. of Skye, Inner Hebrides ; par. 34,251 ac. Pop. 1489. Ham. on N.W. coast, 5 m. N. of Uig. A cross, erected 1871, in the churchyard marks the grave of Flora Macdonald (1721–1790). P.O., T.O.

Kilmuir, ham., on Moray Firth, par. of Knockbain, S.E. Ross and Cromarty, and 3 m. N. of Inverness.

Kilmuir, ham. 1 m. S.E. of Dunvegan, Skye, Inner Hebrides.

Kilmuir Easter, par., on Cromarty Firth, 5 m. S. of Tain, E. Ross and Cromarty ; par. 11,008 ac. Pop. 705. Church 1 m. N.E. of Delny sta.

Kilmuir Wester, another name for Knockbain par., S.E. Ross and Cromarty.

Kilmun, q.s. par. and vil. with quay on N. side of Holy Loch, Argyllshire, and 7½ m. W.N.W. of Greenock. P.O., T.O. Regular steamer communication with Greenock, etc. Vil. pop. 780.

Kilnaughton, pl. 2 m. W. of Port Ellen, Islay, Inner Hebrides.

Kilnave, pl. on W. side of Loch Gruinnard, N.W. Islay, Argyllshire, and 8 m. N.W. of Brigend.

Kilncadzow, ham. 2½ m. S.E. of Carluke, Lanarkshire.

Kilneuair, pl. with ancient chapel, 1½ m. E. of Ford, and at head of Loch Awe, Argyllshire.

Kilninian and Kilmore, united pars. on Sound of Mull, N.W. of Mull Isl., Argyllshire; par. 74,267 ac. Pop. 1508. Includes isls. of Calve, Gometra, Little Colonsay, and Ulva. The two par. churches are very old, Kilninian being 8 m. S.W. of Tobermory, and Kilmore 7 m. N.E. of Kilninian; stone circle in vicinity.

Kilninver and Kilmelfort, par. on Firth of Lorne, mid Argyllshire; par. 31,791 ac. Pop. 298. Contains ham. of Kilninver, 8 m. S. of Oban and on S. side of Loch Feochan. P.O., T.O. Nearby is Kilninver Lodge.

Kilournan, pl., 6½ m. N.N.W. of Helmsdale, Sutherland; K. Hill, 1 m. S.E., is 1043 ft.

Kilpatrick, ry. sta., L.N.E., in Dunbartonshire, and 10½ m. N.W. of Glasgow.

Kilpatrick Hills, 1313 ft., Dunbartonshire and Stirlingshire, 5 m. S. of Killearn; extend from Vale of Leven to Strathbane.

Kilpartick, New, or **East,** par., Dunbartonshire; par. 11,974 ac. Pop. 11,568. Contains Bearsden, or New K., a town 5 m. N.W. of Glasgow. P.O., T.O.

Kilpatrick, Old, or **West,** par. and vil., on the Clyde, Dunbartonshire; par. 12,609 ac. Pop. 55,641. Vil., 5 m. S.E. of Dumbarton. Pop. 2494. Ry. sta., L.N.E. P.O., T.O. Ferry communication with Erskine across r. Clyde.

Kilphedirleg, pl., 8 m. N.W. of Brora, Sutherland.

Kilpurnie Hill (alt. 1134 ft.), in par. of, and 1½ m. N.E. of, Newtyle, S.W. Angus. K. Castle is 1 m. S.S.W. of Newtyle.

Kilquhanity, ham., 2 m. W .of Kirkpatrick-Durham, Kirkcudbrightshire.

Kilravock (pron. "Kilrawk") **Castle,** seat, on r. Nairn, in co. of, and 7 m. S.W. of, Nairn. Was visited by Queen Mary in 1562. Prince Charles Edward was entertained here two days before battle of Culloden. Burns was a guest at the castle in September 1787.

Kilrenny, par., royal and mun. bur. on the Firth of Forth, E. Fife; par. 3777 ac. Pop. 2357; mun. bur. pop. (including Anstruther Easter and Wester) 3325.

Kilrie, seat, 3 m. N.W. of Kinghorn, Fife.

Kilry, q.s. par. in pars. of Glenisla and Lintrathen, W. Angus; 4 m. N.W. of Alyth is K. Lodge. P.O., T.O.

Kilspindie, par. and vil. in S.E. Perthshire; par. 6257 ac. Pop. 433. Vil. 2 m. N.W. of Errol sta.

Kilspindie, ruins of mansion, 1 m. W. of Aberlady, E. Lothian. Very old church at Aberlady. Golf course nearby.

Kilsyth, par. and town (police bur.) in S. Stirlingshire, near r. Kelvin and Forth and Clyde Canal, 38 m. W. of Edinburgh and 12¾ m. N.E. of Glasgow; par. 13,123 ac. Pop. 10,047; bur. pop. 7551; ry. stas., L.M.S. and L.N.E. P.O., T.O. Working of coal, ironstone, and whinstone throughout the par. N. of the town are the ruins of K. Castle; 2 m. N.W. of the town are K. Waterworks.

Kiltarlity, ham., 4½ m. S. of Beauly, Inverness-shire. P.O., T.O.

Kiltarlity and Convinth, par. in N. Inverness-shire. Contains ancient castles, stone circles, and vitrified forts; par. 64,947 ac. Pop. 1670.

Kiltearn, par. on Cromarty Firth, S.E. Ross and Cromarty, and 9 m. S.W. of Invergordon; par. 27,929 ac. Pop. 1012. Par. church dates from pre-Reformation days with modern additions. In the N.E. of the par., near Novar sta., is a gorge called the Black Rock of Kiltearn.

Kiltie Burn, stream, with waterfall, Perthshire, flows 4½ m. S. to Shaggie Burn, 1¾ m. N. of Crieff.

Kilt Rock, 2 m. S.E. of Staffin Bay, on E. coast of Skye, Inner Hebrides. The perpendicular fissures resemble the folds of a kilt, hence the name.

Kiltyrie, pl. on N. side of Loch Tay and 4½ m. W.N.W. of Killin, Perthshire.

Kilvaree, seat, 6½ m. N.E. of Oban, Argyllshire.

Kilvaxter, ham., 1 m. S. of Kilmuir church, Skye, Inner Hebrides.

Kilvickeon, Argyll, see KILFINICHEN AND KILVICKEON.

Kilwinning, par. and town (police bur.), in N.W. Ayrshire; par. 10,991 ac. Pop. 8531; town pop. 5324. Stands on r. Garnock, 3½ m. N.N.W. of Irvine; ry. sta., L.M.S. P.O. T.O. Has large engineering, iron, and fireclay works.

Kimmerghame House, seat, on the Black-adder and 3 m. S.E. of Duns, Berwick-shire.

Kinaldie, seat, on r. Don, in co. of, and 10 m. N.W. of, Aberdeen. P.O., and T.O. at ry sta., L.N.E. A burn of the same name flows into the Don.

Kinaldy, seat, 4 m. S. of St Andrews, Fife.

Kinardochy, Loch, 6 m. N.W. of Aber-feldy, Perthshire, and on road from Kenmore to Tummel Bridge.

Kinbattock, pl., with many ruins and antiquities, 1 m. S.W. of Towie, W. Aberdeenshire.

Kinbeachie, seat and loch, near E. side of Cromarty Firth and 7½ m. N.E. of Conon Bridge, E. Ross and Cromarty.

Kinblethmont, seat, 4¼ m. N. of Arbroath, Angus.

Kinbrace, ry. sta., L.M.S., 16¾ m. N.W. of Helmsdale, Sutherland. The third Duke of Sutherland spent large sums of money trying to reclaim land for agriculture in this district. P.O., T.O.

Kinbrae, seat, Newport, N. Fife.

Kinbroom, seat, 1 m. W.S.W. of Rothie-Norman, Aberdeenshire.

Kinbuck, vil. on Allan Water, and 2¾ m. N of Dunblane, S. Perthshire ; ry. sta. (322 ft.), L.M.S. P.O., T.O.

Kincaid House, seat in S. Stirlingshire, near Milton of Campsie, and 9½ m. N.E. of Glasgow.

Kincairney House, 4 m. N.E. of Dunkeld, Perthshire.

Kincaldrum, seat in co. of Angus and 4 m. S.W. of Forfar. K. Hill, in neigh-bourhood, is 911 ft. P.O., T.O.

Kincaple, vil. and seat, 3 m. N.W. of St. Andrews, Fife.

Kincardine, ancient par., now in par. of Abernethy, E. Inverness-shire. The par. church is on the r. Spey, 6½ m. S.W. of Nethy Bridge.

Kincardine, name of former co. town of Kincardineshire ; stood in Fordoun par., 4¼ m. N.W. of Laurencekirk and 1½ m. N.E. of Fettercairn.

Kincardine, par. on rs. Teith and Forth, S. Perthshire ; par. 10,207 ac. Pop. 1126. Contains several tumuli and a monolith. Church is 2 m. S. of Doune.

Kincardine, or **Kincardine-on-Forth,** small seapt. town on N. shore of Firth of Forth, Tulliallan par., Fife. Formerly

a detached portion of Perthshire ; stands 5½ m. S.E. of Alloa by rail ; ry. sta., L.N.E. P.O., T.O. On E. pier is a fixed red light and on ferry pier two fixed red lights. Has ferry to Higgins' Neuck, Stirlingshire, and bridge across Forth in course of con-struction.

Kincardine, par. in N. Ross and Cromarty ; par. 153,041 ac. Pop. 962. Contains Ardgay, a vil. near Bonar Bridge ry. sta. and 13¼ m. N.W. of Tain.

Kincardine Castle, seat, and ruins of ancient castle, S. Perthshire. K. Glen is on the Ruthven Water, and stretches 2½ m. N.E. to near Auchterarder.

Kincardine O'Neil, par. and vil. in S. Aberdeenshire ; par. 18,185 ac. Pop. 1847. Vil. on r. Dee, 24 m. W. of Aberdeen. P.O., T.O. Vil. pop. 182. Nearby is K. Lodge.

Kincardineshire, or **Mearns,** small co. on E. coast, bounded N. and N.W. by Aberdeenshire, E. by North Sea, S. and S.W. by Angus ; area, 382 sq. m. ; 244,482 ac. of land and 1206 of water. Pop. 39,865. Co. may be divided into four sections, namely, the Coast, the Howe o' Mearns, the Grampians, Dee-side ; the coastal area, Deeside and the " Howe," which is a continuation of the valley of Strathmore, are very fertile and productive, while the mountainous region to the W. is occupied chiefly by game preserves. The chief r. is the Dee, which flows through the N.W. corner of the co. ; area is served by L.M.S. and L.N.E. Co. almost exclusively agricultural, with fishing along the coast. Aberdeen-shire with Kincardine send three members to Parliament.

Kinchurdy House, on r. Spey ; 2½ m. S. of Boat of Garten sta. ; E. Inverness-shire.

Kinclaven, par., on the Tay. S.E. Perth-shire ; par. 6140 ac. Pop. 488. Op-posite confluence of Isla and Tay stand the ruins of K. Castle.

Kinclune, seat, 5 m. W.N.W. of Kirrie-muir, Angus.

Kincorth, seat, 2¼ m. N.W. of Forres, Morayshire.

Kincraig, seat, on r. Spey, 5¾ m. N. of Kingussie, Inverness-shire ; ry. sta. L.M.S. P.O., T.O.

Kincraig House, 1½ m. N.W. of Invergordon, Ross and Cromarty.

Kincraigie, vil., on r. Tay, 1¼ m. N. of Dalguise sta., Perthshire.

Kincraig Point, prom. on Largo Bay, and 4½ m. W.S.W. of St. Monance ; affords shelter to Largo Bay during easterly gales. Many caves, in one of which Macduff is said to have hidden from Macbeth.

Kincurdy House, near Fortrose, E. Ross and Cromarty.

Kindallachan, vil., on the Tay, par. of Dunkeld, and 2 m. S.E. of Ballinluig, Perthshire.

Kindar, Loch, in E. Kirkcudbright, 8 m. S. of Dumfries ; depth, 41 ft., alt. 88 ft. ; nearby is K. House.

Kindeace House, 3½ m. N. of Invergordon, Ross and Cromarty.

Kindrochit House, near Struan sta., Perthshire.

Kindrochit Youth Hostel, Ardtalnaig, Perthshire, on S. shore of Loch Tay, 11 m. N.E. of Killin, 12 m. S.W. of Aberfeldy. A Scottish Youth Hostel.

Kindrogan, seat, at head of Strath Ardle, and 10 m. N.E. of Pitlochry, Perthshire.

Kinellan, Loch (16 ft. deep, alt. 500 ft.), 1 m. S.W. of Strathpeffer, S.E. Ross and Cromarty. Contains a lake dwelling ; nearby is K. Lodge.

Kinfauns, par. on the Tay, S.E. Perthshire ; par. 3885 ac. Pop. 510. Ry. sta. L.M.S., 3¾ m. S.E. Perth. K. Castle is an imposing Gothic structure, now a home ; State forest nearby.

Kingairloch, ham. and deer forest at head of Loch Corrie (on W. side of Loch Linnhe) and 14 m. S.W. of Ardgour, N. Argyllshire. P.O., T.O. Nearby is K. House.

Kingarth, par. in S. of Bute Isl., Buteshire, on Firth of Clyde ; par. 8990 ac. Pop. 951. P.O. Par. church (1826) is 5½ m. S. of Rothesay.

Kingask, seat, 2½ m. S.E. of St. Andrews, Fife.

King Bank Head, mt., alt. 2282 ft., 1 m. N.E. of Culter Fell, S.W. Peebleshire.

Kingcausie, seat, on the Dee, par. of Maryculter, N. Kincardineshire, and 1½ m. S.E. of Culter sta.

King Edward (" Kineddar " or " Kinedart "), par. on the Deveron, N.W. Aberdeenshire ; par. 17,529 ac. Pop.

1955. P.O., T.O, at King Edward ry. sta. L.N.E. The sta. is 6½ m. N. of Turriff. 1 m. S. of sta. is the ruined Castle of King Edward, which was a stronghold of the Comyns.

Kingencleuch House, seat, 1¼ m. from Mauchline, Ayrshire.

Kingennie, ry. sta., L.M.S., 7 m. N.E. of Dundee, Angus. P.O.

Kinghorn, par., seapt., royal and contributory parl. bur. on N. shore of Firth of Forth, Fife ; par. 5333 ac. Pop. 2708 ; pop. bur. 2001. Town lies 3 m. S. of Kirkcaldy and 2½ E. by N. from Burntisland ; ry. sta. L.N.E ; harb. now fallen into decay ; isl. of Inchkeith forms part of the par. K. forms one of the Kirkcaldy dist. of parl. burs., which returns one member. P.O., T.O. Hotel.

Kingie, short stream in S.W. Invernessshire flowing N.E. through a glen of the same name to join the Garry, 3½ m. W. of Tomdown Inn, at K. Pool.

Kingillie House, seat, 7 m. W. of Inverness, in the N. of that county.

Kinglands, pl. in Logiealmond par., in co. of, and 8½ m. N.W. of, Perth.

Kinglassie, par. and vil. on the Lochty, S. Fife ; par. 8414 ac. Pop. 2427. Vil. is 6 m. N.W. of Kirkcaldy. Pop. 1022. P.O., T.O. Coal-mining centre.

Kinglas Water, Argyllshire. See GLEN KINGLAS.

Kingledoors, seat, 5 m. S. of Broughton, Peebleshire.

Kingledoors Burn, trib. of Tweed. 3 m. N. of Tweedsmuir, Peeblesshire.

Kingoldrum, par. and vil. (Kirkton of K.), on the Crombie, W. Angus ; par. 9620 ac. Pop. 294. The vil. is 4 m. W. of Kirriemuir. P.O., T.O.

Kingoodie, vil., on Firth of Tay, S.E. Perthshire, and 4 m. W. of Dundee. Has a small harbour.

Kingsbarns, coastal par., with vil., on the Kenly, E. Fife ; par. 3883 ac. Pop. 534. Vil. is 3 m. N.W. of Crail and 8 m. S.E. of St. Andrews. P.O., T.O. Pop. 245.

Kingsburgh House, ancient mansion which stood on E. side of Loch Snizort Beag, 9 m. N.W. of Portree, Skye, Inner Hebrides. Prince Charles Edward found refuge here, 1746, and Dr. Johnson and Boswell were entertained by Flora Macdonald in 1773.

King's Cave, on the W. coast of Arran, Buteshire. Tradition has it that Robert the Bruce sheltered here. Other caves in the neighbourhood have such names as K. Cellar, K. Stable, K. Kitchen.

Kingscavil, vil., 2 m. E. of Linlithgow, W. Lothian.

King's Cross, pl., 3 m. S.E. of Lamlash, Arran, Buteshire. P.O., T.O. Said to mark the spot where Robert the Bruce embarked for the invasion of Carrick. Kingscross Pt. is a prom. on Lamlash Bay, to which it affords some shelter.

Kingsdale, seat, 1 m. N.W. of Cameron Bridge, S. Fife.

Kingseat, mining vil., 3 m. N.E. of Dunfermline, Fife. P.O.

Kingseat, asylum of Aberdeen Lunacy Board, 8½ m. N.N.W. of Aberdeen.

Kingsford, seat, 1¼ m. S.W. of Alford, Aberdeenshire.

Kingsford, ham., 2 m. N.E. of Stewarton, Ayrshire.

King's Forest, wood, remains of a royal hunting-forest, 19 m. W. of Dunscore, N. Kirkcudbrightshire.

Kingshouse, pl. and hotel, 5 m. from head of Glencoe and 17 m. E. of Ballachulish; on road from Loch Lomond to Fort William, Argyllshire. Originally erected as barracks for royal troops, 1745.

Kingshouse, pl., with hotel and ry. sta., L.M.S., 1 m. S.W. of Balquhidder ry. sta.

Kingside, small loch on borders of Roxburghshire and Selkirkshire, and 5 m. E. of Ettrick church.

Kingskettle, ry. sta., L.N.E., 1 m. S. of Ladybank Junction, Fife. P.O. See KETTLE.

Kingsknowe, ry. sta. (K. Halt), L.M.S., 3 m. S.W. of Princes Street sta., Edinburgh. Serves a sub. of the city.

King's Law, hill, 3 m. N.E. of Carluke, Lanarkshire.

Kingsmills, mansion in co. of, and in S.E. neighbourhood of, Inverness.

Kingsmuir, seat, 4 m. N.W. of Anstruther, Fife.

Kingsmuir, vil., 2 m. S.E. of Forfar, Angus; ry. sta. (384 ft.), L.M.S. P.O.

King's Park, Edinburgh. See HOLYROOD.

King's Park, Fife. See LUCKLAW HILL.

King's Park, public park, on S.W. side of Stirling.

King's Seat, one of summits of Ochils, alt. 2111 ft., 2 m. N.W. of Dollar, Clackmannanshire.

King's Seat, one of summits of Sidlaws, alt. 1235 ft., 14 m. S.W. of Alyth, E. Perthshire.

King's Seat, hill, near Dunkeld, Perthshire; alt. 1250 ft.

King's Seat, alt. 1521 ft., one of the Pentland Hills, in N.W. Peeblesshire.

Kingston, q.s. par. and dist. in Glasgow. P.O., T.O. Contains K. Dock on S. side of the Clyde.

Kingston, small seapt. on W. side of mouth of r. Spey, ¾ m. N. of Garmouth, N.E. Moray. Pop. 288.

Kingston, seat in co. of Angus and 1½ m. S.E. of Forfar.

Kingston, pl., 2½ m. S. of North Berwick, E. Lothian.

Kingswells, ham. and seat in co. of, and 5 m. W. of, Aberdeen.

Kingussie, small town and police bur. on the r. Spey, 44 m. S.E. by road from Inverness. Capital of the dist. of Badenoch, with important annual fairs, and is a noted holiday resort. Hotels. Ry. sta. (745 ft.), L.M.S. P.O., T.O. Pop. of bur. 1067.

Kingussie and Insh, par., at the head of Strathspey, in S.E. Inverness-shire; 114,381 ac. Pop. 2363. Contains K. and the vils. of Newtonmore and Dalwhinnie.

Kinharrachie Lodge, on r. Ythan, 2 m. W. of Ellon, Aberdeenshire.

Kinharvie, a seat of the Duchess of Norfolk, in Kirkcudbrightshire and 9 m. S.W. of Dumfries.

Kininmonth, q.s. par. in pars. of Old Deer, Crimond, Longside, Lonmay, and Strichen, N. Aberdeenshire 4 m. N. of Mintlaw sta.

Kininvie House, 3 m. N.W. of Dufftown, Banffshire.

Kinkell, ancient par., Aberdeenshire. See KEITH-HALL AND KINKELL.

Kinkell, seat, 2½ m. S.E. of St. Andrews, Fife. On the nearby coast are K. Cave, K. Ness, and K. Castle.

Kinkell, vil. on the Earn, 2½ m. N. of Auchterarder, Perthshire.

Kinkell Castle, ancient baronial tower, 1½ m. S.E. of Conon Bridge, S.E. Ross and Cromarty. Was a stronghold of the Mackenzies of Gairloch.

Kinleith Paper Mill, on Water of Leith, Currie, Midlothian.

Kinloch, pl. at S.E. end of Loch Teacuis, Morven dist., Argyllshire.

Kinloch, seat, on S. side of Loch Scridain, near Pennyghael, Mull Isl., Argyllshire.

Kinloch, ham. and seat (K. House), 1½ m. W. of Meigle, E. Perthshire.

Kinloch, par., ham., and seat, 2 m. W. of Blairgowrie, Perthshire ; par. 4000 ac. Pop. 181.

Kinlochailort, ham., with inn and ry. sta. (Lochailort), L.N.E., at head of Loch Ailort, S.W. Inverness-shire, and 26¼ m. by ry. W. of Fort William. P.O., T.O. at sta.

Kinlochaline Castle, ancient square tower, at head of Loch Aline, 15 m. N.W. of Oban, and in par. of Morven, N. Argyllshire.

Kinlochard, ham. at head of Loch Ard, S.W. Perthshire, 12 m. N.W. of Buchlyvie. P.O. Hotel.

Kinlochbeg Lodge and **Deer Forest,** at head of Loch Leven, 9 m. N.E. of Ballachulish, Argyllshire. Forest covers about 27,500 ac.

Kinlochbervie, q.s. par. and ham. on Loch Inchard, par. of Eddrachillis, W. Sutherland ; ham. on N. side of the loch and 3½ m. N.W. of Rhiconich. P.O., T.O. Pier.

Kinloch Castle, seat, on Loch Scresort, E. coast of Rhum Isl., Inner Hebrides.

Kinloch Distillery, Campbeltown, Kintyre, Argyllshire.

Kinlocheil, vil. on Loch Eil, 12 m. N.W. of Fort William, Argyllshire. P.O., T.O.

Kinlochewe, ham., with hotel, in par. of Gairloch, W. Ross and Cromarty, 10 m. N.W. of Achnasheen sta. and 1½ m. S.E. of S.E. end of Loch Maree. P.O., T.O. Nearby is K. Lodge. R. Kinlochewe flows into Loch Maree.

Kinlochewe Deer Forest, see FLOWERDALE.

Kinloch Hourn, ham., at head of Loch Hourn, W. Inverness-shire.

Kinloch House, 2½ m. N.W. of Ladybank, Fife.

Kinloch House, at head of Kyle of Tongue and 4 m. S.W. of vil. of Tongue, N. Sutherland.

Kinlochkerran, ancient name for Campbeltown, Kintyre, Argyllshire, when the property of the Lords of the Isles.

Kinlochlaggan, pl. at head of Loch Laggan, Inverness-shire, 13 m. N.W. of Dalwhinnie. P.O., T.O.

Kinlochlaich, seat, ¾ m. W. of Appin sta., Argyllshire.

Kinlochleven, vil. at head of Loch Leven, N. Argyllshire, and 7 m. E.N.E. of Ballachulish. P.O., T.O. Has large aluminium works. At end of jetty and on N.W. corner of wharf are fixed red lights. Hotel. Pop. 1060 dist.

Kinlochlochy, pl. on Loch Lochy, mid Inverness-shire, 18 m. N.E. of Banavie. Here, in 1544, Clan Ronald defeated the Frasers. P.O.

Kinloch Lodge, in Strath Bran, 8 m. S.W. of Dunkeld, Perthshire.

Kinlochluichart, q.s. par. in par. of Contin, mid Ross and Cromarty, about 17 m. N.W. of Dingwall ; par. church 1 m. W. of head of Loch Luichart sta. K. Deer Forest covers about 43,000 ac.

Kinlochmoidart, seat and deer forest, at head of Loch Moidart, par. of Ardnamurchan, S.W. Inverness-shire. P.O., T.O. called Moidart.

Kinlochmorar, ham., at head of Loch Morar, W. Inverness-shire.

Kinlochmore, pl., on Loch Leven, par. of Lismore and Appin, and 10 m. S.E. of Fort William, Argyllshire. Fine falls where the r. enters the loch.

Kinloch Rannoch, q.s. par. and vil., par. of Fortingall, N.W. Perthshire. Vil. on the Tummel, at the E. end of Loch Rannoch and 21 m. W. of Pitlochry. P.O., T.O. Hotels.

Kinlochspelvie, q.s. par. on Loch Spelvie, par. of Torosay, S.E. of isl. of Mull, Argyllshire.

Kinloch Water, stream, flowing N. to the Kyle of Tongue, Sutherland.

Kinloss, coastal par., vil., and seat, on Moray Firth, Morayshire ; par. 5199 ac. Pop. 778. Vil. 3 m. N.E. of Forres ; ry. sta., L.M.S. Vil. pop. 164. P.O. ; T.O. at sta. K. Abbey was founded for Cistercians by David I., and now only the foundations can be traced.

Kinloss House, ½ m. N.N.W. of Cupar, Fife.

Kinmonth, East, West, and **Upper,** three hams. in Kincardineshire, from ½ to 1½ m. N.W. of Drumlithie.

Kinmount, seat, 4 m. W.N.W. of Annan, S. Dumfriesshire.

Kinmuck, or **Kinmuick,** ham., 3¾ m. E.S.E. of Inverurie, mid Aberdeenshire. P.O.

Kinmundy, seat, par. of Skene, in co. of, and 6½ m. N.W. of, Aberdeen.

Kinmundy, seat, par. of Old Deer, 3 m. S.S.E. of Mintlaw, N.E. Aberdeenshire.

Kinmundy, seat, 1½ m. S. of New Machar sta., Aberdeenshire.

Kinnaber Junction, junc. of L.M.S. and L.N.E., 2½ m. N. of Montrose, Angus.

Kinnabus, loch, 2 m. E.N.E. of Mull of Oa, Islay, Argyllshire.

Kinnaird, par. and vil., S.E. Perthshire; par. 3495 ac. Pop. 183. Vil. 2½ m. W. of Inchture. P.O. K. Castle, a square tower with modern additions, is said to date from the fifteenth century.

Kinnaird, ham. in par. of Moulin and 1½ m. N.E. of Pitlochry, Perthshire.

Kinnaird, vil. in E. Stirlingshire and 3½ m. N. of Falkirk. K. House was repaired and enlarged by James Bruce (1730–1794), the African traveller, who met his death, by an accident, in this house.

Kinnaird Castle, on r. S. Esk and 3½ m. S.E. of Brechin, Angus. Seat of the Earl of Southesk.

Kinnaird House, mansion, on the Tay, 2 m. S. of Ballinluig, Perthshire.

Kinnairds (or **Kinnaird Head**), low prom. on the Moray Firth, N.E. Aberdeenshire. Lighthouse, erected 1787, is 76 ft. high and shows a quarter-minute flashing light, visible 17 m.

Kinnairdy Castle, ancient mansion in Banffshire, on a prom. where the Deveron and Burn of Achintoul meet, 2½ m. S.W. of Aberchirder.

Kinnardochy, loch, 7 m. N.W. of Aberfeldy, Perthshire.

Kinnedar, pl., Morayshire, 1 m. S. of Lossiemouth.

Kinneddar and **Nether Kinneddar,** seats, 5½ m. N.W. of Dunfermline, Fife.

Kinneff and Catterline, coastal par. in Kincardineshire; par. 7130 ac. Pop. 713. Contains the ham. of Kinneff, 7½ m. S. of Stonehaven. P.O., T.O.

Kinneil, vil., on Firth of Forth, par. of Borrowstounness, and in S.W. vicinity of Bo'ness, W. Lothian; ry. sta., L.N.E. (closed). Ironworks. K. House is the property of the Duke of Hamilton.

Kinnelhead, pl. on the Kinnel, 8 m. S.E. of Durrisdeer, N. Dumfriesshire. An ancient stronghold formerly stood here.

Kinnell, par. on the Lunan, E. Angus; par. 6968 ac. Pop. 530. Par. church is on the Lunan and 6 m. S.W. of Montrose.

Kinnellar, par. on the Don, S.E. Aberdeenshire; par. 4218 ac. Pop. 449. P.O., T.O. (spelt Kinellar). K. Lodge, 3½ m. S.E. of Kintore.

Kinnell House, on the E. side of Killin, Perthshire.

Kinnel Water, stream in mid Dumfriesshire; rises on border of Lanarkshire, near source of Clyde, flows 20 m. S.S.E. and falls into the Annan 1¼ m. N.E. of Lochmaben.

Kinnernie, ham., 5 m. S.E. of Tillyfourie sta., Aberdeenshire.

Kinnernie Burn, flowing to Loch Skene, S.E. Aberdeenshire.

Kinnesswood, vil. in co. of, and 5 m. N.E. of, Kinross. Michael Bruce (1746–1767), the poet, was a native. P.O.

Kinnethmont, see KENNETHMONT.

Kinnettles, par., ham., and seat, mid Angus; par. 3143 ac. Pop. 353. Ham. on the Dean and 3½ m. S.W. of Forfar.

Kinneuchar, local pronunciation of Kilconquhar, q.v.

Kinning Park, q.s. par. and dist. in S. of Glasgow; ry. sta., Glasgow District Subway. P.O., T.O.

Kinning Point, pier, 1½ m. W. of Charlestown, Fifeshire, on Firth of Forth. Two fixed red lights.

Kinnoir, pl., 4 m. N.E. of Huntly, Aberdeenshire. P.O.

Kinnordy, seat, 1½ m. N.W. of Kirriemuir, Angus. Was birthplace of Sir Charles Lyell, Bart. (1797–1875), the geologist.

Kinnoull, par., on left bank of Tay, S.E. Perthshire; par. 3331 ac. Pop. 4119. Contains the Brigend sub. of Perth. K. Hill, 729 ft.; from the top a magnificent view of the Tay is obtained.

Kinord Loch, Aberdeenshire. See CANNOR.

Kinrara Lodge, on r. Spey and 3 m. S.W. of Aviemore sta., Inverness-shire. Adjacent hill (Tor of Alvie) crowned by Waterloo Cairn and Duke of Gordon's Monument.

Kinrive Hill, alt. 1063 ft., E. Ross and Cromarty, 5 m. N.W. of Invergordon.

Kinross, par., police bur., mkt.-town, and capital of co. of Kinross-shire; par. 7257 ac. Pop. 2991; pop. of bur. 2525. Town stands on W. side of Loch Leven and is 13¾ m. N. by E. of Dunfermline ; ry. sta. (K. Junction, 400 ft.), L.N.E. P.O., T.O. Hotel. On the peninsula between the town and the loch, is K. House. Wool spinning and linen mfrs. Angling on Loch Leven.

Kinrossie, vil. in co. of, and 8 m. N.E. of, Perth. P.O.

Kinross-shire, the smallest Scottish co. after Clackmannanshire, lies between the cos. of Fife and Perth ; area 81·9 sq. m., or 52,392 ac. of land, 3439 ac. of water. Most of the drainage is into Loch Leven, from which the surface rises to encircling hills—the Ochils in the N.W., the Lomond Hills in the E., Benarty Hill in the S., and the Cleish Hills in the S.W. The central plain is well cultivated, and limestone and sandstone are abundant in the hills. Woollens and linens are manufactured. Co. served by L.N.E. Has no parl. bur., but joins with Perth in returning two members to Parliament in two divisions, namely, Kinross and Western Perthshire and Perth.

Kinsteary House, in co. of, and 3½ m. S.E. of, Nairn.

Kintail, par., seat, and vil., S.W. Ross and Cromarty ; par. 77,440 ac. Pop. 376. Vil. on Loch Duich and 13 m. S.E. of Strome Ferry sta. K. Deer Forest is 27,000 ac. in extent.

Kintessack, vil., 3 m. N.W. of Forres, N.W. Moray.

Kintillo, vil. in par. of Dunbarney, ½ m. S.W. of Bridge of Earn, S.E. Perthshire.

Kintore, par., royal and police bur., E. Aberdeenshire ; par. 9107 ac. Pop. 2202. Town on r. Don, 11¼ m. N.W. of Aberdeen. Pop. 755. Ry. sta. (165 ft.), L.N.E. P.O., T.O.

Kintra, vil., on the Sound of Iona, 6 m. W.N.W. of Bunessan, isl. of Mull, Argyllshire.

Kintradwell, pl., 3 m. N. of Brora, Sutherland. Remains of a large broch and site of an ancient chapel.

Kintraw, pl.; at head of Loch Craignish, Argyllshire, 8 m. S. of Kilmelford.

Kintrockat House, 2 m. S.W. of Brechin, Angus.

Kinture, ham., on E. side of Islay, near Ardmore Pt.

Kintyre, peninsula in S. Argyllshire, between the Firth of Clyde and Atlantic

Ocean ; it is 40 m. long, with an average width of 7 m., and is connected with the mainland by the isthmus of Tarbert. Includes the isls. of Gigha, Cara, and Sanda. Has a mountainous backbone, with slopes of varied steepness on each side ; area of co. dist. 195,361 ac. Pop. 6111. Fishing, farming, and quarrying are the main industries. Mull of K., the S.W. extremity of the peninsula and the nearest pt. of Scotland to Ireland (13 m.). Lighthouse, with group-flashing light, 297 ft. above high water, and visible 24 m. It contains the town of Campbeltown.

Kintyre Distillery, Campbeltown, Argyllshire.

Kinuachdrach, landing-place at N.E. end of isl. of Jura, Argyllshire.

Kinveachie Lodge, 2½ m. W.S.W. of Boat of Garten sta., Inverness-shire.

Kip, stream, N.W. Renfrewshire ; rises near Ravenscraig and flows 4 m. S.W., falling into Firth of Clyde near Inverkip.

Kip, East, and **Kip, West,** two summits of the Pentlands, 4 m. W. of Penicuik, Midlothian. West K. is 1806 ft.

Kippen, par. and vil. on the Forth, N. Stirlingshire ; par. 11,268 ac. Pop. 1356. Vil., 1 m. S.W. of sta., which is 9 m. W. of Stirling ; ry. sta., L.N.E. (passenger service withdrawn). P.O., T.O., and T.O. at sta. Vil. pop. 426.

Kippendavie, seat in N.E. vicinity of Dunblane, on the Allan, Perthshire.

Kippen House, 1 m. S. of Dunning, S.E. Perthshire.

Kippenross, seat, 1 m. S. of Dunblane, Perthshire.

Kippford, ham., on the Urr, 4 m. S. of Dalbeattie, S. Kirkcudbrightshire. P.O., T.O.

Kippilaw, seat, 3 m. S.W. of St. Boswells, N.W. Roxburghshire.

Kipps, pl., N. Lanarkshire, ½ m. N.E. of Coatbridge.

Kip, The, hill, alt. 750 ft., 4 m. S.E. of Kelso, Roxburghshire.

Kirbister, pl., 6 m. W.S.W. of Kirkwall, Orkney.

Kirbister, Loch of, 10 m. S.W. of Kirkwall, Mainland, Orkney ; alt. 52 ft., depth 6 ft.

Kirbuster, pl., 4 m. N. of Stromness, Mainland, Orkney.

Kirdels, pl. in Morayshire, 2½ m. N.E. of Ballindalloch sta.

Kirk, ham., 3 m. N.E. of Watten sta., Caithness. P.O.

Kirkabister, vil., on W. coast of Bressay Isl., Shetland, and 2½ m. S.E. of Lerwick. K. Ness, with Bressay lighthouse, nearby.

Kirkaig, stream, border of Sutherland and Ross and Cromarty; rises in Loch Fewin, flows 3¼ m. N.W., and falls into Loch K., 16 m. N.W. of Ullapool. About a mile below its source there are the Falls of K.

Kirkandrews, ham., on K. Bay, 7½ m. S.W. of Kirkcudbright. Ruined church.

Kirkapoll, pl., 1 m. N.W. of Gott Bay, isl. of Tiree, Argyllshire.

Kirkbank, seat, on the Teviot and 4¾ m. S.W. of Kelso, Roxburghshire; ry. sta., L.N.E.

Kirkbean, par. and vil., on Solway Firth, S.E. Kirkcudbrightshire; par. 8938 ac. Pop. 560. Vil. is 12 m. S. of Dumfries. P.O., T.O.

Kirkbride, seat, 4½ m. S.E. of Maybole, Ayrshire.

Kirkbuddo, seat in co. of Angus, 6 m, S.E. of Forfar; ry. sta. (486 ft.). L.M.S. P.O.

Kirk Burn, trib. flowing N. to Tweed at Cardrona ry. sta., 3 m. E.S.E. of Peebles, Peeblesshire.

Kirkcaldy, par., royal, parl., and mun. bur., manufacturing town and seapt. of Fife, on N. shore of Firth of Forth, 5¾ m. N.E. of Burntisland and 10 m. N. of Edinburgh; par. (K. and Dysart) 9581 ac. Pop. 46,019; parl. and mun. bur. pop. 43,874. Ry. sta. (105 ft.), L.N.E. P.O., T.O. Hotel. The main street of the town is about 4 m. long, hence the name "The Lang Toon." Principal mfrs.: wax-cloth and linoleum, for which it is the chief seat in the kingdom; spinning of flax, tow, and jute, and weaving cloth; engineering, brass- and iron-founding, pottery making, brewing, etc. On the E. pier is a 10-seconds flashing light and on the S. pier a fixed light. K., Burntisland, Dysart, Kinghorn, Buckhaven, Methil, and Innerleven together return one member to Parliament.

Kirkcarsewell, loch, 1¼ m. N.E. of Dundrennan, S. Kirkcudbrightshire.

Kirkchrist, ancient par., on the Dee, now included in par. of Twynholm, S. Kirkcudbrightshire. Convent formerly stood here.

Kirkclaugh, seat, 6 m. S.W. of Gatehouse-of-Fleet, Kirkcudbrightshire.

Kirkcolm, par. and vil. (also called Stewarton), on W. side Loch Ryan, N.W. Wigtownshire; par. 13,330 ac. Pop. 1388. Vil., 6 m. N.W. of Stranraer. Pop. 249. P.O., T.O.

Kirkconnel, par. and vil., Nithsdale, N.W. Dumfriesshire; par. 26,701 ac. Pop. 3962. Vil. on r. Nith, 3¼ m. N.W. of Sanquhar. P.O., T.O. Mineral springs to W. of vil. Ry. sta., L.M.S. Vil. pop. 3329.

Kirkconnel, ancient par., on the Kirtle, now in par. of Kirkpatrick-Fleming, Annandale, S.E. Dumfriesshire. "Fair Helen of K." buried in churchyard.

Kirkconnel Hall, near Ecclefechan, Dumfriesshire.

Kirkconnell, seat, on r. Nith, E. Kirkcudbrightshire, and 7 m. S. of Dumfries.

Kirkconnell Moor, par. of Tongland, 9 m. W. of Dalbeattie, Kirkcudbrightshire.

Kirkcormac, ancient par., on the Dee, now in par. of Kelton, S. Kirkcudbrightshire. Church in ruins.

Kirkcowan, par. and vil., N. Wigtownshire; par. 35,801 ac. Pop. 977. Vil. on Tarf Water, 7¼ m. N.W. of Wigtown; ry. sta., L.M.S. P.O., T.O.

Kirkcudbright, royal and mun. bur., mkt.-town, and capital of co. of K. par. 12,346 ac. Pop. 3188. Town on r. Dee where it flows into K. Bay. Pop. 2311. Ry. sta., L.M.S. P.O., T.O. Hotel. Harbour suitable only for small craft. Principal industry, agricultural produce. On Little Ross Isl., at the mouth of the r., is a lighthouse with flashing light, visible 18 m.; coastguard and lifeboat sta.

Kirkcudbright-Innertig, old name for Ballantrae par., Ayrshire.

Kirkcudbrightshire, also known as Stewartry of Kirkcudbright and East Galloway; a co. of S.W. Scotland, washed on the S. for 50 m. by the Solway Firth, and elsewhere bounded by cos. of Wigtown, Ayr, and Dumfries; area 575,832 ac. of land, 6686 ac. of water. Pop. 30,341. Coast irregular and contains numerous caves, in former times the storehouses of smugglers. Co. is mountainous, especially in the N.W., and from that quarter undulates towards the Solway. Chief heights: Mt. Merrick, Criffel, Caernsmore of Carsphairn, Caernsmore of Fleet. Principal rivers are: Dee, Ken, Cree, and the Ure. Great attention is paid to raising cattle and sheep, but less to agriculture ; mineral wealth of the co. chiefly granite. The co. is served by the L.M.S. Galloway, which comprises the cos. of Kirkcudbright and Wigtown, inclusive of burs. therein, returns one member to Parliament.

Kirkdale House, seat, 5½ m. S.E. of Creetown, S.W. Kirkcudbrightshire. Church in ruins, but burial-ground still in use.

Kirkden, par. in S.E. Angus, 6 m. N.W. of Arbroath ; par. 5406 ac. Pop. 1080.

Kirk Domina, ancient chapel, now in ruins, 1½ m. S.S.W. of Barr, Carrick, S. Ayrshire.

Kirkennan House, on Urr Water, 3 m. S. of Dalbeattie, Kirkcudbrightshire.

Kirkfieldbank, or **Kirkland,** q.s. par. and vil. on the Clyde, in co. of, and 1 m. W. of, Lanark. P.O., T.O. Vil. pop. 892.

Kirkfield House, seat in co. of, and 1¾ m. S.W. of, Lanark.

Kirkford, pl., W. of, and adjoining, Cowdenbeath, Fife. P.O.

Kirkgunzeon, par. and vil. in S.E. Kirkcudbrightshire ; par. 11,929 ac. Pop. 487. Vil. on K. Lane (or Burn), 10 m. S.W. of Dumfries ; ry. sta., L.M.S. P.O., T.O. ; T.O. at K. sta.

Kirk Hill, alt. 850 ft., 6 m. N.E. of Girvan, Ayrshire.

Kirkhill, a State forest, 1¾ m. W. of Bucksburn, Aberdeenshire.

Kirkhill, par. on Beauly Firth, in co. of Inverness ; par. 11,429 ac. Pop. 1155. Vil., 6½ m. W. of Inverness. P.O., T.O. Pop. 582.

Kirkhill, seat, N.W. of Broxburn, W. Lothian.

Kirkhill, ry. sta., L.M.S., 7 m. S.E. of Glasgow, Lanarkshire.

Kirkhill, vil. on N. Esk, ½ m. N.E. of Penicuik, Midlothian. Pop. 658.

Kirkhill, alt. 1293 ft., 3½ m. S.E. of Ettrick, Selkirkshire.

Kirkhill Castle, seat, ½ m. W.N.W. of Colmonell, S. Ayrshire.

Kirkhill House, seat, on S. Esk, 1 m. W.N.W. of Gorebridge, Midlothian.

Kirkholm, isl., in par. of Sandsting, 13¼ m. W.N.W. of Lerwick, Shetland.

Kirkhope, par. on the Ettrick, E. Selkirkshire ; par. 22,734 ac. Pop. 352. K. Tower, an old border keep. Church at Ettrick Bridge, 7 m. S.W. of Selkirk.

Kirkhope Law, alt. 1758 ft., in co. of, and 4 m. S.E. of, Peebles.

Kirkiboll, see TONGUE.

Kirkibost, isl., 6 m. N. of Benbecula, Outer Hebrides. Pop. 6.

Kirkigarth, loch, ½ m. N.W. of Walls, mainland, Shetland.

Kirkinch, ham., 2 m. E. of Meigle, W. Angus.

Kirkinner, par. and vil. on Wigtown Bay, E. Wigtownshire ; par. 15,494 ac. Pop. 1004. Vil. 2½ m. S. Wigtown ; ry. sta., L.M.S. P.O., T.O.

Kirkintilloch, par. and police bur., in detached portion, S.E. Dunbartonshire ; par. 7156 ac. Pop. 17,308 ; pop. bur. 11,817. Town situated on the Luggie Water, where it joins the Kelvin, and on the Forth and Clyde Canal, 8 m. N.E. of Glasgow by rail. P.O., T.O. Iron foundries, chemical works, and extensive collieries.

Kirkland, vil. on the Leven and 1 m. W. of that town, Fife.

Kirkland, vil., 2 m. E. of Moniave, W. Dumfriesshire ; ry. sta. (½ m. from vil.), L.M.S. P.O.

Kirkland Hill, 1670 ft., 2½ m. N. of Kirkconnel, N.W. Dumfriesshire.

Kirklands, seat, on Ale Water, ¾ m. N.W. of Ancrum, Roxburghshire.

Kirklands, mansion in Berwickshire, 2½ m. E. of Melrose.

Kirklands Lunatic Asylum, 1 m. N. of Bothwell, Laharkshire.

Kirklandside, Riccarton, Kilmarnock ,Ayrshire. Town and County hospital.

Kirklee, ry. sta., L.M.S., in N.W. of Glasgow. P.O.

Kirkliston, vil. and par. on the Almond, in N.E. West Lothian and Midlothian ; par. 8010 ac. Pop. 4784. Vil., 10 m. W. of Edinburgh by rail, on r. Almond ; ry. sta., L.N.E. (passenger service withdrawn). Pop. 848. P.O., T.O. Has a distillery.

Kirk Loch, ½ m. S.W. of Lochmaben, Dumfriesshire ; 25 ft. deep and 157 ft. above sea-level.

Kirkmabreck, par., on Wigtown Bay, S.W. Kirkcudbrightshire ; 23,324 ac. Pop. 1294. Contains Creetown. K. Granite Quarries are 1½ m. S.E. of Creetown.

Kirkmadrine, ancient par., now included in par. of Sorbie, E. Wigtownshire. The ruined church stands in the burial-ground, 1 m. N. of Garlieston.

Kirkmadrine, pl. and ruined chapel, in par. of Stoneykirk, S.W. Wigtownshire, and 2 m. S.W. of Sandhead.

Kirkmahoe, par. and vil. on the Nith, S.W. Dumfriesshire ; par. 12,568 ac. Pop. 1075. Vil., 4 m. N. of Dumfries. P.O. called Kirkton.

Kirkmaiden, peninsular par. (forming the extreme S. extremity of Scotland), on Luce Bay, S.W. Wigtownshire ; 13,687 ac. Pop. 1554. Includes the Mull of Galloway, the most southern point of Scotland. The church (1638) is 1 m. W. of Drumore.

Kirkmaiden, ancient par. on Luce Bay, now in par. of Glasserton, S.E. Wigtownshire. Ancient church, now a ruin, stands on Monreith Bay.

Kirkmay, seat, ½ m. S.W. of Crail, Fife.

Kirkmichael, par. on the Ae Water, Annandale, N. Dumfriesshire, and 9 m. N.E. of Dumfries ; 17,023 ac. Pop. 726. 1 m. S.E. of the church (1815) is K. House.

Kirkmichael, par. and vil. on the Dyrock Burn, Carrick, mid Ayrshire ; par. 15,932 ac. Pop. 1575. Vil., 3 m. E. of Maybole. Pop. 255. P.O., T.O. On the S. of the vil. is the seat of K. House.

Kirkmichael, par., on the Avon, S.W. Banffshire ; par. 75,337 ac. Pop. 792. The church (1807) is 4 m. N.W. of Tomintoul.

Kirkmichael, par. and vil. on the Ardle, N.E. Perthshire ; par. 61,818 ac. Pop. 762. Vil., 13 m. N.W. of Blairgowrie. P.O., T.O.

Kirkmichael, par., E. Ross and Cromarty. See RESOLIS.

Kirkmichael Hostel, in Strath Ardle, on the W. side of the r. Ardle, 13 m. N.W. of Blairgowrie. A Scottish Youth Hostel.

Kirkmuirhill, mining vil. in Lesmahagow par., E. Lanarkshire, and 5½ m. S.E. of Larkhall. P.O., T.O.

Kirkness House, seat, near S.E. end of Loch Leven, Kinross-shire, 4 m. N. of Lochgelly.

Kirknewton, par., vil., and seat in W. of Midlothian ; par. 9363 ac. Pop. 2910. Vil., ½ m. S.E. of Mid-Calder sta. P.O., T.O. Vil. pop. 354. The Oakbank shale works are here.

Kirkney Water, burn in N.W. Aberdeenshire. Flows N.E. and falls into the Bogie, near Gartly.

Kirk of Ness, Shetland. See CULLAVOE.

Kirkoswald, par. and vil. on the coast, Carrick, Ayrshire ; par. 14,906 ac. Pop. 1802 ; vil., 203 ; 4 m. S.W. of Maybole. P.O., T.O. Ancestors of Robert Burns and Douglas Graham and John Davidson, the "Tam o' Shanter" and "Souter Johnnie" of his famous poem, are buried in the churchyard.

Kirkpatrick, ham., 3½ m. S.E. of Thornhill, Dumfriesshire.

Kirkpatrick, ry. sta., L.M.S. P.O., T.O., 7 m. S.E. of Ecclefechan, S.E. Dumfriesshire. See KIRKPATRICK-FLEMING.

Kirkpatrick-Durham, par. and vil. on the Urr, N.E. Kirkcudbrightshire ; par. 18,249. Pop. 738. Vil., 5 m. N.E. of Castle-Douglas. P.O., T.O. Vil. pop. 241.

Kirkpatrick-Fleming, par. and vil., S.E. Dumfriesshire ; par. 11,362 ac. Pop. 1067. Vil. on Kirtle Water, 7 m. S.E. of Ecclefechan ; ry. sta. (Kirkpatrick), L.M.S. P.O., T.O. Nearby is K. House.

Kirkpatrick-Irongray, par. on the Cluden, Nithsdale, N. Kirkcudbrightshire ; 13,655 ac. Pop. 602. In burial-ground is a monument to Helen Walker, the original of Scott's "Jeannie Deans."

Kirkpatrick-Juxta, par. on the Annan, N. Dumfriesshire ; 22,365 ac. Pop. 994. Church is 1 m. S.E. of Beattock sta.

Kirkside House, 1¾ m. S.W. of Lauriston sta., S. Kincardineshire.

Kirkstead Burn, N. Selkirkshire ; flows 4 m. S.E. and falls into St. Mary's Loch.

Kirkstyle, ham. on Ewes Water, E. Dumfriesshire, 4 m. N. of Langholm.

Kirkstyle, former bur. of barony, now represented by Carluke, Lanarkshire.

Kirkton, ham., 1½ m. N. of Dumfries. P.O.

Kirkton, seat, Carluke, Lanarkshire.

Kirkton, vil. in par. of Glenelg, Inverness-shire. Pop. 150.

Kirkton Bridge, ry. sta., L.N.E., between Fraserburgh and Cairnbulg, Aberdeen-shire.

Kirkton Burn, flows 4½ m. to the Leven at Barrhead, S. Renfrewshire.

Kirktonhill, vil. on the Esk and 6 m. N.W. of Langholm, Dumfriesshire.

Kirktonhill, seat, 4 m. S.W. of Laurence-kirk, Kincardineshire.

Kirkton Manor, P.O., Peebles. See MANOR.

Kirktown, pl. in co. of, and 3 m. W. of, Inverness.

Kirkurd, par. in W. Peeblesshire; 5704 ac. Pop. 221. Par. church 3 m. S.E. of Dolphinton.

Kirkville House, in co. of, and 8½ m. N. of, Aberdeen.

Kirkwall, royal and mun. bur., seapt. and co. town of Orkney, on the N. shore of a narrow neck of land which divides the Mainland of Orkney into two very unequal portions, 51 m. N. of Wick. Cathedral, dedicated to St. Magnus, was founded in 1137, and was built by Rognvald, one of the Orkney jarls. Fishing centre and coastguard sta. On pier-head a fixed white light, visible 10 m. P.O. Pop. of bur. 3517. Regular steamer communication with Aberdeen, Leith, Lerwick, Stromness, Wick, etc. Scrabster to Stromness and back daily.

Kirkwall and St. Ola, par., Mainland, Orkney; 11,074 ac. Pop. 4398. Contains Kirkwall.

Kirkwood, pl., Dumfriesshire, 1 m. N.W. of Dalton. Pop. Dalton and K. special water dist., 99.

Kirkwood Collieries, Lanarkshire, ½ m. N.E. of Old Monkland.

Kirkwood House, 5 m. S. of Lockerbie, Dumfriesshire.

Kirk Yetholm, see YETHOLM.

Kirn, q.s. par., and sub. of Dunoon, on W. side of Firth of Clyde, Argyllshire. Hotels. P.O., T.O. K. lies N. of Dunoon and is a popular watering-place. The pier has a fixed light. Steamer communication with Greenock, etc.

Kirnan, seat, near vil. of Kilmichael, Argyllshire.

Kirriemuir, par. and mkt.-town (police bur.) on the Gairie, W. Angus; par. 15,028 ac. Pop. 4755. Town 8½ m. by rail N.W. of Forfar. Pop. 3326. Ry. sta., L.M.S. P.O., T.O. Hotel. Birthplace of Sir James Barrie (b. 1860), the author of "Peter Pan," etc. Has linen mfrs.

Kirriereoch Hill, 2562 ft., on borders of Ayrshire and Kirkcudbrightshire and 5 m. S.W. of Loch Doon. K. Loch, 3 m. W. of K. Hill.

Kirroughtree, seat, 1 m. N.E. of Newton-Stewart, Kirkcudbrightshire. State forest nearby.

Kirtle, q.s. par. on the Kirtle, 3 m. S.S.E. of Ecclefechan, Dumfriesshire.

Kirtlebridge, vil. on the Kirtle, 3 m. S.E. of Ecclefechan, Dumfriesshire; ry. sta. (207 ft.), L.M.S. P.O., T.O.

Kirtle Water, stream, flowing 20 m. S. and S.E. and falls into Solway Firth at Kirtlefoot, 2 m. S.W. of Gretna Green.

Kirtomy, ham. at mouth of burn of same name, 3 m. N.E. of Bettyhill of Farr, N. coast of Sutherland.

Kishorn, or **Courthill,** ham. on r. K., where it flows into Loch K., in par. of Apple-cross, and 5 m. W. of Lochcarron. P.O., T.O. At the mouth of the loch is K. Isl.

Kismull Castle, ruined stronghold in K. Bay, par. of Barra, Outer Hebrides.

Kittock, trib. of the White Cart, which it joins near Busby, Lanarkshire.

Kittockside, ham. on the Kittock, 1¾ m. N.W. of E. Kilbride, Lanarkshire.

Kittybrewster, N.W. sub. of Aberdeen; ry. sta., L.N.E. P.O., T.O.

Klett (A'Chleit) Island, 4 m. S.W. of Lochinver, S.W. Sutherland.

Klibreck, Ben, mt., atl., 3154 ft., S. side of Loch Naver, Sutherland.

Klibrick Lodge, at S.W. end of Loch Naver, Altnaharra, Sutherland.

Knab, Lerwick, Shetland. Co. hospital.

Knaik Water, stream in S. Perthshire; flows S.E. and falls into Allan Water near Greenloaning.

Knapdale, dist. in Argyllshire; on the S. it is bounded by E. and W. Loch Tarbert, which separates it from Kin-tyre; on the N. it is bounded by the Crinan Canal, which separates it from the dist. of Lorne.

Knapdale, North, par. on the Sound of Jura, Knapdale, Argyllshire ; 26,286 ac. Pop. 550. Contains Port Crinan and the isls. of Danna and Ulva.

Knapdale, South, par. on Loch Fyne, Knapdale, Argyllshire ; 73,850 ac. Pop. 2188. Contains Ardrishaig at S.E. end of Crinan Canal.

Knapdale State Forest, near Ardrishaig Pier, N. Knapdale, Argyllshire.

Knap of Trowieglen, hill, alt. 1308 ft., Hoy, Orkney.

Knap Point, prom. on W. coast of Knapdale and between Lochs Swin and Killisport.

Knaven, pl., 3½ m. S.W. of Maud Junction, Aberdeenshire.

Knightswood Rows, mining vil., borders of Lanarkshire and Dunbartonshire, 1½ m. W. of Maryhill.

Knipe, The, hill, alt. 1885 ft., in par. of, and 3 m. S.E. of, New Cumnock, Ayrshire.

Knock, ham., 5 m. W. of Whithorn, Wigtownshire.

Knock, ham. in co. of, and 12½ m. S.W. of, Banff ; ry. sta., L.N.E. P.O. ; T.O. at sta. K. Hill (1409 ft.) is 1½ m. N.W.

Knock, vil. on r. Esk, Dumfriesshire, 6½ m. N.W. of Langholm.

Knock (or Eye or Uie), q.s. par. on Broad Bay, par. of Stornoway, Lewis, Outer Hebrides. Comprises Eye Peninsula on E. side of Lewis, and contains Swordle and K. vil., 5 m. S.E. of Stornoway. P.O.

Knockally, ham., 1 m. S.W. of Dunbeath, S. Caithness.

Knockan, ham., 10 m. N.E. of Ullapool, Sutherland.

Knockan, hill, alt. 1219 ft., 6 m. S.W. of Keith, Banffshire.

Knockando, par. on the r. Spey, S.E. Moray ; par. 28,128 ac. Pop. 1334. Ry. sta., L.N.E. P.O., T.O., and P.O. at Upper K. Distilleries. K. House is a seat.

Knockandu, vil., 4 m. N.E. of Tomintoul, Banffshire.

Knockapple, hill, alt. 1118 ft., in co. of, and 5 m. N.E. of, Dumbarton.

Knockard and Erropie, vil. in isl. of, and 1 m. S. of Butt of, Lewis, Outer Hebrides.

Knockbain, par. on N. side of Beauly Firth, S.E. Ross and Cromarty ; 11,635 ac. Pop. 1155. P.O. Church, 5 m. N.W. of Inverness (by ferry across the Firth).

Knockbain, pl., 2½ m. S.E. of Beauly, Inverness-shire.

Knock Bay, 2 m. N.W. of Portpatrick, W. Wigtownshire.

Knockbreac, loch, 2½ m. N. of Feolin Ferry, Jura, Argyllshire.

Knockbreck, pl. on W. side of Loch Snizort, Skye, Inner Hebrides.

Knockbreck, seat, on Dornoch Firth, ¾ m. S.E. of Tain, Ross and Cromarty.

Knockbrex, seat, on Fleet Bay, S. Kirkcudbrightshire, 5½ m. S.W. of Gatehouse-of-Fleet.

Knock Castle, picturesque ruin, 1½ m. S.W. of Ballater, Aberdeenshire ; at confluence of Dee and Muick.

Knock Castle, baronial stronghold and modern mansion on Firth of Clyde, 2 m. N. of Largs, N.W. Ayrshire.

Knock Castle, ruined stronghold on K. Bay, Skye, 3½ m. S.W. of Oronsay Isl., Inner Hebrides. Looks over to Knoydart on the mainland, and was used to defend the Sound of Sleat.

Knockchoilum, pl., 9 m. N.E. of Fort Augustus, Inverness-shire.

Knockdavie Castle, ruin, 2 m. N.W. of Burntisland, Fife.

Knockdaw Castle, ruin, 2½ m. N. of Colmonell, S. Ayrshire.

Knockderry Castle, seat, on E. side of Loch Long and 1 m. N.W. of Cove, Dunbartonshire. Nearby is K. Lodge.

Knockdhu, or Knockdow, seat, 2 m. N.W. of Toward, Cowal, Argyllshire.

Knockdhu Distillery, Knock, Banffshire.

Knockdolian, ruined castle and modern mansion on the r. Stinchar, 1½ m. S.W. of Colmonell, Ayrshire. K. Hill is a conical hill (869 ft.), and known to mariners as "False Crag."

Knockendoch Hill, peak (over 1700 ft.) of the Criffel ridge, par. of, and 2 m. S. of, New Abbey, E. Kirkcudbrightshire.

Knock-en-hair, seat, near Dunbar, E. Lothian.

Knockentiber, seat, 2 m. S.W. of Kilmaurs, Ayrshire.

Knockespock, seat, 4 m. S. of Kennethmont sta., Aberdeenshire.

Knockfarril, conical hill (579 ft.), 1¼ m. E. of Strathpeffer, S.E. Ross and Cromarty. Crowned by a large and complete vitrified fort with ramparts.

Knockfin, or Fingall's Fort, see DUN FIONN.

Knockfin Heights, alt. 1416 ft., 4 m. N.E. of Kinbrace sta., border of Caithness and Sutherland.

Knockfin House, 4 m. S.W. of Inver-cannich, Inverness-shire.

Knockgray, seat in par. of, and 1 m. E. of, Carsphairn, Kirkcudbrightshire.

Knockhall Castle, ruin in par. of Foveran and 1 m. N.N.W. of Newburgh, S.E. Aberdeenshire; dates from 1565.

Knock Head, prom. on Moray Firth, in co. of, and 2½ m. N.W. of, Banff.

Knockhill, seat, 1½ m. W. of Ecclefechan, Dumfriesshire.

Knock Hill, Banffshire. See KNOCK.

Knockhillie, or **Knockhooly,** ham. in S.E. Kirkcudbrightshire, 8 m. S.E. of Dalbeattie.

Knockie Lodge, 8 m. N.E. of Fort Augustus, between Lochs Knockie and Ness, Inverness-shire.

Knockieshee, seat, par. of Glenluce, Wigtownshire.

Knockinaam Lodge, 3 m. S.E. of Port-patrick, Wigtownshire.

Knockinan, pl., with ruins of ancient stronghold, 1½ m. N.E. of Dunbeath, E. Caithness.

Knockleith, seat, 3½ m. S.W. of Auchter-less sta., Aberdeenshire.

Knocknaha, ham., 3 m. S.W. of Campbel-town, Kintyre, Argyllshire.

Knocknair Reservoir, water supply for Port Glasgow, lying 1 m. S. of the town; capacity of 26½ million gallons.

Knocknalling, seat, on the Ken and 6 m. N.W. of New Galloway, Kirkcudbright-shire.

Knock of Braemory, 1493 ft., 2 m. N. of Dava, in par. of Edinkillie, E. Moray.

Knock of Crieff, 911 ft., par. of, and 1½ m. N. of, Crieff, Perthshire; commands a fine view.

Knock of Fyrish, see HILL OF FYRISH.

Knockomie, seat, 1 m. S.W. of Forres, Moray.

Knockshandoch, seat, 7 m. N. of Alyth, Angus.

Knocksting, loch, 6 m. N.E. of Dalry, E. Kirkcudbrightshire.

Knock, The, 1017 ft., 1½ m. N.E. of Bath-gate, W. Lothian.

Knockton, hill, 1605 ft., Angus, 8½ m. N. of Blairgowrie.

Knockvennie, pl., 2 m. N.W. of Kirk-patrick-Durham, E. Kirkcudbrightshire. P.O.

Knowe, pl., par. of Penninghame, 7 m. N.W. of Newton-Stewart, N.E. Wig-townshire. P.O.

Knowegreens, pl., 4 m. N. of Fettercairn, Kincardineshire.

Knowehead, ham., 3½ m. S.E. of Cars-phairn, Kirkcudbrightshire. P.O.

Knowes Hill, 1222 ft., Selkirkshire, 3½ m. N.W. of Galashiels.

Knoweside, ry. sta., L.M.S. (passenger service withdrawn), in co. of, and 11 m. S.W. of, Ayr.

Knowesouth, seat, on Teviot, 2½ m. W. of Jedburgh, Roxburghshire.

Knownoble Colliery, ½ m. S.W. of Clelland, Lanarkshire.

Knowsie, seat, near Lonmay sta., N.E. Aberdeenshire.

Knoxland, q.s. par. in par. of Dumbarton, S. Dunbartonshire.

Knox, Upper, Middle, and **Nether,** 3 hams. in S. Kincardineshire, 1½ m. N.W., S.W., and S.S.W. of Gourdon. K. Hill, 523 ft.

Knoydart, dist. and q.s. par., between Loch Hourn and Loch Nevis, W. Inverness-shire. P.O., T.O. K. Deer Forest embraces 50,000 ac.

Kyle, dist. in mid Ayrshire, between the Irvine and the Doon. Robert Burns was born in the district.

Kyleakin, vil. in the par. of Strath and on Kyle Akin, 8 m. N.E. of Broadford, Skye, Inner Hebrides. Pop. 232. P.O., T.O. Hotel. Castle Maoil is a ruined stronghold. There is a ferry across Kyle Akin to the mainland to Kyle of Lochalsh. K. Lighthouse on Gillean Isl., 1 m. N. of the vil. and at W. entrance to Loch Alsh, is 70 ft. high and shows a fixed light, visible 12 m. Steamer communication to Portree, Stornoway, Oban, Glasgow.

Kyle Castle, ruined castle on the Glen-more, par. of Auchinleck, and 5 m. E. of Cumnock, E. Ayrshire.

Kylemore, channel between Raasay and Scalpa, Outer Hebrides.

Kyle of Durness, sea loch in N. Suther-landshire, 15 m. N.W. of Tongue; receives the Dionard and Grudie Rivers.

Kyle of Laxford, see LAXFORD.

Kyle of Lochalsh, vil. on Loch Alsh, 10½ m. S.W. of Strome Ferry, S.W. Ross and Cromarty; ry. sta., L.M.S. Hotels. P.O., T.O. called Kyle. Steam ferry to Kyleakin. A fixed red light and a fixed green light on pier-head. Pop. 440.

Kyle of Sutherland, continuation of Dornoch Firth, extending 2½ m. N.W. from Bonar Bridge, Ross and Cromarty and Sutherland.

Kyle of Tongue, narrow sea loch, running inland for 7 m., in N. Sutherland, 8 m. E. of Loch Erriboll; receives Kinloch Water. Ferry across it from Melness to Tongue.

Kyle Rhea, narrow strait between E. end of Skye and the mainland, at the S. entrance to Loch Alsh; ⅓ m. wide, with strong current. On W. shore is a fixed white light, visible 9 m. A ferry crosses it.

Kylesku, sea loch and strait, W. Sutherland, 7 m. S. of Scourie; at meeting of Loch Glendhu, Loch Glencoul, and Loch Cairnbawn; is sometimes called Kyle of Assynt. Ferry across strait, with an inn on the S. side.

Kylesmorar, ham., 11 m. E.S.E. of Mallaig, on S. shore of Loch Nevis, W. Inverness-shire. P.O.

Kyles of Bute, strait; narrow arm of Firth of Clyde, between the isl. of Bute and Argyllshire, semicircular in shape and 16 m. long, and presents a constantly changing view of great beauty. Regularly navigated by tourist steamers. Occulting light on Loisgt Isl.

Kyles Scalpay, ham., between N. and S. harbours, Scalpay Isl., E. Loch Tarbert, Harris, 42 m. S. of Stornoway. P.O., T.O.

Kylestrome, pl. on N. side of Kylesku 15 m. N. of Lochinver, W. Sutherland. Remains of a fort.

Kyle Taransay, channel between Taransay and Harris, Outer Hebrides.

Kyllachy Lodge, 2½ m. S.W. of Tomatin, Inverness-shire.

Kymah Burn, headwater of the Livet, Banffshire.

Kynachan, seat, on the Tummel, 10 m. W. of Pitlochry, Perthshire.

Kypes Rig, hill ridge (1173 ft.), 3½ m. S.E. of Strathaven W. Lanarkshire.

Kype Water, rises on Goodbush Hill, flows N., and falls into Avon 1 m. S.E. of Strathaven. There is a reservoir 2 m. from its source.

L

Ladder Burn, stream, Angus, flows 2½ m. S.E. to the Water of Mark, N.W. of Lochee.

Ladders, The, rude stair in precipitous bank of Trossachs, forming the only means of access to foot of Loch Katrine, Perthshire.

Ladhar Beinn, mt., Inverness-shire, on Loch Hourn; alt. 3343 ft.

Ladhope, q.s. par., Roxburghshire, partly in Melrose but chiefly in Galashiels par. L. House, a seat, in vicinity.

Ladshaw Hill, Dumfriesshire, 13 m. N.W of Langholm; alt. 1527 ft.

Lady, par., Orkney, N.E. Sanday; 5361 ac. Pop. 487. P.O.

Ladybank and Monkston, q.s. par. and police bur., with ry. sta. (Ladybank), L.N.E., Fife, 5½ m. S.W. of Cupar. Bur. pop. 1128. P.O., T.O. Linen mfrs., malting, etc.

Ladyburn, seat, Ayrshire, 4 m. S. of Maybole.

Ladyburn, q.s. par., Renfrewshire, Greenock and Port Glasgow pars.

Lady Glenorchy's, q.s. par., E. Edinburgh.

Lady Isle, lighthouse in Firth of Clyde, 5 m. N.W. of Ayr. Shows a group-flashing twenty-seconds light, seen 12 m.

Ladykirk, par. and ham., S.E. Berwickshire; par. 3382 ac. Pop. 305. Ham. on r. Tweed, 6 m. N. of Coldstream. P.O. L. House in vicinity.

Ladykirk, ancient par., Orkney, now in Stronsay and Eday pars., S.W. Stronsay Isl.

Ladykirk, or **Northkirk,** ancient par., Orkney, now in Westray and Papa Westray pars., in N. of Westral Isl.

Ladykirk House, seat, Ayrshire, 4½ m. N.E. of Ayr.

Ladyland, seat, and remains of old tower, Kilbirnie par., Ayrshire, 2 m. S.W. of Lochwinnoch.

Ladylands Siding, ry. sta., L.N.E., on border of Perthshire and Stirlingshire, 2¼ m. E. of Port of Menteith sta.

Ladyloan, q.s. par., Angus, Arbroath and St. Vigeans pars; embraces part of town of Arbroath.

Ladysbridge, ry. sta., L.N.E., 2½ m. W. of Banff. The Banffshire Mental Asylum is here.

Lady's Rock, between Lismore Isl. and Mull Isl., Argyll. Shows a flashing six-seconds light, seen 10 m. In 1523 Maclean of Duart left his wife on this rock to be drowned by the rising tide, but she was rescued by a passing vessel.

Ladyurd Hill, Peeblesshire, 3½ m. N.E. of Broughton ; alt. 1724 ft.

Lady Voe, landing-place, Yell Isl., Shetland.

Ladywell, ham., Kirkcudbrightshire, 12 m. S. of Dumfries.

Lady Yester's, eccl. par., Edinburgh.

Lael State Forest, nearest ry. sta., Garve, 12 m. W. of Dingwall, E. Ross and Cromarty.

Laga, pl. and bay, Argyll, 9 m. N.E. of Tobermory, on Loch Sunart. In vicinity, L. Hill (679 ft.) and L. Loch.

Lagafater Lodge, S. Ayrshire, 6 m. S.E. of Ballantrae.

Lagavulin Distillery, Islay, Argyll, 2½ m. E. of Port Ellen.

Lagg, Bay of, N. of Eigg Isl., Argyll.

Lagg, ham., E. coast of Jura. P.O. Has ferry to N. Knapdale.

Laggan, par. and ham., Badenoch dist., Inverness-shire ; par. 151,916 ac. Pop. 621. Ham. 10 m. S.W. of Kingussie, on r. Spey. P.O., T.O.

Laggan, Loch, Inverness-shire, extends about 7 m. to the S.W. boundary, and has a mean breadth of about 1 m ; greatest depth 174 ft. ; alt. 819 ft. Is now one of the reservoirs for Fort William. Aluminium works.

Laggan, pl., on the Caledonian Canal, Inverness-shire, at N.E. end of Loch Lochy, 9 m. S.W. of Fort Augustus.

Laggan, r., bay, and prom., Islay, Argyll, on E. side of Loch Indail.

Laggan, pl. and seat, Mull Isl., Argyll, on Loch Buie.

Laggan, loch, S.E. Sutherlandshire, 5 m. N.E. of Bonar Bridge.

Laggan, seat, 7 m. S.W. of Inverness.

Laggan, seat, Morayshire, on r. Spey, 1 m. E.N.E. of Carron sta.

Laggan, loch, Stirlingshire, 2 m. S.W. of Kippen.

Laggan, seat, S. Ayrshire, ½ m. E. of Ballantrae.

Laggangarn, pl., with standing-stones, Wigtownshire, New Luce par.

Laggan, South, a State forest, Inverness-shire, 9 m. S.W. of Fort Augustus.

Laggan, The, name given to r. Forth, between Loch Ard and Aberfoyle, Perthshire.

Lag, Lower, ruined massive tower, Dumfriesshire, 1½ m. N.E. of Dunscore. Formerly the seat of the Griersons of Lag.

Lag, Point of, cape, Wigtownshire, E. coast of Luce Bay.

Lahill House, seat, Fife, 1¼ m. E. of Upper Largo.

Laid, ham., N.W. Sutherland, 8 m. S.E. of Durness, on Loch Eriboll. P.O., T.O.

Laide, pl., N.W. Ross and Cromarty, 11 m. N.E. of Gairloch, on Gruinard Bay. P.O.

Laide, loch, Inverness-shire, 4 m. N.E. of Drumnadrochit; alt. 860 ft., depth 9 ft.

Laidlawstiel, seat, Selkirkshire, 5 m. W. of Galashiels.

Laidnagullen, pl., N. Sutherland, 2½ m. W. of Strathy.

Laidon, loch on Rannoch moor, Perthshire and Argyll. It is 7 m. long and about 1 m. broad, and it lies amid a tract of bog, heath, and rock, 5 m. E. of Kingshouse. Greatest depth 128 ft., alt. 923 ft.

Laigh Dalmore, pl., Ayrshire, 3¼ m. N. of Coylton. Plumbago was formerly mined here.

Laighland Wood, at Auchencruive, Ayrshire. 3½ m. N.E. of Ayr.

Laighwood, ham., Perthshire, 5 m. N.E. of Dunkeld.

Laingedail, loch, isl. of Islay, Argyll, 7 m. N.W. of Bridgend.

Lainshaw, seat, N. Ayrshire, ¾ m. S.W. of Stewarton.

Lairdmannoch, Linn of, waterfall, on Tarf Water, S. Kirkcudbrightshire, 5 m. W. of Castle-Douglas.

Laird's Hill, a summit of the Kilsyth Hills, Stirlingshire, 2 m. N.W. of Kilsyth ; alt. 1393 ft.

Laird's Seat, hill, Renfrewshire, 5 m. S.S.E. Greenock ; alt. 1084 ft.

Lairg, par., vil., and ry. sta., L.M.S., Sutherland ; par. 121,390 ac. Pop. 1065. Vil., on r. Shin, 2 m. from ry. sta. and 9 m. N. of Bonar Bridge. Hotel. P.O., T.O. Pop. 305.

Lairg Ghru, the mt.-pass over the Cairngorms from Aviemore to Braemar, a distance of 19 m. without a house; the walk is easy, except about 2 m. at the summit of the pass (2750 ft.), which is strewn with boulders.

Lairig Eala, loch, Perthshire, 3 m. N.W. of Lochearnhead, in Glenogle.

Lairige, Loch na, S.E. Inverness-shire, 7 m. N.E. of Drumgask.

Lairige, Loch Na, loch, 2½ m. S.W. of Ben Lawers, Perthshire; greatest depth 39 ft., alt. 1595 ft.

Laithers House, Aberdeenshire, on r. Deveron, 4 m. W. of Turriff.

Lakehead, pl., 12 m. N. of Dumfries.

Lakenbuie, pl., 3 m. S.E. of Nairn.

Lamancha, pl. and seat, with ry. sta., L.N.E. (passenger service withdrawn), N. Peeblesshire, 3 m. S. of Leadburn, 20¼ m. S.W. of Edinburgh by rail. P.O., T.O. Grange of Romanno, the former name of seat.

Lamanchan Hill, N.W. Kirkcudbrightshire, 7 m. N. of Newton-Stewart; alt. 2349 ft.

Lamba Island, in Yell Sound, Shetland, 1½ m. N.E. of Ollaberry.

Lamba Ness, headland at N.E. extremity of Unst Isl., Shetland.

Lambden, seat and burn, Berwickshire, 3½ m. S.E. of Greenlaw.

Lamberton, ancient par., now part of Mordington par., S.E. Berwickshire. Its church was the marriage-place of King James IV. and the scene of an international convention in 1573, and its toll-bar was long a place of runaway marriages, like Gretna, in the W. The church stands 3½ m. N.W. of Berwick-on-Tweed. L. House, a seat.

Lamb Head, headland, with large Broch, Stronsay, Orkney.

Lambhill, pl. on Forth and Clyde Canal, 2 m. N. of Possilpark, Glasgow. P.O.

Lamb Holm, isl., Orkney, between Burray and the Mainland, in Holm Sound.

Lamb Island, in Firth of Forth, E. Lothian, 1½ m. N.W. of Berwick.

Lamford Hill, Kirkcudbrightshire, 2½ m. E. of L. Doon; alt. 1307 ft.

Lamington, vil., with ry. sta., L.M.S., Lanarkshire, on r. Clyde, 10 m. S.S.E. of Carstairs Junction. L. barony belonged to the wife of Sir William Wallace. L. Tower, once a splendid ancient edifice, is now a small ruin not far from the vil. L. House, the seat of Lord Lamington.

Lamlash, vil., with pier, Arran Isl., Firth of Clyde, on L. Bay, 4 m. S.E. of Brodick. P.O., T.O. A popular summer and sea-bathing resort. L. Bay forms an excellent harb., very large and deep, protected from the sea by Holy Isle. L. lighthouse, S.W. pt., with fixed light, seen 12 m. On Pillar Rock pt., an alternating flashing half-minute white and red light, seen 17 m. Steamer communication with Ardrossan.

Lammermuir Hills, E. Lothian and Berwickshire; highest summit, Meikle Says Law, 1750 ft., and Lammer Law, 1733 ft., 8½ m. S.S.E. and 7½ m. S. of Haddington.

Lamont House, Argyll. See ARDLAMONT.

Lanark, royal and mun. bur., par., and co. town, with ry. sta., L.M.S., S. Lanarkshire, 8½ m. S.E. of Wishaw, 32¼ m. S.E. of Glasgow by rail; par. 10,560 ac. Pop. 9133; bur. pop. 6178. Hotel. P.O., T.O. Has a large cattle and sheep market and has tanning and hosiery industries. In a niche in the church is a colossal statue of the patriot Wallace. Nearby are the celebrated falls of Clyde. L. Moor and L. Loch in vicinity.

Lanark County Asylum, see HARTWOOD.

Lanark Division, parl. division of Lanarkshire. Pop. 66,844.

Lanark, New, vil., 1 m. S. of Lanark, on r. Clyde. P.O., T.O.

Lanarkshire, inland co., comprising the greater part of the basin of the Clyde above Renfrew. The principal tribs. of the Clyde within the co. being the Douglas, the Nethan, the Avon, the Cadder, and the Kelvin. The surface rises towards the S., where the Lowther or Lead Hills reach an alt. of 2403 ft. The co. is divided into three wards—Upper, Middle, and Lower. The Upper Ward is nearly twice the size of the Middle and eight times the size of the Lower, and is chiefly hill or moorland, and has large dairy and sheep farms. Lead is mined. The Middle Ward, lowland in character, is renowned for its orchards. Strawberries, currants, gooseberries, and tomatoes are produced extensively. The Ward is closely identified with the mining and ironworking industries. The Lower Ward is more level, and is the busiest and most populous. It contains the chief seats of the cotton, linen, and woollen industries; the coal and ironstone mines, fireclay beds, blast furnaces, rolling mills, shale mines, and oil works. The principal manufacturing towns are Glasgow, Hamilton, Airdrie, Motherwell, Coatbridge, and Wishaw. Area 882 sq. m., or 564,567 ac. Pop., inclusive of Glasgow, 1,586,047. For parliamentary purposes is divided into seven divisions; one member for each division.

Landends, pl., Kincardineshire, 3½ m. N.W. of Laurencekirk.

Landheads, ham., Dumfriesshire, 2 m. N. of Annan.

Lanergill, pl., Caithness, 4½ m. S.E. of Halkirk.

Lanfine House, seat, Ayrshire, 1½ m. S.E. of Newmilns.

Langa, isl., Shetland, 8 m. W. of Lerwick.

Langalcorad, pl., Bute Isl., 5 m. S. of Rothesay.

Langass, shooting-lodge, N. Uist, 5 m. W.S.W. of Lochmaddy.

Langavat, loch, S. Benbecula, Outer Hebrides, 5 m. S.E. of Balivanich; greatest depth 34 ft., alt. 16 ft.

Langavat, loch, S. Harris, Outer Hebrides, 3¾ m. N. Renish Pt.

Langavat Loch, Lewis Isl., Outer Hebrides, on border of Uig and Lochs pars.; length 7½ m., depth 98 ft., alt. 108 ft.

Langbank, q.s. par. and vil., with ry. sta., L.M.S., Renfrewshire, on r. Clyde, 4¼ m. E. of Port Glasgow. P.O., T.O. Vil. pop. 499.

Langbar, vil., Ayrshire, 1½ m. S.E. of Kilbirnie.

Lang Craig, headland, Angus, S. end of Lunan Bay, 2½ m. E.S.E. of Inverkeilor.

Langdyke, ham., Fife, 2 m. N.W. of Kennoway.

Langholm, town (police bur.) and par., with ry. sta., L.N.E., E. Dumfriesshire; par. 16,995 ac. Pop. 2770. Town on r. Esk, 16½ m. N.E. of Annan and 91¼ m. S. of Edinburgh by rail. Pop. 2448. P.O., T.O. Industries: tweed mills and tan works. L. Castle, a ruined border peel, and L. Lodge, a seat of the Duke of Buccleuch.

Langhope Birks, pl., with traces of encampment, Berwickshire. See CAMPMUIR.

Langhope Burn, trib. of Ale Water, E. Selkirkshire.

Langhouse, seat, Renfrewshire, 2½ m. N.E. of Wemyss Bay.

Langlands, seat, Hawick, Roxburgh.

Langlee, seat, Roxburghshire, 2 m. S. of Jedburgh.

Langlee House, hunting seat, Roxburghshire, near Galashiels.

Langley Park, seat, Angus, 1½ m. S.W. of Dubton Junction.

Langloan, ry. sta., L.M.S., Lanarkshire, sub. of Coatbridge.

Langour, Clackmannanshire. See COALSNAUGHTON.

Langraw, seat, E. Fife, 2 m. S. of St. Andrews.

Langrig, seat, Berwickshire, 7 m. N. of Coldstream.

Langshaw, seat, Dumfriesshire, 1 m. E. of Kirtlebridge sta.

Langshaw, pl. and seat, Roxburghshire, 3 m. N.E. of Galashiels.

Langside, q.s. par. in S. of Glasgow, Cathcart par; ry. sta. (L. and Newlands), L.M.S. P.O., T.O. The L. Battlefield Memorial marks the site of the battle between the troops of Mary, Queen of Scots, and those of the Regent Murray in 1568.

Langton, par. and vil., Berwickshire; 7140 ac. Pop. 363. L. Estate, 2 m. S.W. of Duns. Vil. pop. 169.

Langtonlees, seat, Berwickshire, 4 m. S.W. of Duns.

Langwell, ham., S.E. Sutherlandshire, 15 m. N.W. of Dornoch. Granite quarries.

Langwell, seat of the Duke of Portland, Caithness, at the confluence of Berriedale and L. Waters, 5 m. S.W. of Dunbeath.

Langwell Water, stream, S.E. Caithness, flows 10 m. E. to the sea at Berriedale.

Lann Hall, seat, Dumfriesshire, 2 m. N.E. of Moniaive.

Lanrick Castle, seat, Perthshire, 3 m. W. of Doune.

Lanrick Mead, meadow, W. end of Loch Vennachar, Perthshire. Figures in Scott's "Lady of the Lake."

Lanrig, W. Lothian. See LONGRIDGE.

Lanton, vil., Roxburghshire, 2 m. N.W. of Jedburgh. P.O., T.O. Has old Border peel.

Laoghal, Loch, Sutherland. See LOYAL.

Laorben (Ladhar Bheinn), mt., Inverness-shire, on S. side of Loch Hourn ; alt. 3343 ft.

Laphroaig Distillery, Islay, Argyll, near Port Ellen.

Lappock Rock, in Firth of Clyde, off Ayr coast, 2½ m. N.W. of Troon.

Larachbeg, pl., Argyll, near head of Loch Aline. P.O.

Larbert, par. and town, with ry. sta., L.M.S., E. Stirlingshire ; par. 3971 ac. Pop. 12,959. Town 2½ m. N.W. of Falkirk, 28¼ m. W.N.W. of Edinburgh. Pop. 12,959. P.O., T.O. An important ry. junc. In vicinity, the Stirling District Asylum. L. House, a seat.

Larbrax, pl., Wigtownshire, 5 m. W. of Stranraer.

Larchgrove House, N. Lanarkshire, 1 m. N.E. of Shettleston.

Larg Hill, mt., W. Kirkcudbrightshire, 7 m. N.N.E. of Newton-Stewart ; alt. 2159 ft.

Larg Hill, S.W. Kirkcudbrightshire, 2 m. S.E. of Creetown ; alt. 969 ft.

Largie, pl., Aberdeenshire, 2½ m. N.E. of Wardhouse sta.

Largie Castle, seat, on W. Kintyre, Argyll-shire, near Tayinloan.

Largiemore, pl. and pier on Loch Fyne, Argyll, 1 m. N.E. of Otter Ferry. Ferry communication 1½ m. to W. Otter Ferry.

Larglanlee Hill, Kirkcudbrightshire, 1¼ m. N.W. of Crocketford ; alt. 898 ft.

Largo, par. and ry. sta., L.N.E., Fife, on L. Bay ; 7378 ac. Pop. 2299 ; also the vils. of Upper L. (P.O., T.O.), and the fishing vil. of Lower L. (P.O., T.O.). Pop. 1667. Lower L. was the birthplace of Alexander Selkirk, the prototype of " Robinson Crusoe," 1676–1723. L. House, a seat. L. Law (965 ft.), 1 m. N. of Upper L.

Largoward, vil., Fife, also q.s. par., 4½ m. N.E. of Largo sta. P.O., T.O. Pop. 172.

Largs, mkt.-town (police bur.) and par., with ry. sta., L.M.S., N. Ayrshire ; par. 21,831 ac. Pop. 8470. Town, on L. Bay, Firth of Clyde, 11 m. N. of Ardrossan by rail and 18 m. S.W. of

Greenock. Pop. 6115. Hotels. At N. and S. end of pier are fixed lights. Gives name to a famous battle fought adjacent to it, in 1263, between Alexander III. of Scotland and Haco of Norway. A well-known watering-place and holiday resort. P.O., T.O.

Larguna House, Kinclaven par., Perth-shire.

Largybeg, pl. and headland, Isle of Arran, 1¼ m. S. of Whiting Bay.

Larkfield, Greenock, Renfrewshire. Hospital.

Larkfield, colliery vil., Lanarkshire, 2 m. W.N.W. of Hamilton.

Larkfield House, in co., 1 m. S. of Nairn.

Larkhall, mining town and q.s. par., with ry. stas. (Central and East), L.M.S., Lanarkshire ; town 3½ m. S.E. of Hamilton. P.O., T.O. Pop. 13,121.

Laroch Burn, Argyll, flows N.E. to Loch Leven. See BALLACHULISH.

Laro, Loch, S. Sutherland, 5 m. S.E. of Lairg.

Lassintullich, seat, N.W. Perthshire, 2½ m. S. of Kinloch Rannoch.

Lassodie, colliery vil., Fife, 4 m. N.E. of Dunfermline. P.O., T.O. Vil. pop. 705.

Lasswade, small town (police bur., united with Bonnyrigg, 1929) and par., with ry. sta., L.N.E., Midlothian ; par. (containing Loanhead and parts of Bonnyrigg and Penicuik) 9964 ac. Par. pop. 11,714. Town on r. N. Esk, 9 m. S.E. of G.P.O., Edinburgh, by road, and 9¾ m. by rail. P.O., T.O. Has paper mills and a carpet factory. Bur. pop. 4481.

Lathaich, loch on N. side of Ross of Mull.

Lathallan, seat, E. Stirlingshire, 1 m. E.S.E. of Polmont.

Lathallan House, seat, Fife, 2 m. N.W. of Colinsburgh.

Latheron, par. and fishing vil., S. Caithness ; par. 119,506 ac. Pop. 3276. Vil. (also called Janetstown and Lather-onwheel) 18 m. S.W. of Wick. Pop. 95. P.O., T.O. Latheronwheel House in vicinity.

Lathockar, seat, Fife, 4½ m. S. of St. Andrews.

Lathones, vil., Fife, 5½ m. S. by W. of St. Andrews. Pop. 26.

Lathrisk House, seat, Fife, 1½ m. N.E. of Falkland.

Lauchope House, Lanarkshire, 1½ m. N.E. of Holytown.

Lauder, royal and mun. bur. and par., ry. sta., L.N.E. (passenger service withdrawn), W. Berwickshire, near Leader Water, 7 m. N.N.W. of Earlston ; par. 33,616 ac. Pop. 1214 ; pop. of bur. 628. Hotels. P.O., T.O. Was the scene of the execution of James III.'s favourites by the Earl of Angus, " Bell the Cat."

Lauderdale, ancient dist. in W. of Berwickshire.

Laudle, ham. in Morven dist., Argyll, on S. shore of Loch Sunart. P.O., T.O. called Laudale.

Laurencekirk, small mkt.-town (police bur.) and par., with ry. sta., L.M.S., S. Kincardineshire ; par. 5613 ac. Pop. 1713 ; pop. of bur. 1316. The town stands 10 m. N.E. of Brechin and 30½ m. S.W. of Aberdeen. P.O,, T.O. Hotel. Grain and cattle markets.

Laurence Park, seat, Stirlingshire, 1 m. S.E. of Polmont.

Laurieknowe, pl., Dumfriesshire, near Maxwelltown. P.O.

Laurieston, vil., Stirlingshire, 1½ m. E. of Falkirk. P.O., T.O. Dist. pop. 3147.

Laurieston, q.s. par., Govan par., in S. of Glasgow.

Laurieston, vil., Kirkcudbrightshire, 6 m. N.W. of Castle-Douglas. P.O., T.O. Pop. 122. L. Hall, a seat, 1 m. N.W.

Lauriston, dist. in S. of city of Edinburgh. P.O., T.O.

Lauriston, ry. sta., L.N.E., S. Kincardineshire, 6½ m. N.E. of Montrose. L. Castle, a seat.

Lauriston Castle, formerly a seat, near Davidson's Mains, 4½ m. N.W. of Edinburgh, now national property.

Law, q.s. par., vil., and ry. sta. (L. Junc.), L.M.S., Lanarkshire, 2½ m. S.E. of Wishaw. P.O.; T.O. at sta.

Law Castle, ruined tower, W. Kilbride, Ayrshire.

Lawers, pl., with ferry and hotel, Perthshire, on Loch Tay, at S. skirt of Ben Lawers, 8 m. N.E. of Killin. P.O., T.O. L. Burn flows S.E. to Loch Tay.

Lawers House, seat, Perthshire, 2 m. N.E. of Comrie.

Lawgrove House, seat, Perthshire, Tibbermore par.

Lawhill House, E. Perthshire, ¾ m. N.E. of Dollar.

Law Junction, q.s. par., vil., and ry. junc., L.M.S., Lanarkshire, 2½ m. S.E. of Wishaw. Vil. pop. 1752. P.O.; T.O. at sta.

Lawmuir House, Lanarkshire, 2½ m. W. of East Kilbride.

Lawson's Rock, cliff, N.W. Sutherland, on road from Scourie to Durness.

Laws, The, seat, Angus, 2 m. N. of Monifieth. Vestiges of large vitrified fort.

Laws, The, seat, Berwickshire, 3½ m. S.E. of Duns.

Lawston, pl., with mineral spring, Roxburghshire, 4 m. S.W. of Newcastleton.

Lawting, isl., Shetland, in Tingwall Loch, 4 m. N.E. of Lerwick. Once the seat of the Norse courts.

Lawton, seat, Angus, on Lunan Water, 4 m. N. of Arbroath. A reservoir for the Dundee water supply.

Lawton, ham. and seat, Perthshire, 4 m. S.W. of Coupar-Angus.

Laxadale, loch, N. Harris, Outer Hebrides, 2 m. N.E. of Tarbert.

Laxay, ham. and burn, Lewis Isl., Ross and Cromarty, 10 m. S.W. of Stornoway. P.O., T.O. The burn flows 3 m. W.S.W. to Loch Erisort.

Laxdale, vil., Lewis Isl., Ross and Cromarty, 1¼ m. N.W. of Stornoway. P.O.

Lax Firth, Shetland, in E. of Tingwall par.

Laxfirth Voe, sea loch, S. Shetland, 4 m. N.W. of Lerwick.

Laxford, r., N.W. Sutherland, flows 5 m. N.W. to Loch L. L. Loch, an arm of the sea, extends 3½ m. N.W; sometimes called the Kyle of L.

Laxo, pl. and burn, E. side of Mainland, Shetland.

Leacann, a loch, W. Argyll, 3 m. N.W. of Furnace.

Leachkin, ham., 1½ m. S.W. of Inverness. P.O., T.O.

Leadburn, ham., with junc. ry., L.N.E. (880 ft.), Midlothian, 17½ m. S. of Edinburgh, by rail. P.O.

Leaderfoot, vil., Roxburghshire, 2 m. N.E. of Melrose.

Leadervale House, Roxburghshire, 1 m. N.E. of Earlston.

Leader Water, a small r. running 21 m. S. and S.E. to the Tweed at Leaderfoot; rises on Lammer Law, E. Lothian.

Leadhills, vil. and ry. sta., L.M.S. (alt. 1405 ft.), Crawford par.; q.s. par.; vil. 5 m. S.W. of Elvanfoot sta. P.O., T.O. Derives its name from adjacent lead veins. Is the highest vil. in Scotland. Pop. 670.

Leagarth, seat, Fetlar Isl., Shetland.

Lealt, ham., Isle of Skye, Inverness-shire, 10 m. N. of Portree.

Lealt, r., Isle of Skye, Inverness-shire, runs 4 m. N.E. to Raasay Sound.

Lealt Burn, stream, N. isl. of Jura, Argyll.

Lealty House, Ross and Cromarty, 4 m. N.W. of Alness.

Leanchoil, Forres, Morayshire. Hospital.

Leasgaman, loch, Jura, Argyll, 2 m. S.W. of Lagg.

Leathad an Taobhain, mt., S. Invernessshire, 11 m. E. of Dalwhinnie; alt. 2994 ft.

Leathan, loch, Argyll, 6 m. N. of Lochgilphead.

Leathan, loch, Isle of Skye, 5 m. N. of Portree.

Leckie Burn, stream, N. Stirlingshire, flows N. to the Forth, 2½ m. E.N.E. of Kippen.

Leckie House, seat and glen in Gargunnock par., Stirlingshire. In vicinity, Leckie Castle, visited by Prince Charles Edward, 1745.

Leckmelm, ham. and seat, N.W. Ross and Cromarty, 3¾ m. S.E. of Ullapool.

Leckmelm Deer Forest, N.W. Ross and Cromarty, Lochbroom par; 4740 ac.

Lecropt, eccl. par., adjoining Bridge of Allan, Stirlingshire.

Ledaig, ham., Argyll, 6 m. N.E. of Oban. P.O., T.O.

Ledard, Falls of, two waterfalls, 15 and 20 ft., adjacent to the W. end of Loch Ard, Perthshire.

Ledard Hostel, Kinlochard, Perthshire, 4½ m. W. of Aberfoyle. A Scottish Youth Hostel.

Ledart, seat, Perthshire, 4 m. W. of Aberfoyle, on Loch Ard.

Ledbeg, pl., S.W. Sutherland, 5 m. S. of Inchnadamph. An effort to use the marble found here was not successful.

Leddriegreen House, Stirlingshire, ½ m. N. of Strathblane.

Ledmore, pl., S.W. Sutherland, 6 m. S. of Inchnadamph. L. Lodge, a seat.

Lednock, r., Perthshire, flowing 11 m. S.E. to the Earn at Comrie.

Lee Castle, ancient seat of the Lockharts, 3 m. N.W. of Lanark.

Lee, Loch, a headstream of the N. Esk, Angus, Lochlee par.

Lee Pen, peaked mt. of left side of Tweed, Peeblesshire, 1 m. N.W. of Innerleithen; alt. 1647 ft.

Lees, seat, Berwickshire, in vicinity of Coldstream.

Lees, ham., Shetland, 1 m. S.W. of Mossbank.

Lee, The, seat, 3 m. N.W. of Lanark. Among the many historical heirlooms is the "Lee Penny," the "Talisman" of Sir Walter Scott's romance.

Leetown, vil., Perthshire, 2½ m. W.S.W. of Errol.

Leet Water, stream, Berwickshire, flows 14 m. S.W., S., and S.E. to the Tweed at Coldstream.

Lefnoll Point, headland, Wigtownshire, 3 m. N.E. of Stranraer, on Loch Ryan.

Leftshaw Hill, on borders of Lanark and Dumfries, 6 m. S.E. of Muirkirk; alt. 1513 ft.

Legbrannock, colliery vil., Lanarkshire, 3 m. S.E. of Airdrie.

Legerwood, par. and ham., Berwickshire; 8778 ac. Pop. 332. Ham. 3½ m. N.E. of Earlston.

Leidle, glen and stream, isl. of Mull, Argyll.

Leim Uilleim, mt., Inverness-shire, 3 m. S.E. of south end of Loch Treig; alt. 2971 ft.

Leir, loch, N.E. Sutherland, 4½ m. N.E. of Forsinard.

Leira Ness and **Voe,** cape and bay, Shetland, W. of Bressay.

Leiravay, isl., N. Uist, E. of Lochmaddy. Pop. 8.

Leitfie, ham., E. Perthshire, 2 m. N.W. of Meigle.

Leith, parl. bur., and since 1920 incorporated in the city of Edinburgh, 1½ m. N.E. of Edinburgh G.P.O.; ry. stas., (L. Central) L.N.E. and (L.) L.M.S. On the S. side of the Firth of Forth, is the chief seapt. on the E. coast of Scotland. The harbour works are extensive.

Industries: shipbuilding, flour mills, roperies, chemical works, and sawmills. Imports: grain, flour, sugar, chemicals, esparto, and timber; exports include coal, iron, whisky, and paper. Its history is varied; it was twice seized and burned by the Earl of Hertford, 1544–1547; besieged by the Protestants, 1559–1560; and held by one of Cromwell's generals in 1650. Has a green fixed light on E. pier, seen 8 m.; on the W. pier is a quarter-minute white flashing fixed light, seen 10 m.; at S. entrance to Imperial Dock a fixed red light, seen 8 m. With Portobello and Musselburgh, returns one member to Parliament. Pop. parl. bur. 82,934.

Leithen Lodge, seat, Peeblesshire, 4 m. N. of Innerleithen. Leithen Water flows 9 m. S.E. to the Tweed at Innerleithen.

Leith Hall, Aberdeenshire, 1 m. N.W. of Kennethmont sta.

Leitholm, vil., Berwickshire, 5 m. N.W. of Coldstream. P.O., T.O.

Leith Walk, goods sta., L.N.E.; also connecting thoroughfare, Edinburgh with Leith. P.O., T.O.

Leith, Water of, Midlothian; rising in Mid-Calder par., flows 23 m. N.E. to the Firth of Forth at Leith Harbour.

Leith, Water of, old vil. in W. of Edinburgh, on Water of Leith, W. of the Dean Bridge and opposite the old vil. of Dean.

Lemno Burn, trib. of the S. Esk, Angus, at 6½ m. W.S.W. of Brechin.

Lempitlaw, vil., Roxburghshire, 4 m. S.E. of Kelso. P.O.

Lemreway, pl., Lewis Isl., Ross and Cromarty, on N. side of Loch Sealg. P.O.

Lenach-Gluthin, waterfall on Aray R., Argyll, 3 m. N. of Inveraray.

Lendalfoot, coast vil., Ayrshire, 6 m. S.W. of Girvan. P.O., T.O.

Lendal, Water of, Ayrshire, flows to Firth of Clyde at Lendalfoot.

Lendrick, seat, S. Perthshire, 5½ m. W. of Callander.

Lendrum, Aberdeenshire, 3 m. S.E. of Turriff. Scene of battle between Thane of Buchan and Donald of the Isles (eleventh century).

Lennel House, Berwickshire, 1 m. N.E. of Coldstream; vil. of Lennel Newtown adjoining.

Lennie, pl., Midlothian, 3 m. E. of Kirkliston.

Lennock, The, small trib. of the Lossie, Morayshire.

Lennox, an adjacent co. of Scotland. Comprised Dunbartonshire, a large part of Stirlingshire, and parts of Perthshire and Renfrewshire. It gives the Scottish title of Duke to the English Duke of Richmond.

Lennox Castle, or **Lennox Tower** ancient ruined strong tower of the Earls of Lennox, Currie par., Midlothian.

Lennox Castle, seat, Stirlingshire, 1½ m. W. of Lennoxtown.

Lennox Castle State Forest, 1½ m. N.W. of Lennoxtown, Stirlingshire.

Lennox Hills, hill range, extending 23 m. N.E. from the vicinity of Dumbarton to near Stirling, including the Kilpatrick, Killearn, Campsie, Kilsyth, Fintray, Gargunnock, and Dundaff Hills, and reach a height of 1894 ft. at Earl's Seat.

Lennoxlove, seat, E. Lothian, 1¼ m. S. of Haddington.

Lennoxtown, town, with ry. sta., L.N.E., Stirlingshire, 3 m. N. of Kirkintilloch. P.O., T.O. Pop. 2577. Has print works and bleachfields.

Lennymore, or **North Thundergay,** isl. of Arran, Bute, 4 m. S.W. of Lochranza.

Lentran, seat and ry. sta., L.M.S., 5¾ m. W. of Inverness. P.O.; T.O. at L. sta.

Leny, Pass of, Perthshire, traversed by r. Leny and ry., and overhung by Ben Ledi. R. Leny flows from Loch Lubnaig to the Teith, near Callander. Described by Sir Walter Scott in the "Legend of Montrose" and the "Lady of the Lake." In vicinity of Callander, Falls of L., and L., a seat.

Lenzie, q.s. par. and sub. of Kirkintilloch, with ry. sta., L.N.E., Lanarkshire, 6¼ m. N.E. of Glasgow. P.O., T.O. Pop. of S. Lenzie 1109. Here the Barony Lunatic Asylum, the Glasgow Convalescent Home, and the Broomhill Home for Incurables.

Leochel-Cushnie, par., Aberdeenshire; 12,858 ac. Pop. 814. The church is 5 m. S.W. of Alford. L. Burn flows 9½ m. N. to the Don, near Alford.

Leoid, loch, Argyll, 2½ m. N.W. of Taychreggan; greatest depth 84 ft., alt. 602 ft.

Lephenstrath, S. Kintyre, Argyll, 1½ m. W. of the vil. of Southend.

Lephinmore, pl., Argyll, Loch Fyne, 6 m. N.E. of Otter Ferry.

Lerags, seat, Argyll, 4 m. S.W. of Oban.

Lerevagh, loch, Benbecula, Outer Hebrides, 3 m. N. of Creagorry.

Lerwick, seapt. town (police bur.), par., and capital of the Shetland Isls., on Bressay Sound, E. coast of Mainland, 115 m. N.E. of Kirkwall, Orkney, and 266 m. by steamer from Leith. It is a large fishing centre with extensive herring-gutting and packing stas. Exports: fish, ponies, sheep, cattle, and knitted goods. Par. 10,923 ac. Pop. 6506; police bur. pop. 4221. P.O., T.O. Regular steamer communication with Aberdeen, Wick, Leith, etc. Twageos shows a six-second flashing light, visible 8 m., and fixed lights on pier on N. Ness.

Leslie, par. and ham., Aberdeenshire; par. 4446 ac. Pop. 282. Ham. 4 m. S.W. of Insch. P.O., T.O. L. Castle, a ruin.

Leslie, town (police bur.) and par., with ry. sta., L.N.E. (passenger service withdrawn), Fife, on r. Leven, 4½ m. W. of Markinch; par. 4892 ac. Pop. 3983; town pop. 2477. P.O., T.O. Flax-spinning, bleaching, paper-making, and vulcanite works. L. House, seat of the Earl of Rothes.

Lesmahagow, par. and vil., with ry. sta., L.M.S., Lanarkshire; par. 41,270 ac. Pop. 11,661. Vil. (sometimes called Abbey Green), 6 m. S.W. of Lanark, on r. Nethan. P.O., T.O. Vil. pop. 154. Chief industry is mining.

Lesmore, pl., Aberdeenshire, 2 m. W.N.W. of Rhynie. L. Castle, a ruined ancient castle of the Gordons.

Lesmurdie Cottage, shooting-box, Banffshire, 2½ m. N. of Dufftown.

Lessendrum House, Aberdeenshire, 4½ m. N.E. of Huntly.

Lessudden, vil., N.W. Roxburghshire, 1½ m. S.E. of St. Boswells sta. P.O., T.O. at St. Boswells. L. House, the seat of the Scotts of Raeburn.

Leswalt, par. and vil., Wigtownshire (containing the Sheuchan sub. of Stranraer); 12,529 ac. Pop. 2376. Vil. 3 m. N.W. of Stranraer. P.O., T.O.

Letham, vil., Fife, 4 m. W. of Cupar. P.O., T.O.

Letham, vil., Angus, 5 m. E. of Forfar. P.O., T.O.

Lethame House, Lanarkshire, 1¼ m. W. of Strathaven.

Letham Grange, seat and ry. sta., L.N.E., Angus, 3 m. N. of Arbroath.

Lethamhill House, Lanarkshire, ½ m. S.W. of Millerston.

Letham House, seat, Perthshire, 1½ m. N.E. of Glenfarg.

Letham House, E. Lothian, 1½ m. W. of Haddington.

Letham House, 2 m. N.W. of Perth.

Lethangie, seat, 1¼ m. N.E. of Kinross.

Lethanhill, vil., Ayrshire, 4 m. N.W. of Dalmellington. P.O., T.O. See BURN-FOOTHILL.

Lethendry, ham., Morayshire, 1½ m. E. of Cromdale sta.

Lethendy, par. and vil., E. Perthshire; 1742 ac. Pop. 206. Vil. 4½ m. S.W. of Blairgowrie. Has an interesting old tower.

Lethen House, seat, Nairnshire, 5 m. S.E. of Nairn.

Lethenty, ry. sta., L.N.E., Aberdeenshire, 2¾ m. N. of Inverurie.

Lethington, old name of Lennoxlove, q.v.

Lethnot, ham., N. Angus, 6½ m. N.W. of Brechin. P.O., T.O.

Lethnot and Navar, par., N. Angus; 26,487 ac. Pop. 178.

Letterewe, seat and shooting-lodge, Ross and Cromarty, on N.E. side of Loch Maree, 8½ m. S.E. of Poolewe.

Letterewe and Fisherfield Deer Forest, Ross and Cromarty, on N.E. side of Loch Maree. 45,000 ac.

Letterfearn, dist., Ross and Cromarty, 6 m. S. of Strome Ferry. P.O., T.O.

Letterfinlay, pl., Inverness-shire, on Loch Lochy, 8 m. N. of Spean Bridge.

Letterfourie, seat, Banffshire, 3 m. S.E. of Buckie.

Letter Morar, shooting-estate, Arisaig dist., S.W. Inverness-shire.

Lettermore, pl., with quarries, Argyll, 1½ m. N.W. of Ballachulish.

Letterwalton, seat, Argyll, 2 m. N. Ledaig, on Loch Creran.

Leuchar Burn, Aberdeenshire, flows 7½ m. S.E. from Loch Skene.

Leuchars, par. and vil., with ry. sta. (L. Junction), L.N.E. Fife; par. 13,357 ac. Pop. 3221. Vil., 5 m. N.W. of St. Andrews. P.O., T.O. Pop. 656. Army Air Service sta.

Leuchars House, Morayshire, 3½ m. N.E. of Elgin.

Leuchars Junction, ry. sta., L.N.E., Fife, 1 m. S.W. of Leuchars.

Leuchold, The, mansion, W. Lothian, in Dalmeny Park.

Leum a Chiamhain, Loch, Sutherland, 6 m. N.W. of Kinbrace; alt. 770 ft., depth 51 ft.

Leurary, pl., Caithness, 4½ m. S.W. of Thurso.

Leurbost, vil., Lewis Isl., Ross and Cromarty, 8 m. S.W. of Stornoway.

Levanne House, N.W. Renfrewshire, 1¾ m. W.S.W. of Gourock.

Leven, town (police bur.), with ry. sta., L.N.E., Fife, on W. side of Largo Bay, Firth of Forth, at mouth of r. Leven, 10¾ m. N.E. of Kirkcaldy, 5¾ m. N.E. of Thornton Junction. Pop. 7411. Hotel. P.O., T.O. Has paper-making, engineering, flax-spinning, collieries. A coastguard sta.

Levenbank, pl., with printworks, Dunbartonshire, near Balloch.

Levenhall, vil., Midlothian, E. of Musselburgh. P.O.

Leven, Loch, sea loch, deflecting from head of Loch Linnhe, and extending 12 m. E. between Argyll and Inverness-shire.

Leven, Loch, Kinross-shire, measures 8½ m. in circuit. Contains the famous isl. in which Mary Queen of Scots was imprisoned. Famous for a trout of a special species; maximum depth 83 ft., alt. 350 ft.

Leven River, Argyll, flows 3 m. W. to the head of Loch L. The upper end is now formed into a dam, 10 m. long, to supply power for the Kinlochleven Aluminium Works.

Leven River, Kinross-shire and Fife, flows 16 m. E. to Largo Bay, at L.

Leven Seat, hill, Midlothian, 6 m. S.W. of W. Calder; alt. 1168 ft.

Levenwick, vil., Shetland, 4 m. S.W. of Sandwick. P.O., T.O.

Leverburgh, vil., S.W. coast of Harris, Inverness-shire, the centre for communication with the S., having a regular steamer service from Glasgow. The vil. was formerly known as Obe. P.O., T.O.

Levern, q.s. par. and stream, Renfrewshire; stream enters at White Cart, 2¾ m. E. of Paisley.

Levernbank, vil., Renfrewshire, near Neilston.

Levishie, seat, Inverness-shire, 7 m. N. of Fort Augustus.

Lewis, northern part of Outer Hebrides. Consists of Harris and adjacent isls. in the S. belonging to Inverness-shire, and L. proper and adjacent isls. belonging to Ross and Cromarty. L. proper is divided from Harris by Lochs Seaforth and Reasort and an intervening isthmus; extends 42½ m. N.W. to the Butt; is widest near the S.W. end. Measures 27 m. in extreme breadth and comprises 404,413 ac. About one-fourth of it is bleak high upland, and much of the rest consists of moors, swamps, and lochs. Pop. 29,672. L. Castle, a seat. Chief town, Stornoway.

Lewis, Butt of, the most northerly point of the isl. of Lewis and of the Hebrides.

Lewiston, vil., Inverness-shire, 1 m. S.E. Drumnadrochit.

Leylodge, pl., Aberdeenshire, 2½ m. S.W. of Kintore.

Leys, Upper L., and **Lower L.,** hams., Kincardineshire, Banchory-Ternan par.

Leys Castle, seat, 3½ m. S.W. of Inverness.

Leysmill, vil., ry. sta., L.M.S., Angus, 5 m. N.W. of Arbroath. P.O., T.O. at sta.

Leys of Halliburton, ham., Angus, 2½ m. S.E. of Coupar-Angus.

Lhanbryde, vil., with ry. sta., L.M.S., Morayshire, 3½ m. E. of Elgin. P.O., T.O. Pop. 316.

Liathach, mt. range, 2 m. N.E. of Upper Loch Torridon, Ross and Cromarty; alts. 3456 and 3358 ft.

Libberton, par. and vil., Lanarkshire; par. 8220 ac. Pop. 497. Vil. 2½ m. S.E. of Carnworth.

Liberton, eccl. par. in S. of city of Edinburgh. P.O., T.O.

Liberty, part of the police bur. of Elie, L., and Williamsburgh, Fife. See ELIE.

Libo, Loch, Renfrewshire, 3 m. S.W. of Neilston.

Licklyhead, old mansion, Premnay par., Aberdeenshire, 3 m. S. of Insch sta.

Lickprivick, site of ancient castellated castle, Lanarkshire, 1½ m. S. of East Kilbride.

Liddelbank, estate on the Scottish side of the r. Liddle, Roxburghshire, 4½ m. S.S.W. of Newcastleton.

Liddel Castle, site, Old Castleton, Roxburghshire.

Liddel Water, running 12½ m. in Roxburghshire and 7½ m. on boundary between Scotland and England, all S.S.W., to the Esk, 7 m. S. of Langholm.

Liddesdale, consists of the Liddel's basin down to the point of that river's leaving Roxburghshire, and is the country of "Dandie Dinmont" in Sir Walter Scott's *Guy Mannering.*

Liddesdale, ham., Argyll, on Loch Sunart, S. shore.

Liernish, pl., N. Uist, Outer Hebrides.

Liff and Benvie, par., with ry. sta. (Liff), L.M.S., Angus; 5549 ac. Pop. 2141. Vil., 1¾ m. N.W. of sta., is 4½ m. N.W. of Dundee. Muirhead of Liff, 1 m. N.E. of Liff.

Lightburn, vil., Lanarkshire, in E. end of Cambuslang par.

Lightburn, pl., N. Lanarkshire, ½ m. N. of Shettleston. Hospital for infectious diseases.

Lilliard's Edge, battlefield of Ancrum Moor (1545), 5 m. N. by W. of Jedburgh, Roxburghshire.

Lilliesleaf, par. and vil., Roxburghshire; par. 6668 ac. Pop. 490. Vil. 8 m. S. of Melrose and 5 m. S.E. of Selkirk. P.O., T.O. Pop. 238.

Lillyburn, pl., Stirlingshire, near Milton of Campsie. Has print works.

Lilly Loch, Angus, Fetteresso par. See DEWS LOCH.

Lilly Loch, Lanarkshire, ½ m. S. of Caldercruix.

Limecraigs House, near Campbeltown, Argyll.

Limefield, pl. and seat, Midlothian, 1 m. N.E. of W. Calder.

Limehillocks, pl., Banffshire, 1½ m. N.E. of Grange. Limeworks.

Limekilnburn, pl., Lanarkshire, 3 m. S.W. of Hamilton.

Limekilns, coast vil., Fife, 3¼ m. W. of Inverkeithing. P.O., T.O. Pop. L. and Charleston, 1061.

Limekilns, seat, Lanarkshire, E. Kilbride par.

Limerigg, vil., Stirlingshire, 1 m. S. of Slamannan. P.O.

Linay, Altars of, two reefs off N.E. coast of N. Ronaldshay, Orkney.

Linburn House, Midlothian, Kirknewton par., 1½ m. N.E. of Mid-Calder sta.

Lincluden, seat and ruined ancient collegiate church on the Nith, 1½ m. N. of Dumfries.

Lindean, ry. sta., L.N.E., Roxburghshire, 2 m. N.E. of Selkirk. Ancient par., embraced by Galashiels par.

Lindertis, seat, Angus, 3 m. W.S.W. of Kirriemuir.

Lindores, vil. and sta., L.N.E., Fife, 2 m. S.E. of Newburgh. P.O., T.O. Contains vestiges of castle thought to have belonged to Macduff. In vicinity, L. House, a seat, on L. Loch.

Lindsaylands, estate, Lanarkshire, 1 m. S.W. of Biggar.

Lindsay Tower, or **Crawford Castle,** ruined stronghold of the Lindsays, Lanarkshire, on r. Clyde, opposite Crawford vil.

Linfairn, loch, Ayrshire, 8¼ m. S.E. of Maybole.

Linga, isl., Shetland, 2½ m. S. of Yell Isl.

Linga, isl., Shetland, 2½ m. W. of Scalloway.

Linga, isl., Shetland, 1 m. S. of Walls.

Linga, one of the Treshnish isls., near N.W. of Mull, Argyll.

Linga Holm, isl. and sound, in Stronsay par., Orkney.

Linga Sound, between Whalsey and W. L. Isl., Shetland.

Lingay, isl., Barra par., Outer Hebrides, 3½ m. S. of Vatersay.

Lingay, isl., Outer Hebrides, off N. coast of N. Uist.

Lingore Linn, burn at E. end of Cambusnethan par., Lanarkshire, flows 2 m. to Breich Water.

Lingrow Farm, pl., with remains of "broch," Orkney, 1½ m. S.W. of Kirkwall.

Linhead, pl., 6 m. S. of Banff.

Linhouse Water, Midlothian, flows 10 m. N. to the Almond, near Mid-Calder.

Linicro, ham., Isle of Skye, 2 m. N. of Uig.

Linklet Bay, Orkney, E. side N. Ronaldshay isl.

Links Ness, headland in N. of Stronsay Isl., Orkney.

Links of Forth, serpentine windings of the r. Forth among alluvial lands, from vicinity of Stirling to vicinity of Alloa.

Linktown, part of town of Kirkcaldy, Fife. P.O., T.O.

Linkwood, pl., with extensive distillery, Morayshire, 2 m. S.E. of Elgin.

Linlithgow, royal and mun. bur., par., and co. town of W. Lothian, with ry. sta., L.N.E., 17½ m. W. of Edinburgh; par. 10,264 ac. Pop. 7157; bur. pop. 3666. P.O., T.O. On an eminence bordered by the loch is the ruin of L. Palace, which dates from the time of David I., but was largely rebuilt by James IV. and James V. It was the birthplace of James V. and Mary, Queen of Scots. Chief industries: tanning and currying, glue making, distilling, and paper making.

Linlithgow Bridge, vil., 1 m. W. of Linlithgow, on r. Avon. P.O. Pop. 704.

Linlithgowshire, see West Lothian.

Linmill, vil., Clackmannanshire, 2½ m. E. of Alloa.

Linmill Burn, W. Lothian, flows to Firth of Forth 1 m. W. of Queensferry.

Linmoor, near Culter, Aberdeenshire, Children's Fresh Air Home.

Linne, loch, Knapdale, Argyll, 5 m. N.W. of Ardrishaig.

Linne Mhuirich, arm of Loch L. Sween, Knapdale, Argyll.

Linnhe, Loch, sea loch, Argyll and Inverness-shire. It is 22 m. long, and averages about 5 m. wide. It separates Appin on one side from Morven and Ardgour on the other. Contains the isls. of Lismore, Shuna, and other smaller isls.

Linn House, seat, Renfrewshire, near Cathcart.

Linn Mill, pl., Clackmannanshire, 2½ m. E. of Alloa.

Linplum, seat, E. Lothian, Garvald par.

Linridge, pl., Lanarkshire, ½ m. S.E. of Newhouse ry. sta. Collieries.

Lintalee, seat, Roxburghshire, 1½ m. S. of Jedburgh.

Linthill, seat, Roxburghshire, 1 m. N. of Lilliesleaf.

Linthouse, pl., W. Govan, Glasgow. P.O., T.O. Shipbuilding yards.

Lintlaw, pl., Berwickshire, 4 m. N.E. of Duns.

Lintmill, ham., N. Banffshire, 1 m. S. of Cullen. P.O.

Lintmill Halt, pl., near Campbeltown, Argyll.

Linton, par. in N.E. of Roxburghshire; 6396 ac. Pop. 406. Church, 6 m. S.S.E. of Kelso.

Linton, East, vil. (police bur.), with ry. sta., L.N.E., Prestonkirk par., E. Lothian. Pop. 882. 5¾ m. W. of Dunbar.

Linton House, Aberdeenshire, 3½ m. S.S.E. of Monymusk.

Linton, West, par. and vil., with ry. sta. (Broomlee), L.N.E., in Peeblesshire, 24 m. S.S.W. of Edinburgh; par. 23,257 ac. Pop. 1200. P.O., T.O.

Lintrathen, par. and vil., Angus; par. 22,402 ac. Pop. 375. Vil. (Bridgend of L.) 6½ m. W. of Kirriemuir. P.O., T.O. Loch of L., a Dundee water supply.

Lintrose, seat and cavern, Angus, 1½ m. S.E. of Coupar-Angus.

Linwood, q.s. par., vil., Renfrewshire, 3½ m. N. by W. of Paisley. P.O., T.O. Pop. 1532. Large paper mill, engineering works, etc.

Lionel, pl., Lewis Isl., Ross and Cromarty, 2 m. S. of Butt of Lewis.

Lismore, isl. in N. Argyll, extends along Loch Linnhe from a point 6 m. N.W. of Oban; measures 9½ m. in length and about 1½ m. in mean breadth. P.O., T.O. L. Lighthouse on Mousedale Isl., off S.W. coast. Pop. 280; area 6014 ac.

Lismore and Appin, par., Argyll; 94,725 ac., of which 829 is water. Pop. 3757.

Little Cairney, Arbroath, Angus. Town hospital.

Little Clyde, pl., S.E. Lanarkshire, ⅔ m. N. of Beattock Summit. Roman camp.

Littledean Tower, ruined ancient baronial tower, Roxburghshire, on r. Tweed, 1½ m. N.E. of Maxton.

Littleferry, pl. at mouth of Loch Fleet, Sutherlandshire, with ferry, Golspie-Dornach road.

Little France, ham., Midlothian, 3 m. S.E. of Edinburgh G.P.O.

Littlemill, ham., 4½ m. S.E. of Nairn. P.O.

Littlemill, pl., Ayrshire, 3 m. S.E. of Coylton.

Littlemill, vil., Dunbartonshire, Old Kilpatrick par., on r. Clyde. L. Distillery here.

Little Roe, small isl., Shetland, in Yell Sound.

Little Ross, small isl., with lighthouse, Kirkcudbrightshire, at mouth of Dee estuary. Pop. 5.

Little Scatwell, seat, Ross and Cromarty, Contin par.

Littlester, loch, Yell Isl., Shetland ; alt. 34 ft., depth 9 ft.

Littleton, seat, Angus, 3½ m. S.W. of Kirriemuir.

Little Vantage, ham., Midlothian, 3 m. S. of Mid-Calder ry. sta.

Little Water, trib. of the r. Ythan, Aberdeenshire, 1½ m. N.W. of Methlick.

Liver Glen, and Stream, Argyll, flows through Glen Liver to Loch Etive.

Livet Water, Banffshire, flows 11 m. N.W. to the Avon, 8 m. N. of Tomintoul.

Livilands, Easter and **Wester,** two seats in co., and adjoining Stirling.

Livingstone, par. and vil., with ry. sta., L.N.E., West Lothian ; par. 5368 ac. Pop. 4415. Vil., on r. Almond, 1½ m. E. of ry. sta., 3 m. E. of Bathgate. P.O. (Livingston) ; P.O. at ry. sta.

Livister, ham., Whalsey, Shetland.

Loan, ham., S.E. Stirlingshire, 2 m. S.E. of Polmont Junction sta.

Loanan, r., Sutherland, flows 4 m. to Loch Assynt, at Inchnadamph.

Loanhead, q.s. par. and town (police bur.), with ry. sta., L.N.E. (passenger service withdrawn), Midlothian, 5 m. S.E. of Edinburgh by road. P.O., T.O. Pop. 3940. Colliery.

Loanhead, vil., Angus, 8½ m. N.E. of Dundee.

Loanhead, pl., Angus, 2 m. N.W. of Montrose.

Loanhead, pl., Stirlingshire, 6½ m. S.S.W. of Falkirk.

Loaningfoot, ham., Kirkcudbrightshire, 10 m. S.E. of Dalbeattie.

Loans, vil., Ayrshire, 2 m. N.E. of Troon. P.O., T.O.

Loans, ham., Fife, 4 m. N.N.W. of Kilconquhar.

Loanstone, special water dist., Lasswade and Penicuik pars., Midlothian. Pop. 76.

Lochaber, dist., S. Inverness-shire. The surface is an assemblage of deep glens, broad moors, and lofty mts. Length 33 m., greatest breadth 21 m.

Lochaber Loch, Kirkcudbrightshire, 5 m. S.W. of Dumfries ; depth 55 ft., alt. 298 ft.

Loch a' Bhaid-Luachraich, W. Ross and Cromarty, 4 m. N.E. of Poolewe ; depth 143 ft., alt. 309 ft.

Loch a' Bhaillidh, Knapdale, Argyll, 8 m. W.S.W. of Tarbert ; alt. 326 ft.

Loch a' Bhealaich, Sutherland, 5 m. S.E. of Altnaharra ; depth 80 ft., alt. 572 ft.

Loch a' Bhealaich, W. Ross and Cromarty, 7 m. S.E. of Gairloch ; depth 92 ft.

Loch a' Bhruthaich, Inverness-shire, 5 m. N.W. of Drumnadrochit.

Loch a' Chlair, Sutherland, 6 m. W.N.W. of Kinbrace ; alt. 394 ft., depth 32 ft.

Loch a' Chnuic Bhric, Jura, Argyll, 3 m. N. of Feolin Ferry.

Loch a' Choire, sea loch, Argyll, 5½ m. S.E. of Strontian.

Loch a' Chroisg, Ross and Cromarty, 1¾ m. S.W. of Achnasheen ; depth 168 ft., alt. 508 ft.

Loch a' Chroisg, Ross and Cromarty, 8 m. N.E. of Ullapool.

Loch a' Chuilinn, 8 m. N.E. of Achnasheen, Ross and Cromarty ; depth 43 ft., alt. 361 ft.

Loch a' Ghlinne, N.W. Harris, Outer Hebrides, Inverness-shire.

Loch a' Ghobhainn, S.W. Ross and Cromarty, 6½ m. S.E. of Gairloch ; depth 28 ft., alt. about 1000 ft.

Loch a' Ghriama, Sutherland, adjoining Loch Shin ; alt. 303 ft., depth 64 ft.

Lochailort, pl., with ry. sta., L.N.E., Inverness-shire, 26¼ m. W. of Fort William. P.O., T.O.; T.O. at sta.

Lochaline, vil., with pier, at mouth of Loch Aline, Morvern par. P.O., T.O. called Morvern. L. House, a seat, 4 m. W. Steamers from Glasgow and Oban call.

Loch Allt an Fhearna, loch, 6½ m. W. of Kinbrace ry. sta., Sutherlandshire ; 36 ft. deep, alt. 433 ft.

Loch Allt na h-Airbhe, loch, 60 ft. deep, alt. 119 ft., 5½ m. S.E. of Scourie, Sutherlandshire.

Lochalsh, par., S.W. of Ross and Cromarty ; 49,591 ac. Pop. 1525. P.O., T.O. at Balmacara, on Loch Alsh, 8 m. S.W. of Strome Ferry. Hotel. See KYLE OF LOCHALSH.

Loch a' Mhuilinn, Ross and Cromarty, 8 m. W. of Struy ; depth 94 ft., alt. 418 ft.

Lochan a' Chlaidheimh, N.W. Perthshire, 1¾ m. N.W. of Rannoch.

Loch an Add, Argyll, 4 m. N.W. of Ardrishaig.

Loch an Aircill, S. of Jura Isl., Argyll, 2 m. S. of Loch Tarbert.

Lochan Balloch, Perthshire, 3 m. S.W. of Callander.

Loch an Eilean, Inverness-shire, 3 m. S. of Aviemore sta. ; depth 66 ft., alt. 840 ft. The loch contains an isl. with the ruins of a castle where the ospreys used to nest.

Loch an Eilein, isl. of Tiree, Argyll.

Lochan Fada, loch, Ross and Cromarty, 4½ m. N. of Kinlochewe ; depth 248 ft., alt. about 1005 ft.

Lochan Fada, Sutherland, 7½ m. S.E. of Lochinver.

Loch an Gead, Kintail par., Ross and Cromarty, near Loch Tachdaidh ; alt. 830 ft., depth 30 ft.

Lochanhead, ry. sta., L.M.S., Kirkcudbrightshire, 6 m. S.W. of Dumfries. P.O.

Lochan House, seat, Perthshire, near Amulree.

Loch an Lagain, Sutherland, 4 m. N.E. of Bonar Bridge ; alt. 446 ft., depth 18 ft.

Loch an Leathaid Bhuain, W. Sutherland, 2 m. N.E. of Unapool.

Loch an Leothaid, W. Sutherland, 6 m. N.E. of Lochinver.

Lochan na h-Earba, Inverness-shire, flowing into Loch Laggan ; alt. 1152 ft., depth 81 ft.

Lochan na Lairige, Perthshire, 4½ m. N.E. of Killin ; alt. 1595 ft., depth 39 ft.

Loch an Nostarie, W. Inverness-shire, 1¼ m. S.E. of Mallaig ; alt. 89 ft., depth 35 ft.

Loch an Ruathair, Sutherland, 2½ m. N. of Kinbrace sta. ; alt. 414 ft., depth 26 ft.

Lochans, vil., Wigtownshire, 2½ m. S. of Stranraer. P.O., T.O.

Loch an Staca, Inverness-shire, 6 m. N.W. of Invermoriston ; alt. 1600 ft., depth 51 ft.

Loch an Torran, Knapdale, Argyll, 7 m. W. of Tarbert.

Loch an Tuirc, S.E. Sutherland, 3 m. N. of Lochinver ; alt. 200 ft., depth 39 ft.

Locharbriggs, vil., with ry. sta., L.M.S., 2½ m. N.E. of Dumfries. P.O.

Lochardill, seat, 1½ m. S. of Inverness.

Lochard Lodge, Perthshire, 3 m. W. of Aberfoyle, on Loch Ard. State forest nearby.

Lochar Moss, S. Dumfriesshire, extends 10 m. S.E. from Locharbriggs to Solway Firth. L. Water flows into the Solway Firth 2½ m. N.E. of Caerlaverock.

Lochassynt Deer Forest, N.W. side of Loch Assynt, Sutherland ; 19,000 ac.

Lochassynt Lodge, seat, Sutherland, 6 m. N.W. of Inchnadamph.

Lochavich, seat, Argyll, 14 m. S.W. of Lochawe sta. P.O., T.O.

Lochawe, pl., with ry. sta., L.M.S., and steamboat pier, Argyll, on Loch Awe, 2½ m. W. of Dalmally. P.O., T.O. Hotel.

Lochay, r., Perthshire, flows 11 m. through Glen L. to Loch Tay at Killin.

Lochbay, pl., Isle of Skye, Durinish par.

Loch Bennach, S.E. Sutherland, 4 m. N.E. of Lochinver ; alt. about 250 ft., depth 38 ft.

Loch Beoraid, Inverness-shire, 9 m. E. of Arisaig ; maximum depth 159 ft., alt. 168 ft.

Loch Bog, E. Perthshire, 2½ m. N.W. of Coupar-Angus. Sometimes called Stormont.

Lochboisdale, ham., S. Uist Isl., on Loch Boisdale. Pop. 80. P.O., T.O. Glasgow steamers call regularly.

Lochbroom, coast par., N.W. Ross and Cromarty ; 261,000 ac. Pop. 2004. P.O., T.O. Church stands at the head of salt-water Loch Broom, 6 m. S.S.E. of Ullapool.

Lochbuie, ham., Mull Isl., Argyll, 12 m. S.W. of Auchnacraig. P.O., T.O. L. Castle, seat of the Maclaines; their old castle was at Moy.

Lochburn, dist., Glasgow, 1 m. E. of Maryhill ry. sta.

Loch Cairnbawn, sea loch, W. Sutherland.

Loch Calam, W. Caithness, 4 m. N.N.E. of Altnabreac sta.

Lochcarron, coast par. and fishing vil., S.W. Ross and Cromarty ; par. 85,051 ac. Pop. 967. Vil., also called Jeantown, on W. side of Loch Carron, 5 m. N.E. of Strome Ferry. P.O., T.O. Vil. pop. 259.

Loch Choir Deer Forest, Sutherland, 13 m. N.E. of Lairg ; 53,000 ac.

Loch Connan, Isle of Skye, 6 m. E. of Portree.

Lochcote House, W. Lothian, 4 m. N. of Bathgate.

Lochcraig Head, mt., S. Peeblesshire, 4½ m. N.E. of Hart Fell; alt. 2625 ft.

Loch Crunachun, Inverness-shire, 11 m. W.S.W. of Newtonmore.

Lochdhu, seat, 1½ m. S.W. of Nairn.

Lochdochart House, seat, Perthshire, 3½ m. N.E. of Crianlarich sta.

Lochdonhead, vil., Mull Isl., Argyll, at head of Loch Don. P.O., T.O. at Auchnacraig.

Loch Druim Suardalain, 1 m. S.E. of Lochinver, Sutherland; alt. 134 ft., depth 31 ft.

Lochead, seat, W. Argyll, 9 m. S.W. Ardrishaig.

Lochearnhead, vil. and ry. sta., L.M.S., Perthshire, at head of Loch Earn, 13 m. N.W. of Callander. P.O., T.O. Hotel.

Lochee, q.s. par. and dist., with ry. stas. (L. and L. West), L.M.S., in N.W. Dundee, Angus. P.O., T.O.

Loch Eilde Mor, Inverness-shire, 10 m. N.E. of Ballachulish; depth 100 ft., alt. about 1110 ft.

Locheilside, ry. sta., L.N.E., Inverness-shire, 10 m. W. of Fort William.

Lochenbreck, loch, Kirkcudbrightshire, 4 m. S.W. of New Galloway. Inn. Loch 15 ft. deep, alt. 651 ft.

Lochend, seat, Caithness, 4½ m. E. of Castleton.

Lochend, loch, city of Edinburgh, 1 m. N.E. of G.P.O. Once the water-supply for Leith.

Lochend, ham., 6½ m. S.W. of Inverness.

Lochend, pl., Kirkcudbrightshire, ½ m. S.E. of Killywhan sta. L. Loch (or Loch Arthur) in vicinity; alt. 240 ft., depth 50 ft.

Lochend, loch, Lanarkshire, 2 m. N.W. of Coatbridge.

Lochend, pl., E. coast of Mainland, Shetland, on Colla Firth. P.O., T.O.

Lochend House, E. Lothian, in S. vicinity of Dunbar.

Loch Eport, sea loch and ham., N. Uist Isl., Outer Hebrides, Inverness-shire, 3 m. S. of Lochmaddy. P.O., T.O. Glasgow steamers call.

Locherfield, pl., Renfrewshire, 1 m. S.E. of Bridge of Weir.

Loch Ericht Lodge, Inverness-shire, 5 m. S.W. of Dalwhinnie.

Locher Water, Renfrewshire, flows 8 m. N.E. to Gryfe Water, 5 m. W. of Renfrew.

Loch Etchachan, loch on the E. shoulder, and 1 m. N.E. of the summit, of Ben Macdhui, Aberdeenshire. It is 3058 ft. above sea-level.

Loch Etive, Argyll. See ETIVE.

Lochfauld, pl., Lanarkshire, 1½ m. N.W. of Bishopbriggs.

Loch Fell, mt., N. Dumfriesshire, 5½ m. E. of Moffat; alt. 2256 ft.

Lochfoot, vil., Kirkcudbrightshire, 5¼ m. S.W. of Dumfries. P.O., T.O.

Lochgair, ham., Argyll, 15 m. S.S.W. of Inveraray, on Loch Gair. P.O., T.O. Nearby L. House, a seat.

Lochgarry House, Perthshire, 2½ m. E. of Kinloch Rannoch, on r. Tummel.

Lochgelly, town (police bur.) and q.s. par., with ry. sta., L.N.E., Fife, 7½ m. N.E. of Dunfermline. Town pop. 9297. P.O., T.O. In vicinity, Loch Gelly, and L. House, a seat. Town has ironworks and nearby collieries.

Lochgilphead, police bur. and q.s. par., Argyll, 2 m. N. of Ardrishaig, at head of Loch Gilp. Pop. 974. P.O., T.O. In vicinity, the Argyll and Bute Lunatic Asylum. Steamer communication with Glasgow and W. Highlands.

Lochgoilhead and Kilmorich, par., Argyll; 69,711 ac. Pop. 716. The vil. Lochgoilhead at the head of salt-water Loch Goil, 12½ m. S.W. of Arrochar and 18¼ m. N.N.W. of Greenock by water. P.O., T.O. Has a pier and hotel. Vil. pop. 149.

Lochgoin, pl., Ayrshire, 9 m. N.E. of Kilmarnock. Famous in the history of the Covenanters.

Lochgoin Reservoir, Renfrewshire. A Kilmarnock water supply.

Loch Grennoch Lodge, Kirkcudbrightshire, 8 m. N.E. of Newton-Stewart.

Lochhead Distillery, Campbeltown, Argyll.

Loch Hourn Deer Forest, Inverness-shire, N. side of Loch Hourn.

Lochhouse Tower, ancient edifice, Dumfriesshire, 1¼ m. S. of Moffat.

Lochieheads, seat, Fife, 1½ m. N.E. of Auchtermuchty.

Lochills, ham., Morayshire, 3 m. W. of Garmouth. P.O.

Lochinch, seat of the Earl of Stair, Wigtownshire, on the W. side of Castle Kennedy, 2¾ m. E. by S. of Stranraer.

Loch Indail, sea loch, Islay, W. side ; extends 12 m. inland. Loch Indail Lighthouse, on Dun Pt., shows fixed light, visible 12 m. In vicinity, Lochindail Distillery.

Lochindorb, loch, Morayshire, 3 m. by road S.W. of Dava sta ; alt. 969 ft., depth 51 ft. L. Lodge, a seat.

Lochinvar, loch (depth 10 ft., alt. 736 ft.), Kirkcudbrightshire, 6 m. N.N.E. of New Galloway. Vestiges of the ancient baronial fortalice of the Gordons, Knights of L., and one of them the theme of Lady Heron's song in *Marmion,* "Young Lochinvar." L. Lodge, 6 m. N.E. of New Galloway.

Lochinver, vil., with pier, Sutherland, on Loch Inver. P.O., T.O. A popular tourists' resort. Pop. 216. Hotel.

Loch Kennard Lodge, Perthshire, 8 m. N.W. of Dunkeld.

Lochlaggan, pl., Inverness-shire, 16 m. S.W. of Kingussie, on Loch Laggan. Hotel.

Loch Laich, sea loch, Argyll, 1½ m. N.E. of Port Appin.

Lochlane House, seat, Perthshire, 2 m. W. of Crieff.

Lochlea, pl., Ayrshire, 2½ m. N.W. of Tarbolton.

Lochlee, par., N. Angus ; 58,382 ac. Pop. 198. Contains Tarfside.

Loch Leiter Easaich, Sutherland, 6 m. N.E. of Lochinver ; alt. 217 ft., depth 70 ft.

Lochletter, seat, Inverness-shire, 5 m. W. of Temple pier, L. Ness.

Loch Leven Station, ry. sta., L.N.E., for goods, adjoining Kinross.

Loch Loch, N. Perthshire, 9 m. N.E. of Blair - Atholl ; alt. about 1450 ft., depth 81 ft.

Lochloy, seat, 3½ m. N.E. of Nairn.

Lochloyal, seat, Sutherland, 5 m. S.E. of Tongue.

Lochluichart, ham. and ry. sta., L.M.S., Ross and Cromarty, 17 m. W. of Dingwall. P.O., and P.O. at ry. sta.

Lochlyock Hill, a Tinta summit, Lanarkshire ; alt. 1734 ft.

Lochmaben, royal and mun. bur. and par., with ry. sta, L.M.S., Dumfriesshire ; par. 10,853 ac. Pop. 2460; police bur. pop. 1014. 9 m. N.E. of Dumfries. P.O., T.O. Ruins of L. Castle, in Castle Loch. Sanatorium.

Lochmaddy, sea loch and fishing vil., N.E. North Uist, Outer Hebrides. The loch cuts the land into a multitude of isls. and peninsulas, and forms many safe natural harbours. The vil. stands on the W. shore of Loch Maddy. Pop. 215. P.O., T.O. Steamer communication with Glasgow, Oban, Portree, etc.

Lochmalony, seat, Fife, 4 m. N. of Cupar.

Loch Maol a' Choire, 11 m. E. of Lochinver, Sutherland ; depth 8 ft., alt. about 850 ft.

Lochmaree, Ross and Cromarty, 9 m. N.W. of Kinlochewe. P.O., T.O. Hotel.

Loch Mhaim nan Carn, S.W. Perthshire, 2 m. S. of Stronachlachar.

Lochmill, loch, Fife, 1½ m. S. of Newburgh.

Lochmill Paper Mill, W. Lothian, 1 m. W. of Linlithgow.

Lochmore Lodge, seat, Sutherland, near N.W. end of Loch More, 28 m. N.W. of Lairg.

Loch na Beinne Baine, Inverness-shire, 8 m. W. of Invermoriston ; depth 67 ft.

Loch na Beiste, Ross and Cromarty, 3½ m. N.N.E. of Aultbea ; depth 35 ft.

Loch na Bhaile Mhargaidh, Jura, Argyll, 2 m. S.W. of Craighouse.

Loch na Bharp, S. Uist, 1 m. N. of Lochboisdale.

Loch na Bile, inlet, S.E. of coast of Jura, Argyll.

Loch na Bo, loch, Morayshire, 4½ m. S.E. of Elgin. Lochnabo, a seat.

Loch na Breack More, S.W. Sutherland, 4 m. S.E. of Loch Inver.

Loch na Cairidh, sea loch, between Isle of Skye and Scalpa.

Loch na Caorach, Sutherland, 4 m. S. of Melvich.

Loch na Ceste, N. Harris, 6 m. N. of Tarbert.

Loch na Claise Carnaich, N.W. Sutherland, 1 m. E. Rhiconich.

Loch na Clar, N.E. Sutherland, 6½ m. N.W. of Kinbrace ; depth 82 ft., alt. 394 ft.

Loch na Cloiche Bearga, Jura, Argyll, 7 m. N. of Craighouse.

Loch na Craige, Perthshire, 3 m. S. of Aberfeldy ; depth 13 ft., alt. 1297 ft.

Loch na Criche, N.W. Inverness-shire, 5 m. N.W. of Invermoriston sta.

Loch na Cuaich, S.E. Inverness-shire, 4½ m. N.E. of Dalwhinnie ; depth 85 ft., alt. about 1296 ft.

Loch na Daimb, Ross and Cromarty, 9 m. E. of Ullapool; depth 40 ft., alt. 672 ft.

Loch na Dal, sea loch, Isle of Skye, in Sound of Sleat.

Loch na Dealachd, Sutherland, 9½ m. W. Altnaharra; alt. 805 ft.

Loch na Dearg, isl. of Mull, Argyll, 3 m. S. of Tobermory.

Lochnagar, mt., Aberdeenshire, 9 m. S.W. of Ballater; alt. 3768 ft. It is cut by frightful corries; it has on its shoulder a gloomy tarn, overhung by tremendous precipices. It is the subject of Lord Byron's well-known poem.

Loch nah Oidhche, W. Ross and Cromarty, 7 m. S.E. of Gairloch; depth 121 ft., alt. about 1250 ft.

Loch na Houigh, W. Ross and Cromarty, 3 m. S. Lochmaree Hotel.

Loch na Keal, sea loch, bisecting W. side of Mull Isl., Argyll, penetrating 13½ m. E. by N.

Loch na Larig, S. Inverness-shire, 7 m. N.E. of Dalwhinnie.

Loch na Larige, W. Perthshire, 5 m. N.E. of Killin; depth 39 ft., alt. 1595 ft.

Loch na Lathaich, sea loch, N. coast of Ross of Mull, Argyll.

Loch na Leitreach, Ross and Cromarty, 7 m. N.E. of Invershiel; depth 88 ft., alt. 275 ft.

Loch nam Bonnach, Ross and Cromarty, 3 m. N.W. of Beauly; alt. 934 ft.

Loch nam Braveavach, Caithness, 10 m. W. of Dunbeath.

Loch na Meal, Mull Isl., Argyll, 2 m. S. of Tobermory.

Loch na Meide, Sutherland, 10 m. S.W. Tongue; depth 63 ft., alt. 488 ft.

Loch na Moine, E. Sutherland, 9 m. N.W. of Kildonan.

Loch na Moine Buige, Ross and Cromarty, 4 m. E.N.E. of Poolewe; 60 ft. deep.

Loch nan Ciltean, W. Inverness-shire, S. Morar.

Loch nan Cuinne, Sutherland, 8 m. W.N.W. of Kinbrace ry. sta.; alt. 395 ft., depth 28 ft.

Loch nan Eun, N. Perthshire, 10 m. S.E. of Braemar; depth 50 ft., alt. 2575 ft.

Loch nan Eun, N. Uist, Outer Hebrides; depth 31 ft.

Loch nan Gad, Kintyre, Argyll, 1½ m. N.E. of Clachan.

Loch nan Geireann, N. Uist, Outer Hebrides, 5 m. N.W. of Lochmaddy; depth 31 ft., alt. 7 ft.

Loch nan Lann, Inverness-shire, 7 m. N.E. of Fort Augustus; depth 109 ft., alt. 645 ft.

Loch nan Uamh, sea inlet, W. Inverness-shire, 4 m. N.E. from Sound of Arisaig.

Loch na Seilge, N.E. Sutherland, 4½ m. S.E. of Melvich.

Loch na Shellag, Ross and Cromarty, 9 m. S.W. of Ullapool; depth 217 ft., alt. 279 ft.

Loch na Stainge, W. Perthshire, 9 m. S.W. of Rannoch sta.; alt. 970 ft., depth 14 ft.

Lochnaw Castle, Wigtownshire, 4½ m. N.W. of Stranraer, on S. side of Lochnaw (depth 6 ft., alt. 255 ft).

Lochnawean Hill, mt., W. Kincardine-shire, 9 m. N.W. of Fettercairn; alt. 1906 ft.

Lochnell Castle, seat of the Earl of Dun-donald, Argyll, 7 m. N.N.E. of Oban, on Ardmucknish Bay.

Loch of Cliff, Unst Isl., Shetland.

Loch of Harray, Mainland, Orkney, 7½ m. W. of Kirkwall; depth 14 ft.

Loch of Lows, Perthshire, 1½ m. N.E. of Dunkeld; depth 53 ft., alt. 327 ft.

Loch of the Clans, 4 m. S.W. of Nairn. Has examples of lake dwellings.

Lochore, colliery vil., Fife, 2 m. N.W. of Lochgelly. P.O., T.O. Pop. (L. and Glencraig) 6907. L. House, ½ m. N.W.

Loch Ossian Hostel, on Corrour Estate, on the N.W. corner of the Moor of Rannoch, 9 m. N. of Rannoch sta.; 1300 ft. above sea-level. Closed during shooting season. A Scottish Youth Hostel. Loch depth 132 ft., alt. 1269 ft.

Loch o' the Lowes, Ayrshire, 1 m. N.W. of New Cumnock.

Loch Park, loch, Banffshire, 3 m. N.E. of Dufftown.

Lochportan, Lochmaddy, isl. of N. Uist. P.O., T.O.

Lochranza, q.s. par. and vil., with pier, on the N. coast of Arran, Buteshire, at the head of Loch Ranza, 13 m. N.W. of Brodick. P.O., T.O. Ruins of a fourteenth-century castle, to which Bruce came from Rathlin.

Lochridge, seat, Ayrshire, in vicinity of Stewarton.

Lochriecaur, loch, S. Ayrshire, 8 m. S. of Dalmellington.

Lochrosque, seat, Ross and Cromarty, at end of Loch Rosque, near Achnasheen sta. Also deer forest.

Lochruan Distillery, Campbeltown, Argyll.

Loch Ruel, shooting-lodge, Argyll, 6 m. from Colintraive.

Lochrutton, par., Kirkcudbrightshire ; 7740 ac. Pop. 453. See LOCHFOOT.

Lochryan, q.s. par., Wigtownshire. L. Lighthouse on Cairn Ryan Pt., 5½ m. N. of Stranraer, has fixed light, visible 12 m. L. House, a seat.

Lochs, par., Lewis Isl., Outer Hebrides, Ross and Cromarty ; 114,601 ac. Pop. 3849. Its church stands on the N. shore of Loch Erisort, 11½ m. S.S.W. of Stornoway.

Lochs, shooting-lodge, Perthshire, 10 m. N.N.W. of Killin.

Lochside, pl., Morayshire, 2 m. N.W. of Elgin. Tileworks.

Lochside, seat and ry. sta., L.M.S., Renfrewshire, 1½ m. S.E. of Lochwinnoch.

Lochside, seat, Roxburghshire, on Yet- holm Loch, 1½ m. W. of Yetholm.

Lochside, vil., Stirlingshire, part of Lime- rigg, 1 m. S. of Slamannan sta.

Lochside, seat, Ayrshire, 1¼ m. N.W. of New Cumnock.

Loch Skipport, pl. on E. of South Uist Isl., Inverness-shire. P.O.

Lochslin, an ancient castle and modern mansion in the N.E. corner of Fearn par., Ross and Cromarty, 4½ m. E. of Tain.

Loch Strivenhead, seat, Argyll, at head of Loch Striven.

Loch Tay, ry. sta., L.M.S., Perthshire, at Killin pier.

Lochton House, seat, Perthshire, 3½ m. N.W. of Inchture.

Lochtower, formerly a baronial fortalice of N.E. Roxburghshire, at the foot of Yetholm Loch, 2 m. W. of Yetholm vil.

Lochtummel Lodge, Perthshire, 8 m. W. of Pitlochry.

Lochty, pl., E. Fife, 5½ m. W. of Crail.

Lochty Burn, Morayshire, affl. of r. Lossie, 2 m. W.S.W. of Elgin.

Lochty Burn, stream, Fife; rising on Benarty Hill, flows 10 m. E. to the Ore, near Thornton Junction.

Lochtyside, ham., Fife, ¾ m. N.N.W. of Thornton.

Lochview, pl., Lanarkshire, 1¼ m. E. of Wishaw.

Lochwinnoch, par. and town, with ry. sta., L.M.S., Renfrewshire ; par. 19,449 ac. Pop. 3868. Town stands on the left bank of the r. Calder, at the S.W. end of Castle-Semple Loch, 1 m. N.W. of L. sta., 8¾ m. S.W. of Paisley. P.O., T.O. Pop. 2045. Industries : silk weaving and printing, cabinet and chair making.

Lochwood Tower, a ruined baronial fort- alice, Dumfriesshire, 5½ m. S. of Moffat.

Lochy, r., Argyll; rising W. of Tyndrum, flows 8¾ m. W.S.W. to the Orchy, 1½ m. above Dalmally.

Lochy, loch and r., S.W. Inverness-shire. Is the south-westernmost of the chain of fresh-water lochs in the Great Glen, and forms part of the navigation of the Caledonian Canal. Length 10 m., depth (maximum) 531 ft., alt. 93 ft. The r., issuing from the foot of the loch, flows 8 m. S.W. to Loch Linnhe, near Fort William.

Lochyside, ham., Argyll, on r. Lochy, 2½ m. N. of Fort William.

Lockerbie, mkt.-town and police bur., with ry. sta., L.M.S., Dumfriesshire, 25¾ m. N.W. of Carlisle, and 14½ m. E.N.E. of Dumfries. Pop. 2574. Hotel. P.O., T.O. Noted for sheep, lamb, and cattle fairs. L. House, a seat.

Lodge, The, a seat of the Duke of Buc- cleuch, Dumfriesshire, ¾ m. N. of Langholm.

Logan, seat, Wigtownshire, 2 m. S. of Ardwell.

Loganbank, seat, Midlothian, Glencorse par., 1 m. N.W. of Auchendinny sta.

Logan Burn, Midlothian; rising in the Pentland Hills, flows 7¾ m. to the N. Esk in the vicinity of Auchendinny.

Logan House, a mansion in Old Cumnock par., Ayrshire, on the left bank of the Lugar Water, associated with the mem- ory of the witty "Laird of Logan."

Logan House, old mansion, Midlothian, on Logan Burn, 5 m. N.W. of Penicuik.

Logan House, mansion, on headstream of Logan Water, Lanarkshire, 6 m. S.W. of Lesmahagow. L. Water, affl. of the Nethan.

Loganlea Reservoir, in Pentland Hills, 3½ m. N.W. of Penicuik, Midlothian. Edinburgh water supply

Logan, Mull of, Wigtownshire, N.W. extremity of Port Nessock, 10 m. S.E. of Portpatrick.

Logan Water, affl. of Wauchope Water, Dumfriesshire, 3 m. S.W. of Langholm.

Logie, a seventeenth-century baronial mansion, Morayshire, on rt. bank of Findhorn, 2½ m. N.N.W. of Dunphail sta. P.O.

Logie, par., vil., and estate, Fife ; 3603 ac. Pop. 325. Vil. 3 m. N.N.W. of Dairsie sta. and 5 m. N.E. of Cupar.

Logie, seat, Pitcaple, Aberdeenshire.

Logie, seat, Fife, 1¼ m. S.W. of Dunfermline. Pop. 325.

Logie, q.s. par., Dundee, Angus. P.O.

Logie, seat, Angus, 1½ m. S. of Kirriemuir.

Logie, par., Stirlingshire ; 9096 ac. Pop. 4579. Contains the police bur. of Bridge of Allan and part of the royal bur. of Stirling. Church 1 m. E. of Bridge of Allan.

Logiealmond, par., Perthshire, on N. bank of r. Almond, 10 m. N.W. of Perth ; 14,416 ac. Pop. 277. P.O., T.O. Church 2¾ m. N.N.W. of Methven sta.

Logie-Buchan, par., E. Aberdeenshire ; 6685 ac. Pop. 636. Church 2 m. E.S.E. of Ellon.

Logie-Coldstone, par., S.W. Aberdeenshire, 4 m. W. of Tarland ; 13,978 ac. Pop. 688. P.O., T.O.

Logie-Crimond, united estate, Aberdeenshire, 4 m. S.E. of Lonmay. Scene of Jacobite song, " Logie o' Buchan."

Logie-Easter, vil. and par., N.E. Ross and Cromarty ; 10,018 ac. Pop. 597. 5¼ m. N.E. of Invergordon. Vil. pop. 427.

Logie-Elphinstone, seat, Aberdeenshire, on left bank of Ury, 5 m. N.W. of Inverurie.

Logie Head, E. extremity of Cullen Bay, Banffshire.

Logie House, Perthshire, 3 m. N. of Methven.

Logie-Murdoch, ancient name of Logie par., Fife.

Logie-Pert, par. and vil., N.E. Angus ; 5739 ac. Pop. 905. Vil. 2 m. W. by S. of Craigo, 5½ m. N.W. of Montrose. P.O.

Logierait, par. and vil., N. central Perthshire ; 30,671 ac. Pop. 1452. Vil. on N. bank of Tay, near the influx of the Tummel and ¾ m. W. of Ballinluig sta.

Logierieve, ry. sta., L.N.E., Aberdeenshire, 16¼ m. N. by W. of Aberdeen. P.O.

Logie-Wester, ancient par., Ross and Cromarty ; annexed to Urquhart.

Loin, stream and glen, S.W. Banffshire ; stream affl. of r. Avon.

Loirston, Loch, Kincardineshire, 3½ m. S. of Aberdeen. L. House, a seat.

Lomond Hills, isolated hill-group, on border of Kinross-shire and Fife. Extends 6 m. eastwards from N.E. shore of Loch Leven, rises steeply and ruggedly to the N. but gradually and smoothly on the S., and culminates in two conical peaks—W. Lomond Hill (1713 ft.) and E. Lomond Hill (1471 ft.)

Lomond, Loch, Dunbartonshire and Stirlingshire. The largest loch in Scotland. It covers an area of 2709 sq. m., is 21 m. long, and in breadth varies from 5 m. to 1 m., the S. portion being wide and island-studded. Its greatest depth is 623 ft. and alt. 27 ft. It is almost entirely surrounded by lofty hills. Ben L. overlooks it from the E., and the double-peaked Ben Vorlich rises from its N.W. shore. It drains to the Clyde, by the r. Leven, at the S. end. There is a regular service of steamers. Loch L. Public Park, at S. end, belongs to Glasgow Corporation.

Lonan, stream, Argyll, flowing 6 m. W. to Loch Nell, in Lorn, 3 m. E.S.E. of Oban.

Lone, loch, Ross and Cromarty, 6 m. S.W. of Strathcarron.

Longa, isl., Ross and Cromarty, at mouth of Gair Loch.

Longannet Point, on Firth of Forth, S.W Fife, opposite Grangemouth.

Longart State Forest, near Garve ry. sta., 12 m. W. of Dingwall, E. Ross and Cromarty.

Longay, isl., 6 m. N.W. of Kyle Akin, Skye Isl., Inverness-shire.

Longcastle, ancient par. and remains of old keep of the M'Doualls, 6 m. S.W. of Wigtown.

Longdyke, vil., Stirlingshire, 3½ m. N.W. of Grangemouth.

Longfaugh, well-preserved remains of Roman camp, Midlothian, 1¼ m. E. of Crichton.

Longforgan, par. and vil., with ry. sta., L.M.S., E. Perthshire ; 8526 ac. Pop. 2106. Vil. 1 m. N.W. of ry. sta., 5½ m. S.W. of Dundee. P.O., T.O. Pop. 399.

Longformacus, par. and vil., N. Berwickshire ; 21,068 ac. Pop. 260. Vil. on Dye Water, 6½ m. N.W. of Duns. P.O., T.O. L. House, a seat.

Longhaugh, pl., Renfrewshire, on r. Clyde. Shows a ten-seconds group-flashing light.

Longhaugh and Posso, a shooting, 5½ m. S.W. of Peebles.

Longhaugh Hostel, in Manor Valley, Peeblesshire, 8 m. S.W. of Peebles. A Scottish Youth Hostel.

Longhaugh House, seat, S. Angus, 1½ m. N.E. of Dundee.

Longhaven, seat and ry. sta., L.N.E. (passenger service withdrawn), Aberdeenshire, 2 m. S.W. of Boddam. P.O. ; T.O. at sta.

Longhill, pl., N.W. Aberdeenshire, 4 m. from Huntly. P.O.

Long John's (Ben Nevis and **Nevis) Distilleries,** Fort William, Inverness-shire.

Longleys, vil., Perthshire, 1½ m. S.W. of Meigle.

Long, Loch, sea loch, Dunbartonshire and Argyll. It strikes from the Firth of Clyde at 5 m. W.N.W. of Greenock, is 17 m. long, and from 2 to ¾ m. wide. Sends off Loch Goil north-westward from middle of its W. side.

Long, Loch, sea loch, Ross and Cromarty, projecting 5 m. N.E. from Loch Alsh.

Long, Loch, Benbecula, Outer Hebrides, 4 m. N. of Creagorry.

Long, Loch, S.W. Angus, 1½ m. S. of Newtyle ; alt. 724 ft., depth 42 ft.

Longmanhill, vil. and hill, Banffshire, 2½ m. S.E. of Macduff. P.O.

Longmorn, pl., with ry. sta., L.N.E., Morayshire, 3 m. S. of Elgin. P.O. ; T.O. at sta. A distillery.

Longnewton, ham., E. Lothian, Yester par., 6 m. S. of Haddington.

Longniddry, vil., with ry. sta., L.N.E., E. Lothian, 3½ m. N.E. of Tranent and 13¼ m. E. of Edinburgh by rail. P.O., T.O. Pop. 721.

Longridge, or **Lanrig,** vil., W. Lothian, 1¾ m. S. of Whitburn. P.O., T.O.

Longrigg, East and **West,** 2 vils., New Monkland par., Lanarkshire, 1 m. E.N.E. of Longriggend and 6 m. N.E. of Airdrie.

Longriggend, mining vil. and ry. sta., L.N.E. (passenger service withdrawn), Lanarkshire, 4½ m. N.E. of Airdrie. P.O., T.O.

Longside, par. and vil., with ry. sta.. L.N.E., Aberdeenshire ; 16,844 ac. Pop. 2321. Vil. 6 m. W. by N. of Peterhead. P.O., T.O. Vil. pop. 380. See LINSHART.

Longstone, dist. in W. of Edinburgh city, adjoining Slateford.

Longyester, ham., E. Lothian, 2 m. S. of Gifford.

Lonmay, par. and ry. sta., L.N.E., Aberdeenshire ; 11,233 ac. Pop. 1812. 5 m. S.S.E. of Fraserburgh. P.O., T.O.

Lonmore, pl., W. Isle of Skye, 1½ m. S.E. of Dunvegan.

Loofa Baa, shows white three-seconds flashing light, Lerwick, Shetland.

Lord Arthur's Cairn, hill summit, Aberdeenshire, 5 m. N.W. Alford ; alt. 1699 ft.

Lord Macdonald's Forest, E. Isle of Skye, S.E. from Sligachan.

Lord Reay's Green Table, flat summit, Sutherland, 12 m. N.W. of Lairg and close to Crask House, on the road from Lairg to Tongue. Alt. 945 ft.

Loretto, famous pre-Reformation chapel and hermitage, E. end of Musselburgh. L. School, conducted similar to the great public schools of England.

Lorn, dist., Argyll. It is bounded on the N. by Loch Leven, on the S. by Lochs Awe, Avich, and Melfort, together with arbitrary lines between them ; it measures 33 m. in length and 32 m. in extreme breadth ; 503,975 ac. Gives the titles Baron and Marquess to the Duke of Argyll.

Lorn, Firth of, sea belt, southward from Loch Linnhe, between Lorn dist. and Mull Isl.

Lorn, Lynn of, strait, between Lismore Isl. and Argyll mainland.

Lornty, vil., Perthshire, near Blairgowrie. L. Burn flows 7 m. S.E. from Loch Benachally.

Losgainn (Loch an Losgainn Mòr), loch, W. Argyll, 1¾ m. S.E. of Kilmelford ; alt. 508 ft., depth 29 ft.

Lossie, r., Morayshire, flows 31 m. N.E. to the sea at Lossiemouth.

Lossiemouth, seapt. town (police bur.), with ry. sta., L.N.E., Morayshire, 5½ m. N.E. of Elgin, 92¾ m. N.W. of Aberdeen. Pop. 3914. Hotels. P.O., T.O. Has a coastguard sta., also a good bathing beach and a celebrated golf course. The birthplace of the Rt. Hon. J. Ramsay MacDonald.

Lossit, loch, Islay, Argyll, 3 m. S.W. of Port Askaig.

Lossit Bay, S.W. Islay, Argyll, 2½ m. N.E. of Portnahaven.

Lossit Park, seat, Argyll, 5½ m. W. of Campbeltown.

Loth, coast par., ham., and ry. sta., L.M.S., Sutherland; 17,370 ac. Pop. 343. P.O., T.O. at sta., which is 5½ m. S.W. of Helmsdale. L. Water flows 5½ m. S. to the sea.

Lothbeg, ham., Sutherland, ½ m. S.W. of Loth.

Lother Rock, Pentland Skerries, Caithness. Shows a ten-second white group-flashing light, visible 10 m.

Lothian Bridge, pl., with bridge over S. Esk, Midlothian, 2 m. S.W. of Dalkeith.

Lothianburn, stream, golf course, and park, Edinburgh city, 4½ m. S. of G.P.O.

Lothians, The, dist. on S. side of Firth of Forth, divided into East, Mid, and West Lothian, but anciently more extensive. It gives the title of Marquess to the family of Kerr.

Lothrie Burn, stream, Kinross and Fife, flows 4½ m. S.E. to the Leven, near Leslie.

Lotus, seat, Kirkcudbrightshire, 7 m. N.E. of Dalbeattie. L. Hill, 1050 ft.

Loubcroy, seat, Ross and Cromarty, 16 m. W. Invershin.

Loudoun, par., with ry. sta., L.M.S. (Loudounhill), Ayrshire, on r. L.; 15,509 ac. Pop. 7570. L. Castle, seat of the Earl of L. L. Hill, 1034 ft.

Louisburgh, sub. of Wick, Caithness.

Lounsdale, pl., Renfrewshire, 1½ m. S.W. of Paisley.

Lousiewood Law, a summit of the Lowther Hills, S. Lanarkshire, 3 m. E. of Leadhills; alt. 2028 ft.

Lovat, ham., also site of L. Castle, an ancient fortalice (1230) of the Frasers, Lords Lovat, Inverness-shire, on r. Beauly, ½ m. S.E. of Beauly. L. Bridge, 1½ m. S.W.

Low Banton, ham., Stirlingshire, ½ m. S. of Banton.

Low Coats and Coatbank, Lanarkshire, forms part of Whifflet and Rosehall.

Lower Cabrach, pl., Aberdeenshire, 6 m. S.E. Dufftown. P.O., T.O.

Lower Largo, fishing vil., Fife, in Largo par. P.O., T.O.

Lowes, Loch of the, N.W. Selkirkshire, separated from St. Mary's Loch by a narrow neck of land; depth 58 ft., alt. 810 ft.

Lowes, Loch of, Perthshire, ½ m. N.E. of Dunkeld. The largest of a chain of five lochs, expansions of Lunan Burn; depth 53 ft., alt. 328 ft.

Lowlandman's Bay, E. side of Jura Isl., Argyll, 4 m. S. of Lagg.

Lowlands, all the Scottish mainland not included in the Highlands. Generally applied to the country S. of the Firths of Clyde and Tay.

Lowood, seat, Roxburghshire, ½ m. S. of Melrose.

Lowrans Law, Lammermuir Hills, Berwickshire and E. Lothian, 5 m. S. of Gifford; alt. 1631 ft.

Low Roughrigg, pl., Lanarkshire, 2 m. S.W. of Slamannan.

Lowson Memorial, q.s. par., Angus.

Lowthers, The (or **Leadhills**), mt. range in S. Lanarkshire and N. Dumfriesshire. Chief heights, L. Hill (2377 ft.) and Green Lowther (2403 ft.).

Low Valleyfield, coast vil., Fife, 1 m. E. of Culross.

Low Waters, vil., Lanarkshire, 1 m. S. of Hamilton. P.O.

Loy, r., Argyll, flows 7 m. through Glen Loy to E. side of Caledonian Canal, 3½ m. N.E. of Banavie Hotel.

Loyal, seat, E. Perthshire, 1 m. N. of Alyth ry. sta.

Loyal (or **Laoghal**) **Loch,** Sutherland, 5 m. S.E. of Tongue; maximum depth 217 ft., alt. 369 ft.

Loyne, or **Luine,** stream, affl. of r. Moriston, Inverness-shire. Flows through Glen and Loch Loyne.

Lubnaig, Loch, Perthshire, extending 4 m. S.S.E. to Leny Pass in the vicinity of Callander; depth 146 ft., alt. 398 ft.

Luce Bay, S.W. Wigtownshire penetrates 16 m. inland, and is 18½ m. across from Mull of Galloway to Barrow Head.

Luce, New, par. and vil., with ry. sta., L.M.S., Wigtownshire; 28,811 ac. Pop. 423. Vil. on Luce R., 5 m. N.W. of Glenluce. P.O.; T.O. at sta.

Luce, Old (or **Glenluce**), par., Wigtownshire, 8½ m. S.E. of Stranraer; 31,649 ac. Pop. 2051. Glenluce vil. pop. 750.

Luce, Water of, r., Wigtownshire, formed by the Main and the Cross Water of Luce; flows 6 m. S. to Luce Bay.

Luchie, seat, E. Lothian, 2 m. S.E. of North Berwick.

Luckieslap, vil., Angus, 7 m. N.E. of Dundee; sometimes called Husbandtown.

Lucklaw Hill, Fife, 2 m. W. of Leuchars; 626 ft.

Lude, seat, Perthshire, 1 m. N.E. of Blair-Atholl.

Luffness, seat, E. Lothian, ½ m. N.E. of Aberlady. Has a good golf course.

Lugar, q.s. par. and vil., with ry. sta., L.M.S., Ayrshire, on L. Water, 2¾ m. E. of Auchinleek. Pop. 1148. P.O., T.O. Has ironworks.

Lugar Water, Ayrshire, flows 12½ m. N.W. to r. Ayr.

Lugate Water, stream, flowing 7 m. E. to the Gala, 1 m. S. of Stow, Midlothian.

Luggie Water, stream, Lanarkshire and Dunbartonshire, flowing 11 m. N.W. to the Kelvin at Kirkintilloch.

Lugton, vil., with ry. sta., L.M.S., Ayrshire, 5 m. E. of Beith. P.O., T.O.

Lugton, vil., N.W. sub. of Dalkeith, Midlothian.

Lugton Water, Ayrshire, trib. of the r. Garnock, which it joins 2¼ m. N.W. of Irvine.

Luib, ry. sta., L.M.S., Perthshire, 7 m. S.W. of Killin. P.O.; T.O. at sta. L. Hotel, 1 m. E.

Luib, summit, on L.M.S., Ross and Cromarty, 3 m. S.W. of Achnasheen; alt. 646 ft.

Luibeg, glen, Aberdeenshire, descending from Ben Macdhui to Glen Derry.

Luichart, Loch, see LOCHLUICHART.

Luing, isl., Argyll, off S. side of Seil Isl., 12 m. S. of Oban. Pop. 312. L. Sound between L., Lunga, and Scarba.

Luinga Mhor, isl., off W. coast of S. Morar, Inverness-shire, 4 m. W. of Arisaig.

Luinne Bheinn, mt., Inverness-shire, 2 m. S. of Barrisdale Bay, on Loch Hourn; alt. 3083 ft.

Lui Water, affl. of the Dee, Aberdeenshire, near Linn of Dee.

Lumbister, loch, Yell Isl., Shetland.

Lumgair Law, Kincardineshire, 2 m. S.W. of Stonehaven; 492 ft.

Lumphanan, par. and vil., with ry. sta., L.N.E. Aberdeenshire; par. 8755. ac. Pop. 830. Vil. 5½ m. N.E. of Aboyne. P.O., T.O. Pop. 189.

Lumphinnans, mining town, 1 m. W.S.W. of Lochgelly, Fife. P.O. Pop. 2599.

Lumsden, vil., Aberdeenshire, 11½ m. S.S.W. of Huntly. P.O., T.O. Pop. 315.

Lunan, par., with ry. sta. (L. Bay), L.N.E., Angus, 5 m. S.W. of Montrose, 8½ m. N.E. of Arbroath; 2857 ac. Pop. 306. L. Water flows 13 m. E. to L. Bay. L. House, a seat.

Lunan Burn, stream, Perthshire, flows E. through Caputh and Clunie Lochs to the Isla, 2 m. E. of Meikleour.

Lunanhead, vil., Angus, and 1½ m. N.E. of Forfar. P.O.

Luncarty, vil., with ry. sta., L.M.S., 4 m. N.W. of Perth. P.O., T.O. L. House, a seat.

Lunda Wick, bay, S.W. Unst Isl., Shetland.

Lunderston Bay, Firth of Clyde, 5 m. W.S.W. of Greenock, Renfrewshire.

Lundie, par. and vil., Angus, 9 m. N.W. of Dundee; 4185 ac. Pop. 237. P.O. L. Castle, a seat.

Lundie Loch, close to Loch Clunie, in Glen Moriston, Inverness-shire; alt. 681 ft., depth 25 ft.

Lundie, Loch, near Loch Garry, Perthshire; alt. 445 ft., depth 54 ft.

Lundie, Loch, N.W. Ross and Cromarty, 8 m. N.W. of Lochcarron.

Lundie, Loch, 2½ m. W. of Golspie, Sutherland.

Lundin Links, a summer resort, with good seaside golf course, with ry. sta., L.N.E., Fife, 7¼ m. E. of Thornton Junction, 1 m. W. of Largo, 38 m. N.E. of Edinburgh, by rail. P.O., T.O. Hotel.

Lundin Mill and Drumochie, vil., with ry. sta. (L. Links), L.N.E., Fife, in W. vicinity of Largo. L. House, a seat.

Lundy, stream, Inverness-shire, flows 6 m. W. to Lochy R., 1 m. N.E. of Inverlochy Castle.

Lunga, isl., Argyll, off N. coast of Scarba Isl. Pop. 5.

Lunga, one of the Tresnish Isles, Argyll, off W. coast of Mull.

Lunga, seat, Argyll, 1¾ m. N.W. of Ardfern.

Lungard Loch, Ross and Cromarty, 15½ m. S.W. of Strome Ferry ; depth 129 ft., alt. 761 ft.

Lungawater, loch, Shetland, 3 m. N. of Walls.

Lunna, coast vil., Shetland, 6 m. N.E. of Nesting.

Lunna Firth, Shetland, W. side of L. Ness, Mainland. L. Ness, prom., 4 m. N.N.W. of L.

Lunnasting, ancient par., now united to Nesting, Shetland.

Lunnister, pl. and loch, Mainland, Shetland, 7 m. S. of Collafirth.

Luran, loch, Argyll, 1 m. W.S.W. of Lochgair.

Lure, loch, S.E. Ayrshire, 15 m. S.E. of Maybole ; depth 7 ft., alt. about 990 ft.

Lurgain Loch, Ross and Cromarty, 8 m. N. of Ullapool ; greatest depth 156 ft., alt. 173 ft.

Lurg Hill, Banffshire, 6 m. S. of Cullen ; alt. 1029 ft.

Lurg Mhòr, mt., Ross and Cromarty, 12 m. S.W. of Achnasheen alt. 3234 ft.

Luscar House, seat, Fife, 3 m. N.W. of Dunfermline.

Luskentyre, shooting-lodge, S. Harris, 5 m. W. of Tarbert.

Luss, par. and vil., Dunbartonshire ; par. 24,236 ac. Pop. 517. Vil. on W. side of Loch Lomond, 9 m. N.N.E. of Helensburgh, 12 m. N. of Dumbarton. P.O., T.O. Hotel.

Lussa, r., Jura, Argyll, flows to Jura Sound at L. Point.

Lussa, r., Mull Isl., Argyll, flowing N.E. and S.E. to Loch Spelve.

Lussa, glen and stream, Argyll ; stream flows E. to Ardnacross Bay, S. Kintyre.

Lussa, The, stream, Argyll, N. Knapdale ; flows to Loch Sween.

Luss Water, Dunbartonshire, flows 7 m. E.S.E. to Loch Lomond, at Luss.

Luthermuir, vil., Kincardineshire, on L. Water, 4½ m. S.W. of Laurencekirk. P.O.

Luther Water, Kincardineshire, flows 13 m. S.W. to the N. Esk, 1½ m. N.W. of Marykirk.

Luthrie, vil. and ry. sta., L.N.E., Fife, 4½ m. N.W. of Cupar. P.O., T.O.

Lybster, q.s. par., vil., with ry. sta., L.M.S., Caithness, 13 m. S.W. of Wick. P.O., T.O. Pop. 519. Fishing centre.

Lymphoy, seat, Midlothian, 1 m. W.S.W. of Currie.

Lynchat, ham., Inverness-shire, 2 m. N.E. of Kingussie.

Lyne, par. and ry. sta. (Halt), L.M.S., Peeblesshire, 3 m. W. of Peebles ; 2781 ac. Pop. 98. T.O. at sta. L. Water flows from borders of Midlothian for 20 m. and falls into the Tweed 3 m. above Peebles.

Lyne Dale Lodge, seat of Lord Napier of Magdala, Isle of Skye, 10 m. N.W. of Portree.

Lynedoch, estate, Perthshire, on r. Almond, 7 m. N.W. of Perth.

Lynedoch, ry. sta., L.M.S., Greenock.

Lynegar, seat, Caithness, on N. shore of Loch Watten.

Lynn House, Ayrshire, 1 m. S.W. of Dalry.

Lynturk, pl. and ruined castle, Aberdeenshire, 2¼ m. S. of Alford.

Lynwilg, ham., Inverness-shire, on Loch Alvie, 2½ m. S.W. of Aviemore sta. Inn.

Lyon, r. and loch, Perthshire ; r. flows 34 m. E. through Glen L. to the Tay, 2½ m. N.E. of Kenmore. Loch L., near head of r., 2 m. long, depth 100 ft., alt. 1050 ft.

Lyth, vil., Caithness, 8½ m. N.W. of Wick. P.O., T.O. Burn of L. flows S.E. to Loch of Wester.

M

Maberry, Loch, borders of S. Ayrshire and N.E. Wigtownshire, 5 m. S.E. of Barrhill. Remains of a castle on a small isl. in the loch ; depth 14 ft., alt. 388 ft.

Mabie, seat, 4½ m. S.W. of Dumfries, par. of Troqueer, E. Kirkcudbrightshire.

Macallan, formerly a par., now amalgamated with Knockando, Moray. M. Distillery is near Craigellachie Junction.

Macarthur Head, headland on E. coast of isl. of Islay, and 7 m. S.E. of Portaskaig, Argyllshire. Has a lighthouse 42 ft. high, with a fixed light 128 ft. above high water, visible 17 m.

Macaterick, loch, 9 m. S. of Dalmellington, Ayrshire. At the S. end of the loch is M. Hill ; alt. 1637 ft.

Macbeth's Cairn, a cairn, said to mark the spot where Macbeth was slain (1057), near Lumphanan, Aberdeenshire.

Macbeth's Hillock, Elgin. See HARDMUIR WOOD.

Macbie Hill, seat in N. Peeblesshire ; ry. sta., L.N.E. (passenger service withdrawn), 2½ m. E. of West Linton and 21¾ m. S.S.W. of Edinburgh by rail.

MacCaskin, isl. in Loch Craignish, N. end of Sound of Jura, Argyllshire.

Macdonald Deer Forest, 10,000 ac., Skye, Inner Hebrides. Post town, Broadford.

Macduff, q.s. par., police bur., and seapt. on Moray Firth, Banffshire. Bur. pop. 3276. Town at E. side of mouth of r. Deveron, opposite Banff, with which it is connected by bridges ; ry. sta., L.N.E. P.O., T.O. On N. pier is a fixed light, visible 6 m. Herring and salmon fisheries.

Macduff's Castle, ruin on coast of Fife, near E. Wemyss.

Macduff's Cave, Kincraig Pt., Largo Bay, 1 m. W. of Earlsferry, Fife.

Macduff's Cross, famous antiquity, 1 m. S.W. of Newburgh, Fife.

Machany, ham. and seat, on burn of same name, Perthshire. The stream flows 13 m. E. to the Earn.

Macharioch, seat, 3½ m. N.E. of Southend, Kintyre, Argyllshire.

Macharioch Lodge, 3 m. E. of Southend, Kintyre, Argyllshire.

Macharmore Castle, ancient square tower on left bank of the r. Cree, 1½ m. S.E. of Newton-Stewart, Kirkcudbrightshire.

Machar, New, par. in E. Aberdeenshire ; 9306 ac. Pop. 2113. Ry. sta. (Newmachar), L.N.E. ; the sta. is 11½ m. N.N.W. of Aberdeen. P.O., T.O.

Machar, Old, par., on North Sea coast, adjoining city of Aberdeen ; 7039 ac. Pop. 1438.

Machars, dist. in S.E. Wigtownshire, forming the peninsula between Wigtown Bay and Luce Bay ; 149,713 ac. Pop. 7451.

Machir Bay, W. Islay, Argyllshire, ½ m. W. of Kilchoman.

Machrie, pi. ou bay of same name, on W. side of Arran, Buteshire. P.O., T.O.

Machrie, pl., with golf course, in S.W. Islay, Argyllshire, and 3½ m. N.W. of Port Ellen.

Machrie Water, flows into M. Bay on W. side of Arran, Buteshire.

Machrihanish, holiday centre, with golf course, on W. side of Kintyre, Argyllshire, 6 m. W. of Campbeltown ; ry. stas. (M. and M. Farm). Hotel. P.O., T.O. Lifeboat sta. Light ry. to Campbeltown, now closed.

Machry, stream, 4 m. N.W. of Port Ellen, Islay, Argyllshire.

Mackinnon's Cave, large cave on S. shore of Loch-na-Keal, W. coast of Mull, and 12 m. S. of Tobermory, Argyllshire.

Mackinnon's Cave, Staffa. See CORMORANTS.

Maclachlan, shooting-lodge, near Strachur, Loch Fyne, Argyllshire.

Macleod, q.s. par. in Glasgow.

Macleod's Castle, remains of ancient fortress at mouth of Stornoway Bay, Lewis, Outer Hebrides.

Macleod's Maidens, three insulated basaltic rocks in S. of Duirinish par., off W. coast of Skye, Inner Hebrides.

Macleod's Tables, two flat-topped mts.— Healaval More (1538 ft.) and Healaval Beg (1601 ft.)—par. of Duirinish, Skye, Inner Hebrides.

Macmerry, mining vil., 2 m. E. of Tranent, E. Lothian ; the ry. sta. is now closed. P.O., T.O. Pop. 346.

Madderty, par. in Strathearn dist. in S. Perthshire ; 4860 ac. Pop. 426. Ry. sta., L.M.S., 6¼ m. E. of Crieff. P.O., T.O.

Maddiston and Sooty Hill, vil. in par. of Muiravonside, Stirlingshire, and 1½ m. S.E. of Polmont Junction. P.O. Pop. 616.

Maddy, Loch, Outer Hebrides. See LOCH MADDY.

Maddy Moss, morass, in par. of, and 3 m. N.W. of, Dollar, Clackmannanshire.

Mad Stream, small stream flowing into Loch Scavaig, S. Isle of Skye, Inner Hebrides.

Maeshowe Tumulus, or **The Maidens' Mound,** a conical mound, 36 ft. high and 300 ft. in circumference at base. It was explored in 1861 and found to contain a central chamber with three smaller communicating chambers. The Runic inscriptions on the walls date from the twelfth century. It stands near the head of Loch Harray, 9 m. W. of Kirkwall, Orkney.

Magdalene Bridge, the bridge crosses the Brunstane Burn, and the ham. lies $\frac{3}{4}$ m. E. of Joppa, E. Edinburgh.

Magdalen Green, ry. sta., L.M.S., in W. of Dundee, Angus.

Maggieknockater, ham. in N.W. Banffshire, 3 m. N.E. of Craigellachie. P.O.

Maggotland, seat, 1 m. S. of Inchture, Perthshire.

Magillie, loch, 2½ m. E.S.E. of Stranraer, Wigtownshire ; depth 14 ft., alt. 43 ft.

Magus Moor, wild tract, now reclaimed, in par. of, and 3 m. W. of, St. Andrews, Fife. Scene of murder of Archbishop Sharp, 1679.

Mahaick, Loch ($\frac{3}{4}$ m. long by $\frac{1}{4}$ m. wide), 5 m. E. by S. of Callander, S. Perthshire.

Maich Water, stream in S.W. Renfrewshire. Rises in Misty Law and flows in a S.E. direction along the Ayrshire border to Kilbirnie Loch, a course of 5 m.

Maidenhall, seat in Berwickshire, 2 m. N.E. of St. Boswells.

Maidenhead Bay, 7 m. N. of Girvan, Ayrshire.

Maidenkirk, another name for Kirkmaiden (q.v.).

Maiden Pap, hill, alt. 1587 ft., par. of Latheron, 6 m. N.W. of Berriedale, S.W. Caithness.

Maiden Paps, hill, alt. 1677 ft., par. of Cavers, 8 m. S. of Hawick, Roxburghshire.

Maidens, vil., with ry. sta., L.M.S. (passenger service withdrawn), in par. of, and 1½ m. N.W. of, Kirkoswald, W. Ayrshire. P.O., T.O. Pop. 297.

Maiden Stack, isolated rock in the sea, near the Mull of Eswick, Shetland.

Maiden Stone, sculptured stone, in par. of Chapel of Garioch, and ½ m. S.W. of the church, mid Aberdeenshire. Stone is 10 ft. high and 3 ft. broad, and carved on all sides. Eastern face divided into three compartments, containing representations of a dog, an elephant, and a comb and mirror.

Main, seat, 1¾ m. S.W. of Elgin, Moray.

Maines House, 2 m. E. of Chirnside ry. sta., Berwickshire.

Mainhill, cottage, the home of Thomas Carlyle from 1814–1826, 3 m. N.W. of Ecclefechan, Dumfriesshire.

Mainhouse, seat, par. of Eckford, 4 m. S.E. of Kelso, Roxburghshire.

Mainland, largest of the Shetland Isls., containing the town of Lerwick; 54 m. long from N. to S., and has an extreme breadth of 21½ m., but in general is much narrower, and the "voes" or inlets extend far inland. Surface is irregular, and the loftiest pt. is Ronas Hill (1475 ft.). Pop. 15,172.

Mainland, or **Pomona,** the largest of the Orkney Isls., containing the towns of Kirkwall and Stromness ; about 25 m. long from N.W. to S.E., and an extreme breadth of 15 m. W. coast for most part wild, lofty, and rugged. Towards the E. the various bays and sounds penetrate far inland. Pop. 13,352.

Mainnir nam Fiadh, mt., alt. 2483 ft., Mull Isl., 3½ m. N.W. of Lochdonhead, Argyllshire.

Mains and Strathmartine, united par., S. Angus, partly in Dundee ; 5390 ac. Pop. 1643. Contains Downfield, 2 m. N.W. of Dundee. Baldovan, in the par., is a seat of the Ogilvies. In the neighbourhood is the Baldovan Institution for Imbeciles.

Mains Castle, ruined fortress, 1 m. N.W. of E. Kilbride, Lanarkshire.

Mains House, 1¼ m. W. of Milngavie, Dunbartonshire.

Mains of Park House, par. of Glenluce, Wigtownshire.

Mainsriddle, vil. in par. of Southwick, 7½ m. S.E. of Dalbeattie, Kirkcudbrightshire.

Mair Etive, loch, 9 m. S.W. of Rannoch sta., Argyllshire.

Makerston, par. in N. Roxburghshire ; 2867 ac. Pop. 288. P.O. M. House is on the Tweed, 4 m. S.W. of Kelso.

Malert, small stream draining Loch-a-Choire into the Naver, near E. end of Loch Naver, Sutherland.

Maligar, scattered ham., 13 m. N. of Portree, Skye, Inner Hebrides. M. Burn is a short stream flowing into the Kilmartin.

Mallaig, vil., with harb. and pier, at mouth of Loch Nevis, W. Invernessshire, 41¾ m. W. of Fort William and 164¼ m. N.W. of Glasgow by rail ; ry. sta., L.N.E. P.O., T.O. Pop. 797. Hotels. On Red Rocks is a fixed white light and on White Point is a fixed green light. Calling-place for W. coast steamers. Ferry communication with Armadale, Sleat, Isle of Skye.

Malleny, old seat, 1½ m. S.W. of Currie, near Balerno, Midlothian. Nearby are rifle ranges and the ham. of M. Mills.

Malletsheugh, drainage dist. in par. of Mearns, Renfrewshire. Pop. 79.

Mallie, river, flowing N.E., to enter Loch Arkaig near its E. end, before it drains into Loch Lochy, Inverness-shire.

Malma, loch, 6 m. W. of Arisaig, S. Morar, S.W. Inverness-shire; 359 ft. above sea-level, and maximum depth 44 ft.

Malt Mill Distillery, isl. of Islay, Argyll-shire.

Malzie, pl. on M. Water, in co. of, and 4 m. W. of, Wigtown.

Mambeg, pl., with pier, on W. shore of Gare Loch, W. Dunbartonshire.

Mamlorn, Forest of, formerly a forest in W. Perthshire, at head of Glen Lochay and Glen Lyon.

Mamore Deer Forest, 35,000 ac. ; post town, Fort William, Inverness-shire ; it lies between Glen Nevis and the head of Loch Leven.

Mam Rattachan, or **Mam Ratagain,** hill (1072 ft.) and pass on borders of Inverness-shire and Ross and Cromarty ; pass leads from Glenshiel to Glenelg and is 2½ m. W. of Shiel Inn. Ratagan, a State forest, on the S.W. shore of Loch Duich, S.W. Ross and Cromarty.

Mam Soul (Sodhail), mt., alt. 3862 ft., borders of Ross and Cromarty and Inverness-shire, 3 m. N.W. of Loch Affric.

Mam Suim, mt., alt. 2394 ft., 8 m. E.S.E. of Aviemore, Inverness-shire.

Manar, seat, 3½ m. S.W. of Inverurie, Aberdeenshire.

Manderston, seat, 1¾ m. E. of Duns, Berwickshire.

Mangaster Voe, bay in Mainland, Shetland, 6 m. E. by S. of Esha Ness.

Mangerton Tower, ruined fortress on left bank of the Liddel Water, 1 m. S. of Newcastleton, Roxburghshire. Was a residence of the Armstrongs.

Manish, ham., 2 m. S. of Stockinish, S.E. of Harris, Outer Hebrides. P.O.

Manish, pl. on isl. of N. Ensay, in Sound of Harris, Outer Hebrides.

Mannel, ham., on Hynish Bay, Tiree Isl., Argyllshire.

Mannoch Hill, pl. in par. of Rothiemay, m. N. of Milltown, Banffshire.

Mannofield, q.s. par. and vil., S.E. Aberdeenshire ; vil. 2 m. S.W. of Aberdeen. P.O. Pop. 439. In the par. are two service reservoirs for the Aberdeen water supply.

Man o' Hoy Distillery, Stromness, Mainland, Orkney.

Manor, par. in S. Peeblesshire ; 16,628 ac. Pop. 265. Par. church on r. Tweed, 3 m. S.W. of Peebles. Par. rich in objects of historical and antiquarian interest. P.O. called Kirkton M. M. Water flows 10 m. N. to the Tweed, near Manorfoot Bridge, 1¾ m. S.W. of Peebles.

Mansewood, drainage dist. in par. of Eastwood, Renfrewshire. It is now within the confines of the city of Glasgow.

Mansfield, pl., 1 m. E. of New Cumnock, Ayrshire. Nearby are the seats M. House and M. Hall.

Manuel, pl. in par. of Muiravonside, S.E. Stirlingshire, 2 m. W. of Linlithgow ; ry. sta. and junc., L.N.E. T.O. at sta. M. House is a seat, and near it are the ruins of M. Priory.

Maoile Lunndaidh, mt., alt. 3294 ft., 8 m. S. of Achnasheen, E. Ross and Cromarty.

Maolachy, pl., 10 m. S. of Oban, W. Argyllshire.

Maol Breac, mt., alt. 2115 ft., in N.W. corner of Dunbartonshire, 4 m. due W. of Ardlui.

Maol Cheann Dearg, mt., alt. 3342 ft., on S. border of Ross and Cromarty, 6 m. E. of Loch Hourn Beg.

Maol Mhor, mt., alt. 1420 ft., 4 m. N. of Loch Scridain, Mull, Argyllshire.

Maol na h-Ordaig, headland, S.E. coast of S. Uist, Outer Hebrides.

Maol Odhar, mt., alt. 2550 ft., 3 m. E.S.E. of Strontian, Morvern, Argyllshire.

Mar, ancient dist. in Aberdeenshire, between the Dee and the Don, consisting of Braemar, Cromar, and Midmar. Gives the title of Earl to the family of Erskine. M. bounded with Badenoch on the W.

Marble Lodge, in Glen Tilt, 6 m. N. of Blair-Atholl, Perthshire.

Mar Burn, par. of Durisdeer, N.W. Dumfriesshire ; flows 5 m. E. to the Nith, which it joins about 1 m. below Drumlanrig Castle.

Marcaonach, mt., alt. 3185 ft., 6 m. S.W. of Dalwhinnie, S.E. Inverness-shire.

March, ancient name of Merse (q.v.).

Marchbank, seat, 1½ m. S. of Balerno sta., Midlothian.

Marchbank Wood, seat, 1¾ m. S. of Beattock, Dumfriesshire.

Marchmont, ry. sta., L.N.E., 3¾ m. S.W. of Duns, Berwickshire. Nearby is M. House, a seat. P.O., T.O.

Marcus Lodge, seat, on r. Esk, 6 m. N.E. of Forfar, Angus.

Maree, Loch, par. of Gairloch, 19 m. N.W. of Achnasheen, Ross and Cromarty; 12½ m. long from N.W. to S.E., with a breadth varying from less than ½ m. to 2¼ m.; receives the Kinlochewe, and gives off the Ewe from its N.W. extremity; towards the centre there is a large cluster of isls.; greatest depth 367 ft., alt. above sea-level 29 ft.

Margaretsfield, vil., 8 m. W. by N. of Annan, Dumfriesshire.

Mark, ham., 1¼ m. S. of Castle Kennedy sta., Wigtownshire.

Markie, glen and stream, descending to Strathspey, 2 m. W. of Laggan, Inverness-shire.

Markinch, par., small town and police bur., mid Fife; par. 9686 ac. Pop. 8477. Town 11¼ m. S.W. of Cupar and 33¼ m. N. of Edinburgh by rail. Pop. 1988. Ry. sta., L.N.E. P.O., T.O. Paper mills, bleachfields, and woollen mfrs.

Mark of Shennanton, seat, 3 m. N. of Kirkcowan, Wigtownshire.

Mark, Water of, headstream of N. Esk, Angus.

Marlee House and **Loch,** 1½ m. S.W. of Blairgowrie, Perthshire. See DRUMELLIE.

Marlfield, seat, 2 m. E.S.E. of Eckford, Roxburghshire.

Mar Lodge, seat of Duchess of Fife; stands on left bank of r. Dee, 3 m. S.W. of Braemar, Aberdeenshire. M. Lodge Deer Forest covers 110,000 ac.

Marnoch, par. in N.E. Banffshire, about 8 m. S.W. of Macduff; 14,961 ac. Pop. 2087. P.O., T.O. at Bridge of M., on the Deveron, 2 m. S.W. of Aberchirder.

Marrel and Gartymore, part of W. Helmsdale, Sutherland.

Marrister, ham. in W. of Whalsay, Shetland.

Marscalloch Hill, alt. 1249 ft., 3 m. E. of Carsphairn, N. Kirkcudbrightshire.

Marsco, conical mt., 2414 ft., 3 m. S.E. of Sligichan, Skye, Inner Hebrides.

Marshadder, vil. in par. of Kilmuir, 18 m. N. of Portree, Skye, Inner Hebrides.

Marshall Meadows, pl. and bay on Anglo-Scottish boundary, 2½ m. N.N.W. of Berwick-on-Tweed.

Marsh House, Canonbie, Dumfriesshire.

Marshyland, estate, ½ m. S. of Beith, Ayrshire.

Martin, isl. in par. of Lochbroom, W. Ross and Cromarty. Pop. 3.

Martnaham, Loch, loch, between pars. of Coylton and Dalrymple, in co. of, and 4½ m. S. of, Ayr; it is about 1½ m. long, with an extreme breadth of ¼ m. On a wooded islet are the remains of an old manor house. 1½ m. N.W. of Hollybush sta. is M. House. Loch 29 ft. deep, alt. 269 ft.

Martour Hill, alt. 1649 ft., 6 m. N.W. of Moniaive, on border of cos. of Kirkcudbright and Dumfries.

Martyrs, q.s. par. in Glasgow.

Martyrs, eccl. par., Paisley, Renfrewshire.

Marvig, ham. on E. coast of Lewis Isl. and 10 m. S. of Stornoway, Outer Hebrides. P.O.

Marwhirn Lodge, 2 m. N. of Kirkpatrick-Durham, Kirkcudbrightshire.

Marwick Head, headland on N.W. of Mainland, Orkney. Crowned by cairn in memory of Lord Kitchener.

Marybank, pl. in par. of, and 2 m. N.N.W. of, Urray, Ross and Cromarty.

Maryburgh, an old name for Fort William, Inverness-shire.

Maryburgh, vil. in par. of Cleish, in co. of, and 4 m. S. of, Kinross.

Maryburgh, vil. on r. Conon, 1½ m. S.W. of Dingwall, Ross and Cromarty. P.O. Pop. 275.

Maryculter, par. and ham. in N.E. Kincardineshire; par. 7842 ac. Pop. 924. Ham. 7 m. S.W. of Aberdeen and 2½ m. S.E. of Culter sta. P.O. M. House stands 1 m. N.W., on the r. Dee.

Marydale, pl. on the r. Glass, near Invercannich, 17 m. S.W. of Beauly, Inverness-shire. Here is a Roman Catholic church and a school.

Maryfield, q.s. par. of Dundee, Angus.

Maryhill, a mun. ward in N.W. Glasgow, 3½ m. from the centre of the city ; ry. stas., L.M.S. and L.N.E. P.O., T.O. Dawsholm Gasworks (supplying Glasgow), and M. Barracks.

Maryhill, parl. div. of Glasgow. Pop. 82,393.

Marykirk, eccl. par. in Stirling.

Marykirk, par. and vil., S. Kincardineshire ; par. 9854 ac. Pop. 1163. Vil. on the N. Esk, 6 m. N.W. of Montrose ; ry. sta. L.M.S. (1¼ m. N. of vil.). P.O., T.O.

Marypark, pl., 3 m. N.E. of Ballindalloch, Banffshire. P.O.

Maryport, ham. and bay, 3 m. N.N.W. of Mull of Galloway, Wigtownshire.

Mary's Loch, Ross and Cromarty. See MORIE, LOCH.

Maryston, West, vil. in par. of Old Monkland, 3 m. W. of Coatbridge, N. Lanarkshire.

Maryton, par. in E. Angus, skirted by S. Esk and 3 m. W. of Montrose ; 2293 ac. Pop. 208.

Maryville, pl., 5¼ m. N.W. of Hamilton, Lanarkshire.

Marywell, vil. in par. of St. Vigeans, 2 m. N. of Arbroath, Angus. P.O.

Marywell, vil., 4½ m. S.E. of Aboyne, S. Aberdeenshire.

Mashie Water, burn in S.E. Invernessshire ; rises near Loch Ericht and flows 9 m. N. to the Spey, which it joins 1 m. W. of Laggan Bridge.

Masonhall, vil., 4 m. W. of Strathmiglo, Fife.

Mason's Cave, in S.E. Angus, 2 m. E.N.E. of Arbroath. The cliff, pierced by the cavern, is nearly perpendicular.

Massan, r. in E. Argyllshire ; flows 6 m., partly through Glen M., to the Eachaig.

Masterton, vil., 2 m. N.W. of Inverkeithing, Fife.

Mathers, Kaim of, remains of an old fort, on coast in par. of St. Cyrus, Kincardineshire. The vil. of M. was overwhelmed by the sea.

Mauchline, par. and town in N. Ayrshire ; par. 8915 ac. Pop. 2484. Town 9½ m. S.E. of Kilmarnock ; ry. sta., L.M.S. P.O., T.O. The town is intimately connected with Burns. Long noted for horse and cattle markets. Pop. 1789.

Maud, q.s. par. and vil., N.E. Aberdeenshire ; vil. 13 m. W. of Peterhead and 31¼ m. N. of Aberdeen ; ry. sta., L.N.E. P.O., T.O. Pop. 722.

Maud, Hill of, alt. 900 ft., 4¼ m. S.W. of Cullen, Banffshire.

Mauldslie Castle, seat, with extensive grounds, on r. Clyde, 2½ m. W. of Carluke, Lanarkshire, now demolished.

Maulesden, seat, on S. Esk, 1¾ m. S.W. of Brechin, Angus.

Maulside, seat, 2 m. S.W. of Beith, Ayrshire.

Mauricewood, seat, 1¼ m. N. of Penicuik, Midlothian.

Mavis Grind, isthmus between the Northmaven and Delting portions of Mainland, Shetland ; only about 100 yds. wide, and in great part submerged during spring tides.

Mavis Grove, seat, on the Nith, Kirkcudbrightshire, and 2 m. S. of Dumfries.

Maw, alt. 1753 ft., one of the Pentland Hills, Peeblesshire, 2½ m. N.N.W. of W. Linton.

Mawcarse Junction, ry. sta., L.N.E., Kinross-shire, 4 m. N.E. of Kinross Junction. Nearby is M. House, a seat.

Maxpoffle, seat, 1½ m. S.W. of St. Boswells, Roxburghshire.

Maxton, par., vil., and seat, on S. side of the Tweed, N. Roxburghshire ; par. 4429 ac. Pop. 316 ; ry. sta., L.N.E. ; vil. is 3 m. S.E. of Newtown St. Boswells and 7 m. S.W. of Kelso. P.O. A Roman road passes through the par. Littledean Tower, an ancient fortress of the Kers, is situated on a lofty rock beside the Tweed.

Maxwell, q.s. par., Govan par., Glasgow.

Maxwellheugh, S. sub. of Kelso.

Maxwell Park, public recreation ground in S.W. sub. of Glasgow ; ry. sta., L.M.S., Pollokshields.

Maxwelltown, q.s. par., town, and police bur., on rt. bank of the Nith, E. Kirkcudbrightshire ; bridges over the Nith connect it with Dumfries ; ry. sta., L.M.S. T.O. at sta. To the S. of the town is Corbelly Hill, with an observatory. Has tweed mills, dyeworks, and other industries. Pop. of M. and Dumfries bur. 22,795.

Maxwelton, vil. in par. of, and ½ m. E. by N. of, East Kilbride, N.W. Lanarkshire.

Maxwelton House, seat, near Cairn Water, 3 m. E. of Moniaive, Dumfriesshire.

Mayar, mt., 3043 ft., 6 m. W. of Milton of Clova, Angus.

Maybole, par., mkt.-town, and police bur., W. Ayrshire; par. 21,929 ac. Pop. 6180. Town and bur., 9 m. S.W. of Ayr and 50¼ m. S.W. of Glasgow by air ; ry. sta., L.M.S. P.O., T.O. Pop. 4210. Hotel. Mfrs. boots and shoes. There are also agricultural implement works. M. Castle is a seat of the Marquess of Ailsa.

Mayen House, 5 m. N.E. of Rothiemay sta., Banffshire.

Mayfield, q.s. par. and dist. in S.E. Edinburgh.

Mayfield, seat, 1 m. E. of Dundee, Angus.

May, Isle of, isl. in the N. part of the Firth of Forth, 5½ m. S.E. of Crail ; it is 1 m. long from N.W. to S.E., and has an extreme width of ¼ m. ; area 126 ac. It is included in the co. of Fife, but does not form part of any par. Contains some remains of a thirteenth-century chapel, dedicated to St. Adrian. On the N.E. side is a lighthouse with electric group-flashing light 240 ft. above high water, visible 21 m.

Mayshiel, a shooting, 9 m. S.E. of Haddington, E. Lothian.

May, Water of, in S.E. Perthshire ; rises on John's Hill, 4 m. S. of Dunning, and flows 11 m. N.E. and N.W. to join the Earn near Forteviot.

Meaddie, loch, 10¼ m. S.W. of Tongue, N. Sutherland ; depth 63 ft., alt. 488 ft.

Meaddie, loch, 3 m. S.E. of Bettyhill, N. Sutherland ; alt. 405 ft.

Meadhoin, one of the Mishnish lochs, 2 m. S.W. of Tobermory, Mull, Argyllshire.

Meadowfield, eccl. par. in par. of New Monkland, Lanarkshire.

Meadowmill, vil., 1 m. N. of Tranent, E. Lothian.

Meadowside, Kingussie, Inverness-shire. Co. hospital.

Meal a Bhealaich, mt., alt. 2771 ft., 3 m. S.W. of Ben Alder, on borders of cos. of Perth and Inverness.

Mealabost, ham. on N.W. coast of Lewis, Outer Hebrides.

Mealfourvonie (Meall Fuarmhonaidh), dome-like peak (2284 ft.) on W. side of Loch Ness, 10 m. N.E. of Fort Augustus, Inverness-shire.

Mealista, or **Eilean Mhealastadh,** isl. off W. coast of Lewis, on N. side of mouth of Loch Resort, Outer Hebrides.

Mealisval, mt., alt. 1885 ft., W. Lewis, Outer Hebrides.

Meall a Bhuachaille, mt., alt. 2654 ft., 6 m. E. of Aviemore, Inverness-shire.

Meall a' Bhuiridh, mt., alt. 2448 ft., 5½ m. S.E. of Ballachulish, Argyllshire.

Meall a Chinn Deirg, mt., alt. 3060 ft., 10 m. S.W. of Kinlochewe, Ross and Cromarty.

Meall a Chocaire, mt., alt. 2294 ft., 4 m. N. of Kingussie, Inverness-shire.

Meall a Choire Bhuidhe, mt., alt. 2846 ft., 3 m. W.N.W. of Spittal of Glenshee, Perthshire.

Meall a Choire Chreagaich, mt., alt. 2177 ft., 2 m. S.E. of Kenmore, Perthshire.

Meall a Choire Leith, mt., alt. 3033 ft., 7 m. N.N.E. of Killin, Perthshire.

Meall a Chrasgaidh, mt., alt. 3062 ft., 9 m. N. of Achnasheen, Ross and Cromarty.

Meall a Chuilinn, mt., alt. 2228 ft., 4½ m. E. of Strontian, Argyllshire.

Meall an Aodainn, mt., alt. 2225 ft., 3 m. S. of Ballachulish, Argyllshire.

Meall an Fhudair, mt., alt. 2508 ft., 4 m. N.W. of Ardlui, Argyllshire.

Meall ant Seallaidh, mt., alt. 2792 ft., 3 m. W. of Lochearnhead, Perthshire.

Meall ant Sluichd, mt., alt. 2771 ft., 4 m. S.E. of Braemar, S.W. Aberdeenshire.

Meall a Phubuill, mt., alt. 2535 ft., between Loch Arkaig and Loch Eil, Inverness-shire.

Meall a Phuill, mt., alt. 2882 ft., 6 m. S. of the W. end of Loch Rannoch, W. Perthshire.

Meall Ban, mt., alt. 2148 ft., 7 m. S.W. of Ballachulish, Argyllshire.

Meall Beag, mt., alt. 1559 ft., 2½ m. N. of Loch Tulla, Argyllshire.

Meall Bhalach, mt., alt. 2291 ft., 3 m. N. of Kingshouse Inn, N.E. Argyllshire.

Meall Buidhe, mt., alt. 2043 ft., 6 m. N. of Dalmally, Argyllshire.

Meall Buidhe, mt., alt. 3107 ft., 2 m. N. of E. end of Loch Nevis, Inverness-shire.

Meall Buidhe, mt., alt. 2508 ft., 4 m. S.E. of Kingussie, Inverness-shire.

Meall Buidhe, mt., alt. 2813 ft., 11 m. N.W. of Killin, Perthshire.

Meall Cala, mt., alt. 2203 ft., 4 m. S.W. of Strathyre, Perthshire.

Meall Chuirn, mt., alt. 3007 ft., 7 m. W. of Killin, Perthshire.

Meall Coire nan Saobhaidh, mt., alt. 2695 ft., 4 m. S.W. of Loch Garry, Inverness-shire.

Meall Coire nan Saobhaidhe, mt., alt. 3191 ft., a summit of Lochnagar, S.W. Aberdeenshire.

Meall Crumach, mt., alt. 2217 ft., 4 m. N.W. of Kenmore, Perthshire.

Meall Cuanail, mt., alt. 3004 ft., one of Ben Cruachan summits, Argyllshire.

Meall Daill, mt., alt. 2858 ft., 11 m. N.W. of Killin, W. Perthshire.

Meall Dearg, mt., alt. 3118 ft., 5½ m. E. of Ballachulish, Argyllshire.

Meall Dearg, mt., alt. 2258 ft., 3 m. N. of Amulree, Perthshire.

Meall Dubh, mt., alt. 1935 ft., 1 m. N.W. of Kinloch Rannoch, Perthshire.

Meall Dubh-Achadh, mt., alt. 3268 ft., 8m. E.S.E. of Kingussie, Invernessshire.

Meall Dun Dhomhnuill, mt., alt. 2061 ft., 4 m. S.E. of Kenmore, Perthshire.

Meall Garbh, mt., alt. 2299 ft., 5½ m. S.W. of Kingshouse Inn, Argyllshire.

Meall Garbh, mt., alt. 3661 ft., 8 m. N.E. of Killin, Perthshire.

Meall Ghaordie, mt., alt. 3407 ft., 6 m. N.W. of Killin, Perthshire.

Meall Glas, mt., alt. 2547 ft., 4½ m. N. of Killin, Perthshire.

Meall Gorm, mt., alt. 3109 ft., 8 m. N.E. of Achnasheen, Ross and Cromarty.

Meall Gruaidh, mt., alt. 3280 ft., 9 m. N.E. of Killin, Perthshire.

Meall Horn, mt., alt. 2548 ft., 17 m. S.W. of Tongue, Sutherland.

Meall Luaidhe, mt., alt. 2558 ft., 7 m. N. of Killin, Perthshire.

Meall Mhor, mt., alt. 2457 ft., 3½ m. N.N.W. of Stronachlacher Hotel, Stirlingshire.

Meall Mhor, seat, on Loch Fyne, 4 m. N. of Tarbert, Argyllshire.

Meall Mor, mt., alt. 2215 ft., 2½ m. S.W. of Ballachulish, Argyllshire.

Meallmore Lodge, 3 m. N.W. of Moy, N.E. Inverness-shire.

Meall na Creige, mt., alt. 2683 ft., 8 m. E. of Killin, Perthshire.

Meall na Cuaich, mt., alt. 3120 ft., 3 m. N.E. of Dalwhinnie, Inverness-shire.

Meall na Guaille, mt., alt. 2550 ft., 6 m. N.W. of Braemar, Aberdeenshire.

Meall na h-Aisre, mt., alt. 2825 ft., 6 m. N. of Loch Laggan Hotel, Inverness-shire.

Meall na Leitreach, mt., alt. 2544 ft., 2 m. S. of Dalnaspidal ry. sta., Perthshire.

Meall nan Caora, mt., alt. 2568 ft., 4 m. N.W. of Ardlui, on borders of cos. of Perth and Argyll.

Meall nan Eun, mt., alt. 3039 ft., 5 m. E. of the head of Loch Etive, Argyllshire.

Meall nan Ruadhag, mt., alt. 2120 ft., 3 m. N.E. of Kingshouse Inn, Argyllshire.

Meall nan Tarmachan, mt., alt. 3421 ft., 4 m. N. of Killin, Perthshire.

Meall nan Tigearn, mt., alt. 2423 ft., 5 m. E.S.E. of Dalmally, on borders of cos. of Perth and Argyll.

Meall na Suiramach, see QUIRAING.

Meall Odhar, mt., alt. 2875 ft., 6½ m. S.W. of Kingshouse Inn, Argyllshire.

Meall Odhar, mt., alt. 3019 ft., 8½ m. S. of Braemar, on borders of cos. of Perth and Aberdeen.

Meall Odhar, mt., alt. 2150 ft., 2 m. W. of Tyndrum, on borders of cos. of Perth and Argyll.

Meall Tairneachan, mt., alt. 2559 ft., 4½ m. N.W. of Aberfeldy, Perthshire.

Meall Tionail, mt., alt. 3338 ft., 11 m. S.E. of Kingussie, S.E. Inverness-shire.

Meall Uaine, mt., alt. 2600 ft., 5 m. N.E. of Kirkmichael, Perthshire.

Mealt, loch, 2 m. S.E. of Staffin Bay, Skye, Inner Hebrides.

Mearns, par. and vil. in S.E. Renfrewshire ; par. (which includes the major part of the town of Busby) 10,238 ac. Pop. 4635. Vil., Newton M., is 3 m. S.E. of Busby. P.O., T.O. M. Castle, an ancient fortress of the Maxwells, is 1 m. E. of the vil. In the manse of the par. church Professor John Wilson (" Christopher North ") spent part of his boyhood.

Mearnskirk Sanatorium, Renfrewshire, adjoining Newton Mearns.

Mearns, The, ancient name of Kincardineshire (q.v.).

Measach, Falls of, picturesque waterfall on a headstream of the Broom, 11 m. S.S.E. of Ullapool, N.W. Ross and Cromarty.

Meaul, mt., alt. 2280 ft., 12 m. N.W. of New Galloway, Rhinns of Kells, N.W. Kirkcudbrightshire.

Meavaig, ham., Isle of Lewis, Inverness-shire, 5 m. N.W. of Tarbert. P.O., T.O. called Miavaig.

Medwin, trib. of Clyde, E. Lanarkshire, formed by union of N. and S. Medwins, and joins the Clyde 1½ m. S. of Carnwath.

Medwin House, seat, on Lyne Water, ½ m. N.W. of W. Linton, Peeblesshire.

Meet Hill (181 ft.), appears to be partly artificial, on coast of Aberdeenshire, 1¼ m. S.W. of Peterhead.

Meggernie Castle, on r. Lyon, 9 m. N.W. of Killin, Perthshire. Older part consists of a fifteenth-century square tower.

Megget, dist. in Selkirkshire, 2½ m. W. of St. Mary's Loch.

Megget Water, stream, rising on border of Roxburghshire, flowing in a southerly direction to meet the Esk 7 m. N.W. of Langholm, Dumfriesshire.

Megget Water, stream, 7 m. long, flowing to W. side of St. Mary's Loch, Peeblesshire.

Megginch Castle, seat, par. of Errol, in co. of, and 8½ m. N.E. of, Perth. Mansion was erected in 1575.

Meg Shank, mt., alt. 1537 ft., 5 m. N. of Langholm, Dumfriesshire.

Meig, r. in S.E. Ross and Cromarty; issues from E. end of Loch Benachran and flows to the Conon, which it joins about 10 m. W. by S. of Dingwall.

Meigle Hill, alt. 1387 ft., 2 m. W. of Galashiels, Selkirkshire.

Meigle, par. and vil. in N.E. Perthshire; par. 3984 ac. Pop. 735. Vil. is near the Isla, 5½ m. N.E. of Coupar-Angus and 1¼ m. N. of Alyth Junction; ry. sta., L.M.S. P.O., T.O. M. House is a seat.

Meikle Aikenhead, pl., 1 m. W.S.W. of Rutherglen, in S.E. of Glasgow.

Meikle Auchinstilloch, mt., alt. 1609 ft., 6 m. S.W. of Lesmahagow, Lanarkshire.

Meikle Balloch, hill, alt. 1199 ft., near Keith, Banffshire.

Meikle Bin, one of the summits in the Kilsyth Hills, Stirlingshire, 4 m. N.W. of Kilsyth; alt. 1870 ft.

Meikle Black Law, alt. 803 ft., 2¼ m. N.E. of Grantshouse, Berwickshire.

Meikledale, pl., 5 m. N. of Langholm, Dumfriesshire.

Meikle Earnock, vil., 1½ m. S.W. of Hamilton, Lanarkshire; ry. sta., L.M.S. See EARNOCK.

Meikle Ferry, ferry across Dornoch Firth between Ross and Cromarty and Sutherland, 4½ m. S.W. of Dornoch and 4 m. N.W. of Tain.

Meiklefolla, vil. in par. of Fyvie, 8 m. N.W. of Inverurie, N.E. Aberdeenshire.

Meikle Geal Charn, mt., alt. 2633 ft., 4 m. S.W. of Cock Bridge, borders of cos. of Aberdeen and Perth.

Meikle Law, 1518 ft., 9 m. N.N.E. of Lauder, Berwickshire.

Meikle, Loch, loch, through which flows the Enrick, Glen Urquhart, 4½ m. W. of Drumnadrochit, Inverness-shire; depth 45 ft., alt. 365 ft.

Meikle Millyea, 2246 ft., a summit of the Kells Range, 8 m. W. of New Galloway, Kirkcudbrightshire.

Meikle Mosside, pl., 2½ m. N. of Kilmarnock, Ayrshire.

Meikle Mulltaggart, mt., alt. 2000 ft., 6 m. N.E. of Newton-Stewart, Kirkcudbrightshire.

Meikleour, vil. in E. Perthshire, 4 m. S. of Blairgowrie. P.O., T.O. Here is an old cross dated 1698. At confluence of Tay and the Isla is M. House, with magnificent beech hedges.

Meikle Pap, mt., alt. 3211 ft., near Lochnagar, S.W. Aberdeenshire.

Meikleriggs, Renfrewshire, Paisley. Gleniffer Home for Incurables.

Meikle Says Law, alt. 1750 ft., one of the Lammermuirs, 8½ m. S.E. of Haddington, E. Lothian.

Meikle Seggie, pl., 1½ m. W.N.W. of Milnathort, Kinross-shire.

Meikle Trochry, Perthshire. See TROCHRY.

Meikle Warble or **Warthill,** vil., 6½ m. W.N.W. of Inverurie, Aberdeenshire; ry. sta. (Wartle), L.N.E. P.O.

Meiklewood House, seat, on r. Forth, Stirlingshire, 1½ m. N.E. of Gargunnock sta.

Meil, Bay of, 2 m. N.E. of Kirkwall, Mainland, Orkney.

Meiller (Meall Odhar), mt., alt. 1794 ft., W. side of Loch Tay, 1½ m. W. of Lawers, Perthshire.

Mein Water, stream in S.E. Dumfriesshire, joining the Annan on its left bank, 1½ m. S.W. of Ecclefechan.

Melbost, ham. in par. of, and 3 m. E. of, Stornoway, Lewis, Outer Hebrides.

Melby, seat, 32 m. N. of Lerwick, Shetland.

Meldrum, par., 19 m. N.W. of Aberdeen and 5¾ m. N.E. of Inverurie, Aberdeenshire; 8105 ac. Pop. 1589. See OLD MELDRUM.

Meldrum Hill, ham., 1½ m. S. of Doune, S. Perthshire.

Meldrum House, 1 m. N. of Old Meldrum, mid Aberdeenshire.

Melfort, Loch, sea loch, 11 m. S. of Oban, Argyllshire; stretches about 4 m. inland in an E.N.E. direction. At the head of the loch is the seat, M. House; N. of the House, on the road to Oban, is Pass of M.

Melgam Water, stream in W. Angus, flowing in a south-easterly direction to the Isla opposite Airlie Castle.

Melgum Lodge, shooting-lodge, 7 m. N.W. of Aboyne, Aberdeenshire.

Melgund Castle, ruin, 4 m. S.W. of Brechin, Angus. Said to have been built by Cardinal Beaton.

Melldalloch Loch, 2½ m. S. of Kilfinan, Argyllshire.

Mellerstain, seat of Earl of Haddington, Berwickshire, 6 m. N.W. of Kelso; stands on ground rising from W. bank of Eden Water.

Mell Head, S.W. pt. of Stroma Isl., N.E. Caithness.

Mellock Hill, alt. 1573 ft., one of the Ochils, N.W. Kinross, 5 m. N. of Rumbling Bridge.

Melloncharles, vil. on E. side of Loch Ewe, 2½ m. N.W. of Aultbea, N.W. Ross and Cromarty.

Mellonudrigle, vil. on W. side of Gruinard Bay, 6 m. N. of Aultbea, N.W. Ross and Cromarty.

Melmannock, hill, alt. 1260 ft., Kincardineshire, 4½ m. W.S.W. of Banchory.

Melness, vil. on W. side of Kyle of Tongue, N. Sutherland.

Melrose, par. and mkt.-town (police bur.), Roxburghshire; par. 26,652 ac. Pop. 4518. Town is 3¾ m. S.E. of Galashiels and 37¼ m. S.E. of Edinburgh by rail. Pop. 2052. Ry. sta. (332 ft.), L.N.E. P.O., T.O. In the W. of the town is a large hydropathic and nearby is an asylum for the cos. of Roxburgh, Selkirk, and Berwick. Hotels. M. Abbey, the finest and largest of the great border abbeys, was founded by King David in 1136; burnt and pillaged in many border frays, it suffered so

much in 1322 at the hands of Edward II. that it had to be almost rebuilt, and it is from this period that the present remains mostly date. The heart of Robert Bruce is believed to be buried under the high altar of the abbey.

Melrose, Old, farm (formerly a seat), on a prom. formed by a loop of the Tweed, 2½ m. E. of Melrose, Roxburghshire.

Melsetter, pl. and seat at head of Long Hope, Hoy Isl., and 18 m. S.W. of Kirkwall, Orkney. P.O., T.O.

Meluncart, mt., alt. 1725 ft., 7 m. W.N.W. of Fettercairn, W. Kincardineshire.

Melvaig, vil. on W. coast of Gairloch par. and 10 m. N.W. of Gairloch, N.W. Ross and Cromarty. P.O., T.O.

Melvich, vil., with hotel, on left side of mouth of the r. Halladale, N.E. Sutherland. Adjoins and forms part of the vil. of Portskerra and M., 17 m. W. of Thurso. P.O., T.O. See PORTSKERRA.

Melville, q.s. par. in par. of Montrose, Angus.

Melville Castle, seat of Viscount Melville, on the left bank of the N. Esk, 1 m. N.E. of Lasswade, Midlothian. Castellated three-storeyed building of the eighteenth century.

Melville House, seat, 3 m. N. of Ladybank, Fife.

Melville Monument, obelisk to Henry Dundas (1740–1811), first Viscount Melville, on Dumore Hill, 1 m. N.W. of Comrie, Perthshire.

Memsie, vil., 3½ m. S.S.W. of Fraserburgh, Aberdeenshire. P.O., T.O. The ancient mansion is now a farmhouse.

Memus, pl., 5 m. N.E. of Kirriemuir, Angus. P.O., T.O.

Mendick Hill, alt. 1480 ft., N.W. Peeblesshire, 2 m. S.W. of W. Linton.

Menie House, 3 m. S.W. of Newburgh, Aberdeenshire.

Menmuir, par. and ham. in N.E. Angus; par. 10,695 ac. Pop. 612. Ham. 5 m. N.W. of Brechin. P.O., T.O. Nearby is Hill of M. (881 ft.).

Mennock Bridge, vil. at confluence of M. Water and the Nith, 2 m. S.E. of Sanquhar.

Mennock Pass, road over the Lowther Hills from Wanlockhead to M. Bridge; alt. 1149 ft.

Men of Mey, dangerous reef in the Pentland Firth, separated from the mainland by a narrow channel known as Orkney Ferry. Rocks are submerged except at low tide, and lie off St. John's Pt., Caithness.

Menslaws, seat, 4 m. W. of Jedburgh, Roxburghshire.

Menstrie, q.s. par. and vil. on the M. Burn, at foot of Ochil Hills, Clackmannanshire; vil. 4 m. N.W. of Alloa; ry. sta. (M. and Glenochil), L.N.E. P.O., T.O. Pop. 779. Has a large distillery.

Menteith, dist. in S.W. Perthshire, comprising the pars. of Aberfoyle, Callander, Port of M., Kilmadock, Kincardine, and Lecropt, and part of Kippen, Logie, and Dunblane; dist. is about 28 m. long and 15 m. broad, and embraces much picturesque scenery.

Menteith, Lake of, S.W. Perthshire, 6 m. S.S.W. of Callander; 1½ m. long (E.–W.) and 1 m. broad). Contains three isls.—Inchmahone, Inchtalla, and Dog's Isle. From the lake, the Goodie issues to join the Forth; depth 77 ft., alt. 55 ft.

Menteith, Port of, par. and ham. in S.W. Perthshire; par. 22,238 ac. Pop. 940. Ham. on N. side of Lake of M., 4 m. N.W. of Port of M. sta., 6 m. S.W. of Callander, and 13 m. W. of Stirling; ry. sta. (3½ m. S.E.), L.N.E. P.O., T.O., and T.O. at sta. Hills of M. lie in the N.W. of the par. and rise to a height of 1289 ft.

Menzion Burn, small trib. of Tweed, which it joins near Tweedsmuir Church, Peeblesshire.

Meoble, deer forest (35,000 ac.), 8 m. N.E. of Arisaig, W. Inverness-shire.

Meoble, short stream draining Loch Beoraid into Loch Morar, W. Inverness-shire.

Merchiston, dist. in S.W. of Edinburgh; ry. sta., L.M.S. M. Castle, an ancient baronial fortress, was the birthplace of John Napier (1550–1617), the inventor of logarithms.

Merchiston Castle School, Colinton, Edinburgh. Boys' preparatory school.

Merchiston Hall, in the N.W. of Falkirk, Stirlingshire.

Merchiston House, 1 m. N.W. of Johnstone, Renfrewshire.

Merkland Cross, 4 m. S.E. of Ecclefechan, S.E. Dumfriesshire; ancient cross, said to have been erected in memory of the Master of Maxwell, Warden of the Marches, who was murdered here in 1484.

Merkland Loch, 360 ft. above sea-level and 85 ft. deep, 3 m. N. of N.W. end of Loch Shin, Sutherland.

Merklands, seat, 8½ m. N.W. of Blairgowrie, Perthshire.

Merkland Street, ry. sta., Glasgow Dist. Subway, in Partick.

Merlindale, seat in Peeblesshire, 2 m. S.E. of Broughton sta.

Merrick, mt., alt. 2764 ft., 13 m. N. of Newton-Stewart, Kirkcudbrightshire; highest mt. in S. of Scotland.

Merrylea, q.s. par. in par. of Cathcart, Glasgow.

Merryston West, vil., 3 m. W. of Coatbridge, Lanarkshire. See MARYSTON, WEST.

Merryton, seat, 1 m. N. of Larkhall, Lanarkshire. Collieries in, the neighbourhood.

Merse, the name is variously applied to include the whole of Berwickshire, or the S. part of the co. only, or that part lying between the Lammermuirs and the Tweed, including the part of Roxburghshire N. of the Tweed. As an old political term, it designated all the country between the Lammermuirs and the Cheviots, and was called a March from its position between England and Scotland, and the title Earl of March is still held by the Earl of Wemyss. It is the largest and richest tract of agricultural land in Scotland.

Mersington House, 1 m. E.S.E. of Greenlaw, Berwickshire.

Mertoun, par. in S.W. Berwickshire; 6377 ac. Pop. 522. M. House, 2 m. N.E. of St. Boswells, is the seat of the Earl of Ellesmere.

Methil, q.s. par. and seapt. in par. of Wemyss; vil. 1 m. S.W. of Leven, 5½ m. E. of Thornton Junction sta., and 36 m. N.E. of Edinburgh by rail; ry. sta. L.N.E. P.O., T.O. One of chief coal-exporting ports on Firth of Forth. Red fixed light, visible 9 m., on pier-head. See also BUCKHAVEN, Pop. with Buckhaven 17, 643.

Methilhill, vil., 1 m. N.W. of Methil, Fife. P.O. Pop. 2559.

Methlick, par. and vil., N.E. Aberdeen-shire; par. 13,969 ac. Pop. 1481. Vil. on r. Ythan, 8 m. N.W. of Ellon. P.O., T.O.

Methven, par. and vil., Perthshire; par. 10,848 ac. Pop. 1670. Vil., 7¾ m. W. of Perth by rail; ry. sta., L.M.S. P.O., T.O. Pop. 643. Hotel. M. Castle, a seat, is 1 m. E. of the vil.

Mey, ham. in par. of Canisbay, 13 m. N.E. of Thurso, N.E. Caithness. P.O., T.O. 2 m. W. of the ham. is Loch of Mey. Off the coast is the dangerous reef, Men of M. (q.v.).

Meyrick, Kirkcudbrightshire. See MERRICK.

Mhaim nan Càrn, small loch, S. Perth-shire, 3 m. S. of Stronachlacher.

Mhic Phail or **Mhicfail,** sea loch, N. Uist, 9 m. S.W. of Renish Point, Outer Hebrides. It opens from Harris Sound, and is 1¼ m. wide at its mouth and 3 m. long. Separated at its inner point by a very narrow isthmus from Loch Maddy.

Mhorgay, small isl., par. of N. Uist, Outer Hebrides, 3 m. N.W. of Carinish.

Mhor, Loch, formed by union of Lochs Farraline and Garth, 3 m. E. of Foyers, Inverness-shire.

Miavaig, ham. on N. side of W. Loch Tarbert, par. of Harris, Outer Hebrides. P.O., T.O.

Midbea, ham., par. of Westray, 3 m. S. of Pierowall, Orkney.

Midbrake, seat, 2 m. from Cullavoe, Yell Isl., Shetland.

Mid-Calder, par. and vil., Midlothian; ry. sta., L.M.S. P.O., T.O., and T.O. at sta. See CALDER, MID-. Par. pop. 2793. Vil. pop. 587. Paraffin oil and other chemical products works, also large collieries.

Mid Clyth, ham., 4 m. N.E. of Lybster, Caithness; ry. sta., L.M.S. P.O., T.O. The Stack of Mid Clyth is close to the shore.

Middlebank, seat, 1 m. N. of Inver-keithing, Fife.

Middlebie, par. and vil. in S.E. Dum-friesshire; par. 17,542 ac. Pop. 1440. Vil. is 6 m. N.E. of Annan P.O.

Middlebridge, vil., 1¼ m. N. of Blair. Atholl, Perthshire.

Middle Church, eccl. par. in Greenock, Renfrewshire.

Middle Church, eccl. par. in Paisley, Renfrewshire.

Middledean, vil., 1¾ m. N.W. of Inver-keithing, Fife.

Middle Essie, ham., 6¾ m. N.N.W. of Peterhead, E. Aberdeenshire. P.O. called Mid Essie.

Middlefield Law, mt., alt. 1528 ft., 2 m. N.W. of Muirkirk, E. Ayrshire.

Middlefield, Upper, seat, 1 m. from Wood-side, in co. of, and 2 m. N.W. of, Aberdeen.

Middlefoodie, pl., 3 m. N.E. of Cupar, Fife.

Middleshaw, ham., 4½ m. S. of Lockerbie, near Water of Milk, Dumfriesshire.

Middleton, estate, now called Fettercairn (q.v.).

Middleton, pl., 1½ m. N. of Milnathort, Kinross-shire.

Middleton, ham. and seat, 1¼ m. S. of Borthwick, Midlothian.

Middleton, seat, ½ m. S. of Friockheim ry. sta., S.E. Angus.

Middleton and Neids, a shooting, 4 m. N. of Alyth, Angus.

Middleton Hall, seat at Uphall, W. Lothian.

Middleton House, 6 m. N.W. of Arbroath, Angus.

Midfield House, seat, near Hawthornden, Midlothian.

Midgard, seat, 3 m. N.E. of Hawick, Roxburghshire.

Midhope Tower, ancient castle on M. Burn, 4 m. W. of South Queensferry, W. Lothian. Was once the seat of Earls of Linlithgow. M. Burn flows 2 m. N.E. and N. into Firth of Forth.

Midlem, or **Midholm,** vil. in par. of Bow-den, N.W. Roxburghshire, 5 m. S.W. of St. Boswells. P.O. Pop. 99.

Midleys, gravel mound in co. of, and 2½ m. S.E. of, Inverness, surmounted by ancient Caledonian monument con-sisting of two concentric circles of large stones, a couple of large flat slabs inside the circles and one large upright stone outside.

Mid Locharwoods, farm of Scottish Labour Colony Association, in co. of, and 8 m. S.E. of, Dumfries.

Midlock Water, small stream in S.E. Lanarkshire; rises on borders of Peeblesshire and flows 5½ m. N.W. to the Clyde, opposite Crawford.

Midlothian, co. in southern division of Scotland, bounded on N. by Firth of Forth, W. by West Lothian, S. by cos. of Lanark, Peebles, Selkirk, and Roxburgh, and on E. by East Lothian and Berwickshire; area 234,325 ac., or 366 sq. m. The most prominent hills are the Pentlands in the centre of the co., the highest peak being Scald Law, 1898 ft; Moorfoot Hills are in the S.E., rising, in Blackhope Scar, to 2136 ft. The co. does not possess any large rivers, and most run to Firth of Forth. Co. well served by L.M.S. and L.N.E. rys. Agriculture well developed, and the principal mfrs. are paper, books, whisky, beer, printing ink, candles, soap, rubber. The two divisions of Midlothian and Peebles each return one member to Parliament. Pop. 526,296.

Midlothian and Peebles District Mental Hospital, near Roslynlee ry. sta., Midlothian.

Midmar, par. and vil. in S. Aberdeenshire; par. 10,561 ac. Pop. 767. Vil. 9 m. S.W. of Kintore. P.O., T.O. State forest in neighbourhood.

Midmar Castle, Aberdeenshire, 6 m. N. of Banchory.

Midstone, seat, 2 m. S.W. of Hamilton, Lanarkshire.

Midtown, pl., with bleachfield, 2½ m. S.W. of Johnstone, Renfrewshire.

Mid Yell, eccl. par. and ham. on E. side of Yell Isl., Shetland. Ham. stands at head of Mid Yell Voe. P.O., T.O.

Migdale, ham. at N.E. end of Loch M., and 3 m. N.E. of Ardgay, S.E. Sutherland. Loch M. (alt. 114 ft., depth 49 ft.).

Miglo, trib. of r. Eden, Fife.

Migvie, Aberdeenshire. See TARLAND-MIGVIE.

Milaid, lighthouse on E. side of Lewis, Outer Hebrides; quarter-minute flashing light, visible 11 m.

Milbuie, pl., Ross and Cromarty. See ARDMEANACH.

Milden Lodge, 8 m. N.W. of Edzell, N. Angus.

Mile Dorcha (*dark mile*), between Loch Arkaig and Loch Lochy, Invernessshire. Prince Charlie's tree on roadside.

Milesmark, vil., 1¾ m. N.W. of Dunfermline, Fife.

Milk, Water of, stream in S.E. Dumfriesshire; rises 7½ m. N.W. of Langholm and flows 18 m. S.S.W. to join the Annan, 4 m. S.W. of Ecclefechan.

Mill Bav, inlet on N.E. side of Stronsay Isl., N.E. Orkney, 17 m. N.E. of Kirkwall.

Millbrex, q.s. par. and vil. in N.E. Aberdeenshire, 6½ m. N.E. of Fyvie ry. sta. P.O.

Mill Buie, two heights, alts. 1100 ft. and 1216 ft., 2½ m. E. and S.W. respectively of Dallas, Morayshire.

Millburn, pl., 2½ m. N.N.W. of Strathy, N. Sutherland.

Millburnbank, ham., 8 m. S.E. of Dalbeattie, Kirkcudbrightshire.

Millburn Distillery, in town of Inverness.

Millburn House, Lanarkshire, 7 m. S.E. of Hamilton.

Millburn Tower, seat, 5 m. W. of Edinburgh, near Gogar ry. sta., Midlothian.

Milldewan Hill, 1650 ft., 3½ m. E.N.E. of Kirkton of Glenshee, Angus.

Milldoe, hill, alt. 734 ft., Mainland, Orkney.

Milldriggan, ham. in co. of, and 2 m. S.S.W. of, Wigtown.

Millearne House, seat, on r. Earn, 3 m. N. of Auchterarder, Perthshire.

Millerhill, vil. in Midlothian, 2 m. N.W. of Dalkeith and 6¼ m. S.E. of Edinburgh by rail; ry. sta., L.N.E.

Miller's Acre, pl. in par. of Forteviot and 6 m. S.W. of Perth. Edward Balliol encamped here before the battle of Dupplin (1332).

Millerston and Stepps, vil. on N. side of Hogganfield Loch, N.W. Lanarkshire; partly in the bur. and 3 m. N.E. of Glasgow. P.O., T.O.

Milleur Point, headland on W. side of entrance to Loch Ryan, 8½ m. N.W. of Stranraer, Wigtownshire.

Millfield, seat, ¼ m. N. of Polmont Junction sta., Stirlingshire.

Millfield Paper Mills, Leven, Fife.

Millfire, mt., alt. 2350 ft., 5 m. E. of Merrick, W. Kirkcudbrightshire.

Millfore, mt., alt. 2146 ft., 7 m. N.E. of Newton-Stewart, W. Kirkcudbrightshire.

Millheugh, vil., ½ m. S.W. of Larkhall, Lanarkshire.

Millhill, seat in par. of Longforgan, Perthshire.

Millhills, seat near Crieff, Perthshire.

Millholm Paper Mills, Cathcart, Glasgow.

Millhouse, ham., 3 m. S.W. of Tighnabruaich, S. Argyllshire. P.O., T.O.

Millhouse, pl., par. of Applegarth, 3 m. N.E. of Lochmaben, mid Dumfriesshire. P.O., T.O. called Millhousebridge.

Millifiach, ham., 3 m. S.E. of Beauly, N. Inverness-shire.

Milliken Park, vil., 1½ m. S.W. of Johnstone, Renfrewshire ; ry. sta., L.M.S. P.O., T.O. M. House is 1 m. N. of sta.

Millisle, ry. sta., L.M.S., 1 m. W. of Garliestown, S.E. Wigtownshire.

Millknowe Burn, trib. of Blackadder, 3 m. N.W. of Greenlaw, Berwickshire.

Mill Loch, in par. of, and 1 m. N.W. of, Lochmaben, Dumfriesshire ; depth 55 ft., alt. 172 ft.

Mill Maud, hill, alt. 1250 ft., par. of Lumphanan, 6 m. S. of Alford, S. Aberdeenshire.

Millmore Beacon, a six-seconds flashing light, visible 9 m., Campbeltown, Kintyre, Argyllshire.

Mill of Conveth, estate in par. of Laurencekirk, Kincardineshire.

Mill of Haldane, see HALDANE'S MILL.

Millport, holiday resort and police bur. on bay of same name on S. coast of the Great Cumbrae Isl., Firth of Clyde, Buteshire, 11 m. S.W. of Wemyss Bay. P.O., T.O. Pop. 2083. At Keppel Pier is a marine biological sta. Has two fixed red lights, visible 5 m.

Millseat, ham., 6 m. N.E. of Turriff, Aberdeenshire.

Milltimber, pl. on N. bank of the Dee, in co. of, and 6½ m. S.W. of, Aberdeen ; ry. sta., L.N.E. P.O. ; T.O. at sta.

Millton, vil., 2 m. S. of Glamis, Angus.

Millton, vil. in Glen Urquhart, Invernessshire, 1 m. W. of Drumnadrochit sta.

Milltown, vil. in Banffshire. See ROTHIEMAY.

Milltown, ham., 2 m. N.E. of Elgin, Moray.

Milltown, Jura. See CRAIGHOUSE.

Milltown, pl., 5 m. N.E. of Kirkpatrick, S.E. Dumfriesshire.

Milltown of Cushnie, pl., 8 m. S.W. of Alford, mid Aberdeenshire.

Milnathort, mkt.-town on N. Queich Water and near Loch Leven, N. Kinross-shire, 1⅛ m. N.E. of Kinross Junction and 31½ m. N. of Edinburgh by rail ; ry. sta., L.N.E. P.O., T.O. Has important live stock sales. Pop. 1153. Hotel.

Milne Graden, seat, on the Tweed, 3½ m. N.E. of Coldstream, Berwickshire.

Milngavie (pronounced **"Milguv"**), q.s. par. and town (police bur.), S.E. Dunbartonshire, 7 m. N.W. of Glasgow. Pop. 5056. Ry. sta., L.N.E. P.O. T.O. Has calico-printing, dyeing, and bleaching works.

Milnhead, seat in co. of, and 4 m. W. of, Dumfries.

Milnholm Cross, pl., with ancient cross, near Newcastleton, Roxburghshire.

Milntown, or **Milton of New Tarbat,** vil., 5 m. S. of Tain and near Kildary ry. sta., E. Ross and Cromarty.

Milntown (Milton), ham., ½ m. S. of Applecross, W. Ross and Cromarty.

Milnwood, part of vil. of Clydesdale, adjoining Mossend, Lanarkshire.

Milrig, seat, 2 m. S. of Galston, in par. of Riccarton, Ayrshire.

Milsey Bay, on E. side of North Berwick, E. Lothian.

Milton, ham., 2½ m. S. of Cullen, N. Banffshire.

Milton, ham. and seat, 1 m. N. of Dalmuir ry. sta. and close to Duntocher, Dunbartonshire.

Milton, vil. in co. of, and 2½ m. S.E. of, Dumbarton. M. House is a seat.

Milton, ham., 6 m. N.W. of Arbroath, Angus.

Milton, fishing vil., 2 m. S.W. of Johnshaven, Kincardineshire.

Milton, ham., near a loch of the same name, 7 m. N. of Dalbeattie, E. Kirkcudbrightshire.

Milton, q.s. par. in Glasgow.

Milton, ham., 3 m. N.E. of Crieff, Perthshire.

Milton, ham., 6 m. S. of Dingwall, Ross and Cromarty.

Milton, vil. on the Glazert Water, 1½ m. W. of Kirkintilloch, Stirlingshire ; ry. sta. (M. of Campsie), L.N.E. P.O., T.O. called M. of Campsie. Printing works in the neighbourhood. Pop. 829.

Milton, ham., 1½ m. S.W. of Dunscore, Dumfriesshire. P.O.

Milton, ham., 1 m. S. of Kilmacolm, Renfrewshire.

Milton, seat, 1 m. N. of Straiton, Ayrshire.

Milton, ham., 1 m. W. of Drumnadrochit, N.W. Inverness-shire.

Milton, ham., 2½ m. S.E. of Glenluce, Wigtownshire.

Milton Bridge, ham., 8 m. S. of Edinburgh and 1 m. N.W. of Auchendinny ry. sta., Midlothian. P.O., T.O.

Milton Brodie House, 5 m. N.E. of Forres and 2 m. N.E. of Kinloss ry. sta., N. Morayshire.

Miltonduff, pl., with distillery, 3 m. S.W. of Elgin, Morayshire. P.O., T.O.

Milton Eonan, ham. on r. Lyon, 9 m. N. of Killin, Perthshire.

Milton House, seat, 5½ m. S.W. of Haddington, E. Lothian.

Miltonise, seat, 2 m. N.E. of Glenwhilly, Wigtownshire.

Milton Keith Distillery, Keith, Banffshire.

Milton Loch, ¼ m. S.E. of Crocketford, Kirkcudbrightshire ; depth 15 ft., alt. 410 ft.

Milton Lockhart, seat, 2 m. S.W. of Carluke, Lanarkshire.

Milton Ness, prom., 1½ m. E. of St. Cyrus, Kincardineshire.

Milton of Balgonie, P.O., T.O. See BALGONIE, MILTON OF. Pop. 517.

Milton of Campsie, ry. sta., L.N.E. P.O., T.O. See MILTON.

Milton of Clova, Angus. See CLOVA.

Milton Park, mansion, 2 m. S.E. of Halkirk, Caithness.

Milton Park, Kincardineshire, 1 m. N.N.W. of Dalry, a seat of Lord Sinclair.

Milton, Whins of, vil. in co. of, and 2½ m. S. of, Stirling. Here are remains of Beaton's Mill, where James III. was murdered (1488). Pop. 477.

Minard, ham. and seat (M. Castle) on W. shore of Loch Fyne and 8 m. S.E. of Lochgilphead, Argyllshire. P.O., T.O.

Minchmoor, mt., alt. 1856 ft., 3 m. S.E. of Innerleithen, cos. of Peebles and Selkirk.

Minch, The, channel between the mainland of Ross and Cromarty and Sutherlandshire on the E. and the north part of the Outer Hebrides on the W.; varies from 24 m. to 45 m. wide.

Minch, The Little, channel between the Inner and Outer Hebrides, dividing Skye. on the E., from Benbecula, N. Uist and part of Harris on the W.; varies from 14 m. to 20 m. in width.

Mindork Castle, ruin, in par. of Kirkcowan, Wigtownshire.

Mingary, sea loch, 4½ m. W. of Tobermory, N.W. of Mull, Argyllshire.

Mingary Castle, ancient stronghold of the MacIans, on the N. side of the entrance of Loch Sunart, Ardnamurchan, and 6 m. N. of Tobermory, Argyllshire.

Minginish, dist. in the S.W. of the Isle of of Skye, Inner Hebrides.

Mingulay, small isl., 2½ m. by 1¾ m., 10 m. S. of Barra Isl., Outer Hebrides. Pop. 3.

Minishant, ham. in par. of, and 3½ m. N.E. of, Maybole, Ayrshire. P.O.

Minnick, Water of, stream, in S. Ayrshire and N.W. Kirkcudbrightshire ; rises in Ayrshire and flows 15 m. S. to the Cree, 6½ m. above Newton-Stewart.

Minnigaff, par. and vil. in W. Kirkcudbrightshire ; par. 87,916 ac. Pop. 1144. The vil. is on the r. Cree, nearly opposite Newton-Stewart. Pop. (M. and Creebridge) 405.

Minnoch, Loch, 8 m. N.W. of New Galloway, Kirkcudbrightshire.

Minnygap Height, alt. 1309 ft., 5 m. W.N.W. Wamphray sta.; Dumfriesshire.

Minnyshant, ham., 1 m. N. of Cassillis sta., Ayrshire.

Mintlaw, vil., 9 m. W. of Peterhead, Aberdeenshire ; ry. sta., L.N.E. P.O. ; P.O. and T.O. at M. sta. Pop. 395.

Minto, par. and vil. on r. Teviot, N. Roxburghshire ; par. 5601 ac. Pop. 379. Vil. is 6 m. N.E. of Hawick and 1½ m. E. of Hassendean ry sta. P.O. M. House is the seat of the Earl of M. M. Hills, 905 ft. and 836 ft., overlook the Teviot, and M. Crags are 729 ft.

Misery, Mount, 576 ft., 2 m. N.E. of Balloch, Dunbartonshire.

Mishnish, shooting-lodge, near Tobermory, Mull, Argyllshire.

Mishnish Lochs, chain of three lochs, 2 m. S.W. of Tobermory, Mull, Argyllshire.

Misty Glen, ravine near the head of Glen Lochay, W. Perthshire.

Misty Law, alt. 1663 ft., in par. of, and 6 m. N. of, Lochwinnoch, W. Renfrewshire.

16

Mither Tap, alt. 1698 ft., highest peak of the Bennachie Hills, mid Aberdeenshire.

Miulie, Loch, 16 m. W.S.W. of Beauly, N.W. Inverness-shire. See LOCH A MHUILINN.

Moan, Loch, source of r. Cree, 8 m. N.E. of Barrhill sta., N.W. Kirkcudbright-shire; alt. 674 ft.

Mochrum, par. and vil. on E. shore of Luce Bay, Wigtownshire; par. 24,101 ac. Pop. 1477. Vil. (Kirk of M.) 4½ m. S.W. of Whauphill ry. sta. P.O., T.O. at Elrig. Par. contains M. Loch and M. Fell (646 ft.). The Old Place of M. is a seat of the Marquess of Bute. Loch M. is 12 ft. deep, alt. 247·7 ft.

Mochrum, Loch, par. of Kirkoswald, 1½ m. S.W. of Maybole, Ayrshire.

Moffat, police bur., par., and holiday pl. and seat, on r. Annan, Dumfriesshire, 2 m. N.E. of Beattock sta. and 63¼ m. S.W. of Edinburgh by rail; ry. sta. (335 ft.), L.M.S. Hotels. P.O., T.O. Par. 43,008 ac. Pop. 2522; bur. pop. 2006. M. House and M. Well are nearby.

Moffat, Easter and **Wester,** adjoining Airdrie, Lanarkshire. Town hospital at W. Moffat.

Moffat Mills, vil. in par. of New Monkland, Lanarkshire, 1 m. S.E. of Clarkston ry. sta. Paper is manufactured here.

Moffat Water, burn in N. Dumfriesshire; rises near N. border of Selkirkshire and enters the Annan 2 m. below Moffat.

Moidart, coastal dist. in par. of Ardna-murchan, S.W. Inverness-shire. Separated by Loch Shiel from Argyllshire and indented by the sea lochs Ailort and M.; measures 18 m. long by 7 m. broad. P.O., T.O. (Kinlochmoidart).

Moidart, short stream, 8 m. long, flowing S.W. into Loch M.

Moin, The, extensive tract of moorland between Loch Hope and Kyle of Tongue, N. Sutherland.

Molinburn, vil., 4 m. N.W. of Coatbridge, N.W. Lanarkshire.

Moll, pl., with ruins of a castle, 10 m. S.S.E. of Portree, Skye, Inner Hebrides.

Mollance, seat, 2½ m. N.E. of Castle-Douglas, Kirkcudbrightshire.

Moll's Cleuch Dod, hill, 2571 ft., S. Peeblesshire, 8½ m. N.E. of Moffat.

Molmont Hill, alt. 618 ft., par. of Galston, 7 m. E.S.E. of Kilmarnock, Ayrshire.

Monach Islands, 6 m. S.S.W. of N. Uist, Outer Hebrides; on Shillay is a light-house with flashing light, visible 18 m. Pop. 3.

Monachyle Hostel, near Lochs Voil and Doine, Perthshire, 4 m. W. of Balquhidder. A Scottish Youth Hos-tel.

Monadhliath Mountains, highland area in Inverness-shire, extending N.E. along the W. side of Upper Strathspey; highest pt., Carn Mairg, is 3631 ft., 10 m. W. of Kingussie.

Monadh Mor (Mona More), mt., alt. 3631 ft., 2½ m. S.W. of Cairn Toul, on borders of Inverness-shire.

Monaltrie House, 1 m. N.E. of Ballater, Aberdeenshire.

Monamenach, mt., alt. 2649 ft., on border of Angus and Perthshire, 16 m. N. of Blairgowrie.

Monamore Glen, on E. side of Arran, near Lamlash, Buteshire.

Monar Deer Forest, 20,000 ac., on W. side of Loch M., Ross and Cromarty. Post-town, Beauly.

Monar Loch, on borders of Inverness-shire and Ross and Cromarty, 25 m. S.W. of Beauly, at head of Glen Strathfarrar; 4 m. long E. to W. and about ½ m. wide. M. Lodge is near E. end of the loch; deepest part of loch 260 ft., mean depth 98¼ ft., alt. 664 ft.

Monaughty, an extensive State forest, 5 m. E. of Forres, Morayshire.

Monboddo House, seat, 2 m. N.W. of Fordoun ry. sta., Kincardineshire.

Moncrieffe House, Perthshire, at base of M. Hill (725 ft.), 3 m. S.E. of Perth. M. Hill is on the peninsula between the Earn and the Tay.

Moncur Castle, ruin, ½ m. N.E. of Inchture, S. Perthshire.

Mondynes, pl., with ancient monolith, 1½ m. S.W. of Drumlithie, Kincardineshire.

Monega Hill, alt. 2917 ft., 9 m. N. of Glenisla, N.W. Angus.

Moness Falls, three falls on the M. Burn, which joins the Tay near Aberfeldy, Perthshire. They are described in Burns's poem "The Birks o' Aberfeldy." M. House is close to Aberfeldy.

Moneydie, par. and seat in E. Perthshire; par. 4562 ac. Pop. 248. Par. church is 2 m. W. of Luncarty ry. sta.

Moneypool Burn, stream, 7 m. long, flowing S.W. to Wigtown Bay at Creetown, W. Kirkcudbrightshire.

Mongarrie, special water dist., Tullynessie and Forbes pars., Aberdeenshire. Pop. 143.

Mongour, hill, alt. 1232 ft., 7 m. S.E. of Banchory, Kincardineshire.

Moniack, seat, 4 m. S.E. of Beauly, Inverness-shire.

Moniaive, vil. at junc. of Dalwhat and Craigdarroch Waters, 7 m. S.W. of Thornhill, W. Dumfriesshire; ry. sta., L.M.S. P.O., T.O. Pop. 527.

Monifieth, par. and town (police bur.) on N. shore of Firth of Tay, Angus; par. (containing part of Broughty Ferry) 5171 ac. Pop. 3921. Town, 2 m. N.E. of Broughty Ferry and 6 m. N.E. of Dundee. Pop. 2986. Ry. stas., L.N.E. and L.M.S. P.O., T.O. Golf course. Spinning and weaving machinery.

Monikie, par., ham., and loch in S. Angus. P.O., T.O. Par. 8869 ac. Pop. 1059. Ham. and ry. sta., L.M.S., 8 m. S.W. of Arbroath and 11½ m. N.E. of Dundee. Here are four large reservoirs for Dundee water supply. M. Burn reaches the sea at Carnoustie.

Monimail, par. and vil. in N. Fife; par. 6548 ac. Pop. 687. Vil. is 1 m. N.E. of Collessie ry. sta. and 5½ m. W. of Cupar.

Monkland, ancient barony in Middle Ward of Lanarkshire. It nearly all belonged to the monks of Newbattle, and was divided into the pars. of Old and New M.

Monkland Canal, N. Lanarkshire; extends between the Forth and Clyde Canal at Port Dundas, in N. of Glasgow, and the N. Calder; sends off four branches to works in the neighbourhood.

Monkland, New, par. and vil. in N.E. Lanarkshire; par. (containing Airdrie) 19,814 ac. Pop. 38,499. Vil. on the N. Calder, 1½ m. N.W. of Airdrie.

Monkland, Old, par. in N. Lanarkshire on the N. Calder; 11,006 ac. Pop. 62,824. Contains Coatbridge. Old M. church is 1½ m. S.W. of Coatbridge.

Monklaw, pl., with Roman camp, 1 m. N. of Jedburgh, Roxburghshire.

Monk Myre, loch, par. of Bendochy; depth 12 ft., alt. about 150 ft.; 2 m. S.E. of Blairgowrie, Perthshire.

Monkrigg, seat, 1 m. S.E. of Haddington, E. Lothian.

Monks Burn, trib. to N. Esk, which it reaches 4 m. S.W. of Penicuik, Midlothian.

Monkstadt, ancient mansion, 2½ m. N.W. of Uig, Skye, Inner Hebrides. Was the seat of the Macdonalds and thither Prince Charles Edward Stuart, in the disguise of a maid-servant, was conducted by Flora Macdonald in 1746.

Monkston, part of Ladybank (q.v.).

Monks Water, flows into Douglas Water, near Inches ry. sta., Lanarkshire.

Monkton, vil. in co. of, and 4 m. N. of, Ayr; ry. sta., L.M.S. P.O., T.O. Pop. 653.

Monkton and Prestwick, par. on Firth of Clyde, W. Ayrshire; 3786 ac. Pop. 9427. Contains St. Nicholas, Prestwick, and part of town of Ayr; Monkton vil.

Monktonhall, ham., 1 m. S.W. of Musselburgh, Midlothian.

Monluth Hill, alt. 1231 ft., 6 m. N. of Fordoun, Kincardineshire.

Monquhitter (Hill of the Mist), par. and vil. in N. Aberdeenshire; par. 17,436 ac. Pop. 1804. Contains Cuminestone, 6 m. E. of Turriff.

Monreith, ham., 5½ m. W. of Whithorn, on M. Bay, Wigtownshire. M. House is 1 m. to the N. P.O., T.O.

Mons Hill, in Dalmeny Park and near the Firth of Forth, 1 m. N.E. of Dalmeny ry. sta., W. Lothian; alt. 387 ft.

Monteoffer House, seat of Duchess of Fife, in co. of, and 2 m. S. of, Banff.

Montfode Castle, ruined stronghold, 1½ m. N.W. of Ardrossan, Ayrshire.

Montgarrie, ham., Aberdeenshire, 1¼ m. of Alford. P.O.

Montgarswood, ham., 2 m. E. of Mauchline, Ayrshire.

Montgomerie (formerly called Coilsfield), seat, 1 m. N.E. of Tarbolton ry. sta., Ayrshire. Celebrated in Burns's "Highland Mary."

Montgreenan, ry. sta., L.M.S., and seat, 3 m. N.E. of Kilwinning, Ayrshire. P.O.; T.O. at sta.

Montrave, seat, 4 m. N. of Leven, Fife.

Montreathmont Moor, 5 m. S. of Brechin, Angus. Partly a State forest.

Montrose, par., parl., royal, and mun. bur. and port in N.E. Angus, at mouth of r. Esk; par. 4232 ac. Pop. 11,889. The town is 10 m. E. of Brechin, 30¾ m. N.E. of Dundee, and 90 m. N.E. of Edinburgh by rail; ry. stas., L.M.S. and L.N.E. P.O., T.O. Pop. 10,196. Hotels. M. Basin is a tidal bay, crossed by ry. and road bridges, to the W. of the town. Coastguard and lifeboat sta. Sawmills, linen yarn manufactories, and fruit canning. M. Burs. return one member to Parliament.

Monybuie, ham. and seat, 8 m. S.W. of Moniaive, Kirkcudbrightshire.

Monymusk, par. and vil. in mid Aberdeenshire; par. 10,727 ac. Pop. 855. Vil. on r. Don, 8 m. W. of Kintore and 21 m. N.W. of Aberdeen; ry. sta., L.N.E. P.O., T.O. Vil. pop. 114. 1 m. N. of the sta. is the seat, M. House.

Monynut Water, stream, E. Lothian and Berwickshire; rises among Lammermuirs, flows 8 m. S.E. to join Whiteadder at Abbey St. Bathans. A ridge, along the left side of the valley, is M. Edge; highest point, Heart Law, 1283 ft.

Monzie (pronounced "Mon-ee"), eccl. par. and vil. in Perthshire. Par. in Crieff and Fowlis Wester pars. Vil. 3 m. N. by E. of Crieff. Near the vil. is M. Castle.

Monzie Law, mt., alt. 1812 ft., borders of Roxburghshire and Northumberland, 1½ m. W. of Windygate Hill.

Monzievaird and Strowan, par. in mid Perthshire, on r. Earn; par. 31,718 ac. Pop. 525. Par. church 3½ m. N.W. of Crieff. Close by is M. Loch; depth 39 ft., alt. about 200 ft.

Moodiesburn, vil. in par. of Cadder, 2 m. N.E. of Garnkirk ry. sta., Lanarkshire, and 7 m. N.E. of Glasgow.

Moodlaw Loch, small loch in N.E. of Eskdalemuir par., at meeting-point of cos. of Roxburgh, Dumfries, and Selkirk.

Moonen Bay, inlet on W. coast of Skye, 26 m. W. of Portree, Inner Hebrides.

Moonzie, par. in N. Fife; 1258 ac. Pop. 120. The par. church is 3 m. N.W. of Cupar.

Moorbeck, ham., 2½ m. S.W. of Annan, Dumfriesshire.

Moorbrock Hill, alt. 2136 ft., 4½ m. N.E. of Carsphairn, Kirkcudbrightshire.

Moorfoot Hills, hill range, chiefly in S.E. Midlothian, but partly on border of Midlothian and Peeblesshire; extends S.W. from valley of Gala Water in two masses, which, gradually running farther apart, form a triangle about 10 m. long, with an average breadth of 6 m.; summits mostly rounded and pasture covered. Blackhope Scar is the highest point (2136 ft.).

Moorhouse, farm, 3 m. W. of Eaglesham, Renfrewshire.

Moorpark, pl. in co. of, and 1 m. S.W. of, Renfrew.

Moorpark House, ½ m. N. of Kilburnie, N. Ayrshire.

Moors, The, general name for the greater part of N. Wigtownshire, comprising the whole of the co. which is not included in the Rhinns and the Machars.

Moracha, loch, 1½ m. N. of Obbe, Harris, Outer Hebrides.

Morag, glen, on S. side of Dunoon, Firth of Clyde, Argyllshire.

Moraig, loch, 3 m. N.E. of Blair-Atholl, N. Perthshire; depth 14 ft., alt. 1105 ft.

Moral Fall, waterfall in the Enrick rivulet, par. of Urquhart and Glenmoriston, N. Inverness-shire.

Morar, dist. on W. of Inverness-shire; on the N. is Loch Levis, W., Sound of Sleat, S., Sound of Arisaig, and to the S.E. is Loch Eil; it is 4 to 9 m. broad and 19 m. long; ry. sta., L.N.E., is 2¼ m. S. of Mallaig. P.O., T.O. Loch M. divides the dist. in N. and S. sections; the loch is 12 m. long, from ½ to 2 m. broad, and is the deepest loch in Scotland (1017 ft.); alt. 30 ft. M. Lodge is a seat.

Moray Firth, arm of the North Sea in the N.E. of Scotland. In its wider significance it forms a triangular area, with its apex at the mouth of the Beauly and the extremities of the base at Duncansbay Head, Caithness, and Kinnaird's Head, Aberdeenshire, a distance of 78 m. In a restricted sense, the base runs from Tarbert Ness to Lossiemouth (Moray), a distance of 21 m.; on the N.W. side there are many inlets, the principal being Dornoch Firth and Cromarty Firth. Fish are extremely plentiful in the M. Firth, and along its shores there are six fishery dists.—Banff, Buckie, Findhorn, Cromarty, Helmsdale, and Lybster.

Morayshire (formerly Elginshire), one of the northern cos. of Scotland ; bounded on N.E. by North Sea, on N.W. by Moray Firth, E. and S.E. by Banffshire, on S. and S.W. by Inverness-shire, and on W. by Nairnshire ; area 476·7 sq. m., or 304,931 ac. Pop. 40,805. Upper part of co. is mountainous and well wooded ; highest peak, Carn Eachie (2328 ft.) ; lower part rich, except the Culbin sanddunes. Chief industries : rearing and breeding of cattle, brewing, distilling, and fishing. Chief town, Elgin. With Nairnshire, the co. returns one member to Parliament.

Mordington, par. and ham. on the coast of N.E. Berwickshire ; par. 2940 ac. Pop. 319. Ham. on the Whiteadder and 4 m. W.N.W. of Berwick-on-Tweed. M. House is a seat in the neighbourhood.

Morebattle, par. and vil. in N.E. Roxburghshire ; par. 22,429 ac. Pop. 719. Vil. is on the Kale Water, 7 m. S. by E. of Kelso. P.O., T.O. Pop. 195.

Moredun, Midlothian, near Gilmerton. Animal Diseases Research Association Institute.

Moredun Hill, Perthshire. See MONCREIFFE HILL.

More Head, W. end of Gamrie Bay, N.E. Banffshire.

Moreland, seat in Kinross-shire, 3 m. S.E. of Rumbling Bridge.

More, Loch, 5 m. E. of Altnabreac ry. sta., Caithness ; depth 7 ft., alt. 381 ft.

More, Loch, 6 m. S.E. of Laxford Bridge, Eddrachillis par., W. Sutherland ; depth 316 ft., alt. 127 ft.

Morenish, seat and landing-place on Loch Tay, 2 m. E. of Killin, Perthshire.

Morgay, isl., Inverness-shire, see MHORGAY.

Mornam, par., 3 m. S.E. of Haddington, E. Lothian ; 2086 ac. Pop. 206. Here is a boys' orphanage.

Morie, Loch, or **Mary's Loch,** 9 m. N.W. of Alness, Ross and Cromarty ; source of r. Alness ; depth 270 ft., alt. 622 ft.

Morinsh, pl., Banffshire, 5 m. S.E. of Ballindalloch.

Mor Island, Argyllshire. See EILEAN MORE.

Moriston, r. in N. Inverness-shire ; issues from Loch Clunie and runs 19 m. N.E. along Glen M., entering Loch Ness at Invermoriston, 6 m. N.E. of Fort Augustus.

Morlich Loch, loch, 1 m. by ½ m., 4 m. E. by S. of Aviemore ry. sta., N.E. Inverness-shire ; depth 49 ft., alt. 1045 ft.

Mormond, ry. sta. (300 ft. high), L.N.E., 8 m. S. by W. of Fraserburgh, N.E. Aberdeenshire. T.O. M. Hill, with two summits, 769 ft. and 749 ft. high, is in vicinity. M. House is a seat, 1 m. N.W. of Lonmay ry. sta.

Morningside, q.s. par. and dist. in S. side of city of Edinburgh ; ry. sta., L.N.E., suburban line. P.O., T.O. In the par., and near the church, "The Bore Stane," in which the Scottish Standard was fixed at the gathering before Flodden, is fixed into the wall. See BOROUGH MUIR.

Morningside, vil., 3 m. E. of Wishaw and 5½ m. by rail S.E. of Holytown, N. Lanarkshire ; ry. sta. (511 ft.), L.M.S. (passenger service withdrawn). T.O. at sta. Situated in a mining dist.

Morningside, pl., 2 m. N.E. of Annan, Dumfriesshire.

Mornish, pl., 7 m. W.S.W. of Tobermory, Isl. of Mull, Argyllshire.

Morphie, seat, 4 m. N. of Montrose, Angus. Hill of M., 486 ft., is nearby.

Morrinton, ham., in co. of, and 7 m. W.N.W. of, Dumfries.

Morrishill, seat, near Beith ry. sta., N. Ayrshire.

Morrison's Haven, harbour, 1 m. S.W. of Prestonpans, Firth of Forth, E. Lothian. Nearby are collieries and a brick and tile work.

Morriston, r. in Inverness-shire, see MORISTON.

Morrone Hill, alt. 2819 ft., 2 m. S.W. of Braemar, Aberdeenshire.

Morsgail Deer Forest, 24,000 ac., 24 m. S.W. of Stornoway, Lewis, Outer Hebrides.

Mortlach, par. in mid Banffshire ; 34,222 ac. Pop. 2548. Contains Dufftown, at which is M. Distillery.

Mortlich, hill, alt. 1248 ft., 2 m. N. of Aboyne, S. Aberdeenshire.

Morton, par. in upper part of r. Nith, N.W. Dumfriesshire ; 8047 ac. Pop. 1661. Contains Thornhill, 2½ m. N.E. of which is M. Castle.

Morton Hall, seat and golf course in city, and 3¼ m. S. of G.P.O., Edinburgh.

Morton Hill, alt. 1546 ft., 3½ m. S.W. of Moniaive, Dumfriesshire.

Morusig, mt., alt. 3026 ft., 6 m. S.W. of Achnasheen, Ross and Cromarty.

Morven, mt., alt. 2862 ft., 5 m. N. of Ballater, S.W. Aberdeenshire. M. Lodge lies 2½ m. W.S.W. of the hill.

Morven, mt., alt. 2313 ft., 8 m. N.W. of Berriedale, S. Caithness.

Morven, or **Morvern,** par. in N.W. Argyllshire ; occupies the larger part of a peninsula of triangular form, bounded on the N. by Loch Sunart, on the S.W. by the Sound of Mull, and on the S.E. by Loch Linnhe ; includes also the isls. of Oronsay and Carna ; par. 90,736 ac. Pop. 459. P.O., T.O.

Morvich House, 5½ m. W. of Golspie, S.E. Sutherland.

Moscow, pl., 3 m. S.E. of Fenwick, N.E. Ayrshire.

Mosgail, or **Caen Loch,** small stream, flowing into Caen Loch, Lewis, Outer Hebrides.

Mossat, ham., on M. Burn, 7 m. N.W. of Alford, Aberdeenshire. P.O. See INVERMOSSAT.

Mossband Colliery, 1 m. E.S.E. of Newhouse ry. sta., Lanarkshire.

Mossbank, industrial school, near Millerston, in N.E. of Glasgow, ½ m. S. of Robroyston ry. sta.

Mossbank, vil. in par. of Delting, 29 m. N. by W. of Lerwick, Mainland, Shetland. P.O., T.O.

Mossblown, drainage dist. in Tarbolton par., Ayrshire. Pop. 1674.

Mossburnford, pl., on rt. bank of Jed Water, 3 m. S.S.E. of Jedburgh, Roxburghshire. Here is an ancient fortress which was occupied down to about 1750.

Mossdale, pl. and loch, 4 m. S. of New Galloway, Kirkcudbrightshire. P.O.

Mossend, vil., ½ m. N. of W. Calder, Midlothian.

Mossend, town in par. of Bothwell and 2 m. N.W. of Motherwell, Lanarkshire ; ry. sta. (265 ft.), L.M.S. P.O. Extensive iron and steel works. Pop. of M. and Bellshill, 18,094.

Mossfennan House, 3 m. S. of Broughton ry. sta., Peeblesshire.

Mossgiel, farm, 1 m. N. by W. of Mauchline, mid Ayrshire. Was occupied by Robert Burns from 1784 to 1788.

Mossgreen, q.s. par. and vil., 3 m. N.E. of Dunfermline, Fife.

Mosshope Fell, mt., alt. 1567 ft., 6 m. N.W. of Beattock, N. Dumfriesshire.

Moss House, seat on Blane Water, S.W. Stirlingshire, 1¼ m. S.E. of Killearn ry. sta.

Mossknowe, seat, 1 m. S.S.E. of Kirkpatrick Fleming, S.E. Dumfriesshire.

Mosspark, ry. sta. (M. West), L.M.S., S.W. Glasgow. P.O., T.O.

Mosspaul, pl., with hotel, 12½ m. S.W. of Hawick, on borders of cos. of Roxburgh and Dumfries. Here is the highest point (800 ft.) on the road from Hawick to Langholm.

Moss Road Halt, ry. sta., Machrihanish light ry., Kintyre, Argyllshire, now closed.

Moss-side, vil., 2 m. S.W. of Nairn. Pop. M. and Tradespark, 365.

Mosstodloch, vil. in Morayshire, and 1½ m. N.W. of Fochabers. P.O. Pop. 88.

Mosstower, seat, in par. of Eckford, Roxburghshire.

Mosstowie, ry. sta., L.M.S., 3½ m. W. of Elgin, Morayshire.

Mossy Paper Mills, on Water of Leith, Colinton, Edinburgh.

Mote of Urr, Kirkcudbrightshire. See URR.

Motherwell (bur. of M. and Wishaw), town and police bur. in Middle Ward of Lanarkshire ; ry. sta., L.M.S. P.O., T.O. Pop. of bur. 64,708. Has large iron, steel, and engineering works, also collieries.

Motherwell Division, parl. div. of Lanarkshire. Pop. 64,212.

Motray Water, stream in N.E. Fife ; rises 5 m. N.N.W. of Cupar, and has a winding course of 12 m. E. and S. to the Eden at Guardbridge.

Moulin, par. and vil. in N.E. Perthshire ; par. 44,862 ac. Pop. 2775. In the S.W. of the par. the Garry flows through Killiecrankie Pass and joins the Tummel. Vil. 1 m. N. of Pitlochry. Hotel. P.O. "Howe of M." is a fertile tract.

Moulinalmond, seat, Almondbank, Methven, Perthshire.

Mound, The, ry. junc., L.M.S., 3½ m. S.W. of Golspie, Sutherland. P.O., T.O. at the M. sta.

Mounie Castle, 2½ m. N.W. of Old Meldrum, Aberdeenshire.

Mount, hill, alt. 1385 ft., 4½ m. N.E. of Biggar, Peeblesshire.

Mountaincross, ham., 3½ m. S. of West Linton, Peeblesshire.

Mount Annan, seat, 2 m. N. of Annan, Dumfriesshire.

Mount Benger, farm in co. of, and 12 m. S.W. of, Selkirk. Occupied for some time by James Hogg, "The Ettrick Shepherd."

Mountblair Lodge, 5 m. N.E. of Kirkmichael, Glen Shee, Perthshire.

Mountblairy, seat, in Banffshire, 4 m. N.W. of Turriff.

Mount Blow, Dunbartonshire, 1 m. N. of Clydebank. Child Welfare Home.

Mount Cameron, seat, 1 m. E. of East Kilbride, Lanarkshire.

Mount Charles, seat, in co. of, and 2½ m. S. of, Ayr.

Mountfair, seat, 5 m. N.N.E. of Coldstream, Berwickshire.

Mount Florida, southern sub. of Glasgow; ry. sta., L.M.S. P.O., T.O.

Mountgarrie, vil., 1 m. N. of Alford, Aberdeenshire.

Mountgerald House, seat, 3 m. N.E. of Dingwall, Ross and Cromarty.

Mountherrick Hill, alt. 1400 ft., 4 m. W. of Abington, S. Lanarkshire.

Mount Hill, Fife. See Mount, The.

Mount Kedar, pl. in co. of, and 6½ m. S.E. of, Dumfries.

Mount Lothian, ham., 2½ m. S.E. of Pomathorn sta., S. Midlothian.

Mount Melville, seat and ry. sta., L.N.E., 1½ m. S.W. of St. Andrews, Fife.

Mount Oliphant, farm in co. of, and 2½ m. S.E. of, Ayr. Associations with boyhood of Robert Burns.

Mount Pleasant, part of town of Newburgh, Fife.

Mountquhanie House, seat, 5 m. N. of Cupar, Fife.

Mount Stuart, seat of Marquess of Bute, 5 m. S.E. of Rothesay, on E. side of isl. of Bute, Firth of Clyde.

Mount Stuart, hill, alt. 1567 ft., 6 m. S.E. of Muirkirk. On borders of cos. of Lanark and Dumfries.

Mount Teviot, seat of Marquess of Lothian, 3 m. N. by W. of Jedburgh, and 1¼ m. by rail from Jedburgh, Roxburghshire.

Mount, The, seat, 3½ m. N.W. of Cupar, Fife. Was at one time residence of the poet, Sir David Lindsay. M. Hill is surmounted by a column erected to the Earl of Hopetoun (1766–1823).

Mount, The, seat, near Kilmarnock, Ayrshire.

Mount Vernon, vil., 5 m. E. by S. of Glasgow, Lanarkshire; ry. stas., L.M.S. and L.N.E. Pop. of M. V. and Carmyle, 4435.

Mousa, isl. off E. coast of Mainland, Shetland, 11 m. S. of Lerwick. A channel, 1 m. wide, separates it from Sandwick in Mainland. Contains a Scandinavian fort.

Mousaford, loch, 3½ m. N.W. of Walls, Mainland, Shetland.

Mousalies, loch, 1½ m. N.E. of Walls Mainland, Shetland.

Mousedale, isl. in Argyllshire. See Mus-DALE.

Mouse Water, stream in E. Lanarkshire; rises on Midlothian border and flows 14 m. to the Clyde, which it joins, 1 m. W. by N. of Lanark.

Mouswald, par. and vil. in S. Dumfriesshire, on Lochar Moss; par. 5887 ac. Pop. 412. Vil. is 7 m. S.E. of Dumfries and 2½ m. S.E. of Racks ry. sta. M. Place is a seat nearby. P.O.

Mowhaugh, pl., 4½ m. S. of Yetholm and on Bowmont Water, S.E. Roxburghshire. P.O.

Mowtie, pl., 3 m. N.W. of Stonehaven, Kincardineshire.

Moy, vil. in par. of same name and on Loch M., in co. of, and 14¾ m. S.E. of, Inverness, by rail; ry. sta., L.M.S. P.O., T.O. M. Hall is the seat of The Mackintosh, 1 m. N.W. of the vil.

Moy, Morayshire. See DYKE AND MOY. Pop. of M. Croft 206.

Moy and Dalarossie, par., chiefly in N.E. Inverness-shire, but partly in Nairnshire; 105,968 ac. Pop. 566. The church of M., an eighteenth-century building, stands on Loch M.

Moy Castle, ancient seat of the McLeans, at head of Loch Buy, S. coast of Mull, Argyllshire.

Moydart, dist., Inverness-shire. See MOIDART.

Moy House, N.W. Morayshire, 3½ m. N.W. of Forres.

Moy, Loch, in Glen Moy, ½ m. E. of Moy ry. sta., Inverness-shire; depth 50 ft., alt. 880 ft.

Moy Lodge, shooting-lodge, near S.W. end of Loch Laggan, Inverness-shire, 6 m. N.E. of Tulloch.

Moyness, vil., par. of Auldearn, in co. of, and 5 m. S.E. of, Nairn. Traces of an ancient castle.

Muasdale, vil., 5 m. S. of Tayinloan, Kintyre, Argyllshire. P.O., T.O.

Muchalls, vil. and seat, par. of Fetter-cairn, 4½ m. N.E. of Stonehaven, Kin-cardineshire; ry. sta., L.M.S. P.O. Hotel. Coastguard sta.

Muck, isl., 4 m. N. of Ardnamurchan Point, Inverness-shire; 1586 ac. Pop. 48. Measures 2 m. in length and rises to a height of 451 ft.

Muck, loch, 4 m. S.E. of Dalmellington, Ayrshire; depth 22 ft., alt. 992 ft.

Muckairn, q.s. par. in dist. of Lorne, Argyllshire. Par. church at Taynuilt. M. Castle is on S. side of Loch Etive, 8 m. E.N.E. of Oban.

Muckersie Linn, fall on Water of May, Forteviot, S.E. Perthshire.

Muckhart, par. on r. Devon, and 10 m. S.E. of Crieff, Perthshire; 4932 ac. Pop. 482. P.O., T.O.

Muckle Burn, stream in N.E. Nairnshire, flowing 19 m. N.E. to Findhorn Bay.

Muckle Flugga, rocky islet off N.W. coast of Unst, N. Shetland. On the N. of the isl. is N. Unst lighthouse, with fixed light, visible 21 m.

Muckle Loch, 3 m. S.W. of Banks, Rousay, Orkney.

Muckle Lunga, loch, 4 m. N.W. of Colla-forth, Mainland, Shetland; maximum depth 27 ft., alt. about 450 ft.

Muckle Roe, isl. in St. Magnus Bay, off W. coast of Mainland, Shetland. Pop. 154.

Muckle Skerry, one of the Pentland Skerries. Group-flashing light, visible 19 m.

Muckrach House, seat, 4 m. S.W. of Grantown-on-Spey, Inverness-shire. Nearby is the ruin of M. Castle.

Muckross, see St. Andrews.

Mucomir Falls, on r. Spean, ¼ m. from Gairlochy, near S. end of Loch Lochy, S. Inverness-shire.

Mudale, stream in mid Sutherland, flow-ing 7½ m. S.E. to Loch Naver, near Altnaharra.

Mudell, loch, 4 m. N.E. of Kilchoan, N.W. Argyllshire.

Mudlee Brackshills, alt. 2259 ft., borders of cos. of Aberdeen and Angus.

Mugdock Castle, ancient seat of the Montrose family, in Stirlingshire; 2 m. N. of Milngavie. 1 m. S. is a reser-voir connected with Glasgow water supply.

Mugdrum House, 1 m. N.W. of Newburgh, Fife.

Mugdrum Island, in Firth of Tay, Perth-shire, nearly opposite Newburgh, Fife.

Mugeary, ham., 4 m. S.W. of Portree, Skye, Inner Hebrides.

Mugiemoss Paper Mills, Bucksburn, 4½ m. N.W. of Aberdeen.

Muick, stream in S.W. Aberdeenshire, flowing 3½ m. S.E. through Loch M., and thence 9½ m. N.E. to join the Dee near Ballater. The Linn of M. is a fine waterfall, 6 m. S.W. of Ballater. Loch M. depth 256 ft., alt. 1308 ft.

Muiravonside, par. on the r. Avon, S.E. Stirlingshire; 7964 ac. Pop. 5560. Ancient fortifications on the hilltops and ruins of a pre-Reformation chapel. M. House is 3½ m. S.W. of Linlithgow.

Muirburn, seat, 2½ m. N.E. of Strathaven, W. Lanarkshire.

Muircambus, seat, 1 m. S.W. of Colins-burgh, Fife.

Muirdrum, vil., 5½ m. S.W. of Arbroath and 2½ m. N. of Carnoustie sta., Angus. P.O.

Muirend, pl., 1 m. S. of Cathcart, Renfrew-shire; ry. sta., L.M.S. P.O., T.O.

Muiresk House, seat, 2 m. W. of Turriff, Aberdeenshire.

Muirfield, seat, now convalescent home of the Edinburgh Royal Hospital for Sick Children, 3 m. W. of Drem ry. junc. Golf course.

Muirhead, vil., 4 m. S.E. of Falkland, Fife.

Muirhead or **West Benhar,** vil., 3 m. E.N.E. of Kirk of Shotts, Lanarkshire. P.O.

Muirhead, Lanarkshire. See Chryston and Muirhead.

Muirhead of Liff, vil., 3½ m. N.W. of Dundee, Angus. Pop. Birkhill and M. 440.

Muirhouse, seat, on S. shore of Firth of Forth, 3½ m. E.N.E. of Edinburgh G.P.O.

Muirhouselaw, seat, 1½ m. S.E. of Maxton ry. sta., Roxburghshire.

Muirhouses, vil., 2 m. S.E. of Bo'ness, W. Lothian. Pop. 228.

Muirkirk, par. and town in E. Ayrshire; par. 30,282 ac. Pop. 4358. Town stands on r. Ayr, 26 m. E. by N. of Ayr by rail. P.O., T.O. Ry. sta. Coal mining and iron mfrs. Town pop. 3438.

Muirmadzean, part of Mossend, Lanarkshire.

Muirnag Hill, alt. 808 ft., 11 m. N. of Stornoway, Lewis, Outer Hebrides. The highest land N. of Stornoway in Lewis.

Muir of Fowlis, ham., 3¼ m. S.S.W. of Alford, mid Aberdeenshire. P.O., T.O.

Muir of Ord, vil. in E. Ross and Cromarty, 3 m. N. of Beauly and 13 m. N.W. of Inverness; ry. junc., L.M.S. P.O., T.O. Pop. 307.

Muir of Rhynie, vil. in Aberdeenshire. See Rhynie.

Muirpark Rows, vil., 1 m. N.W. of Bellshill, Lanarkshire.

Muirshiels House, on r. Calder, 4½ m. N.W. of Lochwinnoch, Renfrewshire.

Muirtack, pl., 7 m. S. of Mintlaw sta., Aberdeenshire.

Muirton, vil., 3 m. N.W. of Marykirk, S. Kincardineshire.

Muirton, ham., 1 m. S. of Blairgowrie, Perthshire.

Muirtown, seat, in co. of, and 1 m. N.W. of, Inverness. P.O., T.O.

Muirtown, pl., 2½ m. W. of Forres, Moray.

Muiryhill, pl., 5 m. N. of Thornhill, Dumfriesshire.

Mulben, ry. sta., L.M.S., Morayshire, 5 m. W. of Keith. P.O., T.O.

Muldoanich, small isl., 2 m. S.E. of Barra, Outer Hebrides.

Mulindry, pl., 3 m. E. of Bowmore, Islay, Argyllshire.

Mull, isl. in N.W. Argyllshire, 7 m. W. of Oban; the Sound of M. on the N.E. and the Firth of Lorne on the S.E. divide it from the mainland; on the N.W. and S.W. it is washed by the Atlantic. On a highly indented coastline, the chief inlets are Loch Scridain and Loch-na-Keal on the W. The surface is mountainous, attaining a height of 3185 ft. in Ben More. Tobermory in the N.W. is the chief town. Pop. of isl. 2903, and its area, including Iona and Gometra, is 351¼ sq. m., or 224,802 ac.

Mullach-an-Langa, mt., alt. 2012 ft., N. Harris, Outer Hebrides.

Mullach-an-Rathain, mt., alt. 3258 ft., near head of Upper Loch Torridon, W. Ross and Cromarty.

Mullach Coire Mhic Fhearchair, mt., alt. 3250 ft., 7 m. N. of Kinlochewe, Ross and Cromarty.

Mullach Coire nan Dearsag, mt., alt. 2846 ft., 8 m. E. of Dalwhinnie, Inverness-shire.

Mullach Mor, mt., alt. 2521 ft., 6 m. S.E. of Kingussie, Inverness-shire.

Mullach nan Coirean, mt., alt. 3077 ft., 5 m. S.E. of Fort William, Inverness-shire.

Mullagrach, most northerly of the Summer Isles, 15 m. N.W. of Ullapool, N.W. coast of Ross and Cromarty.

Mulla-of-Dheas, mt., alt. 2439 ft., Harris, Outer Hebrides.

Mullardoch, loch, 6 m. S.E. of Loch Monar, in Inverness-shire and Ross and Cromarty; depth 197 ft., alt. 705.

Mullbuie, Ross and Cromarty. See Ardmeanach.

Mull Head, headland forming N. extremity of isl. of Papa Westray, Orkney.

Mull of Galloway, see Galloway, Mull of.

Mull of Kintyre, see Kintyre.

Mull, Sound of, sea-channel in W. Argyllshire, between M. Isl. and dist. of Morven, and connecting Firth of Lorne with Atlantic; varies from 1½ to 3¼ m. wide. Rudha nan Gal, a rock 1½ m. N. of Tobermory, has a lighthouse with a fixed light, visible 12 m.

Mullwharcher, mt., alt. 2270 ft., 3 m. S. of Loch Macaterick in S.E. Ayrshire.

Munches, seat, 2 m. S. of Dalbeattie, Kirkcudbrightshire.

Muncraig, headland in co. of, and 5 m. S.W. of, Kirkcudbright.

Mundy House, seat, in par. of Aberdalgie, Perthshire.

Mu Ness, headland on the S.E. of isl. of Unst, Shetland.

Mungo, Hill of, alt. 650 ft., 2 m. W. of Huntly, Aberdeenshire.

Munlochy, vil. on bay of same name, 5½ m. S.W. of Fortrose, Black Isle, Ross and Cromarty; ry. sta., L.M.S. P.O., T.O. M. Bay, an inlet of the Moray Firth, is a fishing sta. Vil. pop. 161.

Munnoch Reservoir, 2½ m. S.W. of Dalry, Ayrshire. Supplies Irvine.

Murdoch Isle, in Loch Ard, S.W. Perthshire.

Murdostoun Castle, seat, 1 m. N. of Newmains sta., Lanarkshire.

Murie House, seat, 1½ m. S.W. of Errol, Perthshire.

Murieston, estate belonging to Edinburgh Distress Committee, 14 m. W.S.W. of the city and close to Newpark ry. sta., L.M.S.

Murieston Water, trib. to the Almond, W. Lothian.

Murkle, ham., near bay of same name, 3½ m. E. of Thurso, Caithness.

Murlingden, seat 1½ m. N.W. of Brechin, Angus.

Murrayfield, q.s. par. in the W. of Edinburgh; ry. sta. on Barnton and Leith branches of L.M.S. P.O., T.O. Football ground of Scottish Rugby Union.

Murrayfield House, 1 m. S.E. of Lockerbie, Dumfriesshire.

Murrayshall, seat, in co. of, and 2½ m. S.W. of, Stirling.

Murrayshall Hill, alt. 918 ft., 3 m. N.E. of Perth. On the summit is an obelisk, erected 1850, commemorating Thos. Graham, Lord Lynedoch, a famous Peninsular veteran.

Murraythwaite, seat, 4 m. S.W. of Ecclefechan, Dumfriesshire.

Murroch Burn, flows S.W. to r. Leven, in co. of, and 1 m. N. of, Dumbarton.

Murroes, par. in S. Angus, 4 m. N.E. of Dundee; 6088 ac. Pop. 919. A burn of the same name flows to Dighty Water.

Murthly, q.s. par. and pl. in par. of Little Dunkeld, E. Perthshire, 4½ m. S.E. of Dunkeld and 10½ m. N. of Perth; ry. sta., L.M.S. P.O., T.O. Nearby is Perth Dist. Asylum. 2 m. N.W. of the sta. is the modern seat of M. House, on r. Tay.

Murtle, pl. in co. of, and 5 m. S.W. of, Aberdeen. P.O.; ry. sta, L.N.E. Nearby, on the r. Dee, is M. House.

Musdale, or **Musdile,** islet off S.W. end of isl. of Lismore, 5 m. N.W. of Oban. Lismore lighthouse is on this isl., with fixed light, visible 16 m. Pop. 4.

Muselee, seat, 7 m. S.W. of Hawick, Roxburghshire.

Musselburgh, town, parl. and mun. bur. in Midlothian, at mouth of r. Esk, 6 m. E. of Edinburgh by rail; ry. sta., L.N.E. P.O., T.O. Racecourse and golf links. Forms part of the eastern parl. division of Edinburgh. Pop. 17,007. Industries: net, twine, wire, and paper making, market gardening, and seed growing.

Muthill, par. and vil. on r. Earn, Perthshire; par. 26,181 ac. Pop. 1212. Vil., 3 m. S. of Crieff; ry. sta. (1¾ m. E. of vil.), L.M.S. P.O., T.O. Pop. 605.

Mybster, ham., 4½ m. S.S.E. of Halkirk, Caithness.

Mylnefield Fens, vil. in Perthshire, 3 m. W. of Dundee. The seat stands 1 m. W. of vil.

Myot Hill, alt. 696 ft., 4½ m. N.E. of Kilsyth, S.E. Stirlingshire.

Myrehead, pl. in Dryfesdale par., Dumfriesshire.

Myres Castle, ½ m. S.E. of Auchtermuchty, Fife.

Myres Hill, alt. 1093 ft., in extreme S.E. of Renfrewshire, 4 m. S. of Eaglesham.

Myreton Castle, ruin, 1½ m. W. of Port William, Wigtownshire.

N

Na Bo, Loch, Morayshire, 4 m. S.E. of Elgin.

Nackerty, ham., Lanarkshire, Bothwell par.

Na-h-Oidhche, loch, Ross and Cromarty, 8 m. S.E. of Gairloch; greatest depth 121 ft., alt. about 1250 ft.

Nairn, royal and mun. bur., seapt., par., and co. town of Nairnshire, with ry. sta., L.M.S., on r. Nairn and on the Moray Firth, 9½ m. W. by S. of Forres by rail, and 15½ m. E.N.E. of Inverness; par. 9146 ac. Pop. 5282; bur. pop. 4201. P.O., T.O. Hotels. On W. pier-head is a fixed light. Exports: corn, potatoes, timber, and fish. For its sandy beach and pleasant climate, is a favourite bathing resort.

Nairnshire, a small co. in N.E. of Scotland, on the S. shore of the Moray Firth. It is bounded on the N. by that arm of the sea, on the S.W. by Inverness-shire; area 163·17 sq. m., or 104,452 ac. Pop. 8294. The co. gradually rises from an undulating fertile plain to mountains on the S.; highest point, Cairn Glas, 2162 ft. The Findhorn drains the E. and the r. Nairn the W. Of historical interest is Auldearn, the scene of one of Montrose's victories, and Culloden, where the Highlanders under Prince Charles Edward were defeated by the royal troops under the Duke of Cumberland. Cawdor and Kilravock Castles are in the co. Agriculture and fisheries are the chief industries. With Morayshire, returns one member to Parliament.

Nairnside, seat, in co., and 5 m. E.S.E. of Inverness.

Nakerty, vil., Lanarkshire, 1 m. N. of Uddingston.

Nal Moor, seat, S. Perthshire, 3½ m. S.E. of Glendevon.

Na Moine, loch, Sutherland, 3 m. W. of Kinbrace sta.; alt. 377 ft., depth 8 ft.

Nan Eun, loch, Inverness-shire, 4½ m. E. of Fort Augustus; depth 21 ft., alt. 915 ft.

Nan Eun, loch, on borders of Ross and Cromarty and Inverness-shire, 4 m. N.W. of Beauly; alt. 1051 ft.

Nan Lann, loch, Inverness-shire, 4½ m. N.E. of Fort Augustus; greatest depth 109 ft., alt. 645 ft.

Nan Losganon, loch, Inverness-shire, 8 m. N.E. of Fort Augustus; depth 7 ft.

Nant, stream and loch, Argyll; loch, 4 m. N.W. of Port Sonnachan, on Loch Awe, sends off r., which flows 5½ m. N. through Glen N. to Loch Etive, near Bonawe.

Narrows, The, Loch Leven, Argyll. Shows two-seconds flashing lights at E. and W. end, seen 5 m.

Nathro, seat, Angus, 8 m. N.W. of Brechin.

Naughton, seat and old ruined castle, Fife, 1 m. E. of Balmerino.

Nave, or **Nomah,** islet near entrance to Loch Gruinart, Islay Isl., 10 m. N. by W. of Bowmore. Ancient ruined church and burying ground.

Naver, loch and r., Sutherland; loch, 21 m. N. by W. of Lairg sta., is 6 m. long, and has a maximum depth of 108 ft.; r. issues from the loch and flows 19 m. through Strath N. to Torrisdale Bay, 7 m. E.N.E. of Tongue. A good salmon stream.

Navidale, seat and site of ancient chapel, Sutherland, near Helmsdale.

Navity, seat, Ross and Cromarty, 2½ m. S. of Cromarty.

Neap, ham., Shetland, 10 m. N.N.E. of Lerwick.

Near Eyes, loch, Ayrshire, 2¾ m. N.E. of Barrhill.

Neaty, loch, Inverness-shire, 9 m. S.S.W. of Beauly.

Neave, isl., W. Sutherland, ½ m. N. of Invernaver; alt. 822 ft.

Nedd, Loch, sea loch, W. Sutherland, 6 m. N.E. of Lochinver.

Needle's Eye, cavern, Banffshire, on Gamrie coast, near Troup Head.

Needle's Eye, natural arch, 40 ft. high, on coast of Kirkcudbrightshire, 5 m. S.E. of Dalbeattie.

Needle's Eye, natural archway, S.E. Angus, on coast of St. Vigean's par.

Needles Law, hill, S. Roxburghshire, 2 m. N.W. of Peel Fell; alt. 1457 ft.

Neidpath Castle, massive structure, partly peel-tower, of twelfth century; thick-walled quadrangular castle of fifteenth century, on r. Tweed, 1 m. W. of Peebles. Surrendered to Cromwell in 1650 after a long siege. The property of the Earl of Wemyss.

Neilston, town and par., with ry. stas., L.M.S., Renfrewshire; town on the r. Levern, 2 m. S.W. of Barrhead, 5½ m. S. of Paisley; par. 12,483 ac. Pop. 38,499; town pop. 2509. P.O., T.O. Has print works and bleachfields. In vicinity, N. Pad, 854 ft.

Neish Island, in Loch Earn, Perthshire, near St. Fillans.

Neisland, seat, Hamilton par., Lanarkshire.

Neist Point, N.W. Skye Isl. Lighthouse, showing a half-minute group-flashing light, seen 18 m.

Neldricken, Loch, Kirkcudbrightshire, 11 m. N.E. of Newton-Stewart; alt. 1146 ft.

Nell Loch, Argyll, 3 m. S.E. of Oban; depth 115 ft., alt. 49 ft.

Nemphlar, vil. and braes, on rt. side of Clyde, 2 m. N.W. of Lanark. P.O.

Nenthorn, par. and vil., S.W. Berwickshire, on Eden Water; par. 3443 ac. Pop. 349. Vil., 4 m. N.W. of Kelso. P.O., T.O.   N. House, a seat.

Neptune's Staircase, series of eight lochs on S.W. extremity of Caledonian Canal, Inverness-shire.

Nereabolls, pl., with ancient burying-ground, S.W. Islay, Argyll, on Loch Indaal.

Nerston, vil., Lanarkshire, 1½ m. N.E. of E. Kilbride.

Ness, loch and r., N. Inverness-shire. Loch occupies a considerable portion of the bottom of the N.E. reach of Great Glen. On the line of the Caledonian Canal, extends from Fort Augustus to Bona Ferry, 23 m. in length, and about 1¼ m. in mean breadth. R. issues from the N. end, and flows 7 m. N.E. to the Moray Firth, below Inverness. Maximum depth of loch 754 ft., alt. 52½ ft.

Ness or **Cross,** vil., Lewis Isl., Ross and Cromarty, 18 m. N. of Stornoway. P.O., T.O.

Ness Castle, seat, 2½ m. S.W. of Inverness, on rt. bank of r. Ness.

Ness Glen, narrow ravine, traversed by r. Doon, Ayrshire, 1 m. S. of Dalmellington.

Ness Head, N.E. Caithness, Freswick Bay.

Ness Islands, in par. and co. of Inverness. Pop. 1.

Ness-side House, 2¾ m. S.S.W. of Inverness, on r. Ness.

Ness, The, headland, E. Angus, 1¼ m. S.E. of Montrose.

Nesting, par., Shetland. Church stands 8 m. N. of Lerwick; 30,995 ac. Pop. 1832.

Nethan, rivulet, Lanarkshire, flows 13 m. N. to the Clyde, at Crossford.

Nethanfoot, vil., Lanarkshire, at mouth of Nethan R.

Netherburn, vil., with ry. sta., L.M.S., Lanarkshire, Dalserf par., 3¾ m. S.S.E. of Larkhall. P.O., T.O.

Netherbyres, seat, Berwickshire, 2½ m. N.E. of Ayton.

Nethercleuch, ry. sta., L.M.S., Dumfriesshire, 3 m. N.W. of Lockerbie.

Nethercraigs, pl., Renfrewshire, 2 m. S.S.W. of Paisley.

Nether Croy, seat, Dunbartonshire, 1 m. N.W. of C. sta.

Netherdale House, seat, Banffshire, 5 m. W. of Turriff.

Nether Dallachy, ham., N.E. Morayshire, 3½ m. N.E. of Fochabers.

Netherfield House, Lanarkshire, 1 m. E. of Strathaven.

Netherhall, seat, Ayrshire, 1½ m. N. of Largs.

Netherhill, seat, Renfrewshire, near Paisley.

Nether Kilrenny, vil., Fife. See KILRENNY.

Netherlaw, seat, 7 m. S.E. of Kirkcudbright.

Netherlee, vil., Renfrewshire, 2 m. S.E. of Pollokshaws.

Netherley, ham. and seat, Kincardineshire, 6 m. N.W. of Stonehaven. P.O.

Nether Liberton, dist., S. of city of Edinburgh.

Nethermill, vil., Dumfriesshire, 4 m. N.W. of Lochmaben.

Nethermuir House, Aberdeenshire, New Deer par., 1½ m. N.N.W. of Auchnagatt sta.

Netherplace, seat, Mauchline par., Ayrshire, W. of the town of Mauchline.

Netherplace, pl., Renfrewshire, 3 m. S.E. of Barrhead.

Netherton, dist., Dunfermline, Fife. P.O.

Netherton, vil., Dunbartonshire, 1 m. W. of Maryhill, Glasgow.

Netherton, ham., Angus, 3¾ m. S.W. of Brechin. P.O.

Netherton, mining locality, Lanarkshire, 1¼ m. N. of Wishaw. P.O.

Netherton, seat, E. Perthshire, 2 m. S.E. of Alyth.

Netherton, Strathblane par., Stirlingshire. Special water dist. Pop. 193.

Netherton, seat, Stirlingshire, ½ m. S. of Campsie Glen ry. sta.

Netherurd House, seat, W. Peeblesshire, 2½ m. S. of Dolphinton ry. sta.

Netnerwood, pl., Dumfriesshire, 2 m. S. of Dumfries. Pop. Netherwood, Kelton, and Craig's special water dist. Pop. 340.

Nethy, rivulet, Inverness-shire, running about 14 m. N. from Cairngorm Mts. to the Spey, at Abernethy Church.

Nethy, burn, Perthshire, flows 2½ m. N. to the Earn.

Nethy Bridge, vil., with ry. sta., L.N.E. (alt. 702 ft.), Inverness-shire, 5 m. S.W. of Grantown. P.O., T.O. Pop. 387. Hotel.

Neuck House, Stirlingshire, 4 m. N.E. of Larbert ry. sta.

Nevay, see EASSIE AND NEVAY.

Nevis, Bridge of, ham., S.W. Inverness-shire, 1 m. N.E. of Fort William. In vicinity N. and Ben N. Distilleries.

Nevis, Loch, arm of the sea, in Glenelg par., Inverness-shire, 7 m. N.E. of Arisaig; length 14½ m., and from 1 to 4 m. wide.

Nevis State Forest, near Fort William, Inverness-shire.

Nevis, Water of, traverses Glen Nevis, flows N.W. to Loch Eil, at Fort William.

New Abbey, par. and vil., E. Kirkcudbrightshire; par. 13,177 ac. Pop. 728. Vil., 7½ m. S. of Dumfries. P.O., T.O. Pop. 173. Sweetheart, or New Abbey, after which the par. is named, lies E. of the vil. This Abbey was founded in 1275 by Devorgoil, who also founded Baliol College, Oxford.

New Abbey Pow, or Glensome Burn, Dumfriesshire, rises 5½ m. S.W. of Dumfries and flows to the Solway Firth.

New Aberdour, coast vil., Aberdeenshire, 8 m. W.S.W. of Fraserburgh. P.O., T.O. Pop. 426.

New Alyth, vil., E. Perthshire, 1 m. S. of Alyth.

Newark, pl., with well-preserved ancient Caledonian stone circle, Lonmay par., Aberdeenshire.

Newark, restored old baronial fortalice, 4 m. S. of Ayr.

Newark, q.s. par., adjacent to Port Glasgow, Renfrewshire. Contains ruins of N. Castle; government property.

Newark Castle, ruin, 4 m. W. by N. of Selkirk, scene of Sir Walter Scott's "Lay of the Last Minstrel."

Newark Castle, ruins of the ancient seat of the Leslies, St. Monans, Fife.

Newarthill, town, Lanarkshire, 3 m. N.E. of Motherwell. P.O., T.O. In rich mining area. Pop. 2487.

Newbarns, seat, Angus, 3½ m. N.E. of Kirriemuir.

Newbattle, par. and vil., Midlothian; par. 5186 ac. Pop. 7455. Vil. on r. South Esk, 1 m. S. of Dalkeith. N.

Abbey, formerly a seat of the Marquess of Lothian, now being taken over as an adult educational centre for working men.

Newbigging, sub. of Musselburgh, Midlothian.

Newbigging, ham., adjacent to Auchtertool, Fife.

Newbigging, vil., Angus, 7½ m. E.N.E. of Dundee. P.O.

Newbigging, ham., Angus, 1 m. N.W. of Newtyle.

Newbigging, ham., Angus, 4 m. N. of Dundee.

Newbigging, pl., with ancient stone circle, Angus, 3 m. W. of Edzell.

Newbigging, vil., with ry. sta., L.M.S., Lanarkshire, 3 m. S.E. of Carnwarth. P.O.; T.O. at sta.

Newbigging, pl., N.E. West Lothian, ¾ m. S. of South Queensferry.

Newbigging, ham., Perthshire, 1 m. S.E. of Cargill.

Newbridge, ham., Midlothian, on rt. bank of Almond Water, 8 m. W. by S. of Edinburgh.

Newbridge, ham., Kirkcudbrightshire, 3 m. N.W. of Dumfries. P.O.

Newburgh, royal and mun. bur., seapt., and par., with ry. sta., L.N.E., Fife, on Firth of Tay, 7½ m. N.W. of Ladybank, 11¼ m. E.S.E. of Perth; par. 2017 ac. Pop. 2019; mun. bur. pop. 2152. P.O., T.O. Has linen and floorcloth mfrs., and salmon fishing.

Newburgh, seapt. vil., Aberdeenshire, on the r. Ythan, 5 m. S.E. of Ellon ry. sta. P.O., T.O. Pop. 414.

Newburn, par., S.E. Fife, on Firth of Forth; 3049 ac. Pop. 238. Church, 1½ m. N.E. of Largo.

Newby Glen, 2 m. S. of Peebles.

Newbyres Castle, ruined castle, Midlothian, near Gorebridge.

Newbyth, seat, East Lothian, 2 m. N. of East Linton.

New Byth, vil., N. Aberdeenshire, 6½ m. N.E. of Turriff. P.O., T.O. Pop. 291.

New Calder Paper Mill, Mid-Calder, Midlothian.

Newcastleton, vil., with ry. sta., L.N.E., Roxburghshire, on Liddel Water, 21 m. S.W. of Hawick and 24 m. N.E. of Carlisle. Pop. 833. P.O., T.O. State forest, 1½ m. S.E. of town.

New Cathcart, locality in S. of Glasgow. See CATHCART.

New Clunie, pl., Perthshire, 3 m. N. of Caputh. P.O.

New Craighall, q.s. par. and mining dist., in city of Edinburgh and partly in Midlothian, 2 m. S.W. of Musselburgh. P.O.

New Cumnock, par. and town, with ry. sta., L.M.S., Ayrshire ; par. 48,165 ac. Pop. 6419. Town at confluence of Afton Water and the Nith, 5½ m. S.E. of Cumnock. P.O., T.O. Pop. 2022.

New Deer, vil., Aberdeenshire, 6 m. W. of Old Deer. P.O., T.O. See DEER, NEW.

Newe, seat, Aberdeenshire. See CASTLE NEWE.

Newfield, seat, Ayrshire, 4 m. S.W. of Kilmarnock.

Newfieldyke, part of Binniehill vil., Stirlingshire, 1 m. S. of Slamannan.

New Galloway, royal and mun. bur. and small town, Kirkcudbrightshire, on Water of Ken, 5 m. N.W. of its sta. and 19 m. N. of Kirkcudbright. Pop. of bur. 307. P.O., T.O. Hotel.

New Galloway, ry. sta., L.M.S., Kirkcudbrightshire, 9 m. N.W. of Castle-Douglas by rail and 5 m. S.E. of New G. T.O.

New Gilston, vil., Fife, 2½ m. S.E. of Ceres. P.O.

New Grange, seat, West Lothian, 1 m. N. of Linlithgow.

New Hailes, seat and ry. sta., L.N.E., Midlothian, 1 m. W. of Musselburgh.

Newhall, q.s. par. in Glasgow.

Newhall, seat, Roxburghshire, 1½ m. N.W. of Belses ry. sta.

Newhall, seat, Midlothian, 4½ m. S.W. of Penicuik, on r. North Esk.

Newhall, seat, Ross and Cromarty, on Cromarty Firth, 2 m. S. of Invergordon.

Newhalls, vil., W. Lothian, in the eastern vicinity of S. Queensferry.

Newhaven, q.s. par., with ry. sta., L.M.S., in city of Edinburgh, on Firth of Forth, 2¼ m. N. by W. of Edinburgh G.P.O. P.O., T.O. Has a large fish market. East pier shows a fixed white light, visible 12 m.

Newhill, seat, N.W. Fife, 2 m. N.W. of Auchtermuchty.

Newhills, par., S.E. Aberdeenshire, on r. Don; 10,286 ac. Pop. 6722. Church, 4 m. N.W. of Aberdeen. Vil. pop. 809.

Newholme, seat, Lanarkshire, ½ m. S.E. of Dunsyre ry. sta.

Newhouse, pl., with ry. sta., L.M.S., Lanarkshire, 3¾ m. S.E. of Airdrie. Has a reservoir of the Lanarkshire County Council.

Newhouses and Westerton, vil., West Lothian, 1 m. S.E. of Broxburn.

Newington, q.s. par., dist., and ry. sta., L.N.E., in S.E. part of city of Edinburgh. P.O., T.O. N. House is a home for blinded soldiers.

New Kelso, ham. and deer forest, Ross and Cromarty, near Strathcarron ry. sta. Deer forest, 13,800 ac.

Newkirk, vil., S. Aberdeenshire, 3 m. W. of Tarland.

New Lanark, vil., 1 m. S. of Lanark, on r. Clyde. P.O.

Newland House, seat, Borthwick par., Midlothian.

Newlandrigg, pl., Midlothian, 4 m. S.S.E. of Dalkeith.

Newlands, par., Peeblesshire ; 12,531 ac. Pop. 613. Church, 3 m. S. of West Linton.

Newlands, q.s. par. and ry. sta. (Langside and N.), L.M.S., in Glasgow. P.O., T.O.

Newlands, shooting-lodge, Roxburghshire, 4½ m. N. of Newcastleton.

Newlands of Forse, ham., Caithness, 2½ m. N.E. of Latheron.

Newlaw, hill, Kirkcudbrightshire, 1 m. N.W. of Dundrennan ; 599 ft.

New Leeds, vil., Aberdeenshire, 8 m. S. of Fraserburgh. P.O.

Newliston, seat and pl., West Lothian, 1¼ m. S.W. of Kirkliston. Shale-oil works.

New Logans, vil., Lanarkshire, on N.W. side of Motherwell.

New Luce, ry. sta., L.M.S., Wigtownshire, 5 m. N.W. of Glenluce. P.O., T.O. See LUCE, NEW.

Newmachar, vil. and ry. sta., L.N.E., 11½ m. N.W. of Aberdeen. P.O., T.O. Pop. 229.

Newmains, town, with ry. sta., L.M.S., Lanarkshire, 2 m. N.E. of Wishaw. P.O., T.O. Pop. 5104. Has ironworks.

Newmanswalls House, Angus, 1 m. N. of Montrose.

New Maude, vil., Aberdeenshire, 13 m. W. of Peterhead.

New Mill, q.s. par. and vil., Banffshire, on r. Isla, 1¼ m. N. of Keith. P.O. Pop. 632.

Newmill, pl., Kincardineshire, 2 m. N.E. of Drumlithie.

Newmill-on-Teviot, pl., Roxburghshire, 5 m. from Hawick. P.O., T.O.

New Mills, vil., Banffshire, Fordyce par., near Cornhill ry. sta.

Newmills, pl., Fife, 4 m. W.S.W. of Dunfermline. P.O., T.O.

Newmilns and Greenholm, town, police bur., with ry. sta., L.M.S., Ayrshire, on r. Irvine, 7 m. E. of Kilmarnock. Pop. 3979. P.O., T.O. Has lace and muslin manufactures.

Newmore, seat, Ross and Cromarty, 3½ m. N.W. of Invergordon.

New North, or **West St. Giles,** eccl. par., Edinburgh.

Newpark, ry. sta. (alt. 526 ft.), L.M.S., Midlothian, 4 m. S.W. of Mid-Calder.

New Pitsligo, vil., N. Aberdeenshire, 4 m. W. of Strichen ry. sta. P.O., T.O. Pop. 1284.

Newport, seapt. town, mun. bur., and q.s. par., with ry. stas. (E. and W. Newport), L.N.E., N.E. Fife, on Firth of Tay, opposite Dundee, at S. end of the Tay Bridge. Bur. pop. 3275. P.O., T.O. Regular ferry service with Dundee.

New Rattray, vil., Perthshire, on r. Ericht, opposite Blairgowrie.

New Sauchie, part of Alloa, Clackmannanshire.

New Saughton Hall, Midlothian, ½ m. S.E. of Loanhead. Private lunatic asylum.

Newseat, ry. sta., L.N.E., N.E. Aberdeenshire, 3½ m. W. of Peterhead.

Newshot Island, marshy isl., in r. Clyde, 2 m. N.W. of Renfrew.

Newstead, Doune, Perthshire. Boy's Private School.

Newstead, vil., Roxburghshire, 1 m. E. of Melrose, on r. Tweed. Adjoins ground where numerous Roman relics have been found. P.O. Pop. 215.

New Stevenston, vil., N. Lanarkshire, adjoining Holytown. P.O., T.O. Pop. of N. S., Holytown, and Carfin, 12,011.

Newton, ham., Argyll, on Loch Fyne, 4 m. S.W. of Strachur.

Newton, seat, adjoining town of Nairn.

Newton, ruined seat, 8 m. W. of Inverness.

Newton, ham., W. Lothian, 2 m. W.S.W. of S. Queensferry. P.O.

Newton, vil., Caithness, 1 m. S.W. of Wick.

Newton, par. and vil., N.E. Midlothian; par. 2312 ac. Pop. 1122. Vil. pop. 1047. N. House, a seat.

Newton, mining vil., East Lothian, 3 m. S.E. of Tranent.

Newton, vil. and ry. sta., L.M.S., Lanarkshire, 5½ m. S.E. of Glasgow. Heavy steel-casting works. N. House, a seat. Pop. of N. and Flemington 4176.

Newton, vil., 1 m. S.W. of Renfrew.

Newton, pl., Dumfriesshire, adjoining Wamphray sta.

Newton, pl., S.E. Dumfriesshire, 1½ m. N.W. of Kirkpatrick.

Newton, seat, Roxburghshire, 3 m. W. of Jedburgh.

Newtonairds, seat and ry. sta., L.M.S., 7 m. N.W. of Dumfries.

Newton and West Haven, part of Carnoustie, Angus.

Newton Dee, seat, 5 m. S.W. of Aberdeen.

Newton Don, seat, Berwickshire, 2 m. N.W. of Kelso.

Newton Ferry, N. Uist Isl., Outer Hebrides, 8 m. N.N.W. of Lochmaddy. P.O., T.O.

Newtongrange, mining vil. and ry. sta., L.N.E., Midlothian, 2½ m. S. of Dalkeith. P.O., T.O. Pop. (N. and Easthouses) 6171.

Newton Hall, seat, Fife, 2½ m. E. of Markinch.

Newton Hall, seat, East Lothian, 2 m. S.W. of Gifford.

Newtonhead, pl., adjoining Ayr. P.O.

Newtonhill, vil., with ry. sta., L.M.S., N.E. Kincardineshire, 5½ m. N.E. of Stonehaven. P.O., T.O. Pop. 194.

Newton House, seat, Morayshire, 3½ m. W. of Elgin.

Newton House, seat, Lanarkshire, on r. Clyde, opposite Elvanfoot sta.

Newton House, Aberdeenshire, 2¼ m. E.N.E. of Insch ry. sta.

Newtonloan, pl., Midlothian, 3 m. S. of Dalkeith. County hospital.

Newton Mearns, vil., Renfrewshire, 7 m. S. by W. of Glasgow. P.O., T.O. Pop. 1548.

Newtonmill House, Angus, 2½ m. N. of Brechin.

Newtonmore, vil., with ry. sta., L.M.S., Inverness-shire, on r. Spey, 3 m. S.W. of Kingussie. P.O., T.O. Pop. 655. Hotel.

Newton of Balcanquhal, ham., Perthshire, 1½ m. N.E. of Glenfarg.

Newton of Balcormo, pl., E. Fife, 2 m. N.W. of Pittenweem.

Newton of Falkland, vil., Fife, 1 m. E. of Falkland. P.O. Pop. 143.

Newton of Ferintosh, ham., Nairnshire, 2 m. S.E. of Conon Bridge.

Newton of Pitcairns, vil., Perthshire, 4½ m. N.E. of Auchterarder.

Newton on Ayr, q.s. par. and ry. sta., L.M.S., Ayr.

Newton State Forest, near Mosstowie ry. sta., 3½ m. W. of Elgin, Moray.

Newton-Stewart, police bur. and ry. sta., L.M.S., Wigtownshire, on r. Cree (including also sub. of Creebridge, in Minnigaff par., Kirkcudbrightshire), 7 m. N. by W. of Wigtown and 23½ m. E. by N. of Stranraer. Pol. bur. pop. 1914. Hotels. P.O., T.O.

Newtown, vil., W. Lothian, ½ m. S. of Bo'ness. Pop. 975.

Newtown, ham., Stirlingshire, 5 m. S.E. of Balfron.

Newtown, pl., Morayshire, 1 m. S. of Garmouth.

Newtown, S.E. sub. of Dumbarton. P.O., T.O.

Newtown St. Boswells, vil., with ry. sta. (St. B.), L.N.E., Roxburghshire, 3¾ m. S.E. of Melrose. Pop. 669. P.O., T.O. Important ry. junction.

New Trows, ham., Lanarkshire, 1 m. S. of Lesmahagow.

Newtyle, par. and vil., with ry. sta., L.M.S., Angus; par. 5195 ac. Pop. 831. Vil., 16¾ m. N.W. of Dundee. P.O., T.O. Pop. 831.

Newtyle Hill, Perthshire, in eastern vicinity of Dunkeld; alt. 996 ft.

Newtyle State Forest, near Dunphail ry. sta., 8½ m. S. of Forres, Moray.

Nibon, isl., Shetland, in St. Magnus Bay, 5¼ m. S.E. of Hillswick.

Niddrie, colliery dist., city of Edinburgh, 1½ m. S.W. of Portobello. P.O.

Niddry, vil., West Lothian, 1 m. S.E. of Winchburgh. N. Castle, a ruined baronial fortalice. Received Mary, Queen of Scots, on her escape from Loch Leven.

Niddry Marischal, seat, city of Edinburgh, 3½ m. S.E. of Edinburgh G.P.O.

Nielsland, colliery, Lanarkshire, ½ m. N.E. of Meikle Earnock ry. sta.

Nigg, par., Kincardineshire, on r. Dee; 3982 ac. Pop. 1506. Church, 2 m. S.E. of Aberdeen. P.O. Situated near N. Bay is the Scottish Fishery Board's Experimental Station.

Nigg, par. and vil., N.E. Ross and Cromarty, on the N. side entrance to the Cromarty Firth; par. 7832 ac. Pop. 725. Vil., on Nigg Bay, 4 m. S. of Nigg ry. sta., and 4 m. by ferry N.E. of Cromarty. P.O., T.O. Pier shows fixed red light, seen 5 m. Hill of N., 1 m. E. of Nigg; alt. 666 ft.

Nigg Station, ry. sta., L.M.S., Ross and Cromarty, 4 m. N. of Nigg vil. and 5 m. S.E. of Tain. P.O., T.O.

Ninecairn Edge, mt., N.W. Berwickshire, 6 m. E. of Tynehead sta.; alt. 1479 ft.

Ninemileburn, vil., Midlothian, near Habbie's Howe, 4 m. S.W. of Penicuik.

Ninestone Rig, ridge, Roxburghshire, 7 m. N. of Newcastleton; alt. 943 ft.

Ninewells, seat, Berwickshire, 1 m. S.E. of Chirnside ry. sta.

Nisbet, a seat of Lord Sinclair, 2 m. S. of Duns, Berwickshire.

Nisbet, ham., with ry. sta., L.N.E., Roxburghshire, on r. Teviot, 3 m. N.N.E. of Jedburgh sta. P.O., T.O.

Nith, r., mainly of Dumfriesshire, but partly of Ayrshire and Kirkcudbrightshire; rising near Enoch Hill, Dumfriesshire, flows 70¾ m. in a S.S.E. course to the Solway Firth, 14¼ m. S. by E. of Dumfries.

Nith Lodge, shooting, 5 m. N.E. of Dalmellington, Ayrshire.

Nithsdale, the western of the divisions of Dumfriesshire. Takes its name from its being drained and traversed by the r. Nith.

Nitshill, vil., with ry. sta., L.M.S., Renfrewshire, 3¼ m. S.E. of Paisley. P.O., T.O. Chemical works.

Nivingston, seat, 3 m. S.W. of Kinross.

Noblehall, ham., Peeblesshire, 3 m. S.E. of West Linton.

Nochty, Water of, rivulet, Aberdeenshire, flows 5 m. S.E. to the Don, at Innernochty.

Noddsdale Water, Ayrshire, flows 7 m. S.W. to Largs Bay. Sometimes called Noddle Burn.

Noe, rivulet, Argyll, flows into N.W. of Loch Etive, 4 m. N.E. of Taynuilt.

Noltland Castle, ruin, Orkney, on N.W. coast of Westray.

Noness, coast vil., Shetland, Dunrossness par., 2 m. S.S.E. of Sandwick. Shows a ten-second flashing light, seen 15 m.

Noranside Sanatorium, Fern par., Angus, 7½ m. W. of Brechin.

Noran Water, stream, Angus, flows 10½ m. S.S.E. to the r. S. Esk, 4½ m. S.W. of Brechin.

Norby, vil., Walls par., Shetland, 32 m. W.N.W. of Lerwick.

Nordrach-upon-Dee, sanatorium, Kincardineshire, 1 m. N.W. of Banchory.

Normandykes, vestiges of a Roman camp, in S.W. corner of Peterculter par., Aberdeenshire.

Norman's Law, hill, with ancient fort, Fife, 4 m. N.E. of Newburgh; alt. 936 ft.

Norrie's Law, tumulus, Fife, 2½ m. N.W. of Largo. Many ancient relics have been found here.

Norrieston, q.s. par., S. Perthshire. Contains Thornhill vil., 3 m. S.W. of Doune.

North Ballachulish, vil., S. Invernessshire, N. side of Loch Leven, opposite Ballachulish. P.O., T.O.

Northbay, ham., Barra Isl., Outer Hebrides. P.O., T.O.

North Berwick, royal and mun. bur. and par., and a popular seaside resort, E. Lothian, at the S. entrance of the Firth of Forth, 22½ m. E.N.E. of Edinburgh by rail; ry. sta., L.N.E.; par. 5073 ac. Pop. 4083; mun. bur. pop. 3473. It has noted golf links, fine bathing pool, and an open-air swimming pond. P.O., T.O. Hotels. A coastguard sta. Shows a fixed light on N. pier head.

North British Distillery, in W. of city of Edinburgh.

North Broomage, pl., Stirlingshire, ½ m. from Stenhousemuir. The Scottish National Institution for the Education of Imbecile Children in vicinity.

North Bute, shooting-box, Buteshire, 8 m. N.W. of Rothesay.

North Church, q.s. par., Dunfermline, Fife.

North Church, eccl. par., Greenock, Renfrewshire.

North Church, eccl. par., Paisley, Renfrewshire.

North Connel, ry. halt sta., L.M.S., Argyll, ½ m. from Connel Ferry.

Northcraig Reservoir, Ayrshire, 3 m. S.E. of Stewarton.

North Craigs, ham., Dumfriesshire, 5½ m E. of Ecclefechan.

North Easdale, Argyll, isl. of Seil, 4 m. S.E. of Mull. P.O.

North Esk, q.s. par., Inveresk par., Midlothian. Church in Fisherrow.

North Esk, r., Angus. See ESK, NORTH.

North Esk, r., Midlothian. See ESK.

North Esk Distillery, Marykirk, S. Kincardineshire.

Northfield, seat, Annan par., Dumfriesshire, on left bank of r. Annan, 1 m. N. of the town.

Northfod, seat, Fife, 2 m. E. or Dunfermline.

North Haven, inlet, Fair Isl., between Orkney and Shetland Islands.

North Haven, pl., N.E. Aberdeenshire, 5 m. S.S.W. or Peterhead.

North Kirk, eccl. par., Aberdeen.

North Leith, ry. sta., L.N.E., Leith, city of Edinburgh.

Northmaven, par., N. Mainland, Shetland; 49,045 ac. Pop. 1343. Church at Hillswick.

Northmuir, vil., Kirriemuir par., Angus, 1 m. N. by W. of Kirriemuir.

North Ness, Lerwick, Shetland. Has fixed light.

Northouse, loch, mainland, Shetland, 6 m. N.E. of Walls; depth 13 ft., alt. 79 ft.

North Port Distillery, Brechin, Angus.

North Queensferry, see QUEENSFERRY, N.

North Roe, dist. and ham., in N. of Northmaven par., Shetland. Ham., W. of Yell Sound. P.O., T.O.

North Rotten Burn, stream, partly forming the boundary between Ayrshire and N.W. Renfrewshire, 4 m. long, and flowing to Shaw's reservoir.

Northshield Rings, hill-foot, 5¼ m. N. of Peebles.

Northton, ham., S. Harris, Invernessshire, 3 m. N.W. of Leverburgh. P.O., T.O.

North Water Bridge, ham., near mouth of N. Esk, with ry. sta., L.N.E., on border of Kincardineshire and Angus, 3 m. N. of Montrose. P.O., T.O.

North West Castle, Stranraer, Wigtownshire. At one time the home of Sir John Ross, the Arctic explorer, where his father was parish minister.

Norton, seat, Midlothian, 1 m. N.W. of Ratho.

17

Nor Wick, bay, N.E. coast of Unst Isl., Shetland.

Norwick, ham., Unst Isl., Shetland, 2½ m. N.E. of Haroldswick.

Norwood, seat, Clackmannanshire, near Alloa.

Norwood House, seat, 2 m. S.W. of Aberdeen.

Noss, isl., Bressay par., Shetland ; length 1½ m., on E. side Noss Head ; and a rock called the Noup of N., which attains on one side a precipitous and almost perpendicular height above sea-level of 592 ft. On the S.E. side the islet, Holm of N., one of the greatest curiosities of Shetland.

Noss Head, a bold prom., Caithness, 3¼ m. N.N.E. of Wick. Lighthouse showing a half-minute flashing light, seen 18 m.

Nostarie, loch, W. Inverness-shire, 1¼ m. S.E. of Mallaig.

Nostie, ham., 5 m. E. of Kyle of Lochalsh, W. Ross and Cromarty. P.O. called Nostle.

Nothland Castle, see NOLTLAND.

Noth, Tap o', hill, with remains of vitrified fort, Aberdeenshire, 7 m. S.W. of Huntly ; alt. 1851 ft.

Notman Law, mt., on borders of Peebles and Selkirk, 6 m. S.E. of Drumelzier ; alt. 2408 ft.

Noup Head, N.W. coast of Westray, Orkney. N. Hill, or the Stack of N., a little to the S.E. N. H. lighthouse shows a one-minute group-flashing light, seen 22 m.

Novar, ry. sta., L.M.S., E. Ross and Cromarty, 6½ m. N.E. of Dingwall. N. House, a seat. T.O.

Nungate, part of town of Haddington, E. Lothian.

Nunlands House, Berwickshire, 3½ m. S. by E. of Ayton.

Nunraw Castle, E. Lothian, ½ m. E.S.E. of Garvald. Built in the fifteenth century as a peel-tower or fortalice ; modernised.

Nunraw Reservoir, E. Lothian, 6½ m. S.E. of Haddington. North Berwick water supply.

Nunton, vil., Benbecula, S. Uist par., Outer Hebrides, 18 m. S.S.W. of Lochmaddy. P.O., T.O.

Nunwood, seat, N.E. Kirkcudbrightshire, 2½ m. N.W. of Dumfries.

Nutberry Hill, Lanarkshire, 6 m. S.W. of Lesmahagow ; alt. 1712 ft.

O

Oa, headland and q.s. par., Kildalton par., Islay, Argyll. Contains Port Ellon. The headland, Mull of Oa, 400 ft., has monument to U.S. soldiers who were temporarily buried here.

Oakbank, vil., with oil works, Midlothian, 1 m. S. of Mid-Calder. P.O.

Oakendean House, Roxburghshire, 1 m. E. of Melrose.

Oakfield, vil., S.W. Fife, 5 m. N.E. of Dunfermline.

Oakley, seat, S.E. vicinity of Kirkcudbrightshire.

Oakley with **Comrie,** vil., with ry. sta. (O.), L.N.E., Fife, 4½ m. W.N.W. of Dunfermline. P.O., T.O.

Oakwood Tower, ruined baronial fortalice, Selkirkshire, rt. bank of Ettrick Water, 4½ m. S.W. of Selkirk. Belonged to the famous wizard, Sir Michael Scott, later (sixteenth cent.) to Wat Scott of Harden.

Oathlaw, par., mid Angus. Church stands 4½ m. N.N.E. of Forfar ; 5273 ac. Pop. 362.

Oatlands, q.s. par., Govan, S.W. Glasgow. P.O.

Oatridge, pl., W. Lothian, 1½ m. N.N.W. of Uphall.

Oban, mun. bur., seapt., and q.s. par., with ry. sta., L.M.S., Argyll, 101 m. N.W. of Glasgow and 123½ m. N.W. of Edinburgh by rail. P.O. T.O. Great tourist and yachting centre, and one of the most fashionable watering-places in Scotland. Commands one of the grandest views of isl. and sea-coast scenery on W. of Scotland. Among the antiquities are the ruins of Dunolly and Dunstaffnage castles. Has regular steamboat communications with Glasgow, Inverness, and the Western Isles. Has a distillery. Mun. bur. pop. 5759. Hotels. Shows a fixed light (Dunollie), visible 11 m.

Obbe Point, S.W. coast of Harris, Outer Hebrides. Shows a five-seconds' occulting light, visible 11 m.

Obbe, see LEVERBURGH.

Obisary, loch, N. Uist, Outer Hebrides, 4 m. S. of Loch Maddy ; depth 151 ft., alt. 8 ft.

Obney Hills, E. Perthshire, 2¼ m. S. of Dunkeld ; alt. 1323 ft.

Occumster, ham., with ry. sta., L.M.S., Caithness, 1½ m. E.N.E. of Lybster. P.O.

Ochil Hills, a range extending for 24 m. over parts of the Counties of Clackmannan, Kinross, Fife, but chiefly in S.E. Perthshire. Chief summits are Bencleugh (2363 ft.), Dunnyat (1375 ft.), and King's Seat (2111 ft.). Ochil Hills Sanatorium, 2¼ m. N.W. of Milnathort.

Ochiltree, par. and vil., with ry. sta., L.M.S., Ayrshire, on Lugar Water; par. 18401 ac. Pop. 2023. Vil. 1½ m. N. of sta., and 11½ m. E. of Ayr. P.O., T.O. Vil. pop. 642. O. Castle, 1½ m. N.W. Convalescent Home.

Ochiltree, vil., 3 m. S.E. of Linlithgow. Shale-oil works.

Ochiltree, Loch, Wigtownshire, 8½ m. N.W. of Newton-Stewart; depth 34 ft., alt. 341 ft.

Ochterlony, seat, Angus, 1 m. S.W. of Guthrie sta.

Ochtertyre, seat, Perthshire, on r. Leith, 4 m. N.W. of Stirling.

Ochtertyre, seat, Perthshire, 3 m. N.W. of Crieff.

Ockle, ham., Argyll, 10 m. E.N.E. of Ardnamurchan Point; O. Point, 1 m. N.W.

Ockran Head, Shetland, N.W. coast of Mainland.

Octafad, ham., Islay, Argyll, on W. coast of Loch Indail, 3 m. S.W. of Port Charlotte.

Octomore, ham., Islay, Argyll, on W. coast of Loch Indail, 1 m. N.W. of Port Charlotte.

Odin Ness, headland, Orkney, on E. side of Stronsay.

Odin's Cave, coast of Mull Isl., Argyll, E. side of Loch Buy.

Ogilface, ruins of fortalice, W. Lothian, 3¼ m. W. of Bathgate.

Ogil House, Easter, Angus, Tannadice par.

Ogilvie, ruined castle, S. Perthshire, 1 m. S.E. of Blackford.

Ogilvie Glen, Angus, 2 m. S. of Glamis.

Oich, loch, Inverness-shire. Lies in the centre of the Great Glen, and forms the summit level of the Caledonian Canal. It is about 4 m. in length, and varies from ½ furlong to ¼ m. in breadth. It is fed by the Garry and sends off the r. Oich which flows 6½ m. toward Loch Ness, at Fort Augustus. Alt. 106 ft., depth 133 ft.

Oigh-sgeir, islet, W. of the isl. of Rhum, Inverness-shire. Lighthouse (Hyskier) with group-flashing light, seen 17½ m.

Oisinneach Mòr Lochan, Perthshire, 6 m. E.S.E. of Pitlochry. Lochan O Beag is 1 m. N.E.

Old Aberdeen, P.O., T.O. See ABERDEEN.

Oldany, isl., W. Sutherland, Assynt par. on S. entrance to Kylesku, 7 m. N. of Lochinver.

Old Bridge-of-Urr, ham., Kirkcudbrightshire, on Urr Water, 4 m. N. of Castle-Douglas.

Old Carsehead, Ayrshire, part of Dalry.

Old Craighall, mining vil., Midlothian, 1¼ m. S. of Musselburgh.

Old Dailly, ham., Ayrshire, 2 m. N.E. of Girvan.

Old Deer, par. and vil., Aberdeenshire. See DEER, OLD.

Old Hall, seat, Caithness, 5 m. S.E. of Halkirk.

Oldhamstocks, par. and vil., E. Lothian; par. 7397 ac. Pop. 405. Vil. on O. burn, 3 m. S. of Innerwick sta. P.O., T.O. Vil. pop. 51.

Old Head, headland, Orkney Isl., S.E. extremity of S. Ronaldshay.

Old Meldrum, police bur., with ry. sta., L.N.E., Aberdeenshire; vil., 5¾ m. N.E. of Inverurie. Bur. pop. 980. P.O., T.O. See MELDRUM.

Oldmill Poorhouse, 2¼ m. W. of Aberdeen.

Old Patrick Water, aff. of Black Cart Water, Renfrewshire, 2½ m. W.S.W. of Paisley.

Old Place, seat, Wigtownshire, 4½ m. S.W. of Whauphill ry. sta. See MOCHRUM.

Old Rayne, vil., Aberdeenshire, on r. Ury, 2 m. N. of Oyne ry. sta. P.O.

Old Rome, mining vil., Ayrshire, 2 m. W.S.W. of Kilmarnock.

Oldshore, ham., Sutherland, 3 m. N.W. of Kinlochbervie.

Old Water of Cluden, Kirkcudbrightshire, flows 6½ m. N.E. to Cairn Water.

Olginey, loch, N.W. Caithness-shire, 3 m. S.W. of Halkirk.

Oliver, seat and site of ancient baronial fortalice, S.W. Peeblesshire, nearly opposite Tweedsmuir Church.

Ollaberry, vil., E. coast of Northmaven par., Shetland, 17 m. N.W. of Voe. P.O., T.O.

Ollach pl., Skye Isl., 4½ m. S.E. of Portree.

Ollayht, Loch, S. Uist, Outer Hebrides, connected with L. Eynort.

Olna Firth, bay, Delting par., Shetland, penetrating 3 m. to the vil. of Voe, 19 m. N.N.W. of Lerwick ; ham. of Olnafirth at head of bay.

Olrig, coast par., N. Caithness, 10,025 ac. Pop. 1198 ; containing Castletown, 5 m. E. by S. of Thurso. O. House, a seat, 1½ m. S.W. O. Hill, 463 ft.

Omoa, ry. sta., L.M.S., and part of Cleland, Lanarkshire, 3 m. N.E. of Motherwell.

Onich, vil., Inverness-shire, at entrance to Loch Leven, 2½ m. W.N.W., of Ballachulish ferry. P.O., T.O. Hotel.

Opinan, ham., W. Ross and Cromarty, 3 m. S.W. Badachro, Gairloch. P.O.

Orangefield, seat, Ayrshire, 1 m. S.E. of Monkton ry. sta.

Orbliston, pl. and ry. junc., sta. L.M.S., N.E. Morayshire, 2½ m. W. of Fochabers. T.O. at sta.

Orbliston, New and **Old,** pl., with collieries, Lanarkshire, ½ m. S. Bellshill.

Orbost, seat, Skye Isl., 4 m. S.E. of Dunvegan.

Orcades, see ORKNEY.

Orchard, seat, Stirlingshire, 3 m. N.E. of Kilsyth.

Orchard Rig, mt., Peeblesshire, 2 m. W. of Traquair ; alt. 1463 ft.

Orchardton, seat, S. Kirkcudbrightshire, near the head of O. Bay, 6½ m. S.S.E. of Castle-Douglas.

Orchardton Tower, the only round tower in Galloway, 4 m. S.W. of Dalbeattie.

Orchill, seat, Perthshire, 3½ m. N.W. of Blackford sta.

Orchy, stream, Argyll, flows 16 m. S.W. through Glen Orchy to Loch Awe at Kilchurn Castle.

Orchy, Bridge of, see BRIDGE OF ORCHY.

Ord, q.s. par., N. Banffshire. Church, 3 m. S.S.E. of Tillynaught sta.

Ord, or **Muir of Ord,** pop. 307. See MUIR OF ORD.

Ord, or **Ord-of-Caithness,** an abrupt broad granite hill, on the border of Sutherland and Caithness, 4 m. N.E. of Helmsdale. Rises from the sea to 652 ft.

Ordain Hill, Inverness-shire, 9 m. E.N.E. of Kingussie ; alt. 1250 ft.

Ord Distillery, at Muir of Ord, Ross and Cromarty.

Ordens Halt, ry. sta., L.N.E., Banffshire, 1 m. E. of Tillynaught.

Ord Fundlie, hill, Aberdeenshire, 1 m. S.W. of Torphins ; alt. over 800 ft.

Ordhead, pl., Aberdeenshire, 3 m. S.S.W. of Monymusk. P.O., T.O.

Ord Hill, Ross and Cromarty, 2 m. N. of Inverness ; alt. 633 ft.

Ord House, Ross and Cromarty, 1¼ m. W. of Muir of Ord sta.

Ordie, vil., Aberdeenshire, 6¼ m. N.E. of Ballater. P.O.

Ordie Burn, rivulet, Perthshire, flows 8½ m. E.S.E. to the Tay a little above Luncarty.

Ordie Loch, Perthshire, 5 m. N. of Dunkeld ; depth 69 ft., alt. 946 ft.

Ordiequish, ham., N.E. Morayshire, 1 m. S. of Fochabers.

Ordiequish, hill with state forest, 1½ m. S. of Fochabers, Moray ; alt of hill, 401 ft.

Ordiquhill, par., N.E. Banffshire ; 4756 ac. Pop. 536. Church, 2 m. S.S.W. of Carnhill sta.

Ord, The, headland, in S. of Bressay, Shetland ; alt. 542 ft.

Ore, stream, S. W. Fife, flows 17 m. E. of r. Leven, near Cameron Bridge.

Oreval, mt., Harris, Outer Hebrides ; alt. 2165 ft.

Orgill Lodge, seat in N.E. of Hog isl., Orkney.

Orka Voe, bay, mainland, Shetland, 4 m. S.E. of Collafirth.

Orkney Islands, a group of 67 islands separated from Caithness by the Pentland Firth (6½ to 8 m. broad), and extending 50 m. N.N.E. Area, 376·3 sq. m., or 240,847 ac. Pop. 22,075. About thirty of the islands are inhabited ; the largest, known as Mainland, contains Kirkwall and Stromness, the only two towns. The country, except in Hoy, is low lying, being a continuation of the Old Red Sandstone plain of Caithness. Fishing forms an important industry ; the staple manufacture is knitted goods. The Orkneys were subdued by Harold Haarfager in 875 A.D., and were ruled by Scandinavian jarls down to 1231, and in 1468 were pledged to the Scottish Crown. Archæological remains include the standing stones of Stenness, and the cairn at Malshowe, also numerous bracks (fortresses) and underground houses. There is regular steamer communication with Leith, Aberdeen, Lerwick, and Thurso. Unites with Shetland in returning one member to Parliament.

Ormiclate, pl., S. Uist Isl., Outer Hebrides, 4 m. S. of Howmore.

Ormidale, seat, Argyll, at the head of Loch Riddon, 5½ m. N.E. of Tighnabruaich.

Ormidale Pier, on W. side of Loch Riddon, Argyll, 3 m. N.N.E. of Tighnabruaich.

Ormiston, seat, Midlothian, near Mid-Calder.

Ormiston, par. and vil., with ry. sta., L.N.E. (passenger service withdrawn), E. Lothian ; 3443 ac. Pop. 2032. Vil. on the E. bank of the Tyne, 2¾ m. S.S.E. of Tranent, 12 m. E.S.E. of Edinburgh by rail. P.O., T.O. Vil. pop. 1498. O. Hall, a seat of the Marquis of Linlithgow.

Ormiston House, Roxburghshire, 5 m. S.S.W. of Kelso.

Ormsaigbeg, Kilchoan Bay, Ardnamurchan, W. Argyll.

Ormsaigmore, Kilchoan Bay, Ardnamurchan, W. Argyll.

Ormsary, seat, Knapdale par., Argyll, 12 m. S.W. of Ardrishaig. P.O., T.O.

Ornsay, off N. coast of N. Uist Isl., Outer Hebrides.

Oronsay, rocky isl., N.W. Argyll, at mouth of Loch Sunart.

Oronsay, isl., Argyll, at S. end of Colonsay Isl., see COLONSAY and O. Pop. 6.

Orphan Homes of Scotland, Renfrewshire, 1½ m. N.W. of Bridge of Weir. P.O. called O. Homes.

Orphir, (including **Cava Isl.**), par. and vil., S. Mainland, Orkney ; 12,763 ac. Pop. 684. Vil. near W. end of Scapa Flow, 9 m. W.S.W. of Kirkwall. P.O., T.O.

Orrin, r., S.E. Ross and Cromarty ; rises 2¼ m. N. of Loch Monar, flows 26 m. E.N.E. to the Conon opposite Brahan Castle, 4 m. S.S.W. of Dingwall. O. Falls, 3 m. above the confluence with the Conon.

Orrock Hill, with basaltic columns, Fife, 1¾ m. N.N.W. of Burntisland.

Orroland, seat, 6½ m. S.E. of Kirkcudbright

Orsay, isl., S.W. Islay, Argyll. See OVERSAY.

Orton, pl., with ry. sta., L.M.S., Morayshire, 4 m. S.W. of Fochabers. P.O., T.O. T.O. at sta. O. House in vicinity.

Orval, mt., W. of Rum Isl., : alt. 1869 ft.

Orwell, par., N.W. Kinross-shire, on Loch Leven ; 13,113 ac. Pop. 1997.

Ose, glen and stream, N.W. Isle of Skye, Inverness-shire, 2½ m. N.W. of Bracadale.

Osnaburgh, or **Dairsiemuir,** vil., N.E. Fife, 2⅓ m. N.E. of Cupar.

Ossian Hostel, Loch, on Corrour Estate, on the N.W. corner of the Moor of Rannoch, 9 m. N. of Rannoch sta. ; 1300 ft above sea-level. A Scottish Youth Hostel, closed during shooting season.

Ossian, Loch, S. Inverness-shire, 15 m. E. of Ben Nevis ; depth 132 ft., alt. 1269 ft.

Ossian's Grave, large stone, near r. Almond, Perthshire, 5 m. S. of Amulree, 6 m. N.E. of Crieff.

Ossian's Hall, pl., Perthshire, 1½ m. W. of W. Dunkeld.

Otterburn, seat, Roxburghshire, 1 m. W.S.W. of Morebattle.

Otter Burn, trib. of Eye Water, Berwickshire, 1½ m. S.W. of Grant's House ry. sta.

Otterden House, Ayrshire, 3 m. N. of Maybole.

Otter Ferry, vil., Argyll, on E. shore of Loch Fyne, 5 m. S.E. of Lochgilphead. P.O., T.O. Ferry, 1½ m. broad, across Loch Fyne. O. House, seat, 5½ m. N.W. of Tighnabruaich.

Otter Rock, S. Islay, Argyll, 5 m. S.E. of Port Ellen. A light-ship showing a quarter-minute group-flashing light, visible 10 m.

Otterston, seat and loch, Fife, 1½ m. N.W. of Aberdour.

Otters Wick, bay, N. coast of Sanday, Orkney.

Otters Wick, pl. and bay, E. coast Yell Isl., Shetland.

Oude, stream, Lorne, Argyll, flows 4 m. S.W. to Loch Melford.

Ouplaymoor, vil., with ry. sta., Renfrewshire, 3½ m. S.W. of Neilston. P.O., T.O. called Uplawmoor.

Ourahag, small loch, 11 m. N.W. of Stornoway, Leurs', Outer Hebrides ; depth 33 ft., alt. 89 ft.

Outh, seat, S.W. Fife, 4 m. N. of Dunfermline.

Out Skerries, group of isles, 5 m. N.E. of Whalsay. Lighthouse (on Bound Skerry) showing a one-minute flashing light, visible 18 m. See HOUSIE SKERRIES.

Overbrae, hill, N.W. Aberdeenshire, 5 m. W.N.W. of New Pitsligo; alt. 701 ft.

Oversay Island, S.W. Islay, Argyll, near Rhynns Point. Lighthouse, showing a five-seconds' flashing light, seen 18 m. Pop. 14.

Overscaig Hotel, on N. side of L. Shin, Sutherland, 17 m. N.W. of Lairg.

Overstone, Fife, 1½ m. N. of Pittenweem. County Hospital.

Overton, vil. Dreghorn par., Ayrshire, 3¼ m. E. of Irvine.

Overton Paper Mills, Upper Greenock, Renfrewshire.

Overtoun, seat, 2 m. E. of Dumbarton.

Overtown, q.s. par. and vil., with ry. sta., L.M.S., N. Lanarkshire, 1½ m. S.E. of Wishaw. P.O., T.O. Pop. of O. and Waterloo, 3629.

Overtown, pl., Renfrewshire, 2½ m. W. of Howwood.

O Wood Row, pl., N. Lanarkshire, 1 m. N.E. of Holytown.

Owskeich, loch, N.W. Ross and Cromarty, 2 m. N.E. of Achiltibue Hotel, greatest depth, 153 ft, alt. 72 ft.

Oxcar, a rock in the Firth of Forth, S.E. of Inchcolm, 5 m. N.W. of Leith. Lighthouse, shows a ten-seconds' occulting light (white and red sectors) seen 12 m.

Oxendean, seat, Berwickshire, 2 m. N. of Duns.

Oxenfoord Castle, formerly a seat of the Earl of Stair, Midlothian, on r. Tyne, 4 m. E.S.E. of Dalkeith. Now a private school for girls.

Oxhill, seat, Stirlingshire, 1 m. E. of Buchlyvie sta.

Oxna, isl., Tingwall par., Shetland, 3¼ m. S.W. of Scalloway.

Oxnam, par. and ham., S.E. Roxburghshire; par. 20,459 ac. Pop. 522. Ham. on Oxnam Water, 4½ m. S.E. of Jedburgh, which flows 9¾ m. from the Cheviots to the Teviot near Crailing.

Ox Rocks, rocky islets, W. Wigtownshire, 1 m. S.W. of Corsewall Lighthouse.

Oxton, vil. and ry. sta., L.N.E. (passenger service withdrawn), Berwickshire, near the r. bank of Leader Water, 5 m. N.N.W. of Lauder. P.O., T.O. Pop. 92.

Oykell, r., Sutherland and Ross and Cromarty, flows 35¼ m. from near Ben More to the Dornoch Firth at Bonar Bridge, passing through Loch Ailsh.

Oykell Bridge, pl., Ross and Cromarty, 17 m. S.W. of Lairg Inn.

Oyne, par. and ry. sta., L.N.E., Aberdeenshire; 10,131 ac. Pop. 630. Sta., 4 m. W. by N. of Inveramsay sta., and 24½ m. N.W. of Aberdeen. P.O., T.O.

P

Pabay or **Pabaidh,** isl. in W. Loch Roag, W. coast of Lewis, Outer Hebrides. Pop. 3.

Pabay, isl. off Broadford Bay in S.E. of Skye, Inner Hebrides.

Pabbay, isl. in Sound of Harris, Outer Hebrides. Pop. 3.

Pabbay, isl., 6½ m. S. of Barra in Outer Hebrides, rises to 560 ft. at the Hoe.

Padanaram, vil., 2 m. N.W. of Forfar, Angus.

Paddy Rock, Upper Loch Fyne, opposite Lachlan Bay, Argyllshire; 6-seconds' flashing light, visible 9 m.

Paible, ham., 11 m. W. of Lochmaddy, North Uist isl., Outer Hebrides.

Pairney Burn, flows 5½ m. N.W. to Ruthven Water, S. Perthshire.

Paisley, par., mun. and parl. bur., 53 m. W. by S. of Edinburgh and 7 m. S. by W. of Glasgow; par. 16,665 ac. Pop. 120,268. The town has two stations on the L.M.S. (St. James and Gilmour Street). P.O., T.O. Pop. 86,441. The most important industry is the manufacture of cotton thread, but there are numerous bleaching and printing works, also carpet works; the bur. returns one member to Parliament.

Palace, seat, 4 m. N.E. of Jedburgh, Roxburghshire.

Palace Brae, pl., where formerly stood a mansion belonging to the ancestors of the Duke of Atholl, in S. Perthshire, 2½ m. S.E. of Dollar.

Palace Craig, see FASKINE AND PALACE CRAIG.

Palacerigg, estate belonging to Glasgow Distress Committee, 2 m. S.E. of Cumbernauld ry. sta., in the detached portion of Dunbartonshire.

Paldy, ancient chapel founded in honour of St. Palladius ; Fordoun par. churchyard, Kincardineshire.

Palmer's Cross, seat, 1 m. S.W. of Elgin, Morayshire,

Palnackie, small port, at mouth of the Urr Water, 4 m. S. of Dalbeattie, Kirkcudbrightshire. P.O., T.O. Pop. 177.

Palnure, vil., near junction of Palnure Burn and the r. Cree, W. Kirkcudbrightshire, and 3 m. S.E. of Newton-Stewart. P.O., T.O. Ry. sta., L.M.S.

Pananich Wells, pl. with chalybeate springs, 2 m. E. of Ballater, Aberdeenshire ; Pananich Hill (1896 ft) is 1 m. S.W.

Panbride, par. and fishing ham. in S.E. Angus ; par. 6104 ac. Pop. 1584. the ham. is 1 m. N.E. of Carnoustie.

Pan Hope, bay on E. side of isl. of Flotta, S. Orkney.

Panmure House, seat of Earl of Dalhousie, 3 m. N.W. of Carnoustie, Angus.

Pannanich Wells, see PANANICH WELLS.

Papa, isl., 2 m. S.W. of Scalloway, Shetland.

Papa, Holm of, Shetland. See HOLM OF PAPAL.

Papa, Little, isl. at entrance to Aith Voe, W. coast of Mainland, Shetland.

Papa Sound, between Stronsay and Papa Stronsay, in N.W. of Orkney.

Papa Sound, between Westray and Papa Westray, N. Orkney.

Papa, Sound of, between Papa Stour and Mainland, W. Shetland.

Papa Stour, isl. ($2\frac{3}{4}$ m. by $2\frac{1}{4}$ m.) in W. of Shetland, on S. side of St. Magnus Bay and 11 m. N.W. of Walls. P.O. Pop. 100.

Papa Stronsay, small isl., $\frac{1}{4}$ m. off N.E. of Stronsay Isl., N.E. Orkney. Pop. 18.

Papa Westray, small isl. in N. Orkney, $1\frac{1}{4}$ m. N.E. of Westray Isl. and 25 m. N. of Kirkwall. P.O. The northern extremity is a headland called the Mull of Papa. Pop. 237.

Papdale House, seat in the eastern neighbourhood of Kirkwall, Orkney.

Paphrie Burn, trib. of the West Water which joins the N. Esk, 6 m. N.W. of Brechin, Angus.

Papigoe, ham., $1\frac{1}{2}$ m. N.E. of Wick, Caithness.

Paps of Jura. See JURA.

Park, small loch, $2\frac{1}{2}$ m. N.E. of Dufftown, Banffshire.

Park, seat and ry. sta., L.N.E., in co. of and 11 m. S.W. of Aberdeen.

Park, seat in par. of Inchinnan, Renfrewshire.

Park, seat, $1\frac{1}{2}$ m. S.E. of Cornhill ry sta., Banffshire.

Park, dist. between Loch Erisort and Seaforth Loch, in S.E. of Isl. of Lewis, Outer Hebrides.

Park, q.s. par in Glasgow.

Park Burn, stream flowing 5 m. S.E. to join the Nith, 2 m. N.W. of Thornhill, N.W. Dumfriesshire.

Parkend, seat, 3 m. S.W. of Lockerbie, S. Dumfriesshire.

Parkfoot, part of the town of Falkirk, Stirlingshire.

Parkfoot and Longcroft, vil., in par. of and $2\frac{1}{2}$ m. S.W. of Denny, Stirlingshire.

Parkgate, ham., in co. of and 8 m. N.E. of Dumfries ; P.O., T.O.

Parkgate, ham., in co. of and $2\frac{1}{2}$ m. S. of Kirkcudbright.

Parkhead, q.s. par. and ry. stas. L.M.S and L.N.E., in E. of Glasgow. P.O., T.O., Heavy steel casting works.

Parkhill, seat, 1 m. N.E. of Blairgowrie, Perthshire.

Park Hill, ry. sta., L.N.E., $7\frac{1}{2}$ m. N.W. of Aberdeen, $\frac{3}{4}$ m. S.E. of the station and on R. Don is Parkhill House.

Parkhill, seat, 3 m. N. of Arbroath, Angus.

Parkhill, seat, $\frac{1}{2}$ m. N. of Polmont sta., E. Stirlingshire.

Parkhouse, seat, in par. of Neilston, Renfrewshire.

Park House, seat in the S. side of town of Kirkcudbright.

Park, Loch, Banffshire. See LOCH PARK.

Park, Loch, 2 m. N.W. of Park sta., N. Kincardineshire.

Park Lodge, seat, Lochs par., Lewis Isl., Outer Hebrides.

Parkmore Distillery, $\frac{1}{2}$ m. N.E. of Dufftown, Banffshire.

Parknook, Fife. See MILESMARK and PARKNOOK.

Park Place, par. of St. Ninian, Stirlingshire ; special drainage dist. Pop. 305.

Park Place, ancient Castle near Glenluce, Wigtownshire.

Park Terrace Rows, part of Den vil., $2\frac{1}{4}$ m. N.E. of Dalry, N. Ayrshire.

Parlan Hill, alt. 2001 ft., 2½ m. N.E. of Ardlui Pier, N.E. extremity of Dunbartonshire.

Parney Burn, Perthshire. See PAIRNEY BURN.

Partick, parl. division of Glasgow. Pop. 55,370.

Partick, East and **West,** two wards and q.s. par., in Govan par., on the N. bank of the Clyde and the W. bank of the Kelvin in the West of Glasgow; ry. stas., (Partick) L.N.E., (Partick Central and Partick West) L.M.S. P.O., T.O. Ferry to Govan; has ship-building, sawmilling, and engineering.

Parton, par. and ham., in mid Kirkcudbrightshire; par. 15,813 ac. Pop. 548. The ham. stands on the r. Dee, 7 m. N.W. of Castle-Douglas; ry. sta., L.M.S. P.O., T.O. P. House is a seat, 1½ m. E. of sta.

Pass of Awe, between Loch Awe and Loch Etive in Argyllshire; is a continuation of Pass of Brander.

Pataig, see r. PATTACK.

Pathhead, part of the vil. of New Cumnock, E. of mid Ayrshire.

Pathhead, vil. on r. Tyne, 5 m. S.E. of Dalkeith, Midlothian, and 11 m. S.E. of Edinburgh. Pop. 422.

Pathhead, eccl. par. in par. of Kirkcaldy and Dysart, Fife, in royal bur. of Kirkcaldy. P.O., T.O.

Path of Condie, ham. on the Water of May, 5½ m. S.E. of Forteviot ry. sta., S.E. Perthshire. P.O. called Pathstruie.

Pathstruie, Perthshire. P.O. See PATH OF CONDIE.

Patiemuir, vil. in Fife. See PETTYMUIR.

Patna, q.s. par. and vil. in mid Ayrshire; vil. stands on r. Doon, 10 m. S.E. of Ayr; ry. sta., L.M.S. P.O., T.O. Pop. 465. Coal and iron industries.

Pattack, r. flowing from Loch Pattack, 10 m. N. to Loch Laggan, Invernessshire.

Pattack, Loch Pattack, 6½ m. S. of Loch Laggan Hotel; depth 69 ft., alt. 1419 ft.

Patt Deer Forest, 1 m. S.W. of Loch Monar, S. Ross and Cromarty.

Patterton, pl. in Renfrewshire, 7½ m. S.W. of Central sta., Glasgow; ry. sta. (for Darnley Rifle Ranges), L.M.S.

Pavilion, adjacent to Ardrossan, N. Ayrshire.

Pavilion, seat, on r. Tweed, 1½ m. N.W. of Melrose, Roxburghshire.

Paxton, vil. on the Whitadder, S.E. Berwickshire, 5 m. W. of Berwick-on-Tweed. P.O., T.O. 1 m. S.W. is P. House.

Peacock Cross, dist. in the W. of Hamilton, Lanarkshire. P.O., T.O.

Peaks of the Castles, mt. alt. 2735 ft., in N. of the Isl. of Arran, Buteshire.

Peallach, one of the Mishnish lochs, 1½ m. S.W. of Tobermory, Mull, Argyllshire.

Pean, small stream flowing through Glen Pean into W. end of Loch Arkaig. S.W. Inverness-shire.

Pearsie, seat on the Prosen Water, 4 m. N.W. of Kirriemuir, Angus.

Pearson (Loch a'Phearsain), loch, ½ m. N.E. of Kimelfort, dist. of Lore, Argyllshire; depth 53 ft., alt. 226 ft.

Peartree, pl. in Kirkcudbrightshire, 7½ m. W. of Dumfries.

Pease Dean, deep wooded ravine, 1¼ m. S.E. of Cockburnspath, Berwickshire. Pease Bridge (1786) crosses the ravine and is 300 ft. long and 127 ft. high.

Peat Inn, pl. 3½ m. E. of Ceres, Fife. P.O.

Peat Law, alt. 1209 ft., 6 m. S.W. of Oldhamstocks, S.E. E. Lothian.

Peaton, ham. and seat on Loch Long, 3 m. N.W. of Roseneath, Dunbartonshire.

Peddieston, pl. in par. of, and 4 m. S.W. of, Cromarty, Black Isle, E. Ross and Cromarty.

Peebles, par., ancient royal bur. and county town, 22 m. S. of Edinburgh, and 22 m. W. by N. of Selkirk, at confluence of Tweed and Eddleston Water; par. 16,603 ac. Pop. 6402; pop. bur. 5853. P.O., T.O. Ry. stas., L.N.E. and L.M.S. A holiday resort with hydropathic. Hotels; golf course; and there are woollen mills.

Peeblesshire, otherwise Tweeddale, in the S. part of Scotland, is bounded on the N. and N.E. by Midlothian and on the E. and S.E. by Selkirkshire, on the S. by Dumfriesshire and on the W. by Lanarkshire; extreme length over 30 m., extreme breadth 20 m.; area, 347 sq. m. or 222,240 ac. of land, and 690 ac. of water; the county is an uninterrupted series of hill and mountain ranges and there is scarcely a plain of moderate dimensions to be found; chief heights, Hartfield (2916 ft.), Broad Law (2850 ft.), Dollar Law (2840 ft.). Chief crop oats, but large quantities of turnips are raised as winter fodder for the large number of sheep raised. Principal r. is the Tweed; county served by L.N.E. and L.M.S. rys. The counties of Midlothian and Peebles return two members to Parliament. The county contains one royal bur., Peebles, and one police bur., Innerleithen. Pop. 15,051.

Peel, mansion on r. Tweed, in co. of, and 4½ m. N.W. of, Selkirk.

Peel Bog, ancient circular fort, just S.W. of Lumphanan church, 4 m. N.E. of Aboyne, S. Aberdeenshire.

Peel Fell, alt. 1975 ft., 13 m. S. of Jedburgh on border of Roxburghshire and Northumberland.

Peelwalls, seat, 1 m. S. of Hyton, E. Berwickshire.

Peesweep Row, ham. close to Roughrigg, Lanarkshire.

Peesweep Sanatorium, Neilston, Renfrewshire.

Peffer Burn, East and **West,** two streams in E. Lothian; the former flows 8 m. N.E. to the North Sea, 2½ m. below Whitekirk; the latter flows 7 m. W. to Aberlady Bay in the Firth of Forth.

Peffermill, pl., ¼ m. W. of Duddingston sta., Edinburgh. University athletic grounds.

Peffery, small stream flowing 9¼ m. through Strathpeffer to Cromarty Firth, near Dingwall.

Peinchorran, ham. on N. shore of Loch Sligachan, Skye, Inner Hebrides.

Peirceton, seat, Ayrshire. See PERCETON.

Pencaitland, par. and vil. in E. Lothian; par. 5056 ac. Pop. 1398. Vil. on r. Tyne, 2 m. E.S.E. of Ormiston and 13¾ m. E.S.E. of Edinburgh by rail; ry. sta. L.N.E. (passenger service withdrawn). P.O., T.O. Vil. pop. 601.

Penchrise, shooting-lodge, 5 m. S. of Hawick, Roxburghshire.

Penicuik, par. and town (police bur.) on the N. Esk, in S. Midlothian, 10 m. by road and 15¾ m. by rail S. of Edinburgh; par. 19,429 ac. Pop. 5198; pop. bur. 2750. The town has paper mills; in the dist. are coal, shale, and iron mines. Ry. sta. (550 ft.), L.N.E. P.O., T.O. P. House is 1¼ m. S.W. of town.

Penielheugh, hill, alt. 774 ft., 4 m. N.E. of Jedburgh, Roxburghshire; crowned by Waterloo Column (150 ft. high).

Penifiler, ham., 3 m. S.E. of Portree, Skye, Inner Hebrides. P.O.

Peninver, ham., 2½ m. N.E. of Campbeltown, Kintyre, Argyllshire. P.O.

Penkill Burn, stream in W. Kirkcudbrightshire, joining the r. Cree, after an 8 m. S.W. course, near Newton-Stewart.

Penkill Castle, seat, 3 m. N.E. of Girvan, Ayrshire.

Pennan, fishing vil. in N. Aberdeenshire, close to the Banffshire boundary; coastguard sta. P.O., T.O. Has two fixed lights.

Pennick, ham. in co. of, and 4 m. S.E. of, Nairn; the deans of Moray formerly resided here.

Penninghame, par. on the r. Cree, N.E. Wigtownshire; 33,765 ac. Pop. 2924. Contains Newton-Stewart, 4 m. N.W. of which stands P. House.

Pennyghael, seat and ham. on Loch Scridain, and 8 m. N.E. of Bunessan, Mull, Argyllshire. P.O., T.O.

Pennygown, pl. on Sound of Mull and Firth of Lorne, Mull Isl., Argyllshire. Contains the ruins of Duart Castle and of Lochbury Castle.

Penpont, par. and vil. in N.W. Dumfriesshire; par. 21,998 ac. Pop. 783. Vil. stands on Scar Water, 2 m. S.W. of Thornhill. P.O., T.O. Vil. pop. 358.

Penrioch, pl. on N.W. of isl. of Arran, Buteshire.

Penshiel Hill, alt. 1395 ft., 7 m. S.E. of Gifford, E. Lothian.

Penston, colliery vil., 1 m. S.W. of Gladsmuir, E. Lothian.

Pentland Firth, strait between Caithness and Orkney, connecting the North Sea and Atlantic Ocean. It is from 6 to 8 m. broad, and 14 m. long. Navigation is very difficult and dangerous owing to the rapidity of the tidal current (6 to 10 m.p.h.), and the dangerous nature of eddies.

Pentland Hills, a broken hill-range, stretching from 3 m. S.W. of Edinburgh in a S.W. direction for some 16 m. through Midlothian, Peebles and Lanarkshire to near Carnwath. The highest point is Scald Law (1898 ft.). The hills are generally rounded, affording good pasture for sheep, and many streams, with which are connected several reservoirs, have their source here.

Pentland, New, ham., 1 m. W. of Loanhead ry. sta., Midlothian, and 4½ m. S. of Edinburgh G.P.O.

Pentland Skerries, group of isls., Orkney, at E. entrance to the Pentland Firth and 4½ m. N.E. of Duncansbay Head; on Muckle Skerry is a lighthouse 118 ft. high with an 18-seconds' group-flashing light, visible 19 m. Pop. 5.

Perceton, seat, 2½ m. N.E. of Irvine, N. Ayrshire.

Persie, q.s. par. on the Black Water, in pars. of Alyth and Kirkmichael, N.E. Perthshire; par. church, 9 m. N. of Blairgowrie.

Persley, ry. sta., L.N.E., in co. of, and 3¼ m. N.W. of, Aberdeen.

Perth, (formerly St. John's Town or St. Johnstoun), par., ancient city and royal bur. and co. town, on r. Tay, 10 m. from the Firth, 46 m. N. of Edinburgh, 22 m. S.W. of Dundee, and 62 m. N.E. of Glasgow; par. 2653 ac. Pop. 28,613; bur. pop. 34,807. Ry. sta. used jointly by L.M.S. and L.N.E. P.O., T.O. The city stands in Strathmore, bounded on the N. and S. by two fine meadows known as the N. and S. Inches, and used as public recreation grounds. The city has long been noted for its dyeworks, but also makes inks, linen and jute, carpets, beer, and whisky; a centre of live-stock trade in Scotland.

Perthshire, one of the most extensive cos. in Scotland, lying nearly in the centre of the country, and forming the S.E. dist. of the Highlands; bounded on the E. by Angus, on the S.E. by Kinross-shire and Fifeshire, further on the S. its boundaries are Stirlingshire and Clackmannanshire, on the S.W. Dumbarton touches it for a short distance, on the W. Argyllshire, and on the N.W. and N. it is bounded by the cos. of Inverness and Aberdeen; area, 2493·7 sq. m. or 1,595,802 ac. Pop. 120,793. The co. abounds in lakes and rivers. In the S.E. the Ochil and Sidlaw Hills are separated by the Firth of Tay, while the N. and N.W. are occupied by the Grampians; between them lies the fertile vale of Strathmore, while the Carse of Gowrie is that fertile tract between the Sidlaws and the Firth of Tay. A little coal, copper, and ironstone are found in the Ochils. There are extensive deer forests in the co., and the r. fisheries are valuable. The principal industries in the towns and vils. of the co. are the weaving of woollens, linens, and cottons; also dye works. Fruit growing and the raising of animals are important in the country dists. For parl. purposes, Perth and Kinross are divided into two divisions, each returning one member.

Peterculter, par. on left bank of lower r. Dee, S.E. Aberdeenshire; 12,784 ac. Pop. 6309. P.O., T.O. Church is near Culter sta., 7½ m. S.W. of Aberdeen.

Peterhead, par., mun. bur., and seapt. in the dist. of Buchan, N.E. Aberdeenshire; par. 9449 ac. Pop. 15,285. Town is situated on a peninsula (the most easterly point of Scotland), about a mile S. of the mouth of the Ugie R., and is 141 m. N. by E. of Edinburgh, and 33 m. N.E. of Aberdeen; ry. sta. L.N.E. P.O., T.O. Pop. 12,545. Centre of herring fishing and coastguard and lifeboat sta. On the W. pier of the S. harbour is a fixed red light, visible 10 m., and a similar light is on the N. harbour, while on the breakwater is an 11-seconds' group-flashing light, visible 8 m. In the dist. are red granite quarries and a large convict prison.

Peter Hill, alt. 2025 ft., 7 m. S.E. of Aboyne, S. Aberdeenshire.

Petershill Reservoir, W. Lothian, ¾ m. N.E. of Bathgate, to which town it supplies water.

Petmathen House, seat, near Oyne ry. sta. and 7½ m. N.W. of Inverurie, Aberdeenshire.

Pettinain, par. and vil. on the Clyde, Lanarkshire; par. 3889 ac. Pop. 277. Vil. is 3 m. S. of Carstairs Junction. P.O.

Petty, par. and vil. on the Moray Firth, N.E. Inverness-shire; par. 9757 ac. Pop. 1223. Vil. 1½ m. S.W. of Dalcross ry. sta. and 5 m. N.E. of Inverness.

Pettycur, small watering-place on bay of same name, 1 m. W. of Kinghorn, Fife; small harbour and quay.

Pettymuir, vil., 2½ m. S. of Dunfermline, Fife.

Phantassie, seat, ½ m. E. of E. Linton, E. Lothian.

Pharay (**Fara,** or **South Pharay**), isl., ½ m. E. of isl. of Hoy, Orkney. Pop. 28.

Pharay (Fara, or **North Pharay),** isl., 1 m. W. of isl. of Eday, Orkney; near its N. end is the Holm of Pharay, a pastoral isl. Pharay Sound separates the isl. from Eday. Pop. 40.

Phesdo, seat, 5 m. N.W. of Laurencekirk, Kincardineshire.

Philiphaugh, seat on Yarrow Water, in co. of, and 3 m. S.W. of, Selkirk. In 1645 Montrose was here defeated in battle by Leslie.

Philorth and **Philorth Bridge Halt,** ry. stas., L.N.E., 2 m. S. of Fraserburgh, Aberdeenshire. P., the seat of Lord Saltoun, is nearby.

Philpstoun, vil., 3½ m. E. of Linlithgow, W. Lothian; ry. sta., L.N.E. P.O., T.O. 1 m. N.E. is the seat P. House.

Phoines, seat, 4 m. S. of Newtonmore, Spey Valley, Inverness-shire.

Physgill, seat, 2 m. S.W. of Whithorn, S.E. Wigtownshire.

Pickerstonhill, part of Newarthill vil., 3 m. N.E. of Motherwell, Lanarkshire.

Picketlaw Reservoir, ½ m. S. of Eaglesham, S.E. Renfrewshire.

Piehills, ham., 9½ m. N.W. of Annan, Dumfriesshire.

Pierceton, seat, Ayrshire. See PERCETON.

Pierowall, vil. and bay in the N.E. of Westray Orkney; fixed white light, visible 5 m.

Piershill, dist. in the E. of Edinburgh; ry. sta. and depot, L.N.E.

Piersland, seat, Troon, Ayrshire.

Pike Fell, alt. 1637 ft., 7 m. N.E. of Langholm, Dumfriesshire.

Pike, The, mt., alt. 1454 ft., 4½ m. S.E. of Ettrick Church, Selkirkshire.

Piketlaw Hill, alt. 1844 ft., 8 m. N. of Langholm, border of cos. of Dumfries and Roxburgh.

Pikey Hill, alt. 1164 ft., 5 m. N.W. of Rothes, Moray.

Pillar Rock Point, on Holy Isl., mouth of Lamlash Bay, E. coast of Arran, Buteshire; has half-minute alternating flashing light, visible 17 m.

Pilmour Links, pl., with a golf course, 1 m. N.W. of St. Andrews, Fife.

Pilmuir, ham., 3½ m. S.W. of Haddington, E. Lothian.

Piltanton Burn, rises near Stranraer and flows 3 m. E. to Luce Bay, Wigtownshire.

Pilton, dist. in the Granton area, N. Edinburgh; has a public health hospital; ry. halt on Leith Branch, L.M.S.

Pinkerton, pl., 3 m. S.E. of Dunbar, E. Lothian.

Pinkhill, ry. sta., L.N.E., for the Corstorphine area of Edinburgh; 3 m. W. of Waverley sta.

Pinkie House, in E. neighbourhood of Musselburgh, Midlothian; battle of Pinkie, 1547.

Pinmore, pl., 5 m. S. of Girvan, Ayrshire; ry. sta., L.M.S. T.O. at sta. The seat, P. House, is 1½ m. S.E.

Pinnacle, seat, 4½ m. N.W. of Jedburgh, Roxburghshire.

Pinnaclehill, seat, on the Tweed, near Kelso, Roxburghshire.

Pinwherry, pl. on r. Stinchar, 8 m. S. of Girvan, Ayrshire; ry. sta., L.M.S. P.O., T.O.

Piperhall, pl., 4 m. S.E. of Rothesay, Bute Isl., Buteshire.

Pirn, seat on Gala Water, 8 m. N.W. of Stow, Midlothian.

Pirnie, pl., Roxburghshire, 2½ m. S.E. of Maxton.

Pirnmill or **Alltgobhlach,** vil., with steamboat pier on N.W. of Arran, Kilbrennan Sound, Buteshire, 6 m. S.W. of Loch Ranza. P.O., T.O.

Pirntaiton, pl., 4 m. N.W. of Stow, Midlothian; has traces of a Roman camp.

Pitbladdo, seat, 2 m. N.W. of Cupar, Fife.

Pitcairlie, seat, 2½ m. N. of Auchtermuchty, Fife.

Pitcairnfield, seat in co. of, and 3 m. N.W. of, Perth.

Pitcairngreen, vil. on r. Almond, in co. of, and 4½ m. N.W. of, Perth.

Pitcairns, Newton of, part of vil. of Dunning, in co. of, and 10 m. S.W. of, Perth; nearby is the seat P.

Pitcaithly Wells, see PITKEATHLY.

Pitcalnie, pl., 1½ m. N.E. of Nigg, N.E. Ross and Cromarty.

Pitcaple, vil. on the Ury, 5 m. N.W. of Inverurie, mid Aberdeenshire; ry. sta., L.N.E. P.O., T.O. Nearby is P. Castle.

Pitcarity, vil. in Glen Prosen, 8 m. N.W. of Kirriemuir, Angus.

Pitcarmick, seat, 9 m. N.W. of Blairgowrie, Perthshire.

Pitcastle, seat, near r. Tay, 5½ m. N.W. of Ballinluig Junction, Perthshire.

Pitchroy, seat, ½ m. S.W. of Blacksboat ry. sta., Morayshire.

Pitcorthie, former seat, now a Church of Scotland Holiday Home, 1½ m. N.E. of Colinsburgh, Fife.

Pitcox, pl. in par. of Stenton, 4 m. S.W. of Dunbar, E. Lothian.

Pitcrocknie Platform, ry. sta. L.M.S., 1 m. from Alyth, Perthshire.

Pitcruvie, see BALCRUVIE.

Pitcullo, seat, 4½ m. N.E. of Cupar, Fife.

Pitcur, vil., par. of Kettins, 3 m. S.E. of Coupar-Angus, Angus ; ruins of a castle here.

Pitfichie Castle, ruin, 2 m. N.W. of Mony-musk, Aberdeenshire.

Pitfirrane Park, seat, 2¼ m. S.W. of Dunfermline, Fife.

Pitfodels, ry. sta., L.N.E., in co. of, and 3 m. S.W. of, Aberdeen ; a high service reservoir here.

Pitfour Castle, seat, in co. of, and 6 m. E. of, Perth.

Pitfour House, seat, 2 m. N.W. of Mint-law, Aberdeenshire.

Pitgaveny House, seat, 2¼ m. N.E. of Elgin, Morayshire.

Pitheavlis, seat, in co. of, and 1 m. S.W. of, Perth.

Pitkeathly, pl., with mineral wells, 1 m. S.W. of Bridge of Earn, Perthshire.

Pitkellony, seat near Muthill, Perthshire.

Pitkennedy, pl., 5¼ m. N.E. of Forfar, Angus.

Pitkerro House, seat, near Kingennie sta., Monifieth, Angus.

Pitkierie, Easter and **Wester,** two seats, 1½ m. N.W. of Anstruther, Fife.

Pitleoch Burn, stream, joining the Ballin-loan and flowing to the Bran, 3½ m. W.S.W. of Dunkeld, Perthshire.

Pitlessie, vil. in par. of Cults, 4 m. S.W. of Cupar, and 2 m. E. of Ladybank, Fife. P.O., T.O. Pop. 328.

Pitliver, seat, 2 m. S.W. of Dunfermline, Fife.

Pitlochry, vil., delightfully situated in the wooded valley of the Tummel, 28½ m. N. of Perth ; ry. sta., L.M.S. P.O., T.O. Has two hydropathics and two distilleries ; manufactures woollens. Hotels. Pop. 1942.

Pitlour House, seat, 1¼ m. N.W. of Strath-miglo, Fife.

Pitlurg, seat, 6 m. N.E. of Ellon, Aberdeen-shire ; ry. sta., L.N.E. (passenger service withdrawn). T.O. at sta.

Pitlyal, or **Round Loch,** Tay Basin, S. Angus, 5 m. E. of Coupar-Angus ; alt. 606 ft., greatest depth 19 ft.

Pitmain, seat, 1 m. S.W. of Kingussie, E. Inverness-shire.

Pitmedden, pl. in co. of, and 8¼ m. N.W. of, Aberdeen ; ry. sta., L.N.E. ; nearby is P. House, a seat.

Pitmedden, ham., 4 m. S.W. of Ellon, Aberdeenshire ; P. House is 1 m. N.W.

Pitmiddle, vil., ¾ m. N. of Kinnaird, E. Perthshire.

Pitmilly, seat, 1 m. E. of Boarhills ry. sta., Fife.

Pitmudie, ham., 7½ m. W. of Kirriemuir, Angus.

Pitmuies, seat, ¼ m. S. of Guthrie Junction, S. Angus.

Pitmurchie House, 1½ m. S.E. of Lum-phanan ry. sta., Aberdeenshire.

Pitmurthly House, in co. of, and 4 m. N.W. of, Perth.

Pitnacree, seat, on r. Tay, 4 m. N.E. of Aberfeldy, Perthshire.

Pitreavie, seat, 2¾ m. S.E. of Dunfermline, Fife.

Pitrodie, vil., 2 m. N.W. of Errol, Perth-shire. P.O.

Pitscandly, seat, 2 m. N.E. of Forfar, Angus.

Pitscottie, ham., par. of Ceres, 3 m. S.E. of Cupar, Fife. P.O.

Pitsligo, coastal par. in N. Aberdeenshire ; 4646 ac. Pop. 2087. The parish church is 1 m. S. of Rosehearty, near which is the ruin of P. Castle.

Pitsligo, New, q.s. par. and vil. in N. Aberdeenshire ; vil. stands 4 m. W. of Strichen ry. sta. P.O., T.O.

Pittarthie Castle, ruin, 6 m. S. of St. Andrews, Fife.

Pitteadie Castle, ruin, near Kinghorn, Fife.

Pittenane, par. in Lanarkshire. See PETTINAIN.

Pittencrieff, public park and glen in Dunfermline, Fife.

Pittenweem, par., royal and mun. bur., and seapt. on Firth of Forth, 1½ m. W. of Anstruther and 3½ m. N.E. of Elie, Fife ; par. 666 ac. Pop. 1644 ; pop. bur. 1619. P.O., T.O. Shows two red fixed lights ; ry. sta., L.N.E.

Pittodrie, seat, 1½ m. S.W. of Pitcaple ry. sta., Aberdeenshire.

Pittormie, pl., 3½ m. N.E. of Cupar, Fife.

Pittrichie, seat, 3 m. S.E. of Old Meldrum, Aberdeenshire.

Pittulie and Sandhaven, conjoint vil., par. of Pitsligo, N. Aberdeenshire, 3 m. W. of Fraserburgh; inhabited chiefly by fisherfolk. P.O., T.O. at Sandhaven. P. Castle is a ruin.

Pitversie, ham., ½ m. S.W. of Abernethy, Perthshire.

Pityoulish, loch (674 ft. above sea-level and 74 ft. deep), 2 m. N.E. of Aviemore, Spey Valley, Inverness-shire.

Place, seat in N. Ayrshire, ¾ m. W. of Kilbirnie.

Place of Tilliefourie, Aberdeenshire. See TILLIEFOURE.

Pladda, islet at mouth of Firth of Lorne, Argyllshire, 1 m. E. of Lismore Isl.

Pladda, isl. off the S.E. of Arran, Buteshire; lighthouse with half-minute group-flashing light, visible 17 m. Pop. 7.

Plaidy, ry. sta. L.N.E., 4½ m. N. of Turiff, Aberdeenshire. T.O.

Plains, vil. in par. of New Monkland, on N. Calder Water, and 2½ m. N.E. of Airdrie, Lanarkshire. P.O.

Plantation, q.s. par. in Govan par., S.W. Glasgow. P.O., T.O.

Plantation, ry. halt, 1 m. W. of Campbeltown, Machrihanish ry., Kintyre, Argyllshire.

Plean, q.s. par. and vil. (East Plean), and ry. sta., L.M.S. in par. of St. Ninian's, E. Stirlingshire. Vil. 3 m. S.E. of Bannockburn. P.O., T.O. Mining centre. Nearby is the seat P. House. Vil. pop. 1434.

Pleanmeanoch, ham., 1 m. S.W. of Lochailort ry. sta., Inverness-shire.

Pleasance, ham., 1¼ m. N.W. of Auchtermuchty, Fife.

Plockton, q.s. par. and fishing vil. on Loch Carron, S.W. Ross and Cromarty; vil. 5 m. S.W. of Strome Ferry; ry. sta., L.M.S. P.O., T.O. Vil. pop. 250.

Plotcock, colliery vil., 5 m. S.E. of Hamilton, Lanarkshire.

Plunton Castle, in co. of, and 5 m. W. of, Kirkcudbright.

Pluscarden, pl., 6 m. S.W. of Elgin, Morayshire. P.O. Pluscardine Abbey is a well-preserved ruin of a Cistercian monastery founded in 1230.

Pogbie, seat, 3 m. S. of Humbie sta., E. Lothian.

Point of Ayre, headland on S.E. coast of Mainland, Orkney. T.O. at lighthouse; four-seconds' flashing light, visible 8 m.

Poitna-h-I, loch (Ross of Mull), in isl. of Mull, Argyllshire, 1 m. E. of Iona Ferry.

Polbae, seat, 11 m. N.W. of Newton-Stewart, Wigtownshire.

Polbain, ham., 2 m. N.W. of Achiltibuie, N.W. Ross and Cromarty.

Polbeth, pl., 1 m. N.E. of W. Calder, Midlothian.

Polchar, seat, 1½ m. S. of Aviemore, Spey Valley, Inverness-shire.

Poldar House, Perthshire, 1½ m. N.W. of Kippen.

Polerscaig, ham., in q.s. par. of Strathy, Sutherland.

Polglass, ham., 1 m. S.E. of Achiltibuie, N.W. Ross and Cromarty.

Polkemmet, seat, 1½ m. W. of Whitburn, W. Lothian.

Poll, loch, 5 m. N. of Lochinver, W. Sutherland.

Pollo, seat, in par. of Easter Kilmuir, Delny, Ross and Cromarty.

Polloch, pl., on r. Polloch which runs from Loch Doilate to Loch Shiel, N. Argyllshire.

Pollok Castle, seat, 2½ m. S.E. of Barrhead, Renfrewshire.

Pollok Division, parl. division of Glasgow.

Pollok House, seat, on the White Cart, in Renfrewshire, 1 m. N.W. of Pollokshaws and within the Glasgow mun. bur.

Pollokshaws, q.s. par., on the White Cart, in the par. of Eastwood, in S.W.Glasgow; ry. sta., L.M.S. P.O., T.O. Weaving and dyeing, bleachfields, paper, iron, and other industries.

Pollokshields, q.s. par., in the par. of Govan, in S.W. Glasgow; ry. sta., L.M.S. P.O., T.O. The dist. is almost entirely a residential area; contains Maxwell Park.

Pollokshields East, ry. sta., L.M.S., in S.W. of Glasgow.

Pollokshields West, ry. sta., L.M.S., in Pollokshields, Glasgow.

Poll Uidhe a Chro, loch, 2½ m. S.E. of Poolewe, W. Ross and Cromarty.

Polmadie, hill; alt. 1852 ft., 2 m. S.E. of Barr, Ayrshire.

Polmadie, S. sub. of Glasgow. P.O. T.O.

Polmaily House, seat, on r. Endrick, in co. of, and 16 m. S.W. of, Inverness.

Polmaise Castle, seat in co. of, and 1½ m. S.W. of, Stirling.

Polmaise, Old, seat, on r. Forth, in co of, and 3 m. E. of, Stirling.

Polmont, q.s. par. and town in S. Stirlingshire; town 3 m. E. of Falkirk and 22¼ m. W. of Edinburgh by rail; ry. sta., L.N.E. P.O., T.O., and P.O. and T.O. at Old Polmont. P. Institution is a Borstal one. Seats nearby at P. Park and P. Bank. Vil. pop. 6204.

Polmood, pl., with seat, 6 m. S. of Broughton, S.W. Peeblesshire. P.O., T.O.

Polmood Burn, stream, flowing N.W. to the Tweed near Crook Inn, Tweedsmuir, Peeblesshire.

Polnish, ham., 1 m. W. of Lochailort ry. sta., W. Inverness-shire.

Polnoon Castle, traces of an ancient seat, near Eaglesham, Renfrewshire.

Poltalloch, ancient seat of the Malcolms, on Loch Craignish, 10 m. N.W. of Lochgilphead, Argyllshire.

Polton, vil. on the r. North Esk, 1½ m. S.W. of Lasswade, Midlothian, 7 m. S. by road from Edinburgh; ry. sta., L.N.E. P.O., T.O. Has paper mills. P. House is a seat.

Polton Farm Colony, Midlothian, ½ m. S. of Lasswade. Hospital for consumptives.

Poltonhall (and **Dalhousie Colliery**), vil., ¾ m. S. of Bonnyrigg, Midlothian. Pop. (P. and Sherwood), 819.

Polvaird, small loch, near summit of Cruffel, 6 m. S. of Sanquhar, Dumfriesshire.

Polwarth, par. and vil. in mid Berwickshire; par. 2999 ac. Pop. 160. Vil. stands 4 m. S.W. of Duns.

Pomathorn, pl., with ry. sta., L.N.E., Midlothian, 15 m. S. of Edinburgh.

Pomona, Orkney. See Mainland.

Poneil Water, stream, flowing 9½ m. N.E. to the Douglas Water near Ponfeigh, Lanarkshire.

Ponfeigh, colliery vil. in co. of, and 6 m. S. of, Lanark; ry. sta., L.M.S. T.O. at sta. Pop. 1342.

Pool, vil. in S. Perthshire, 3½ m. N.E. of Dollar.

Poolewe, q.s. par. and vil. in par. of Gairloch, W. Ross and Cromarty; vil. at head of Loch Ewe, 6 m. S. of Aultbea and 27½ m. W. of Achnasheen ry. sta.

P.O., T.O. Motor communication with Achnasheen. Frequent communication by steamer with Portree, Oban, Glasgow, etc.

Pool House, seat, Poolewe, W. Ross and Cromarty.

Pooltiel, sea loch on W. coast of Skye, 3 m. S. of Dunvegan Head, Inner Hebrides.

Porridge Cairn, alt., 2484 ft., on borders of counties of Peebles and Selkirk, 2 m. N.E. of Talla Reservoir.

Portachoillan, pl. at S.W. end of W. Loch Tarbert, Kintyre, Argyllshire; a ferry to Ardpatrick on the opposite shore.

Portacree, ham. on W. shore of Luce Bay, 10 m. S.E. of Port Patrick, Wigtownshire.

Port Allan, small bay in S.E. Wigtownshire, part of Wigtown Bay, 3½ m. S. of Garliestown.

Port Allen, on Firth of Tay, 1 m. S. of Errol, Perthshire.

Port Appin, vil. on E. shore of Loch Linnhe, opposite N. end of Lismore Isl., and 2½ m. S.W. of Appin, Argyllshire. P.O., T.O. Pop. 74.

Port Askaig, small vil. on N.E. coast of Islay, on Sound of Islay, Argyllshire. P.O., T.O. spelt Portaskaig.

Port Bannatyne, or **Kamesburgh,** vil. on Kames Bay on E. side of isl. of Bute, 2¼ m. N.W. of Rothesay with which it is connected by tramway. P.O., T.O. Pop. 1161.

Port Charlotte, vil. on W. shore of Loch Indail, on S. side of isl. of Islay, Argyllshire; nearly opposite Bowmore. P.O., T.O. Pop. 312.

Portclair, State and deer forest, lying between Glen Moriston and Loch Ness, Inverness-shire, 4 m. N.E. of Fort Augustus.

Port Crinan, or **Crinan,** vil., with hotel, on Loch Crinan, near W. end of Crinan Canal, Knapdale, Argyllshire. P.O., T.O. called Crinan Pier.

Port Dearg, ham. in S.W. extremity of isl .of Mull, on Iona Sound, Argyllshire.

Port Dundas, dist. in the N. of Glasgow. P.O., T.O. Situated at junc. of Forth and Clyde and Monkland Canals. Has distilleries, iron foundries, oil mills, etc.

Port Edgar, small harb,, 1 m. W. of S. Queensferry, on Firth of Forth, W. Lothian.

Port Ellen, seapt. vil. in S.E. of isl. of Islay, 12 m. S.E. of Bowmore, Argyllshire. Has a distillery and fishing industry. P.O., T.O. Regular steamer communication with Glasgow. Pop. 632.

Port Elphinstone, vil. on r. Don, 1 m. S. of Inverurie, Aberdeenshire. P.O. Has meal and paper mills.

Portencross, pl., 3 m. N.W. of W. Kilbride, Ayrshire. P.O., T.O. The ruin of P. Castle, said to have been the residence of Stuart kings, stands at the base of a cliff.

Port Errol (or **Ward of Cruden**), fishing vil. at the mouth of the Water of Cruden, 11 m. N.E. of Ellon and 8½ m. S.W. of Peterhead. P.O., T.O. called Cruden Bay. Coastguard sta.

Portessie, fishing vil. in bur. of, and 1¼ m. N.E. of, Buckie, Banffshire; ry. sta., L.N.E. P.O., T.O.

Port Glasgow, par., town and police bur., and seapt. on the S. bank of the Clyde, 20 m. N.W. of Glasgow and 3 m. E. of Greenock; par. 1119 ac. Pop. 19,616; pop. bur. 19,580; ry. sta., L.M.S. P.O., T.O. Contiguous to the harbours is a large graving dock; shipbuilding is carried on, and there are also sawmills, roperies, and engineering works.

Port Gordon, fishing vil., 2 m. S.W. of Buckie, Banffshire; ry. sta., L.N.E. P.O., T.O. Salmon fishery. Pop. 1183.

Port Gower, fishing vil., 2 m. S.W. of Helmsdale, E. Sutherland.

Port Henderson, 4 m. W.S.W. of Gairloch, W. Ross and Cromarty.

Portincaple, pl. on S. side of Loch Goil, where it meets Loch Long, ½ m. W. of Whistlefield ry. sta., Argyllshire.

Portincross Castle, see PORTENCROSS.

Portinisherrich, ham. on E. side of Loch Awe, 7½ m. W. of Inveraray, Argyllshire. P.O., T.O.

Portinleck, pl., with hotel, 4 m. N.W. of Bonar Bridge, Sutherland. See INVERSHIN.

Portkil, small bay in W. Dunbartonshire, 1¼ m. E. of Kilcreggan; nearby is a fort for Clyde defences.

Portknockie, police bur. and fishing vil., 5 m. N.E. of Buckie, Banffshire. Pop. 1619. Ry. sta., L.N.E. P.O., T.O.

Portlethen, q.s. par. and fishing vil., with ry. sta., L.N.E., in N.E. Kincardineshire; vil., 8 m. S.W. of Aberdeen. P.O., T.O.

Port Logan, or **Port Nessock,** coastal vil., par. of Kirkmaiden, on Portnessock Bay, 10 m. S.E. of Stranraer, Wigtownshire. P.O., T.O. Lifeboat sta. Pop. 65.

Portmahomack, fishing vil. on Dornoch Firth, N.E. Ross and Cromarty, 8 m. N.E. of Tain. P.O., T.O. Hotel.

Port Maluag, on N.E. coast of isl. of Lismore, Loch Linnhe, Argyllshire.

Port Mary, seat in co. of, and 7½ m. S.E. of, Kirkcudbright.

Port Mary, pl. in N. of Luing Isl., Firth of Lorne, W. Argyllshire.

Portmoak, par. on Loch Leven, Kinross-shire; 9833 ac. Pop. 871.

Portmore, seat, Peeblesshire, 1½ m. N.E. of Eddleston. P. Loch, in vicinity, 1000 ft. above sea-level; depth 41 ft.

Portnacraig, pl. on the Tummel opposite Pitlochry, Perthshire.

Portnacroish, vil. on E. side of Loch Linnhe, 14 m. N.E. of Oban, Argyllshire.

Port-na-Curaich, supposed landing-place of St. Columba in 563; on S. of Iona, off S.W. of Mull, Argyllshire.

Portnaguran, vil. on Eye Peninsula, and 9½ m. N.E. of Stornoway, Lewis, Outer Hebrides. P.O., T.O.

Portnahaven, q.s. par. and vil. in par. of Kilchonan, S.W. end of Islay, Argyllshire, 10 m. W.S.W. of Bowmore. Pop. of P. and Port Wemyss, 341.

Portnalong, ham., 3½ m. N.W. of Carbost, Skye, Inner Hebrides. P.O.

Portnellan, seat, 2 m. E. of Crianlarich, Perthshire.

Port Nessock Bay, or **Port Logan,** 10 m. S.E. of Portpatrick, Wigtownshire.

Portobello, watering pl. and q.s. par., on Firth of Forth within city of, and 3 m. E. of G.P.O., Edinburgh; ry. sta., P.O., T.O. Sea-water baths. Industries: brick, stoneware, paper, and bottle making.

Port of Menteith, par. and ry. sta., L.N.E., (passenger service withdrawn), Perthshire. Pop. 940; 3 m. N.E. of Buchlyvie. P.O., T.O., and P.O., T.O. at sta.

Port of Ness, ham., 1½ m. S.E. of .Butt of Lewis, N. Lewis, Outer Hebrides. P.O., T.O.

Portpatrick, par. and seapt. vil., on the North Channel, W. Wigtownshire; par. 9073 ac. Pop. 1109. Vil. 7½ m. S.W. of Stranraer; ry. sta., L.M.S. P.O., T.O., and T.O. at wireless sta. Hotel. Coastguard and lifeboat sta. Pop. 569.

Portrack, seat, on r. Nith, 5 m. N. of Wadebridge, Dumfriesshire.

Portramsay, vil., with good harbour, on N.W. side of Lismore Isl., Loch Linnhe, Argyllshire. Is 10 m. N. of Oban.

Portree, par., vil., and seapt. in the E. centre of Skye; par. 56,915 ac. Pop. 2008. P.O., T.O. It is the capital of the isl. and lies about 120 m. N.W. of Oban and 74 m. N. of Tobermory; reached by steamer from Oban or from Kyle of Lochalsh by ry. steamer; has daily steamer communication with Stornoway, Glasgow etc. On pier is a red fixed light, visible 4 m. Makes tweeds and plaids. Vil. pop. 802. Hotel. P. Lodge is a seat.

Port Seton, part of Cockenzie, E. Lothian; a small watering-place with an open-air bathing pool.

Portskerra (and **Melvich**), vil. at mouth of r. Halladale, N.E. Sutherland, 17 m. W. of Thurso.

Portsonachan, North and **South,** two hams., with pier and ferry, on Loch Awe, 8½ m. S.W. of Dalmally, Argyllshire. P.O., T.O. Hotels. Ferry to Taychreggan, no cars.

Portsoy, q.s. par., police bur., and seapt. in par. of Fordyce, N. Banffshire. Town is 8½ m. W. of Banff; ry. sta., L.N.E. P.O., T.O. Pop. 1651. Hotel. Fishing centre with a rocket apparatus; has a distillery.

Port Tannachy, on W. of Port Gordon (q.v.) N. Banffshire.

Port Vasco, 4 m. N.W. of Tongue, N. Sutherland.

Port Wemyss, coastal vil., 15 m. S.W. of Brigend, isl. of Islay, Argyllshire. Pop. of P.W. and Portnahaven 341.

Port William, small port and holiday resort on E. shore of Luce Bay, 7 m. N.W. of Whithorn, Wigtownshire. P.O., T.O. Pop. 539.

Portyerrock, coastal ham., on bay of same name, Wigtown Bay, 2½ m. S.E. of Whithorn, Wigtownshire.

Possilpark, q.s. par. in N. of Glasgow; ry sta. (Possil), L.N.E. P.O., T.O.

Posso, seat in co. of, and 5¼ m. S.W. of, Peebles. Posso Craigs, a hill nearby, is 1817 ft.

Potarch, Bridge of, vil. on the r. Dee, in S. Aberdeenshire, 2 m. S.E. of Kincardine O'Neil; has fairs three times a year.

Potrail Water, one of the headstreams of the Clyde in upper Lanarkshire; flows 9 m. N.E. to join the Daer Water, 2½ m. S. of Elvanfoot sta.

Potterhill, vil. in par. of Paisley, Renfrewshire; ry. sta., L.M.S. Pop. 1186.

Poulary, loch, 11½ m. W. of Invergarry, N.W. Inverness-shire; alt. 320, depth 47 ft.

Poundland, vil. on the r. Stinchar, 8 m. S. of Girvan, Ayrshire.

Pourie, ruins of castle, 3 m. N.E. of Dundee, Angus.

Pow, any one of the numerous sluggish streams or stagnant burns in marshy or alluvial districts of Scotland.

Pow Burn, short stream flowing S. of Airth Castle, Stirlingshire, to Firth of Forth, opposite Kincardine-on-Forth.

Powderhall, dist. in N. side of city of Edinburgh; has athletic ground.

Powfoot or **Queensberry,** ham. on Solway Firth, 4 m. S.W. of Annan, Dumfriesshire. P.O.

Powfoulis House, 6 m. N. of Falkirk, Stirlingshire.

Powis, q.s. par. in city of Aberdeen.

Powis, seat in co. of, and 2 m. N.E. of, Stirling.

Powis House, seat near Old Aberdeen.

Powmill, ham., 1¾ m. S.S.W. of Crook of Devon ry. sta., Kinross-shire.

Powmillion Burn, stream in N.W. Lanarkshire, flowing 7 m. chiefly in a S.E. direction to Avon Water, 1 m. E. of Strathaven.

Powtrail Water, Lanarkshire. See Potrail Water.

Poyntzfield, vil. on S. side of Cromarty Firth, E. Ross and Cromarty, 2 m. S.S.E. of Invergordon by ferry. P.O., T.O. Vil. also called Jemimaville.

Premnay, par. in E. Aberdeenshire; 5433 ac. Pop. 713. Par. church 2 m. S.E. of Insch ry. sta.

Preshome, pl., 3 m. S.S.E. of Port Gordon, Banffshire; has a Roman Catholic church.

Press Castle, seat, $4\frac{1}{2}$ m. W. of Eyemouth, Berwickshire.

Pressmennan Loch, $4\frac{1}{2}$ m. S.S.W. of Dunbar, E. Lothian.

Preston, ham., par. of Cranston, 4 m. E. of Dalkeith, Midlothian. P. Hall is a seat.

Preston, ham., $2\frac{1}{2}$ m. N.E. of Duns, Berwickshire.

Preston, vil., $\frac{1}{2}$ m. S.E. of Prestonpans, E. Lothian. It contains an ancient pillar-cross. N. of the vil. is a ruined ancient historical tower supposed to have been originally a fortalice of the Earls of Home. To the E. are the remains of the ancient manorial residence of Lord Grange.

Preston, or **Prestonmill,** ham. in Kirkcudbrightshire, 14 m. S. of Dumfries. P.O.

Prestonfield, q.s. par. in S.E. of Edinburgh. Golf course.

Prestongrange House, near Prestonpans, E. Lothian, 9 m. E. of Edinburgh, now the home of the Royal Musselburgh Golf Club.

Prestonhall, seat, 1 m. N.E. of Cupar, Fife.

Prestonholm, ham. on the S. Esk, $3\frac{1}{2}$ m. S.E. of Lasswade, Midlothian.

Preston House, seat, 14 m. S. of Linlithgow, W. Lothian.

Prestonkirk, par. on r. Tyne in mid East Lothian, 5 m. W. of Dunbar ; 7076 ac. Pop. 1628. Near the S. side of the par. are the ruins of Hailes Castle. P.O. T.O. called Prestonkirk or East Linton.

Prestonmill, see PRESTON, Kirkcudbrightshire.

Prestonpans, par. and coast town (police bur). in E. Lothian, on the Firth of Forth, $9\frac{1}{2}$ m. E. of Edinburgh by rail and 9 m. by road ; par. 1299 ac. Pop. 5986. Bur. pop. 2426 ; ry. sta., L.N.E. P.O., T.O., and T.O. at sta. In the neighbourhood Prince Charles Edward Stuart defeated Sir John Cope (21st Sept. 1745); has manufactures of bricks, tiles, fireclay goods, salt, soap, and brewing ; coal mines in vicinity.

Prestwick, small town. Pop. 8538. In par. of Monkton and Prestwick, and in co. of, and $2\frac{1}{2}$ m. N. of, Ayr ; ry. sta., L.M.S. P.O., T.O. Its charter as a bur. of barony dates from 1600 ; holiday place and golfing centre.

Prestwick, New, vil. in co. of, and $1\frac{1}{2}$ m. N. of, Ayr. P.O.

Prett's Mill, pl. in co. of, and $2\frac{1}{2}$ m. N.E. of, Lanark.

Priest Craig, mt., alt. 2215 ft., near N. boundary of Dumfriesshire and 2 m. S.E. of Hart Fell.

Priesthill, farm, 4 m. E. of Muirkirk, Ayrshire.

Priesthill Height, alt. 1615 ft., on S.W. border of Lanarkshire, 4 m. N.E. of Muirkirk in E. Ayrshire.

Priest Island, near the entrance of Gruinard Bay and Loch Broom, N.W. Ross and Cromarty.

Priestside Bank, shoal in Solway Firth, 3 m. S. of Annan.

Primrosehill, ham., $2\frac{1}{2}$ m. N. of Duns, Berwickshire.

Primside Loch, 1 m. W. of Yetholm, Roxburghshire.

Prince Charles's Cave, on S.E. side of Loch Ericht, S. Inverness-shire.

Prince Charles's Cave, 4 m. N.E. of Portree, Skye ; hiding place of the Prince for some time after Culloden.

Princeland, seat on N. side of Coupar-Angus, Perthshire.

Prince's Pier, pier and ry. sta., L.M.S., at Greenock, Firth of Clyde ; starting point for Clyde steamers.

Prinlaws, q.s. par. in par. of Leslie and Kinglassie, mid Fife.

Prior Bank, seat, near Melrose, Roxburghshire.

Priorhill, pl., near Canonbie sta., Dumfriesshire.

Proaig Bay, on the E. coast of Islay, Argyllshire.

Prosen Water, stream in N.W. Angus, flowing S.E. for 18 m. to join the S. Esk, 3 m. N.E. of Kirriemuir.

Prospecthill, pl. with colliery, $\frac{1}{2}$ m. S.W. of Morningside ry. sta., N. Lanarkshire.

Protstownhill, ham., $11\frac{1}{2}$ m. E. of Banff, E. Banffshire.

Provanmill, pl., 3 m. N.E. of Glasgow G.P.O.

Ptarmigan, mt., alt. 2395 ft., 2 m. N. of Rowardennan pier (Loch Lomond), Stirlingshire ; there is also a seat of the same name.

Pulteneytown, business and manufacturing part of the bur. of Wick, Caithness. P.O., T.O. Has a distillery.

Pulwhirrin Burn, short stream, 7 m. long, flowing into Kirkandrews Bay, S.W. Kirkcudbrightshire.

Pumpherston, vil. on the r. Almond, 1½ m. N. of Mid-Calder, Midlothian. P.O., T.O. Shale-oil works in the neighbourhood.

Purves Hall, seat, 3½ m. E. of Greenlaw, Berwickshire.

Purvis Hill Terrace, pl., 1½ m. E. of Innerleithen, Peebles.

Q

Quaich, stream, flowing 7 m. E. to head of Loch Freuchie, 10 m. W. of Dunkeld, Perthshire.

Quair, stream, flowing 7 m. N.E. to the Tweed, which it joins a little below Innerleithen, Peeblesshire.

Quanterness, pl., with Broch, 2 m. N.W. of Kirkwall, Orkney. The structure is 140 ft. in circuit and 12 ft. high.

Quarff, q.s. par., in par. of, and 6½ m. S.W. of, Lerwick, Shetland. P.O., T.O.

Quarrelton, mining vil., 4 m. S.W. of Paisley, Renfrewshire.

Quarrelwood, ham., par. of Kirkmahoe, in co. of, and 3½ m. N.W. of, Dumfries.

Quarry Hill, alt. 1089 ft., 4 m. N. of Darvel, Ayrshire.

Quarrywood, pl., with sandstone quarries, 2½ m. N.W. of Elgin, Moray.

Quartalehouse, Stuartfield and Q. special water and lighting dist., Aberdeenshire. See STUARTFIELD.

Quarter, seat, 2½ m. N.W. of Largs, Ayrshire.

Quarter, seat, 2 m. S.W. of Broughton sta., Peeblesshire.

Quarter, q.s. par., in Lanark. P.O., T.O. See QUARTER IRONWORKS.

Quarter House, seat, 1½ m. N.W. of Denny, Stirlingshire.

Quarter Ironworks and Darngaber, vil., 3 m. S. of Hamilton by road and 7¾ m. by rail, Lanarkshire; ry. sta. (Quarter, alt. 573 ft.), L.M.S. See DARNGABER.

Quarter, West, vil., 2 m. N.E. of Strathaven, Lanarkshire; nearby is the seat, Westquarter House.

Queensberry, mt., alt. 2285 ft., 7 m. S.W. of Moffat, Dumfriesshire.

Queensferry, or **South Queensferry,** q.s. par., mun. and royal bur., and small port on S. side of Firth of Forth, W. Lothian, 9½ m. W. of Edinburgh by rail; ry. sta. (Dalmeny), L.N.E.; also a halt on the Ratho line; ferry to N. Queensferry; bur. pop. 1798. P.O., T.O.

Queensferry, North, vil., opposite S. Queensferry, on N. shore of Firth of Forth, Fife, and at N. end of Forth Bridge, 2 m. S. of Inverkeithing; ry. sta., L.N.E. Pop. 1353. P.O., T.O. Ferry to Queensferry.

Queenshill, seat, 2 m. N. of Tarff ry. sta., Kirkcudbrightshire.

Queenside, loch, 5 m. N.W. of Lochwinnoch, Renfrewshire.

Queen's Park, q.s. par., in par. of Govan and Cathcart, in S. of Glasgow; ry. sta., L.M.S.

Queen Victoria School, Dunblane (q.v.).

Queich, North, short stream rising in Ochils and flowing 6½ m. S.E. to Loch Leven, 1 m. S.E. of Milnathort, Kinross-shire.

Queich, South, short stream, partly in Perthshire and Kinross-shire, and flowing 8½ m. S.E. to Loch Leven at Kinross.

Quendale, pl., S. Mainland, Shetland, near Bay of Q. P.O.

Quendale Bay, sea inlet near S. end of mainland, Shetland, 1½ m. N.W. of Sunburgh Head; Q. House stands at the head of the bay.

Quickningar Hill, alt. 1601 ft., 6 m. N.N.E. of Eskdalemuir, on border of N.E. Dumfriesshire and Selkirkshire.

Quien, loch, 4 m. S.W. of Rothesay, isl. of Bute, Firth of Clyde.

Quinag, mt., alt. 2653 ft., on N. side of Loch Assynt and 8 m. N.E. of Lochinver, Sutherland.

Quinish House, 5 m. W. of Tobermory, Mull, Argyllshire.

Quiraing, mt., alt. 1779 ft. par. of Kilmuir, Skye, Inner Hebrides; consists of a number of wonderful rock formations, including "Needle Rock" and the "Table."

Quithelhead, seat, 3 m. S.E. of Banchory, Kincardineshire.

Quivals, ham., 1½ m. N.E. of Broughtown, Sanday, Orkney. P.O.

Quoich, Loch, (5¼ m. long by 3 m. broad and 281 ft. deep), at head of Glen Q., S.W. Inverness-shire. It receives the Q. Water, and from it issues the Garry; depth 281 ft., alt. 555 ft.

Quoich Water, stream, in par. of Crathie and Braemar, S.W. Aberdeenshire; rises between Ben a' Bourd and Ben Avon and flows 8½ m. to the Dee, 2 m. N.W. of Braemar. 1 m. above its mouth is a beautiful waterfall, the Linn of Q.

Quothquan, vil., 2 m. N.E. of Thankerton, Lanarkshire.

Quoyburray, ham., 5½ m. S. of Kirkwall, Orkney. P.O.

Quoyloo, pl., 8 m. N. of Stromness, Orkney. P.O.

R

Raa, loch, N.W. Ross and Cromarty, 2 m. N. of Achiltibuie.

Raasay, isl. on the W. of Skye and separated from it by Sound of Raasay, Inverness-shire; measures 13 m. by 3½ m. and has a total area of 15,704 ac., most of which is moorland; highest summit Dun Caan, 1456 ft., and Boswell says he danced on the top on 10th Sept. 1773. P.O., T.O. Pop. 354. R. House is a seat.

Rabbit Islands, three isls., Sutherland, in the entrance to the Kyle of Tongue.

Rachan House, seat, Peeblesshire, on Holms Water, 1 m. S.E. of Broughton sta. P.O. at Rachan Hill.

Racks, vil., with ry. sta., L.M.S., near Lochar Water, in co. of, and 4 m. S.E. of, Dumfries.

Rackwick, pl., Hoy Isl., Orkney, 2 m. E. of Rora Head.

Raddery House, seat, Ross and Cromarty, 4 m. N.W. of Fortrose.

Radernie, ham., Fife, 6 m. S.W. of St. Andrews.

Radnor Park, pl. adjoining Kilbowie, Dunbartonshire. P.O.

Rae, loch (Tay Basin), 1½ m. S.W. of Blairgowrie, Perthshire; alt. 195 ft., depth 16 ft.

Raeberry, remains of a stronghold of the Maclellans in co. of, and 6 m. S.E. of, Kirkcudbright.

Raehills, seat, Dumfriesshire, near Kinnel Water, 10 m. N.W. of Lockerbie.

Raemoir, seat, Kincardineshire, 3 m. N. of Banchory.

Rafford, par. and vil., 2½ m. S.E. of Forres, Morayshire. P.O., T.O. Par. 12,456 ac. Pop. 803.

Rahane, ham., with steamboat pier, Argyll, on W. shore of Gareloch opposite Shandon.

Rahoy, pl., N.W. Morven, Argyllshire, on the N.E. side of Loch Teacuir.

Rait, vil., Perthshire, 3 m. N.W. of Errol. P.O.

Rait Castle, ruined seat in co. of, and 3 m. S. of, Nairn; formerly a seat of the Cumyns; late of the Macintoshes of Rait.

Raith, q.s. par. and loch, Abbotshall, Fife, Raith House, 2½ m. W. of Kirkcaldy.

Raith, Little Raith, lighting dist., Auchtertool par., Fife. Pop. 259.

Raith Colliery, vil., Fife, in Auchtertool par.

Ralston, seat, Renfrewshire, 2 m. E. of Paisley.

Ramasaig, ham. and bay, N.W. isl. of Skye, 8 m. S.W. of Dunvegan.

Rammerscales, seat, Dumfriesshire, 3½ m. S. of Lochmaben.

Ramornie, seat, Fife, 1 m. E. of Ladybank Junction.

Ramoth, vil., Perthshire, 3 m. S.E. of Thornhill.

Ramsaycleugh, ham., Selkirkshire, on Ettrick Water, 5 m. S.E. of St. Mary's Loch. P.O.

Ranfurly Castle, ruin, Renfrewshire, ½ m. S. of Bridge of Weir; belonged to the forebears of John Knox.

Rangag Loch, 5 m. N.N.W. Latheron, Caithness.

Range Castle, ancient British camp, Dumfriesshire, on Almagill hill, 2 m. N.W. of Dalton.

Rankeillour Nether, seat, Fife, 3½ m. S.W. of Cupar.

Rankeillour Over, seat, Fife, 1 m. N. of Nether Rankeillour.

Rankinston, vil., with ry. sta., L.M.S., Ayrshire, 8 m. S.W. of old Cumnock. P.O., T.O.

Rankle Burn, Selkirkshire, flows 9 m. N. to Ettrick Water, opposite Tushielaw Tower.

Rannoch, dist. of the West Highlands mostly moorland and lochs, Argyll and Perth cos.; ry. sta., L.N.E., is 5 m. W. of Loch R.; R. Deer Forest, Braes of R., and Loch R. (668 ft. alt. and 440 ft. depth) form part of the dist. R. Lodge is at the W. end of the loch, and there are three hotels at the E. end of the loch. P.O.

Ranza, see GLEN RANZA and LOCHRANZA.

Rapness, the S.E. extremity of Westray, Orkney Islands.

Rarnish, vil., isl. of Lewis, Ross and Cromarty, 5 m. S.S.W. of Stornoway.

Rasay, or **Blackwater,** r., Ross and Crom-arty, rising in the Derrymore mts. and flowing 19 m. to the Conon at Moy.

Rasay Island, Inverness-shire. See RAASAY.

Rashfield, ham., Argyll, on r. Eachaig, 1 m. N.W. of Kilmun.

Ratagan, a State forest on S.W. shore of Loch Duich, S.W. Ross and Cromarty. See MAM RATTACHAN.

Ratagan Hostel, on S. side of Loch Duich, 1 m. W. of Shiel Bridge, Ross and Cromarty. A Scottish Youth Hostel.

Rathen, par. ham. and ry. sta., L.N.E., Aberdeenshire; par. 9592 ac. Pop. 2399. Ham., 4 m. S. of Fraserburgh. P.O. and T.O. at sta.

Rathillet, vil. and seat, Fife, 4 m. N. of Cupar. P.O.

Ratho, par., vil., and ry. junc., L.N.E., Midlothian; par. 6142 ac. Pop. 1672. Vil. 1¼ m. S. of sta. and 8¼ m. W. of Edinburgh. Pop. 1392. P.O., T.O., and P.O., T.O. at sta. R. Hall a seat, and R. Park, a golf course.

Rathobyres, estate and seat, Midlothian, near Ratho.

Rathven, coast par. and vil., with ry. sta., L.M.S., Banffshire; par. 23,165 ac. Pop. 15,099. Vil. 1½ m. N.E. of Buckie.

Ratter, ham., Caithness, 7 m. N.E. of Castletown.

Rattray, vil. and seat, Aberdeenshire, 7½ m. N.W. of Peterhead. There is a coastguard sta. on Rattray Head and an important Lighthouse, light visible 15 m.

Rattray, town (police bur.) and par., Perthshire; par. 4534. Pop. 2004. Has flax and jute spinning. See BLAIRGOWRIE.

Ravelston, seat, S.W. Edinburgh and dist.

Ravenscraig, ruined castle, 2½ m. N.W. of Peterhead, Aberdeenshire, on r. Ugie.

Ravenscraig, ruined castle in par. and 1 m. S.W. of Dysart, Fife.

Ravenscraig, ry. sta., L.M.S., Renfrew-shire, 3½ m. S.W. of Greenock. Convalescent Home.

Ravenshall, ham., 1½ m. N. of Falkland, Fife.

Ravenshall Point, headland, Kirkcud-brightshire, on E. side of Wigtown Bay and 6 m. S.E. of Creetown; Scott mentions this coast in *Guy Mannering.*

Ravenstone, seat of Lord Borthwick, Wigtownshire, 5 m. N.W. of Whit-horn.

Ravenstruther, vil., in co. of, and 3 m. N.E. of, Lanark. P.O.

Ravenswood, seat, 2 m. E. of Melrose, Roxburghshire.

Rawyards, vil., with ry. sta., L.N.E. (passenger service withdrawn), N.E. sub. of Airdrie, Lanarkshire. P.O.

Rayne, par., Aberdeenshire; 7885 ac. Pop. 921. Contains Old Rayne vil. on r. Ury, 1½ m. E. of Oyne ry. sta.

Reawick, seat, Shetland, on Scalloway Bay, 6 m. N.W. of Scalloway. P.O., T.O.

Reay, par. and vil., Caithness; par. 46,326 ac. Pop. 719. Originally the par. was partly in the cos. of Suther-land and Caithness, but the Sutherland portion was transferred by the Commissioners to the par. of Farr. Vil. near Sandside Bay, 10½ m. W. of Thurso. P.O., T.O.

Reay Deer Forest, Sutherland; 45,000 ac.; post town, Lairg, 35 m.

Recar Loch, 8½ m. S.W. of Dalmellington, S.E. Ayrshire.

Redcastle, vil., with ry. sta., L.M.S., E. Ross and Cromarty, 3¼ m. E. of Muir of Ord.

Redcastle, seat, Ross and Cromarty, 6½ m. N.W. of Inverness; said to be the oldest inhabited castle in Scotland, built originally in 1179.

Red Castle, Angus, a ruin on Lunan Bay, 4½ m. S.W. of Montrose.

Red Head, a conspicuous headland, Angus, on S. side of Lunan Bay; alt. 267 ft.

Red Point, headland, Ross and Cromarty, 12 m. S.W. of Gairloch.

Red Row, vil., Midlothian, ½ m. N.W. of Millerhill.

Redding, colliery vil., Stirlingshire, 2½ m. S.E. of Falkirk. P.O.

Redford, large military barracks, Edin-burgh, ½ m. N.E. of Colinton or 3½ m. from Edinburgh G.P.O.

Redford, vil., Fife, on r. Ore, 3¼ m. N.W. of Kirkcaldy.

Redford, seat, Stirlingshire, near r. Avon, 4 m. S.W. of Linlithgow.

Redhall Mills, formerly a seat, now an industrial dist. of Edinburgh, Lanark Road, Slateford.

Redhouse, vil., Perthshire, 2½ m. S. of Doune.

Redhouse Castle, ruin, 1 m. E. of Longniddry sta., E. Lothian.

Redhythe Point, headland, Banffshire, 1½ m. N.W. of Portsoy.

Rednock Castle, seat, 4 m. N. of Port of Menteith sta., Perthshire.

Redpath, ham., 2 m. S. of Earlston, Berwickshire. Pop. 50.

Redscar Law, mt., N. Selkirkshire, 4 m. N.N.E. of Walkerburn ; alt. 1837 ft.

Redswire, scene of the " Raid of the Redswire " (1575), Roxburghshire, 9 m. S.E. of Jedburgh near source of the r. Rede.

Redwells, hill, with tower, Kinglassie, Fife.

Ree, or **Righ,** isl. in Loch Craignish, Argyll.

Reekie Linn, waterfall on r. Isla, Angus, 3½ m. N. of Alyth.

Reichip House, seat, 5 m. N.E. of Dunkeld, Perthshire.

Reiss, vil., Caithness, 4 m. N.W. of Wick. P.O.

Relugas, ham. and seat, Morayshire, 6 m. S.W. of Forres. P.O., T.O.

Rendall, q.s. par., and ham., Orkney, 12 m. N.W. of Kirkwall. P.O., T.O.

Renfrew, parl., royal, and mun. bur., par., and co. town of Renfrewshire, with ry. sta., L.M.S., 3 m. N.E. of Paisley and 5 m. W. of Glasgow ; par. 4840 ac. Pop. 40,816 ; Pop. of town 14,971. P.O., T.O. ; made a royal bur. in 1396. Shipbuilding and engineering ; has ferry to Yoker.

Renfrewshire, co. in S.W. Scotland, bounded on the N. by the Clyde and Dunbartonshire, E. by Lanarkshire, S. by Ayrshire, and W. by the Firth of Clyde. Surface flat in N.E., rising to high moorland in W. and S. ; principal streams are Black Cart, White Cart, and Gryfe, all flowing to the Clyde. Agriculture, dairying, and stock-raising are carried on ; produces coal, iron, and shale ; manufactures thread, cotton and chemicals ; shipbuilding, engineering, print and bleach works. Chief towns : Renfrew, Paisley, Greenock, Gourock, and Port Glasgow ; area 151,431 ac. Pop. 288,586. Returns two members to Parliament—one each for Eastern and Western Divisions.

Renton, town and q.s. par., Dunbartonshire, on r. Leven, 2 m. N. of Dumbarton; ry. sta., L.N.E. and L.M.S. P.O., T.O. Town pop. 4562. Has calico-printing, dyeing, and bleaching works.

Renton, pl., Berwickshire, 1 m. S.E. of Grant's House ry. sta. R. House, a seat in vicinity.

Renton Hall, seat, E. Lothian, 2½ m. S.E. of Haddington.

Repentance Tower, a tower of the fifteenth century, Dumfriesshire, 4 m. S.W. of Ecclefechan.

Rerigonium, said to be the site of an ancient Caledonian town on the E. shore of Loch Ryan, Wigtownshire.

Rerrick, coast par., Kirkcudbrightshire ; 20,010 ac. Pop. 1228. The church is in Dundrennan vil., 5 m. S.E. of Kirkcudbright.

Rescobie, par., Angus ; 6559 ac. Pop. 559. The church is on the N. side of R. Loch, 3 m. N.E. of Forfar.

Reservoirs, The, applied to lochs, 5 m. N.W. of Muirkirk in E. Ayrshire.

Resipol, pl., Ardnamurchan, Argyll, 2 m. E. of Salen Hotel and 3 m. N.E. of Ben R. ; alt. 2774 ft.

Resolis, or **Kirkmichael,** par., Ross and Cromarty, on S. side of Cromarty Firth, and opposite Invergordon ; par. 12,449 ac. Pop. 846. The par. contains the vils. of Gordon's Mills and Jemimaville. There are numerous tumuli and ancient camps within the par.

Resort, or **Reasort,** sea loch on border of Lewis and Harris, on W. side of Lewis Isl., Ross and Cromarty.

Rest and be Thankful, stone seat in Glen Croe, Argyll.

Rest and be Thankful, a resting pl. on E. side of Corstorphine Hill, Edinburgh ; commands an excellent view of Edinburgh and the surrounding country.

Restal Loch, 5½ m. N.W. of Arrochar, E. Argyll.

Restalrig, originally a vil. N.E. of Edinburgh, now forming part of the city. The ecclesiastical history goes back to the eighth century.

Restenneth, ancient par in Angus, N. of Forfar ; has traces of a priory founded by David I.

Reston, vil., with ry. sta. (Reston junc.), L.N.E., Berwickshire, on Eye Water, 7 m. N.E. of Duns. P.O., T.O. Hotel.

Reston Junction, sta., L.N.E., Berwickshire, 7 m. N.E. of Duns.

Reswallie, seat, Angus, 3 m. N.E. of Forfar.

Retreat, seat, Grant's House, Berwickshire.

Revack, seat, 1½ m. S. of Grantown, N.E. Inverness-shire.

Rha Hostel, at R. Bridge, on N. side of Uig Bay, Skye. A Scottish Youth Hostel.

Rhiconich, originally a hotel and fishing centre, now a shooting-lodge only; at the head of Lochinchord, Sutherland, 15 m. S. of Durness. P.O., T.O.

Rhidorroch Deer Forest, Ross and Cromarty, 5 m. N.E. of Ullapool; 45,000 ac.

Rhifail, seat, 15 m. N.W. of Kinbrace, Sutherland.

Rhilochan, pl., Sutherland, 5 m. N. of Rogart. P.O.

Rhinns, The, peninsula on the S.W. of the isl. of Islay, Argyll. On the small isl. of Oversay, at the S. end, is a lighthouse 150 ft. above high water.

Rhinns, The, in Wigtownshire; forms the western of the three divisions of the county.

Rhu, par. and vil., also a favourite watering pl., Dunbartonshire; with ry. sta., L.N.E. (Rhu); par. (containing Helensburgh) 20,120 ac. Pop. 11,373. Vil. with pier on Gare Loch, 2 m. N.W. of Helensburgh. P.O., T.O. Pop. 929.

Rhubana, seat, 9 m. N.E. of Arisaig, W. Inverness-shire.

Rhudunan, Isle of Skye, Inverness-shire; headland S. of Loch Brittle; also known as Rudh an Dunain. P.O.

Rhum, or **Rum, Island,** 15 m. N.W. of Ardnamurchan Pt., Inverness-shire; large isl. measuring 8½ m. by 8 m., and its area 26,786 ac., of which only 300 ac. are arable. The S. is hilly, rising at Haskeval to 2667 ft. Most of the isl. is a deer forest. P.O., T.O.

Rhunahaorine, pl., 17 m. N.W. of Campbeltown, Argyll, on the W. of Kintyre.

Rhu Stoer, a high prom. jutting out into the sea, Sutherland, 9 m. N.W. of Lochinver; has a lighthouse with a range of 20 m.

Rhynd, par. and ham., Perthshire; par. 2468 ac. Ham. 4¾ m. S.E. of Perth. P.O.

Rhynie, par. and vil. in Aberdeenshire; par. 12,879 ac. Pop. 781. Vil. also known as Muir of Rhynie on Water of Bogie. P.O., T.O. Vil. pop. 388.

Riccarton, par., Ayrshire; 7649 ac. Pop. 8354. Contains part of Hurlford.

Riccarton, seat, Midlothian, 6 m. S.W. of Edinburgh.

Riccarton, Kincardineshire. See RICKARTON.

Riccarton, vil. with ry. sta., L.N.E. (R. Junction), Roxburghshire, 13 m. S. of Hawick. P.O., T.O.

Rickarton, q.s. par., Fetteresso and Glenbervie pars., Kincardineshire. R. House, a seat on Cowie Water, is 3 m. W. of Stonehaven.

Riddon Loch, sea loch, Cowal dist., Argyll; extends 4 m. N. from the Kyles of Bute.

Riechip, seat, 5½ m. N.E. of Dunkeld, Perthshire.

Riemore, seat, 6½ m. N.E. of Dunkeld, Perthshire.

Riff Bank, sandbank in Moray Firth, 3 m. N.E. of Fort George.

Rigg, vil., with ry. sta., L.M.S., Dumfriesshire, on Kirtle Water, 6 m. E. of Annan. P.O., T.O.

Rigg Bay, Argyll, 1 m. S. of Garliestown.

Riggend, vil., Lanarkshire, 3½ m. N. of Airdrie.

Righ, isl. in Kilmartin par., W. Argyllshire. Pop. 7.

Rigside, vil., 4 m. N.E. of Douglas, Lanarkshire.

Ringford, vil. in co. of and 5 m. N. of Kirkcudbright. P.O., T.O.

Rinns, The, W. district of Wigtownshire; extends from Carsewall Point to the Mull of Galloway; also called Rhinns (q.v.).

Risdale, or **Isle Ristol,** isl., Ross and Cromarty, 14 m. N.W. of Ullapool, between Rhu More and Summer Isles.

Rispond, ham. and harbour near the mouth of Loch Eriboll, and 4 m. E. of the vil. of Durness, Sutherland.

Roadmeetings, vil., 1½ m. E. of Carluke, Lanarkshire.

Roag Loch, sea loch on the W. of Lewis, Outer Hebrides.

Roan, isl. at entrance to the Kyle of Tongue, N. Sutherland. Pop. 61.

Roan Fell, mt. on borders of Roxburghshire and Dumfriesshire, 4½ m. N.W. of Castleton ; alt. 1862 ft.

Robertland, seat and site of old castle, 2 m. N.E. of Stewarton, Ayrshire.

Roberton, vil., Lanarkshire, near r. Clyde, 3 m. S.E. of Lamington sta. P.O. Pop. 98.

Roberton, par., Roxburghshire, on Borthwick Water ; 29,456 ac. Pop. 406. Church, 5 m. W. of Hawick. P.O., T.O.

Roberton Hostel, on Borthwick Water, ½ m. up from Roberton, between Eskdale and Teviotdale, 6 m. W. of Hawick. A Scottish Youth Hostel.

Robroyston, pl., with ry. sta., and former seat, Lanarkshire, 3½ m. N.E. of Buchanan Street ry. sta., Glasgow. In the garden there is a memorial Cross to Sir Wm. Wallace.

Rochsoles, seat, Lanarkshire, 2 m. N.W. of Airdrie.

Rockcliff, ham., Kirkcudbrightshire, 7 m. S.E. of Dalbeattie. P.O. Pop. 95.

Rockfield, par. and vil., Ross and Cromarty, 8 m. E. of Tain.

Rockhall, seat, Dumfriesshire, 6 m. E. of Dumfries.

Rockvale, Saltcoats, Ayrshire. Child Welfare Home.

Rodel, pl., with pier, extreme south end of Harris, Outer Hebrides, 1 m. N. of Renish Point. R. House, a seat. On the west shore of R. Loch is a church ruin.

Rodono, hotel, on St. Mary's Loch, in co. of, and 18 m. S.W. of, Selkirk. Formerly called Summerhope.

Roebank, pl., N. Ayrshire, 1 m. N. of Beith. Silk printing works.

Roeness Hill, Shetland, 8 m. N. of Sandwick ; alt. 1475 ft.

Roerwater, stream and loch, Shetland, Northmaven par.

Rogart, par. and ry. sta., L.M.S., Sutherland ; par. 63,400 ac. Pop. 732. P.O., T.O. Vil. 7½ m. W. by N. of Golspie.

Rogie Falls, on r. Rasay, Ross and Cromarty, 3 m. W. of Strathpeffer.

Romach Loch, Morayshire, 5 m. S.E. of Forres.

Roman Camp, pl., with shale oilworks at Uphall, W. Lothian.

Roman Camp, seat at Callander, Perthshire.

Romanno Bridge, ham., Peeblesshire, on Lyne Water, 3½ m. S.E. of W. Linton. On a steep ground are ancient cultivation terraces. R. House, a seat, nearby. Hotel. P.O. and T.O.

Ron Rock, Aberdeenshire. See RATTRAY HEAD.

Rona, isl., Inverness-shire, between Skye Isl. and the mainland, 11½ m. N.E. of Portree. On N.E. point there is a lighthouse, alt. 42 ft., seen for 21 m. ; alt. of isl., 404 ft. Pop. 16.

Rona, small isl. at an equal distance of 44 m. from the Butt of Lewis amd Cape Wrath. Was once inhabited.

Ronachan, seat, 1½ m. S.W. of Clachan, Kintyre, Argyll.

Ronaldshay North, isl., one of the Orkney group forming the extreme N.E. of the Orkneys, 3 m. long and 2 m. broad. The shallow firth which divides it from Sanday is dangerous to navigation. There is a lighthouse on the N.E. point, 139 ft., and can be seen for 18 m. Pop. 298.

Ronaldshay, South, one of the Orkney group of islands forming the S.E. boundary of the Orkneys and the E. of the Pentland Firth ; extends to 15,062 ac. From N. to S. it measures 7¾ m. and varies from 1½ to 6 m. in breadth. The Bay of St. Margaret's Hope is a safe anchorage. Pop. 1312.

Ronal Loch, 5 m. N.W. of Kirkcowan, Wigtownshire.

Ronas Hill, Shetland, on N. side of Ronas Voe and 8 m. N. of Sandwick ; alt. 1475 ft.; its summit formed one of the stations of the principal triangulation of the British Isles.

Ronay, isl., Inverness-shire, 1 m. S. of S.E. point of N. Uist. Pop. 6.

Roome Bay, ½ m. E. of Crail, Fife.

Rora, ham., 6 m. N.W. of Peterhead, Aberdeenshire. P.O.

Rora Head, headland, in W. of Hoy Isl., Orkneys.

Ros, see ROSE and ROSS.

Rosa Glen, Buteshire. See GLEN ROSA.

Roscobie, pl., with lime works, 4 m. N. of Dunfermline, Fife.

Roseacre, seat, ½ m. S. of Portsoy, in N. Banffshire.

Rosebank, seat near Wick, Caithness.

Rosebank, vil., Lanarkshire, on r. Clyde, 3 m. S.E. of Larkhall. P.O.

Rosebank Distillery, Falkirk, Stirlingshire.

Roseberry, seat, Midlothian, on r. South Esk, 4 m. S.W. of Gorebridge. There is a reservoir (Edinburgh Water Works) here.

Rosehall, pl., forming part of Whifflet, Lanarkshire, 1 m. S.E. of Coatbridge.

Rosehall, seat, estate, and ham. at the junc. of the rs. Oykell and Cassley, Sutherland. P.O., T.O.

Rosehaugh, seat, Ross and Cromarty, 4 m. S.W. of Fortrose.

Rosehearty, fishing vil. (police burgh), Aberdeenshire, 4½ m. W. of Fraserburgh. P.O., T.O. Pier and lighthouse. Pop. 1079.

Roseisle, ham. and seat, Morayshire, 2 m. S.E. of Burghead. State forest nearby.

Rosemarkie, par. and vil., Ross and Cromarty, on Moray Firth. Contains the royal burgh of Fortrose; par. 6675 ac. Pop. 1203. Pop. of Fortrose, 875. P.O., T.O.

Rosemount, seat, Symington, Ayrshire, 2 m. N.E. of Monkton sta.

Rosemount, seat, Angus, 1 m. N.W. of Dubton sta.

Rosemount, vil., seat, and ry. sta., L.M.S., Perthshire, 1½ m. S.E. of Blairgowrie. P.O. Pop. 405.

Rosemount, q.s. par., Aberdeen. P.O., T.O.

Roseneath, par. and vil., Dunbartonshire, on the promontory between Loch Long and Gare Loch; par. 8467 ac. Pop. 2227. Vil., with pier on W. shore of Gare Loch, 2½ m. W. of Helensburgh. P.O., T.O. R. House, a seat, 1 m. S.E. of vil.

Roseneath Point, prom., Dunbartonshire, at W. entrance to Gare Loch, 2 m. S.E. of Roseneath. Fixed white light, seen 9 m.

Rose Ness, S.E. extremity of Mainland, Orkney. Shows a six-seconds white flashing light, visible 14 m.

Rosetta, seat in co. of, and 1 m. N.W. of, Peebles.

Rosewell, q.s. par. and mining vil., Midlothian; vil. 4 m. S.W. of Dalkeith. P.O., T.O.

Roshven, seat, W. Inverness-shire, on S. side of Loch Ailort.

Roslin, q.s. par., with ry. stas., Roslin (passenger service withdrawn), Rosslyn Castle, and Rosslynlee, L.N.E., Midlothian, on N. Esk. Vil. 6½ m. S. of Edinburgh. Pop. 1497. P.O., T.O. Hotels. Has an explosives factory. Rosslyn Castle (P.O.), a ruin, a former seat of the St. Clairs, Earls of Orkney. The chapel, founded in 1445, is full of exquisite stone carving, including the famous 'prentice pillar. The Midlothian and Peebles Mental Hospital is at Rosslynlee.

Ross, see ROSS AND CROMARTY.

Ross, isl. of Mull, Argyll. See ROSS OF MULL.

Ross, a seat, 1¼ m. N.E. of Hamilton, at meeting of the rs. Avon and Clyde.

Ross, seat, 12 m. N. of Arisaig, Argyll.

Ross and Burnmouth, fishing vil. and ry. sta. (Burnmouth), L.N.E., Berwickshire, 5½ m. N.W. of Berwick-on-Tweed. See BURNMOUTH.

Ross and Cromarty, maritime co. in N.W. Scotland, bounded on the N. by Sutherland and on the S. by Inverness. It extends 67 m. N. and S., and 75 m. E. and W. Contains 1,976,707 ac. or 3088·6 sq. m. Pop. 62,802. The co. consists of a mainland and insular portion of which the isl. of Lewis forms the major part. On the E. side the Dornoch Firth provides harbours. Of sea lochs, Loch Broom, Little Loch Broom, Gruinard Bay, and Loch Ewe are the most important. The land lochs are Lochs Ewe, Maree, Fannich, Luichart, Sheallag, and Rosyne. Chief mts. are Bens Mam Soul (3862 ft.), Wyvis (3429 ft.)., Ben Fhada (Attow)*long mt.*—Ben More Coigach (2438 ft.). The chief towns and vils. are Tain, Dingwall (co. town), Invergordon, Cromarty, Stornoway (Lewis), Alness, Ullapool, Fortrose, and Muir of Ord. The export is mainly sheep, wool, cattle, grain, and fish. The making of Harris tweed and the distilling of whisky are the chief mfrs. The Ross and Cromarty Division and the Western Isles Division of Inverness and Ross and Cromarty each return one member to Parliament.

Rossdhu, seat, 3¼ m. S. of Luss, Dunbartonshire, on W. side of Loch Lomond.

Rossend Castle, ancient mansion adjacent to Burntisland, Fife.

Rossie, seat, Perthshire, 3½ m. W. of Bridge of Earn.

Rossie, Angus. See INCHBRAYOCK.

Rossie Castle, seat, Angus, 2¼ m. S.W. of Montrose.

Rossie House, seat, Fife, 1½ m. N.E. of Auchtermuchty.

Rossie Island, in mouth of S. Esk, Angus. Pop. 334. Connected by a suspension bridge with Montrose, and to the mainland by a stone bridge. Included in the parl. bur. of Montrose. Formerly Inchbrayoch Isl.

Rossie Priory, Inchture, Perthshire, seat of Lord Kinnaird, 8 m. W. of Dundee.

Rosskeen, par., Ross and Cromarty, on Cromarty Firth; 33,725 ac. Pop. 3580. Contains Invergordon. P.O., T.O. Ry. sta. L.M.S.

Ross, Little, isl., Kirkcudbrightshire. See LITTLE ROSS.

Rosslyn Castle, ry. sta., L.N.E., Midlothian, 1¼ m. S. of Roslin vil. P.O.

Rosslynlee, ry. sta., L.N.E., Midlothian, 1¾ m. S. of Roslin vil.

Ross Point, prom., Fife, near Burntisland.

Ross Point, headland on E. side of Loch Lomond, 2 m. S. of Rowardennan, Dunbartonshire.

Ross Priory, seat, Dunbartonshire, on the S.E. side of Loch Lomond, 4½ m. N.E. of Balloch sta.

Rosyth Castle, ruin in Inverkeithing par., Fife, 1¾ m. N.W. of Queensferry, close to Rosyth Naval Base. Mentioned by Sir Walter Scott in *The Abbot.*

Rosyth Naval Base, on N. shore of Firth of Forth, and included in the town of Dunfermline, Fife, 30 m. from the mouth of the Forth. The Base, created in 1909, is one of the largest in the world, capable of docking and repairing all classes of warships. Ry. sta., L.N.E. (Rosyth Halt). P.O., T.O.

Rothes, par. and police bur., with ry. sta., L.N.E., Morayshire; par. 19,816 ac. Pop. 1854. Vil. 9¾ m. S.E. of Elgin. Pop. 1292. P.O., T.O. Hotel. Has distilleries.

Rothes, vil., Fife, 1 m. W. of Markinch. P.O., T.O. Has a paper mill.

Rothesay, royal and mun. bur., par., and co. town of Buteshire, Firth of Clyde, 19 m. S.W. of Greenock; par. 2439 ac. Pop. 9346 (par. and bur.). A summer resort, situated in a sandy bay, good harb. with pier; has a mild climate. Hotels. Steamer communication with Dunoon, Greenock, and Glasgow. Headquarters of the Royal Northern Yacht Club. P.O., T.O. R. Castle, a ruin, gives the title of Duke to the heir-apparent of the British throne.

Rothiemaise Hill, 4 m. N.E. of Insch ry. sta., Aberdeenshire; alt. 854 ft.

Rothiemay, par. and vil., with ry. sta., L.N.E., Banffshire; par. 9365 ac. Pop. 926. Vil. is on r. Deveron, 2¼ m. N.E. of Rothiemay sta. (in Aberdeenshire, 5 m. N. of Huntly). P.O., T.O., and T.O. at sta. Near the vil. is R. House.

Rothiemurchus, q.s. par., Inverness-shire. Church is on the Spey, 2 m. S.W. of Aviemore sta.

Rothiemurchus Deer Forest, Inverness-shire; 22,000 ac. Post town, Aviemore.

Rothie-Norman, ry. sta., L.N.E., Aberdeenshire, 7½ m. N. of Inveramsay. P.O., T.O. Pop. 128.

Rothney, vil., Aberdeenshire, near Insch ry sta. Pop. (R. and Insch) 914.

Rottal Lodge, 11 m. N.W. of Kirriemuir, N. Angus. R. and Glenmoye shooting, 14,500 ac.

Rotten Calder, The, name given to lower course of r. Calder, Lanarkshire, after joining with the Rotten near Torrance.

Roucan, vil., Dumfriesshire, 3½ m. N.E. of Dumfries.

Rouchan, Low, seat, S.E. Wigtownshire, 2½ m. S.W. of Whithorn.

Rough Castle, seat, with antiquities and Roman remains, 3 m. W. of Falkirk, Stirlingshire.

Rough Firth, sea loch, Kirkcudbrightshire, 4 m. S. of Dalbeattie.

Roughrigg, colliery vil., Lanarkshire, 2½ m. S.W. of Slamannan.

Roughsware, hill, S.E. Midlothian, 3 m. E. of Gladhouse Reservoir; alt. 1394 ft.

Rouken Glen, public park, with house, Pollokshaws; now part of Glasgow.

Round Loch, Tay Basin, S. Angus. See PITLYAL.

Round of the Dungeon Loch, 8½ m. S.W. of Carsphairn, N.W. Kirkcudbrightshire.

Roundy Hill, ham., near Kirriemuir, Angus.

Rousay, isl., Orkney, 10 m. N. of Kirkwall. Pop. 468. P.O., T.O. Has a hilly surface; measures 5½ m. by 4½ m.

Rousay, Egilshay, and Wyre, isls., Orkney, forming one par; 13,761 ac. Pop. 597.

Rovie Lodge, 1 m. W. of Rogart, Sutherland.

Row, Dunbartonshire. See RHU.

Rowadill, Outer Hebrides. See RODEL.

Rowanburn, vil., 1½ m. S.E. of Canonbie, Dumfriesshire, on Liddel Water. Pop. 237.

Rowantree Hill, Barr par., S.E. Ayrshire; alt. 1811 ft.

Rowardennan Hotel and **Pier,** on E. side of Loch Lomond, Stirlingshire, 9½ m. N. of Balloch Pier; usually the starting-place to ascend Ben Lomond. P.O., T.O. R. Lodge, a seat. Ferry to Inverbeg; no cars.

Rowbank, reservoir, Paisley water supply, Renfrewshire, 8 m. S.W. of Paisley.

Rowchester, seat, 3 m. S.E. of Greenlaw, Berwickshire.

Rowchoish, pl. on E. side of Loch Lomond, opposite Tarbet, shortest path to summit of Ben Lomond.

Roxburgh, par. and vil., with ry. sta., L.N.E.; par. 7714 ac. Pop. 698. Vil. near r. Teviot, 3 m. S.W. of Kelso, Roxburghshire. P.O., T.O. R. Castle, a ruin, near the confluence of Teviot and Tweed.

Roxburghshire, border co., S. of Scotland, bounded by Berwickshire, Northumberland, Cumberland, Dumfriesshire, Selkirk, and Midlothian; its length N. to S., 42 m.; breadth E. to W., 30 m.; area 665·7 sq. m. or 426,060 ac. Pop. 45,787. The chief part of the co. lies within the drainage area of the Teviot, and the co. is often referred to as Teviotdale. The three Eildons in the N.W. of the co. are an interesting feature, and rise to an alt. of 1385 ft. The S. portion is high upland ground suitable for pastoral farming. The principal streams are: Teviot, Tweed, Borwick, Ale, Slitrig, Rule, Jed, Oxnam, and Kale. The co. is divided into 30 pars., and the chief towns are Kelso, Hawick, Jedburgh, Melrose, with the vils. of Castleton, Yetholm, Roxburgh. The industry consists chiefly of the manufacture of woollens, tweeds, and blankets. Combines with Selkirkshire in returning one member to Parliament.

Roy, r., S. Inverness-shire, rises 8 m. S. of Fort Augustus and flows to the Spean, near Roybridge.

Roy Bridge, or **Roybridge,** pl., with ry. sta., L.N.E., Inverness-shire, near the meeting of Roy and Spean. P.O., T.O. Hotel.

Roy's Cairn, hill, 1 m. S.W. of Knockando, Morayshire; alt. 1691 ft.

Royston, dist., Edinburgh, between Granton and Cramond, on Firth of Forth.

Rozelle, seat, 2 m. S. of Ayr.

Ruadh Sgeir, in Sound of Jura, Argyll. Shows a six-second white flashing light, visible 11 m.

Ruaig, pl., N.E. Tiree Isl., Argyll, 3 m. N.E. of Scarinish. P.O., T.O.

Ruard Loch, 8 m. N.W. of Dunbeath in S.E. Caithness.

Ruathair An, loch, Sutherland, 2½ m. N. of Kinbrace sta.; alt. 418 ft., depth 26 ft.

Rubers Law, hill, Roxburghshire, 4¾ m. N.E. of Hawick; alt. 1392 ft.

Rubislaw, q.s. par., W. side of bur. of Aberdeen. Large granite quarry.

Ruchill, public park, N. suburbs of Glasgow.

Ruchill Water, Perthshire, flows 10 m. along Glen Artney to r. Earn, which it enters at Dalginross opposite Comrie.

Ruchlaw, seat, E. Lothian, 3½ m. S.E. of East Linton.

Ruddons Point, prom. on E. side of Largo Bay, Fife.

Rudha Geall, prom., on E. of Lewis, Outer Hebrides, 5 m. S.E. of the Butt of Lewis.

Rudha Hunish, headland at N. end of the Isle of Skye, Inverness-shire.

Rudha Mhail Lighthouse, on N. pt. of the isl. of Islay, Argyll; alt. 147 ft., seen for 18 m.

Rudha na h-Ordaig, headland, S.E. coast of S. Uist, Outer Hebrides.

Rudh' an Dunain, headland, Skye Isl., S. of Loch Brittle. P.O. called Rhudman.

Rudh' Re, headland, 10 m. N.W. of Gairloch, Ross and Cromarty.

Ruel River, salmon and trout stream flowing through Glendaruel to the head of Loch Riddon, Argyll.

Rueval, sound between N. Uist and Benbecula, Outer Hebrides, Inverness-shire; sound 9 m. long and often called Loch Rueval.

Ruffell, seat, Perthshire, 2 m. N. of Murtly sta.

Ruff Reef, off Cantick Head, Walls, Orkney. Shows a ten-second group-flashing light, seen 10 m.

Ru Hunish, Skye. See RUDHA HUNISH.

Rule Water, Roxburghshire, rises in Hobkirk par. and flows 13 m. to join the Teviot 2 m. N.E. of Denholm.

Rullion Green, pl., Midlothian, 1¾ m. N.W. of Penicuik. Scene of the defeat of the Covenanters in 1666.

Rumbling Bridge, pl. and ry. sta., L.N.E., on border of Kinross and Perth shires, 4 m. N.E. of Dollar. P.O., T.O. Hotel.

Rumbling Bridge, bridge over a deep chasm on the r. Bran, Perthshire, 2¾ m. S.W. of Dunkeld.

Rumford, vil., Stirlingshire, 1 m. S. of Polmont. Pop, 1498.

Rum Island, Inner Hebrides, Inverness-shire, 15 m. N.W. of Ardnamurchan Point. Pop. 32. P.O., T.O. called Rhum. A mountainous isl., measuring 8½ m. by 8 m. Greatest alt., Huskeral, 2667 ft. Mostly deer forest.

Rumness, headland, Angus, 4½ m. N.E. of Arbroath.

Rumsdale, seat, 1 m. S. of Altnabreac in W. Cathness.

Runnan eun Point, S.E. Bute Isl. Shows flashing light, visible 10 m.

Ru Rea, Ross and Cromarty. See RUDH' RE.

Ruskie, ham., Perthshire, on Loch Ruskie, 3½ m. S. of Callendar. P.O.

Rusko, seat, 4 m. S.E. of Dromore, E. Kirkcudbrightshire.

Rusness, ham., N.E. of Sanday Isl., Orkney. P.O.

Ru Stoer, Sutherland, West Coast. See RHU STOER.

Rutherford, pl. and ry. sta., L.N.E., Roxburghshire, 6 m. S.W. of Kelso.

Rutherglen, royal and mun. bur. and par. Lanarkshire; par. 2158 ac. Pop. 34,223. Town on r. Clyde, 2 m. S.E. of Glasgow; ry. sta., L.M.S. Bur. pop. 25,157. P.O., T.O. In the neighbourhood are collieries, chemical and dye works, paper mills, and ropeworks.

Ruthreie, Linn of, cascade on Aberlour Burn, 1 m. S. of Aberlour, Banffshire.

Ruthrieston, ry. sta. (R. Halt), L.N.E., on border of Aberdeen and Kincardine Shires, 2 m. S. of Aberdeen.

Ruthven, vil., Aberdeenshire, 2 m. N.W. of Rothiemay. P.O. Pop. 33.

Ruthven, par. and ham., Angus; par. 2049 ac. Pop. 226. Ham on Alyth Burn, 4 m. N.E. of Meigle. P.O. Nearby is R. House.

Ruthven, loch in co. of, and 11 m. S. of, Inverness; alt. 701 ft., depth 42 ft.

Ruthven, pl., E. Inverness-shire, 1 m. S. of Kingussie.

Ruthven Castle and **Ruthvenfield,** Perthshire. See HUNTINGTOWER.

Ruthven Road, ry. sta., L.M.S., in co. of, and 2½ m. N.W. of, Perth.

Ruthven Water, Perthshire, flows 9 m. E. to r. Earn, 1 m. N. of Aberuthven.

Ruthwell, coast par., and ry. sta., L.M.S., Dumfriesshire; par. 8686 ac. Pop. 646. Vil. 1½ m. S.E. of sta. and 9 m. S.E. of Dumfries. P.O., T.O. Pop. R. and East Raffles, 615.

Rutton Loch, 2 m. N.W. of Lochanhead, E. Kirkcudbrightshire.

Ryan Loch, sea loch in N.W. Wigtownshire, 8 m. long and 1½ to 2 m. broad. Affords safe anchorage. See LOCHRYAN.

Ryedale, seat, Kirkcudbrightshire, on r. Nith near Dumfries.

Ryefield House, Ayrshire, 1 m. N.W. of Dalry.

Ryehill House, Ross and Cromarty, 2 m. R. of Conon Bridge.

Ryeland, ry. sta., L.M.S., Lanarkshire, 3¾ m. S.W. of Strathaven.

S

Saasaig, ham., in par. of Sleat, 3 m. N.E. of Armadale pier, Skye, Inverness-shire.

Sabhal Beag and **S. Mor,** (1) mt., alt. 2393 ft., 6 m. S.E. of Foinne Bheinn ; (2) mt., alt. 2280 ft. 5 m. S.E. of Foinne Bheinn, Sutherlandshire.

Sabiston, loch, depth 3 ft., 6½ m. N.W. of Firstown, Orkney.

Sackville, seat, 4½ m. N.E. of Watten in E. Caithness.

Sacquoy Head, N.W. extremity of Rousay Isl., Orkney.

Saddell, q.s. par. Argyll, on E. side of Mull of Kintyre. P.O., T.O. S. Abbey, a ruin of twelfth century. S. House at the foot of S. Glen on S. Bay, 9 m. N.N.E. Campbeltown.

Saddell and Skipness, par. Argyll, on E. side of Kintyre; 46,888 ac. Pop. 946.

Saddle, The, mt., Inverness-shire, 3 m. S. of Shiel Hotel; alt. 3317 ft.

Saddle Yoke, mt., 6 m. N.E. of Moffat, Dumfriesshire; alt. 2412 ft.

Saighe, loch, N.W. Caithness, 6½ m. S.S.W. of Reay.

Sail Mhor, mt., Ross and Cromarty, S. of Little Loch Broom and 6½ m. S.W. of Ullapool; alt. 2508 ft.

Saint, loch, 4 m., N.W. of Blairgowrie, E. Perthshire.

St. Abbs, vil., Berwickshire, 1½ m. N.E. of Coldingham. P.O., T.O. Pop. 278.

St. Abb's Head, rocky prom., Berwickshire, 4 m. N.W. of Eyemouth; alt. 310 ft., named after St. Ebba, who founded a monastery here about the middle of the seventh century. There is a lighthouse, a Lloyd's signal sta., and coastguard sta.

St. Andrews, par., royal and mun. burgh, watering-place and port in Fife, 13½ m. S.E. of Dundee; par. (St. A. and St. Leonards), 12,594 ac. Pop. 9984, burgh pop. 8269. Ry. sta., L.N.E.; P.O., T.O. The university, the oldest in Scotland, was founded in 1411, and unites with those of Aberdeen, Glasgow, and Edinburgh in returning three members to Parliament. The town was for long the ecclesiastical centre of Scotland and witnessed the martyrdom of Patrick Hamilton (1527) and the assassination of Cardinal Beaton (1546); the cathedral, now a ruin, was founded in 1159; the church of St. Regulus or St. Rule is a well-preserved ruin. St. A. is the "Mecca" of golf and the headquarters of the "Royal and Ancient Club," founded 1754.

St. Andrews and Deerness, par., on E. side of the Mainland of Orkney; par. 12,830 ac. Pop. 1155. Includes the islands of Copinshay and Kirkholm. The church of St. Andrews is 6 m. S.E. of Kirkwall, on Deer Sound.

St. Andrews Lhanbryde, par., Morayshire, on r. Lossie; 9197 ac. Pop. 1165. Contains part of Elgin.

St. Ann's, ham., 8¼ m. N.W. of Lockerbie, Dumfriesshire. P.O.

St. Anthony's Chapel, a ruin above St. Margaret's loch on north base of Arthur's Seat, Edinburgh, ¼ m. S.E. of Holyrood House.

St. Baldred's Cradle, rock, on N. side of the mouth of the r. Tyne, E. Lothian, 2½ m. N.E. of Tynninghame.

St. Bernard's Well, mineral spring on the Water of Leith and in the Stockbridge dist. of Edinburgh.

St. Boswells or **Lessuden,** par. and vil., Roxburghshire (containing part of Newtown St. Boswells); par. 3466 ac. Pop. 592. Vil. 1½ m. S.E. of St. Boswells ry. sta., L.N.E., and 2½ m. from Melrose. Hotel. P.O., T.O. Pop. 564. Important live-stock sales.

St. Boswells Junction, ry. sta., L.N.E., Roxburghshire, 3½ m. S.E. of Melrose.

St. Bride's Chapel, ruin at S. end of Loch Lubnaig, Perthshire. Mentioned by Scott in the "Lady of the Lake."

St. Catherine's, pl., with hotel and ferry on E. shore of Loch Fyne, Argyll, opposite Inveraray. P.O., T.O.

St. Columba, q.s. par., Kilmore and Kilbride par., Oban, Argyll.

St. Combs, fishing vil., with ry. sta., L.N.E., Aberdeenshire, 5¾ m. S.E. of Fraserburgh. P.O., T.O. Pop. 599.

St. Cyrus, par. and vil., with ry. sta., L.N.E., Kincardineshire; par. 8249 ac. Pop. 1173. Vil. 5 m. N.N.E. of Montrose. P.O., T.O. Vil. pop. 356.

St. David's, seapt. vil., Fife, on Firth of Forth, 1½ m. E. of Inverkeithing. P.O.

St. David's, vil., Perthshire, 5½ m. E. of Crieff.

St. Enoch, terminus of Glasgow ry. sta., L.N.E.

St. Fergus, coast par. and vil., Aberdeenshire; par. 8856 ac. Pop. 841. Vil. 5 m. N.N.W. of Peterhead. P.O., T.O.

St. Fillans, vil., with ry. sta., L.M.S., Perthshire, at E. end of Loch Earn, 11½ m. W. of Crieff. P.O., T.O. Pop. 170. Hotel.

St. Fink, seat, 2½ m. N.E. of Blairgowrie, Perthshire.

St. Fort, ry. sta., L.N.E., Fife, 3½ m. N.W. of Leuchars Junction, 1 m. N.N.E. is St. Fort House.

St. Helena, islet, Wigtownshire, at head of Luce Bay. Pop. 1.

St. Germains, seat, 2 m. N.E. of Tranent, E. Lothian.

St. Katherine's, pl., N. Aberdeenshire, 2½ m. S.S.E. of Fyvie. P.O.

St. Kilda, the most Western group of isls. in Scotland, being 40 m. W. of Harris, 95 m. S.W. of the Butt of Lewis, and 100 m. W. of the mainland of Scotland. The group consists of 4 isls., St. Kilda (N. Kilder, a well), Dun, Soay (sheep's isle) and Boreray (N. isle). These isls. appear to have been inhabited for over 2000 years, as indicated by archæological remains. The pop. in 1697 was 180, but it went gradually down to 43 in 1927, and in 1929 the remaining inhabitants were taken to the mainland. The area of the group is 2117 ac. and the highest land is 1397 ft. Originally the property of McLeod, it was recently sold to the Earl of Dumfries.

St. Madoes, par., Perthshire, on Firth of Tay : 1561 ac. Pop. 274. The church is near Glencarse ry. sta., 6 m. E. of Perth.

St. Magdalene, distillery in the town of Linlithgow, W. Lothian.

St. Magnus Bay, on W. coast of Mainland of Shetland.

St. Margaret's Hope, fishing vil. and harbour in N. of South Ronaldshay, isl. of Orkney, 13 m. S. of Kirkwall. P.O., T.O.

St. Margaret's Hope, anchorage in Firth of Forth, 1 m. N.W. of North Queensferry. Steamers from Leith, Aberdeen, etc., call regularly.

St. Margaret's Loch, small artificial lake at the foot of Arthur's Seat and ¼ m. E. of Holyroodhouse, Edinburgh.

St. Martins, par., Perthshire, on r. Tay ; 7017 ac. Pop. 801. Church 4¾ m. N.E. of Perth. St. Martin's Abbey, a seat.

St. Mary's, q.s. par., S. Ronaldshay and Burray par., Orkney.

St. Mary's Coast, vil. in S. of Mainland of Orkney, 7 m. S. of Kirkwall. P.O., T.O. Called St. Mary's Holm.

St. Mary's Croft, pl. on W. side of Loch Ryan, Wigtownshire, 4 m. N.W. of Stranraer.

St. Mary's Isle, seat in co. of, and 1¼ m. S. of, Kirkcudbright.

St. Mary's Loch, Selkirkshire, 16 m. S.W. of Selkirk. Loch is 814 ft. above sea-level and measures 3 m. long, ½ m. broad. The beauty of the loch and its surrounding scenery has been sung by Scott, Wordsworth, Hogg, and others ; several large streams flow into, and the r. Yarrow issues from, its N. end, and at its S. end is the well-known Tibbie Shiels Inn, while at its W. side is the Rodono Hotel.

St. Monans, par. and fishing vil., with ry. sta. (St. Monans), L.N.E., Fife ; par. 1205 ac. Pop. 1819. Vil. 2 m. E. of Elie. Pop. 1700. P.O., T.O. Good harbour. Favourite resort of artists.

St. Mungo, par., Dumfriesshire, on the Water of Milk and the r. Annan ; 4901 ac. Pop. 534. The church is 3 m. S. of Lockerbie.

St. Ninians, or **St. Ringans,** par. and suburb of Stirling, on r. Forth ; par. 37,543 ac. Pop. 14,757. P.O., T.O. Manufactures nails, screws, bolts, etc.

St. Ninian's Cave, Wigtownshire, 3 m. N.W. of Burrow Head.

St. Oran's Chapel, ruin, Iona isl., Argyll.

St. Quivox, former par., Ayrshire, 3 m. N.E. of Ayr is the par. church ; is now part of the par. of Ayr.

St. Ringans, Stirlingshire. See ST.NINIANS.

St. Rollox, dist. and parl. div. ; ry. sta., L.M.S., on N. side of Glasgow. P.O., T.O. Workshops of L.M.S.

St. Ronan's Well, Peeblesshire. See INNERLEITHEN.

St. Serf's Island, islet in Loch Leven, Kinross-shire. Has ruins of a priory.

St. Vigeans, q.s. par., Arbroath and St. Vigeans and Panbride pars., Angus.

Salachan, pl., N. Argyll, on W. side of Loch Linnhe, 2½ m. W. of Ardgour. Hotel.

Salachie Loch (Loch an-t-Salachaidh) in par. 5 m. W. of Golspie, Sutherland.

Salen, q.s. par., vil. and bay, on E. side of isl. of Mull, Argyll. Vil. pop. 119. P.O., T.O. Called Aros. 8 m. S.E. of Tobermory.

Salen, pl. on N. shore of Loch Sunart, 10 m. N.W. of Strontian, Argyll. P.O., T.O., and hotel.

Salen State Forest, near Salen Pier on Loch Sunart, Argyllshire.

Saligo Bay, Argyll, on W. of Islay. Saligo R., which drains Loch Gorm, flows into this bay.

Saline, par. and vil. in Fife ; par. 8757 ac. Pop. 1510. Vil., 2½ m. N. of Oakley sta. P.O., T.O. Pop. 1095. 1 m. N. of vil. is S. Hill ; alt. 1178 ft.

Salisbury Crags, a high cliff extending to ¼ m. on the N.W. of Arthur's Seat, City of Edinburgh.

Sallachy, seat, 5 m. N.W. of Lairg, Sutherland.

Sallachy, vil., 3½ S.E. of Strome Ferry, S.W. Ross and Cromarty. P.O.

Salsburgh, vil., Lanarkshire, 3¼ m. N.W. of Shotts; ry. sta. P.O., T.O. Pop. 1130.

Saltburn, vil., Ross and Cromarty, 1¼ m. N.E. of Invergordon. P.O.

Saltcoats, police bur., seapt., and watering-pl., with ry. sta., L.M.S., Ayrshire, 1¼ m. S.E. of Ardrossan; pier, hotels. P.O., T.O. Pop. 10,173.

Salton, or **Saltoun,** par., with ry. sta. (Saltoun), L.N.E. (passenger service withdrawn), 4 m. S.E. of Ormiston, E. Lothian, on r. Tyne; par. 3641 ac. Pop. 419. Church at E. Salton, vil. 6¼ m. S.W. of Haddington. W. Salton, vil. with P.O., T.O. 1 m. W. of East Salton. Vil. pop. 141. S. Hall, a seat.

Salzcraggie, seat and ry. platform, L.M.S., Sutherland, 3 m. N.W. of Helmsdale.

Samphrey, isl., Shetland, in S.E. entrance to Yell Sound; measures ½ m. by 1 m.

Samson's Lane, pl., 1 m. from Stronsay, Orkney. P.O.

Samson's Ribs, high cliff at S.W. of Arthur's Seat, city of Edinburgh, showing twisted basaltic columns.

Samuelston, vil., E. Lothian, on r. Tyne, 3½ m. S.W. of Haddington.

Sanaighmore, ham. and bay on N.W. coast of Islay, Argyll, 5 m. S.W. of Ardnave Pt.

Sanda, isl., Argyll, at W. side of entrance to the Firth of Clyde, 2 m. S. of Kintyre Pt.; measures ¾ m. by 1¼ m.; alt. 405 ft. Has lighthouse, 165 ft., with range of 18 m. Pop. 14.

Sandaig, ham. and bay, Knoydart, Argyll, on N. side of the mouth of Loch Nevis.

Sanday, isl., Inner Hebrides, Inverness-shire, off E. side of Canna Isl. and 4 m. N.W. of Rhum Isl.; measures 1¾ m. by ½ m., and is connected with Canna at low water. Pop. 20.

Sanday, one of the largest of the North Isles of the Orkney. It lies 2¼ m. N. of Stronsay and measures N.E. to S.W. 13 m.; the breadth varying from

¼ m. to 5 m.. It is very flat, rising to only 173 ft. The harbours are at Kettletoft on S.E., and Otterswick on the N.E. Lighthouse at Start Pt. There are eleven small lakes, the largest 7½ m. in circumference. Pop. 1160. P.O., T.O.

Sandbank, q.s. par. and vil., Dunoon and Kilmun par., Argyll, on S. side of Holy Loch. Vil., 2¾ m. N.W. of Dunoon P.O., T.O. Pop. 942. Hotels and pier where Glasgow steamers call.

Sandend, fishing vil., Banffshire, 2½ m. W. of Portsoy. Pop. 259.

Sanderay, isl., Outer Hebrides, Inverness-shire, 3½ m. S. of Barra Isl; hill of gneiss, alt. 800 ft.

Sandford, vil., Lanarkshire, 1½ m. S.E. of Stenhouse. P.O.

Sandford Bay, Peterhead, Aberdeenshire.

Sandhaven, vil., Aberdeenshire, 2 m. W. of Fraserburgh. Pop. 598. P.O., T.O.

Sandhead, vil., Wigtownshire, on Luce Bay, 7 m. S.E. of Stranraer. P.O., T.O.

Sandilands, ry. sta., L.M.S., Lanarkshire, 3 m. S. of Lanark. P.O., T.O.

Sandness, pl., 8 m. N. of Walls, Shetland, P.O., T.O. S. Hill, alt. 817 ft.

Sandray, one of the Barra isls., Outer Hebrides, 3½ m. S. of Barra.

Sands, seat, on Firth of Forth, 1½ m. S.E. of Kincardine, Fife, in Tulliallan par.

Sandside House, seat, Caithness, 1 m. W. of the vil. of Reay, near Sandside Bay.

Sand Sound, voe or bay, Shetland, on N. side of Scalloway bay.

Sandsting, par. in Mainland of Shetland, 13 m. N.W. of Lerwick. The surface is hilly and rises to 457 ft. at Ward of Scollan. There are many lakes, the larger being Clousta, Vaara, Hulma, Gossa, Sulma, and Voxterby. 39,870 ac. Pop. 1541.

Sandwick, par. on mainland of Orkney; 11,827 ac. Pop. 901. The Church is 5 m. N. of Stromness. P.O., T.O.

Sandwick, vil., 1¼ m. S.E. of Stornoway, Lewis Isl., Ross and Cromarty.

Sandwick, q.s. par. and vil., Dunrossness par., Shetland. Ham. 13 m. S.W. of Lerwick. P.O., T.O., and P.O., T.O. at West S.

Sandwick, pl., N.E. coast of Yell Isl., Shetland.

Sandwood, pl. and lake on the W. of Sutherland, 7 m. S. of Cape Wrath.

Sandyknowe Crags, 1½ m. S.W. of Smailholme, Roxburghshire.

Sannick Bay, small bay, ¾ m. W. of Duncansbay Head, Caithness.

Sannox, ham. on E. side of Arran isl., Buteshire, at foot of Glen Sannox, 7 m. N. of Brodick.

Sanquhar, par. and royal mun. bur. in Dumfriesshire; par. 40,905 ac.; pop. 3346, which includes, 1753 the pop. of the town. Town stands on the r. Nith, 26¼ m. N.W. of Dumfries, and the name is derived from the Gaelic scam Cathair—*old fort*; was made a royal bur. 1598. Stands in the midst of a coalfield and has iron and steel works, spade and shovel factory, brick and tile works. Ry. sta., L.M.S. P.O., T.O.

Sanquhar House, seat, 1 m. S. of Forres, Morayshire.

Sark, small r. in the extreme S.E. of Dumfriesshire; its source lies in the Eskdale Hills, and after a run of 11 m. falls into the head of the Solway Firth.

Sauchen, vil., Aberdeenshire, 2 m. S. of Monymusk ry. sta. P.O., T.O.

Sauchie, vil., with ry. sta., L.N.E., and q.s. par. Alloa, Clackmannanshire. P.O., T.O. Pop. 2782.

Sauchie, seat and ruined fortalice, Stirlingshire, 3 m. S.W. of Stirling, close to Sauchieburn where James III. was defeated in 1488 and murdered near Beaton's Mill while trying to escape.

Sauchie Law, mt., Selkirkshire, 4 m. S.E. of Ettrick Church; alt. 1449 ft.

Sauchie, New, part of Alloa in co. of Clackmannan.

Saughs, Water of, Angus. See WEST WATER HILL OF SAUGHS, mt. on N. border of Kincardineshire and Aberdeenshire; 8 m. S.E. of Ballater; alt. 2141 ft.

Saughton, dist. and ry. sta., L.N.E.; formerly an estate, now part of the city of Edinburgh; pleasure gardens and golf course.

Saughtree, ry. sta., L.N.E., Roxburghshire, 2¾ m. E. of Riccarton Junction.

Saulseat, or **Soulseat,** ancient par. and site of Abbey of the twelfth century, Wigtownshire, 3 m. S.E. of Stranraer.

Savalmore, ham., 1 m. N. of Lairg ry. sta., E. Sutherland.

Savoch, q.s. par., now part of Ellon, New Deer, Old Deer, Aberdeenshire.

Saxa Nord, hill, near N. of Unst Isl., Shetland; alt. 934 ft.

Scaddon Lighthouse, see FAIR ISLE.

Scald Law, highest summit of the Pentland Hills, Midlothian; alt. 1898 ft.

Scallastle, seat and bay, N. coast of Mull, Argyll, 2 m. N.W. of Craignure.

Scalloway, seapt. vil. on W. coast of Mainland of Shetland, 6 m. S.W. of Lerwick. P.O., T.O. Pop. 602. Has important fisheries. Old castle of S. is a ruin (1600).

Scalpay, isl. Outer Hebrides, Inverness-shire, at the entrance of Loch Tarbert, Harris; measures 1½ m. × 2¾ m.; mostly clad with heath. On E. point there is a lighthouse, alt. 130 ft., can be seen for 17 m. Pop. 636. P.O., T.O., at Kyles, Scalpay.

Scalpay, isl., Inner Hebrides, Inverness-shire, 4 m. N.W. of Broadford, Skye. P.O., T.O. Pop. 27. Measures 3 m. by 4 m.; 6489 ac. The highest point is 1298 ft.

Scalpsie Bay, on S.W. of Bute Island.

Scalp, The, mt., Banffshire, 3½ m. S.E. of Dufftown; alt. 1599 ft.

Scamadale Loch, Argyll, 7 m. S. of Oban is 1½ m. long.

Scamodale, pl., N. Argyll, at the mouth of Allt Scamodale, on the E. bank of Loch Shiel.

Scaniport, in co. of, and 4 m. S.W. of, Inverness-shire. P.O., T.O.

Scapa, pier for Kirkwall, Orkney, 2 m. S. of Kirkwall on Scapa Bay. A safe anchorage in Scapa Flow.

Scapa Flow, a wide expanse of water enclosed by islands, Orkney. Extends from the Mainland of Orkney for 15 m. to the Pentland Firth. Contains many good harbours.

Scarba, isl., Argyll; lies 1½ m. N. of Jura Island and 3½ m. W. of Craignish; it measures 3½ m. by 2¾ m., has an area of 3676 ac., and rises to 1476 ft. The pasture is heathy and is used as a deer forest. Pop. 5.

Scar, Big and **Little,** two rocks in entrance to Luce Bay, Wigtownshire.

Scarfskerry, coast ham. 4 m. N.E. of Dunnet, Caithness. P.O., T.O.

Scarmclate Loch, Caithness, 2¼ m. E. of Halkirk; alt. 89 ft., depth 5 ft.

Scarnish, ham., with harbour, Tiree Isl., Argyll, between Got Bay and Hynish Bay. P.O., T.O. Called Scarinish.

Scar Nose, headland, on W. side of Cullen Bay, Banffshire.

Scarp, isl. in N.W. Harris, Inverness-shire; measures 2¼ m. by 3 m. and rises to 1011 ft. Pop. 95.

Scarsburgh, ancient camp, near Jedburgh, Roxburghshire.

Scar, The, sandbank on the W. shore of Loch Ryan, Wigtownshire.

Scar Water, in N.W. Dumfriesshire; flows 19 m. S.E. to the r. Nith, 2 m. S.W. of Thornhill.

Scatsta, ham. N. Mainland of Shetland, 1 m. S.W. of the Voe of Scatsta on E. side of Sullom Voe.

Scatwell, seat, Ross and Cromarty, on r. Conon, 10 m. N.W. of Muir of Ord; ry. sta.; P.O., T.O. S. and Cabaan Deer Forest covers 10,000 ac.

Scaut Hill, mt., Banffshire, 5½ m. S.W. of Dufftown; alt. 1987 ft.

Scavaig Loch, sea inlet on S.W. coast of Skye, S. of the Cuillin Hills, Inverness-shire.

Scaven, loch, 3 m. N.E. of Glencarron, Ross and Cromarty.

Scawa Law, mt. on borders of Lanark and Dumfriesshire, Upper Clydesdale; alt. 2166 ft.

Schiehallion, mt., Perthshire, 4 m. S.E. from Kinloch Rannoch; alt. 3547 ft. This mt. was used by Dr. N. Maskelyne in 1774 to ascertain the mean density of the earth. The name means the mountain of the Caledonians.

Schil, The, alt. 1985 ft., 2 m. N.W. of The Cheviot (2676 ft.), Roxburghshire.

Schivas, ancient mansion, 2¼ m. N.E. of Haddo House, Aberdeenshire, now restored and occupied.

Schoolhill, ry. sta., L.N.E., in city of Aberdeen.

Sclaterford, the scene of a battle between Scots and English, 1513, between Hawick and Jedburgh, Roxburghshire.

Sclattie, part of Auchmull, Aberdeenshire.

Scoly, Coch (Tay Basin), Perthshire, 4 m. E.S.E. of Aberfeldy; depth 12 ft.

Scone, par. and vil., Perthshire; par. 6963 ac. Pop. 2559. Vil. (New Scone) 2 m. N.E. of Perth. Pop. 1858. P.O., T.O. 1½ m. W., on r. Tay, is Scone Palace, occupying the site of an Abbey founded in 1411 and destroyed in 1559. Old Scone, now a decayed ham. near the palace, was the capital of the kingdom of Scotland, and was long the place where the Scottish kings were crowned. The Coronation Stone or Stone of Destiny was removed to Westminster by Edward I. in 1296, where it is now. S. Palace, seat of the Earl of Mansfield.

Sconser, ham., 10 m. S.E. of Portree, Skye Isl., Inverness-shire, on Loch Sligachan. P.O.

Scoonie, par. on S. coast of Fife; 4107 ac. Pop. 7953; contains Leven. Pop. 7411.

Scootmore State Forest, 1 m. W. of Blacksboat ry. sta., 23 m. N.E. of Boat of Garten, Inverness-shire.

Scoraig, pl., Ross and Cromarty, on N. side of Little Lochbroom.

Scotasay, isl. in E. Loch Tarbert, Lewis Isl., Inverness-shire.

Scotlandwell, vil., Kinross-shire, 5 m. W. of Leslie. P.O., T.O.

Scotsbrig, farm, Dumfriesshire, 3 m. N.E. of Ecclefechan; was the home of the Carlyle family in 1826.

Scotsburn, ham., seat, and wood, 5 m. S.W. of Tain, Ross and Cromarty.

Scotscalder, seat and ry. sta., L.M.S., 2¾ m. S.W. of Halkirk, Caithness. P.O.

Scotscraig, seat, Fife, 1 m. S.W. of Tayport. Now the property of Dundee Corporation.

Scots Dyke, The, old ditch and embankment forming part of the boundary between Scotland and England. Extended from the r. Esk to the Sark.

Scotstarvet Tower, old tower on an eminence, Fife, 2 m. S.W. of Cupar.

Scotstoun, town, with ry. sta., L.M.S. (S. and S. West), in co. of, and 1 m. E. of, Renfrew. P.O., T.O.

Scotstounhill, vil., with ry. sta., L.N.E., in par. and co. of Renfrew.

Scotstoun, vil., Argyll, 2 m. N. of Strontian.

Scotstown Head, headland, N.E. Aberdeenshire, 4 m. N.N.W. of Peterhead.

Scoubar, loch, 6 m. N. of Lochbuie, S.W. isl. of Mull, Argyll.

Scour, see SGURR.

Scour Gairoch, mt., W. Inverness-shire, on S. shore of Loch Quoich; alt. 3015 ft.

Scourie, vil., Sutherland, on Scourie Bay, 43½ N.W. of Lairg. P.O., T.O. Hotel. Pop. 51.

Scour Ouran (Sgurr Fhuaran), mt., Ross and Cromarty, 2 m. S.W. of Beinn Fhada (Ben Attow); alt. 3505 ft.

Scourst, loch, 9 m. N.W. of Tarbert, N. Harris, Outer Hebrides.

Scousburgh, ham., on bay of same name, 5½ m. S.W. of Sandwick, Dunrossness, Mainland, Shetland. P.O.

Scrabster, seapt. vil., 1½ m. N.W. of Thurso, Caithness, on W. side of Thurso Bay pier. P.O., T.O. The scant ruins of S. Castle are close by. Regular steamer communication with Kirkwall and Stromness.

Scrape, The, mt., in Peeblesshire, 7½ m. S.W. of Peebles; alt. 2347 ft.

Scresort Loch, sea loch on E. side of Rhum Isl., Argyll.

Scridain Loch, sea loch on W. side of Mull, Argyll. 9 m. long.

Scriden Rocks, at N. end of Arran Isl., Buteshire.

Scrishven (Sgribhis Bheinn), mt., on the coast of N.W. Sutherland, 4 m. E. of Cape Wrath. It has a great scar on its E. side, hence the name.

Scroggiehill, vil., Perthshire, 4 m. N.W. of Perth.

Scroof, ham., N.E. Selkirkshire, 3½ m. W.S.W. of Stow.

Scuir of Eigg. This remarkable hill, made of up columnar basalt almost as picturesque as Staffa, rises to 1289 ft. on the S. side of the isl. of Eigg, Inverness.

Scuir Vuillin (Sgurr a Mhuilinn), mt., Ross and Cromarty, on N. side of Strath Conon; alt. 2778 ft.

Scullamie, or **Skullomie,** a bay on the E. side of the Kyle of Tongue, Sutherland.

Seacliff House, seat, E. Lothian, 4½ m. S.E. of North Berwick.

Seafield, S. suburb Ayr. County hospital.

Seafield, q.s. par. in Rathven par., Banffshire. Gives the title of Earl to the family of Ogilvie-Grant.

Seafield, suburb of Leith, Edinburgh.

Seafield, · vil., near Livingstone, W. Lothian. P.O.

Seafield House, Broughty Ferry, Angus. Boys' preparatory school.

Seafield Tower, ruined seat in par. and 1 m. N.E. of Kinghorn, Fife.

Seaforth Loch, sea loch on E. side of Lewis Isl., Ross and Cromarty.

Seagate Castle, Ayrshire. See IRVINE.

Seal Island, Sutherland. See EILEAN-NAN-ROAN.

Sealy Loch, Ross and Cromarty. See SHELL LOCH.

Seamab Hill, a summit of the Ochil Hills, 3 m. N.E. of Dollar, Perthshire; alt. 1442 ft.

Seamill, vil., 3¼ m. N.W. of Ardrossan, N.W. Ayrshire. P.O., T.O. Hydro.

Sea Park, seat, Morayshire, 1 m. W. of Kinloss ry. sta.

Seathope Law, mt., on N. border of Selkirkshire, 3 m. N.W. of Thornielee sta.; alt. 1778 ft.

Seaton House, a seat, 1½ m. N.E. of Arbroath, Angus.

Seenes Law, mt. on N. border of Berwickshire, 5½ m. S. of Gifford; alt. 1686 ft.

Seggieden, seat, Perthshire, on r. Tay, 4 m. S.E. of Perth.

Seil, isl., Argyll, 3½ m. S. by W. of Kerrera and 4 m. S.E. of Mull, separated from the mainland by a very narrow strait. It measures 4¼ m. long by 2¼ broad and has an area of 3820 ac. Pop. 367.

Seilich Loch-an-t, 1 m. long, drains into the r. Tromie, S.W. Inverness-shire, 9 m. S. of Kingussie; alt. 1500 ft.

Selkirk, royal and mun. bur., and capital of the county. It dates back to 1113. Of the old castle little is known beyond that David I. occasionally resided here. The town stands on a pleasant slope near the banks of the Ettrick. In the old Town Hall there is the court-room where Sir W. Scott, as sheriff of the county, used to preside. The old market cross was rebuilt in 1898 and stands in the market-place. There is a Masonic Hall and Free Library. The principal manufactures carried on consist of various kinds of woollen goods; there is one weekly newspaper, *Southern Reporter*. There are monuments to Scott and Mungo Park. The area of the civil par. is 17,854 ac. Pop. 7075. The bur. is 544 ac. Pop. 5667. P.O., T.O. Ry. sta., L.N.E.

Selkirkshire, the county was origina-ally designated Sheriffdom of Ettrick Forest. The Romans occupied part of it during their 350 years in Britain. There are a number of ramparts and ditches through several parts of the county known as the " Catrail." The county is mostly hill and dale. The chief dales are the Ettrick, Yarrow, and Tweed, with many " cleughs " and " hopes." The hills rise at Ettrick Pen to 2269 ft.: the Wiss 1932 ft., Elibank Law 1715 ft. Sheep farming is of the greatest importance. There are no minerals except igneous rocks and limestone. The chief manufactures are `in the towns of Selkirk and Galashiels, where woollen tweeds and other material are manufactured. The rivers are the Tweed, Ettrick, and Yarrow, all with good fishing. There are many lochs, but small, St. Mary's Loch, 3 m. by $\frac{1}{2}$ m., being the largest. The L.N.E. railway passes through the N. of the county, with a branch to Selkirk from Galashiels. The county comprises seven pars. and 2 towns—Selkirk and Galashiels. Its area is 266·8 sq. m., 170,793 ac. Pop. 22,608. United with Roxburghshire in return-ing one member to Parliament.

Sellafirth, vil., Yell Isl., Shetland, 5 m., S.S.W. of Cullivoe. P.O.

Seton, ancient par., now in Tranent par., E. Lothian. The ruin of a fourteenth-century church stands $1\frac{1}{2}$ m. N.E. of Tranent, near to which S. palace, built in the time of James V., stood. S. House now occupies the site of the ancient palace.

Setter, loch, $2\frac{1}{4}$ m. W. of Lerwick, Shetland Islands.

Sgarsoch, An, mt., on S.W. border of Aberdeenshire, 16 m. W.S.W. of Brae-mar ; alt. 3300 ft.

Sgeir Maoile, lighthouse, sound of Jura, Argyll. See SKERVUILE.

Sgor an Fhuarain, mt., Inverness-shire, on S. side of Loch Quoich ; alt. 2961 ft.

Sgor Mor, mt., Inverness-shire, on S. shore of Loch Quoich ; alt. 3290 ft.

Sgor na Ciche, mt., W. Inverness-shire, 3 m. N. E. of head of Loch Nevis ; alt. 3410 ft.

Sgor nan Coireachan, mt., S.W. Inverness-shire, at the head of Glenfinnan ; alt. 3133 ft.

Sgor nan Gillean, mt., S.- of Rhum Isl., Argyll ; alt. 2503 ft.

Sgurr, see also SCUIR.

Sgurr an Airgid, mt., S.W. Ross and Cromarty, on N.E. shore of Loch Duich ; alt. 2757 ft.

Sgurr Mhòr, hill in N. of Skye Isl., Inver-ness-shire ; alt. 1460 ft.

Sgurr Mhòr, mt., Ross and Cromarty, 4 m. N. of Loch Fannich ; alt. 3637 ft.

Sgurr na Banachdich, peak of the Cuillin Hills, Skye, Inverness-shire ; alt. 3167 ft.

Sgurr na Lapaich, mt., Inverness-shire, 2 m. N. of Loch Affric ; alt. 3401 ft.

Sgurr na Lapaich, mt. Inverness-shire 3 m. S. of Loch Monar ; alt. 3773 ft ; observed on the principal triangulation of Great Britain.

Sgurr nan Ceathreamhnan, mt., on border of Inverness-shire, 8 m. E.N.E. of the head of Loch Duich ; alt. 3771 ft.

Sgurr nan Gillean, a peak of the Cuillin Hills, Skye, Inverness-shire ; alt. 3167 ft.

Sgurr Ruadh, mt. on N. border of Inver-ness-shire, 7 m. W.N.W. of Struy ; alt. 3254 ft.

Sgurr Ruadh, mt., S.W. Ross and Cromarty, 3 m. N.W. of Achnashellach sta., Glen Carron ; alt. 3141 ft.

Shade Mount, N.W. Kincardineshire, 8 m. S.W. of Banchory ; alt. 1662 ft.

Shader, vil., isl. of Lewis, Ross and Cromarty, $3\frac{1}{2}$ m. N.E. of Barvas. P.O.

Shaggie Burn, stream, Perthshire, flows 6 m. to Turret Burn, $1\frac{1}{4}$ m. N.W. of Crieff.

Shalloch on Minnoch, mt., W. Ayrshire, 2 m. W.S.W. of Loch Macaterick ; alt. 2520 ft.

Shambellie, seat, Kirkcudbrightshire, 7 m. S.W. of Dumfries.

Shandon, vil., with ry. sta., L.N.E., Dunbartonshire, on E. shore of Gare Loch, 5 m. N.W. of Helensburgh. P.O., T.O. Has steamboat pier and a hydropathic establishment. Agnes Miller Wilson Children's Home.

Shandwick, vil., Ross and Cromarty, on Moray Firth, 5 m. E. of Nigg ry. sta. S. House, a seat.

Shankend, ry. sta., L.N.E., Roxburghshire, 7 m. S.E. of Hawick.

Shankeston, loch, 8 m. N.E. of Maybole, S.W. Ayrshire.

Shannochie, ham., Arran Isl., 4 m. W. of Kildonan. P.O., T.O.

Shanwell House, seat, Kinross, 3 m. W. of Milnathort.

Shapinsay, isl. and par., Orkney, 1 m. N. of the Mainland, and 4 m. N.E. of Kirkwall; 6761 ac. Pop. 584. Daily steamer communication with Kirkwall.

Sharpitlaw House, seat, N.E. Roxburghshire, 1 m. N.E. of Kelso.

Shawbost North, vil., on N.W. side of Lewis, Ross and Cromarty, 19 m. N.W. of Stornoway. P.O., T.O. called Shawbost.

Shawhead, vil., Kirkcudbrightshire, 7 m. W. of Dumfries. P.O., T.O.

Shawhill, seat, Ayrshire, on r. Irvine, near Hurlford.

Shawlands, S.W. suburb and ry sta., L.M.S., Glasgow.

Shaws Water Reservoir, near Greenock. See THOM LOCH.

Sheabost, Ross. See SHAWBOST NORTH.

Shebster, ham., Caithness, 7 m. S.W. of Thurso. P.O.

Shedog, pl., Arran Isl., Buteshire, 6 m. W. of Lamlash.

Sheeoch Burn, Kincardineshire; flows 8½ m. N.E. to join the Dee near Durris Church.

Sheep Craig, headland on E. coast of Fair Isle between Orkney and Shetland.

Sheep Isle, Inner Hebrides, Argyll, in Firth of Lorne, 1 m. N.W. of Seil Isl.

Sheep Isle, Argyll, close to Sanda, S. end of Mull of Kintyre.

Shee Water, stream, Perthshire; flows through Glen Shee; lower it is called Black Water till it joins the Ardle Water to form the r. Ericht.

Sheigra, pl., N.W. Sutherland, 4 m. N.W. of Kinlochbervie.

Sheil Loch, on the borders of Inverness and Argyll; extends from Glenfinnan S.W. for 17 m. and discharges the r. Sheil which flows 3 m. to the sea at Loch Moidart.

Shell Loch, sea loch, Outer Hebrides, Ross and Cromarty, on E. side of Lewis, 15 m. S.W. of Stornoway.

Sheriffmuir, a heathy hill rising 2 m. E. of Dunblane, where an indecisive battle was fought in 1715 between the forces of the Pretender under the Earl of Mar and the Royalists under the Duke of Argyll.

Sherwood, pl., Midlothian, ½ m. S.W. of Bonnyrigg. Pop. (S. and Poltonhall) 819.

Sheshader, vil., par. of Stornoway, Lewis, Ross and Cromarty.

Shetland, or **Zetland,** group of isls. 50 m. N. of the Orkney isl. politically formed into a county which, together with the Orkneys, returns one member of Parliament. The total area is 551·4 sq. m. or 352,889 ac. Pop. 21,421. The group consists of 100 isls., 27 of which are inhabited. The large isl., known as the Mainland, measures 54 m. N. to S. with a maximum breadth of 21½ m. The chief industry is fishing, mainly herring, and the chief centre is the town of Lerwick. Pop. 4221. Steamer communication from Leith and Aberdeen about three times weekly. The two harbours are Lerwick and Scalloway. The group, all except two, are gathered round the Mainland, i.e., Foula, 14 m. W. of the Mainland, and Fair Isle, halfway between the Orkneys and Shetland.

Shettleston, town and q.s. par. and parl. division, with ry. sta., L.N.E., Glasgow par., and forming an eastern suburb of Glasgow. Pop. of Shettleston Division 76,717. P.O., T.O.

Sheuchan, q.s. par. Leswalt and Portpatrick pars., Wigtownshire; includes part of Stranraer.

Shian, ham., Perthshire, near r. Quaich, 13 m. S.W. of Dunkeld.

Shian, North and **South,** two hams., on either side of Loch Creran, W. Argyll; a ferry.

Shiant Isles, a group of islets in the Minch, Outer Hebrides, Ross and Cromarty, 4½ m. S.E. of Lewis.

Shiel, pl., with inn, Ross and Cromarty. at mouth of the r. Shiel, 12 m. S.E. of Lochalsh.

Shieldaig, vil. and q.s. par. Ross and Cromarty; vil. is on Loch S., an arm of Loch Torridon, 18 m. N.W. of Strome Ferry. P.O., T.O. Vil. pop. 86.

Shieldaig Hostel, Strathcarron, Rossshire. A Scottish Youth Hostel.

Shieldhall, dist., Govan Combination, Lanarkshire. Contains works and factories of the Scottish Co-operative Wholesale Society.

Shieldhill, pl., with ry. sta., L.M.S., Dumfriesshire, 7 m. N.W. of Lockerbie.

Shieldhill, seat, Lanarkshire, 4 m. N.W. of Biggar.

Shieldhill and Blackbraes, q.s. par., Grangemouth and Muiravonside pars., Stirlingshire. P.O., T.O. named Shield-hill.

Shieldhill, East, vil., Stirlingshire, near Polmont Junction. Dist. pop. 1714.

Shields, ry. sta., in S.W. suburbs of Glasgow.

Shields Road, ry. sta., Glasgow Subway, in S.W. Glasgow, near Shields sta. and Pollokshields.

Shielhill, seat, 3½ m. N.E. of Kirriemuir, Angus.

Shiels, pl., 1½ m. N.W. of Belhelvie, Aberdeenshire.

Shillay Island, N. Uist. Pop. 3. See MONACH LIGHTHOUSE.

Shillofad, mt., Kincardineshire, 5 m. S.S.E. of Banchory ; alt. 1207 ft.

Shin Loch, one of the largest lochs in Sutherland, extending from Lairg in a N.W. direction for 17 m., with a varying breadth of ½ to 1½ m. ; alt. 270 ft., depth 162 ft.

Shin, r., issuing from Loch Shin. It flows nearly 8 m. S. to join the Oykell. Excellent salmon river.

Shinnel Water, stream, Dumfriesshire ; flows 12½ m. S.E. to Scar Water, near Penpont.

Shiskine, ham., W. coast of Arran isl., Buteshire, 10 m. S.W. of Brodick. P.O., T.O.

Shochie Burn, Perthshire, rises in Glen-shee and flows 6 m. S.W. to the Tay at Luncarty.

Shona, isl., Inverness-shire, in Loch Moidart; measures 1⅓ m. by 3 m. Pop. 62.

Shortwoodend Hostel, on Moffat Water, 5 m. N.E. of Moffat, Dumfriesshire. A Scottish Youth Hostel.

Shotts, par. and vil. (Kirk of Shotts), with ry. sta., L.M.S., Lanarkshire ; 24,835 ac. Pop. 20.537. Vil. 2¾ m. N.W. of Shotts sta. P.O., T.O. Coal and ironstone abound in the par. and form the chief industry in the dist. Pop. of Shotts and Dykehead 10,665.

Shuna, isl., Inner Hebrides, Argyll, between Luing Isl. and the mainland and near the mouth of Loch Melfort; measures 1½ m. by 2½ m. and rises to 291 ft. Pop. 12.

Shuna Island, Argyll, in Loch Linnhe, 14 m. N.N.E. of Oban ; measures 1 m. by ½ m. and rises to 233 ft. Pop. 4.

Shurrery Loch, 9 m. S.W. of Thurso, Caithness ; alt. 322 ft., depth 7 ft.

Shurroch (or **Schurroch**) **Hill,** 1½ m. S.W. of Kingoldrum, Angus ; alt. 798 ft. Has three ancient stone circles.

Siàn, or **Shian,** bay, on W. coast of Jura, Argyll, 4 m. N. of Loch Tarbet. Shian R. flows into it from E.S.E., and Shian Isl. lies on the N. side.

Sibbaldbie. P.O., near Lockerbie. See APPLEGARTH and S.

Sidlaw Hills, a range of hills from Kin-noull Hill at Perth, extending N.E. to the North Sea. They are only detached from the Ochil range by the valley of the Tay. The highest pt. is Craigowl Hill, 1492 ft. They form the S. part of Strathmore.

Sighthill, dist. between St. Rollox and Springburn, N.E. Glasgow.

Silverbanks, part of the town of Cambus-lang, Lanarkshire.

Silverbithall, seat Roxburghshire, 1 m. N. of Hawick.

Silver (or **Alva**) **Glen,** in the locality of Alva, Clackmannanshire, is wooded glen. Silver mine worked here at one time.

Sinclair's Bay, on E. coast of Caithness. Castle Sinclair (in ruins), stands on the S. shore of the bay.

Sinclairstown, ham., 4¾ m. S.W. of Ochiltree, Ayrshire.

Sinclairtown, vil. and ry. sta., L.N.E., Fife, ¾ m. W. of Dysart. Forms part of Kirkcaldy. Ironfoundries, linoleum and engineering works.

Singer, ry. sta., L.N.E., Dunbartonshire, 1 m. W. of Clydebank.

Sinnahard, pl., with ruins of ancient chapel, Aberdeenshire, 7 m. W. of Alford. P.O.

Sionascaig Loch, large loch, 5 m. S. of Lochinver, with an irregular outline ; alt. 248 ft., depth 216 ft.

Siòr Loch, 5 m. W. of Taychreggan, Argyll.

Skae, loch, 6 m. N.E. of New Galloway, Kirkcudbrightshire, depth 35 ft., alt. 864 ft.

Skaill, ham., Mainland of Orkney, 8 m. N. of Stromness. Bay of S. and Loch S. here.

Skares, ry. sta., L.M.S., Ayrshire, 1 m. E. of Ochiltree. P.O., T.O.

Skateraw, vil., E. Lothian, 4 m. S.E. of Dunbar. Lifeboat sta.

Skateraw, fishing vil., with small harb. Kincardineshire, 6 m. N.E. of Stonehaven.

Skea, hill, Westray Isl., Orkneys, Inga Ness ; alt. 257 ft.

Skeabost, ham., Skye Isl. Inverness-shire, on Loch Snizort, 5 m. N.W. of Portree. P.O., T.O. called Skeabost Bridge.

Skeir Maoile Lighthouse, Argyll. See Iron Rock.

Skelbo, ry. sta., L.M.S., 4 m. N.W. of Dornoch, Sutherland. P.O. 1 m. N.W. is S. Castle, ancient seat of the Sutherlands.

Skelda Voe, inlet, 1½ m. long, W. Mainland, Shetland. To the S. lies Skelda Ness, 10½ m. W. of Lerwick.

Skeld, Easter, ham., with pier, on Skelda Voe, 10 m. W. of Lerwick, Mainland, Shetland. Wester S. lies inland 1½ m. S.W.

Skeldon House, seat, Ayrshire, 2 m. S.E. of Dalrymple.

Skelfhill, pl., Roxburghshire, 7 m. S.S.W. of Hawick.

Skellister, ham., E. Mainland, Shetland, 7½ m. E. of Aith. P.O., T.O.

Skelloch, loch, 11 m. S.E. of Maybole, S.E. Ayrshire.

Skelmorlie, coast vil., Ayrshire, and q.s. par., Renfrewshire, on Firth of Clyde, 4½ m. N.W. of Largs. P.O., T.O. Pop. 1105. Has a hydropathic establishment. S. Castle, seat of Earl of Eglinton, nearby.

Skelpick, seat, 4½ m. S.E. of Bettyhill, N. coast of Sutherland.

Skene, par. and loch, Aberdeenshire ; par. 10,246 ac. Pop. 1310. Ham. 9 m. W. of Aberdeen. P.O., T.O. S. House, seat, 2¾ m. N.W. of ham.

Skene Loch, or **Skeen,** 5 m. S.W. of Tibbie Shiels Hotel, St. Mary's Loch, in Dumfriesshire, close to the border of Selkirkshire.

Skerray, ham., Sutherland, 3 m. N.W. of Bettyhill of Farr. P.O., T.O.

Skerries, Out, group of islands, Shetland, 9 m. E. of Mainland. P.O. called Skerries. See Housie Skerries.

Skerries Pentland, see Pentland Skerries.

Skerrington, dist., Ayrshire, 1 m. S. of Cumnock. Pop. 308.

Skerrow Loch, Kirkcudbrightshire, 13 m. N.W. of Castle-Douglas.

Skerryvore Lighthouse, isolated rock, 11 m. S.W. by S. of S. end of the isl. of Tiree, Argyll. It lies within a dangerous reef about 10 m. long. The lighthouse was built in 1844, and can be seen for 20 m.

Skervule, isl., Inner Hebrides, Argyll, near S. entrance to Jura Sound. It has a lighthouse visible 14 m.

Skiach Loch, 3 m. S.W. of Logierait, Perthshire ; alt. 1385 ft., depth 55 ft. Close to it is Little Loch Skiach.

Skiack, or **Skiach,** rivulet, running 8 m. N.E. to Cromarty Firth, near Kiltearn Church, Ross and Cromarty.

Skibo Castle, seat, 4 m. S.W. of Dornoch, Sutherland.

Skinidin, ham., W. Isle of Skye, Inverness-shire, 4½ m. W. of Dunvegan. P.O., T.O.

Skinsdale, stream, Sutherland. Flows 13 m. S.E. to the Blackwater, 3 m. N.W. of Loch Brora.

Skipmyre, ham., 5 m. N.E. of Dumfries.

Skipness, q.s. par. and fishing vil., Argyll, on E. side of Kintyre ; vil. 1½ m. S. of Tarbert. P.O., T.O. S. Castle, a seat.

Skiport, sea loch, in N.E. of South Uist, Outer Hebrides. Good harbourage.

Skirling, par and vil. near W. border of Peeblesshire ; par. 3423 ac. Pop. 200. Vil. 2½ m. N.E. of Biggar. P.O.

Skiros, loch, 1 m. N. of Bridgend, Islay, S.W. Argyll.

Skirsa, pl. and headland, N.E. Caithness, 3½ m. S. of Duncansbay Head.

Skolie Burn, tributary of Breich Water, Midlothian, 1 m. W. of Addiewell.

Skye, the largest, most varied, and most picturesque of the Hebrides. The isl. measures 48¾ m. from Rudha Hunish to Point of Sleat and 44½ m. from Isle of Ornsay to Vaternish Point, its surface area being 411·704 ac. and pop. 9908. The isl. is indented with sea lochs, the most important being Eishort, Slapim, Scavaig, and Brittles on the S. and Bracadale, Dunvegan, and Snizort on W. and N. The chief mountains are the Cuillin Hills, with peaks 3309 ft., 3197 ft., and 3167 ft. These hills, with Loch Coruisk in their midst, form the wildest and most picturesque in Scotland. There is no special industry in the isl. beyond the fishing, agriculture, and cattle and sheep breeding. The pop., which numbered 18,908 in 1861, is now (1931) only 10,490.

Skyreburn, ham., Kirkcudbrightshire, on Skyre Burn, 4 m. S.W. of Gatehouse.

Slaggan, pl., N.W. Ross and Cromarty, between S. Bay., N. of the mouth of Loch Ewe and 9 m. N. of Poolewe, and Loch an-t-Slaggan.

Slains, coast par. and loch, Aberdeenshire ; par. 9154 ac. Pop. 744. Vil. 6 m. E. of Ellon. A prominent object is the ruin of S. Castle, demolished in 1594 by James VI.

Slam, loch, 5 m. S.E. of Tongue, N. Sutherland.

Slamannan, par. and vil., with ry. sta., L.N.E., Stirlingshire ; par. 10,856 ac. Pop. 2959. Vil. near r. Avon, 5½ m. S.W. of Falkirk. P.O., T.O. Coal and iron mining dist. ; vil. pop. 1319.

Slapin, sea loch, S. of the isl. of Skye, Inverness-shire, 5 m. S.W. of Broadford.

Slate Islands, group of islands, Argyll, includes Luing, Shuna, Torsay, Seil, and Easdale.

Slateford, ham., 1½ m. N. of Maybole, Ayrshire.

Slateford, vil., with ry. sta., L.M.S., Midlothian, on Water of Leith and within the City of Edinburgh. P.O., T.O. Has dye-works.

Slattadale, a State forest, on S. side of Loch Maree, Ross and Cromarty.

Sleach Water, stream, 10 m. long, flowing past Altnabreac sta. to Loch More, W. Caithness.

Sleat, par. and vil., Skye, Invernessshire, on Sound of Sleat, 41,550 ac. Pop. 792. Vil. 10 m. S. of Broadford. The ruin of Castle Knock is here.

Sleat Point, extreme S. point of Skye, Inverness-shire.

Sleat Sound, the channel that divides Skye from the mainland of Scotland ; extends for about 20 m. in a N. and S. direction and is from ½ m. to 7 m. wide.

Sliddery, pl., Arran Isl., Buteshire, on Sliddery Water, 9 m. S.W. of Lamlash. P.O., T.O.

Sligachan, pl., with hotel, Isl. of Skye, Inverness-shire. This, a suitable centre to visit the Cuillin Hills, is at the head of Loch Sligachan, 9 m. S. of Portree. P.O., T.O. S. Loch extends for 3 m. S. means *place of shells.*

Slitrig Water, Roxburghshire, rises between Teviotdale and Liddesdale, and flows 11 m. N. to r. Teviot at Hawick.

Slochd, deep hollow on the high ridge between Aviemore and Inverness on the L.M.S. ry ; highest alt. 1315 ft.

Sloy Loch, Dunbartonshire, at S.E. base of Ben Vorlich.

Slugain, stream and glen, 3½ m. long flowing to the valley of the Dee, 1 m. N. of Braemar, Aberdeenshire.

Sluie, pl., on r. Findhorn, Morayshire, 5 m. S.W. of Forres.

Slungie Hill, one of the Ochil Peaks, 5 m. N.W. of Milnathort, Kinross-shire ; alt. 1354 ft.

Smaddy Loch, 2 m. N. of Crossmichael, Kirkcudbrightshire.

Sma' Glen, a picturesque glen through which the r. Almond passes, 5 m. N. of Crieff.

Smailholm, par. and vil., Roxburghshire, on Eden Water par. ; 4194 ac. Pop. 256. Vil. 6 m. N.W. of Kelso. P.O.

Smallholm, vil., near Lochmaben, Dumfriesshire.

Small Isles, par. in cos. Argyll and Inverness ; 37,752 ac. Pop. 281.

Small Lochs, 3 in number, 12 m. N.W. of Tarbert, N. Harris, Outer Hebrides.

Smeaton, ry. sta., L.N.E., Midlothian, 8 m. E. of Edinburgh.

Smeaton Hepburn, seat, 1½ m. N. of East Linton, E. Lothian.

Smoo Cave, a huge limestone cave 67 ft. high and a fair-sized burn falling into an inner cave and lake, in the par. of Durness, Sutherland, 1 m. E. of the par. church.

Snargow, seat, Perthshire, 4 m. E. of Dunkeld.

Snar Water, Lanarkshire, rises near Leadhills and flows 6 m. N. to Duneaton Water.

Snib, The, headland, W. Lothian, ¾ m. N.W. of Cramond.

Snizort, par. and ham., Skye, Invernessshire ; par. 53,280 ac. Pop. 1260. Ham. 7 m. N.W. of Portree. P.O., T.O.

Snizort Loch, sea loch in N.E. Skye, Inverness - shire. Enters from the Little Minch and separates Trotternish and Vaternish dists.

Snout Hostel, on the Borthwick Water, ½ m. from the Hawick-Tushielaw road at Roberton, between Eskdale and Teviotdale, 6 m. W. of Hawick. A Scottish Youth Hostel.

Soay, isl., Skye, Inverness-shire, at entrance to Loch Scavaig, measures 1¾ m by 3 m. and rises to 445 ft. P.O. Pop. 64.

Soay, isl., Sutherland. See SOYEA.

Soay, one of the St. Kilda group of isls., rises to 1225 ft., and is stocked with Mouflon sheep which run wild. The isl. extends to 244 ac.

Sollas, N. Uist Isl., Outer Hebrides, 10 m. W. of Lochmaddy. P.O., T.O.

Solway Firth, an arm of the Irish Sea ; extends inland for 46 m. and varies from 20 to 2 m. in breadth, about 6 m. N.W. of Carlisle, where the r. Eden enters it.

Sonachan House, seat, Argyll, on S.E. shore of Loch Awe, 12 m. N.W. of Inveraray.

Sorbie, par. and vil., with ry. sta., L.M.S., Wigtownshire ; par. on Wigtown Bay ; 9777 ac. Pop. 1096. P.O., T.O. Vil. 6 m. S. of Wigtown. Pop. 145.

Sorn, par. and vil., Ayrshire ; par. 19,195 ac. Pop. 3369. Vil. on r. Ayr, 4½ m. E. of Mauchline. P.O. Pop. 586.

Sourin, ham., Rousay, Orkney, 15 m. N. of Kirkwall. P.O.

South Alloa, ham., Stirlingshire, on r. Forth, opposite Alloa. P.O., T.O.

Southannan, estate, Ayrshire, 4 m. S. of Largs.

Southdean, par., S.E. Roxburghshire ; 27,986 ac. Pop. 525. Church at Chesters, 7 m. S.W. of Jedburgh. P.O.

Southend, par. and vil., in S. extremity of Kintyre peninsula. Par., including Sanda Isl., 30,805 ac. Pop. 640. Vil. 9¼ m. S.W. of Campbeltown. P.O., T.O. A lifeboat sta.

Southfield, Auchtermuchty, Fife. County Hospital.

Southfield and Blackwood, vil., with ry. sta. (Blackwood), 3 m. N.W. of Lesmahagow, Lanarkshire.

Southfield Sanatorium, Liberton, Edinburgh.

South Hall, seat, Argyll, on the Kyles of Bute, 5 m. N.W. of Rothesay.

Southhall, seat, 2 m. S. of Colintraive, Argyll.

Southhook, vil., 7 m. N.E. of Troon, Ayrshire.

Southmuir, part of the town of Kirriemuir, Angus. P.O.

South Queensferry, W. Lothian, 9½ m. W. of Edinburgh. P.O., T.O. See QUEENSFERRY.

South Renfrew, ry. sta., L.M.S., at town of Renfrew.

South Walls, isl., Orkney. See CANTICK.

Southwick, ry. sta., L.M.S., Kirkcudbrightshire, 3 m. N.E. of Dalbeattie. P.O., T.O.

Southwick, q.s. par., Kirkcudbrightshire, 7 m. S.E. of Southwick ry. sta.

Soutra, ancient par. now in Fala and Soutra par., E. Lothian. Near it is Soutra Hill (1209 ft.), the most westerly ridge of the Lammermuir Hills, 8 m. N.N.W. of Lauder.

Soval, seat, 9 m. S.W. of Stornoway, Lewis, Ross and Cromarty.

Soyea, isl., Sutherland, off the mouth of Loch Inver.

Spalander, loch, 6 m. W. of Maybole, Ayrshire.

Spango Water, Dumfriesshire, flows 7 m. E. to join the Crawick Water, 5½ m. N.E. of Sanquhar.

Spar Cave, near S. end of W. side of Loch Slapin, Skye, Inverness-shire.

Spean, river, Inverness-shire. Issues from Loch Laggan and flows through Glen Spean 20½ m. to the r. Lochy.

Spean Bridge, ham. and ry. sta., L.N.E., Inverness-shire, 9 m. (by rail) from Fort William. P.O., T.O. Pop. 135. Hotel.

Spear Head, prom., 2½ m. N.W. of Thurso, Caithness.

Speddoch, seat, Dumfriesshire, 10 m. N.W. of Dumfries.

Spelve Loch, sea loch, isl. of Mull, Argyll, on E. side opposite the isl. of Kerrera.

Spey, r., rises about 1500 ft. on S.E. side of Creag a' Chait, 5 m. from the western shore of Loch Laggan, passes through the cos. of Inverness, Moray, and Banff, and drains 1300 sq. m. ; joins the sea at Garmouth. Its total length is 107 m. and it is the most rapid r. in Scotland.

Spey Bay, pl. and ry. sta., L.N.E., Morayshire, near Portgordon. P.O., T.O. at sta. Hotel.

Speybridge, pl., Morayshire, ½ m. S.E. of Grantown-on-Spey. P.O.

Speyburn-Glenlivet Distillery, Rothes, Morayshire.

Speymouth, par., and q.s. par., Morayshire, near the mouth of the r. Spey; par. 6776 ac. Pop. 611. The par. church is 1½ m. N.W. of Fochabers. Salmon fishing is the most important industry.

Speyside Distillery, Kingussie, Inverness-shire.

Spiggie, pl., Shetland, 17 m. S.W. of Lerwick, near Loch Spiggie and opposite Colsay Isl.

Spinningdale, vil., S.E. Sutherland, on Dornoch Firth, 7¾ m. W. of Dornoch. P.O.

Spital Hill, small hill, 3½ m. S.E. of Halkirk, Caithness; alt. 577 ft.

Spital House, seat, Berwickshire, 5¼ m. W. of Berwick-on-Tweed.

Spitalhaugh, seat, Peeblesshire, on Lyne Water, S.E. of W. Linton.

Spittal, ham., Caithness, 14 m. N.W. of Wick. P.O.

Spittal, pl., with site of ancient hospital, Roxburghshire, on Rule Water, 1½ m. E. of Denholm.

Spittal, vil., 2½ m. S.E. of Cambuslang, Lanarkshire.

Spittalfield, vil., Perthshire, 3 m. N. of Murthly ry. sta. P.O., T.O.

Spittal of Glenshee, ham., with hotel, Perthshire, 20 m. N.W. of Blairgowrie. See GLENSHEE.

Spittal, South, or **Meikle, Port of,** pl. on W. coast of Wigtownshire, 2¼ m. S.E. of Portpatrick.

Spott, par. and vil., E. Lothian; par. 7831 ac. Pop. 325. Vil. 3 m. S. of Dunbar. S. House, a seat, is near.

Spottes, seat, Kirkcudbrightshire, on Urr Water, 4 m. N.W. of Dalbeattie.

Spottiswoode, seat, Berwickshire, 5½ m. N.E. of Lauder.

Spout of Ballaggan, Stirlingshire. See BALLAGGAN.

Springbank, pl., with church, Berwickshire, near Ayton.

Springburn, q.s. par. and parl. div., with ry. sta., L.N.E; a suburb of the city of Glasgow. Public park and Stobhill General Hospital.

Springfield, vil., Dumfriesshire, on r. Sark, ½ m. E. of Gretna Green.

Springfield, vil., 1½ m. S.W. of Lasswade Midlothian, on r. North Esk; has paper mills.

Springfield, vil., with ry. sta., L.N.E., Cupar par., Fife, near the left bank of the Eden, 3 m. S.W. of the town of Cupar. P.O., T.O. Pop. 723.

Springhill, seat, Berwickshire, near r. Tweed, 4 m. W. of Coldstream.

Springholm, vil., 1 m. E. of Kirkpatrick, Kirkcudbrightshire. P.O.

Springkell, seat, Dumfriesshire, on Kirtle Water, 2½ m. N.E. of Kirtlebridge sta.

Springside, vil., with ry. sta., L.M.S., 3 m. N.W. of Kilmarnock, Ayrshire. P.O., T.O.

Springside Distillery, Campbeltown, Argyll.

Springvale, Ardrossan, Ayr. County hospital.

Springwell, part of town of Stonefield, Lanarkshire.

Springwood House, seat, 1 m. S.W. of Kelso, Roxburghshire, near r. Teviot.

Sprouston, par. and vil., with ry. sta., L.N.E., on N.E. border of Roxburghshire; par. 8637 ac. Pop. 720. Vil. near r. Tweed, 2 m. N.E. of Kelso. P.O., T.O.

Spurness, Holms of, three islets in Spurness Sound, Orkney, between Stronsay and Sanday islands.

Spynie, New, par., Morayshire, on r. Lossie; 5861 ac. Pop. 1872. The church is 2½ m. N.W. of Elgin, and in the N.E. of the par. are the ruins of the palace of Spynie. The par. has also Loch Spynie.

Square Point, ham., 3 m. S. of Corsock, Kirkcudbrightshire.

Squod (Sguod), loch, 5 m. N.W. of Poolewe, W. Ross and Cromarty.

Sron Smeur, loch (Tay Basin), NW. Perthshire, 4 m. N.W. of W. end of Loch Rannoch; depth 33 ft.

Sronphadrick, deer forest, Perthshire, part of Dalnacardoch deer forest, q.v.

Stack Loch, Sutherland, on E. side of Ben Stack and 4 m. S.E. of Laxford Bridge; depth 108 ft., alt. 117 ft.

Staffa, an uninhabited isl., well known for the columnar columns and great caves, the principal being Fingal's Cave, 227 ft. long, 42 ft. broad, and 66 ft. high. The island is 6 m. N. of Iona.

Staffin, or **Stenscholl,** ham., Skye, Inverness-shire, in S. Bay, 18 m. N. of Portree. S. House, a hotel. P.O., T.O.

Stagehouse Inn, ½ m. N.W. of the head of Loch Shiel, S.W. Inverness-shire.

Stainrig, seat, 5 m. N.W. of Coldstream, S. Berwickshire.

Stair, par. and vil., Ayrshire, on r. Ayr; par, 5316 ac. Pop. 1034. Vil. 6½ m. E. of Ayr. P.O. S. House, seat, near it.

Stairhaven, fishing harbour, Wigtownshire, on Luce Bay, 2½ m. S. of Glenluce.

Stake, Hill of, on border of cos. Ayr and Renfrew, 4 m. N.E. of Largs; alt. 1711 ft.

Stake Law, in co. of, and 5½ m. S.E. of, Peebles; alt. 1784 ft.

Stake Ness, headland, in co. of, and 3 m. N.W. of, Banff.

Standburn, vil., Stirlingshire, 3 m. S. of Polmont. P.O.

Stane, vil., Lanarkshire, on South Calder Water, near Shotts Ironworks.

Stanely Castle, ruined seat, Renfrewshire, 2 m. S.W. of Paisley.

Stanhope Burn, Peeblesshire, flows 4½ m. N.W. to the Tweed, 2½ m. N. of Crook Inn.

Stanley, Perthshire, vil. and ry. sta., L.M.S., on r. Tay, ½ m. S.W. of Stanley Junction, 7 m. N. of Perth. Pop. 1192. P.O., T.O. S. House close by.

Stanley Reservoirs, 2 m. W. of Paisley, Renfrewshire.

Stanley School, Bridge of Allan. Boys' private school.

Stannochy, seat, Angus, 1 m. S.W. of Brechin.

Stapleton Towers, seat, 3 m. N.E. of Annan, Dumfries.

Star, vil., Fife, 1¼ m. N.E. of Markinch. P.O.

Starlaw. ham., 2 m. S.E. of Bathgate, W. Lothian.

Starleyburn, ham. and stream in par. and 1 m. W. of Burntisland, Fife.

Start Point, headland, at E. end of Sanday Isl., Orkney. Has a lighthouse 75 ft. high, visible 14 m.

Staxigoe, fishing vil., 2 m. N.E. of Wick, Caithness.

Steel Road, ry. sta., L.N.E., Roxburghshire, 3 m. S. of Riccarton Junction.

Steinscholl, Skye. See STENSCHOLL.

Steisavat, loch, 2½ m. N.W. of Rodel House, S.E. Harris, Inverness-shire.

Stemster, ham., 7 m. S.W. of Thurso, Caithness. S. House, seat, 2¾ m. N.W. of Bower sta.

Stemster, loch, 4 m. N. of Latheron, Caithness.

Stenhouse, ham., Midlothian, 1 m. N.W. of Gilmerton.

Stenhouse, seat, 1¼ m. N.E. of Larbert sta., Stirlingshire.

Stenhousemuir, town, Stirlingshire, near Larbert sta., 3 m. N.W. of Falkirk. P.O., T.O.

Stenness, par., Mainland, Orkney; 7766 ac. Pop. 469.

Stenness, q.s. par., on Loch of Stenness, Mainland of Orkney, 10 m. W. of Kirkwall. P.O., T. O. 1 m. W. of Stenness church are the remarkable Standing Stones of Stenness.

Stenscholl, ham. and q.s. par., Skye, Inverness-shire, also known as Staffin, near head of S. Bay, 18 m. N. of Portree.

Stenton, par. and vil., E. Lothian; 7530 ac. Pop. 462. Vil. on Sauchet Water, 4 m. S.E. of E. Linton. P.O., T.O. Pop. 129.

Stenton, seat, Perthshire, on r. Tay, 3¼ m. S.E. of Dunkeld.

Stepford, ry. sta., L.M.S., S.W. Dumfriesshire, 1½ m. S. of Dunscore.

Steps, or **Stepps,** residential dist. and ry. sta. (Stepps), L.M.S., Lanarkshire, 4½ m. N.E. of Buchanan Street ry. sta. P.O., T.O. called Stepps. Pop. 2678.

Stevenson, seat, 2½ m. E. of Haddington, in E. Lothian.

Stevenston, town and par., with ry. sta., L.M.S., Ayrshire, containing Ardeer explosive works and part of the town of Saltcoats; par. 3836 ac. Pop. 11,572. The town is 2½ m. E of Ardrossan. Pop 6633. P.O., T.O. Coal-mining, ironworks, and dynamite.

Stevenston, New, part of Holytown, Lanarkshire.

Stewartfield, vil., Aberdeenshire, 11 m. W. of Peterhead. P.O. named Stuartfield.

Stewarthall, seat, in co. of, and 2½ m. E. of, Stirling.

Stewarton, par., town, and police bur., with ry. sta., L.M.S., on Annick Water, Ayrshire; par. 13,635 ac. Pop. 3700. Town, 5 m. N. of Kilmarnock. P.O., T.O. Bonnet-making, hosiery, lace, and woollens. Burgh pop. 2749.

Stewarton, Argyll. See STUARTOWN.

Stewarton, vil., 5½ m. N.W. of Stranraer, Wigtownshire. See also KIRKCOLM.

Stichill, Roxburghshire. See STITCHEL.

Stinchar, r. Ayrshire, rises near Loch Skelloch and flows 30 m. S.W. to the sea at Ballantrae.

Stirling, par., parl., royal, and mun. bur. and co. town, with ry. stas., L.M.S. and L.N.E., on r. Forth, 36½ m. N.W. of Edinburgh and 20½ m. N.E. of Glasgow; par. 1825 ac. Pop. 22,897. The town, said to have been a Roman sta., is first mentioned in a charter of David I., 1124. It is pleasantly situated on the slope of a rocky hill which is crowned by the Castle, which was one of the favourite residences of the Stuart Kings. The Parliament House, erected by James III., was a magnificent building, especially the great hall 126 ft. in length. The par. church is on a commanding site above the town, and the earliest church dates back to 1153. There is a monument to Sir William Wallace, and 2½ m. to the S. of the town was fought the battle of Bannockburn, 1314. Chief industries in the town are: carpet-weaving, worsted-spinning, rubber works, ironfounding, making of agricultural implements, etc.; also important cattle markets. Area of burgh, 1377 ac. Pop. 22,593. P.O., T.O.

Stirling Bridge, see CHAPEL AND STIRLING BRIDGE.

Stirlingshire, this is a midland county with the Highlands to the N. and the low valleys of the Forth and Clyde E. and W. It is bounded by Dunbarton, Lanark, and W. Lothian, with Perthshire on the N. and Clackmannan on the E. The chief rs. are Forth, Kelvin, Avon, and Endrick. Loch Lomond lies on the W. The figure of the county is extremely irregular; its length from E. to W. is 36 m., while its breadth varies from 12 to 20 m., and its area is 451·3 sq. m. or 288,842 ac. Pop. 166,447. It comprises 22 civil pars. and 35 complete and 4 parts of eccl. pars. Agriculture: corn crops 21,850 ac., turnips 3751 ac., potatoes 3108 ac. (these figures are for 1926). Of the total area of 288,842 ac. mountain and heath land 149,861, permanent grass 11,645 ac., and arable 50,438 ac. The chief manufacture is of iron goods; also tartans and carpets in Stirling. Mineral production is mostly coal, fireclay, igneous rock, iron ore, etc. The land rises towards the W. in Ben Lomond to 3192 ft., Beinn a' Choin, 2524 ft.

Stitchel, par., Roxburghshire and Berwickshire on Eden Water; 2798 ac. Pop. 238. S. vil. is 3¼ m. N.W. of Kelso. Pop. 111. P.O., T.O. S. House, a seat, and S. Linn a picturesque waterfall.

Stob Ban, mt., S. Inverness-shire, 5 m. E. of Ben Nevis; alt. 3217 ft.

Stobcross, ry. sta., L.M.S. near Queen's Dock, Finnieston, Glasgow. P.O., T.O.

Stob Garbh, mt., Perthshire, 3 m. S.E. of Crianlarich; alt. 3148 ft.

Stob Ghabhar, mt., N.E. Argyll, 5 m. N.W. of Inveroran Hotel; alt. 3565 ft.

Stobhall, ancient mansion in co. of, and 8 m. N. of, Perth.

Stobhill, q.s. par., now in pars. of Borthwick, Cockpen, and Newbattle, 4 m. S.E. of Dalkeith, Midlothian.

Stobo, par. and ry. sta., L.M.S., Peeblesshire, 6½ m. S.W. of Peebles; par. 10,309 ac. Pop. 292. P.O., T.O. Stobo Castle is near the sta.

Stobs, pl., with ry. sta., L.N.E., 4 m. S. of Hawick, Roxburghshire. Used as a military camp, with S. Castle as headquarters. P.O.

Stobsmuir, local reservoir for Dundee water supply.

Stockbridge, pl., with church, ¾ m. S. of Cockburnspath, Berwickshire.

Stockbridge, N.E. dist. of Edinburgh.

Stockbriggs, seat, 3 m. S.W. of Lesmahagow, Lanarkshire, on r. Nethan.

Stockinish, ham., isl., and loch, E. coast of Harris, Inverness-shire. P.O., T.O.

Stoer, dist. and q.s. par., Assynt par., Sutherland, on W. coast, 5 m. S.E. of Rhu Stoer, and 6 m. N.W. of Lochinver. P.O.

Stonefield, ham., Argyll, on shore of Loch Etive, 3½ m. W. of Taynuilt sta.

Stonefield, seat, S. Knapdale, Argyll, on W. shore of Loch Fyne, 2 m. N. of Tarbert.

Stonefield, vil., Lanarkshire, 2½ m. N.W. of Hamilton. P.O.

Stonegunn Quarries, 3 m. S.E. of Thurso, N. Caithness.

Stonehaven, seaport, police bur., and co. town of Kincardineshire, on S. Bay, with ry. sta., L.N.E. and L.M.S., 16 m. S.W. of Aberdeen. P.O., T.O. Pop. 4185. Has a good harbour and important fisheries, net-making, distilling, and woollen mfrs. A lighthouse at the harbour, visible 8 m. It is a frequented summer resort. Hotels.

Stonehouse, town and par., with ry sta., Lanarkshire, 7 m. S.E. of Hamilton; par. 6264 ac. Pop. 3704. P.O., T.O. Town pop. 3304.

Stoneridge, seat, Berwickshire, 6 m. N.E. of Coldstream.

Stoneyburn, vil., W. Lothian, 4½ m. S. of Bathgate. P.O., T.O. Pop. 2479.

Stoneykirk, par. and vil., Wigtownshire, on W. side of Luce Bay; par. 19,597 ac. Pop. 2092. P.O., T.O. The par. contains the large vil. of Sandhead.

Stoneywood, vil., with ry. sta., L.N.E., in Newhills par., Aberdeenshire. On the rt. bank of the Don, 1 m. N. of Auchmull. Large paper-works. S. House is near. P.O.

Stoneywood and Fankerton, Denny and Dunipace dist., Stirlingshire. Special drainage dist. Pop. 457.

Stoneywood Paper Mill, Denny, Stirlingshire.

Stony Hill, mt., border of Lanarkshire, 4 m. S.E. of Muirkirk; alt. 1843 ft.

Storm, loch, 6 m. N.W. of Lerwick, Mainland, Shetland Islands.

Stornoway, seapt., police bur., and par., Lewis Isl., Ross and Cromarty; 51 m. N.W. of Ullapool and 180 m. from Oban; par. 63,160 ac. Pop. 12,117. Par. pop. 3771. P.O., T.O. An important fishery sta., comprising the whole of the Outer Hebrides; steamers ply regularly from Aberdeen, Dundee, Glasgow, and Liverpool. S. Castle, a fine building now the property of the town, is on the opposite side of the Bay. S. Lighthouse, on Arnish Point, can be seen for 13 m. The town is well supplied with hotels; good fishing on the r. Cree. Daily mail-steamer from Kyle of Lochalsh.

Stornoway Loch, 1½ m. N.W. of West Loch Tarbert, Argyll.

Storr, mt., in Skye, Inverness-shire, 7 m. N. of Portree; alt. 2360 ft. The Old Man of Storr is a rock pinnacle 160 ft. high. Storr Loch is near.

Stoup, vil., in co. of, and 1 m. N.E. of, Dumfries.

Stour Head, Sutherland, W. coast. See RHU STOER.

Stow, par. and vil., with ry. sta., L.N.E., Midlothian, on Gala Water, 6¾ m. N. of Galashiels; par. 26,138 ac. Pop. 1280. Vil. has town hall and woollen mills. Pop. 462. P.O., T.O.

Straad, pl. Buteshire, 4 m. W.S.W. of Rothesay. P.O.

Stracathro, par., Angus, on West Water; 5304 ac. Pop. 445. Nearly 4000 ac. are under cultivation, 400 woodland. A battle took place in 1130, when David I. defeated the followers of Angus Mormaer of Moray. S. House is a seat on r. Esk.

Strachan, par. and vil., Kincardineshire; par. 41,672 ac. Pop. 546. Vil. on Water of Feugh, 3 m. S.W. of Banchory. P.O., T.O.

Strachur, par. and vil., Argyll, on E. side of Loch Fyne; par. 23,986 ac. Pop. 499. Vil. 1 m. S.E. of Creggans. P.O., T.O. Hotel. Vil. pop. 179. S. Park in the vicinity.

Straiton, par. and vil., Ayrshire; par. 49,855 ac. Pop. 1080. Vil. on Girvan Water, 7 m. S.E. of Maybole. P.O., T.O.

Straiton, vil., 5 m. S. of Edinburgh G.P.O. P.O. Had extensive oil works, now closed.

Stralachlan, par., Argyll, 6 m. S.W. of Strachur; 14,290 ac. Pop. 179.

Straloch, seat, Aberdeenshire, 2½ m. W.N.W. of New Machar sta.. Was once the home of the famous cartographer Sir R. Gordon, who rescued the M.S. maps of Timothy Pont and prepared the sheet for publication in Blaeus Great Atlas (1580–1661).

Stranathrow, fishing vil., Kincardineshire, 5 m. N.E. of Stonehaven.

Strand, loch, 5 m. N.W. of Lerwick, Shetland Isls.

Stranraer, royal and mun. bur., seapt., and par., with ry. stas., L.M.S., and harbour, Wigtownshire, at head of Loch Ryan, $7\frac{1}{2}$ m. N.E. of Portpatrick ; par. 61 ac. Pop. 2802. The town area is 389 ac. Pop. 6527. The bay of Stranraer affords excellent anchorage and a considerable trade is carried on with Ireland. There is a coastguard sta. and a custom house. S. was made a bur. in 1596 ; was made a royal bur. in 1617. The ruin of the old castle stands in the centre of the town. Hotel. P.O., T.O., and T.O. at S. Harbour sta. Steamer communication with Glasgow and Belfast (via Larne). Shortest sea route to Ireland.

Strath, par. in S.E. Skye, Inverness-shire ; 69,765 ac. Pop. 1573, includes the isl. of Scalpay and Pabbay. The church is at Broadford.

Strath, ham., 2 m. N.W. of Gairloch, W. Ross and Cromarty. P.O., T.O.

Strath Affric, Inverness-shire. See GLEN AFFRIC.

Strathairly House, seat, 2 m. E. of Largo, S.E. Fifeshire.

Strath Allan, valley of r. Allan, Perthshire and Stirlingshire.

Strathallan Castle, seat, Perthshire, on Machany Water, 3 m. N.W. of Auchterarder.

Strathallan School, Forgandenny, Perthshire. Boy's private school.

Strathan, ham., Sutherland, $1\frac{1}{2}$ m. S. of Lochinver.

Strath Ardle, valley of Ardle Water, in N.E. Perthshire.

Strathaven, town, with ry. sta., L.M.S., Lanarkshire, on Powmillon Burn, 7 m. S.W. of Hamilton. P.O., T.O. Pop. 4041. The town has a weaving and brewing trade and is a market-town. S. Castle (fifteenth century) a ruin.

Strath Avon, valley of the r. Avon, in S.W. Banffshire.

Strathbeg, Loch of, Aberdeenshire, 6 m. S.E. of Fraserburgh.

Strathblane, par. and vil., with ry. sta., L.N.E., Stirlingshire ; par. 9085 ac. Pop. 1161. Vil. on r. Blane, 4 m. W. of Lennoxtown. P.O., T.O. Calico printing works. S. Hills rise to 1624 ft.

Strath Bogie, seat and valley of Bogie Water, in N.W. of Aberdeenshire.

Strath Bran, valley of the r. Bran, Perthshire. Strath Bran is a seat.

Strath Bran, valley of the r. Bran, Ross and Cromarty between Achnasheen and Loch Luichart.

Strath Brora, the valley through which the r. Brora passes in S.E. Sutherland.

Strathbungo, q.s. par., and ry. sta. in S. suburbs of Glasgow. P.O., T.O.

Strath Carnach, valley in S.E. Sutherland.

Strathcarron, ry. sta., L.M.S., Ross and Cromarty, $7\frac{1}{4}$ m. N.E. of Strome Ferry. P.O., T.O.

Strath Carron, valley of Carron Water in S.W. Ross and Cromarty.

Strath Clyde, at one time a kingdom formed by the northerly Romanised Britons. It extended from the Clyde to the middle of Lancashire, its capital being Dumbarton. It was divided in the tenth century and part of it ceded to Scotland.

Strath Conon, valley of the r. Conon in S.E. of Ross and Cromarty. See STRATHCONON.

Strathconon, pl., Carnoch, q.s. par., Ross and Cromarty, on the r. Conon, 16 m. N.W. of Muir of Ord sta. P.O., T.O.

Strathconon Deer Forest, Ross and Cromarty, extends to 74,000 ac. Post Town, Muir of Ord. Said to be the best deer forest in Scotland.

Strath Dearn, valley, Inverness-shire and Nairnshire, is drained by the r. Findhorn.

Strathdee Distillery, near Aberdeen.

Strath Dionard, valley, Sutherland ; applied to the upper part of the r. Dionard to loch Dionard from which it issues.

Strathdon, par. and vil., Aberdeenshire ; par. 53,681 ac. Pop. 809. Vil. on r. Don., 14 m. N. of Ballater. P.O., T.O.

Strath Earn, valley of the r. Earn, Perthshire, extends over 30 m. from Loch Earn to Bridge of Earn. A wide and flat valley.

Strathendry, seat, 1 m. from Leslie, Fife.

Strathfillan, q.s. par., Killin par., Perthshire.

Strath Fillan, valley of Fillan Water, Perthshire.

Strathfinella Hill, Kincardineshire, $2\frac{1}{4}$ m. W. of Fordoun ; alt. 1358 ft.

Strathgarry House, seat, Perthshire, on r. Garry, 1½ m. S.E. of Blair-Atholl.

Strath Gartney, land along the northern shore of Loch Katrine between Ellen's Isle and Stronachlachar.

Strath Garve, valley extends for 18 m. from Loch Garve, Ross and Cromarty.

Strath Glass, valley of r. Glass in N.W. Inverness-shire.

Strath Gryfe, valley of the Gryfe Water in W. Renfrew.

Strathgyle, hill, Kincardineshire, 2½ m. S.E. of Durris ; alt. 975 ft.

Strathkanaird, pl., 5 m. N. of Ullapool, Ross and Cromarty. P.O.

Strathkinness, q.s. par. and vil., Fife, 3 m. W. of St. Andrews. P.O., T.O. Pop. of vil., 434.

Strathlachlan, Argyll. P.O., T.O. See STRALACHLAN.

Strathleven, seat, in co. of, and 2 m. N. of, Dumbarton.

Strathloch, seat, 8 m. N.E. of Pitlochry, N.E. Perthshire.

Strathmartine, vil., Angus, 3½ m. N.W. of Dundee. P.O.

Strathmashie, seat, 3 m. S.W. of Laggan Bridge, S. Inverness-shire.

Strathmiglo, par. and vil., with ry. sta., L.N.E., Fife ; par. 9162 ac. Pop. 1561. Vil. on r. Eden, 6½ m. W. of Ladybank Junction. P.O., T.O. Pop. 990.

Strathmill Distillery, Keith, Banffshire.

Strathmore, the great valley which partly divides the Highlands from the Lowlands of Scotland. It extends from Ardmore in Dunbartonshire to the North Sea at Stonehaven, but the name is generally applied to the dist. N.W. of the Sidlaw Hills and reaching from Methven in Perthshire to near Brechin in Angus.

Strathmore, valley traversed by the r. Thurso, Caithness.

Strathmore, see also STRATH MORE.

Strath More, valley in N. Sutherland, extending along the Strathmore R. from Gobernuisgach to Loch Hope. Has the antiquity known as the Broch of Donnigil.

Strath Naver, valley extending for 19 m. from Loch Naver to near the sea, N. Sutherland. P.O., T.O.

Strathord, ry. sta., L.M.S., Perthshire, 2 m. S.W. of Stanley, 6 m. N.W. of Perth.

Strath Oykell, valley on border of Sutherland and Ross and Cromarty, extends from Kyle of Sutherland to Bridge of Oykell, along the r. Oykell which forms the boundary between Ross and Sutherland.

Strathpeffer, vil., with ry. sta., L.M.S., Ross and Cromarty, 5 m. W. of Dingwall. P.O., T.O. Noted for its mineral springs. A favourite watering-place ; hotels and baths. Pop. 419.

Strath Polly, small valley in extreme N.W. of Lochbroom par., Ross and Cromarty.

Strath Rusdale, valley along the Alness R., E. Ross and Cromarty.

Strath Shinary, valley descending to Sandwood Loch near the sea, N.W. Sutherland.

Strathtay, ham, Perthshire, 4½ m. N.E. of Aberfeldy. P.O., T.O.

Strath Tirry, flat valley along the r. Tirry from its mouth at Loch Shin. Extends for 10 m. N. to near the Crask, 12 m. N.W. of Lairg, Sutherland.

Strathtummel, pl. in Perthshire, 9 m. N.W. of Pitlochry. P.O., T.O.

Strathtyrum, seat, 1 m. W. of St. Andrews, N.E. Fifeshire.

Strath Ullie, valley, Sutherland, on r. Helmsdale. Is about 15 m. long and has many antiquities along it.

Strath Vagastie, Sutherland. Extends from Crask Inn to Altnaharra at the S. end of Loch Naver.

Strathvaich, shooting, Ross and Cromarty. Post town, Garve.

Strathy, q.s. par. and vil., N. coast of Sutherland. Vil. at mouth of Strathy Water, 18 m. W. of Thurso. P.O., T.O.

Strathyre, seat, vil., and ry. sta., L.M.S., Perthshire, 8 m. N.W. of Callander. Summer resort. P.O., T.O. Hotel.

Strathyre and Balquhidder State Forest, stas., Strathyre or Balquhidder, Perthshire.

Stravithie, ry. sta., L.N.E., 4½ m. S.E. of St. Andrews, Fife. P.O.

Streanach Head, N.E. point of Loch Sealg, E. Lewis, Ross and Cromarty.

Streap, mt., S.W. Inverness-shire, 4½ m. E. of Glenfinnan ; alt. 2988 ft.

Strichen, par. and vil., with ry. sta., L.N.E., Aberdeenshire ; par. 14,409 ac. Pop. 2024. Vil. on North Ugie Water 10 m. S.W. of Fraserburgh. P.O.,

T.O. Vil. pop. 958. S. House, seat, ¾ m. S.W. of sta.

Striven Loch, an arm of the Firth of Clyde, Argyll. Extends from Ardyne Point N. for 10 m.

Stroan Loch, 1 m. W. of New Galloway ry. sta., Kirkcudbrightshire.

Stroma, isl., in the Pentland Firth, Caithness, 3½ m. N.W. of Duncansbay Head, 2 m. N.W. of John o' Groats' House. Measures 2¾ m. by ⅓ m. P.O. Pop. 193.

Stromedearg, loch, 1¾ m. S.W. of Loch-boisdale, S. of South Uist.

Strome Ferry, pl., with ry. sta., L.M.S., and pier, 10 m. N.E. of Kyle of Loch-alsh. P.O., T.O. Hotel. Continuous daily motor ferry boat service across Loch Carron.

Strome, North, Hostel, on N. side of Loch Carron, at Strome Ferry, W. Ross and Cromarty. A Scottish Youth Hostel.

Strome (N. and **S.) State Forest,** near Strome Ferry, 10½ m. N.E. of Kyle of Lochalsh, W. Ross and Cromarty.

Stromness, seapt. town and police bur. and par. S.W. Mainland of Orkney, 14 m. W. of Kirkwall. Par. 7618 ac. Pop. 2116. Town pop. 1592. P.O., T.O. The town is on an excellent natural harbour and has distillery, ropeworks, and shipbuilding yards. The most important industry is the herring fishing. A large concrete pier was built in 1926, which can accommo-date a large number of vessels. Town created a burgh of barony in 1817, now a police burgh. The town is old and very irregular, the main street in some places being too narrow to allow two vehicles to pass each other. There are gas works, established and episcopal churches, hotels, and an interesting natural history museum. The bur. extends to 454 ac. Steamer com-munication with Aberdeen, Kirkwall, Leith, Lerwick, Wick, etc.

Stronachie Distillery, Forgandenny, S.E. Perthshire.

Stronachlachar, pl. on S.W. of Loch Katrine, Stirlingshire, 5 m. N.E. of Inversnaid. Has steamboat pier. Hotel. P.O., T.O.

Stronachullin, seat, 3½ m. S. of Ardris-haig, W. Argyll.

Stronaclachich, mt., 1 m. W. of Killin, Perthshire ; alt. 1708 ft.

Stronardron, seat, 14 m. N.W. of Colin-traive, W. Argyll.

Stronchreggan House, N.E. of Loch Linnhe, Argyll, nearly opposite Fort William.

Strone, watering-place, Argyll. On point which projects between Loch Long and Holy Loch, 6 m. N.W. of Greenock. P.O., T.O.

Stronelairg, seat, 11 m. S.E. of Foyers, Inverness-shire.

Stronend, highest point of Fintry Hills, 2 m. N. of Fintry ; alt. 1676 ft.

Strone of Callie, pl. and seat, Persie q.s. par., Perthshire.

Stronfearnan, or **Fernan** (q.v.).

Stronsay, isl., Orkney, 12 m. N.E. of Kirkwall, meaures 4½ m. × 7¼ m., area 9840 ac. Highest points are 141 and 154 ft. The par. of Stronsay includes Auskerry, Eday, Papa-Stronsay, Pharey, and Stronsay, 16,404 ac. Pop. 975. P.O., T.O.

Stron Smeur, loch, 2½ m. N.E. of Rannoch, Perthshire.

Strontian, vil. and q.s. par., Morven dist. of Argyll. Vil. near head of Loch Sunart, 24 m. S.W. of Fort William. P.O., T.O. S. House, a seat, near vil.

Stronvar, seat, Perthshire, on S. shore of Loch Voil, 5½ m. S.W. of Lochearnhead.

Strowan, seat, Perthshire, near r. Earn, 3 m. W. of Crieff.

Struan, ham. with inn on Loch Bracadale, Skye, Inverness-shire, 11 m. S.E. of Dunvegan. P.O., T.O. Steamers from Glasgow call regularly.

Struan, ham. and ry. sta., L.M.S., Perth-shire, 4¾ m. W. of Blair-Atholl. P.O.

Struie Hill, N.E. Ross and Cromarty, 6 m. S.E. of Bonar Bridge ; alt. 1068 ft.

Struy, ham., Inverness-shire, at meeting of rs. Farrar and Glass, 10 m. S.W. of Beauly. P.O., T.O. S. Deer Forest extends to 18,000 ac.

Stuartfield, vil., 2½ m. S.W. of Mint-law, Aberdeenshire. P.O., T.O. See STEWARTFIELD. Pop 390 (S. and Quartalehouse).

Stuartown, part of vil. of Campbeltown, Inverness-shire.

Stuckgown House, seat, Dunbartonshire, on Loch Lomond, 1 m. S.E. of Tarbet.

Sturdy Hill, borders of Angus and Kincardine, 5½ m. N. of Edzell; alt. 1784 ft.

Suainaval, loch and hill (1404 ft.) on W. of Lewis, 2 m. S.E. of Camas Uig, Ross and Cromarty.

Sugar Loaf, The, name often applied to Suilven, the pillar mountain, 4 m. S.E. of Lochinver (2399 ft.), Sutherland.

Suilven—frequently called the Sugar Loaf —a mt. on the W. of Sutherland, 4 m. S.E. of Lochinver; alt. 2399 ft.

Suisgill Burn, tributary of the r. Helmsdale, Sutherland. It was mainly along this burn that the gold diggings in 1869-1870 were carried on.

Suishnish Point, or **Susnish,** headland separating Loch Slapin and Loch Eishort.

Sula Sgeir, a pillar rock 6 m. S.W. of Rona Isl. and about 50 m. N.W. of Cape Wrath.

Sule Skerry, isl., Stromness par., Orkney, with lighthouse (direction - finding beacon), 113 ft., visible 17 m. The rock is 40 m. N.W. of Stromness, and Stack Skerry is 5½ m. S.W. of Sula Skerry.

Sullam, or **Sullom,** ham., Shetland, on Sullom Voe, 9 m. N.W. of Voe. P.O.

Sumburgh Head, a bold headland at the extreme S. end of the Mainland of Shetland. Has lighthouse 300 ft., seen for 24 m. S. Roost is a strong tidal current which sweeps round this headland. P.O., T.O.

Summer Isles, a group of isls. at the entrance to Loch Broom, Ross and Cromarty. Only one inhabited, namely, Tanera.

Summerhill, pl., with church, Ayton par., Berwickshire.

Summerhill, ham., 1 m. from New Machar, Aberdeenshire. P.O., T.O.

Summerston, ry. sta., L.N.E., Stirlingshire, 6½ m. N. of Glasgow by rail.

Sunadale, pl., E. coast of Kintyre, Argyll, 4 m. N. of Airds.

Sunart, dist., Ardnamurchan, Argyll, along the shore of Loch Sunart for 12 m.

Sunart Loch, sea loch, Argyll, extends E. and W. for 14 m. and divides the dists. of Ardnamurchan and Morven.

Sundhope Height, mt., mid Selkirkshire, 3 m. S.W. of Yarrow church; alt. 1684 ft.

Sundrum, seat, Ayrshire, on Water of Coyle, 5½ m. E. of Ayr.

Sunlaws, seat, Roxburghshire, on r. Teviot, 3 m. S.W. of Kelso.

Sunnybank, Port Glasgow, Renfrewshire. Convalescent home.

Sunnyside, pl., Berwickshire, 4 m. W.N.W. of Coldstream.

Sunnyside, pl., Angus, 2½ m. N.W. of Montrose. Montrose Lunatic Asylum here.

Sunnyside, dist. and ry. sta., L.N.E., Gorebridge, Lanarkshire. P.O.

Sunnyside, seat, Perthshire, 7 m. E.N.E. of Crieff.

Sunwick, seat, Berwickshire, 6¼ m. W. of Berwick-on-Tweed.

Suorach, loch, N. Caithness, 4 m. S.E. of Reay.

Sutherland, maritime co., forming the extreme N.W. of Scotland. Bounded by the Atlantic Ocean on the W., by the North Sea on the N., by Caithness on the E., and by Dornoch Firth, Kyle of Sutherland, and Ross and Cromarty on S.E. and S. Its greatest length is 63 m., breadth 60 m., area 2027·9 sq. m. or 1,297,849 ac. Pop. 16,100 or 7·8 persons to the sq. m. The E. of the co. is served by L.M.S., which enters the co. at the Kyle of Sutherland and follows the E. coast to Helmsdale and through Kildonan, passing into Caithness 4 m. beyond Forsinard ry. sta. and hotel. The surface is chiefly moor and mt.; of the latter, Ben More Assynt, 3273 ft., is the highest, Ben Klibreck 3154 ft., Ben Hope 3040 ft., Foinaven 2990 ft. The only town is Dornoch, the capital, and in the S.E. are the vils. of Lairg, Rogart, Brora, Helmsdale, Tongue, Durness, Scourie, and Lochinver. The chief rs. are Oykell, Brora, Golspie, Helmsdale, Halladale, Naver, Hope, and Dionaird; and the largest lochs, Shin, Assynt, Loyal, Naver, Hope, More, and Sonascaig. Sheep farming is the chief industry. There are woollen mills at Rogart and Brora, with inferior coal mined at the latter. Mail coaches run daily from Invershin to Lochinver, and from Lairg to Scourie, Durness, and Tongue. The co. is divided in 13 pars., and with Caithness it returns one member to Parliament.

Sutors of Cromarty, this name applies to the two outstanding headlands at the entrance of the Firth of Cromarty, rising to over 400 ft. on S. and N. side of the Firth, Ross and Cromarty.

Swannay, ham., Orkney, on Loch of Swannay, 14 m. N.W. of Finstown. P.O.

Swattle Fell, hill in N. Dumfriesshire, 4 m. N. of Moffat ; alt. 2388 ft.

Sween Loch, sea loch, Knapdale, Argyll, opposite Jura Isl., 9 m. long. On E. side, 4 m. N. of Knap Point, is the ham. of Castlesween and ruins of an ancient castle, near which is a ferry to Danna Isl.

Sweetheart Abbey, Kirkcudbrightshire. See NEW ABBEY.

Swiney, vil., mansion, and ruined castle, Caithness, 1½ m. W. of Lybster.

Swinister, pl., Shetland, on S. Voe, 2 m. S. of Mossbank.

Swinridgemuir, seat, 2 m. E. of Dalry, Ayrshire.

Swinton, par. and vil., Berwickshire ; par. 5570 ac. Pop. 676. Vil. on Leet Water, 5½ m. S.E. of Duns. P.O., T.O. S. House, seat, 1 m. W. of vil.

Swinton, vil., Lanarkshire, near Baillieston, 2½ m. W. of Coatbridge.

Swona, isl., Orkney, in Pentland Firth, 4 m. W.of S. extremity of S.Ronaldshay. Pop. 6. Lighthouse, showing an eight-seconds' white-flashing light, visible 12 m.

Swordale, seat, 2¼ m. S.W. of Novar sta., E. Ross and Cromarty.

Swordle and Knock, vil., 6 m. E. of Stornoway, Ross and Cromarty.

Sydenham, seat, 2 m. N. of Kelso, Roxburghshire.

Symington, par. and vil., Ayrshire ; par. 3725 ac. Pop. 651 ; vil. 3½ m. N.E. of Monkton sta. and 6 m. S.W. of Kilmarnock. Pop. 192. P.O., T.O.

Symington, par. and ry. junc., L.M.S., Lanarkshire, on r. Clyde ; par. 3515 ac. Pop. 580. Sta. 3¼ m. W.S.W. of Biggar. P.O., T.O. Vil. pop. 429. Hotel.

Symington House, seat, 3¼ m., N.W. of Stow, Midlothian, on Gala Water.

Synniness, ruined castle, Wigtownshire, 3 m. S.E. of Glenluce.

Syre, loch and seat, 16 m. N.W. of Kinbrace ry. sta., E. Sutherland.

T

Taberon Law, mt., Peeblesshire, 3 m. S.E. of Drumelzier ; alt. 2088 ft.

Tachdaidh, Loch an, Ross and Cromarty, 11 m. E. of Lochcarron, depth 62 ft. ; alt. 831 ft.

Tafts Ness, prom. headland, N. extremity of Sanday Isl., Orkney.

Tail of the Bank, sandbank, Firth of Clyde, off Greenock.

Tain, seapt., royal and mun. bur., and par., with ry. sta., L.M.S., Ross and Cromarty, on S. shore of Dornoch Firth, 44¼ m. N. E. of Inverness, 25½ m. N.E. of Dingwall ; par. 16,843 ac. Pop. 2176 ; mun. bur pop. 1383. P.O., T.O. Hotel. A quaint town with a curious old tower and a fine Gothic church of 1471.

Talisker, ham., Isle of Skye, 1 m. W. of T. bay, Loch Bracadale. T. Point, a bold lofty headland, near E. of entrance to Loch Bracadale.

Talisker Distillery, Carbost, Bracadale, Isle of Skye.

Talla Bheith, deer-forest, Perthshire, between Loch Ericht and Loch Rannoch. Embraces 15,000 ac.

Talladale, pl., Ross and Cromarty, S. side of Loch Maree,9 m.N.W. of Kinlochewe. P.O., T.O. called Lochmaree. Hotel.

Tallant, loch, Islay, Argyll, 2 m. S.E. of Bowmore.

Talla Reservoir, Peeblesshire, in T. Glen, Tweedsmuir. Edinburgh water supply, 299 ac. ; capacity, 443,000,000 cubic ft., maximum depth 79 ft.

Talla Water, Peebleshire, flows 3 m. N.W. to the Talla Reservoir, Tweedsmuir.

Talmine, ham., Sutherland, 4 m. N. of Tongue. P.O., T.O.

Tambowie Distillery, Dunbartonshire,1½ m. N.W. Milngavie.

Tamdhu Distillery, Morayshire, on r. Spey, 8 m. S.W. of Craigellachie.

Tam Head, Aberdeenshire, 4½ m. W. of Fraserburgh. Shows fixed light, visible 5 m.

Tandlemuir, pl., Renfrewshire, 2¼ m. N.W. of Lochwinnoch.

Tanera, largest of the Summer Isles, Loch Broom, Ross and Cromarty.

Taner, Water of, Aberdeenshire, flows 11 m. N.E. through Glen Taner, to the Dee, 1½ m. W. of Aboyne.

Tang Head, prom., N. Caithness, 2 m. W. of St. John's Point.

Tangwick, ham. and bay, W. side of Mainland, Shetland, on N. side of Magnus Bay.

Tangy, loch, Argyll, 5 m. N.W. of Campbeltown.

Tankerness, vil., seat, and loch, Orkney, 6 m. E.S.E. of Kirkwall, on Deer Sound, P.O., T.O.

Tannach, ham., Caithness, 4 m. S.W. of Wick. T. Hill, 1 m. W. ; 457 ft.

Tannachy, seat, Banffshire, 3¼ m. S.W. of Buckie.

Tannachy, ham., S.E. Sutherland, 5 m. N. of Rogart sta.

Tannadice, par. and vil., with sta., L.M.S., Angus ; par. 21,316 ac. Pop. 975. Vil. on the S. Esk, 6 m. N. of Forfar. P.O., T.O. T. House, 1 m. S.E.

Tanna Loch, Arran Isl., Buteshire, 3½ m. E.N.E. of Whitefarland.

Tannochside, mining vil., Lanarkshire, 1½ m. N.W. of Bellshill. Pop. (Aitkenhead and T.) 3060. P.O.

Tannochside, or **St. Enoch's Hall,** seat, Lanarkshire, 2½ m. N.N.E. of Bothwell.

Tantallon Castle, E. Lothian, 2¾ m. E. of North Berwick. Ruined strong famous ancient castle on peninsulated, lofty, precipitous sea-rock. Belonged to the Earls of Douglas till 1455, passed to the Earls of Angus and served them to give defiance to James IV. and James V. Besieged and captured by the Covenanters in 1639. Described by Sir W. Scott in "Marmion." Now the property of the nation.

Taransay, isl., off mouth of W. Loch Tarbert, N.W. Harris, 4½ m. by 2 m. Pop. 33.

Tarbat, par., N.E. Ross and Cromarty ; 6820 ac. Pop. 890. Church ½ m. S. of Portmahomack.

Tarbat House, seat of Countess of Cromartie, Ross and Cromarty, on Cromarty Firth, ¾ m. S.E. of Kildary ry. sta.

Tarbat Ness, between Dornoch Firth and Moray Firth and 11 m. N.E. of Tain. Lighthouse showing a half-minute group-flashing light, seen 18 m.

Tarbert, seapt. vil. and q.s. par., Argyll. Vil. at head of E. Loch Tarbert off Loch Fyne, 13½ m. S. of Lochgilphead. P.O., T.O. Vil. pop. 1184. Hotel.

Carries on extensive herring fishery. Steamers from Glasgow to Ardrishaig land passengers for Islay.

Tarbert, ham., Harris, Outer Hebrides, at the head of E. Loch Tarbert. P.O., T.O. called Harris. Pop. 297.

Tarbert, sea loch, nearly bisecting Jura Isl., Argyll. T. deer forest in vicinity.

Tarbert, stream, Argyll, flowing through Glen T. to Loch Linnhe, 5 m. E. of Strontian.

Tarbert, East and **West,** two sea lochs on mutual boundary of Knapdale and Kintyre, Argyll. Isthmus, 1 m. wide. West loch descends 11 m. S.W. to the sea opposite middle part of Islay.

Tarbert, East and **West,** two sea lochs, nearly bisecting Harris, Outer Hebrides. Isthmus between the lochs a ½ m. wide.

Tarbet, ham., with ry. sta. (Arrochar and T.), L.N.E., Dunbartonshire, on W. shore of Loch Lomond, 1½ m. E. of Arrochar and 42 m. from Glasgow by rail. P.O., T.O. Has a steamboat pier and large hotel.

Tarbolton, par. and vil., with ry. sta., L.M.S., Ayrshire par. (containing Annbank), 12,058 ac. Pop. 5131. Vil. 1½ m. N.N.W. of Tarbolton sta., 4 m. W. of Mauchine. P.O., T.O. Pop. 1226. P.O. at sta. called T. Station. From 1777 to 1784 a habitual resort of the poet Burns.

Tarduf, seat, Stirlingshire, 2 m. S.E. of Polmont Junction.

Tarff, loch, Inverness-shire, 3 m. N.E. of Fort Augustus, depth 89 ft ; alt. 956 ft.

Tarff, ry. sta., L.M.S., Kirkcudbrightshire, 3½ m. N. of Kirkcudbright.

Tarff, stream, Inverness-shire, flowing 12 m., through Glen T. to Loch Ness at Fort Augustus.

Tarff Water, Kirkcudbrightshire, flows 9 m. S. to r. Dee, 2 m. N. of Kirkcudbright.

Tarfside, ham., Angus, Lochlee par., 9 m. N.W. of Edzell. P.O., T.O.

Tarf Water, Wigtownshire, tributary of Bladenoch, 1 m., S.E. of Kirkcowan.

Tarf, Water of, Angus, flows 7½ m. S.E. to the N. Esk, near Tarfside.

Tarf, Water of, Perthshire, flowing 11 m. E. to r. Tilt, near the Bedford Memorial Bridge superseding a dangerous ford.

20

Tarland, par. and vil., Aberdeenshire ; 6300 ac. Pop. 704. Vil. on T. Burn, 7 m. N.W. of Aboyne ry. sta. P.O., T.O. Vil. pop. 357.

Tarland Lodge, Aberdeenshire, 1 m. N.E. of Tarland.

Tarland-Migvie, eccl. par., Aberdeenshire.

Tarlogie House, Ross and Cromarty, 2 m. N.W. of Tain.

Tarner, isl., Isle of Skye, Inverness-shire, in Loch Bracadale.

Tarradale, pl. and seat, Ross and Cromarty, 2 m. S.E. of Muir of Ord. ry. sta.

Tarras Water, Dumfriesshire, flows 11½ m. S.W. to r. Esk, 2½ m. S.E. of Langholm.

Tarry Croys, Miltown of, ham., Banffshire, 3 m. N.W. of Keith.

Tarskavaig, ham., Isle of Skye, 12 m. S.W. of Isle of Ornsay. P.O.

Tartan Lodge, N.W. Argyll, 10 m. N.E. of Ballachulish.

Tarth Water, Peeblesshire, flows 7 m. S.S.E. to the Lyne Water, near Drochil Castle.

Tarty Burn, affl. of r. Ythan, Aberdeenshire, 2 m. N. of Newburgh.

Tarves, par. and vil., Aberdeenshire ; par. 16,300 ac. Pop. 2026. Vil. 5½ m. N.E. of Old Meldrum. P.O., T.O. Vil. pop. 239.

Tarvie, seat, Perthshire, 6 m. N.E. of Pitlochry.

Tarvit House, seat, Fife, near Cupar. T. Hill, 2 m. S. of Cupar, has market-cross of Cupar.

Tavool House, Mull, Argyll, on N. shore of L. Scridain.

Tay, the longest r. in Scotland, rising on the N. slope of Ben Lui (Beinn Laoigh) at an alt. of 2980 ft., and flows under the names of Fillan and Dochart until it joins Loch Tay, and after 14½ m. issues from that loch as the Tay, and flows for 54½ m. until it is joined by the Earn, and flows for 54½ m. to join the North Sea beyond Dundee. Its total is 119¾ m., and it drains 2400 sq.m., the largest area of any of the Scottish rs. It passes through some of the finest scenery in the Highlands and is crossed by the Tay ry. bridge at Dundee, at Perth by a ry. bridge and a stone bridge built in 1766, at Dunkeld by a Telford Bridge built in 1809, and by a General Wade Bridge at Aberfeldy built in 1733. The estuary, r., and loch are among the best salmon waters in Scotland. Its chief tributaries, Ericht, Tummel, Earn, and Lyon.

Tay Bridge, ry. viaduct across the Firth of Tay at Dundee. Opened in 1888 in place of the old bridge blown down in December 1879. Is over 2 m. in length, and rests on 86 piers, which have a span varying from 68 to 245 ft. Cost £650,000.

Taychreggan, ham., Argyll, on Loch Awe, opposite Port Sonachan, with which it is connected by ferry. No cars.

Tayfield, seat, Fife, near Newport.

Tay, Firth of, extends 24½ m. N.E. from the confluence of the Tay and Earn to the North Sea, widens to 3½ m. at its broadest part and separates the cos. of Perth and Angus on the N. from Fife in the S. Has a light-vessel at Abertay Sands, and lighthouses at Buddon Ness and Tayport.

Tayinloan, vil., Argyll, W. coast of Kintyre, 18 m. S.W. of Tarbert. P.O., T.O.

Tayinlone Shooting-lodge, Isle of Skye, 4 m. N.W. of Carbost.

Tay Lighthouse, Angus. See Buddon Ness.

Tay, Loch, Perthshire. A magnificent highland loch (alt. 355 ft.) surrounded by mts., Ben Lawers, 3945 ft., being on its W. side. It receives the rs. Dochart and Lochy at the S.W. and discharges itself by the r. Tay at Kenmore, its N.E. extremity. Length 15 m., breadth ½ to over 1 m., greatest depth 508 ft. ; alt. 349 ft.

Taylorton and **Lower Taylorton,** Stirlingshire, 1½ m. E. of Stirling. County hospital.

Taylun Bay, Coll Isl., W. coast, Argyll.

Taymount, seat, Perthshire, on r. Tay, 2 m. N.E. of Stanley.

Taymouth Castle, formerly a seat of the Marquis of Breadalbane, now a hotel, Perthshire, on r. Tay, and near the fort of Loch Tay, 5 m. S.W. of Aberfeldy.

Taynish, ham. and seat and isl., Knapdale, Argyll, on W. shore of Loch Sween, 2½ m. S. of Tayrallich.

Taynuilt, vil. and ry. sta., L.M.S., Argyll, near the S. shore of Loch Etive, 13 m. E. of Oban. P.O., T.O. Pop. 415. Hotel and 9-hole golf course.

Tayport, or **Ferryport-on-Craig,** town, police bur. and par., with ry. sta., L.N.E., Fife ; par. 2806 ac. Pop. 3270. Police bur. pop. 3164. The town is on the S. side of entrance to Firth of Tay, opposite Broughty Ferry, 3½ m. S.E. of Dundee. P.O., T.O. Has linen and jute and engineering works, also a large sawmill. Maintains a ry. ferry to Broughty Ferry. Two lighthouses, High and Pill, with fixed lights, seen 10 and 12 m.

Tayvallich, ham., Argyll, on Loch Sween, 8 m. S.W. of Crinan Pier. P.O., T.O.

Teacuis, sea loch, extending from Loch Sunart, N.W. Argyll.

Teallach, or **An Teallach,** one of the most picturesque mts. in Scotland, 4 m. S.W. Ullapool, Ross and Cromarty. Two peaks 3483 and 3474 ft.

Tealling, par. and vil., Angus ; par. 6980 ac. Pop. 515. Vil. 6 m. N. of Dundee. P.O., T.O. T. House, a seat.

Teanassie, ham., Inverness-shire, 4 m. S.W. of Beauly.

Teangue, ham., Isle of Skye, on Knock Bay, 3 m. S.S.W. of Isle Ornsay. P.O., T.O.

Teaninich, Seat, Ross and Cromarty, W. of Alness. T. Distillery here.

Teasses, seat, Fife, 3 m. S. of Ceres.

Techmuiry, ham., N.E., Aberdeenshire, 5 m. S.S.W. of Fraserburgh.

Teindland, State forest, in Morayshire, 2 m. S.W. of Orbliston ry. junc.

Teinside, pl., Roxburghshire, 6 m. S.W. of Hawick.

Teith, r. Perthshire, formed by two headstreams which rise 2 m. E.N.E. of the head of Loch Lomond and which unite at Callander, and running 13 m. S.E. to the Forth, 2¼ m. N.W. of Stirling.

Tempar, pl., Perthshire, 1 m. E. of Kinloch Rannoch.

Templand, vil., Dumfriesshire, 2½ m. N. of Lochmaben. P.O.

Temple, par. and vil., Midlothian ; par. 13,630 ac. Pop. 352. Vil. on the S. Esk, 6¼ m. S.W. of Dalkeith. P.O.

Temple, part of Lower Largo, Fife.

Temple, q.s. par., Govan, Glasgow. P.O.

Temple Hall, N.E. Berwickshire, ½ m. W. of Coldingham.

Templeland House, Aberdeenshire, 6 m. N.E. of Huntly ry. sta.

Temple Pier, on W. side of Loch Ness, Inverness-shire.

Tenandry, q.s. par., Perthshire on r. Garry. Church 1 m. S. of Killiecrankie ry. sta.

Tents Muir, a sandy tract on the Fife coast, Leuchars par. T. M. Point its N.E. point. State forest, 3 m. N.E. of Leuchars.

Ternate, loch, Argyll, drained by Ronach Water, into Loch Aline.

Ternemny, pl., Banffshire, 4½ m. N. o Rothiemay sta.

Terpersie, pl., Aberdeenshire, 1 m. N.W. of Tullynessie.

Terraughtie, seat, Kirkcudbrightshire, 2½ m. W. of Dumfries.

Terregles, par., N.E. Kirkcudbrightshire and part in Dumfriesshire, 3828 ac. Pop. 580. P.O. T. House belongs to the Board of Agriculture. Terregleston, a seat.

Terrenzean Castle, ruined ancient seat, Ayrshire, 1¼ m. W.N.W. of Old Cumnock.

Tertowie, seat, Aberdeenshire, 5 m. S.E. of Kintore.

Tervie, Burn of, tributary of Livet Water, Banffshire.

Teviot, r., Roxburghshire, rises on the Dumfriesshire border, flows 36 m. N.E. to the Tweed at Kelso. Tributaries, Allan, Slitrig, Rule, Jed, and Kale Water on r. bank and Borthwick and Ale Water on l. bank.

Teviotbank, seat, Roxburghshire, 5 m. N.E. of Hawick.

Teviotdale, Roxburghshire, vale of the r. Teviot.

Teviothead, par., Roxburghshire, on r. Teviot ; 31,498 ac. Pop. 353. P.O., T.O. Church 9 m. S.W. of Hawick.

Tewel, pl., Kincardineshire, 2½ m. W. of Stonehaven.

Texa, islet, S.E. coast of Islay Island, Argyll, 2 m. E.S.E. of Port Ellen.

Thainstone, seat, Aberdeenshire, 2¼ m. N.N.W. of Kintore.

Thankerton, vil., and ancient par., Lanarkshire, ry. sta., L.M.S., on r. Clyde, 5 m. S.S.E. of Carstairs Junction. P.O., T.O. Vil. pop. 220.

Thankerton Square, pl., N.E. Lanarkshire, 1 m. N.E. of Bellshill.

The Braes, vil., Isle of Skye, 6 m. S.E. of Portree. P.O., T.O. See also BRAES, THE.

The Bush, seat, Midlothian. See BUSH.

The Craigs, pl., Ross and Cromarty, 1 m. from Ardgay. P.O., T.O.

The Glen, Peeblesshire, P.O., T.O. See GLEN, THE.

The Hirsel, seat of the Earl of Home, 1½ m. N.W. of Coldstream, Berwickshire. H. Loch in vicinity.

The Mound, ry. sta., L.M.S., Sutherland, 7¾ m. N. of Dornoch. P.O., T.O. at sta.

The Poles, pl., Sutherland, 2½ m. N.W. of Dornoch.

Thief's Road, old freebooters' mountain-path, from Border through Peeblesshire to Midlothian.

Thirdpart, estate, Fife, 1½ m. S.W. of Crail.

Thirlestane Castle, seat of Earl of Lauderdale, Berwickshire, on Leader Water, N.E. of the town of Lauder.

Thirlestane Castle, seat of Lord Napier and Ettrick, Selkirkshire, 17 m. S.W. of Selkirk, on l. bank of Ettrick Water.

Thirlestane Gardens Hostel, opposite Thirlestane Castle, Ettrick Valley, Selkirkshire, 18 m. W. of Hawick and 18 m. S.W. of Selkirk. A Scottish Youth Hostel.

Thomanean, seat, Kinross-shire, 2¼ m. S.W. of Milnathort.

Thomaston, ruined castle, Ayrshire, 4½ m. W.S.W. of Maybole. Said to have been founded by a nephew of Robert Bruce in 1335.

Thom, Loch, Renfrewshire, 2 m. S. of Greenock. Reservoir of the Greenock Waterworks. Also called Shaws. Depth 42 ft., alt. 632 ft.

Thomshill, pl., Morayshire, 3½ m. S. of Elgin.

Thorlies Hope, pl., S.E. Roxburghshire, 8 m. N.E. of Newcastleton.

Thormaid, loch, N. Caithness, 4 m. S.E. of Reay.

Thorn, sub. of Johnstone, Renfrewshire. P.O.

Thornhill, vil., with ry. sta., L.M.S., Dumfriesshire, on r. Nith, 14¼ m. N.N.W. of Dumfries. Hotel. P.O., T.O. Pop. 1044. Has live-stock sales. Has an interesting museum containing Burns's relics.

Thornhill, vil., Perthshire, 4 m. W.S.W. of Doune. P.O., T.O.

Thornhill House, Stirlingshire, at Grahamston.

Thornice, Lanarkshire, part of vil. of Braidwood, 1¼ m. S.E. of Carluke.

Thornielee, pl., with ry. sta., L.N.E., Selkirkshire, 5 m. E. of Innerleithen.

Thorniewood, vil., Lanarkshire, ¾ m. N.E. of Uddingston.

Thornliebank, q.s. par., with ry. sta., L.M.S., in bur. of Glasgow, 1½ m. S.S.W. of Pollokshaws. Has calico printing and Turkey-red dye works. P.O., T.O. Pop. 1512. T. House in vicinity.

Thornlybank Hill, W. Renfrewshire, 3 m. S.W. of Bridge of Weir ; alt. 908 ft.

Thornton, town and q.s. par., with ry. sta. (Thornton Junction), L.N.E., Fife, 5 m. N. of Kirkcaldy and 30¾ m. N. by E. of Edinburgh by the Forth Bridge. P.O., T.O. Town pop. 2342.

Thornton, ham., Angus, 1 m. E. of Glamis.

Thornton, pl., East Lothian, 5 m. E.S.E. of Dunbar. Ruins of old castle.

Thornton Castle, seat, Kincardineshire, 3 m. W. of Laurencekirk.

Thornton Hall, pl., with ry. sta., L.M.S., Lanarkshire, 1 m. S.E. of Busby. P.O., T.O.

Thornton House, Ayrshire, 3 m. W. of Kilmarnock.

Thorntonloch, look-out sta., East Lothian, 5 m. E.S.E. of Dunbar.

Thornwood, vil., Lanarkshire, 4 m. N. of Hamilton. T. House, a seat.

Thornyhills, pl., 3 m. S.E. of Lanark.

Thornyhive, bay, Kincardineshire, 2 m. S. of Stonehaven.

Threave Castle, ancient castle, on an islet in the r. Dee, Kirkcudbrightshire, 1¼ m. W. of Castle-Douglas. Erected as a palace fort by the third Earl of Douglas near the close of the fourteenth century.

Threepland Burn, headstream of White Cart, Renfrewshire.

Threepwood, seat, N. Roxburghshire, 3½ m. S.W. of Lauder.

Threestone Hill, mt., border of Banffshire and Aberdeenshire, 12 m. S. of Dufftown ; alt. 2065 ft.

Threipmuir Reservoir, Pentland Hills, Midlothian, 2 m. S. of Balerno, Edinburgh water supply, depth 17 ft. ; alt. 831 ft.

Throsk, ry. sta., L.M.S., Stirlingshire, 2¼ m. from Alloa.

Thrumster, ham. and ry. sta., L.M.S. Caithness, 4½ m. S.W. of Wick. P.O. T. House, a seat.

Thulachan Loch, S. Caithness, 8 m. N.W. of Dunbeath.

Thundergay, pl., N.W. Arran, 3 m. S.W. of L. Ranza.

Thura, pl., Caithness, 5 m. S.E. of Castletown.

Thurdistoft, seat, N. Caithness, 6 m. E. of Thurso.

Thurso, par., police bur., and seapt. on bay of same name, at mouth of r. Thurso, and 20¾ m. N.W. of Wick by rail; par. 21,182 ac. Pop. 4095; bur. pop. 2946. Ry. sta., L.M.S. Hotel. P.O., T.O. Flagstones for paving exported; at one time an important industry but now practically ceased owing to the introduction of concrete pavements. T. Castle is the seat of the Sinclairs of Ulbster. Fixed red light, visible 5 m., on breakwater. Steamer communication with Aberdeen, Kirkwall, Lerwick, etc.

Thurso, r., Caithness, rises on Morven mts., flows 27 m. N.E. to Thurso Bay.

Thurston, seat, E. Lothian, 5 m. S.E. of Dunbar.

Tibbermore, par., vil., and ry. sta. (Tibbermuir Crossing), L.M.S., Perthshire; par. 7131 ac. Pop. 3588. Vil. 5 m. W. of Perth. P.O.

Tibbers, vestige of an ancient castle, Dumfriesshire, 2¼ m. N.N.W. of Thornhill.

Tibbie Shiels, a famous hostelry, Selkirkshire, between St. Mary's Loch and the Loch of the Lowes, 19 m. S.W. of Selkirk. On the hillside above the isthmus dividing the two lochs is a monument to James Hogg, the Ettrick shepherd.

Tiel Burn, S.E. Fife, flows 6 m. E. to the Firth of Forth at Kirkcaldy.

Tig, glen and stream, S.W. Ayrshire, flowing W. to r. Stinchar.

Tigerton, ham., Angus, 5 m. N.W. of Brechin.

Tigh, islet, E. Skye, between S. Rona and Raasay. Pop. 10.

Tigharry, ham., N. Uist Isl., Outer Hebrides, 14 m. W. of Lochmaddy. P.O., T.O.

Tigh Mor, mt., Inverness-shire, 4 m. N. of Loch Clunie; alt. 3222 ft.

Tighnabruaich, q.s. par. and vil., on Kyles of Bute, Argyll, 9½ m. N.W. of Rothesay by water. P.O., T.O. Vil.

pop. 935. A favourite seaside retreat with a 9-hole golf course. Has a steamboat pier.

Tighnalinn (Georgetown), ham., Perthshire, at head of L. Rannoch.

Tillanburn, ham., Lanarkshire, on Tillan Burn, 1 m. S.E. of Newhouse ry. sta.

Tillery House, Aberdeenshire, 1½ m. S.E. of Udny ry. sta.

Tillicoultry, town, police bur., and par., with ry. sta., L.N.E., Clackmannanshire; par. 6946 ac. Pop. 4461. Town stands on the r. Devon, 3 m. S.W. of Dollar and 3½ m. N.E. of Alloa. Pop. 2953. P.O., T.O. Has woollen manufactures.

Tilliery Hill, Kinross-shire, a summit of the Ochils, 2 m. N.W. of Milnathort; alt. 1787 ft.

Tillietudlem, pl., with ry. sta., L.M.S., Lanarkshire, near Craignethan Castle, the prototype of T. in Sir Walter Scott's *Old Mortality,* 9 m. S.E. of Hamilton.

Tillyangus, Clatt par., Aberdeenshire, scene of skirmish in 1572 between the clans of Forbes and Gordon.

Tillycairn Castle, ruin, Cluny par., Aberdeenshire.

Tillyduke, pl., W. Aberdeenshire, 2¾ m. S.E. of Strathdon.

Tillyfoure, estate, Aberdeenshire, 4 m. N.E. of Whitehouse ry. sta.

Tillyfourie, ry. sta., L.N.E., Aberdeenshire, 5½ m. S.E. of Alford. P.O.; T.O. at sta.

Tillylair Hill, N.W. Kincardineshire, 4 m. W.S.W. of Banchory; alt. 815 ft.

Tillylodge, pl., Aberdeenshire, 4 m. Tarland. P.O.

Tillymorgan, pl., Aberdeenshire, 10 m. S.E. of Huntly. T. Hill, 1249 ft.

Tillynaught, ry. sta., L.N.E., Banffshire, 6 m. S.W. of Banff. T.O.

Tillyochie, ham., 2½ m. W. of Kinross.

Tillypronie, seat, Aberdeenshire, 10 m. N.W. of Aboyne.

Tillywhallie, estate, Kinross-shire, 1 m. W. of Milnathort.

Tilquhilly Castle, ruins, N. Kincardineshire, 1½ m. S.E. of Banchory.

Tilt, r., N. Perthshire, issues from Loch Tilt (alt. 1653 ft.), and flows 15 m. S.W. through Glen T. to r. Garry at Bridge of Tilt.

Tilt, Bridge of, part of Blair-Atholl, 35¼ m. N.W. of Perth.

Tima Water, Selkirkshire, flows 6½ m. N. to Ettrick Water, 4½ m. S.E. of St. Mary's Loch.

Timsgarry, ham., Isle of Lewis, 2½ m. W. of Miavaig, on W. Loch Tarbert. P.O., T.O.

Tingwall, par., Shetland; 29,217 ac. Pop. 1531. P.O., T.O. Church 4 m. N.W. of Lerwick. T. Loch, depth 60 ft.

Tinnis Hill, border of Roxburgh and Dumfries, 3 m. S.W. of Newcastleton; alt. 1326 ft.

Tinto Hill, in co. and 8 m. S.E. of Lanark; alt. 2335 ft. The immediate peaks are known as Tinto Hills.

Tinwald, par., Dumfriesshire; 10,312 ac. Pop. 671. Vil. 4 m. N.E. of Dumfries. T. House, 1½ m. E. of Lochartriggs.

Tipperty, pl., Kincardineshire, 9½ m. W.S.W. of Stonehaven.

Tirandrish, seat, Inverness-shire, Kilmonivaig par., 1 m. E. of Spean Bridge.

Tirga More, mt., Harris, Outer Hebrides; alt. 2227 ft.

Tiroran, seat and vil., Mull, Argyll, on the N. side and 3 m. from the head of Loch Scridain. P.O.

Tirry, r., Sutherland, flows 14 m. S.E. to Loch Shin, 2 m. N.W. of Lairg.

Tirvine, seat, Argyll, 2 m. S.W. of Loch Awe ry. sta.

Tister House, seat, Caithness, 12½ m. N.W. of Wick and 2 m. N.W. of Bower sta.

Titwood, q.s. par., Glasgow, Govan par.

Tiumpan Head, Lewis Isl., Outer Hebrides, 9½ m. N.E. of Stornoway. Lighthouse with a half-minute group-flashing light, visible 19 m.

Toab, vil., Orkney, 5 m. S.E. of Kirkwall. P.O., T.O.

Tobermory, spt., police bur., and q.s. par., Mull Isl., on Sound of Mull, 28 m. N.W. of Oban. Bur. pop. 771. P.O., T.O. Has regular steamer communication with Glasgow, Oban, Portree, etc. On pier a fixed light, white with red sectors, visible 5 m. Has a distillery. Is a Marconi telegraph sta.

Toberonochy, vil., Argyll, on E. side of Luing Isl., 7½ m. S. of Easdale. P.O., T.O. Pop. 100.

Tochieneal, ry. sta., L.N.E., Banffshire, 2 m. S.E. of Cullen. T. also a seat.

Todden Hill, N. border of Kirkcudbrightshire, 4 m. S.E. of Dalmellington; alt. 1565 ft.

Todhead Point, prom., Kincardineshire, 5¾ m. S. of Stonehaven. Lighthouse with a half-minute group-flashing light, visible 17 m.; also an automatic siren.

Todhills, ham., Angus, 6 m. N. of Dundee.

Toe Head, S.W. Harris, Outer Hebrides, 9½ m. N.W. of Renish Point.

Toftcombs, seat, Lanarkshire, 1 m. N.E. of Biggar.

Toftingall, Loch, Caithness, 5½ m. S.E. of Halkirk sta.

Toll-a'-Mhuic, loch, Ross and Cromarty, 11 m. N.W. of Garve sta.

Toll-an-Lochain, loch, Ross and Cromarty, 7½ m. S.W. of Ullapool.

Toll Creagach, mt., W. Inverness-shire, 2 m. S. of Loch Mullardoch; alt. 3452 ft.

Tollcross, dist., with ry. sta., L.M.S., in city of Glasgow, also q.s. par. P.O., T.O. Has a large public park.

Tollie Loch, Ross and Cromarty, 3 m. N.E. of Gairloch; alt. 388 ft., depth 86 ft.

Tollohill Lodge, Kincardineshire, 2½ m. S.S.W. of Aberdeen.

Tolmount, mt., borders of Angus and Aberdeenshire, 8 m. S.E. of Braemar; alt. 3143 ft.

Tolmuick Deer Forest, Ross and Cromarty. Post town, Garve.

Tolquhon Castle, ruin, sixteenth-century castle, Aberdeenshire, 2½ m. S.E. of Tarves.

Tolsta, vil. and headland, Lewis Isl., Ross and Cromarty, 13 m. N.E. of Stornoway. P.O., T.O. called North T.

Tolsta-Chalois, pl., Lewis Isl. See Tolstachoellsh. P.O., T.O.

Tolstachoellsh, pl., Lewis Isl., Ross and Cromarty, 5½ m. N.W. of Callanish. P.O. and T.O. called Tolsta-Chaolais.

Tom-a-Bhuachaille, mt., Argyll, 5½ m. N.W. of Inveraray; alt. 1414 ft.

Tomacharrich, pl., Inverness-shire, 5½ m. S.W. of Spean Bridge.

Tomachastle, hill, 3 m. W. of Crieff. On summit a monument to General Sir David Baird (1757–1829); alt. 434 ft.

Tom-a-Chòinich, mt., borders of Ross and Cromarty and Inverness-shire, 2 m. S. of Loch Mullardoch; alt. 3646 ft.

Tomatin, pl., with ry. sta., L.M.S. (alt. 1029 ft.), Inverness-shire on r. Findhorn, 19 m. S.E. of Inverness. P.O., T.O. Hotel. Has a distillery.

Tombae, pl., Banffshire, 4½ m. S.E. of Glenlivet.

Tom Breac, mt., Aberdeenshire, 3½ m. N.W. of Balmoral ; alt. 2276 ft.

Tom Buidhe, mt., N.W. Angus, 8 m. N.W. Milton of Clova ; alt. 3140 ft.

Tomdoun, pl., Inverness-shire, 15 m. S.W. of Fort Augustus. P.O., T.O.

Tomich, pl., Inverness-shire, 22 m. S.W. of Beauly. P.O., T.O.

Tomintoul, q.s. par. and vil., Banffshire. Vil., 14½ m. S. of Ballindalloch ry. sta. P.O., T.O. The second highest vil. in Scotland ; alt. 1160 ft. Hotel. Pop. 457. Scottish Youth Hostel here. Motorbus communication with Ballindalloch.

Tomluncart, pl., N.E. Inverness-shire, 3½ m. S.W. of Nairn.

Tomnacross, ham., Inverness-shire, 3½ m. S. of Beauly.

Tomnagrew, seat, Perthshire, 6 m. S.W. of Dunkeld.

Tomnahurich, isolated eminence, 1 m. S.W. of Inverness. Has a beautifully laid-out cemetery.

Tomnavonlin, pl., Banffshire, 8 m. S.E. of Ballindalloch ry. sta. P.O., T.O.

Tomperran, seat, Perthshire, ½ m. W. of Comrie.

Tong, vil., Lewis Isl., Outer Isles., 4 m. N.E. of Stornoway. P.O., T.O.

Tongland, par. and ham., 2¼ m. N.E. of Kirkcudbrightshire ; 9660 ac. Pop. 669. P.O. spelt Tongueland. Here one of the hydro-electric stas. of the Galloway Water Power Co.

Tongue, par. and vil., on N. coast of Sutherland ; par. 81,048 ac. Pop. 1184. Vil., T. or Kirkiboll, on E. side of Kyle of Tongue, 38½ m. N. of Lairg ry. sta. and 44 m. W. of Thurso. P.O., T.O. Hotel. T. House, 1 m. below the vil. Ferry service with Auchenooldroch across Kyle of Tongue.

Tongue, Kyle of, sea loch, N. coast of Sutherland, 9 m. long and 5 m. broad at entrance.

Tongueland, N.E. Kirkcudbrightshire. P.O. See TONGLAND.

Tonley House, seat, Aberdeenshire, 2 m. S.E. of Alford.

Tor Achilty, wooded hill, Ross and Cromarty, 1 m. S. of Contin ; alt. 650 ft.

Tor Aluinn, seat, Argyll, 2 m. S. of Dunoon.

Tor Alvie, conspicuous crag, Inverness-shire, E. of Loch Alvie. Upon it are two monuments—a lofty pillar in memory of the last of the old Dukes of Gordon, and a cairn in memory of Highland soldiers who fell at Waterloo. alt. 1175 ft.

Toravon, seat, Stirlingshire, Muiravonside par.

Torbanehill House, West Lothian, 1 m. N.N.E. of Whitburn.

Torboll Burn, stream, S.E. Sutherland, flowing 9 m. through Loch Buie to Letty R.

Torbraehead, mt., Dumfriesshire, 3½ m. N. of Moniaive ; alt. 1311 ft.

Torbreck, pl., Sutherland, 9½ m. N.W. of Golspie.

Tor Castle, seat, Inverness-shire, 2 m. N.E. of Banavie. Fragment of ancient castle of the Mackintoshes.

Tordarroch Bridge, pl., 7 m. S. of Inverness.

Torduff Point, headland, S. Dumfriesshire, 5 m. S.E. of Annan.

Torduff Reservoir, Pentland Hills, Midlothian, 1 m. S. of Colinton. Edinburgh water supply.

Tore, pl., S.E., Ross and Cromarty, 4 m. W. of Munlochy. P.O., T.O.

Torgyle, pl., Glen Moriston, Inverness-shire, 7 m. N.W. of Fort Augustus.

Torhousemuir, seat, 3 m. N.W. of Wigtown

Torloisk, ham., and seat, N.W. Mull, Argyll, 12 m. S.S.W. of Tobermory. P.O.

Torlum Hill, conical wooded hill, Perthshire, 3 m. S.W. of Crieff ; alt. 1291 ft.

Tormaid, loch, N.W. Caithness, 4 m. S.E. of Reay.

Tormaukin, seat, Perthshire, 3 m. N.N.W. of Rumbling Bridge.

Tormore, seat, Isle of Skye, 2 m. S.W. of Armadale.

Tormsdale, pl., Caithness, 6½ m. S. of Halkirk.

Tormusk, mt., Arran, Buteshire, 3½ m. N.E. of Blackwaterfoot ; alt. 1259 ft.

Tor-Na-Dee Sanatorium, on r. Dee, 5 m. S.W. of Aberdeen.

Tornaveen, seat and ham., Aberdeenshire, 3 m. N.W. of Torphins sta. P.O.

Torness, pl., Inverness-shire, 14 m. S.W. of Inverness. P.O., T.O.

Tor Ness, headland, S. of Hoy, Orkney Islands.

Torosay, par., Mull Isl., Argyll ; 89,584 ac. Pop. 610. Also q.s. par.

Torphichen, par. and vil., West Lothian ; par. 9937 ac. Pop. 4175. Vil. 2½ m. N.W. of Bathgate. P.O., T.O. Vil. pop. 525. Was the Scottish seat of the Knights of St. John of Jerusalem ; retains an interesting portion of their preceptory church.

Torphin Hill, N. slope Pentland Hills, Midlothian, ½ m. S. of Juniper Green. Golf course.

Torphins, q.s. par. and vil., with ry. sta., L.N.E., Aberdeenshire, 7 m. N.W. of Banchory. P.O., T.O. Vil. pop. 473. Hotel.

Torquhan, seat, S.E. Midlothian, 2 m. N.N.W. of Stow.

Torran, pl., Raasay Isl., Inner Hebrides. P.O.

Torrance, vil., with ry. sta., L.N.E., Stirlingshire, 3 m. W. of Kirkintilloch. P.O. Pop. (Balmore and T.) 1101.

Torrance Castle, seat, Lanarkshire, 2 m. S.E. of East Kilbride.

Torran Rocks, off S.W. coast of Mull, Argyll.

Torraston, pl., N. Coll Isl., Argyll.

Torr House, Kirkcudbrightshire, 1 m. N. of Auchencairn.

Torridon, ham., Ross and Cromarty, at head of Loch T., 10 m. S.W. of Kinlochewe. P.O., T.O. T. House, seat, 2 m. to the N.W.

Torridon Deer Forest, Ross and Cromarty ; 11,630 ac. P.O., Torridon.

Torridon, Loch, sea loch, Ross and Cromarty, opens opposite the N.E. of Skye ; penetrates 13 m. E.S.E., but forms three successive reaches separated by narrow straits.

Torrie House, seat, Fife, near Torryburn.

Torrin, ham, Isle of Skye, on E. side of upper part of Loch Slapin, 5 m. S.W. of Broadford. P.O., T.O.

Torrinch, islet, S. end of Loch Lomond, Dunbartonshire.

Torrisdale Castle, seat, Argyll, E. Kintyre, 1¼ m. S.S.W. of Carradale.

Torrish, seat, E. Sutherland, 4 m. N.W. of Helmsdale.

Torr-na-Eighin, loch, W.Sutherland, 1½ m. S. of Drumbeg.

Torroboll, pl., Sutherland, 1½ m. S. of Lairg.

Torry, fishing vil., Kincardineshire, near mouth of r. Dee, and connected with the city of Aberdeen by the Victoria Bridge. The vil. is included within the boundaries of that town. P.O., T.O.

Torryburn, par., vil., and ry. sta., L.N.E., Fife ; par. 3423 ac. Pop. 1910. Vil. on the Firth of Forth, 4½ m. W.S.W. of Dunfermline. P.O. Pop. (T. and Valleyfield) 2932.

Torsay, one of the Slate Islands, Argyll, separated by narrow straits from Seil and Luing. Pop. 6.

Torsonce House, seat, Midlothian, 5 m. N.W. of Galashiels.

Tors Point, headland, Auchenairn Bay, Kirkcudbrightshire.

Torthorwald, par. and vil., Dumfriesshire ; par. 6827 ac. Pop. 718. Vil. 4 m. N.E. of Dumfries. P.O. T. Castle, a ruin.

Torwood, ham., Perthshire, near Dunkeld.

Torwoodlee, seat, Selkirkshire, 2 m. N.W. of Galashiels. Also ruins of large ancient fortalice.

Toryglen, golf course, in S.E. of Glasgow, 1 m. W.N.W. of Rutherglen.

Toscaig, pl., stream, and loch, W. Ross and Cromarty, 3½ m. S. of Applecross.

Totaig, pl., with pier, on Loch Alsh, W. Ross and Cromarty, 7½ m. S.E. of Kyle of Lochalsh. Ferry for Dornie.

Tote, seat, Isle of Skye, 5½ m. N.W. of Portree.

Touchadam, shooting, 5 m. S.W. of Stirling.

Touch Burn, tributary of r. Forth, 2 m. W. of Stirling. At its source T. Moor, with a reservoir.

Touch House, seat, 3 m. S.W. of Stirling.

Touchie Mill, pl., 3¼ m. W.N.W. of Kinross.

Tough, par., Aberdeenshire, 7109 ac. Pop. 487. Church, 1½ m. S.W. of Whitehouse sta. P.O.

Tour, seat, Ayrshire, 2 m. N.W. of Kilmarnock.

Tournaig House, Ross and Cromarty, 2 m. N.E. of Poolewe.

Toward, ham., S. Cowal, near Castle Toward, 2¾ m. N.E. of Rothesay. P.O., T.O.

Toward Point, headland, S.E. extremity of Cowal, Argyll, 6¼ m. S.S.W. of Dunoon. P.O., T.O. Steamboat pier. Lighthouse, with a ten-second flashing light, visible 14 m.

Towerhill House, Ayrshire, 2½ m. N.W. of Kilmarnock.

Tower Lindsay, ruined castle, Lanarkshire, on r. Clyde, opposite Crawford vil.

Towford, pl., Roxburghshire, 3 m. S. of Hounam. Remains of large Roman camp.

Towie, par., Aberdeenshire, on r. Don; 16,736 ac. Pop. 569. Church 8½ m. W.S.W. of Alford. T. Castle, a ruin, anciently the fortified seat of a branch of the Forbes.

Towie Barclay, ancient castle, Aberdeenshire, 4½ m. S.E. of Turriff.

Towie Hill, Banffshire, 5 m. S.W. of Keith; alt. 1108 ft.

Towiemore-Glenlivet Distillery, Banffshire, 5 m. S.W. of Keith.

Townend, a N.E. sub. of Dumbarton.

Townend House, seat, Ayrshire, 4 m. E. of Troon.

Townhead, pl., Ayrshire, 4½ m. S.W. of Maybole.

Townhead, q.s. par., Glasgow.

Townhead, vil., Dumfriesshire, 3 m. N.W. of Penpont.

Townhead, pl., Kirkcudbrightshire, 2 m. N.W. of Castle-Douglas.

Townhead, seat, Dumfriesshire, 1½ m. E. of Moniaive.

Townhead, pl. and reservoir, Stirlingshire, 1 m. E. of Kilsyth.

Townhill, mining vil., Fife, in bur., and 1½ m. N.E. of Dunfermline. P.O., T.O.

Toxside, seat, Midlothian, 7 m. S.W. of Gorebridge.

Trabboch, vil., with ry. sta., L.M.S., 8 m. W. of Ayr.

Tradespark, vil., 1 m. W. of Nairn. Pop. (T. and Moss-side) 365.

Tradeston, parl. division, Glasgow. P.O.

Trahenna Hill, Peeblesshire, 1¾ m. N.E. of Broughton; alt. 1792 ft.

Traigh Bhagh, bay, Tiree Isl., Argyll, 2 m. N. Hynish.

Traigh House, seat, Inverness-shire, 3 m. N. of Arisaig.

Trailflat, ham., Dumfriesshire, 2½ m. N.E. of Lochmaben.

Trailtrow, ham., S.E. Dumfriesshire, 4½ m. N.W. of Annan.

Tralaig, loch, Argyll, 3 m. N.E. of Kilmelford; depth 117 ft., alt. 420 ft.

Tranent, par. and police bur., East Lothian; par. 5918 ac. Pop. 9002. Town 1¼ m. S.E. of Prestonpans and

9½ m. E. of Edinburgh. Pop. 4526. P.O., T.O. Centre of large mining area. Motor bus service with Edinburgh.

Transy, dist., Dunfermline, Fife. P.O.

Traprain Law, conical hill, East Lothian, 4½ m. E. of Haddington; alt. 724 ft. A pit, 2 ft. deep and 2 ft. wide, was discovered here in 1919 filled with a rich collection of fourth-century silver plate.

Traquair, par. and vil., Peeblesshire; par. 18,489 ac. Pop. 551. P.O., T.O. Vil. 1½ m. S. of Innerleithen, with which it is connected by a bridge. T. House and T. Riggs are seats.

Traside, seat, Midlothian, 8 m. S.W. of Gorebridge.

Trearne, seat, N. Ayrshire, 1½ m. E. of Beith.

Treaslane, pl. and stream, Isle of Skye, 8 m. N.W. of Portree.

Tredwall, loch, Papa Westray Isl., Orkney.

Treesbank, seat, Ayrshire, 2¼ m. S.W. of Kilmarnock.

Treig, Loch, Inverness-shire. Lying 784 ft. above sea-level, it extends 5 m. N.E.; sends off the r. Treig, which flows 2½ m. N.E. to the Spean at a point 6 m. E. of the Bridge of Roy. Maximum depth 436 ft., alt. 787 ft. It is now used as the chief reservoir for the Fort William Aluminium Works to which it is connected by a tunnel through Ben Nevis.

Treshnish Isles, a group of Hebridean islets, 5 m. N.N.W. of Staffa. Consist of five principal isles and some intervening rocks.

Treshnish Point, headland, N.W. Mull Isl., Argyll, 3 m. S. of Caliach Point.

Treslaig, pl., near Fort William, Inverness-shire. P.O.

Tressady, seat, 2 m. W. of Rogart, S. Sutherland.

Tresta, ham., Shetland, on T. Voe, 11 m. N.W. of Lerwick. P.O., T.O.

Tresta, ham., Shetland, Fetlar Isl., on Wick of T.

Trinafour, pl., Perthshire, 6¼ m. S.W. of Struan ry. sta.

Trinity, sub. of Edinburgh, between Newhaven and Granton.

Trinity, ham., Angus, 1¼ m. N. of Brechin. P.O.

Trinity, q.s. par., Aberdeen.

Trinity College, see GLENALMOND.

Trinity Gask, par. and vil., Perthshire ; 5691 ac. Pop. 354. Vil. 4 m. N. of Auchterarder.

Trochraigue, seat, Ayrshire, Girvan par.

Trochry, vil., Perthshire, 4 m. S.W. of Dunkeld. P.O., T.O.

Trodday, isl., off N. extremity of Isle of Skye. Lighthouse with a ten-seconds' group-flashing light, visible 18 m.

Trodigal, near Machrihanish, Kintyre, Argyll, was a ry. halt, now closed and buses used.

Trohoughton, ham., 3 m. S.E. of Dumfries.

Troisgeach, mt., W. Perthshire, 3 m. N.W. of Ardlui ; alt. 2395 ft.

Tromie, stream, Inverness-shire, flows 12 m. N. through Glen T. to r. Spey, 1½ m. N.E. of Kingussie.

Tromlie, loch, W. Argyll, 2 m. N. of Taychreggan.

Tronach Castle, ruin, Banffshire, near Portknockie.

Trondra, isl., Tingwall par., Shetland. Pop. 91.

Trool, Loch, W. Kirkcudbrightshire, 14 m. N. of Newton-Stewart ; sends off the Water of T. which flows to the Water of Minnoch ; depth 55 ft., alt. 246 ft.

Troon, seapt. town (police bur.), q.s. par., with ry. sta., L.M.S., Ayrshire, 6⅓ m. N. of Ayr and 35 m. S.W. of Glasgow. Bur. pop. 8544. P.O., T.O. Hotels. A popular holiday resort, with five first-class golf courses, fine sands, and good bathing. There is also an excellent open-air swimming-pool. W. pier shows a fixed light, visible 6 m.

Troqueer, par., Kirkcudbrightshire and part in Dumfriesshire, on r. Nith ; 12,125 ac. Pop. 7594. Contains town of Maxwelltown.

Troquhain, seat, N. Kirkcudbrightshire, 3 m. E.N.E. of New Galloway.

Trossachs, romantic wooded defile, S.W. Perthshire, extending 1 m. W. from the head of Loch Achray to the foot of Loch Katrine and forming a portal of the Western Highlands. From T. Pier on Loch Katrine a steamer plys twice daily in summer to Stronachlachar. T. Hotel, 8 m. W. of Callander, near the N. shore of Loch Achray. A little to the S.E. is the church of the q.s. par. of the Trossachs. P.O., T.O. A new road has recently been made between Aberfoyle and the T.

Trostan, loch, W. Kirkcudbrightshire, 9 m. W. of Dalry.

Trotternish, dist. comprehending all Portree, Kilmuir, and Snizort pars. and small parts of Bracadale and Strath in Isle of Skye. Most of it is peninsular N.N.W. from the isthmus between head of Loch Portree and Loch Snizort.

Troup House, seat, Banffshire, 3 m. N.E. of Gardenstown, near T. Head (300 ft.).

Truderscaig, loch, Sutherland, 9 m. W. of Kinbrace ry. sta. ; depth 12 ft., alt. 426 ft.

Truffhill, Ayrshire, part of Waterside vil., 11½ m. S.E. of Ayr.

Truim, stream, Inverness-shire, flows 16 m. N.E. through Glen T. to the r. Spey at Glentruim House, 6 m. S.W. of Kingussie. The Falls of T., 2½ m. S. of the confluence.

Trumisgarry, q.s. par., North Uist, Outer Hebrides.

Trumland House, seat, S. of Rousay, Orkney.

Trumpan, pl., N.W. Skye Isl., Inverness-shire, 3½ m. S. Vaternish Point.

Trusta, Hill of, Kincardineshire, 5⅓ m. W. of Stonehaven ; alt. 1051 ft.

Tuath, Loch, sea loch, Argyll, between Mull and Ulva Isl.

Tudhope Hill, Dumfriesshire, 2 m. W. of Mosspaul Hotel ; alt. 1961 ft.

Tugnet, pl., Morayshire, 1 m. N.E. of Garmouth.

Tuiteam-Tarbhach, Sutherland. Scene of a clan fight between the Macleods and Mackays in 1397. On the Oykell r. and 9 m. W.N.W. of Invershin.

Tulach Hill, Perthshire, 1 m. S.W. of Blair-Atholl ; alt. 1541 ft.

Tulchan Lodge, seat of the Earl of Airlie, Perthshire, Glenisla, Alyth.

Tulchan Lodge, shooting-box, Moray-shire, 2¼ m. S.W. of Advie ry. sta.

Tulla, stream and loch, Argyll. Stream flows 11 m. S.W. to Loch Tulla, 7 m. N.W. of Tyndrum, depth 84 ft. ; alt. 542 ft.

Tulliallan, par., W. Fife, on r. Forth ; 3685 ac. Pop. 2166. T. Castle, 4⅓ m. S.E. of Alloa. Lighting area pop. 2012. State forest here.

Tullibardine, vil., with ry. sta., L.M.S., Perthshire, 6 m. S.E. of Crieff. Ruins of ancient castle.

Tullibody, vil., Clackmannanshire, 2½ m. N.W. of Alloa. P.O. Pop. 824. T. House, a seat, 1¼ m. W. of Alloa.

Tullich, ham., Aberdeenshire, 1½ m. N.E. of Ballater.

Tullich, pl., Ross and Cromarty, 3½ m. N.N.E. of Invergordon.

Tullichewan Castle, seat, Dunbartonshire, 1 m. N.W. of Alexandria.

Tullich Hill, mt., N.W. Dunbartonshire, 2½ m. S. of Arrochar ; alt. 2075 ft.

Tulloch, pl., with ry. sta. (for Loch Laggan), L.N.E., Inverness-shire, 18 m. N.E. of Fort William. P.O., T.O. at sta.

Tulloch, vil., 2 m. N.W. of Perth.

Tulloch Castle, Ross and Cromarty, 1 m. N. of Dingwall.

Tullochgorum, Inverness-shire, on r. Spey, 6 m. S.W. of Grantown.

Tulloch House, Aberdeenshire, 2 m. N. of Old Meldrum.

Tullochvenus, seat, Aberdeenshire, 6½ m. N.N.E. of Aboyne.

Tulloch, Wester and **Easter,** hams., N.E. Inverness-shire, 2 m. and 3 m. S.S.E. of Boat of Garten.

Tullybeagles, seat, 12 m. N.W. of Perth.

Tullybelton, seat and loch, 3 m. S.W. of Bankfoot, Perthshire.

Tullymet, seat and ham., Perthshire, 2¼ m. N.E. of Ballinluig Junction. P.O.

Tullynessie, vil., Aberdeenshire, 3 m. N.W. of Alford.

Tullynessie and Forbes, par., Aberdeenshire, on r. Don ; 11,283 ac. Pop. 727.

Tummel, loch and r., N. Perthshire. The loch, formed by an expansion of the river, commences 9 m. W.N.W. of Pitlochry, is about 3 m. long, with a breadth of ½ m. and has a depth of 128 ft. ; alt. 454 ft. The r. Tummel, issuing from the foot of Loch Rannoch, flows 29 m. E. and S.E. till, near Ballinluig sta., it forms a confluence with the Tay. Falls of T., 3 m. N.W. of Pitlochry.

Tummel Bridge, ham., Perthshire, on r. Tummel, 7 m. E. of Kinloch Rannoch. Hotel. P.O., T.O.

Tuna, loch, W. Argyll, 6 m. S.E. of Clachan, W. Kintyre.

Tundergarth, par., Dumfriesshire ; 10,486 ac. Pop. 321. Church 3 m. S.E. of Lockerbie.

Tungadal, glen and stream, Isle of Skye, Inverness-shire, 4½ m. S.W. of Portree.

Tuquoy, bay, Westray, Orkney Isl.

Turdalewater, loch, Shetland, 5 m. N.E. of Walls.

Turfholm, ham., Lanarkshire, adjoining Lesmahagow.

Turin Hill, Angus, 4½ m. N.E. of Forfar ; alt. 814 ft.

Turin House, Angus, 2¾ m. N.E. of Letham and 5½ m. N.E. of Forfar.

Turk, r., Perthshire, flows 6½ m. S.E. through Glen Finglass to Brig o' Turk to join the Black Water.

Turnberry, pl., with large hotel and splendid seaside, golf links, with ry. sta., L.M.S., on T. Bay, ·6 m. N. of Girvan, 56 m. S.W. of Glasgow by rail. P.O., T.O. T. Castle said to have been the birthplace of Robert Bruce, who in 1307, mistaking a kiln fire on the neighbouring " Bogle's Brae " for a signal, prematurely came over from Arran to begin his struggle for Scottish independence.

Turner Hall, seat, Aberdeenshire, 3 m. N. of Ellon.

Turnhouse, pl., with golf course and ry. sta., L.N.E., in city and 6½ m. W. of Edinburgh.

Turnlaw Reservoir, Lanarkshire, 1 m. S. of Cambuslang.

Turret Water, Perthshire, flows 8 m. S.E. through Glen T. and Loch T. to r. Earn, near Crieff. Falls of Turret, 3 m. N.W. of Crieff.

Turriff, pol. bur. and par., with ry. sta., L.N.E., Aberdeenshire, 38½ m. N.N.W. of Aberdeen, by rail ; par. 18,397 ac. Pop. 3944. Bur. pop. 2298. P.O., T.O. Prosperous agricultural centre.

Tushielaw, seat and pl. and hotel, Selkirkshire, on Ettrick Water, 15 m. S.W. of Selkirk. Ruined tower, was the stronghold of Adam Scott, " King of Thieves," beheaded by James V. in 1529.

Twatt, pl. and loch, Orkney, Mainland, 11 m. N. of Stromness. P.O.

Twatt, ham., Shetland, 12 m. N.W. of Lerwick.

Twechar, q.s. par. and vil., with ry. sta., L.N.E., Dunbartonshire, on Forth and Clyde Canal, 2 m. S.W. of Kilsyth. P.O.

Tweed, r., chiefly in Scotland, partly on boundary with England, and briefly in England. Rises in the centre of the S. Highlands, at Tweed's Well, near the source of the Clyde and Annan, and flows N.E. through Peeblesshire, E. through Selkirkshire and Roxburghshire, and N.E. to Berwick-on-Tweed. Length 97 m. Its basin comprises about 1870 sq. m. ; and both its own waters and those of its chief tributaries, the Ettrick (which receives the Yarrow), the Teviot, and the Till on the right bank and the Lyne, Biggar, Eddlestone, Gala, Leader and Eden Waters, and the Whitadder on the left bank, afford famous fishing.

Tweeddale, old name of Peeblesshire. Gives the title of Marquess to the family of Hay.

Tweeden Burn, S.E. Roxburghshire, tributary of Liddell water.

Tweedhill, seat, Berwickshire, 4½ m. W.S.W. of Berwick-on-Tweed.

Tweedsmuir, par., Peeblesshire, on r. Tweed ; 32,278 ac. Pop, 177. P.O., T.O. Church 1½ m. S.W. of Croak Inn, and 8 m. S. of Broughton ry. sta.

Twinlaw Cairns, hill, Berwickshire, 3 m. N. of Westruther ; alt. 1466 ft.

Twin Laws, 2 summits, S.W. Pentland Hills, Lanarkshire, 2 m. S.E. of Tarbrax; alts., 1124 and 1026 ft.

Twynholm, par. and vil., Kirkcudbrightshire ; par. 10,528 ac. Pop. 720. Vil. 3 m. N.W. of Kirkcudbright. P.O., T.O. Pop. 240.

Tyndrum, vil. and ry. stas., L.N.E. and L.M.S., Perthshire, 34¼ m. N.W. of Callander, 36½ m. E. of Oban. P.O., T.O. Hotel.

Tyne, r. rises in Midlothian and flows 28 m. N.E. chiefly through E. Lothian to the North Sea, 2½ m. N.W. of Dunbar.

Tynecastle, q.s. par. and dist. in W. city of Edinburgh. Heart of Midlothian football ground.

Tynehead, ry. sta., L.N.E., Midlothian, 16 m. by ry. S.E. of Edinburgh. P.O.

Tyneholm, estate, E. Lothian, ½ m. E. of Pencaitland.

Tynninghame, vil., E. Lothian, on r. Tyne, 2 m. N.E. of E. Linton. P.O. spelt Tyninghame. Vil. pop. 151. T. House seat of the Earl of Haddington, ¾ m. N.E.

Tynron, par. and ham., Dumfriesshire, on Shinnel Water ; par. 15,708 ac. Pop. 294. Ham., 5 m. S.W. of Thornhill. P.O., T.O. T. Doon, 945 ft.

Tyree, or **Tiree,** an isl. and par., Inner Hebrides, Argyll, 19 m. N.W. of Iona. It is treeless and flat ; contains several fresh-water lochs ; has marble quarries, Scandinavian forts and standing stones. Length 14 m. by ¾ to 6 m. in breadth ; Par. (comprising T. and Skerryvore), 18,896 ac., or 30 sq. m. Pop. 1451. P.O. and T.O. called Scarinish. On Gott Bay are two fixed white lights seen 5 m. Pop of isl., 1448 (excluding Skerryvore).

Tyrie, par., Aberdeenshire ; 11,185 ac. Pop. 2195. P.O., T.O. Church 5 m. S.W. of Fraserburgh.

Tyrie, pl., S. Fife, 1 m. S. of Kirkcaldy.

U

Uachdarach, loch. in isl. of Jura, 4½ m. N.E. of Portaskaig (Islay), W. Argyllshire.

Uaine, loch on the r. Turret, which flows to the Earn near Crieff, Perthshire ; alt. 1520 ft., depth 10 ft.

Uamh Bheag, mt., 2 peaks ; alts. 2166 ft. and 2179 ft., 3 m. N.E. of Callander, Perthshire.

Uamh Mhor, or **Uamvar,** mt., 5 m. N.E. of Callander, Perthshire. A cavern here was formerly a haunt of robbers.

Udale Bay, inlet on S. side of Cromarty Firth, 4 m. W. of Cromarty, E. Ross and Cromarty.

Uddingston, q.s. par. and town near the Clyde, and 7 m. S.E. of Glasgow by rail, Lanarkshire ; ry. stas. (Uddingston) L.M.S., and (Uddingston and Uddingston W.) L.N.E. P.O., T.O. Town pop. 8352. Iron foundries and agricultural implement making.

Uddington, vil., 2½ m. N.E. of Douglas, S. Lanarkshire.

Udny, par. and vil., mid Aberdeenshire ; par. 11,242 ac. Pop. 1474. Vil., 14½ m. N.W. of Aberdeen ; ry. sta. (2¼ m. S.E. of the vil.), L.N.E. P.O., and P.O. and T.O. at the sta. U. Castle is a seat.

Udston, vil. and seat, 2 m. S.W. of Hamilton, Lanarkshire.

Ugadale Point, E. side of Kintyre, Kilbrennan Sound, 2 m. S. of Saddell, Argyllshire.

Ugie, r. in N. Aberdeenshire; rises 2 m. S.W. of New Aberdour and flows 21 m. S.E. to North Sea, 1¼ m. N.W. of Peterhead.

Uig, vil., glen, and bay on E. shore of Loch Snizort, N.W. Skye, Inner Hebrides. Vil., 14 m. N.W. of Portree. P.O., T.O. On pier fixed red light, visible 5 m. Steamers from Glasgow call regularly.

Uig, par. and seat, W. of Lewis, Outer Hebrides; par. 129,109 ac. Pop. 3363. Contains isls. of Bernera and Flannen; on shore of Uig Bay, 34 m. W. of Stornoway, stands the par. church.

Uig, glen and small stream descending to Glen Prosen, N.W. Angus.

Uigeadail, loch, 4 m. N.E. of Port Ellen, Islay, Argyllshire.

Uiginish Lodge, seat, 1 m. W. of Dunvegan, Skye, Inner Hebrides. On U. Point is a three-seconds' flashing light, visible 12 m.

Uisg, loch, 1 m. E. of Lochbuie, S. of Mull, W. Argyllshire.

Uisge-Dubh, or Blackwater, headstream of r. Leven which flows into Loch Leven, a branch of Loch Linnhe.

Uisgnaval More, mt.; alt. 2392 ft., Harris, Outer Hebrides.

Uist, North, isl. and par., in Outer Hebrides, S.W. of Harris; par. 75,513 ac. Pop. 2349. P.O., T.O. at Lochmaddy on E. side. Par. includes the isls. of Baleshare and Illeray, Boreray, Grimsay, Heisker, Levara, Kirkibost, Monach, Ronay, and Vallay. The coast is very much indented and the isl. is mainly marsh and moorland. Chief industry is fishing.

Uist, South, isl. and par. in Outer Hebrides and separated from N. Uist by Benbecula; par. 90,093 ac. Pop. 2810. P.O. at Howmore, 36 m. S.W. of Lochmaddy (N. Uist). Par. includes the isls. of Benbecula, Eriskay, Fladda, Grimsay, and Wiay. The northern part of the isl. is flat, but the S. is hilly or mountainous. Chief industries are fishing and the raising of cattle.

Ulaidh, mt., alt. 3258 ft.; 5 m. S.E. of Ballachulish, N. Argyllshire.

Uibster, ham., on the coast of Caithness, 7½ m. S.W. of Wick; ry. sta., L.M.S. P.O.

Ullapool, q.s. par. and vil. on N.E. shore of Loch Broom, N.W. Ross and Cromarty. The vil. has a motor connection with Garve ry. sta., which is 32 m. S.E. P.O., T.O. Coastguard sta. Hotels. Vil. pop. 538. Chief industry, herring fishing. Regular steamer communication with Glasgow, Portree, Stornoway, etc. Ferry service with Aultnaharrie across Loch Broom.

Ullapool Hostel, Ullapool, Ross-shire. A Scottish Youth Hostel.

Ullaval, mt., alt. 2153 ft., Harris, Outer Hebrides.

Ulladale, loch, 12 m. N.W. of Tarbert, Harris, Outer Hebrides.

Ullinish, seat, 2 m. W. of Struan, Loch Bracadale, Skye, Hebrides.

Ulsta, loch and ham. on Ulsta Bay, in S.W. of Yell isl., Shetland. P.O., T.O.

Ulva, small isl. on W. side of Loch Sween, at mouth of Linne Mhuirich, N. Knapdale, Argyllshire. Pop. 9.

Ulva, q.s. par. and isl. off W. coast of Mull, Argyllshire; q.s. par. comprises the isls. of Ulva, Gometra, Little Colonsay, and Staffa, and part of Mull. There is a ferry over Ulva Sound to the Mainland of Mull. P.O., T.O. called Ulva Ferry. Rocks show basaltic colonnades. U. House is situated near U. Ferry. Pop. of isl. 25.

Umma Voe, sea inlet in N. of Mainland, Shetland.

Unapool, ham., at mouth of Glencoal, ¼ m. S.E. of Kylesku Inn, W. Sutherland. The U. Burn enters Loch Glencoal.

Underhough, seat, ½ m. S. of Cloch Point, W. coast of Renfrewshire.

Underwood, seat, 3 m. N.W. of Tarbolton, Ayrshire.

Unich, Water of, stream flowing E. to Water of Lee, near Lochlee, N. Angus.

Union Bridge, suspension bridge across the Tweed between Scotland and England at the vil. of Horncliffe, 31½ m. S.W. of Berwick-on-Tweed. It was built in 1820, and is 449 ft. long, 18 ft. wide, and 27 ft. above the water. Open for vehicular traffic.

Union Canal, canal in the cos. of Midlothian, West Lothian, and Stirlingshire, extending 31½ m. W. from Edinburgh to join the Forth and Clyde Canal at Port Downie, 1½ m. S.W. of Falkirk.

Unst, isl. and par., in the extreme N. of Shetland ; par. 29,874 ac. including the isls. of Muckle Flugga, Unst, and Vyea. Par. pop. 1341. P.O., T.O. at Baltasound, 43 m. N.E. of Lerwick. Fishing is the chief occupation, but, some Shetland hosiery is made. U. Isl. pop. 1247.

Uphall, par. and vil., in W. Lothian ; par., which contains the town of Broxburn, 4529 ac. Pop. 11,119. The vil. stands on the Brox Burn 1½ m. N.W. of Broxburn ; ry. sta., L.N.E., 1 m. S.E. of the vil., is 13¼ m. W.S.W. of Edinburgh. P.O., T.O., and P.O. called Uphall Sta. Pop. 8464. Shale mining and large paraffin works.

Uplawmoor, vil., 3½ m. S.W. of Neilston, Renfrewshire ; ry. sta., L.M.S. P.O., T.O.

Uppat House, seat, 2½ m. S.W. of Brora, Sutherland.

Upper Bighouse, ham., 5 m. S. of Melvich, N.E. Sutherland.

Upper Knockando. P.O. See KNOCKANDO.

Upper Largo, vil., in Fife. P.O., T.O. See LARGO.

Upperton, pl., with Longriggend Fever Hospital, 1 m. N.W. of Longriggend, N.E. Lanarkshire.

Upsettlington, ham. and ancient par., now included in par. of Ladykirk, Berwickshire. Ham. stands 1 m. S.W. of Ladykirk.

Urafirth, ham. and sea inlet in N. of Mainland, Shetland. On Hillswick Ness is a fixed light, visible 15 m.

Uraraidh, loch, 5½ m. N.E. of Port Ellen, Islay, Argyllshire.

Urchany Deer Forest, 4340 acres ; 5 m. W.S.W. of Beauly, Inverness-shire.

Ure, r. in Glen Ure (q.v.), Argyllshire.

Urie, loch, 2¼ m. S.W. of Lamlash, Arran, Buteshire.

Urie, stream, rising 4 m. S.E. of Huntly and flowing 19 m. S.E. to join the Don near Inverurie, Aberdeenshire.

Urie House, 2 m. N.W. of Stonehaven, Kincardineshire. Seat of Lord Stonehaven.

Urie Lingey, small isl. between Fetlar and Unst isls., N. Shetland.

Urigill, Loch, alt. 515 ft. and 40 ft. deep ; on S.E. of Assynt, W. Sutherland.

Urlar Burn, small stream, flowing N.E. to the Tay at Aberfeldy, Perthshire.

Urquhart, par. and vil. in N.E. Morayshire. Par. on Moray Firth ; 12,492 ac. Pop. 1473. Vil. 5 m. E. of Elgin by rail ; ry. sta., L.N.E. P.O., T.O. at station.

Urquhart and Glenmoriston, par. on Loch Ness, Inverness-shire ; 122,708 ac. Pop. 1488. Ruins of Urquhart Castle stand on W. side of Loch Ness, 13 m. S.W. of Inverness.

Urquhart and Logie Water, par. in E. Ross and Cromarty ; 13,603 ac. Pop. 1792. Par. church is 8 m. N. of Beauly.

Urr, par., on the Urr Water and containing the town of Dalbeattie, Kirkcudbrightshire ; 15,429 ac. Pop. 4032.

Urrall, pl. and hill (604 ft.), 9½ m. N.E. of Glenluce, Wigtownshire.

Urrard House, seat, on r. Garry, 2½ m. S.E. of Blair-Atholl, Perthshire. On the spot where the house stands the battle of Killiecrankie was fought, and a mound marks the spot where Claverhouse fell.

Urray, par., partly in N.E. Inverness-shire and partly in S.E. Ross and Cromarty ; 72,513 ac. Pop. 1826. Vil. of Urray stands near confluence of the rs. Conon and Errin, and 3 m. N.W. of Muir of Ord ry. sta. P.O., T.O.

Urr, Bridge of, ham., on Urr Water, 4 m. N. of Castle-Douglas, Kirkcudbrightshire. P.O.

Urr, Haugh of, vil., on Urr Water, 3¼ m. N.W. of Dalbeattie, Kirkcudbrightshire. P.O., T.O.

Urrich, loch, 5 m. N.E. of Foyers, mid Inverness-shire.

Urr, Loch on the borders of Kirkcudbright and Dumfries, 4 m. S.W. of Moniaive ; depth 42 ft., alt. 624 ft.

Urr, Mote of, an earthwork, one of the finest examples of a "mote" in Britain, 3 m. W. of Southwick ry. sta., near Dalbeattie, Kirkcudbrightshire.

Urr Water, issues from Loch Urr on the N.E. border of Kirkcudbrightshire, and flows 27 m. S.E. to the Solway Firth, 3¼ m. S. of Dalbeattie.

Usan, fishing ham. and coastguard sta., 2½ m. S.E. of Montrose, E. Angus.

Usan House, 1½ m. S. of Montrose, Angus.

Ushinish Lighthouse, on a headland on the E. coast of isl. of S. Uist, Outer Hebrides. A twenty-four seconds' flashing light, visible 18 m.

Ussie, Loch, 419 ft. above sea-level and 35 ft. deep, E. Ross and Cromarty. U. Valley is 3 m. N.W. of Conon Bridge.

Uyea, small isl. on S. side of isl. of Unst, N. Shetland, from which it is separated by Uyea Sound. Pop. 12.

Uyea, small isl. off the N.W. extremity of Mainland, Shetland.

Uyeasound, vil., on Uyea Sound, on S. coast of Unst, Shetland, and 37 m. N. of Lerwick. P.O., T.O.

V

Vaa, loch, 3½ m. S. of Carrbridge, S.E. Inverness-shire.

Vaara, loch, 13 m. N.W. of Lerwick, Mainland, Shetland, depth 52 ft. ; alt. 26 ft.

Vach, or **Bhac,** loch, 1½ m. S.E. of Struan, N. Perthshire.

Vadals, loch, 2 m. S.W. of Clousta, Shetland.

Vaila, isl., 1½ m. by 1¼ m., in mouth of Vaila Sound, 15 m. N.W. of Lerwick, Shetland. On Rams Head is a 3¾-seconds' flashing light and on V. Pier is a fixed red light, visible 5 m.

Valdigarth, ham., 3¼ m. S.E. of Pierowall, Westray, Orkney. P.O.

Vale of Leven, lighting, water supply, and drainage dist., in par. of Bonhill, Dunbartonshire. Pop. 14,974.

Vale Paper Works, Denny, Stirlingshire.

Valigan, loch, 6 m. N.E. of Blair-Atholl, N. Perthshire.

Valla Field, hill range along W. coast of isl. of Unst, Shetland ; alt. 703 ft..

Vallay, isl., 2½ m. long and average breadth ½ m., near N.W. side of isl. of N. Uist, Outer Hebrides. Pop. 19.

Valley, loch, 10 m. N. of Newton-Stewart, Kirkcudbrightshire.

Valleyfield, coal-mining and garden vil., 1½ m. N.E. of Culross, Fife. Pop. (V. and Torryburn), 2932.

Valleyfield House, near Tarff ry. sta., Kirkcudbrightshire.

Valleyfield Paper Mills, Penicuik, Midlothian.

Vallich, small loch, 7 m. S.E. of Gairloch, W. Ross and Cromarty. See LOCH A' BHEALAICH.

Valtos, vil., 13½ m. N. of Portree, Skye, Inner Hebrides.

Valtos, vil., on sea loch of Loch Roag, 34 m. W. of Stornoway, Lewis, Outer Hebrides.

Vane Castle, ruin, on Noran Water, 6 m. W. of Brechin, Angus.

Varragill, glen, with stream flowing N. to Portree Loch, Skye, Inner Hebrides.

Varrich (Caisteal Bharraich), prom., with ruined castle, Kyle of Tongue, N. Sutherland.

Vassa, ham. and bay, 7 m. N. of Lerwick, Shetland.

Vatachan, loch, N.W. Ross and Cromarty, 2 m. N.E. of Achiltibuie.

Vaternish, or **Waternish,** peninsular dist. in N.W. Skye, between Lochs Snizort and Dunvegan.

Vatersay, irregularly shaped isl. to the S. of Barra, from which it is separated by Sound of Vatersay, Outer Hebrides. Pop. 240.

Vatsetter, pl. and loch on the E. of isl. of Yell, Shetland. Fixed light (white and red sectors), visible 14 m.

Vatskir, vil. and prom., 6 m. N.E. of Stornoway, Lewis, Outer Hebrides.

Vatster, loch, 5¼ m. N.W. of Lerwick, Shetland.

Vatten, pl., 3½ m. S.E. of Dunvegan, in the W. of Skye, Outer Hebrides.

Vat, The, romantic cave, 5 m. N.E. of Ballater, S. Aberdeenshire. The V. Burn runs through it to flow into the W. end of Loch Kinord.

Vaul, ham., on bay of same name, N. coast of Tiree, W. Argyllshire.

Vawsay, loch, 1½ m. N.E. of Paible, N. Uist, Outer Hebrides.

Vayne Castle, ruin, on Noran Water, 1 m. N.E. of Tannadice ry. sta., Angus.

Vealloch (Loch a' Bhealaich), loch, 6 m. E. of Crask and at S. end of Loch Coire, E. Sutherland ; depth 80 ft. alt. 572 ft.

Veantrow Bay, in N. of Shapinsay, Orkney.

Vellore Castle, seat, in Muiravonside par., Stirlingshire.

Vementry, isl., in S.E. corner of St. Magnus Bay, Sandsting par., W. Mainland, Shetland. Isl. has extreme length and breadth of $2\frac{1}{4}$ m. and 3 m.; alt. 298 ft.

Venchen, seat, 1 m. N. of Yetholm, Roxburghshire.

Venlaw, seat and hill (1066 ft.), in co. of, and 1 m. N. of, Peebles.

Vennacher, Loch, $3\frac{1}{2}$ m. long by $\frac{1}{2}$ m. broad (average) and 111 ft. deep; alt. 264 ft.; lies $2\frac{1}{4}$ m. S.W. of Callander, Perthshire. Formed by the expansion of the r. Teith, a tributary of the Forth. At the end of the loch are the sluices of the Glasgow Waterworks by which at least, by Act of Parliament, 40,000,000 gallons must be sent down the r. daily.

Veyatie, Loch, 6 m. S.E. of Lochinver, and on borders of N.W. Ross and Cromarty and S.W. Sutherland. It is 4 m. long, a little over $\frac{1}{4}$ m. in breadth (average), and 126 ft. deep; alt. 366 ft.

Victoria Falls, $1\frac{1}{2}$ m. N.W. of Loch Maree Hotel, near Garavaig Bridge, W. Ross and Cromarty.

Victoria Park, ry. sta., L.N.E., Whiteinch, W. subs. of Glasgow. Here is Victoria Park, with fossil grove.

Vidlin, ham. on bay of same name, 5 m. E.N.E. of Voe, N. Mainland, Shetland. P.O., T.O.

Viera, isl., to the S.E. of Rousay, from which it is separated by Viera Sound, Orkney. P.O. Isl. is 2 m. long by 1 m. broad. Pop. 44. The land rises to 104 ft.

Viewfield, Selkirk; nursing home.

Viewfield, seat, in S. vicinity of Mauchline, Ayrshire.

Viewfield House, in S. vicinity of Portree, Skye, Inner Hebrides.

Viewfield Rows, part of Cantsdam (q.v.), Fife.

Viewpark, colliery, ham., near Uddingston, Lanarkshire.

Vinny Water, tributary of the Lunan Water, Friockheim, Angus.

Virkie, ham., $2\frac{1}{2}$ m. S. of Dunrossness, Shetland. P.O.

Voe, ham., with pier, at head of Olna Firth, 16 m. N. by W. of Lerwick, Shetland. P.O., T.O.

Vogrie, mental hospital, on Tyne Water, 5 m. S.E. of Dalkeith, Midlothian.

Voil, Loch, 3 m. W. of Balquhidder ry. sta. An expansion of the r. Balvag,

and is $3\frac{1}{2}$ m. long and 98 ft. deep; alt. 414 ft.

Vord Hill, alt. 521 ft., Fetlar Isl., N.E. Shetland.

Voshimid, loch, 12 m. N.W. of Tarbert, Harris.

Voulin (Loch a' Mhuilinn), loch, 4 m. W. of Amulree, mid. Perthshire.

Vow Island, near head of Loch Lomond, Dunbartonshire.

Voxterbey, loch, 3 m. N.E. of Walls, W. Mainland, Shetland.

Vrotachan, loch, 9 m. S.W. of Braemar, Aberdeenshire.

Vrotten (Loch a' Bhrodainn), loch, 10 m. S. of Kingussie, Spey valley, Inverness-shire.

W

Wadbister, vil., on voe or inlet of same name, 6 m. N.W. of Lerwick, Shetland.

Wakefield Burn, tributary to Kenly Burn at Dunino, E. Fife.

Walkerburn, q.s. par., in pars. of Inner-leithen and Traquair, and vil. on the Tweed, E. Peeblesshire. Vil. $1\frac{3}{4}$ m. E. of Innerleithen; ry. sta. (450 ft.), L.N.E. P.O., T.O. Pop. 1076. Has woollen mfrs.

Walkerton, vil. in pars. of Kinglassie and Leslie, and close to town of Leslie, mid. Fife.

Walkinshaw House, $2\frac{1}{2}$ m. N.W. of Paisley, Renfrewshire.

Wallace's Loup, falls on Kinnel Water, Dumfriesshire.

Wallacestone and Standrig, joint village, $1\frac{1}{2}$ m. S.W. of Polmont Junction, Stirlingshire.

Wallacetown, q.s. par., in town of Ayr. P.O.

Wallacetown, q.s. par. in Dundee, Angus.

Wallacetown, ham., 1 m. N. of New Dailly, S. Ayrshire.

Wallhouse, seat, 3 m. N.W. of Bathgate, W. Lothian.

Walls and Flotta, par. in Orkney; 24,030 ac. Pop. 1115. Par. consists of S. or Walls part of the isl. of Hoy, the isls. of Flotta and Pharay, and some un-inhabited isls.

Walls and Sandness, par. and ham. in the W. of Mainland, Shetland; par. 23,668 ac. Pop. 1102. Par. occupies the more W. part of the peninsula, which strikes W. from the centre of Mainland and includes the isls. of Varla, Linga, Foula, and Papa Stour. Ham. of Walls on Varla Sound, 15 m. N.W. of Lerwick. P.O., T.O.

Wallyford, coal-mining vil. in Midlothian, 1½ m. S.E. of Musselburgh and ½ m. E. of Edinburgh. P.O.

Walston, par. and ham. on r. South Medwin, Lanarkshire; par. 4365 ac. Pop. 227. Par. church stands 5½ m. N. of Biggar.

Walton Park, seat, 1½ m. N.W. of Kirkpatrick-Durham, Kirkcudbrightshire.

Walton Reservoir, 2 m. E. of Fintry, mid. Stirlingshire.

Wampherflat, seat in co. of, and near town of, Lanark.

Wamphray, par., on r. Annan and Wamphray Water in N. Dumfriesshire; 13,136 ac. Pop. 394. Ry. sta., L.M.S., 5 m. S.E. of Beattock and 9 m. N. of Lockerbie. P.O., T.O. W. Water rises on W. side of Loch Fell and runs S.S.W. to the Annan, which it joins 7 m. from its source.

Wandell and Lamington, par. on the r. Clyde, Lanarkshire. Contains the vil. of Lamington, with ry. sta., L.M.S., 6 m. S.W. of Biggar. P.O., T.O.; par. 12,793 ac. Pop. 255. L. House is a seat.

Wanlock Dod, mt., alt. 1808 ft., 1¼ m. S.W. of Leadhills, S. Lanarkshire.

Wanlockhead, q.s. par. and vil., with ry. sta., L.M.S., in N.W. Dumfriesshire. Vil. stands on Wanlock Water in Dumfriesshire on border of Lanarkshire, 1½ m. S.W. of Leadhills and 7½ m. S.W. of Elvanfoot ry. sta. P.O., T.O. Lead Mines, opened in 1680, are still being worked. A light ry. connects W. and Leadhills with Elvanfoot. These vils. are the highest in Scotland, the ry. terminus being 1384 ft. above sea-level.

Wanton Walls, pl. near Newcraighall in the extreme E. of city of Edinburgh.

Ward Hill, 881 ft., 7 m. W. of Kirkwall, Mainland, Orkney.

Ward Hill, 1565 ft., in N.W. of Hoy isl., Orkney.

Wardhouse, seat, 10 m. S.E. of Huntly, Aberdeenshire; ry. sta., L.N.E.

Wardie, part of the Granton dist., in the city of Edinburgh.

Ward Law, mt., alt. 1951 ft., 1 m. N. of Ettrick church, Selkirkshire.

Wardlaw Hill, q.s. par., in par. of Rutherglen, Lanarkshire. P.O.

Wardlaw Law, alt. 1630 ft., 2½ m. S. of Muirkirk, E. Ayrshire. P.O.

Ward of Bressay, hill, alt. 742 ft., in S. of Bressay, Shetland.

Ward of Cruden, Aberdeenshire. See PORT ERROL.

Warmanbie, seat, 1½ m. N. of Annan, Dumfriesshire.

Warriston, dist. in the N. of the city of Edinburgh. P.O., T.O.

Warroch Hill, one of the Ochil Hills, 4½ m. W. of Milnathort, Kinross-shire; alt. 1134 ft.

Warroch House, 3 m. W. of Milnathort, Kinross-shire.

Warthill, hill, alt. 412 ft., 3 m. S.W. of Duncansbay Head, Caithness.

Warthill House, Aberdeenshire, 1 m. N.N.W. of Wartle ry. sta. P.O. Wartle, Meikle Wartle.

Wartle, ry. sta., with T.O., L.N.E., 14 m. S. of Turriff, Aberdeenshire.

Warwickhill House, seat, 4 m. E. of Irvine, N. Ayrshire.

Wasbister, ham. and loch in the N. of Rousay Isl., Orkney. P.O.

Wasdale, loch, 8 m. N.W. of Kirkwall, Orkney.

Washington, vil. on borders of Angus and Perthshire, 3½ m. N.E. of Coupar-Angus.

Watch Hill, or **Cnoc-an-Fhreiceadain,** alt. 1009 ft., 2 m. N.E. of Tongue, N. Sutherland.

Watch Water, tributary of the Dye Water, near Longformacus, N. Berwickshire.

Waterbeck, vil. on the Beck Water and 7 m. S.E. of Lockerbie, Dumfriesshire. P.O., T.O.

Wateresk, pl. in par. of Cortachy and Clova, N. Angus.

Waterfoot, harbour at mouth of Annan Water and 1½ m. S. of Annan, Dumfriesshire.

Waterhead, mt , alt. 1549 ft., 6½ m. S.S.W. of Dalmellington, S.E. Ayrshire.

Waterhead, a shooting, 5 m. N.E. of Dalmellington, S.E. Ayrshire.

Waterloo, vil., 1 m. S.E. of Wishaw, N.W. Lanarkshire. P.O. Pop. of W. and Overtown, 3629.

21

Westfield, vil. on the r. Ericht, 2 m. N. of Blairgowrie, N.E. Perthshire. P.O.

Westfield, seat, 5 m. S.W. of Thurso, N. Caithness.

Westfield, ham., 7 m. S.W. of St. Andrews, Fife.

Westfield, vil., 2½ m. E.S.E. of Alloa, Clackmannanshire.

Westfield House, 5 m. S. of Nigg ry. sta., E. Ross and Cromarty.

West Gordon, vil. in Berwickshire. See GORDON.

West Hall, seat, 1½ m. N.E. of Oyne sta., Aberdeenshire.

Westhorn House, ¾ m. S.W. of Tollcross, N. Lanarkshire.

Westing, pl. on W. coast of isl. of Unst, N. Shetland.

West Kilbride, par. and small town, N. Ayrshire; par. 10,161 ac. Pop. 3946. Town, 4¼ m. N.W. of Ardrossan; ry. sta., L.M.S. P.O., T.O. Pop. 3093. Seamill Hydro.

West Kirk, eccl. par., Aberdeen.

West Linga, isl. between Mainland and isl. of Whalsey, Shetland.

West Linton, par. and vil. in N. Peebleshire; par. 23,257 ac. Pop. 1200. Vil. 15 m. S. of Edinburgh; ry. sta., L.N.E. P.O., T.O. Pop. 522.

West Lomond, mt., alt. 1713 ft., 3½ m. W. of Falkland, Fife.

West Longrigg, vil. near Longriggend, Lanarkshire.

West Lothian, formerly Linlithgowshire, bounded on the N. for 17 m. by the Firth of Forth, E. and S. by Midlothian, and W. by the cos. of Lanark and Stirling. Surface varied with knolls and gentle undulations. Chief rs., the Almond and the Avon. Coal is largely mined in the co. around Bathgate; at Armadale and Fauldhouse, iron-ore is also produced. Oil shale for the production of paraffin and other shale products extensively manufactured in the E. of the co. Limestone and freestone are largely worked. There are also large farms, and all the principal crops are grown with success. Co. town, Linlithgow. Area 120·1 sq. m., or 76,861 ac. Pop. 81,426. W. Lothian returns one member to Parliament.

Westmuir, pl., near Carntyne sta. in E. of Glasgow.

Westmuir, vil., 1¾ m. S.W. of Kirriemuir, Angus.

Westness, pl. and seat, on Eynhallow Sound, S. side of Rousay, Orkney.

West Newport, ry. sta., L.N.E., 3 m. W. of Tayport, N. Fife.

Westown, ham., 2½ m. N.W. of Errol, Perthshire.

Westpans, ham., 1 m. E. of Musselburgh, Midlothian.

West Quarter, vil., 2 m. N.E. of Strathaven, W. Lanarkshire; nearby is the seat, Westquarter House.

Westquarter House, 1 m. W. of Polmont, Stirlingshire.

Westray, par., containing Westray and the smaller isl. to the N.W., Papa Westray, in N. Orkney; 11,551 ac. Pop. 1270. The par. church stands in the centre of Westray Isl.

Westray, isl., 10 m. long and varying from ½ m. to 6¼ m. broad, in N. Orkney; Westray Firth separates it on the S. from Rousay Isl. P.O., T.O. Steamer communication with Kirkwall. Pop. 1269.

Westrigg, colliery vil., 2 m. W.S.W. of Armadale, W. Lothian.

Westruther, par. and vil. in W. Berwickshire; par., 14,633 ac. Pop. 421. Vil. 7½ m. N.E. of Lauder and 5½ m. N. of Gordon ry. sta. P.O., T.O. In N.E. extremity of the par. stand the ruins of Ively Castle. Vil. pop. 51.

West Salton, vil. in E. Lothian. See SALTON.

Westsandwick, ham. situated about middle of W. coast of Yell Isl., Shetland. P.O., T.O.

West Sidewood, seat, 3 m. W. of Carnwath, Lanarkshire.

West Street, ry. sta. on Glasgow Dist. Subway, ¼ m. W.S.W. of Bridge Street sta.

West Water, trib. of the N. Esk, Angus.

West Water, tributary of the r. Tyne, 1 m. S.E. of W. Linton, Peeblesshire.

Wether Hill, alt. 1100 ft., westermost point of the Cleish Hills, 3 m. N.E. of Saline, S.W. Fife.

Weydale, ham., 3½ m. S.E. of Thurso, N. Caithness. P.O. Quarries near by.

Whale Firth, sea inlet running S. and S.E. on W. side of Yell Isl., Shetland.

Whaligoe, pl., 7½ m. S.S.W. of Wick, Caithness.

Whalplaw Burn runs to Soonhope Burn, which joins the Leader, 3 m. N.E. of Lauder, Berwickshire.

Whalsay, isl. on E. side of Mainland, Shetland ; measures 5½ m. from S.W. to N.E. and has a maximum breadth of 2⅛ m. P.O., T.O. Pop. 897. Steamer communication with Lerwick. Has fixed white light at S. end, visible 6 m.

Whalsay and Skerries, q.s. par. in par. of Nesting, Mainland, Shetland. Par. church is on isl. of Whalsay.

Wharral, loch, 2 m. N.E. of Clova, N. Angus.

Whatley, loch, about middle of isl. of Unst, Shetland.

Whauphill, vil. in co. of, and 4 m. S. of, Wigtown ; ry. sta., L.M.S. P.O., T.O.

Wheelrig Head, hill, alt. 1465 ft., 1¼ m. N.W. of Peel Fell, S. Roxburghshire.

Whifflet, vil., 1 m. S.E. of Coatbridge, Lanarkshire ; ry. stas., L.M.S. and L.N.E. P.O., T.O. Collieries and ironworks.

Whigstreet, ham., 4½ m. S.E. of Forfar, Angus.

Whim, pl. and seat, ½ m. N.E. of Lamancha ry. sta., Peeblesshire.

Whinfield, seat, in co. of, and adjoining, Kinross.

Whinnyfold, pl., 8½ m. S.S.W. of Peterhead, N.E. Aberdeenshire.

Whinnyleggate, pl. in co. of, and 3 m. N.E. of, Kirkcudbright.

Whins of Milton, vil. in co. of, and 2½ m. S. of, Stirling. Has remains of Beaton's Mill, where James III. was murdered 1488. Special lighting dist. Pop. 477. See MILTON, WHINS OF.

Whinyeon, Loch, in co. of, and 8 m. N.W. of Kirkcudbright ; depth 33 ft., alt. about 715 ft.

Whistleberry, pl., with ruins of ancient castle, 2½ m. N.E. of Bervie, Kincardineshire.

Whistlefield, pl., with inn on E. shore of Loch Eck, 11 m. N. of Dunoon, Cowal, Argyllshire.

Whistlefield, pl. near Loch Long and 1¼ m. N.W. of Garelochhead, W. Dunbartonshire ; ry. sta., L.N.E.

Whitburgh, seat, 3 m. S.E. of Pathhead, E. Lothian.

Whitburn, par. and small town (police bur.) on the Almond in the S. of West Lothian ; par. 9782 ac. Pop. 12,619. Town stands 2½ m. S.W. of Bathgate. Pop. 2440. Ry. sta. (1 m. E.), L.N.E. (passenger service withdrawn). P.O., T.O. Coal and iron dist.

Whitburn, East, vil. in par. of Whitburn, W. Lothian.

Whitchester, cottage hospital, 2 m. N.E. of Longformacus, N. Berwickshire.

Whitadder, stream rising in E. Lothian and flowing S.E. through Berwickshire to the Tweed, which it joins, after a 34 m. course, 2½ m. W.S.W. of Berwick.

Whitebridge, ham. on r. Foyers, 10 m. N.E. of Fort Augustus, mid Inverness-shire. P.O., T.O.

Whitecairns, pl., with inn, 3 m. S.E. of New Machar, E. Aberdeenshire. P.O.

White Cart, stream rising on moors of Eaglesham, S.E. Renfrewshire, and flowing N. and W. for 18½ m,. to join the Black Cart at Inchinnan Bridge. The united stream flows 1 m. N. to the Clyde.

Whitecleuch, pl., 4½ m. S.W. of Crawfordjohn, upper ward of Lanarkshire.

White Coomb, mt., alt. 2695 ft., 8 m. N.E. of Moffat, on N. border of Dumfriesshire.

Whitecraigs, ry. sta., L.M.S. (for Mearns), Renfrewshire, 6½ m. S.W. of Glasgow Central Sta.

Whitecross, pl., Stirlingshire, 2 m. S.E. of Maddiston. P.O.

Whitecrosses House, seat, ½ m. N. of Dunblane, Perthshire.

White Esk, r. in Dumfriesshire. See ESK.

Whiteface, ham., 5¾ m. W. of Dornoch, Sutherland. P.O.

Whitefarland, pl. and headland on Kilbrennan Sound, 6½ m. S.W. of Loch Ranza, Arran, Buteshire.

Whitefarland Point, in Gourock Bay, N.W. Renfrewshire. Fixed green light on pier-head, visible 9 m.

Whitefield Loch, 3 m. S.E. of Glenluce, Wigtownshire ; depth 14 ft., alt. 192 ft.

White Hall, 8 n. S.W. of Eyemouth, E. Berwickshire.

Whitehall, vil., with pier, Stronsay, opposite Papa Stronsay, Orkney.

Whitehall Bay, Papa Sound, isl. of Stronsay, Orkney. Fixed red light, seen 5 m.

Whitehaugh, seat, on r. Don, 1½ m. N.E. of Alford, Aberdeenshire.

Whitehaugh Burn, tributary of the r. Ayr, 4½ m. W. of Muirkirk, E. Ayrshire.

White Hill, alt. 2544 ft., 4½ m. E. of Milton of Clova, Angus.

Whitehill, vil., 2 m. S. of New Pitsligo, N.E. Aberdeenshire.

Whitehill, seat, 1 m. S.E. of Hawthornden ry. sta., Midlothian.

Whitehill, ham., 1½ m. S.E. of Dalkeith, Midlothian.

Whitehill, seat, 1 m. N.W. of Aberdour, Fife.

Whitehills, near Forfar, Angus. County hospital.

Whitehills, fishing vil. in co. of, and 3 m. N.W. of, Banff. P.O., T.O. Lifeboat sta. Pop. 1013.

Whitehills, seat, 1 m. W. of Abernyte, which is 10 m. N.E. of Perth.

White Hope Edge, mt., alt. 1560 ft., 8½ m. W. of Langholm, Dumfriesshire.

Whitehope Law, mt., alt. 2038 ft., 3 m. N. of Innerleithen, N. Peeblesshire.

Whitehouse, pl., with ry. sta., L.N.E., 2¾ m. S.E. of Alford, Aberdeenshire. P.O., T.O.

Whitehouse, ham., on W. Loch Tarbert, 6 m. S.W. of Tarbert, S. Argyllshire. P.O., T.O.

Whitehouse, shooting-lodge, ½ m. S. of Lamlash, E. coast of Arran Isl., Buteshire.

Whiteinch, q.s. par. on N. bank of the Clyde, in par. of Govan, Glasgow; ry. stas. L.M.S. (Whiteinch), L.N.E. (Victoria Park), P.O., T.O. Here is Victoria Park with fossil grove.

Whitekirk and Tynninghame, coastal par. on the Tyne, N.E. of East Lothian; 6333 ac. Pop. 732. P.O., T.O. called Whitekirk. Vil. stands 4½ m. S.E. of North Berwick, and the church—one of the finest specimens of pre-Reformation buildings—was burnt down, but now restored.

Whiteknowe, seat, 5 m. N.E. of Lockerbie, Dumfriesshire.

Whiteknowe Head, mt., in co. of, and 7 m. S.S.E. of, Peebles; alt. 1678 ft.

White Law, alt. 1941 ft., summit in the Lowther Hills, 2¼ m. N.N.E. of vil. of Leadhills, S. Lanarkshire.

Whitelaw Hill, alt. 584 ft., 1 m. N.W. of Garvald, E. Lothian.

Whitelaws, hill, alt. 1664 ft., 4 m. N.W. of Fettercairn, Kincardineshire.

White Loch, or **Loch of Inch,** 3 m. E. of Stranraer, Wigtownshire; depth 38 ft., alt. 54 ft.

White Loch, 5 m. S.E. of Dalbeattie, Kirkcudbrightshire.

White Loch, 2 m. S. of Blairgowrie, Perthshire; depth 32 ft., alt. 154 ft.

White Loch of Myrton, 1 m. E. of Portwilliam, Wigtownshire; depth 40 ft., alt. 98 ft.

Whitemire, vil., 6 m. S.E. of Forres and 3¾ m. S. of Brodie ry. sta., Morayshire. P.O.

Whitemount (or **Whitemonth**) **Deer Forest** (4540 ac.), 5 m. S.E. of Braemar, S.W. Aberdeenshire.

Whiteness, ham., 9 m. N.W. of Lerwick, Shetland. P.O., T.O.

Whiteness and Weisdale, q.s. par. in Tingwall par., Mainland, Shetland.

Whiteness Voe, inlet, 5½ m. W.N.W. of Lerwick, on W. coast of Mainland, Shetland.

Whiten Head, or **Kennageall,** headland (935 ft.) between Loch Eriboll and Kyle of Tongue, N. coast of Sutherland. Many fantastic caverns have been formed by the action of the sea.

Whiterashes, ham., 3½ m. N.W. of New Machar, E. Aberdeenshire. P.O.

Whiterigg, seat, 1 m. S.W. of St. Boswells sta., Roxburghshire.

Whiterigg, vil., 3 m. N.E. of Airdrie, Lanarkshire; ry. sta., L.N.E. (passenger service withdrawn).

Whitesands Bay, E. of Aberdour, Fife.

Whiteside Hill, alt. 1817 ft., 3 m. N. of Queensberry Hill, on borders of cos. of Lanark and Dumfries.

Whitevale, dist. in E. of city of Glasgow. P.O., T.O.

Whithope Burn, flowing into Loch o' the Lowes, Yarrow, Selkirkshire.

Whithorn, par. and royal and mun. bur. in S.E. Wigtownshire; par. 11,848 ac. Pop. 1796; town pop. 951. Is 12¼ m. S. of Wigtown by rail; ry. sta., L.M.S. P.O., T.O. St. Ninian built a church here in 397 and attached a monastery to it; believed to be the first stone-built church in Scotland and the earliest Christian church in the country. The par. church is a spacious building, erected partly on the ruins of a priory founded by St. Ninian of which there are now only remains.

Whithorn, Isle of, vil. in extreme S.E. of Wigtownshire. See ISLE OF WHITHORN.

Whiting Bay, q.s. par., bay, and vil., with pier, in S.E. of Arran Isl., Buteshire. The bay is 3 m. wide at mouth and penetrates about ¾ m. inland. Vil. 4 m. S.E. of Lamlash. P.O., T.O. On pier is a fixed light, visible 6 m. Pop. 314.

Whitletts, vil., 2 m. N.E. of Ayr. P.O. Pop. 991.

Whitliemuir Dam Reservoir, 3 m. S. of Johnstone, Renfrewshire.

Whitslaid, ruined tower, on the Leader Water, 3 m. S.E. of Lauder, Berwickshire.

Whitsome, par. and ham., on the Leet Water, S.E., Berwickshire; par. 4895 ac. Pop. 433. Ham. 4 m. S. of Chirnside. P.O., T.O.

Whittingehame, par. in S. of East Lothian; 15,466 ac. Pop. 493. P.O. W. Water flows eastward through the par. and reaches the sea at Belhaven. On the stream, and 3 m. S. of East Linton, is W. House, the seat of the Earl of Balfour. Nearby are the ruins of W. Castle.

Whitton, pl., with ruins of a castle, 7 m. E. of Jedburgh, Roxburghshire.

Whitton Loch, 3½ m. S.W. of Morebattle, Roxburghshire.

Whytbank, seat in N. Roxburghshire, 3 m. W. of Galashiels.

Wiay, isl., in Loch Bracadale, W. Skye, Inner Hebrides.

Wiay, isl., off S.E. coast of Benbecula, Outer Hebrides; deeply indented coast.

Wick, par., seapt., royal and mun. bur., and co. town of Caithness; par. 47,247 ac. Pop. 10,383. Town stands on Wick Water where it enters Wick Bay, and is 161¼ m. N.E. of Inverness and 279¼ m. N.E. of Perth by rail; ry. sta., L.M.S. P.O., T.O. and T.O. at radio station. The town consists of three parts—Wick proper (the oldest part) and Louisburgh on the N. side of the river and bay and Pulteneytown on the S. The port consists of two harbours and is the centre of an important and extensive fishing district. It is also a coastguard and motor lifeboat sta. Regular steamboat communication is maintained with Aberdeen, Leith, Kirkwall, and Lerwick. On the N. and S. piers are fixed lights, visible 12 m. and 11 m. respectively. An interesting ruin in the vicinity is the Castle of Old Wick, or the Auld Man o' Wick, which, along with the lighthouse on Noss Head, form excellent landmarks to seamen. There are many remains of Pictish dwellings in the dist. Pop. 7548.

Wick of Gruting, bay on N. shore of Fetlar Isl., Shetland.

Wick of Tresta, bay on S. side of Fetlar Isl., Shetland.

Wick Water, stream, issuing from Loch Watten and flowing 8 m. S.E. to Wick Bay, N.E. Caithness.

Wideford Hill, alt. 741 ft., 2 m. W. of Kirkwall, Mainland, Orkney.

Widewall Bay., in N.W. of S. Ronaldshay, Orkney.

Wightman Hill, alt. 1153 ft., 1½ m. S.W. of Oldhamstocks, E. Lothian.

Wigtown, par., seapt., royal and mun. bur., and co. town of Wigtownshire; par. 8206 ac. Pop. 1531; town. pop. 1261. Stands on W. side of Wigtown Bay, 7¼ m. S. of Newton-Stewart by rail; ry. sta., L.M.S. P.O., T.O.

Wigtown Bay, inlet of the Irish Sea, running up for about 9 m. between the cos. of Wigtown and Dumfries, with an average breadth of 3 m. Receives the rs. Cree and Bladenoch, and has large sandbanks.

Wigtownshire, maritime co., S.W. Scotland, with a coast-line of 120 m., bounded S. and W. by the Irish Sea, on the N. by Ayrshire, and on the E. by Kirkcudbrightshire. It is deeply indented on the S. by Luce Bay, and Loch Ryan on the N., thus forming a double-headed peninsula, the " Rinns of Galloway "; and between Luce Bay and Wigtown Bay on E. is formed the prom. called the Machers. The rest of the co. bears the general name of the Moors. The co. consists chiefly of stoney moors, rising in the N. into rocky hills, reaching an alt. of 1000 ft. It has good arable land in the Rinns and the Machers. The principal rs. are the Cree, the Bladenoch, the Tarff, and the Luce. The climate is mild and humid. About 47 per cent. of the area is under cultivation, the chief crops being oats, but the chief industries are cattle raising and dairy farming. There are also woollen works and engineering works. Principal towns are Stranraer, Wigtown, and Newton-Stewart; area 487·4 sq. m. or 311,984 ac. Pop. 29,331. Forms part of Galloway, which returns one member to Parliament.

Wildmanbridge, pl., 2½ m. E.S.E. of Wishaw, Lanarkshire.

Wilkieston, vil. in par. of Kirknewton, and 3 m. E. of Mid-Calder, Midlothian.

Williamsburgh, E. sub. of Paisley, Renfrewshire.

Williamsburgh, Fife. See ELIE.

Williamston House, 3 m. N.E. of Insch ry. sta., Aberdeenshire.

Williamwood, ry. sta., L.M.S., 1½ m. N.N.W. of Busby, Lanarkshire. Dist. pop. 306.

Willies Law, alt. 1626 ft., 5 m. S.E. of Gifford, E. Lothian.

Wilson Burn, stream, flowing from Hellmuir Loch to Ale Water, 3 m. S.W. of Roberton Church, Selkirkshire.

Wilsontown, colliery vil., 9¼ m. N.W. of Carstairs, Lanarkshire; ry. sta. (alt. 917 ft.), L.M.S. Pop. of W. and Forth, 2039.

Wilton, q.s. par., Glasgow.

Wilton, q.s. par. in Hawick, r. Teviot, Roxburghshire. Wilton Dean, Hawick par., special lighting dist. Pop. 155.

Winchburgh, q.s. par. and vil. in Kirkliston par., W. Lothian; vil., 5½ m. E. of Linlithgow and 12 m. W. of Edinburgh; ry. sta., L.N.E. P.O., T.O. Pop. 2142.

Windburgh Hill, alt. 1662 ft., 4 m. N. of Riccarton ry. junc., S.W. Roxburghshire.

Wind Fell, mt., alt. 2180 ft., 6 m. E. of Moffat, Dumfriesshire.

Windlestraw Law, alt. 2161 ft., 8 m. E.N.E. of Peebles, on E. border of the co.

Windy Gate Hill, alt. 2034 ft., 8 m. S. of Yetholm, on borders of cos. of Roxburgh and Northumberland.

Windygates and Cameron Bridge, vil. in par. of Markinch, 3½ m. N.E. of Thornton Junc., Fife; ry. sta. (Cameron Bridge), L.N.E. P.O., T.O. called Windygates. Has a distillery. Pop. of Windygates and Balcurvie 1744.

Windy Standard, hill, alt. 1760 ft., 3 m. E.S.E. of Dalmellington, S.E. Ayrshire.

Windy Standard, mt., alt. 2287 ft., in N. Kirkcudbrightshire, 7 m. S. of New Cumnock, Ayrshire.

Winless, small loch, 4½ m. N.W. of Wick, Caithness.

Winshields Hostel, on the Caldwell Burn, a trib. of the Dryfe Water, Dumfriesshire, 4 m. S.W. of Eskdalemuir Church, 9 m. N. of Lockerbie. A Scottish Youth Hostel.

Wintercleuch Fell, alt., 1804 ft., 5 m. S.E. of Elvanfoot sta., S. Lanarkshire.

Winterfield House, 1 m. W. of Dunbar, E. Lothian.

Winterhope Burn, trib. of the Megget Water, Peeblesshire.

Winthank, seat, 3 m. S.W. of St. Andrews, Fife.

Winton, ry. sta., L.N.E., E. Lothian, 13 m. S.E. of Edinburgh.

Winton Castle, seat, on the r. Tyne, 3 m. S.E. of Tranent, E. Lothian.

Winton Park, seat, 1 m. E.N.E. of Prestonpans, E. Lothian.

Wirren, Hill of, alt. 2220 ft., 7 m. W.N.W. of Edzell, Angus.

Wirren, West, mt. alt. 2060 ft., 7 m. W.N.W. of Edzell, Angus.

Wishaw, q.s. par. and town in par. of Cambusnethan, N. Lanarkshire; town stands 15 m. S.E. of Glasgow by rail; ry. stas. (South and Central), L.M.S. P.O., T.O. Town now amalgamated with Motherwell to form the police bur. of Motherwell and W. Pop. of W. and Motherwell, 64,710; important centre in the coal, iron, and steel trades. W. House, seat of Lord Belhaven and Stenton, 1 m. N.W.

Wisp Hill, alt. 1950 ft., 9 m. N. of Langholm, Dumfriesshire.

Wiston, vil., 1½ m. N.W. of Lamington ry. sta., Lanarkshire.

Wiston and Roberton, par., on the Clyde, E. Lanarkshire; 13,156 ac. Pop. 385. Par. church near vil. of Wiston. Wiston Lodge is a seat.

Witches Cairn, hill, 1045 ft., 5 m. S.W. of Oldhamstocks, E. Lothian.

Witches' Knowe, 649 ft., 2 m. N.E. of Foulden, S.E. Berwickshire.

Woden Law, alt. 1388 ft., 4 m. S. of Hounam, Berwickshire.

Wolf Craig, alt. 2343 ft., between Glen Mark and Glen Lee, N. Angus.

Wolfelee, seat, on Rule Water, 6½ m. S.E. of Hawick, Roxburghshire. Wolfelee Hill, in the neighbourhood, is 1288 ft.

Wolf Hill, alt. 2703 ft., 3 m. N. of Milton of Clova, Angus.

Wolfhill, vil. in co. of, and 6 m. N.E. of, Perth. P.O., T.O. Pop. of W., Guildtown, and Balbeggie, 425.

Woll, The, seat in co. of, and 4 m. S. of, Selkirk.

Woodbank, seat, ½ m. W. of Balloch, Dunbartonshire.

Woodcockdale, seat, 2 m. S.W. of Linlithgow, W. Lothian.

Woodcote Park, seat, 5 m. S.E. of Pathhead, E. Lothian.

Woodend, vil. on the r. Don, in co. of, and 4½ m. N.W. of, Aberdeen.

Woodend, vil., 1 m. N.W. of Armadale, W. Lothian.

Woodend, seat, 1¼ m. S.W. of Madderty sta., Perthshire.

Woodend, seat, 2 m. S.S.W. of Rothesay and on W. side of Loch Fad. It was built for Edmund Kean, the actor.

Woodend, seat, 2½ m. N.W. of Johnstone, Renfrewshire.

Woodend Loch, 2 m. N.W. of Coatbridge, Lanarkshire.

Wooden House, seat, 1¾ m. S.E. of Kelso, Roxburghshire.

Woodhall, seat, on Water of Leith, opposite Juniper Green, Edinburgh; paper mill adjacent.

Woodhall, seat, 1 m. N. of Holytown, Lanarkshire.

Woodhall House, seat, Pencaitland, E. Lothian.

Woodhall Loch, 2 m. S. of New Galloway Station, Kirkcudbrightshire; depth 49 ft., alt. 172 ft.

Woodhead, vil., 1½ m. E. of Fyvie, N. Aberdeenshire. P.O.

Woodhead, vil. in N. Kirkcudbrightshire, and 11 m. S.E. of Dalmellington sta. in Ayrshire.

Woodhill, seat, 4 m. W. of Carnoustie, Angus.

Woodhill House, seat, 7½ m. N.W. of Blairgowrie, Perthshire.

Woodhouse, ham., Peeblesshire, 1 m. S. of Manor Church.

Woodhouselee, seat, 6 m. S. of Edinburgh, Midlothian.

Woodhouselees, pl., 1 m. S.W. of Canonbie, Dumfriesshire.

Woodhouse Tower, ruin, 5 m. N.E. by E. of Annan, S.E. Dumfriesshire. Said to have been the post place which harboured Robert Bruce when he was fleeing from Edward I.

Woodilee, the Glasgow Lunacy District Asylum ; adjoins Lenzie, Dunbartonshire.

Woodland, seat, 2½ m. N.W. of Maxwelltown, E. Kirkcudbrightshire.

Woodlands, pl., ½ m. W. of Old Monkland, N. Lanarkshire.

Woodlandslee, trib. of Leithen Water, 4½ m. N. of Innerleithen, Peeblesshire.

Woodmuir, pl., 3½ m. S.W. of W. Calder, Midlothian.

Woodneuk House, 2½ m. N.W. of Coatbridge, N. Lanarkshire.

Woodside, ward and q.s. par., with ry. sta., L.N.E., in Aberdeen. P.O., T.O.

Woodside, seat, 1 m. N. of Beith, Ayrshire.

Woodside, ham., 7 m. S. of Sanquhar, Dumfriesshire.

Woodside, vil., 3¼ m. N. of Largo, Fife. Pop. 314.

Woodside, vil., with ry. sta. (Woodside and Burrelton), L.M.S., Perthshire, 2½ m. S.W. of Coupar-Angus. P.O., T.O. ; T.O. at Burrelton.

Woodside, vil., 1½ m. S.W. of Markinch, Fife.

Woodside House, in co. of, and 2½ m. N.W. of, Aberdeen.

Woodside of Ardbeadie, part of Banchory, Kincardineshire.

Woodthorpe, seat, Murtle, Peterculter, Aberdeenshire.

Woodville, seat, 2 m. N.W. of Arbroath, Angus.

Wood Wick, pl. and bay, 9 m. N.W. of Kirkwall, Orkney.

Wooplaw, seat in N.W. Roxburghshire, 3 m. N. of Galashiels.

Wormadale, Loch, 12 m. N.W. of Lerwick, Shetland.

Wormiston, seat, 1½ m. S. of Eddleston, Peeblesshire.

Wormistone House, 1½ m. N. of Crail, Fife ; seat of Earl of Lindsay.

Wormit, q.s. par., pl., and bay on Firth of Tay, Fife ; ry. sta., L.N.E., 2 m. W. of Newport with which it is incorporated. P.O., T.O. Vil. pop. 109.

Wrath, Cape, headland in the N.W. extremity of the mainland of Scotland. The headland is 370 ft. high, and is surmounted by a lighthouse with a one-minute alternating flashing light 400 ft. above high water and visible 27 m. Dangerous reefs run out from the base of the cliffs. The Norsemen called it Hvarf, *a turning point.*

Wreath's Tower, ruin, 2 m. S.W. of Kirkbean, S. Kirkcudbrightshire.

Wrightspark, seat, 1½ m. S.S.W. of Kippen, W. Stirlingshire.

Wyre (or **Viera**), isl., Orkney, off S. coast of Rousay. P.O. Pop. 44.

Wyseby, seat, 1 m. S.E. of Kirtlebridge sta., and 4 m. N.E. of Annan, S. Dumfriesshire.

Wyvis, seat, at N. end of Loch Glass, E. Ross and Cromarty.

Wyvis, the largest and most conspicuous mt. in the N. of Scotland, rising with two summits to an alt. of 3429 and 3295 ft. It lies 10 m. N.W. of Dingwall, Ross and Cromarty.

Y

Yair House, in co. of, and 5 m. N.W. of Selkirk.

Yarehouse, loch, 5½ m. S.S.W. of Wick, Caithness.

Yarrow, par., on Yarrow Water, Selkirkshire ; 48,851 ac. Pop. 459. Par. church 9 m. W. of Selkirk. P.O., T.O.

Yarrowfeus, ham., on Yarrow Water, 2 m. S.W. of Yarrow Church, Selkirkshire.

Yarrowford, ham, on Yarrow Water, in co. of, and 5 m. N.W. of, Selkirk. P.O., T.O. Hotel.

Yarrows, Loch of, 5 m. S.W. of Wick, Caithness. At the S. end is a cairn known as Cairn Reian, or *Seat of the King.*

Yarrow Water, stream in Selkirkshire ; rises on border of Dumfriesshire, flows through Loch o' the Lowes and St. Mary's Loch, 14½ m. N.E., to join the Ettrick Water 2 m. S.W. of Selkirk.

Yearngill Head, mt., alt. 1804 ft., 3½ m. E. of Crawford, S. Lanarkshire.

Yell, par. and isl. lying between the isls. of Mainland and Unst, Shetland, and separated from Mainland by Yell Sound ; par. 52,136 ac. Pop. 1883.

The isl. is the second largest of the Shetland group, and is 17 m. long and from ½ m. to 6½ m. broad. The surface is bleak moorland and fishing is the chief occupation. Pop of Yell Isl. 1879.

Yell, Mid, eccl. par. in the par. of Yell, N. Shetland. P.O., T.O.

Yell, North and **South,** eccl. pars. in par. of Yell, N. Shetland.

Yesnaby, pl., 14 m. W. of Kirkwall, Mainland, Orkney.

Yester, par. in the S. of East Lothian ; 9028 ac. Pop. 691. Par. church at Gifford, 4 m. S.S.E. of Haddington. Y. House, seat of the Marquess of Tweeddale, on Gifford Water, is 1 m. S.E. of vil. of Gifford.

Yetholm, par. and vil. in N.E. Roxburghshire ; par. 5869 ac. Pop. 612. Vil. stands athwart Bowmont Water, 7½ m. S.E. of Kelso. Hotel. P.O., T.O. Pop. (Kirk Yetholm) 140, (Town Yetholm) 295. Was for long the head-quarters of the Scottish gipsies.

Yetholm Loch, 1½ m. W. of Kirk Yetholm, Roxburghshire.

Yieldshields, vil. in Lanarkshire. See ROADMEETINGS.

Yoker, dist., with ry. stas., L.N.E. and L.M.S., Renfrewshire, on the Clyde, in bur., and 7¾ m. W.N.W. of Queen Street, Glasgow. P.O., T.O. Has ferry to Renfrew.

Yoolfield, pl., 3 m. E. by N. of Cupar, Fife.

Ythan, r. in N.E. Aberdeenshire. Rises at the Wells of Ythan and flows for 35 m. mainly in a S.E. direction to the North Sea, near Newburgh.

Ythanbank, pl., 5 m. N.W. of Ellon, E. Aberdeenshire. P.O., T.O.

Ythan Wells, q.s. par. in pars. of Forgue and Auchterless, N.E. Aberdeenshire. Par. church 9 m. E. of Huntly. P.O., T.O.

Ythsie, pl., 4½ m. W. of Ellon, E. Aberdeenshire.

Z

Zetland, another name for Shetland (q.v.). Gives the title of Marquess to the Dundas family.

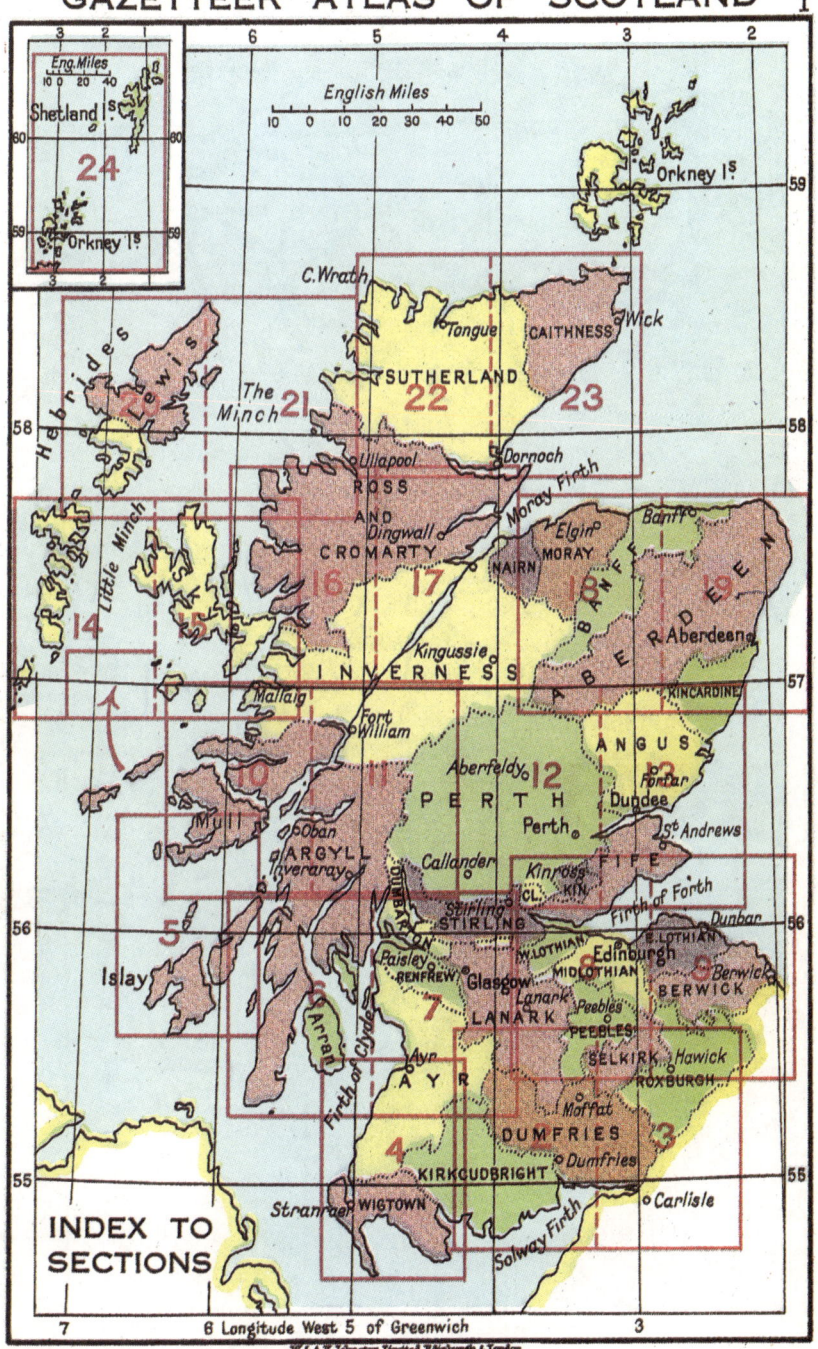

INDEX TO
SECTIONS

LE

3°

30'

Glen Ho. Traquair Abbotsford Ferry Melrose Makerstoun Ch. Sta.
Minchmoor St Boswells Lindean Bowden Maxton Rutherford Roxburgh
Yarrowford Newark Cas. SELKIRK St Boswells Green Ancrum Kirkbank Kale W.
ar Law Yarrow Ch. 1150 Midlem Belses Nisbet Eckford
Quair W. Gordon Arms Hotel Lilliesleaf 1645

Ettrick Fore
Rodon SELKIRK Ashkirk Ale W. Crailing Whitton
Tibbie Shiels Inn St Mary L. Hassendean JEDBURGH Abbey Oxnam Hownam
Chapelhope Tushielaw Inn Kelmuir L. Minto Sta.
Ettrick Ch. Roberton Ch. ROXBURGH Bedrule Ternichirst Cas.
Ettrick Pen Buccleuch HAWICK Ruletownhead Edgerston
2269 1098 Hobkirk Chesters Southdean Carter Bar
Foulbog Teviothead Ch. Stobs Bonchester Br. Dykeraw Tower 1371 Hungry Law
ch Fell Shankend Southdean 1742 1643
2256 Mosspaul 848 Greatmoor Hill Note o' the Gate Carter Fell HILLS
Eskdalemuir 1964 Ltmekilnedge 1250 Peel Fell
1195 Sta. 1975
Riccarton Junc. Hermitage Saughtree
Hermitage W. Larriston Fells
Steele Road Roan Fell 1524
Ewes Ch. Kirkstile Castleton Ch. Kershope W.
Bentpath Newcastleton R. Lyne
Langholm Sta.
Hutton Ch. Bigholm Canonbie Haltwhistle
Middlebie Gilnockie Brampton Sth. Tyne
Half Morton Sta. Sta.
Kirkpatrick Longtown R. Irthing
Gretna Green 55'
Annan Dornock R. Eden
Lt. Ho. Tordoff Pt. CARLISLE Alston
Bowness Sta.
Wigton

N O R T H U M B E R L A N D

C U M B E R L A N D

Edinburgh & London.

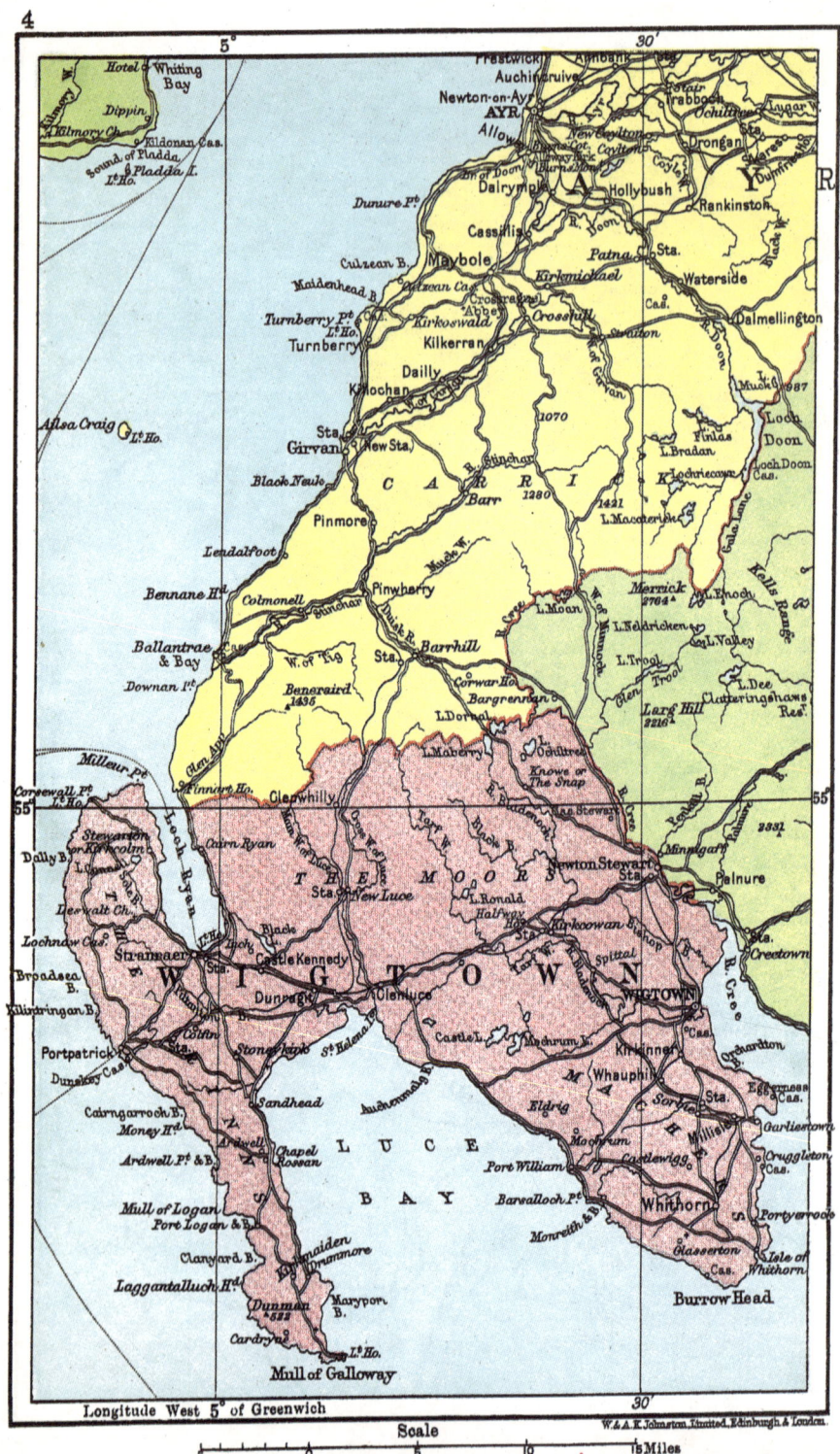

4

Scale

W. & A. K. Johnston Limited, Edinburgh & London.

5   10   15 Miles

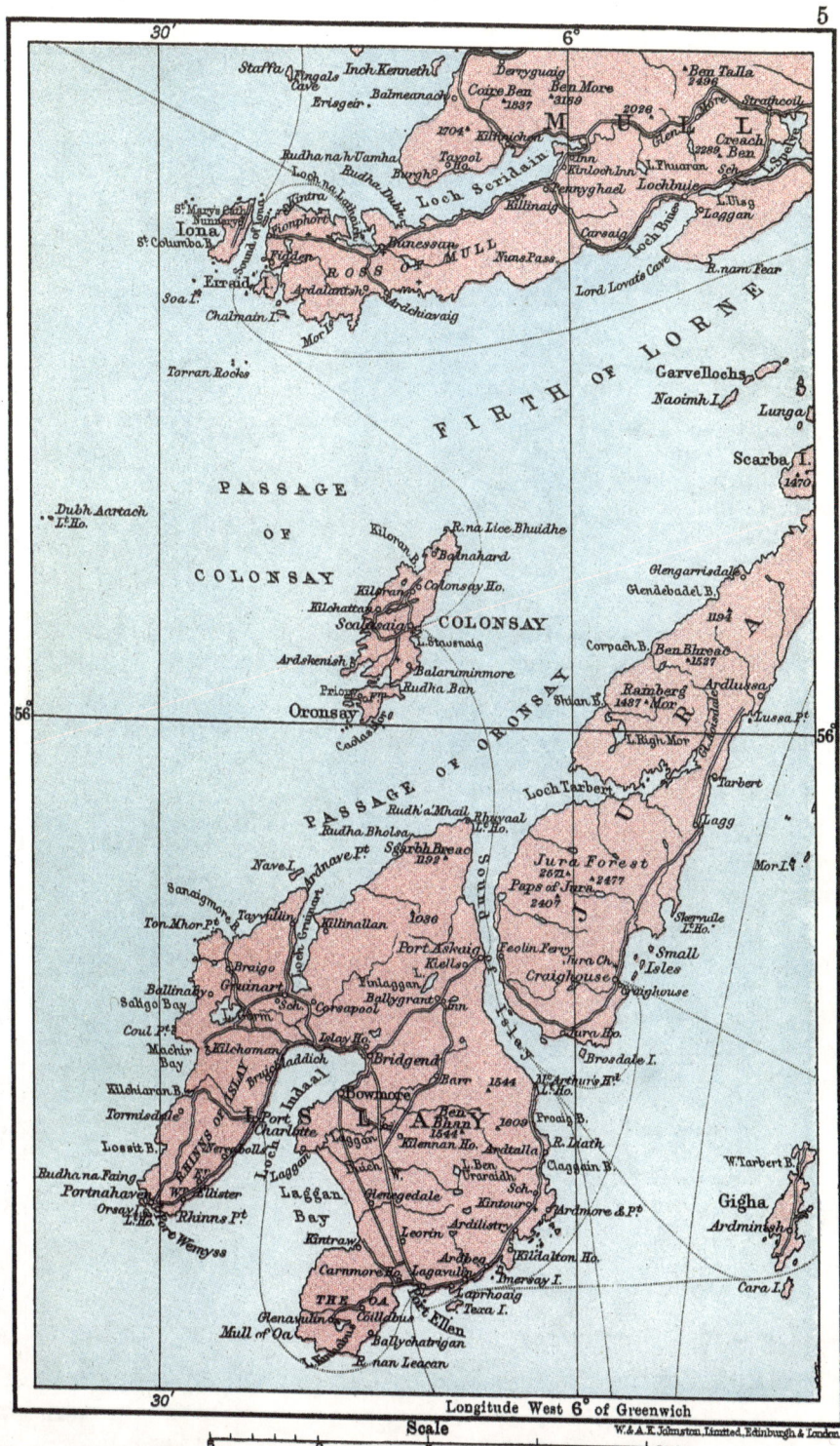

Scale

W. & A. K. Johnston, Limited, Edinburgh & London.

Miles

30'
4°

L. Menteith
Blair Drummond
Sta.
Rowardennan Hotel
Gartmore
Sta.
B. Forth of Menteith
R. Forth
Inverbeg
3409
Gartness
Kelty
Kippen
Carfgunnoc
STIRLING
St. Ninians
Whistlefield Hotel
Luss (Upper)
Balmaha
Buchlyvie
Drymen
Burn
Gakelochhead
Killearn
Earl's Seat
Finry
L. Coulter
Shandon Hydro.
Drymen Sta.
1894
Fintry Hills
Carronbridge
Sta.
Caldarvan
Dumgoyne
CAMPSIE FEELS
1154
1870
R. Carron
Bonnybridge
Dennyloanhead
Cas.
56°
Balloch
Jamestown
Blanefield
Strathblane
Campsie
Kilsyth Hills
Kilsyth
Banknock
Castlecary
Alexandria
Lennoxtown
Campsie Glen
Gavell
Cumbernauld
Rentone
Bonhill
Milton of Campsie
Torrance
PART OF DUMBARTON
Cumbernauld
GOUROCK
DUMBARTON
Kilpatrick Hills
N. Milngavie
Salmons
Summerston
GREENOCK
Bowling
Kilpatrick
Bearsden
Bishopbriggs
Blairs
Stepenboig
Gartcosh
Coatbridge
Caldercruix
Plains
Inverkip
Kilmacolm
Clydebank
AIRDRIE
Duchal Ho.
Houston
GLASGOW
Bridge of Weir
PAISLEY
Johnstone
RENFREW
Bellahill
Holytown Junc.
Lochwinnoch
Howwood
Glenfield
Barrhead
Busby
High Blantyre
Bothwell
Motherwell
Kilbirnie
Neilston Sta.
Thorntonhall
Kilbride
Wishaw
Fairlie
Beith
Uplawmoor
Meikle Earnock
Larkhall
HAMILTON
Dalry
Barrmill
Lugton
Eaglesham
Quarter
Netherburn
Glassford
West Kilbride
Dunlop
Stonehouse
Strathaven
Blackwood
DROSSAN
Kilwinning
Stewarton
Quarry Hill 1089
Ryeland
Lesmahagow
Saltcoats
Waterside
Drumclog
Stevenston
Cunninghame
KILMARNOCK
Darvel
Dungavel Hill 1502
Gleghuck
Inches
Irvine Bay
Crosshouse
Newmilns
Loudounhill
Gailes
Galston
Hart Hill 1294
Markirk
Sta.
Barrassie
Craigie
Lady I.
Troon
Mauchline
Sorn
Cairn Table 1944
Prestwick
Tarbolton
Catrine
Cronberry
Auchincruive
Ochiltree
Cumnock
Newton-on-Ayr
AYR
Drongan
Old
Alloway
Hollybush
Dalrymple
Rankinston
New Cumnock
Kirkconnel
R. Nith
Dunure Pt.
Cassillis
Patna Sta.
Afton Bridgend
Culzean B.
Maybole
Waterside
Enoch Hill 1865
Blackcraig Hill 2298
Kirkmichael
Kirkoswald
Crosshill
Kilkerran
Dalmellington
Dailly
Alhang 2100

Longitude West 30' of Greenwich
4°

Scale

5   0   5   10   15 Miles

W. & A.K. Johnston, Limited

W. & A.K. Johnston, Limited

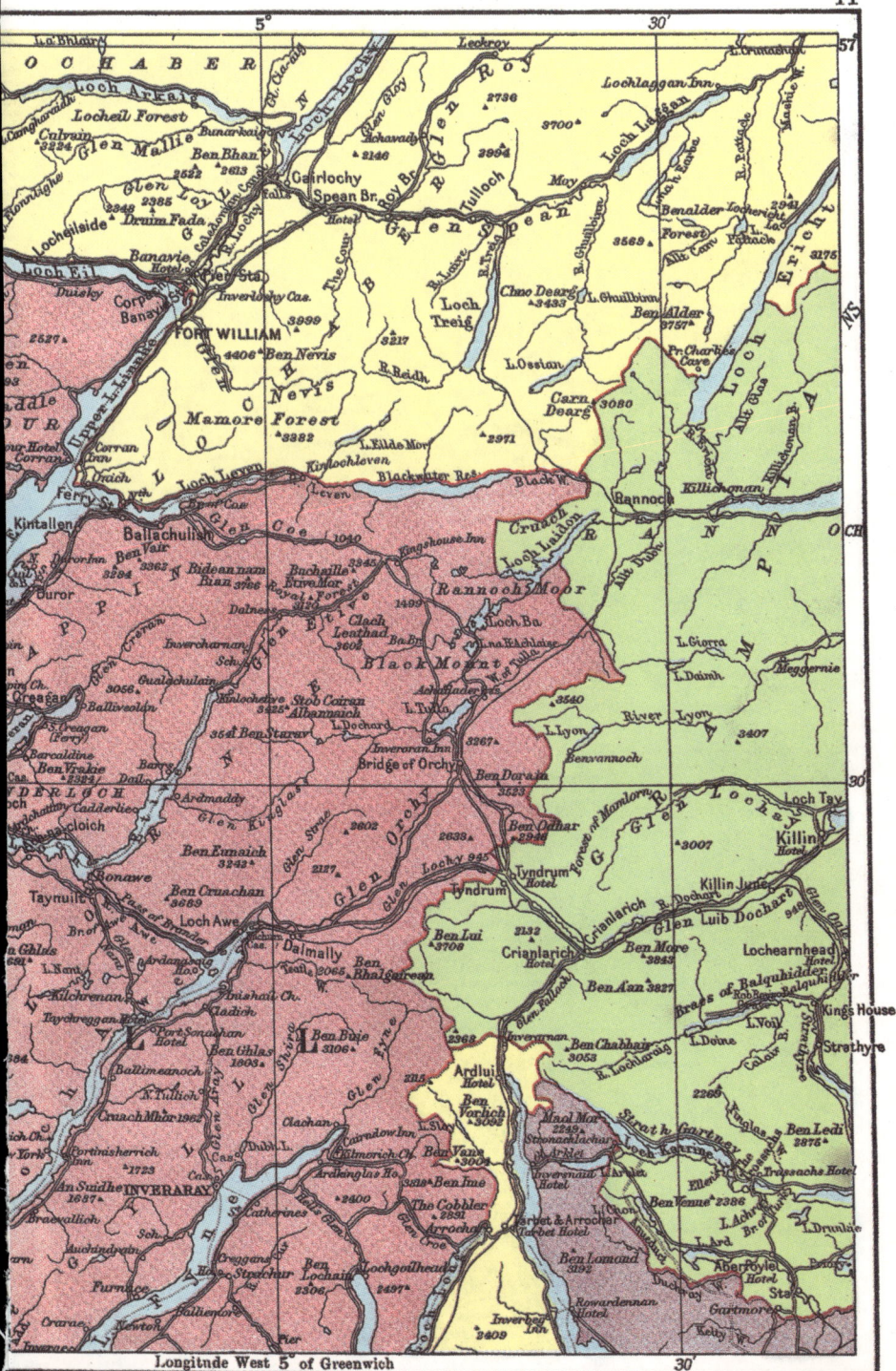

Scale

5   0   5   10   15 Miles

W. & A. K. Johnston, Limited

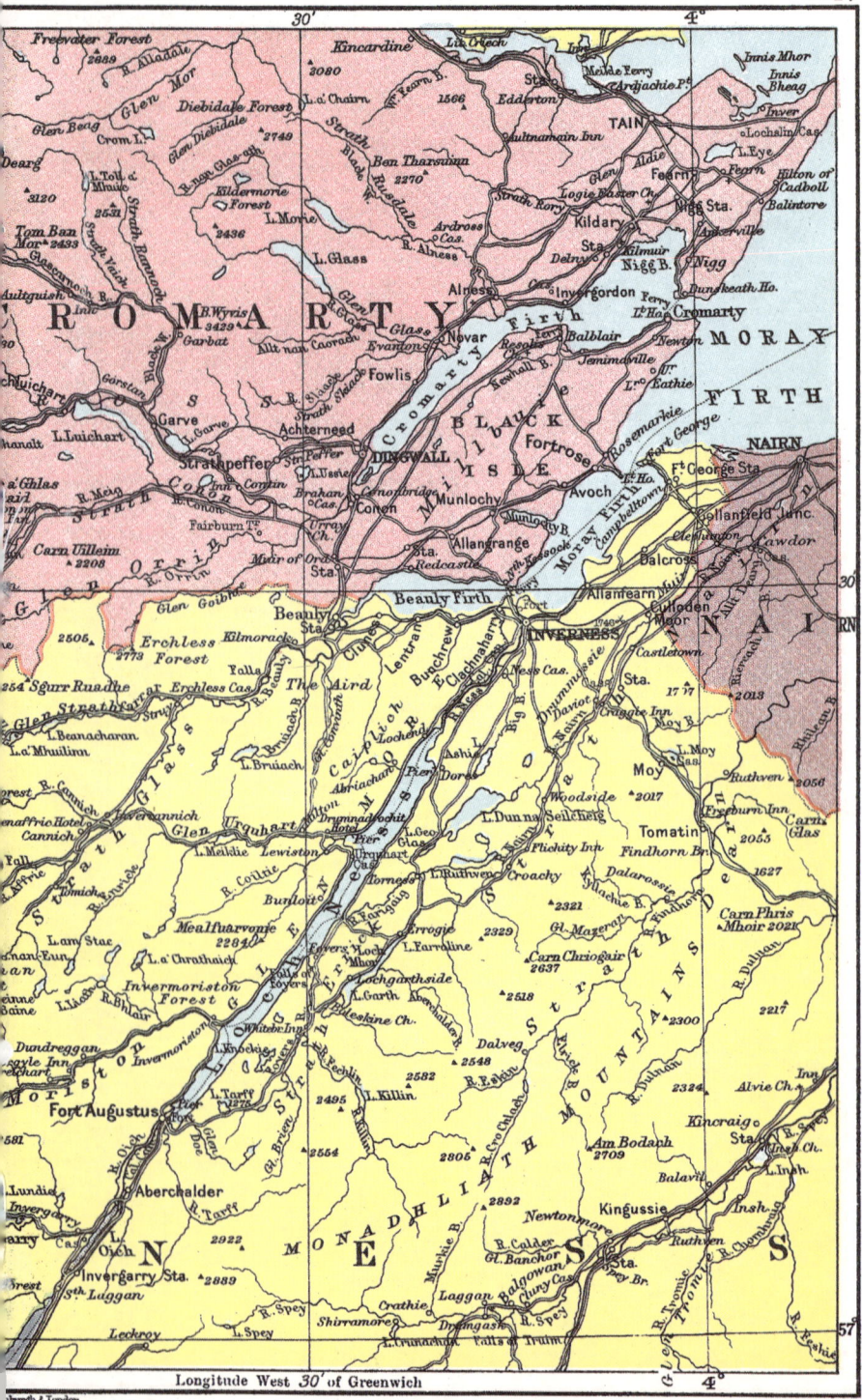

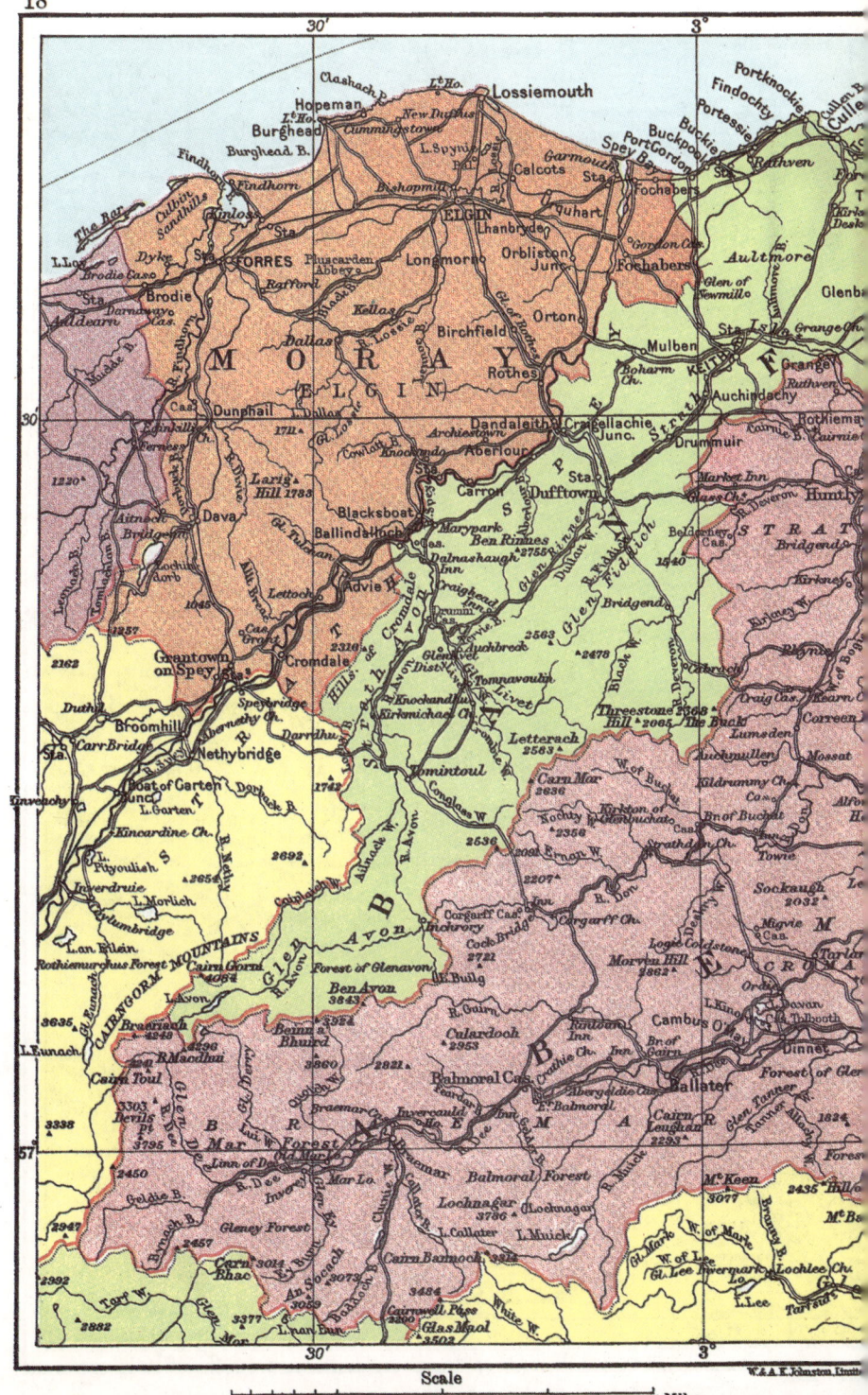

Scale

5   0   5   10   15 Miles

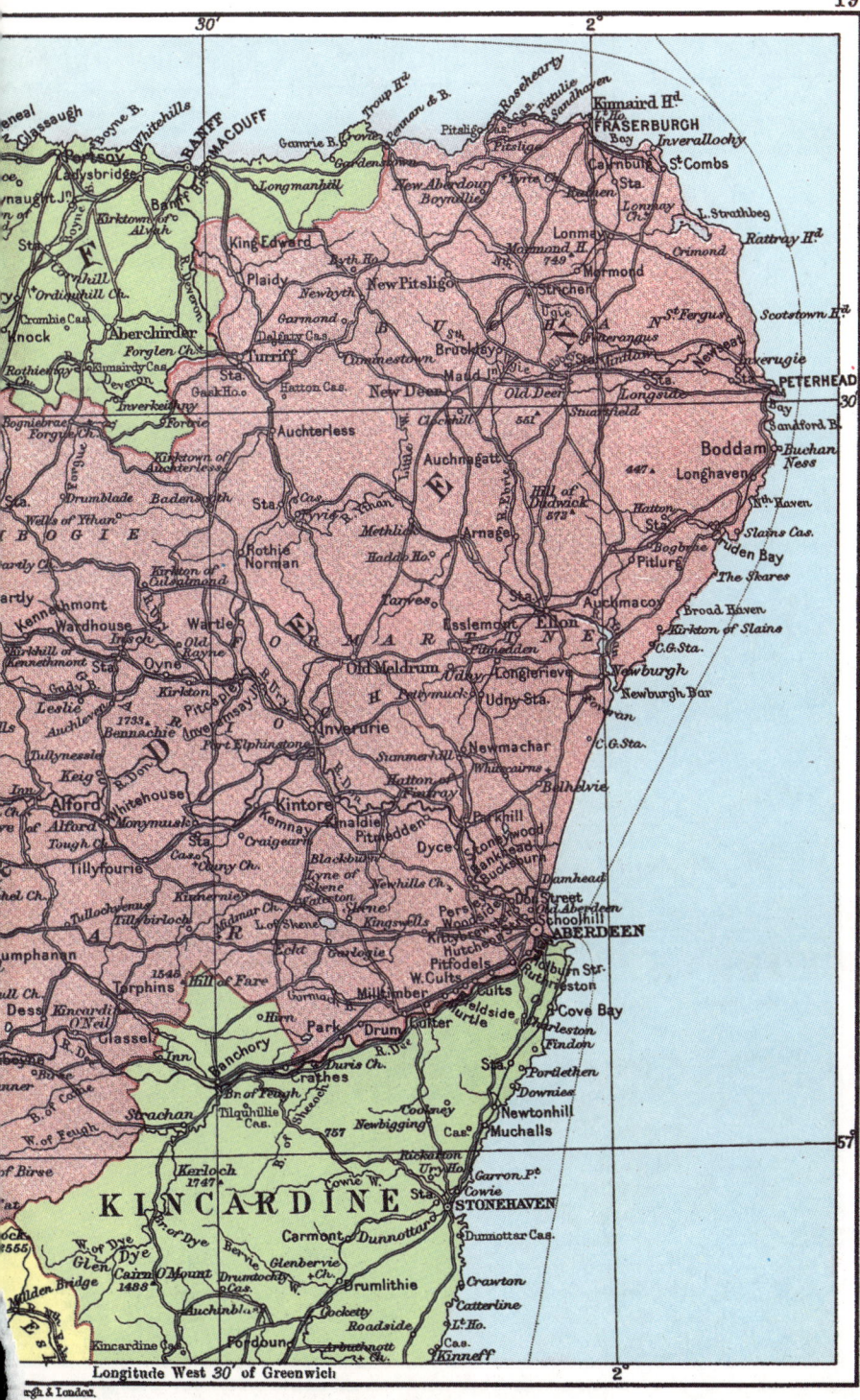

7°   30'

30'

Butt of
Lᵗ Ho.   Eoropi
Lᵗ Ho.   Port
Ness Ch.   Lionel
Aird Dhail
Galson   Dail O'Dheas
Borgh
Borgh   L. Langabhat
L. Ranish
Aird Barvas
L. More Barvas   Shuder
Barvas   El. Shader
L. Gress
Sgeir Mhor   Monsoch 800
L. Shawbost   Glen More Nevis   R. Gress
Shawbost   Bragar   Cress
Bragar   L. Brevat   Sellib
L. Carloway   Arnoll   Back
Lit. Bernera   Carloway   Coll.
West   857   Vatisker P
Gallan Hᵈ   Loch Roag   Glen Bragar
Loch Roag   Breasclete   Broad
Vatios   Bay
Callanish   B. Barvas 815   STORNOWAY
Aird Mhor Mhangursta   Uig B.   Garrynahine   R. Laxdale
Uig Ch.   L. an Oash   Carr
Suainabhal   L. an Tarbert   Bramahuie B.
1250   Grimersta   Chicken
L. Sgastabhat   L. Tungabhat   Ranish Pᵗ
L. Suainabhal   Little   L. Airidh
Aird Breidhnis   Mealasbhat   L. Roag   Sheach   Lurbost
Breidhnis   L. Crurdabhat   L. Airidh   Leurbost
nah Airde   L. nan   L. Leurbost
Criomabhal   Bitheanach   Sandsh
Mhealastadh I.   1500   900   Crossmore
Ceann   Balallan   Calabost
Resort   L. Bhennisbhat   L. Skibacleit   Ojayer   L. Ouin
L. Langabhat   1250   Kebock Hᵈ
Loch Resort   Loch   P   A   R   K   (Cabog)
Scarp I.   Langabhat   Seaforth   Ujmenway
L. Craxadale   Ardvourlie   Oanforth   Shell   Lubhard I.
Craxadale   Ben Mhor
Husinish Pᵗ   Forest of Harris   Crionaig   Sound of Shiant
Husinish B.   Nᵗʰ   H A R R I S   1500
Leosavay   L. Leosaid   Clisham 2622
Soay More   Soay St.   A. Maaruig
W. Loch Tarbert   Ardhasig   Laxadale
Taransay I.   875   2664   L.   Shiant Iˢ
Udh   B. Luskentyre   St. of Scalpay
Aird Vanish   1529   Meavag   Scalpay I.
Toe Head   Borve   Sᵗʰ   H A R R I S   W. Loch Tarbert
Coppay I.   Chaipaval   Geocrab   Flabbay
1092   Sᵗ Harris Ch.   Cluer   Grosebay
Langavat   Ardvey   Stockinish
Ensay I.   L. Stsesavat   Flodabay   Fladda-
Leverburgh   Lingarabay I.   chuain
Killegray   Rodil   Troday I.
Berneray I.   Breast of   Renish Pᵗ   Rudha Hunish
L. Bhuust   Groay   Gilsay I.   L. Hunish
Groay   Scaravay   Duntulm B.   Kilmaluag B.
Torogay I.   Score Bay   Kilmaluag
Foreman Long   Inn
Berneray   Knockbae   Mon. to   Floa
Flora Macdonald
Knockbae

Scale
5   0   5   10   15 Miles

W. & A.K. Johnston, Limᵈ

22

Cape Wrath
L.ᵗ Ho.
Stack Clo Kearvaig
Kyle of Durness
Faraid Hᵈ

Keisgaig B.
976
Fashven
1495
Airidh
na-Beinne
Daill
Durness
Whiten Hᵈ

Sandwood B.
1592
E. Dearg Mhor
Sandwood
1597
Kyoldaig
Hoan I.
Ben Hutig
1340
Midtown
Rabbit I.
Ron I.
Torrisdale B.
Port

Sheigra
Inverhope or Melness
Inveruaver
Farr
Bettyhill

Loch Inchard
Portnancon
Heilan
Kyle of Tongue
Tongue
L. Buidhe

Kinlochbervie
Cranstackie
2630
Erriboll
Loch
Hope
Borie
L. Slaim
L.na Caorach

Achlanish
Whitanich
2530
Foinaven
Str. Dionard
Pollan
3040
Ben Hope
Kinloch Ho.
Ben Loyal
2504
Loch
Laoghal
(Loyal)
Rhifail

Badnaboy
2554
L. Dionard
L. Dubreibh
L. Chalum
Skail
Dalvina Lo.
L. Crocach
1375

Laxford Br.
2580
Arkle
Gt. Gualin
Ben Stack
2364
L. Stack
Forest
L. More
Ben Hee
2864
Loch
Meadie
L.t.E.Geanach
Ben Griam
Mhor
1936
Ben Griam
Bac

Ackfarry
L. Coire
na Sih
Mudale
Loch Naver
L.
Rimsdale
L. Coire
Achnam

Unap
Quinag
2653
L. Glendhu
L. Glencoul
L. Merland
L. Flag
Altnaharra Hotel
Ben Klibreck
3154
L.a'Bhealaich
2773
Ben Armine
2338

SUTHERLAND

Skiag Br.
3597
Gorm
Mor
L. Ghriama
Inn
Over
945 The Crask
1403
1381

Loch Assynt
Ardvreck Cas.
Inchnadamph
B. More
Assynt
3273
2345
Overscaig
Srn an Loin
1428
Strath na Seilga
Gaa-loch
Mor
B. Gheil
L. Beannach

L. Gainmhi
2779
2338
R. Oykel
Glen Cassley
Ben Sgaireach
1561
L. Sgaireach
B. Brora

Cam.
L. Awe
Sgonnan Mor
2028
Authnacealgach Hotel
L. Ailsh
1785
R. Glencassley Cas.
Cruinn D.
Lairg
Hotel
Mun
L.a' Choinn

Loch
Veyatie
Elphin
L. Urigill
1692
L. Craggie
L. Claise Moire
Inveroykell
Invercassley
R. Oykel
Rivehall
Lairg Sta.
L. Cra_sal Mor
L. Buidhe
Rogart Sta.

Cronleh Hills
L.a'Chrois
1398
Luberoy
Strath Oykell
Kyl
L.Laro
Inveran Sta.
L-an-Lagain
The Mound

R. Kanard
Rhidorroch Forest
Rappach W.
Oykell Br. Inn
Glen Einig
1616
Achnahanat
Inveroykell Cas.
Culrain
L. Migdale
Bonar Bridge
Clashmore

L. Achall
Glen Achall
L-an
Daimh
Str. Muivie
Freewater Forest
2559
Croick
Black W.
R. Carron
Craig Ho.
Ardgay
Kincardine
2080
Str. Loch
R. Evelix
Pronay
Cas.
Meikle Ferry
Ardjach

Rhidorroch Lo.
Leckmelm

Scale
5   0   5   10   15 Miles

W. & A.K. Johnston, Limited